53rd Annual Loss Prevention Symposium 2019 (LPS)

Topical Conference at the 2019 AIChE Spring Meeting
and 15th Global Congress on Process Safety

New Orleans, Louisiana, USA
31 March - 4 April 2019

ISBN: 978-1-5108-8688-9

Printed from e-media with permission by:

Curran Associates, Inc.
57 Morehouse Lane
Red Hook, NY 12571

CURRAN ASSOCIATES INC. proceedings.com

Some format issues inherent in the e-media version may also appear in this print version.

Printed by Curran Associates, Inc. (2019)

For permission requests, please contact AIChE
at the address below.

AIChE
120 Wall Street, FL 23
New York, NY 10005-4020

Phone: (800) 242-4363
Fax: (203) 775-5177

www.aiche.org

Additional copies of this publication are available from:

Curran Associates, Inc.
57 Morehouse Lane
Red Hook, NY 12571 USA
Phone: 845-758-0400
Fax: 845-758-2633
Email: curran@proceedings.com
Web: www.proceedings.com

TABLE OF CONTENTS

Analysis of Risk Due to Compound Extreme Events

Patrick A. Harr
Jupiter Intelligence
181 2nd Ave., San Mateo, CA 94401
Email address

Paul Boehm[2], Josh Hacker[1], Bob Haddad[3], Harri Kytomaa[4], John Osteraas[5], and Steve Sain[1]
[1] Jupiter Intelligence, Boulder, CO
[2] Environmental & Earth Sciences, Exponent, Inc., Maynard, MA
[3] Environmental & Earth Sciences, Exponent Inc., Menlo Park, CA
[4] Exponent Inc, Bowie MD
[5] Buildings & Structures, Exponent Inc., Menlo Park, CA

Prepared for Presentation at
American Institute of Chemical Engineers
2019 Spring Meeting and 15th Global Congress on Process Safety
New Orleans, LA
March 31 – April 3, 2019

Analysis of Risk Due to Compound Extreme Events

Patrick A. Harr
Jupiter Intelligence
181 2nd Ave., San Mateo, CA 94401
Email address

Paul Boehm[2], Josh Hacker[1], Bob Haddad[3], Harri Kytomaa[4], John Osteraas[5], and Steve Sain[1]
[1] Jupiter Intelligence, Boulder, CO
[2] Environmental & Earth Sciences, Exponent, Inc., Maynard, MA
[3] Environmental & Earth Sciences, Exponent Inc., Menlo Park, CA
[4] Exponent Inc, Bowie MD
[5] Buildings & Structures, Exponent Inc., Menlo Park, CA

Keywords: Flood Frequency Analysis, Storm Surge, Coastal Flooding

Abstract

Extreme weather events are often comprised of compound, connected events that individually amplify the impacts to process safety in the industrial sector. During Hurricane Harvey (2017) an extreme event triggered by prolonged intense rainfall over the Houston metropolitan area was made more extreme due to compound influences of a large region of coastal storm surge, which was not extreme in itself but contributed to reduced discharge of water over the inland area. Compound effects led to significant operational, safety, financial, and legal ramifications, such as the dozens of chemical release events documented after Harvey.

Often hazard risk assessments focus on one factor rather than the compound nature of multiple forcing mechanisms. It is possible that extreme event analysis in a univariate context may underestimate the probabilities and impacts of extreme events. In this study, a new framework addresses multivariate analysis of risk due to compound hazards. Combinations of observations and simulations are used to identify expected frequencies of annual chance extreme events forced via connected individual events. The framework emphasizes the statistical dependence of multiple physical variables that contribute to extreme compound events when individual events are not extreme.

1 Introduction

The concurrent or consecutive occurrence of multiple extremes is referred to as a compound event[1]. Examples include a heat wave and drought or ocean and fluvial flooding. Compound events can be categorized as several types. One such type is successive or simultaneous extreme events such as extreme precipitation and storm surge. A second is an indirect amplification of an impact from one or more individual extremes (e.g., drought and heat wave). A third type is one in which individual events are not extreme by themselves but the combination of events leads to extreme events or impacts. This may occur when a moderate storm surge occurs at an above average high tide.

Many studies[2],[3] have identified the potential for the occurrence and severity of compound events to increase in the near future. This requires advances in recognition, understanding, and modeling of interacting physical processes to better quantify the risk that compound events pose to infrastructure and society[4]. A primary challenge to assessing risk associated with compound events is that dependencies among environmental factors, which comprise a particular hazard (i.e., flood, drought, fire, etc.), can cause the estimation of event probabilities to be much more difficult than if the factors are independent[5]. Poor representation of the dependencies among physical factors that define compound events can often lead to underestimation of the risk of such events provided that the impacts of dependent factors are larger than the impacts of the independent combination of individual factors[6].

In this presentation, the hazard to be addressed is coastal flooding. The risk of coastal flooding is related to the joint occurrence of heavy precipitation and storm surge. Often, a strong storm surge will prevent the discharge of water, which has made its way into streams and rivers during heavy precipitation, to the open ocean. A methodology to derive the bivariate joint distribution of precipitation and storm surge. Once derived, the bivariate distribution is used to assess the return probabilities of the joint occurrence of precipitation and storm. The joint probability is compared to the univariate probability of the separate occurrence of precipitation and storm surge. Finally, extensions to the assessment of uncertainty and more complete data analytics are defined.

2 Data and Methods

For this presentation, ocean water level data and precipitation data are taken from the region of Charleston, SC (Fig. 1). The ocean data are defined from the Charleston, Cooper River Entrance tidal gauge, which is the National Oceanic and Atmospheric Administration (NOAA) gauge 8665530. Precipitation data are defined from several rain gauge sites, which include Charleston, Charleston International Airport (KCHS), the Charleston Executive Airport (KZJI), and Mount Pleasant (KLRO). The analyzed precipitation data are an average of the three stations.

Water level (Fig. 2) and precipitation data are available hourly from 1930 to the present. From the hourly water level data, the daily maxima are defined and the yearly maxima are defined from the daily values. For each annual maximum water level, the corresponding daily maximum precipitation values are obtained. Precipitation data are scanned to identify the daily maximum that occurred from 2 days prior to 1 day after the day of maximum water level.

The series of annual maximum water level (Fig. 3a) and the corresponding maximum daily precipitation data (Fig. 3b) define pairs of environmental factors that define the annual maximum water level hazard at the water-level gauge in downtown Charleston SC. Most rainfall occurs the day before or on the day of the annual maximum water level (Fig. 4).

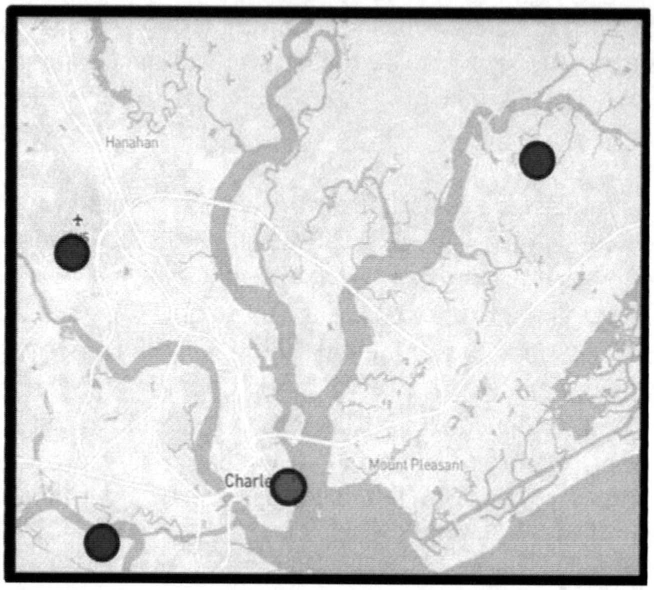

Figure 1. Location of water level gauge (red circle) and precipitation gauges (gray circles) used to define the annual maximum water level and corresponding maximum daily precipitation for the Charleston region from 1930 through 2017.

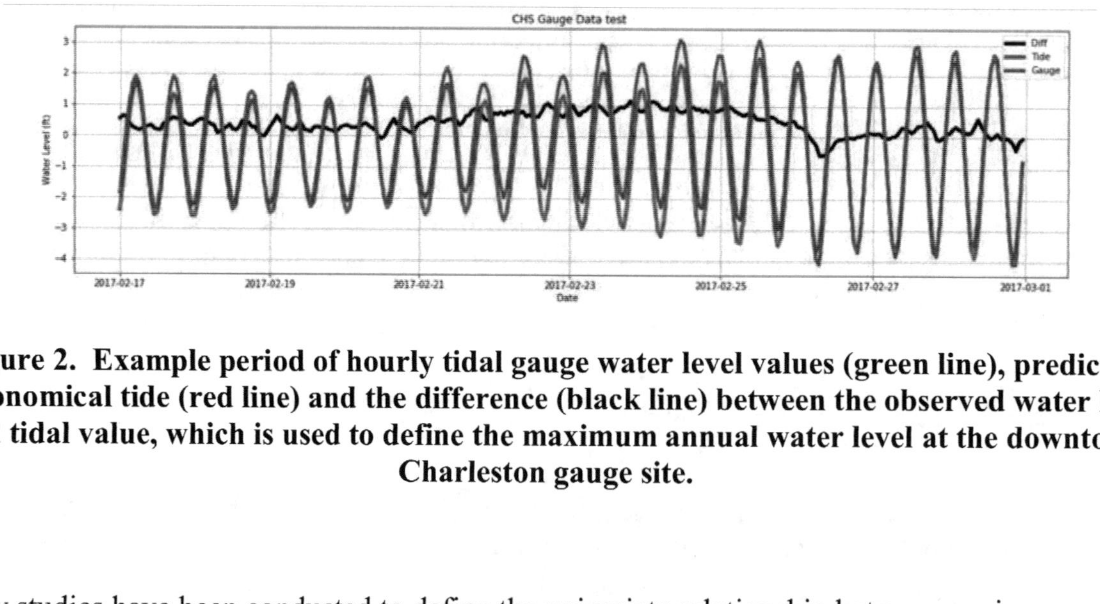

Figure 2. Example period of hourly tidal gauge water level values (green line), predicted astronomical tide (red line) and the difference (black line) between the observed water level and tidal value, which is used to define the maximum annual water level at the downtown Charleston gauge site.

Many studies have been conducted to define the univariate relationship between environmental factors such as rainfall or storm surge and flood frequency[7,8]. In a univariate framework, a probability distribution is fit to the environmental data and extreme value statistical methods can be employed to identify probabilities associated with levels of forcing (i.e., rainfall amount or storm surge level). Often, the univariate analysis is expanded to define joint distributions of environmental forcing factors[10,11,12]. However, these methods are often designed such that the same family of marginal distributions is assumed for all environmental forcing. For example, a

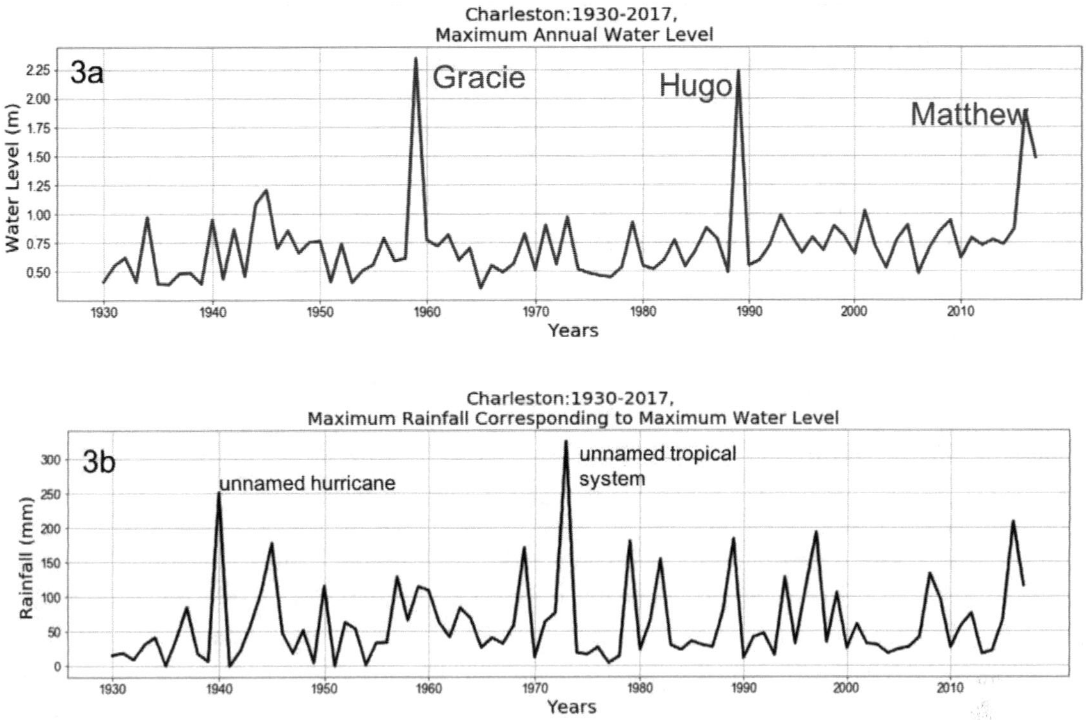

Figure 3 (a) Annual maximum water level (m) at the downtown Charleston tidal gauge, and (b) the maximum daily precipitation (in) corresponding to the annual maximum water level in (a). Key natural events are identified for relative extreme values.

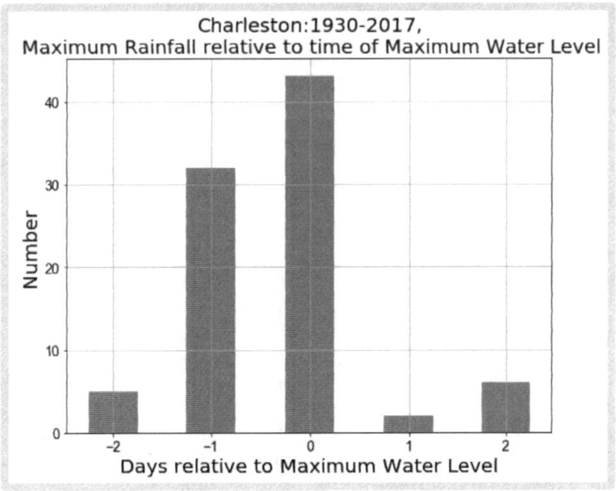

Figure 4. Days relative to the day of the annual maximum water level on which the daily maximum amount of precipitation occurred.

bivariate Gumbel[11], bivariate normal[10], or bivariate gamma[12] have all been used in various flood frequency studies.

However, recent investigations have shown that the best fit marginal distribution of individual environmental forcing often belong to different families of probability distributions[5,13,14,15].

These studies rely on the concept of the copula[16] to model the dependence structure among multiple environmental forcing factors without regard to their respective marginal distributions. It has been shown[15] that copula-based analyses result in a more realistic assessment of the multivariate environmental forcing mechanisms of a compound event because the resulting copula, which defines the dependence structure of the multivariate distribution, fits the empirical distribution of data better than a distribution established around joint parametric distribution and marginal distributions assumed to be of the same family for all forcing factors.

In this extended abstract, the steps taken to compare a univariate and multivariate flood frequency analysis in which water level and precipitation represent environmental pairs are defined as a basis to the formal analysis of compound extreme events.

2.1 Marginal Distributions

Several distributions (Table 1) were fit to the annual maximum water level data and precipitation data. The parameters for each distribution were defined via maximum likelihood approximation and each distribution was assessed using the chi-squared goodness of fit parameter. For the rainfall data (Fig. 5), the generalized extreme value distribution fit the data (Fig 6) with the lowest chi-square statistic. The fit of the distribution became a bit less representative for higher values of precipitation (Fig. 7), however the overall fit was better than any other of the tested distributions.

Table 1. Parametric distributions fit to the annual maximum water level and corresponding daily maximum precipitation data.

Distribution	
Beta	Log-Gamma
Chi-square	Log-Normal
Exponential	Pareto
Generalized Pareto	Pearson3
Gamma	Weibull
Gumbel	Normal
Logistic	Generalized Extreme Value

For the water level data (Fig. 8), the Pearson3 distribution (Fig. 9) resulted in the lowest chi-square goodness of fit statistic. Although the Pearson3 distribution was the best fit of those tested, the percentile plot (Fig. 10) indicates that it does not fit the large values in the right-hand tail of the distribution very well. A major aspect to the definition of marginal distributions for the chosen environmental factors is that the best fit distribution differs for each parameter. However, in the copula-based approach to defining the joint dependencies between water level and precipitation it is not required that both variables follow the same marginal distribution.

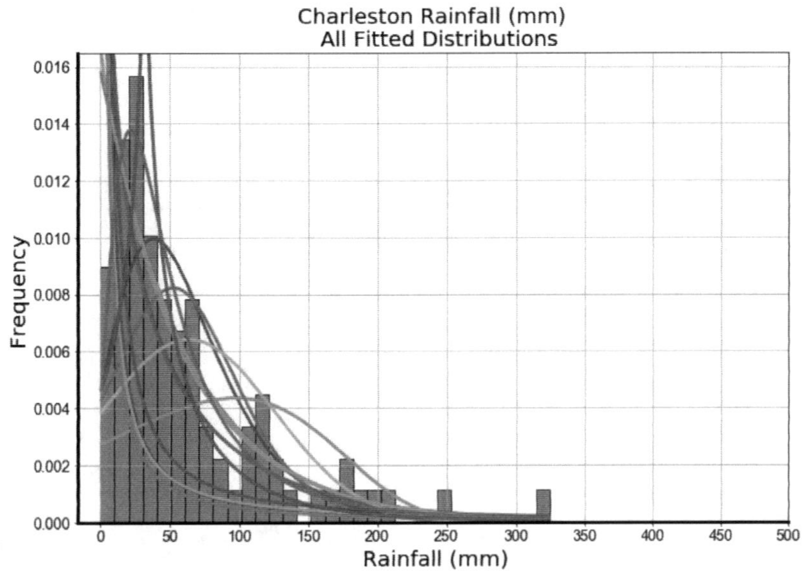

Figure 5. The relative histogram (gray bars) of maximum rainfall that corresponds to the annual maximum water level and the probability distribution functions defined in Table 1 and fit to the precipitation data.

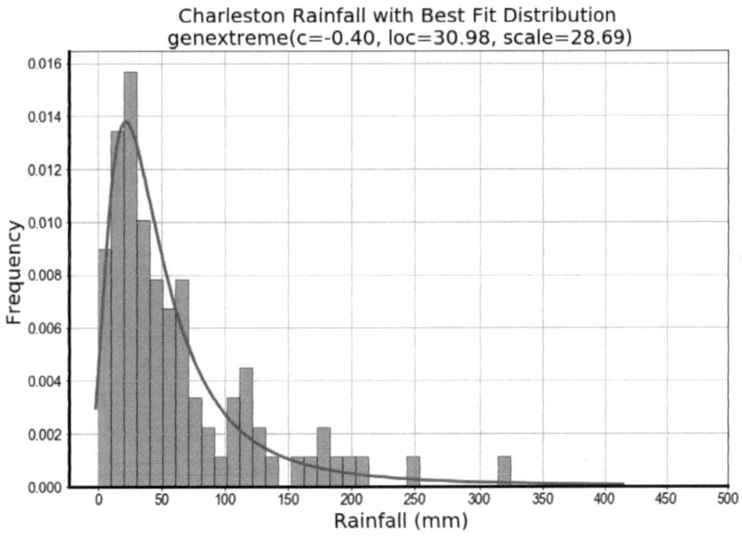

Figure 6. The relative histogram (gray bars) of maximum rainfall that corresponds to the annual maximum water level and the generalize extreme value distribution fit to the precipitation data.

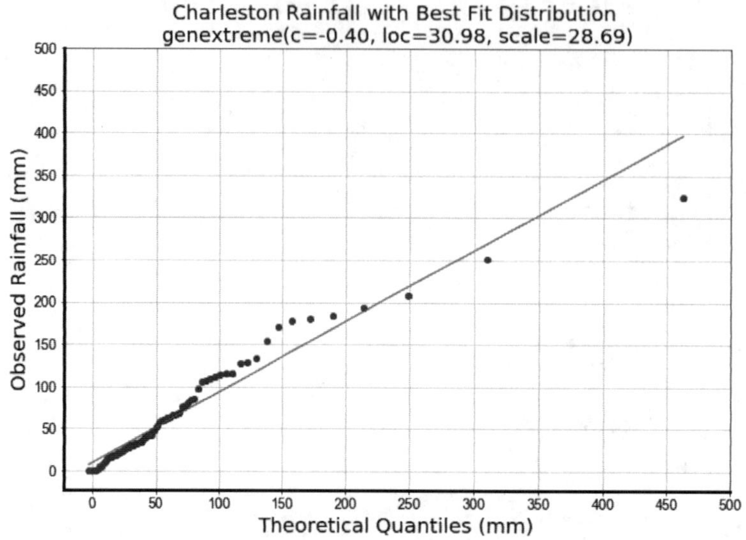

Figure 7. A percentile plot of the observed maximum rainfall values (blue dots) plotted versus the theoretical quantiles of the generalized extreme value distribution contained in Fig. 6.

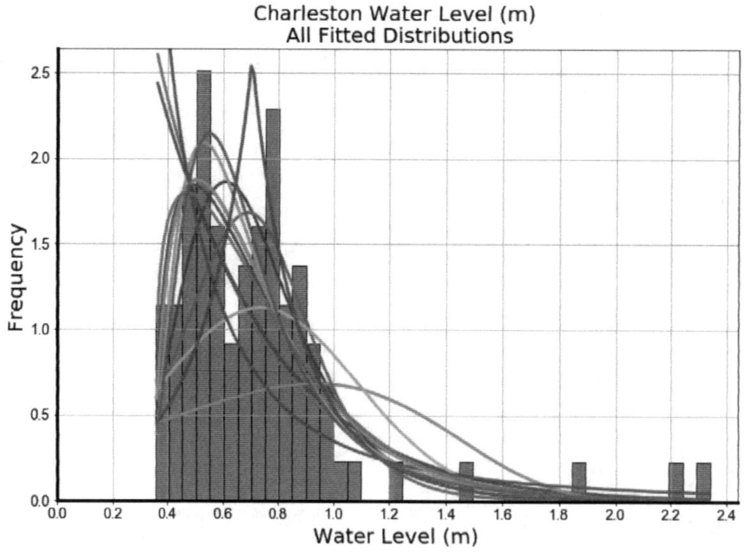

Figure 8. The relative histogram (gray bars) of maximum water level that corresponds to the annual maximum water level and the probability distribution functions defined in Table 1 and fit to the precipitation data.

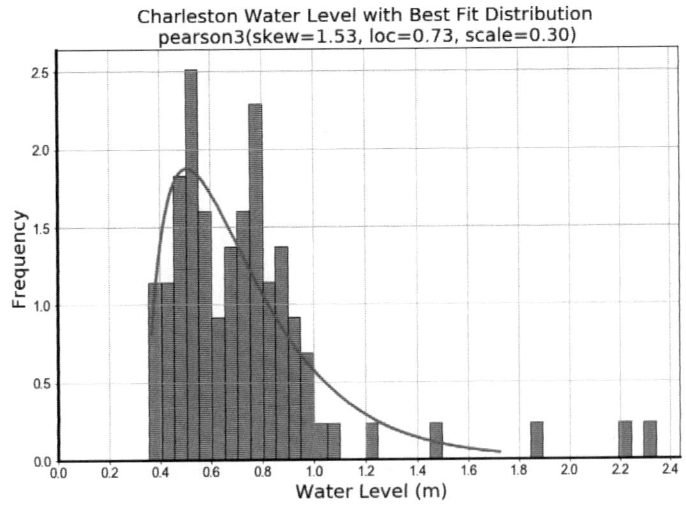

Figure 9. The relative histogram (gray bars) of maximum water level that corresponds to the annual maximum water level and the Pearson3 distribution fit to the water level data.

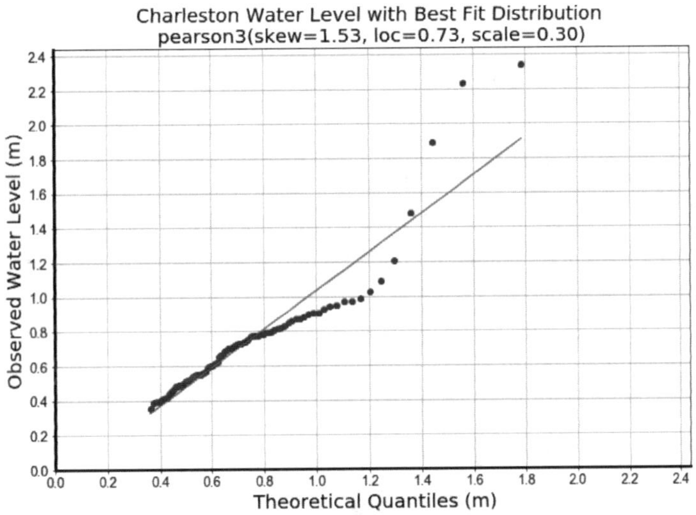

Figure 10. A percentile plot of the observed maximum water level values (blue dots) plotted versus the theoretical quantiles of the Pearson3 distribution contained in Fig. 9.

2.2 *Joint Distribution*

A simple plot (Fig. 11) of annual maximum water level and the corresponding precipitation maximum indicates that there is some relationship between these two factors. However, it is also likely that the marginal distributions of each parameter are different (Figs. 6 and 9). Therefore, the joint distribution of water level and rainfall is modeled using the copula function to avoid the restriction that the marginal distribution of the two variables be of the same family.

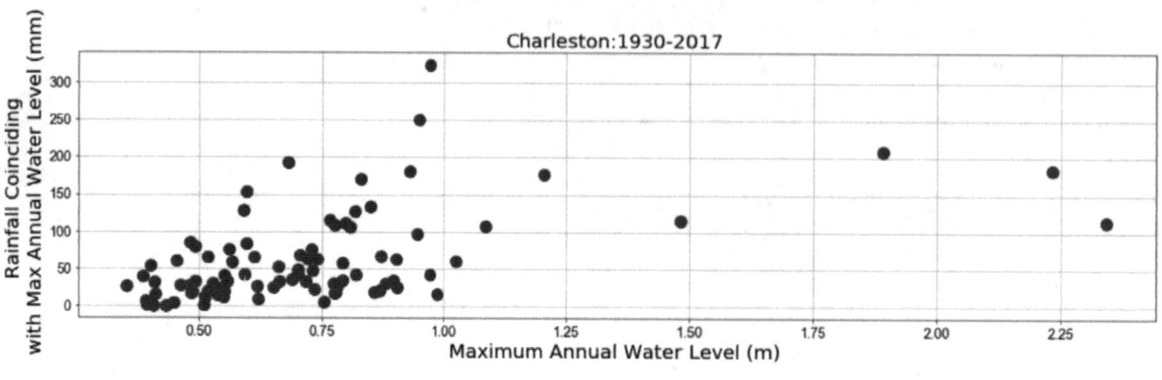

Figure 11. Maximum annual water level (abscissa) and the corresponding maximum rainfall amount (ordinate).

The basis of the use of copulas is contained in Sklar (1959)[16] in which it is showed that an n-dimensional distribution function F and be written as:

$$F(x_1, x_2, \ldots\ldots, x_n) = C\,[F(x_1), \ldots F(x_n)]$$

where the $F_1, \ldots\ldots, F_n$ are marginal distribution functions and continuous. Then, C is a unique function that has the following structure:

$$C(u_1, \ldots\ldots, u_n) = F(F^{-1}(u_1), \ldots., F^{-1}(u_n)), \text{ for } 0 \le u_1, \ldots., u_n \le 1$$

Therefore, if C is a copula function, the $F_1, \ldots, F_n$ are distribution functions and F is a n-dimensional distribution function with marginals $F_1, \ldots, F_n$.

In general, three classes of copula functions are Archimedean, elliptical, and extreme value. Based on Nelsen (1999)[17], the Archimedean family of copula have properties that lend themselves well to use in flood-frequency analysis. These properties include definition of a generating function, ϕ, and copula parameter, θ. The copula parameter, θ, can be defined using the Kendall's rank correlation tau statistic, which measures the dependence between the two variables (water level and precipitation). In this study, three Archimedean copulas (Table 2) are defined and fit to the pairs of water level and precipitation data.

Table 2. Archimedian Copulas employed in this study.

Copula Function
Ali-Mikhail-Haq Family
Cook-Johnson Family
Gumbel-Hougaard Family

The following steps are taken to define each of the three copula functions:

1) Compute Kendall's rank correlation, τ, from the observed data;
2) Determine the copula parameter, θ, from τ;
3) Compute the generating function, ϕ, for each copula; and
4) Obtain the copula function from ϕ and θ.

To identify the copula that best fits the data, a method[16] is used to compare the quantiles of the parametric copula with a nonparametric distribution that is based on the empirical nonexceedance probabilities of the joint occurrences of water level and rainfall. The resulting copula Q-Q plot (Fig. 12) provides evidence as to the copula that best fits the original data, which in this case is the Gumbel-Hougaard copula as the quantiles of the copula function lie nearest the diagonal line that defines the empirical quantiles equal to the copula quantiles. The Cook-Johnson copula does not fit the data well at the more extreme values.

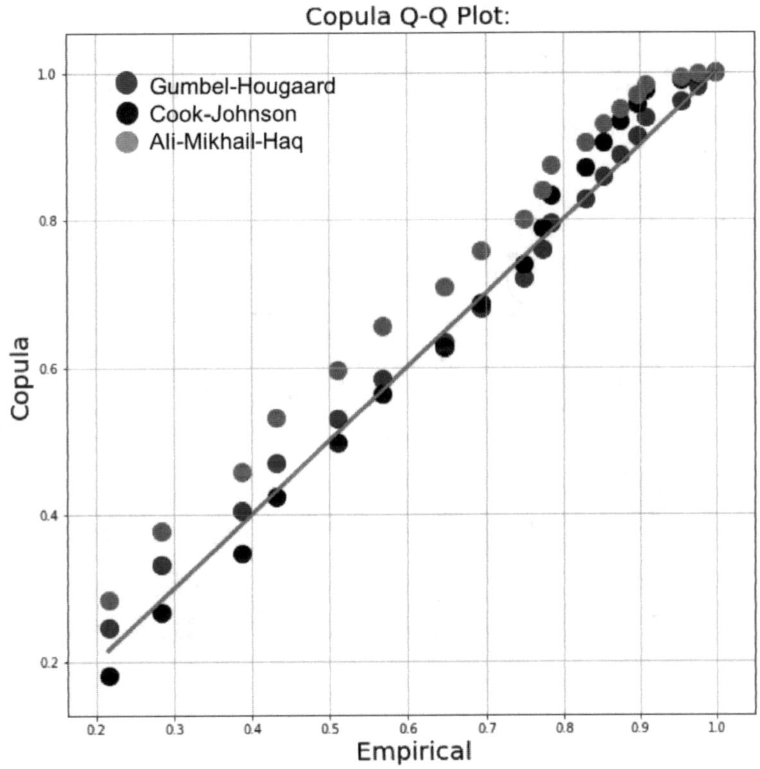

Figure 12. Copula Q-Q plot that defines the quantiles of the three parametric copulas with the empirical joint nonexceedance probabilities.

The choice of the Gumbel-Hougaard copula is also justified by computing the root mean-square error and Akaike Information Criterion (AIC). The Gumbel-Hougaard copula also has the lowest root mean square error (RMSE) and Akaike Information Criterion (AIC) value (Table 3). Finally, the isolines of the joint distribution of annual maximum water level and corresponding precipitation values from the Gumbel-Hougaard copula are defined in Fig. 13.

Table 3. The RMSE and AIC values for three copula functions fit to the annual maximum water level and corresponding precipitation data.

Copula Function	RMSE	AIC
Gumbel-Hougaard	0.05308	-53.78
Cook-Johnson	0.05390	-53.49
Ali-Mikhail-Haq	0.05769	-52.19

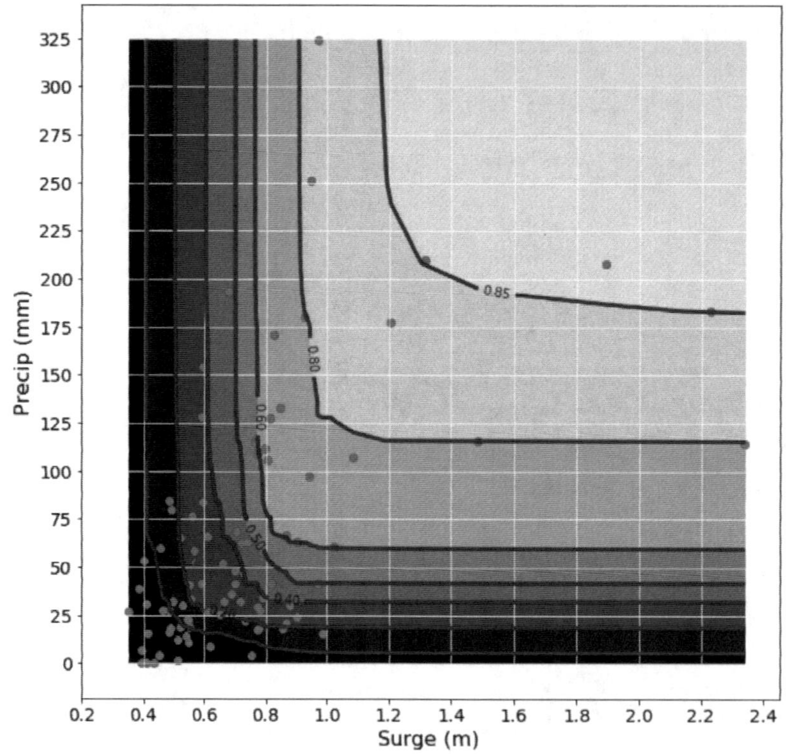

Figure 13. Isolines of the joint cumulative distribution of annual maximum water level and corresponding precipitation defined by the Gumbel-Hougaard copula.

3 Return Periods

In the current framework where the water level data are annual maxima, the return period, T, associated with a given univariate environmental factor is defined as:

$$T_x = 1 / (1 - F_x(x^*)).$$

In the case being examined here, the variable x is water level and F is the distribution of x and ($1-F_x(x^*)$ is the probability that the water level exceeds a value, x^*. A similar univariate return period for y (i.e., rainfall) is defined as:

$$T_y = 1 / (1 - F_y(y^*)).$$

The distributions of water level and rainfall are defined above and the univariate return periods for water level (Fig. 14a) and rainfall (Fig. 14b) can be computed with associated uncertainties defined using the normal approximation and delta method.

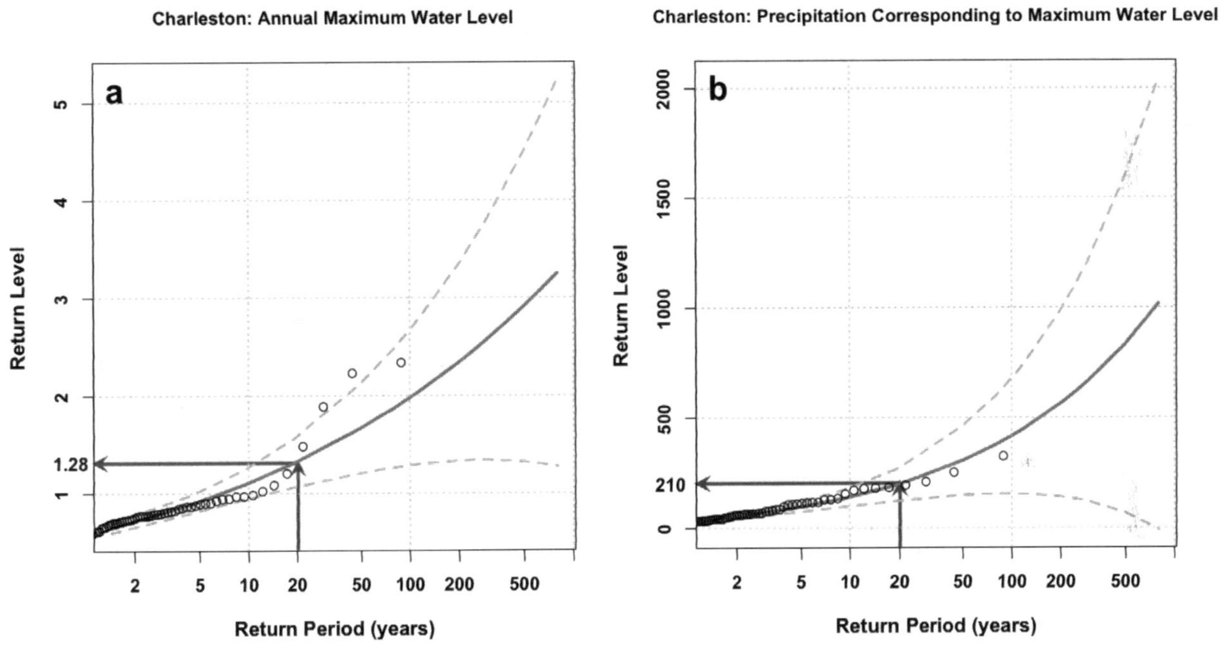

Figure 14. Return period analysis for (a) annual maximum water level (m) and (b) maximum precipitation (mm) corresponding the annual maximum water level. The vertical and horizontal arrows identify the 20-year return period water levels and precipitation amount. The redline defines the computed return period, the dashed lines represent 5%-95% uncertainty region, and the open circles define the observed values.

The schematic in Fig. 15 identifies the bivariate case in which paired observations of water level and precipitation are analyzed as regions in which the probability of the water level exceeds a value x^* and the precipitation value exceeds a value y^*, or water level exceeds x* or precipitation exceeds y^*. This is defined as the bivariate OR case [5] in which the hazard region is related to a critical pair of values (x^*,y^*).

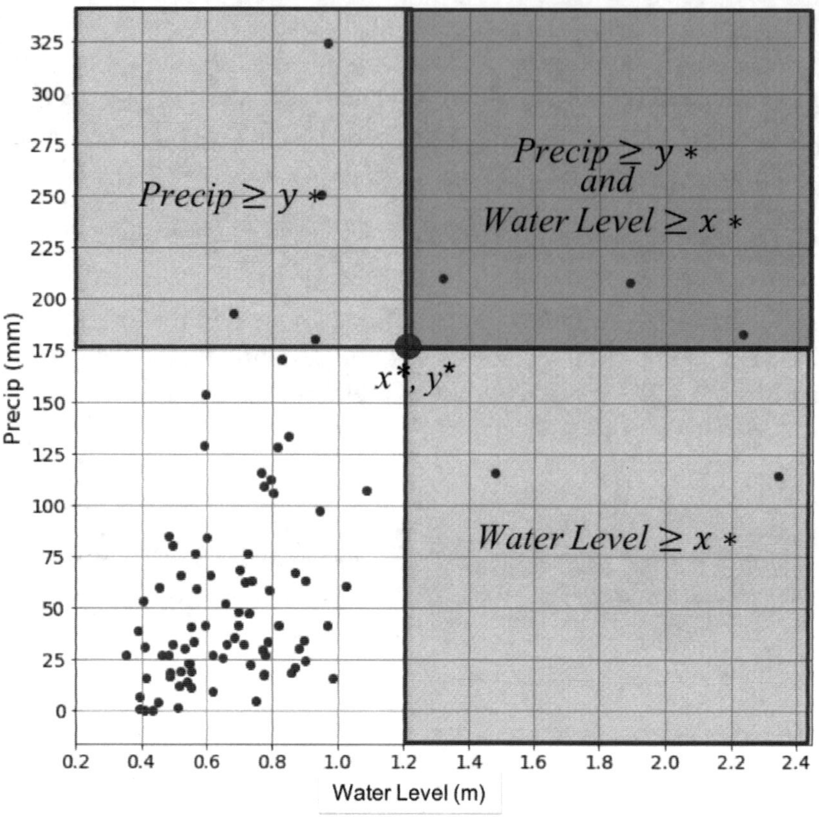

Figure 15. Scatter diagram of water level (m) and precipitation (mm) with the regions labeled that define the bivariate OR case for calculating the joint return period.

Based on the definition of the copula function, the joint return period is calculated in manner similar to the univariate return period. The joint return period is:

$$T_{xy} = 1 / (1\text{-}F_{xy}(x^*,y^*)) = 1 / (1 - C_{xy}(F_x(x^*),F_y(y^*))).$$

In this case, F_{xy} is the joint distribution of x (water level) and y (rainfall) and $1\text{-}F_{xy}$ is the probability that a bivariate pair (x,y) lie in the shaded region of Fig. 15. For all copula functions,[5] Cxy:

$$C_{xy}(F_x(x^*),F_y(y^*)) \leq \min(F_x(x^*),F_y(y^*)).$$

Therefore, pairs of (x,y) are more likely to fall in the bivariate hazard regions (i.e., shaded region in Fig. 15) than in the corresponding univariate regions. As a consequence, univariate estimates of return periods may underestimate the hazard amplitude.

The joint return period is displayed in Fig. 16 for the range of water level and precipitation value pairs contained in the observation record. For the 20-year return period curve, the point nearest the maximum probability density level for that value is highlighted to indicate the joint occurrence of a water level of 1.35 m and a precipitation value of 214 mm for the joint 20-year return level. In the univariate cases, the 20-year return water level is 1.28 m (Fig. 14a) and the precipitation value is 1.28 mm (Fig. 14b). Therefore, the univariate analyses underestimate the level of the 20-year event as the univariate analyses do not account for the dependence structure between water level and precipitation. The differences between univariate and joint return levels increase with increasing return period.

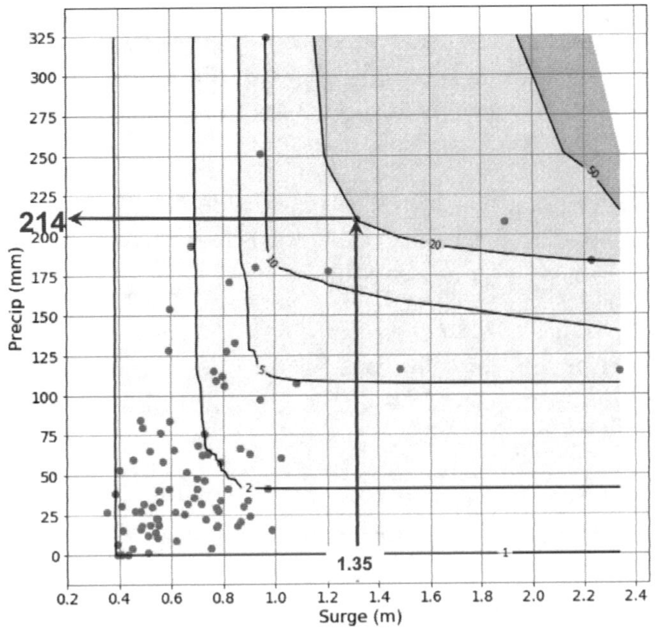

Figure 16. Isolines of joint return period (years) for annual maximum water level (m) and corresponding maximum precipitation (mm). The arrows highlight a 20-year return period coinciding with the maximum probability density value along the 20-year return period curve.

4 Conclusion

The purpose of this analysis is to highlight the motivation and steps to be taken to conduct a joint frequency analysis for a hazard in which multiple dependent factors may contribute to increase the impact of the hazard beyond what could occur without the joint occurrence of the environmental factors. To make the analysis clear, one location is analyzed using both univariate and a joint analysis. The joint analysis is conducted using copula functions to remove the restriction that marginal distributions need come from the same family of probability functions. The primary result is the illustration that the univariate frequency analysis underestimates the magnitude of the extreme events because the univariate analyses do not account for the dependence structure between the paired environmental factors, which in this case are water level and precipitation.

This analysis will be expanded to a larger sample of locations and will also include uncertainty estimates in the definition of return periods.

5 References

[1] T. Wahl, S. Jain, J. Bender, S. D. Meyers, and M. E. Luther, "Increasing Risk of Compound Flooding from Storm Surge and Rainfall to Major U.S. Cities. Nature Climate Change, 5(12), p. 1093, December 2015.

[2] R. Kopp, D. Easterling, T. Hall, K. Hayhoe, R. Horton, K. Kunkel, and A. LeGrande, "Potential surprises – compound extremes and tipping elements," Climate Scinec Special Report: A Sustained Assessment Activity of the U.S. Global Change Research Program, pp 608-635, 2017.

[3] C. B. Field, "Managing the Risks of Extreme Events and Disasters to Advance Climate Change Adaption: Special Report of the Intergovernmental Panel on Climate Change. Cambridge, UK: Cambridge University Press, 2012.

[4] M. Leanard, S. Wesrram A, Ohatak, M. Lambert, B. can den Hurk, K. McInnes, J. Risbey, S. Schuster, D. Jakob, and M. Stafford-Smith, "A Compound Event Framework for Understanding Extreme Impacts," WIREs Climate Change, 5, pp 113-128, January-February 2014.

[5] G. Salvadori, F. Durante, C. De Michele, M. Bernardi, and L. Petrella, "A Multivariate Copula-Based Framework for Dealing with Hazard Scenarios and Failure Probabilities. Water Resources research, 52, pp3701-3721, 2016.

[6] J. Zscheischler, S. Westra, J. J. J. M. van den Hurk, S. I. Seneviratne, P. J. Ward, A. Pitman, A. AghaKouchak, D. N. Bresch, M. Leonard, T. Wahl, and X. Zhang, "Future Climate Risk from Compound Events, Nature Climate Change, 8, pp469-477, 2018.

[7] G. W. Kite, "Frequency and Risk Analysis in Hydrology," Publications in Water Resources. Colorado State University, Ft Collins, CO, 1978.

[8] D. W., Scott, "Multivariate Density Estimation, Theory, Practice, and Visualization. Joh Wiley, NY, NY, 1992.

[9] F. Ashkar, and J. Rousselle, "Some Remarks on the Truncation used in Partial Flood Series Models." Water Resources Research, 19, pp477-480, 1983.

[10] B. Sackl, and H. Bergmann, "A Bivariate Flood Model and its Application." Hydrologic Frequency Modeling. Edited by V. P. Singh, pp 571-582, Dreidel, Dordrecht, Netherlands, 1987.

[11] S. Yue, T. B. M. K. Ouarda, B. Bobe'e, P. Legendre, and P. Bruneau, "The Gumbel Mixed Model for Flood Frequency Analysis. Journal of Hydrology, 246, pp1-18, 2001.

[12] S. Yue, "A Bivariate Gamma Distribution for use in Multivariate Flood Frequency Analysis. Hyrdrological Processes. 15, pp1033-1045, 2001.

[13] A.-C. Favre, S. El Adlouni, L. Perreault, N. Thie' monge, and B. Bobe' e, "Multivariate Hydrological Frequency Analysis using Copulas., Water Resources Research, 40, W01101. Doi:10.1029/2003WR002456, 2004.

[14] S. Grimaldi, and F. Serinaldi, "Asymmetric Copula in Multivariate Flood Frequency Analysis. Adv. Water Resources, 29, pp1115-1167, 2006.

[15] L. Zhang, and V. P. Singh, “Bivariate Flood Frequency Analysis using the Copula Method., Journal of Hydrologic Engineering, ASCE, 11, pp150-164, 2006.

[16] A. Sklar, “Fonction de re’partition a’ n dimensions et leurs marges,” Publications de L’Institute de Statistique, Universite’ de Paris, 8 pp229-231, 1959

[17] R. B. Nelson, “An Introduction to Copulas, Springer, NY, NY, 1999.

19 AIChE Spring Meeting
15TH GLOBAL CONGRESS ON PROCESS SAFETY
NEW ORLEANS • 2019

A Non-Stationary Approach to Conducting Site-Specific Integrative Risk Management Assessments at Industrial Facilities at Risk from Extreme Weather Events

Harri K. Kytomaa, Ph.D., P.E., CFEI
Exponent, Inc.
9 Strathmore Rd., Natick, MA 01760
hkytomaa@exponent.com

Paul Boehm[1], John Osteraas[2], Bob Haddad[3], Josh Hacker[4], Lindsey Gilman[5], Ezra Jampole[6], Patrick Murphy[7], Shukri Souri[8]

(1) Exponent, Inc., Maynard, MA, pboehm@exponent.com, (2) Exponent, Inc., Menlo Park, CA, josteraas@exponent.com, (3) Exponent, Inc., Menlo Park, CA, bhaddad@exponent.com, (4) Jupiter Intelligence, Boulder, CO, josh.hacker@jupiterintel.com, (5) Exponent Inc., Natick, MA, lgilman@exponent.com, (6) Exponent, Inc., New York, NY, ejampole@exponent.com, (7) Exponent, Inc., New York, NY pmurphy@exponent.com, (8) Exponent, Inc., New York, NY ssouri@exponent.com

Prepared for Presentation at
American Institute of Chemical Engineers
2019 Spring Meeting and 15th Global Congress on Process Safety
New Orleans, LA
March 31 – April 3, 2019

A Non-Stationary Approach to Conducting Site-Specific Integrative Risk Management Assessments at Industrial Facilities at Risk from Extreme Weather Events

Harri K. Kytomaa, Ph.D., P.E., CFEI
Exponent, Inc.
9 Strathmore Rd., Natick, MA 01760
hkytomaa@exponent.com

Paul Boehm[1], John Osteraas[2], Bob Haddad[3], Josh Hacker[4], Lindsey Gilman[5], Ezra Jampole[6], Patrick Murphy[7], Shukri Souri[8]
(1) Exponent, Inc., Maynard, MA (2) Exponent, Inc., Menlo Park, CA, (3) Exponent, Inc., Menlo Park, CA, (4) Jupiter Intelligence, Boulder, CO, (5) Exponent Inc., Natick, MA, (6) Exponent, Inc., New York, NY, (7) Exponent, Inc., New York, NY (8) Exponent, Inc., New York, NY

Keywords: Extreme Weather Events, Risk Management

Abstract

The physical forces and environmental stressors that occur during extreme weather events can place facilities at risk for multiple infrastructure failures, loss of operations, product loss, and major impactful chemical releases, all of which affect directly a company's bottom line. Hurricane Harvey (2017) resulted in over 100 such failures and chemical releases. Going forward, non-static (non-stationary) risk management approaches, wherein risk predictions incorporate and account for evolving environmental factors such as continuous sea level rise, will allow us to more accurately predict storm surge flooding as a function of time and provide more realistic long-term (decades) predictions to assist in actionable planning. An integrated three-part approach to assessing risk of infrastructure damage and chemical releases and the business and legal consequences therefrom will be presented. This approach consists of developing: a) temporally variant and spatially localized probabilistic predictions of flooding and forces related to flooding (FloodScore^{TM}) with unprecedented resolution; b) detailed impact predictions on facility infrastructure and structural and supporting elements thereof based on these predictions; and c) a quantitative means of scoring the environmental/financial risk and consequences of chemicals releases as derived from (b). This integrated approach, which assesses risk of losses in the near term and out to 50 years, includes the assessment of ecological and human impact levels and provides actionable information for resiliency and risk mitigation planning.

1 Introduction

Recent extreme weather events have had profound social and economic impacts in the United States. In 2018 alone, there were 14 weather and climate events that each resulted in losses exceeding $1 billion, and the National Oceanic and Atmospheric Administration (NOAA) estimated the total losses in 2018 to be $91 billion. [1] The year 2018 is ranked as the fourth highest in total cost, behind years 2017, 2005, and 2012. Additionally, when evaluating data from 1980-2018, the annual average number of events is 6.2 (Consumer Price Index (CPI)-adjusted), while the annual average from 2014-2018 is 12.6 (CPI-adjusted). While the growing frequency of extreme weather event occurrence is still debatable, experts agree that climate change will only compound the already increasing severity of tropical cyclones, coastal flooding, and wildfires. [1] It is therefore important for corporations and government entities alike to consider a changing, non-stationary set of conditions when assessing and mitigating the risk of infrastructure destruction, business interruption, and consequences of chemical releases resulting from extreme weather events.

Weather-related vulnerabilities to multiple sectors, including the energy sector [2], continue to increase. In response, companies typically adopt one of three approaches to extreme-weather-related risk mitigation. The most common approach builds reactive response plans based on static or stationary data such as historical flood maps and information from past events. [3] While flood maps can provide valuable historical insight, they do not account for other factors such as sea level rise and associated increases in the severity of storm surges that are expected over the next thirty years. A more progressive risk management approach leverages dynamic sets of data that account for both the changing environment and the accelerated pace of data availability. Dynamic data sets include information on rising sea levels, rising ocean surface water temperatures, increasing severity of storms, increasing forces on facilities, and changes in the land and habitat. The third and most proactive approach to extreme-weather-related risk management focuses on using non-stationary data in a longer-term planning horizon. Companies who subscribe to this approach plan for weather-related occurrences with an agreed-upon risk tolerance and leverage probabilistic predictions to build risk mitigation and resiliency around their facilities. This proactive assessment and mitigation approach can help inform not only the risk management of existing facilities, but also the location of new and future facilities, design specifications, and the degree of resiliency that should be built into the design.

Companies wishing to more rigorously assess and mitigate extreme-weather-related risk can benefit from an approach wherein detailed weather, engineering, environmental, and health analyses are integrated into a systematic methodology. Sole reliance on information from past occurrences and use of historical data sets can inadvertently result in plans that are ill-equipped to withstand the future environment. Leveraging dynamic data sets instead, which account for future changes, can greatly improve risk management outcomes. This improvement is especially enhanced when proactive engineering analyses that examine failure modes resulting from extreme weather forces are coupled with assessments of environmental and health risks and consequences of potential chemical releases. To underscore this point, during the 2017 weather event known as Hurricane Harvey, over 100 sites released hazardous pollutants. [4]

An integrated three-part approach to assessing risk of infrastructure damage and chemical releases and the business and legal consequences therefrom consists of developing: a) temporally variant and spatially localized probabilistic predictions of flooding and forces related to flooding (FloodScore™) with unprecedented resolution; b) detailed impact predictions on facility infrastructure and structural and supporting elements thereof based on these predictions; and c) a quantitative means of scoring the environmental/financial risk and consequences of chemicals releases from these predictions. This integrated and localized approach to determining facility-level asset vulnerabilities, quantifying potential impacts, identifying risk management actions, and implementing risk transfer strategies provides actionable information for resiliency and risk mitigation planning.

2 Jupiter's Prediction of Flooding from Extreme Weather Events

As presented in Section 1, the first step in assessing infrastructure risk and potential damage is to develop temporally variant and spatially localized predictions of flooding related to an extreme weather event. In this section, the development of Jupiter's methodology and modeling workstream is presented to illustrate the real-time (operational) prediction of events that appear imminent.

2.1 Methodology

By definition, predicting compound extreme events requires modeling multiple hazards that can contribute to a peril. In coastal zones, the canonical compound event is a storm surge combined with heavy rain. Often antecedent conditions, such as nearly-to-completely saturated soils or a high water table, can contribute as well. While one or the other may not be extreme in isolation, together they can produce an impactful flood.

Predicting floods is accomplished through modeling each of the physical factors that can lead to flooding, and their interactions with the natural and built environment. Those factors combine nonlinearly. The best approach to modeling flooding where multiple factors are involved is to couple the models in flood simulation and forecasting. Here, a hydraulic model forced by coastal ocean and precipitation predictions ultimately provides the flooding prediction. The coastal ocean model, forced by far-field ocean lateral boundary conditions and a bias-corrected wind prediction, is the basis for surge and provides tailwater (downstream) boundary conditions for the hydraulic model. A groundwater model and a land hydrology model provide the over-land sources and sinks that constrain the hydraulics, and a high-resolution meteorological model provides precipitation. Figure 1 gives a schematic of Jupiter's FloodScore Operations (FSO) modeling system release for predicting floods up to five days in advance. Some details are given below.

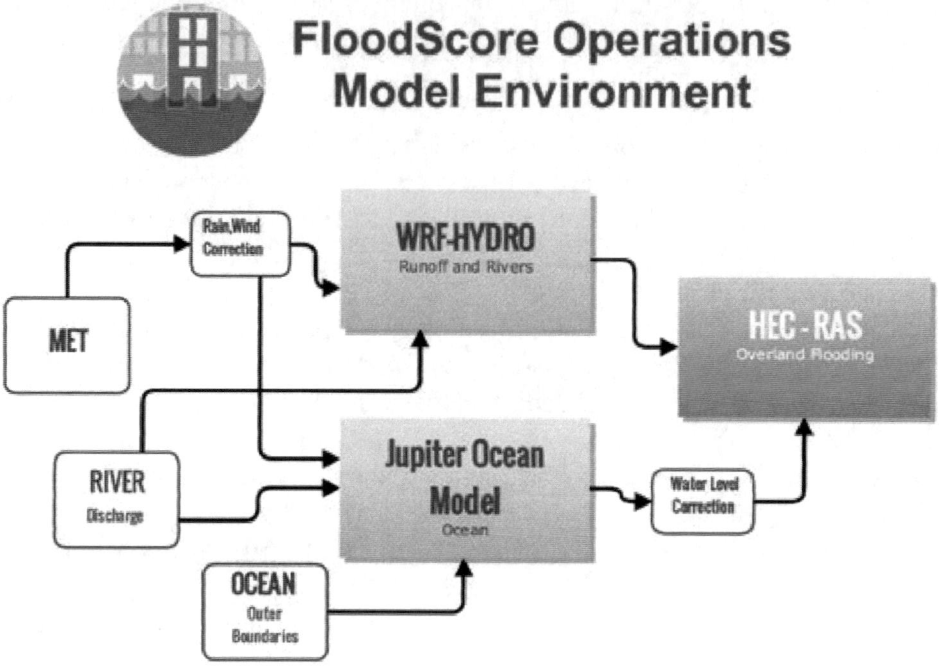

Figure 1. FSO Diagram.

2.2 Jupiter's Model Workstream

Ensemble techniques lead to a probabilistic view of forecasts from zero to 5 days. A multi-model ensemble of 43 global weather forecasts, downscaled to provide finer-scale details in wind, temperature, and rainfall provide boundary conditions to a coastal ocean model (hydrodynamic) as well as a hydrologic and hydraulic (H&H) modeling framework. The ensemble of coastal ocean model forecasts provides a sample from the distribution of forecast water levels along the coast and over the coastal flood plain. A limited set of H&H model executions, which include rainfall from the downscaled atmospheric forecasts, provide bounds on the probabilities of flood depths and velocities, given that the surge or rain levels are sufficient to lead to flooding.

The atmospheric forcing (primarily rain and winds) result from dynamically downscaling global weather forecasts. The Weather Research and Forecasting (WRF) [5] model is the state-of-the-art community regional atmospheric model and is suitable for deriving 1-km gridded forecasts from global forecasts models that range from approximately 9 to 20 km grid spacing. Use of the WRF model allows for regional optimization; the resulting rain and winds are further bias-corrected via a recursive algorithm that minimizes the forecast errors compared to observations, over the recent history of forecasts. The output forces both the hydrodynamic and H&H models.

The Jupiter Ocean Model, based on two open-source ocean models, provides currents, water elevations, and other oceanic variables to indicate the potential for flooding. It is an advanced free-surface, terrain-following, primitive equation ocean model. The model includes wetting and drying in the coastal plain and the ability to accept river and sewer discharges dynamically, enabling accurate simulation of estuarine environments. A global ocean model such as executed by Mercator Ocean (http://marine.copernicus.eu) provides initial and lateral boundary conditions; multiple global atmospheric models provide ocean surface boundary conditions, and a hydrologic and river routing model provides upstream boundary conditions for both the Jupiter Ocean Model and the over-land H&H modeling framework. The Jupiter Ocean Model is currently executed at 50-m grid spacing or smaller, twice daily (0000 UTC and 1200 UTC), with each forecast corresponding to a unique atmospheric forecast from the 43-member ensemble. Resulting water levels are corrected based on the recent history of errors measured by comparison against available water level gages. The forecasts from the ensemble of model runs are sub-selected to create three flooding scenarios for the next 5 days: a low-potential-impact (5%), most-probable-impact (50%), and a high-potential-impact (95%). These three timeseries of water levels are then provided as downstream (tailwater) boundary conditions to force a hydraulic model to predict the water level and velocity over ground.

Various approaches to hydrology are available on Jupiter's platform, and the Hydrologic Engineering Center's River Analysis System (HEC-RAS) is designed to perform one and two-dimensional hydraulic calculations for a full network of natural and constructed channels at DEM-scale. The hydraulic model also incorporates run-off from a hydrology model, which is currently a regionally optimized implementation of the National Water Model WRF-Hydro [6] configuration.

HEC-RAS is triggered for operation in a scalable cloud-based system when the hydrodynamic forecast water level is near a level that could flood over land. Once the model is triggered, the output from HEC-RAS includes: water depth above ground, water surface elevation, and velocity for each of the flood scenarios.

3 Impact Prediction and Risk Evaluation through Fragility Curves

The output from Jupiter's flooding prediction model of water depth and velocity can be used in combination with fragility curves to estimate the risk and impact of various flooding scenarios. Conditional fragility relations, or fragility curves, relate the probability of a damage state (DM) to an engineering demand parameter (EDP), or directly to a hazard intensity measure (IM). [7,8] Damage states are typically defined to capture repair or disruption details. Figure 2 shows example fragility curves relating water height to the functional condition of a transformer. The fragility curves describe the probability of a given damage state as a function of water height. The three damage states are based on the following repair conditions: (1) rust on the transformer not requiring immediate repair for operation; (2) partial repair required for operation; and (3) permanent replacement required for operation. Figure 2a shows, given a water height, the conditional probability of exceeding a damage state; and Figure 2b shows the conditional probability of experiencing a particular damage state.

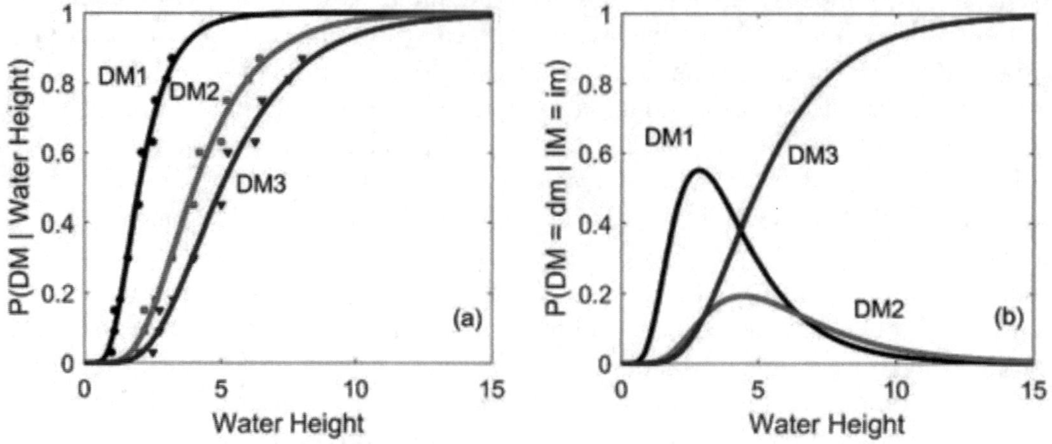

Figure 2. Example fragility relationships for an electrical box for three damage states conditioned on water height: a) shows the conditional probability of exceeding a damage state; and b) shows the conditional probability of experiencing a particular damage state.

For assessing the risk of damage to critical infrastructure and equipment during flood events, it can be necessary to use multiple fragility curves in series. For example, a piece of electrical equipment or a tank containing a hazardous substance may be behind a levee, in which case fragility curves must be used to estimate the probability of levee failure or overtopping, followed by fragility curves to estimate the probability of damage states to the electrical equipment or rupture of the tank, given that water has breached the levee.

Fragility curves for electrical equipment or hazardous substance tanks may be constructed using multiple intensity measures because of the numerous intensity measures that relate to damage states. These include: water depth; velocity on arrival, duration of submersion; and wind speed. Multiple intensity measures can be incorporated into fragility curves by using multinomial logistic regressions. [9] Important additional considerations for developing fragility curves for water-sensitive equipment include whether there is freshwater or saltwater, the nature of flood-borne silt, debris and contaminants, and if the equipment is energized or de-energized when the water arrives. In many cases when flooding is expected, electrical equipment is de-energized in order to prevent catastrophic damage to the equipment including downstream and upstream effects.

Development of detailed impact predictions on electrical infrastructure requires an understanding of the multifarious effects of flooding on individual items of electrical equipment, as well as an understanding of how each item interacts with the facility-wide electrical system and how it in turn interacts with the facility as a whole.

The process for creating fragility curves for equipment in water events can be challenging, namely because of the numerous relevant intensity measures, and because they must be created for each piece of equipment. They are typically created through observations after events, such as for electrical substation equipment performance during earthquakes [10] prior experience with such equipment, or laboratory testing of the equipment at numerous intensities. In case of the latter, the probability of a damage state at each intensity is observed, then a curve is fit through the data (i.e. a lognormal cumulative distribution function using the method of moments) for each damage state. For larger infrastructure, such as a protective levee where laboratory test data does not exist and is impractical to perform, the fragility curves may be determined through numerical simulation.

4 Evaluation of Environmental Risk and Chemical Release Consequences

Section 3 presented the concept of site vulnerability as a function of flood levels. Incorporating an understanding of the human health and environmental risks associated with specific outcomes from the vulnerability analysis provides greater insights regarding the types and the magnitude of potential risks (i.e., the risk profile) arising from the identified vulnerability.

This risk profile can be further explored by assessing specific elements of the nature of the release including factors describing the release (type of chemicals released, volume released, catastrophic vs. slow release, time to leak detection, etc.) as well as factors describing both the potential human health exposure and potential ecological exposure. These include the human health and ecological toxicity of the released material, the mobility of the released material[1], and scenarios describing the various types of exposure (e.g., adult v. child, resident vs. worker, aquatic wildlife vs. terrestrial wildlife, etc.). The final element to consider is the potential financial exposure; including costs of responding to the release, costs of remediation, and any other litigation and/or civil liabilities associated with the release.

Development of the various release, exposure, and financial factors that are related to the vulnerability is followed by a process whereby these factors are scored on a relative ranking scale (Figure 3). For example, the likelihood of a release may be characterized as negligible (0%), low (50%), and high (100%). Similarly, the size of the release can be characterized as a fraction of the maximum: negligible (0%), medium (50%), and high (100%). These release factors are then integrated into a relative ranking scale. The same is done to describe and score relative exposure/toxicity scales and the degree of financial impact (e.g., small scale vs. large scale remediation, statute-driven litigation, etc.). For a given set of conditions identified at a given time, an estimate of relative environmental and financial risk can be determined by combining the scores of the release factors, the exposure/toxicity factors, and the financial exposure factors.

[1] The mobility of released material is also informed by the depth, velocity and direction of the flood water at the time of the release.

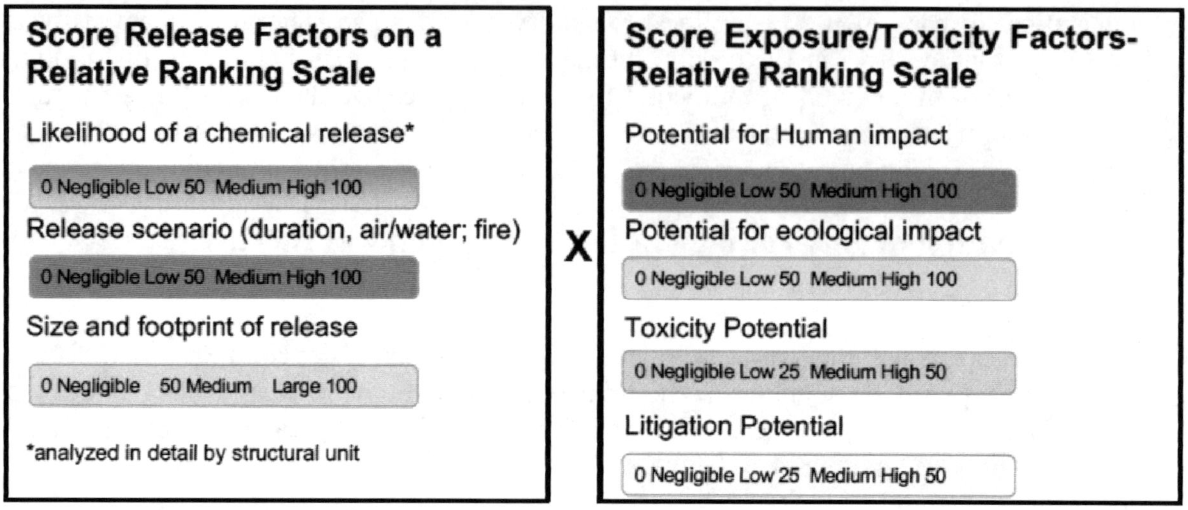

Figure 3. Development of Relative Risk Factors.

Finally, combining the relative risk values determined for a given scenario with the likelihood of a specific outcome from the vulnerability analysis allows for the creation of a matrix which describes the change in risk profile as a function of exposure condition (Figure 4). Further, incorporation of non-stationary data into the vulnerability analysis allows us to determine and assess the evolution of the risk profile at a given site.

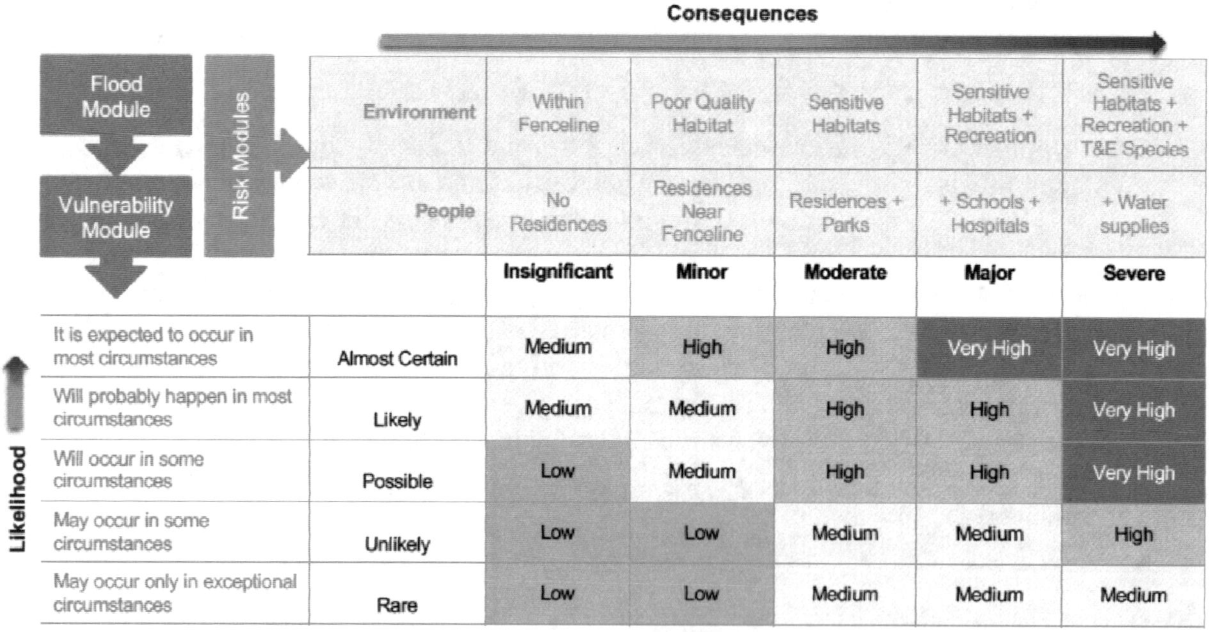

Figure 4. Example Risk Profile.

5 Extreme Weather Impacts and Failures

The physical forces and environmental stressors that occur during extreme weather events can place facilities at risk for multiple infrastructure failures, loss of operations, product loss, and major impactful chemical releases, all of which affect directly a company's bottom line. Some examples of recent impacts from extreme weather events is presented and discussed in this section.

5.1 Levee/Flood Wall Breach

Flood protection systems, such as the levees and floodwalls in New Orleans, can provide the illusion of safety, while risk assessments reveal that there is substantial risk. Figure 5 illustrates a situation in which the storm surge from a hurricane breached floodwalls and overtopped levees, flooding a processing plant at which chemicals were stored in large tanks. A tank failed, releasing chemicals that spread beyond the property line. A risk assessment before the event would have identified that: relatively frequent flood events were enough to breach the levee and flood walls; only a small amount of flood water and debris contacting the tank would likely cause tank failure; and tank failure with even minimal flood water height would transport the tank contents off the property and into the community before a response team could be deployed to mitigate the situation. This type of chemical release could be prevented by any number of measures, such as (1) reducing the probability of failure of the floodwalls or of overtopping the levee, (2) reducing the probability of tank failure, given flooding, or (3) relocating the facility to somewhere more remote.

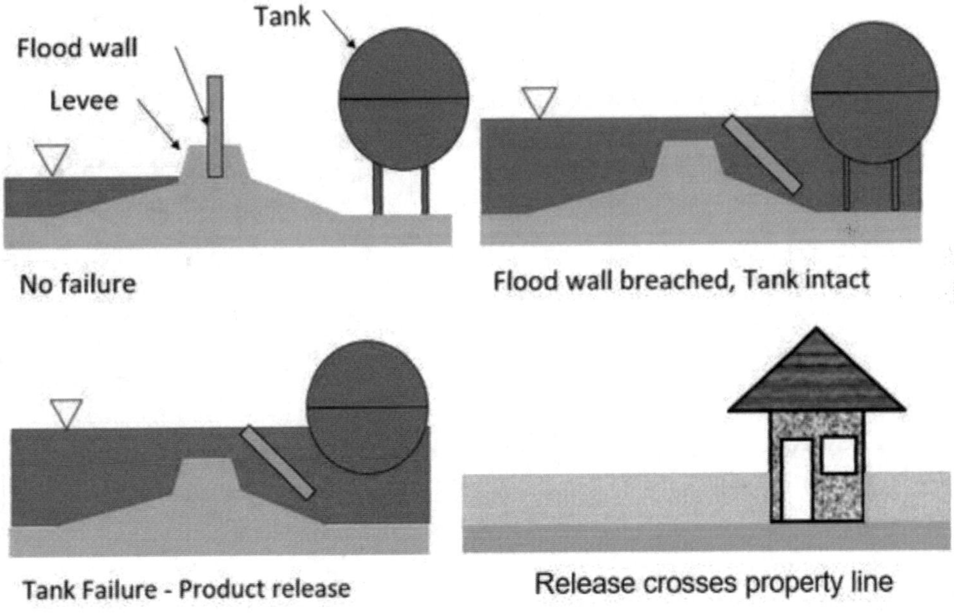

Figure 5. Levee/flood wall breach and subsequent chemical release.

5.2 Generator Dual Tank Air Vent Failure

Failures suffered during past events stress the urgency of risk assessments to identify and quantify potentially unknown hazards. For example, Superstorm Sandy caused flooding in a metropolitan area and an unexpected power outage in a building. As depicted in Figure 6, emergency generators and a suspended fuel tank were located at the top of the building to eliminate the risk of water damage and maintain electricity during a weather event; however, the tank was supplied by pumping fuel from a tank in the basement. Water entered the basement fuel tank through a vent located at the elevation of the "100-year flood", that ultimately was barely below the actual flood water level. When water was pumped to the emergency generators, they malfunctioned causing a power outage. Relocating a vent is among the easiest measures available for preventing flood intrusion. With a proper risk assessment, the risk could have been identified, and the vent moved to even the "10,000-year flood" elevation at minimal cost. Additionally, the notion of a "100-year flood" is only meaningful as an indication of the past now that water levels are non-stationary. It does not provide forward looking probabilities for future flooding events. The present framework is designed to specifically address this inconsistency between the past, present and the future.

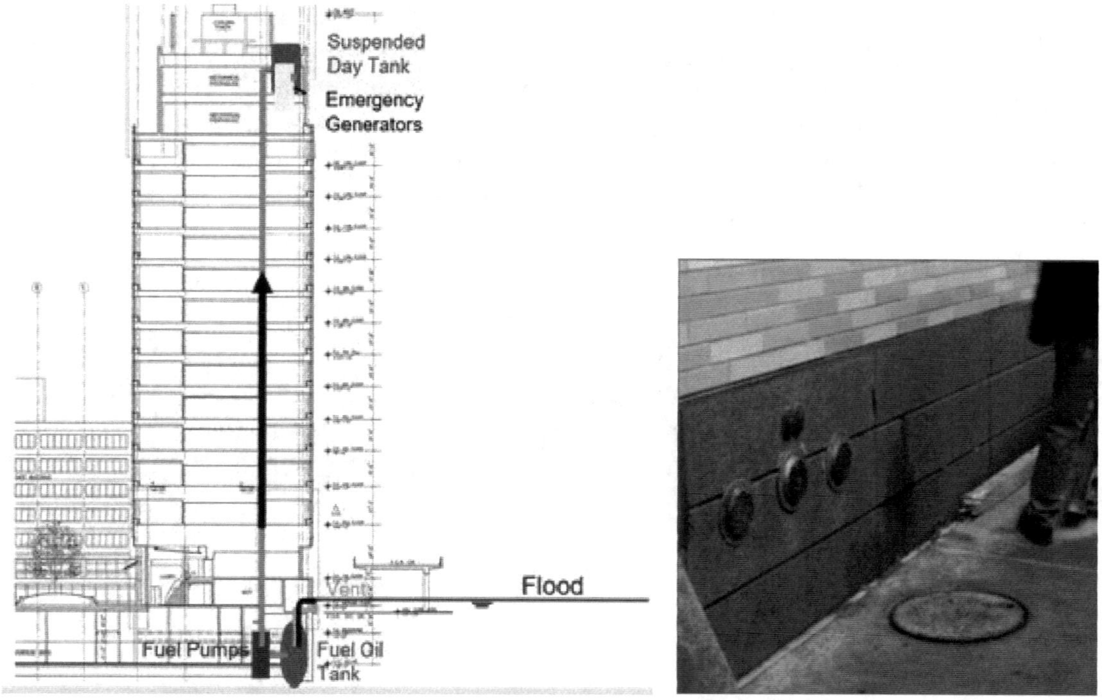

Figure 6. Schematic diagram of water intrusion through a vent, and pumping to emergency generators.

5.3 Electrical Substation Failure

One of many dramatically visible effects of "Superstorm" Sandy, to say nothing of the human costs, was the darkening of the skyline of most of lower Manhattan, which is shown in Figure 7. [11] Flooding at two Con Edison 345 kV transmission substations at the East River Complex (adjacent to the East River at East 14th Street) accounted for the outages of ten electrical distribution networks in lower Manhattan. Separately, three more lower Manhattan Con Edison networks shut down during the storm: two networks were preemptively de-energized by Con Edison and one network shutdown due to flooding at a substation near the East River in the Seaport area. [12, 13, 14]

Figure 7. Aerial photograph published on the cover of New York Magazine of Manhattan during Sandy power outage.

The peak water level as measured at the Battery (at the southern tip of lower Manhattan) was 14.06 feet above mean lower water level ("MLLW"), two feet higher than the maximum National Weather Service forecast and exceeding by several feet any known historical flood level. [15] The peak water level observed in the vicinity of the Con Edison East River Complex was 13.8 feet (MLLW), similarly exceeding forecasts, resulting in several feet of flooding at street level, as shown in Figure 8. [16] Flood waters overtopped or dislodged temporary protective measures at the East River Complex, such as sandbags and water-filled polyethylene berms, that had been installed up to an elevation of 13.6 feet (MLLW) based on forecasts and historical observations. [12, 13]

Figure 8. Photo published by Con Edison showing street-level flooding during Sandy in the vicinity of the Con Edison East River Complex.

Con Edison reported that the typical elevations for the top of foundations at the East River Complex are at 10.6 feet (MLLW) and the lowest elevation of critical equipment at 11.2 feet (MWWL). [12, 13] Figure 9 shows a datum diagram published by Con Edison indicating the elevations of the forecast and observed flooding levels as compared to the general equipment elevation at the East River Complex. [17]

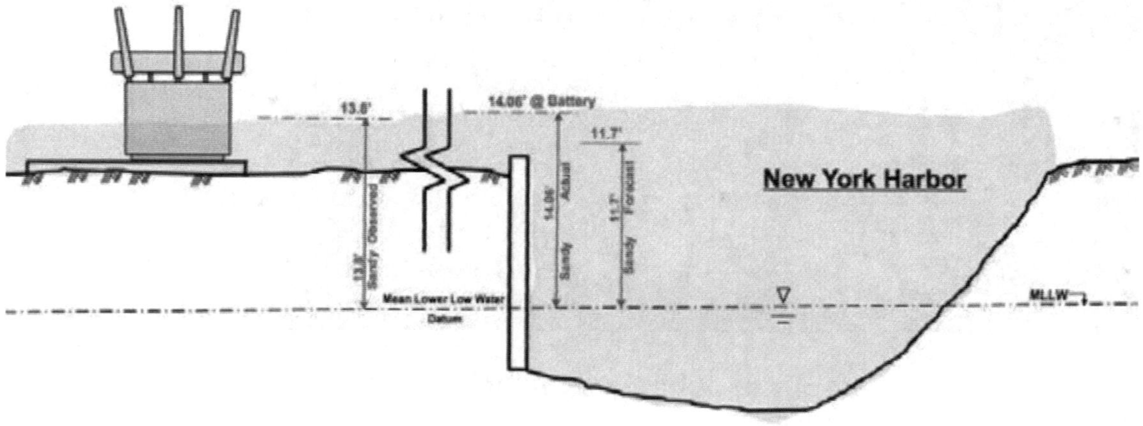

Figure 9. Datum diagram published by Con Edison comparing the 11.7 feet (MLLW) forecast flood height at the Battery to the observed 13.8 feet (MLLW) flooding height observed in the vicinity of the Con Edison East River Complex during Sandy.

Con Edison reported that the outages at the East River Complex were due to flooding of critical components of the low-voltage protective relay system as well as components of the system that maintains flow of pressurized dielectric oil for insulating feeders. [12, 13] Published photographs, shown in Figure 10 [18] and Figure 11 [19] demonstrate the effects of flood waters on relay equipment. Also during Sandy, failures at the East River Complex produced arcing and explosions that were visible from across the East River in Brooklyn. A sequence of still images from a video of the event is shown in Figure 12. [20]

Figure 10. Photo published by Con Edison of protective relay equipment damaged during flooding at East River Complex during Sandy.

Figure 11. Photo published by Con Edison of protective relay equipment damaged by flooding during Sandy at Con Edison facilities. Yellow arrow indicates horizontal water mark suggesting maximum flood water level at this location.

Figure 12. Sequence of still images from a video published on YouTube of an apparent arcing and explosion event at the East River Complex during Sandy.

A typical Con Edison substation layout cross section is shown in Figure 13. [21] The low voltage protective relay and control equipment mainly reside in the Relay House and Control Room and are connected by low voltage cables to terminal boxes on the breakers and other equipment in the facility. The pumping plant typically houses the pumps and associated control system for the pressurized dielectric oil that is fed through oil lines to the transmission and distribution cables.

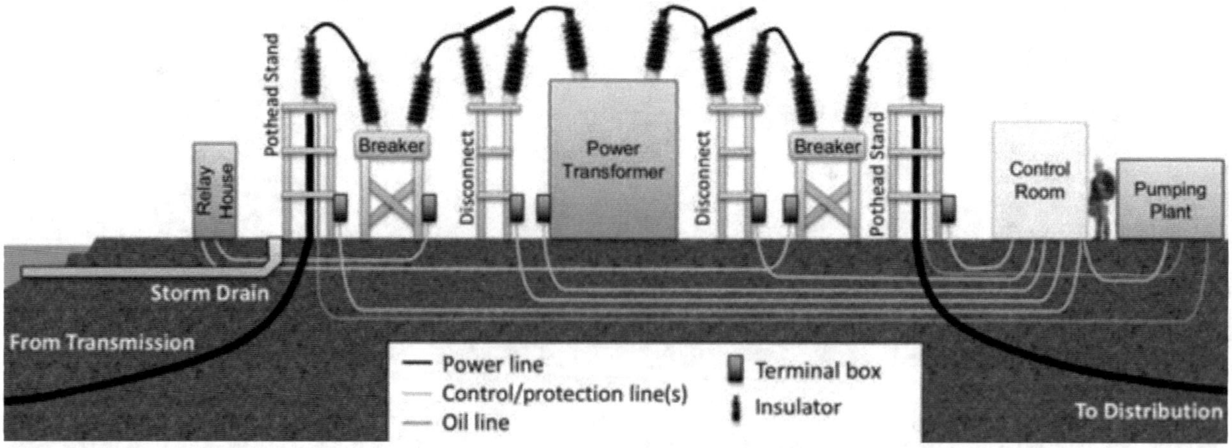

Figure 13. Diagram published by Con Edison showing cross section of typical substation layout.

Protective relay equipment may include electromechanical or electronic relays, cables, current and potential transformers, and metering. Such equipment that is still energized will tend to rapidly fail to perform its intended function when submerged in flood water, unless specifically designed for submersion. For example, if an electromechanical protective relay is submerged while energized, the power supply for the trip coil may short out, resulting in the failure for the circuit breaker to open if a fault occurs. Alternately, if submerged, the relay itself may experience a short at the contacts, causing the circuit breaker to open without an actual fault present.

In the first case (unprotected fault), severe damage can occur to both upstream and downstream transmission and distribution equipment and facilities if a fault occurs and is not detected. Resulting damage can include system disruption due to voltage drop, thermal stress and mechanical stress on equipment due to high currents and associated magnetic fields, and arcing, explosions, and burning due to, for example, failure of insulation. In the second case (open breaker without a fault present), an open breaker may cause other feeders to unnecessarily carry more load or cause other system disruptions.

Protective relay equipment that has been pre-emptively shut down will also tend to be damaged when submerged in flood water, unless specifically designed for submersion. A detailed technical evaluation along industry guidelines is generally required when assessing how flood waters may damage electrical equipment. [22] For instance, the mechanical operation of breakers and switches can be impaired by corrosion, deposited silt, and the removal of lubricants. The dielectric properties of insulating materials and insulators will tend to degrade, depending on the insulating materials and contaminants in the flood water. It is possible for certain electromechanical equipment, breakers, and motors to be reconditioned by properly trained personnel in close consultation with the manufacturer. However, most other electrical equipment such as transformers, batteries, communications systems and electronic protective relays, breakers, and meters are in general severely damaged due to flooding and require replacement.

Switchgear are key components of an electrical power distribution system at an industrial facility. Components of a typical switchgear include the chassis, conductive busbars, circuit breaker, sensors, and protective relay. Electronic protection relays are essentially special-purpose computers, with built in display and touchpad, and which receive input from current and voltage sensors and provide output to activate the circuit breaker. A fragility curve for this type of equipment may include two separate curves: if the switchgear is energized or if it is de-energized. In the case of an energized switchgear, one can assume catastrophic damage if energized high voltage terminals are exposed to floodwater. In the case of de-energized switchgear, depending on the nature of the flood water, a degree of damage would be expected to occur to the chassis, busbars, and other mechanical parts of the circuit breaker with the possibility to recondition these parts in close consultation with the manufacturer. Once the sensors are submerged, however, there is an increased risk of complete damage to the sensors, low-voltage cables, and electronic protection relay due to the potential for flood water to infiltrate the electrical connectors and cables that connect these components.

As demonstrated by the Con Edison experience during Sandy, the failure of energized equipment caused catastrophic damage both to equipment that was directly submerged, as well as damage to upstream and downstream equipment that may not have been directly submerged. The overall damage at the East River Complex was sufficient to cause the outage of ten electrical distribution networks for several days. A similar phenomenon of cascading effects can be expected at industrial sites that experience severe flooding. For example, failure of electrical distribution equipment or cables may stop the operation of powered ventilation equipment in areas where a hazardous classification exists. In this scenario, explosive vapors and/or dust may accumulate, increasing the hazardous classification of the area, and increasing the risk of an explosion.

6 Conclusion

In the United States, recent extreme weather events have resulted in profound social and economic impacts. These events highlight the importance for corporations and government entities alike to consider a changing, non-stationary set of conditions when assessing and mitigating the risk of infrastructure destruction, business interruption, and consequences of chemical releases resulting from extreme weather events. Weather-related vulnerabilities to multiple sectors, including the energy sector [2], continue to increase. Companies wishing to more rigorously assess and mitigate extreme-weather-related risk can benefit from an approach wherein detailed weather, engineering, environmental, and health analyses are integrated into a systematic methodology.

In this paper, an integrated three-part approach to assessing risk of infrastructure damage and chemical releases and the resulting business and legal consequences was presented. This includes a non-stationary methodology to quantify flooding and its related forces, impact predictions on infrastructure based on this flooding model, and a method to quantify environmental and financial risk associated with these predictions. This localized approach to determining facility-level asset vulnerabilities, quantifying potential impacts, identifying risk management actions, and implementing risk transfer strategies provides actionable information for resiliency and risk mitigation planning.

7 References

[1] NOAA National Centers for Environmental Information (NCEI) U.S. Billion-Dollar Weather and Climate Disasters (2019). https://www.ncdc.noaa.gov/billions/, Accessed on February 11, 2019.

[2] U.S. Department of Energy, "U. S. Energy Sector Vulnerabilities to Climate Change and Extreme Weather," July 2013, DOE/PI-0013.

[3] U.S. Department of Homeland Security, Office of Inspector General, "FEMA Needs to Improve Management of Its Flood Mapping Programs," September 2017, OIG-17-110.

[4] F. Bajak and L.Olsen, Silent Spills, Houston Chronicle, https://www.houstonchronicle.com/news/houston-texas/houston/article/In-Houston-and-beyond-Harvey-s-spills-leave-a-12771237.php, accessed February 15, 2019.

[5] Skamarock, W. C., and Coauthors, 2008: *A Description of the Advanced Research WRF Version 3*. NCAR Technical Note NCAR/TN-475+STR, doi:10.5065/D68S4MVH.

[6] Gochis, D. J., M. Barlage, A. Dugger, K. FitzGerald, L. Karsten, M. McAllister, J. McCreight, J. Mills, A. RafieeiNasab, L. Read, K. Sampson, D. Yates, W. Yu, (2018). The WRF-Hydro® modeling system technical description, (Version 5.0). NCAR Technical Note. 107 pages. Available online athttps://ral.ucar.edu/sites/default/files/public/WRF-HydroV5 Technical Description. Source Code DOI:10.5065/D6J38RBJ.

[7] Deierlein, G.G., 2004, June. Overview of a comprehensive framework for earthquake performance assessment. In Performance-Based Seismic Design Concepts and Implementation, Proceedings of an International Workshop (pp. 15-26).

[8] Krawinkler, H. and Miranda, E., 2004. Performance-based earthquake engineering. Earthquake engineering: from engineering seismology to performance-based engineering, 9, pp.1-9.

[9] Jampole E, Hunt J. Structural performance factors and building damage following the 19 September 2017 Puebla, Mexico Earthquake. 17th U.S.-Japan-New Zealand Workshop on the Improvement of Structural Engineering and Resilience, Applied Technology Council, Queenstown, New Zealand, 2018.

[10] Development of an Electrical Substation Equipment Performance Database for Evaluation of Equipment Fragilities (PG&E).pdf, pg. 51.

[11] I. Baan. The City and the Storm, New York Magazine, November 12, 2012. Available at http://nymag.com/nymag/letters/hurricane-sandy-editors-letter-2012-11/ Accessed on February 13, 2019.

[12] Consolidated Edison Company of New York, Inc. Report on 2012 Performance under Electric Service Reliability Performance Mechanism. April 1, 2013. Available at http://documents.dps.ny.gov/public/Common/ViewDoc.aspx?DocRefId=%7BF996CF89-4A2C-4409-8C62-9E69AF5A9B19%7D Accessed on February 13, 2019.

[13] Consolidated Edison Company of New York, Inc. Report on Preparation and System Restoration Performance Sandy October 29 through November 12, 2012. January 11, 2013. Available at http://documents.dps.ny.gov/search/Home/ViewDoc/Find?id=%7B2D1BF3D9-95DC-4C2D-9F24-6DE65926275B%7D&ext=pdf Accessed on February 13, 2019.

[14] Con Edison, Inc., Edison Electric Institute Finance Conference, Southern Manhattan Outage Map November 11-13, 2012. Available at http://phx.corporate-

ir.net/External.File?item=UGFyZW50SUQ9NDg1NDc3fENoaWxkSUQ9NTM2MjYwfFR5cGU9MQ==&t=1 Accessed on February 13, 2019.

[15] A. Freedman. Hurricane Sandy Set to Deliver Massive Blow to East Coast. October 28th, 2012. Available at https://www.climatecentral.org/news/hurricane-sandy-to-deliver-historic-blow-to-mid-atlantic-northeast-15176 Accessed on February 13, 2019.

[16] K. Field. Con Ed Nearing Completion of $1B Post-Sandy Fortification Effort. T&D World, Nov 21, 2017. Available at https://www.tdworld.com/substations/con-ed-nearing-completion-1b-post-sandy-fortification-effort Accessed on February 13, 2019.

[17] F. A. Cuomo, M. Mangan, J. O'Hagan. Consolidated Edison Company of New York Steam Operations Storm Hardening: Impacts and Solutions. Available at https://www.districtenergy.org/HigherLogic/System/DownloadDocumentFile.ashx?DocumentFileKey=4589eb65-3ffd-2761-1fa3-687615efbcae. Accessed on February 13, 2019.

[18] K. Field. Con Ed Nearing Completion of $1B Post-Sandy Fortification Effort. T&D World, Nov 21, 2017. Available at https://www.tdworld.com/substations/con-ed-nearing-completion-1b-post-sandy-fortification-effort. Accessed on February 13, 2019.

[19] P. Di Lillo and A. Martello. Con Edison E. 13th Street Substation - ABB PASS Business Case. Available at http://aieq.net/wp-content/uploads/2018/03/panel_3_PatDilillo_and_Andrea_Martello_Con_Edison.pdf Accessed on February 13, 2019.

[20] J. Mattiuzzi. Hurricane Sandy Causes Huge Explosion NYC at Con Ed. Available at https://www.youtube.com/watch?v=3K2OWYG61GY Accessed on February 13, 2019.

[21] Con Edison Storm Hardening & Resiliency Collaborative, Substation Operations Update, May 20, 2014. Available at http://documents.dps.ny.gov/public/Common/ViewDoc.aspx?DocRefId=%7BC73D6658-896D-4AE0-B909-ADB468421421%7D Accessed on February 13, 2019.

[22] Evaluating Water-Damaged Electrical Equipment, National Electrical Manufacturers Association. August 25, 2016. Available at https://www.nema.org/Standards/Pages/Evaluating-Water-Damaged-Electrical-Equipment.aspx. Accessed on February 13, 2019.

"Yes, You Too Can Be Sued"

or

Legal Liability, Good Engineering Judgment and Adaptation

Mary Ellen Ternes, Esq., FAIChE, FACOEL
Earth & Water Law, LLC
5030 North May, No. 324
Oklahoma City, Oklahoma
Maryellen.ternes@earthandwatergroup.com

NOTE: This paper does not constitute legal advice.
If you have concerns or questions, please consult your attorney

Prepared for Presentation at
American Institute of Chemical Engineers
2019 Spring Meeting and 15th Global Congress on Process Safety
New Orleans, LA
March 31 – April 3, 2019

Legal Liability, Good Engineering Judgment and Adaptation

Mary Ellen Ternes, Esq., FAIChE, FACOEL
Earth & Water Law, LLC
5030 North May, No. 324
Oklahoma City, Oklahoma
Maryellen.ternes@earthandwatergroup.com

Keywords: Process Safety Management, Good Engineering Judgment, RAGAGEP, Hazard Risk Assessment, Adaptation, Risk Management, Negligence, Legal Liability

Abstract

Hurricane Harvey's devastation resulted in flooding and fire, as well as criminal indictments against Arkema's plant manager and CEO, and related claims against Arkema, Inc. This current litigation is just part of a recent trend in civil litigation, from citizen suits to common law actions, that is redefining causes of action and defenses in the context of climate change. This trend is creating greater concern among engineering and management professionals regarding litigation risk and potential liability arising from individual or managerial actions undertaken on behalf of an employer or client. These concerns are exacerbated by greater ability to forecast anticipated impacts from storm events in hurricane prone areas, which lends support to allegations that such events are foreseeable, and that engineers have a duty to prevent harm resulting from such events. In this context, chemical engineers should read the Chemical Safety Board's Arkema Final Report, especially focusing on the CSB's recommendations for further action by AIChE's Center for Chemical Process Safety, and brush up on the basics of professional liability.

This paper, developed specifically for AIChE members by an AIChE Fellow, senior environmental attorney and Fellow of the American College of Environmental Lawyers, provides a brief overview of legal process and concepts regarding professional liability, including classes of defendants such as plant managers, employees, licensed and unlicensed professionals, and independent contractors, types of civil and criminal claims, standing and other threshold issues, general defenses and immunities, and evidentiary issues at trial.

1. Introduction to Legal Liability and Process

Chemical engineers employed as independent contractors maintaining their professional engineering licenses generally anticipate the possibility that they could be sued individually, whether the allegations are meritorious or not, for professional negligence. On the other hand, chemical engineers working as employees for large and economically viable corporations, working pursuant to the "industrial exemption" without professional engineering licenses, may feel insulated from individual liability, particularly where simply performing their duties within the scope of their employment. However, as recognized by the National Society of Professional Engineers, chemical engineers in both employment scenarios may face potential personal legal liability in certain circumstances. [1] However, it becomes more difficult to fully appreciate the full impact of potential legal liability when employers shift more engineers to independent contractor status with varying commitments to indemnify the engineer should litigation arise from the employer or client's activities, which the engineer supports.

Given fuzzier lines between employer and client, and recognition of the scope of the engineering profession regardless of professional licensing status, those who believe they are flying under the radar as unlicensed chemical engineer employees may find themselves named as a party in a lawsuit. The engineer could be named primarily for strategic reasons, such as the plaintiff's attempt to destroy federal court diversity jurisdiction to remain in state court because state courts are generally recognized as a more favorable venue for plaintiff claims than federal court. Thus, even assuming reliance on reasonably adequate professional liability insurance, engineers should understand the nature of individual liability and how it might arise.

To remove some of the mystery from the legal process and provide perspective regarding how a plant engineer's notes might become Plaintiff's Exhibit 1, this paper first lays a brief foundation describing the process by which legal liability is established using a familiar fact pattern for illustration. The paper then discusses common types of personal legal liability alleged against engineers and recent trends impacting this analysis.

While litigation itself feels quite overwhelming when one is named as a party, the procedural and evidentiary rules are actually quite logical. These rules generally allow the development of proof, for and against, legal claims that is fair to both sides, meaning, to both the plaintiff and the defendant. Thus, conscientious engineers competently performing their work consistent with the standards of the industry and good engineering practice, should gain some confidence in the legal process, especially where they have secured adequate professional liability insurance and possibly, for belt and suspenders, contractual indemnification provisions from their client/employee. Of course, given the pure nature of the beast – directly opposing forces steered by sophisticated and zealous professional advocates, who are individually strangers to the facts and likely science as well, yet hired for their debating skills to develop and strategically position their arguments to their client's best advantage, unafraid to file forests of paper pursuant to strict timelines while complying with complex procedure - the process is both chaotic and prohibitively expensive for most defendants. Thus, even defendants with excellent defenses will often consider settlement of meritless claims.

For purposes of illustration in the discussion below, we will loosely adapt facts from a 1988 New Jersey case,[2] to create this Hypothetical:

> A chemical engineer, completing work for a financially viable chemical manufacturer, performed chemical engineering calculations in the process of his work involving a distillation vessel in operation at the manufacturer's facility. The distillation vessel exploded, and someone was killed. The plaintiff files a wrongful death lawsuit alleging claims of negligence against the company and also the chemical engineer, who are both co-defendants. The plaintiff is represented by plaintiff's attorneys, and the defendants, including the chemical engineer, are represented by the defense attorneys.

2. Civil and Criminal Liability

Legal liability can arise when someone is harmed by the actions of others, though of course, not all harms are actionable. The someone harmed can be a natural person, a corporation, or municipality, or any entity recognized by the law, where this someone can support a claim for discrete and quantifiable damages shown to result from actions of the other parties. These actions of other parties can include intentional actions, or unintentional actions, as well as intentional results or unintentional results. Where the intent is to cause harm to someone else, these claims may fall within the criminal realm. However, when the intent is to simply perform one's professional duties, and harm results, the actor may find themselves party to a civil claim for damages, and only very rarely, named as a criminal defendant.

Civil liability usually results in common law claims for damages to compensate for harm, arising from contract or tort, or penalties arising from failure to comply with administrative regulations. Most relevant to the scope of this paper are claims for damages arising from injury, or penalties for noncompliance with statutory or regulatory requirements. For civil liability, in addition to the requirements for threshold criteria (mootness, ripeness and standing), causation and damages, the burden of proof for civil cases rests with the plaintiff to prove their claims by a "preponderance of the evidence," meaning basically, more than fifty percent, or more likely than not. [3]

Violation of the environmental regulations, such as the CAA §112(r) regulations, codified at 40 C.F.R. Part 68 (the RMP regulation), can result in another type of liability: strict liability for failure to comply with regulatory requirements. "Strict liability" simply means "without fault." Thus, an agency's proof of its claims that the regulation was violated does not require a demonstration of fault or negligence, or failure to comply with any standard of care, but simply that the rule was violated. Most often, the entity subject to regulation is the owner or operator of a facility, and thus the employee engineer is not usually penalized. However, intentional violation of environmental requirements can result in criminal liability, and under the Clean Water Act, negligent violation resulting in death (e.g., release of gasoline into a creek that explodes and kills someone) can result in criminal negligence. [4]

Criminal liability results in indictments followed by trial and convictions, where the prosecutor prevails. Criminal liability may arise from facts involving injury or death from exposure to chemical releases or explosion through criminal negligence or intentional violation of standards which were adopted to prevent such harm. For criminal liability, the burden of proof falls on the prosecutor to prove their claims "beyond a reasonable doubt." This is the highest burden of proof, considered met if the facts allow no other logical explanation. [5] As articulated by the U.S. Supreme Court, this standard requires the prosecutor to demonstrate the absence of

> such doubt as would give rise to a grave uncertainty, raised in your mind by reasons of the unsatisfactory character of the evidence or lack thereof ... What is required is not an absolute or mathematical certainty but a moral certainty. [6]

3. Negligence Liability

One of the most common claims faced by engineers responsible for actions that can impact others is the claim of "negligence." Within the scope of "professional negligence," engineers are considered "professionals" regardless of licensure status (where the scope of the term "professionals" is not to be confused with the more limited scope and protections of the term "Professional Engineer" or "P.E."):

> The engineer is 'one who is trained…in a branch of engineering' that has been described as 'the application of scientific and mathematical principles to practice ends such as the design, manufacture, and operation of efficient and economical structures, machines, processes and systems. [T]he engineer [is] involved in planning, designing, supervising, and maintaining safety standards in connection with the construction of …large structures. [7]

And for this reason, engineers are treated like other design professionals, for example architects, under the law "for the purpose of assessing issues involving their duty of reasonable care and liability for negligent conduct." [7] The most relevant wrongful act alleged against licensed, or unlicensed, engineering professionals is the claim of negligence. The legal claim of negligence falls within a class of common law actions known as "torts," or wrongful acts, or infringements upon the rights of another, which are legally actionable.

Negligence is defined as the "failure to exercise reasonable care." [8] Negligence is "conduct which falls below the standard established by law for the protection of others against unreasonable risk." [8] For negligence to be actionable, the risk must be foreseeable.

> [Negligence] necessarily involves a foreseeable risk, a threatened danger of injury, and conduct unreasonable in proportion to the danger. If one could not reasonably foresee any injury as the result of one's act, or if one's conduct was reasonable in the light of what one could anticipate, there would be no negligence, and no liability. [8]

The duty of care is normally that level of care which would used by a reasonable person. However, where a person has "superior leaning and experience," or "special skills, knowledge or training" such as engineering professionals, then these persons are required to exercise that level of care reasonable for a person with this level of understanding.

For engineers, this professional negligence standard has been explained as follows:

> Generally, individuals 'performing… engineering services are performing professional services, and the law imposes upon such persons the duty to exercise a reasonable degree of skill and care, as determined by the degree of skill and care ordinarily employed by their respective professions under similar conditions and like surrounding circumstances.' This standard pertains to skills relative to problems encountered in …engineering …, which exceed the knowledge of the normal layperson. Note that this standard refers to the minimum level of performance, not the average or highest level of performance. Thus, when an engineer, for example, fails to exercise reasonable care under the same or similar circumstances as 'an engineer of ordinary prudence' would have exercised or

> when he does something contrary to this standard, the tortfeasor engineer may be held liable for professional negligence. [7]

The specific elements of this professional standard can be stated as follows:

> for a person who holds themselves out to the public in a professional capacity is of average ability in the profession and will be presumed in law to (1) possess the requisite degree of learning, skill, and experience ordinarily possessed by similarly-situated professionals in the community, (2) use reasonable and ordinary care and diligence in the exercise of this skill to accomplish the purpose for which the professional is employed and (3) use his or her best judgment. [9]

Thus, for professionals, the standard is one of professional negligence, and the reasonable duty of care that applies is that duty applicable to that specific type of professional of average ability using their best judgment. This duty varies with the degree to which the professional is able to foresee harm, and thus, arguably, the degree to which the potential for the alleged harm falls within the realm of foreseeability afforded the engineer by the scope of his engineering tasks. As recognized by the court in *Sykes* [2]:

> [A]lthough all engineers have a professional obligation to see that the work they do is accurate and in conformance with accepted standards of care, the duty to foresee and prevent a particular risk of harm from materializing should be commensurate with the degree of responsibility which the engineer has agreed to undertake.

4. How it Begins: Pleading Threshold Elements

As an introduction to legal claims generally, know that any legal action, to be sustained in court from its initial filing by the plaintiff of their complaint through to the court's final judgment, must first sufficiently plead (allege) facts supporting the plaintiff's claim as properly within the intent of the United States Constitution, Article III "case or controversy," such that it is properly before the court. These facts must show that the legal claims filed are not yet moot, that they are ripe for legal action, and that the person filing the claim is otherwise is an appropriate claimant that has standing to bring the claim. These threshold criteria must be properly plead through allegation of facts, based on information and belief, which are accepted as true to test the sufficiency of the allegations. [3]

The element of "standing" is often challenged. This element requires the plaintiff to articulate how they have been harmed by the person(s) they have sued in a manner that is legally actionable. The three-part test for this element was set by the United States Supreme Court and requires that the plaintiff has suffered "injury in fact," i.e., a legally protected interest which is both "concrete and particularized, as well as "actual or imminent," causally connected to the defendant's actions, for which the court can provide a remedy. [10]

Usually, the defendant files a motion to dismiss the complaint arguing that these threshold elements have not been properly plead ("failure to state a claim"), or that there is a complete defense (claims are barred, waived, estopped), or perhaps argue a jurisdictional deficiency. If the case is not dismissed at this stage, based on the law alone before the development of proof regarding the underlying facts, then discovery commences.

In our Hypothetical, the plaintiff would have to articulate how the actions or failure to act of the company, the plant manager and the chemical engineer caused their injury resulting in damages that are legally recoverable. For example, if alleging that the deceased died from injuries sustained after being struck by a car, following the explosion of the distillation unit but in a manner not directly caused by the explosion, the plaintiff could not likely sustain their lawsuit due to the lack of causal connection, or at least may lose the case due to the potentially prevailing complete defense of an intervening cause.

The engineer defendant in our Hypothetical would have been consulted initially by defense counsel to get a handle on the facts, weigh the relative merits, risk and difficulty that might factor into mounting a defense and generally develop a basic "story of the case." However, discovery has not yet commenced, and the defense attorneys are primarily concerned with ensuring that any failures in the plaintiff's initial pleading are properly addressed with a motion to dismiss. After this initial pleading stage, the defense attorneys will rely on the engineer's in-depth contribution of both the facts, as well as information regarding the applicable engineering standard. But for now, in developing their motion to dismiss, the defense attorneys will be focusing on legal, not factual, arguments, primarily based on the insufficiency of the plaintiff's pleadings as a matter of law. So, based solely on the law alone, the defense will file its motion to dismiss. The plaintiff files its response, and the defense replies. If the court grants the defendant's motion, the plaintiff may be able to amend its complaint to correct any legal deficiencies. However, if the plaintiff cannot correct its legal deficiencies, the case is likely over.

5. How it Continues: Developing Facts for Trial

Following the initial pleading phase, assuming the case is not dismissed, through settlement or otherwise, then the case proceeds to discovery. Discovery is the process through which both parties develop facts to offer as evidence at trial pursuant to the rules of evidence. [3] The discovery phase allows each side, plaintiffs and defendants, to develop the facts of the case through interrogatories (asking questions which require specific written answers), requests for production (for production of documents and other record evidence, including data, hard drives, flash drives, mirrored servers, inspection records, email, texts, phone messages, etc.) and then depositions (sworn testimony, taken by the party issuing the subpoena to the witness to be deposed, i.e., questions asked of the witness on the record to which the witness must give responses which the witness swears are true). Confidential Business Information and other confidential information is subject to this discovery process, unless accepted as legally privileged and excluded, such as confidential attorney client communications or work product developed under the guidance of the attorney in preparation for litigation.

In our Hypothetical, the plaintiff would likely request production of the engineer's calculations for the distillation unit, as well as all data regarding operation, maintenance, and design, in addition to records of all communication relevant to the engineer's hiring, education, and other work that would tend to show the expertise and competency of the engineer. The plaintiff would also ask for records from the company, and plant manager regarding the defendant engineer, including hiring, training, discipline and other documents generally relevant to company decisions and process regarding competency of its engineering staff. The defendants would object but likely produce relevant documents, protecting confidential business information or other sensitive documents with confidentiality agreements or production with court protection under seal. The defendants would ask the plaintiff to provide documents supporting the plaintiff's claims of damages. After production of discovery by both sides, the parties will schedule depositions. The engineer will likely be deposed by the plaintiff's attorney, which deposition is generally limited in time (e.g., 7 hours, if in federal court) and place so as not to impose an unreasonable burden on the deponent. The engineer will be defended during the deposition by the defense attorneys. After the deposition, the engineer should have an opportunity to review the deposition transcript for accuracy before it is finalized.

6. How it Ends (Part 1)—Motion Practice and Settlement

Upon development of the facts, the defendant may file a motion for summary judgment, alleging that, based upon the material facts which are not in dispute, plaintiff cannot prevail on their claims as a matter of law. It is rare for a court to grant such motions, because inevitably, the opposing party will allege the presence of material facts which are in dispute, and because granting of a motion for summary judgment can then be appealed and possibly reversed. When courts do grant these motions, it is usually with a thoughtful opinion crafted upon careful analysis.

However, the process of developing and arguing these motions often allows the parties to see the strength of their own and each other's cases, as well as gauge their likelihood of success before a jury. Note that the strength of the case is not always an indicator of success before a jury. Juries often have difficulties with scientific proof, so much so that courts sometimes appoint special masters to assist juries with scientific evidence. Thus, at this point in the case, after significant discovery has allowed evaluation of the evidence, the scope of the arguments the evidence might support, and whether those arguments are within the grasp of a lay jury to understand, settlements may occur, if settlement is possible.

In our Hypothetical, the defense will review the discovery and deposition statements and develop a list of material facts which they can reasonably argue are not in dispute in preparation to develop and file a motion for summary judgment. Relying on this list of arguably undisputed facts, the defense will file a motion for summary judgment arguing that, based on these facts, as a matter of law the plaintiff could not prevail on its claims regarding "good engineering judgment" as applied to the chemical engineer, the chemical engineer's alleged violation of the standard, or allegations that the violation by the chemical engineer of this applicable standard was a foreseeable cause, and/or cause in fact, of the plaintiff's injuries. The defense may attach to their motion exhibits such as the engineer's calculations or other work product, pages from the engineer's deposition transcript, or other demonstrative evidence to support their arguments. The plaintiff will file a response, arguing that the defense is legally wrong and, of course, there are many facts in dispute, to which the defense will reply. At this point, after the facts are developed, and the legal arguments are aired, any chances of settlement will be fairly apparent, and settlement may be reached if settlement is possible.

7. How it Ends (Part 2)--Trial and Expert Testimony

Assuming the plaintiff has met the threshold requirements and properly plead facts, based on information and belief, sufficient to support the elements of standing, case or controversy and ripeness, and the case has not yet settled even after all the facts have been developed, the case proceeds to trial. At trial, the plaintiff has the burden to prove in court the substantive elements of their civil claim by a preponderance of the evidence.

The first element the plaintiff must prove is the presence of a standard of care applicable to the chemical engineer in the work the plaintiff alleges as the cause of their injuries. For a claim of negligence, because the basis for the negligence case would be failure to exercise the duty of reasonable care, the first element of a claim of negligence that must be supported by evidence (sufficient to demonstrate the claim "more likely than not") before a reviewing court would be evidence of the existence of a duty or obligation recognized by law, "requiring the person to conform to a certain standard of conduct, for the protection of others against unreasonable risks." [8], p. 164. Generally, for engineers, this duty would arise when engaged in engineering work within the scope of the engineer's professional expertise, where the court may recognize as applicable the duty to utilize "good engineering judgment" or "good engineering practices." Considering our Hypothetical above, if the nature of the work is subject to regulatory standards, such as the requirements of EPA's Risk Management Program (RMP) and OSHA's Process Safety Management (PSM), then the applicable duties, or at least a portion of them, are described within the rules themselves. For example, EPA's RMP design provisions, 40 C.F.R. §68.48, require "good engineering practice" where compliance with the standard (and hence evidence arguably sufficient to demonstrate compliance with the applicable due of care) can be proven with evidence of "compliance with Federal or state regulations that address industry-specific safe design or with industry-specific design codes and standards." The RMP maintenance provisions, 40 C.F.R. §68.56, require "recognized and generally accepted good engineering practices," which is also the standard required by OSHA's PSM standard. 29 C.F.R. §1910.119.

The second element the plaintiff must prove is the violation of the applicable duty, or the failure to comply with the allegedly applicable standard of care. An important element relevant to these first and second elements is whether the harm that occurred was reasonably foreseeable - within the applicable standard of care - such as to create a duty for the professional to act within that applicable standard of care. The third element the plaintiff must prove is sufficient demonstration of causation, including both "proximate cause," as well as "cause in fact." Again, both sides may offer expert testimony of causation. The fourth and final element the plaintiff must prove is actual loss, or damages. Where there are no damages, there is no legal claim.

Where evidence regarding proof for or against any of these elements involves complex scientific or engineering judgment, demonstration of this evidence will likely require expert testimony. In this case, the evidence would be offered by the plaintiff as Daubert expert testimony, which would be rebutted by the defendant's own Daubert expert testimony – resulting in a "battle of the experts." This process allowing expert testimony ensures this evidence is sound as well, including a qualification process for experts whose testimony is offered to support and explain complex scientific evidence. This process, important for chemical engineers defending their actions in court, involves development of expert testimony through an expert witness pursuant to the Rules of Evidence 702 [3]:

> If scientific, technical, or other specialized knowledge will assist the trier of fact in issue, a witness qualified as an expert by knowledge, skill, experience, training, or education, may testify thereto in the form of an opinion or otherwise, if (1 the testimony is based upon sufficient facts or data, (2) the testimony is the product of reliable principles and methods, and (3) the witness has applied the principles and methods reliably to the facts of the case.

To assess whether an expert's opinion testimony will meet the requirements of Rule 702, the court will assess its merit following the standard set by the United States Supreme Court in *Daubert v. Merrell Dow Pharmaceuticals, Inc.*, 509 U.S. 579 (1993) (the "Daubert Standard") which requires consideration of: (1) whether the theory or technique can be or has been tested; (2) whether subjected to peer review and publication; (3) the known or potential rate of error; (4) general acceptance in the scientific community. [11]

In our Hypothetical, the case has failed to settle and is proceeding to trial. At trial, the engineer will likely have to testify regarding the engineer's actions related to the applicable standard of care, and the defense will likely need expert testimony to support the engineer's testimony. The plaintiff's counsel will cross-examine the engineer and try to use the engineer's deposition transcript to impeach the credibility of the engineer, especially if the engineer is a credible and persuasive witness. The trial itself may take days or weeks, depending upon the complexity of the case, and the court may not issue an opinion until weeks or months after the trial.

7. Case Study--Arkema Crosby, Texas (2017)

During Hurricane Harvey, the Arkema, Inc. facility in Crosby Texas was inundated by flood waters resulting in a fire that continued for days, evacuations and injuries. Arkema was investigated by the Chemical Safety Board (CSB), and is facing a criminal trial in state court, and a civil trial in federal court.

The CSB's Arkema findings touch on adaptation planning, recommending that chemical engineers responsible for hazard risk assessments consider those "reasonably anticipated external events," consistent with RAGAGEP, including weather events exacerbated by climate change, which are now being forecast with greater reliability. CSB's findings, relevant to both the engineer's applicable standard of care as well as the element of foreseeability, should be considered in evaluating litigation risk for similar facilities.

Most chemical engineers have heard about the Arkema event during Hurricane Harvey. Hurricane Harvey hit the Houston/Crosby area on August 24, 2017, with record 39 inches of precipitation over the following six-day period, and 17.54 inches on August 27, 2017 alone. The Arkema facility in Crosby, Texas stored organic peroxides which required refrigeration to prevent decomposition. Power was interrupted during Hurricane Harvey, and Arkema's emergency generators became flooded as well, thus, Arkema lost emergency generator power, and as a result, lost refrigeration, resulting in self-accelerating decomposition of organic peroxide on August 30, 2017, followed by fume and fire events observed August 31, 2017. Arkema's over-pressurization, accelerating decomposition and fire event lasted four days, with road closures, evacuation of over 200 residents for over a week, and 21 people seeking medical attention. [12]

The CSB completed its investigation pursuant to its express statutory authority to investigate accidental releases resulting in a fatality, serious injury or substantial property damages, provided by CAA § 112(r)(6)(C)(i), codified at 42 U.S.C. § 7412(r)(6)(C)(i). In its final report, the CSB found that Arkema did complete a CAA § 112(r) Risk Management Plan for two chemicals managed at its facility, but did not consider the organic peroxides (i.e., the chemicals that needed to be refrigerated), to be subject to these requirements. Additionally, according to the CSB, Arkema completed the Process Hazard Assessment (PHA) portion of its RMP for the low temperature warehouse where the organic peroxides were stored, but did not expressly consider flooding. Inexplicably, Arkema's common mode failure was flooding, but the PHA did not consider power loss due to flooding of the emergency generators. [12]

The CSB's May 2018 final report recognizes that the CAA § 112(r) RMPs should evaluate "whether a process is vulnerable to damage caused by external events, such as … floods, high winds, and evaluate the potential consequences if such events damaged the integrity of the process" considering, again, "reasonably anticipated external events" utilizing "recognized and generally accepted good engineering practices (RAGAGEP)." While also recognizing "effective guidance from consensus industry codes and standards is lacking with respect to flood risk," CSB found that flood risk could be analyzed under the existing terms of the existing regulatory structure, despite lack of express requirements. [12], § 7.1.1.2. EPA Risk Management Program, ¶¶ 238-240.

The CSB provides specific recommendations including, among others, that AIChE's CCPS develop broad and comprehensive guidance to help companies assess risk from extreme weather events, covering common mode failures, facility susceptibility and involving sufficient expertise, including engineers, to ensure risk assessments and process hazard analyses are sufficiently robust. Moreover, our historical record regarding storm characteristics and recent governmental reports predicting the scope and likely impact of future storms, flooding and sea level rise impact the element of foreseeability as it arises in negligence cases. Thus, the CSB appears to have enlarged the engineering considerations for implementing RAGAGEP in risk management planning where foreseeability of future storms may be accepted by the courts. And for this reason, plaintiffs could attempt to use the CSB's report to enlarge the applicable standard of care where future storms are foreseeable to prove allegations of professional negligence and recover damages arising in future cases with similar facts.

Following the CSB report, Arkema was served with criminal charges. The State of Texas filed separate criminal charges against Richard Rowe, the President and CEO of Arkema, Leslie Comardelle, the Crosby, TX Plant Manager, and Arkema, Inc. By August 2018, these charges became criminal indictments by the Harris County grand jury. [13] The allegations against the CEO and Plant Manager include "Reckless Emission of an Air Contaminant F3," caused by:

> failing to remove temperature sensitive organic peroxides from the Arkema facility…before the arrival of rainfall and/or flooding associated with Hurricane Harvey, recklessly caused the emission of an air contaminant, namely organic peroxides and/or

byproducts of organic peroxides and/or petroleum distillates and/or soot and/or particulate matter on or about 8/31/2017 thereby placing persons in imminent danger of death or serious bodily injury, and said release was not in strict compliance with Chapter 382 of the Health and Safety Code or a permit variance or order issued or rule adopted by the TCEQ.

These three cases are developing and have not yet reached resolution. [13]

But the first action filed against Arkema was the civil action filed on September 1, 2017, by seven civil plaintiffs, including first responders and homeowners adjacent to the Arkema facility, in Harris County, Texas district court. The Plaintiffs alleged claims seeking damages against Arkema, Inc. as well as Arkema representatives Richard Rennard, Richard P. Rowe and Andrew Burdett. Plaintiffs alleged three causes of action including negligence, gross negligence and negligence per se, for failing to: properly store chemicals, have proper procedures for backup refrigeration of chemicals, have adequate procedures in place to protect the safety and welfare of the community in the event of a catastrophe, to provide the public and first responders accurate information on the chemicals at risk of exploding, to implement procedures put in place by Arkema, Inc. as well as governmental agencies regarding the handling of chemicals, and to adequately prepare for a major flood event, having had the knowledge that such an event was foreseeable. The case was removed to the United States District Court for the Southern District of Texas in October 2017 and has since added more parties and more causes of action. On May 15, 2018, Plaintiffs filed their Third Amended Complaint, adding 135 more individual plaintiffs (now numbering 142), two more corporate defendants, and adding nine more causes of action to the original three, including: negligent misrepresentation, nuisance, trespass, conversion, intentional infliction of emotional distress, assault, fraud by non-disclosure, common law fraud and strict liability. [14]

This civil case is developing, and it will be some time before there will be any decision. However, note that in the civil case, the Plaintiffs have specifically and carefully alleged facts to support their negligence causes of action, including the scope of reasonable care and the foreseeability of facts relevant to both causation as well as resulting harm. Plaintiffs' alleged that: "Arkema failed to heed the warnings and ignored the foreseeable consequences of failing to prepare"; that Arkema's plans failed to "call for isolating hazardous materials from high wind or water – two very common hazards of any hurricane or tropical storm that could foreseeably and regularly hit the gulf coast"; and that Arkema failed "to adequately prepare for a major flood event having had the knowledge that such an event was foreseeable."

8.0 Conclusion

With the number of attorneys graduating from law school, and the general lack of science education among the jury pool, chemical engineers should be aware of potential personal liability arising from higher risk chemical plant operations. This is even more true now given the potential for more extreme weather events, the CSB's Arkema findings arguably enlarging the duty of chemical engineers to consider impacts from these more extreme weather events in completing Hazard Risk Assessments and the emphasis on the foreseeability of such events in recent litigation.

References

[1] "Liability of Employed Engineers," Arthur Schwartz, NSPE Deputy Executive Director and General Counsel, NSPE https://www.nspe.org/resources/professional-liability/liability-employed-engineers

[2] *Sykes v. Propane Power Corp.*. 224 N.J. Super. 686, 541 A.2d 271, 1988 NJ. Super. LEXIS 166.

[3] Title 28 of the United States Code, covering Judiciary and Judicial Procedure, providing the Federal Rules of Civil Procedure (Fed. R. Civ. P.), 28 U.S.C. §1651, et seq., and the Federal Rules of Evidence (F.R.E.), 28 U.S.C. §1781, et seq. See also, 28 U.S.C. §2072. For burden of persuasion or proof in a civil action, see F.R.E. 301.

[4] Criminal liability arising where a person negligently or knowingly discharges oil or a hazardous substance into a water of the United States, upon adjoining shorelines or into the contiguous zone in a harmful quantity. 33 U.S.C. §§ 1319(1), (2), 1321(b)(3), 40 C.F.R. pts. 110, 116, 117.

[5] Title 18 of the United States Code, Covering Crimes and Criminal Procedure (Crim. P., 18 U.S.C. § 3001, et seq.

[6] *Victor v. Nebraska*, 511 U.S. 1 (1994).

[7] Fain, *Architect and Engineer Liability*, 35 Washburn L. J. 32, 34 (1995).

[8] See generally W. Paige Keeton Et Al., *Prosser and Keeton on the Law of Torts* 28-43 (5th ed. 1984 & Supp. 1988) (see especially Section 32, footnote 28).

[9] Circo, *When Specialty Designs Cause Building Disasters: Responsibility for Shared Architectural and Engineering Services*, 84 Neb. L. Rev. (2005).

[10] *Lujan v. Defenders of Wildlife*, 504 U.S. 555 (1992).

[11] *Daubert v. Merrell Dow Pharmaceuticals, Inc.*, 509 U.S. 579 (1993).

[12] *Organic Peroxide Decomposition, Release and Fire at Arkema Crosby Following Hurricane Harvey Flooding*, Investigation Report, U.S. Chemical Safety and Hazard Investigation Board (May 2018), available at: https://www.csb.gov/arkema-inc-chemical-plant-fire-/.

[13] *The State of Texas v. Richard P. Rowe*, Case No. 160031101010-3, Harris County District Court, Texas, No. 339; *The State of Texas v. Leslie Comardelle*, Case No. 160031201010-3, Harris County District Court, Texas, No. 339; and *The State of Texas v. Arkema, Inc.*, Case No. 201776961, Harris County District Court, Texas, No. 157 (transferred from Court 339 to 157, as "other civil").

[14] *Graves, et al., v. Arkema, Inc.,* Plaintiffs' Third Amended Complaint, U.S. Dist. Ct. (S. Dist. TX, Houston Div.) (4:17-cv-03068).

For further regarding Hurricane Harvey's implications for chemical process plants, see:

a. *Climate Change Adaptation, by Any Other Name*, Mary Ellen Ternes, NR&E Fall 2018, p. 53;
b. ACC's Summary Report, *Hurricane Harvey Forum: Lessons Learned*, American Chemistry Council (Mar. 5, 2018), available at: https://www.americanchemistry.com/Hurricane-Harvey-Summary-Report.pdf.

Deflagrations by Design

Darren R. Malik
3330 Oakwell Court, Suite 100
San Antonio, TX 78218
DMalik@BakerRisk.com

J.K. Thomas, O.A. Rodriguez, B.J. Horn, and J. Geng
Baker Engineering and Risk Consultants, Inc.

Keywords: DDT, VCE, deflagration, detonation, vapor cloud explosion

Abstract

The consequences of large vapor cloud explosions (VCEs) and the potential for deflagration to detonation transition (DDT) are current topics of considerable interest in the petroleum refining and chemical processing industries. The potential for a DDT is particularly relevant since a detonation not only results in a very high flame speed, but can significantly increase explosion energy and decrease standoff distance, both of which act to increase blast load. A number of potential mitigation options to limit VCE severity have recently been explored and tested (e.g., multiple simultaneous ignition sources to limit flame travel distance, suppressants to limit flame speed). Another mitigation approach is to control the congested volume geometry as a means of limiting the flame travel distance (i.e., minimize flame acceleration within a congested volume) and incorporate open areas between congested volumes (e.g., "safety gaps"). An additional congested volume geometry mitigation approach is to minimize the flame travel distance at which lateral venting becomes effective. (i.e., the minimum distance to free vent). Several simplified VCE blast load prediction methods consider flame travel distance within a congested volume as an input parameter, but do not explicitly consider the distance to free vent.

Decreasing the distance to free vent (D_{FV}) acts to limit flame acceleration and decrease the potential for a DDT for a given congestion/confinement level and fuel mixture. High fuel reactivities and elevated congestion/confinement levels require smaller distances to free vent to preclude a DDT and limit the maximum flame speed to the deflagration regime. This paper describes a test program which takes a first step towards demonstrating this mitigation option. Tests were performed using near-stoichiometric ethylene-air mixtures in an elongated test rig configured with a medium level of congestion. The free vent distance required to prevent a DDT in this configuration was determined. This paper presents the results of the test program, the associated Flame Acceleration Simulator (FLACS) predictions, and potential areas for future research.

1.0 Introduction

The consequences of large vapor cloud explosions (VCEs) and the potential for deflagration to detonation transition (DDT) are current topics of considerable interest in the petroleum refining and chemical processing industries. One mitigation option is to suppress the flame front or decrease the flame speed using a water spray or particulate chemical inhibitor, but this approach requires successful detection and timely system activation. The congested volume can also be controlled as a mitigation approach. One such congested volume mitigation option is to limit the flame travel distance within a congested volume (i.e., minimize flame acceleration) and incorporate open areas between congested volumes (e.g., "safety gaps").

Another possible congested volume mitigation option, and the one explored in this work, is to limit the minimum distance to free vent (D_{FV}), which is the flame travel distance at which lateral venting becomes effective. Minimizing the flame travel distance at which lateral venting becomes effective (i.e., designing plants with the smallest practical D_{FV}) would represent a passive mitigation strategy. Decreasing the D_{FV} would limit flame acceleration and decrease the potential for a DDT for a given congestion/confinement level and fuel mixture. High fuel reactivities and elevated congestion/confinement levels would require smaller D_{FVs} to preclude a DDT and limit the maximum flame speed to the deflagration regime.

It is noted that several simplified VCE blast load prediction methods consider flame travel distance within a congested volume, as well as fuel reactivity, as input parameters [1,2], but do not explicitly consider the D_{FV}. Prior work by BakerRisk had noted a correlation between normalized flame run-up length (Ln_f), distance to free vent (D_{FV}) and the propensity of a VCE to undergo a DDT [3].

Figure 1 depicts an elongated flammable cloud with a length L, a width W and a height H. It is assumed that the flammable cloud covers the entire congested volume (i.e., that none of the cloud occupies an empty uncongested space), so that the cloud and the congested volume have the same dimensions. The parameters shown in Table 1 are utilized to characterize the cloud.

The "*Free Vent Distance*" (D_{FV}) is defined as the minimum of the cloud height and half the cloud width. The "*Cylindrical Distance*" (L_{Cyl}) is introduced to measure the maximum distance a cylindrical flame front can travel before reaching the edge of the cloud. The "*Flame Travel Distance*" (L_f) is the distance from ignition to the flame front, with a maximum value equal to the distance to the edge of the cloud; the ignition source could be located at the cloud center, cloud end or anywhere in the cloud. The "*Normalized Flame Travel Distance*" (Ln_f) is also given in Table 1.

Figure 2 provides an example of the observed deflagration and detonation regimes for elongated cloud geometries [3]; observed flame speeds from three different tests performed as part of the Buncefield Joint Industry Project (JIP) [4] are shown. Two of the tests entered the detonation regime within six normalized flame travel (Ln_f) distances and underwent a DDT. The third test remained below the deflagration regime threshold at $Ln_f = 6$. This test remained a deflagration and did not undergo a DDT.

The purpose of the work described in this paper was to further explore the relationship between D_{FV} and Ln_f. The specific goal of this work was to demonstrate that, for a given congestion pattern and fuel mixture, there is a critical D_{FV} that would preclude a DDT and result in a steady-state (i.e., approximately constant flame speed) deflagration.

A preliminary FLACS analysis (Section 2.0) indicated that the initial test setup (Section 3.0) would support the test program objectives. A discussion of the study results is provided in Section 4.0; conclusions and areas for future work are provided in Section 5.0.

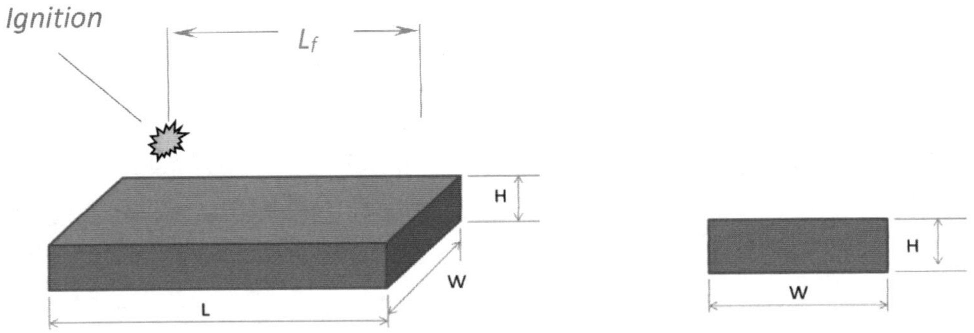

Figure 1. Dimensions of Elongated Vapor Cloud

Table 1. Cloud Geometry and Blast Field Parameter Definitions

Parameter	Symbol	Definition
Free Vent Distance	D_{FV}	$\mathrm{MIN}\left(H, \frac{W}{2}\right)$
Cylindrical Distance	L_{Cyl}	$\mathrm{MAX}\left(H, \frac{W}{2}\right)$
Flame Travel Distance	L_f	*n/a*
Normalized Flame Travel Distance	Ln_f	L_f/D_{FV}

2.0 FLACS Predictions

A preliminary FLACS simulation was performed to determine whether or not a DDT would be expected during the test. The criteria used to evaluate the likelihood of a DDT followed the Gexcon AS guidelines, where the normalized spatial pressure gradient across the flame front parameter (DPDX) is used to assess the potential for a DDT. Gexcon's DPDX DDT criteria [5] is summarized in Table 2.

The FLACS model employed a simplified geometry which accounted for both the pipes and main frame of the test rig. No instrumentation or cross-bracing were modeled in the simplified geometry; as discussed in Section 4, a more detailed geometry including the cross-bracing was evaluated post-test. The FLACS geometry is shown in Figure 3. Note that the vertical piping is shown in multiple colors only for visual purposes.

For both the simplified and higher fidelity FLACS models the normalized spatial pressure gradient predicted by FLACS did not reach the 0.5 threshold, and the probability of a DDT occurring within the test geometry was classified as "unlikely" based on GexCon's DDT criteria.

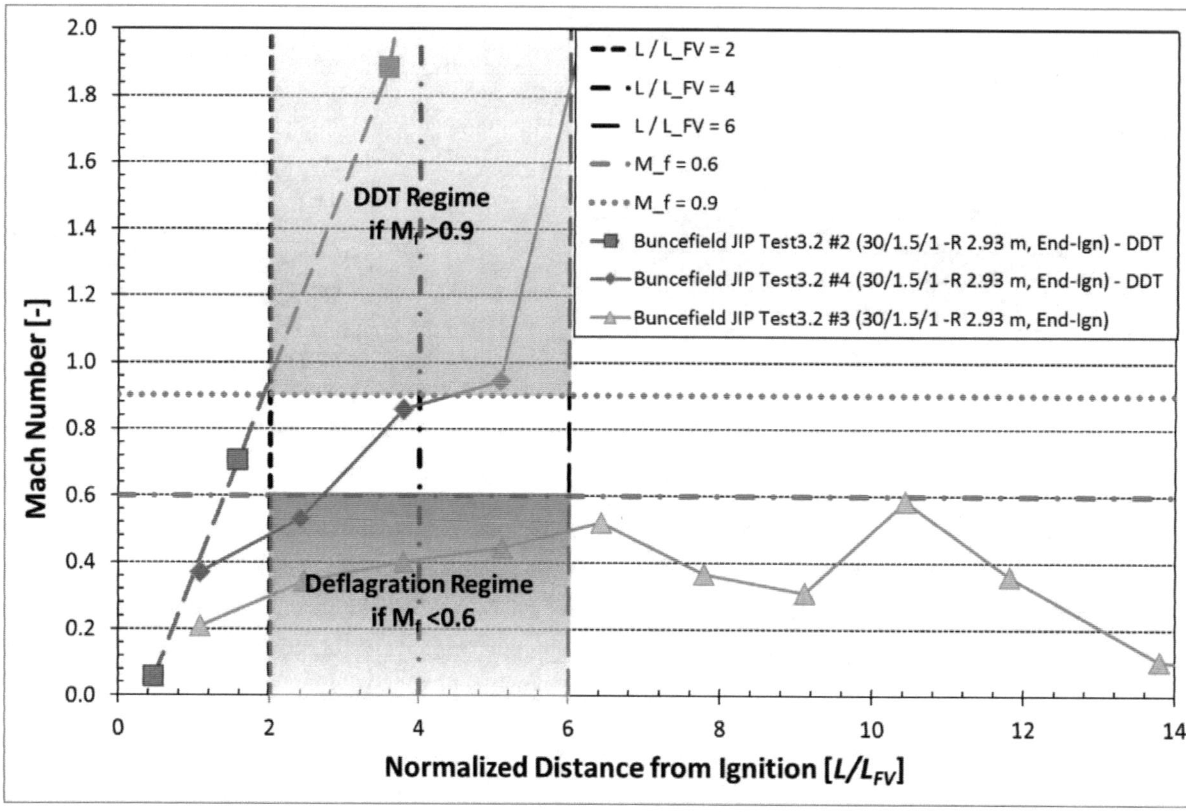

Figure 2 Observed Deflagration and Detonation Behavior Based on Normalized Flame Travel Distance

Table 2 Guidelines for DDT Evaluation using DDT

DPDX Value	DDT Likelihood
DPDX < 0.5	Unlikely
0.5 < DPDX < 2	Possible
2 < DPDX < 5	Possible/Likely
5 < DPDX	Likely

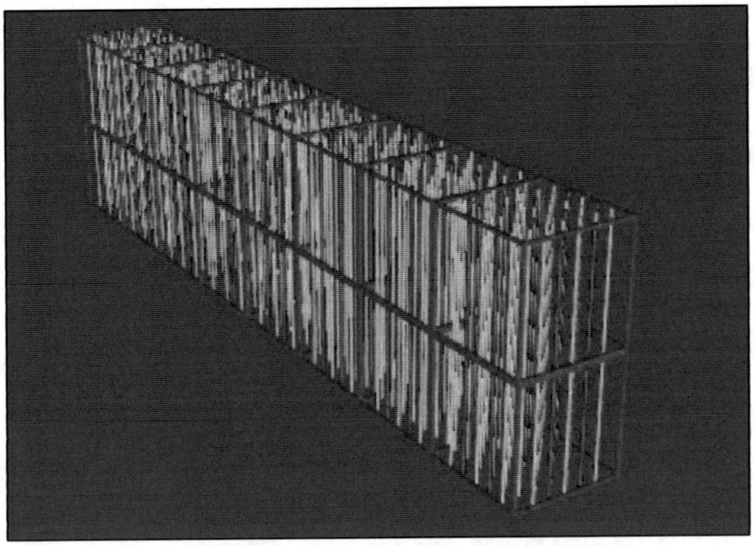

Figure 3 FLACS Geometry

3.0 Test Description

The VCE tests were carried out in a test rig which used vertical pipes to promote flame acceleration. The same medium congestion level used in the creation of the Baker-Strehlow-Tang (BST) flame speed tables [6, 7] was used in these tests. The congestion array was made up of a regular array of vertical circular tubes with a 2.375-inch (6 cm) diameter. For each 6 ft. (1.8 m) cube, a total of 45 vertical tubes were installed, in addition to the 4 cube corner supports. A plan view illustrating this congestion pattern is provided in Figure 4. This congestion pattern is intended to be conservative (i.e., more congested) relative to the congestion in a typical process plant. The area and volume blockage ratios of this congestion pattern are 23% and 3.9%, respectively. The ratio of surface area of congestion per unit volume within an individual cube is approximately 0.8/ft.

The congested portion of the test rig was 48 feet (14.6 m) long, 6 feet (1.8 m) wide and 12 feet (3.7 m) high. Photographs of the test rig are provided in Figure 5.

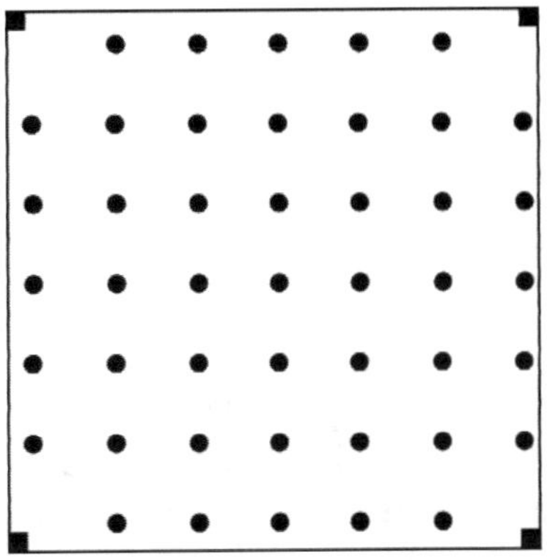

Figure 4. Schematic of Medium Congestion Pattern

Figure 5. Photo of the South (Top) and North (Bottom) Faces of the Test Rig

3.1 *Creation of Fuel-Air Mixture*

Prior to the start of the test, the entire test rig was covered in a 1.5 mil (0.0015 in., 0.038 mm) thick plastic sheet. The sheet was retained by a series of straps placed along the length of the rig during the generation of the fuel-air mixture. A quiescent wait period was followed once the target concentration was achieved. The straps were released via remote actuators just prior to ignition in order to allow the VCE to expand as freely as possible. The acceptable upper and lower fuel concentrations were based on a 1% decrease in peak Laminar Burning Velocity (LBV) for each fuel species; the target ethylene concentration for these tests was 7.4%, with an acceptance range of 7.1% to 7.7%. The fuel delivery, gas sampling, and mixing systems are described below.

3.1.1 Fuel Delivery

The fuel was introduced into the rig through a series of venturis which pull gas mixture in from one end and expel an enriched fuel-air mixture out of the other. The venturis were oriented to expel the enriched fuel-air mixture toward the south edge of the rig (counter flow to the mixing fans). The rig was divided into 2 fuel delivery zones with four venturis each, in order to support the generation of a uniform fuel-air mixture throughout the rig. The locations of the venturis are shown in Figure 6. Each venturi was centered along the long axis of the rig and deployed 3 feet (0.91 m) from the mid-height [i.e., at 3 feet (0.91 m) or 9 feet (2.7 m) from grade].

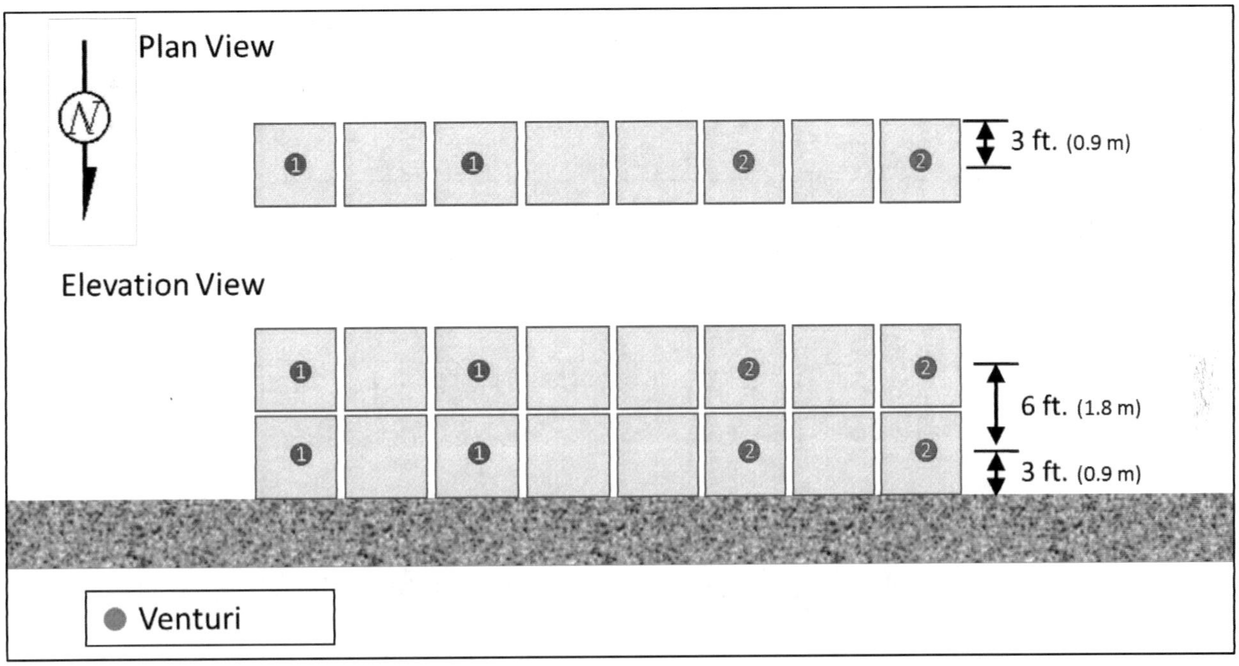

Figure 6. Location of Fuel Delivery Venturis

3.1.2 Fuel Circulation

Fans were used to circulate the fuel-air mixture in order to quickly achieve a uniform mixture throughout the rig. The fans were positioned at the center of each cube along the south side of the rig and oriented to force the fuel-air mixture back to the center of the rig (i.e., provide counter flow to the venturis). Each fan was rated for a 110 cfm flow rate, which is equivalent to turning the rig volume over every two minutes. The fans were shut off prior to ignition in order to allow the mixture to reach a quiescent state.

3.1.3 Gas Sampling

Fuel concentrations were determined indirectly based on oxygen concentration, which was measured using a California Analytical Model 600P Oxygen Analyzer. The analyzer was calibrated prior to each test using a known concentration of a suitable span gas balanced with nitrogen and with nitrogen as the zero gas. The sample points used to monitor concentration during the test were distributed along the long axis of the rig in the pattern shown in Figure 8.

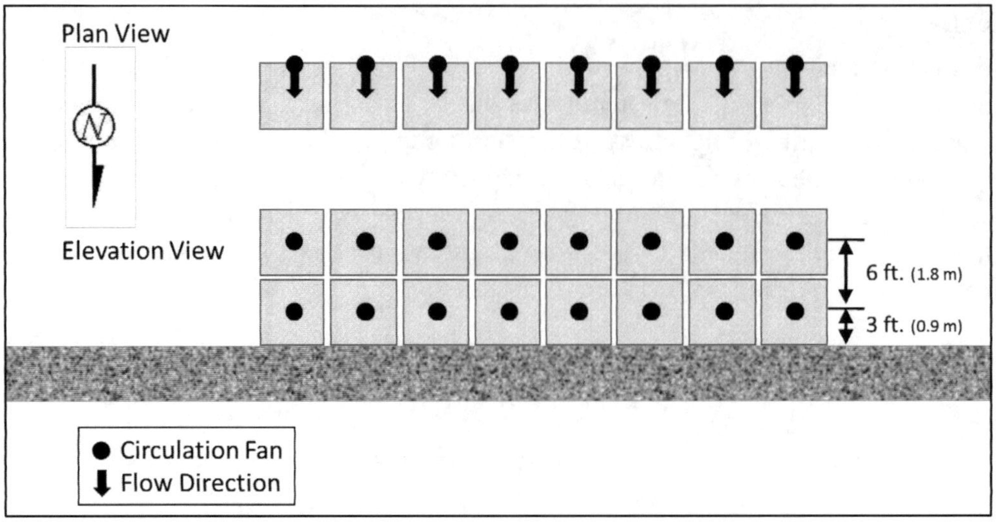

Figure 7. Location of Circulation Fans

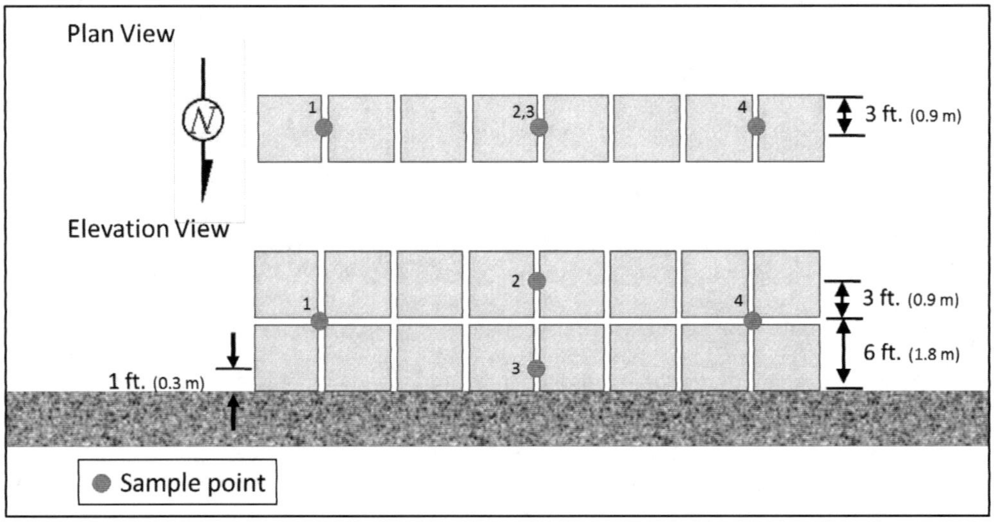

Figure 8. Location of Fuel Sampling Points

Gas samples were pulled through 0.18-in (0.46 cm) inner diameter sample lines, ranging in length from 30 to 60 ft (9 to 24 m), depending on the sample point. The sample pump flow rate was 10 ft^3/hr (17 m^3/min), giving a transit time from the sample point to the analyzer of 2 to 5 seconds. The time required for the analyzer to reach 90% of full scale is 30 seconds. The combined time for a sample to reach the analyzer and for the analyzer to respond was therefore approximately one half minute. This allowed the 4 sample points to be cycled through in approximately 2 minutes.

3.2 *Ignition System*

The location of the ignition source was grade level near the center of the rig. An exploding fuse wire was used for the ignition source in each test. The fuse wire ignition system was composed of equal length wire leads and a Parr exploding fuse wire (part #45C10) approximately 5 cm long. The fuse wire has a heat of combustion of 9.6 J/cm, giving a nominal ignition energy of 50 J.

3.3 *Instrumentation*

Pressure and flame speed data were collected for each of the tests.

3.3.1 Pressure Measurement

Blast pressure histories were measured using PCB® Piezotronics general purpose ICP® dynamic pressure sensors (Model CA102B18) connected to a line-powered ICP® sensor signal conditioner using microdot connectors and BNC cables. Each pressure gauge was mounted on a ½-inch steel plate approximately 12 inches (30 cm) in length and width. The pressure gauges internal to the test rig were installed with the gauges mounted in a recess to reduce the effects of thermal radiation from the fireball. The gauges external to the rig were installed with the gauge face flush with the top of the mount. The signal from each pressure sensor was recorded using a National Instruments PXIE-1082 Chassis.

3.3.2 Flame Speed Measurement

Flame propagation was observed with high definition and high speed video. The high-speed camera was set to record at 3000 frames per second based on the expected flame speed. The high definition video was captured at 30 frames per second.

4.0 Results and Discussion

Preliminary FLACS simulations indicated that the test geometry shown in Figure 9 and Figure 10 would not undergo a DDT; however, a DDT was observed in the first test run in this configuration. Review of the high speed video showed that the DDT occurred just outside the congested array and that interaction with the plastic sheeting at the end of the test rig may have contributed to the DDT.

A second test was therefore run with an additional 6 feet of uncongested volume on either end of the rig. This additional uncongested volume was used to reduce the potential interaction of the plastic sheeting with the flame front as it exited the congested volume. This test rig configuration, with the plastic sheeting installed, is shown in Figure 11. This configuration also resulted in a DDT. Review of the high speed video indicated that the DDT occurred as the flame jetted into the uncongested volume and not due to interaction with the plastic. A sequence of still frames from the second test is provided in Table 3. A second DDT, not shown, was observed on the west end of the rig approximately 2 ms after the DDT on the east end of the rig. Flame speeds for the first two tests that underwent a DDT are provided in Figure 12.

Following the observed DDT in the second test, the D_{FV} was reduced from 3 ft. (0.91 m) to approximately 2.2 ft. (0.66 m) by removing the outer rows of congestion from each side along the long axis. A schematic of the original 7 x 7 array with a 3 ft. D_{FV} and modified 5x7 array with a 2.2 ft. D_{FV} is provided in Figure 13. Three tests were run in this configuration, and a DDT was not observed in any of these tests (i.e., all 3 tests yielded a deflagration). A sequence of still frames from one of these tests is provided in Table 4.

Figure 9. Original Test Rig Configuration (Long Axis)

Figure 10. Original Test Rig Configuration (Short Axis)

Figure 11. Modified Test Configuration with an Additional uncongested Volume

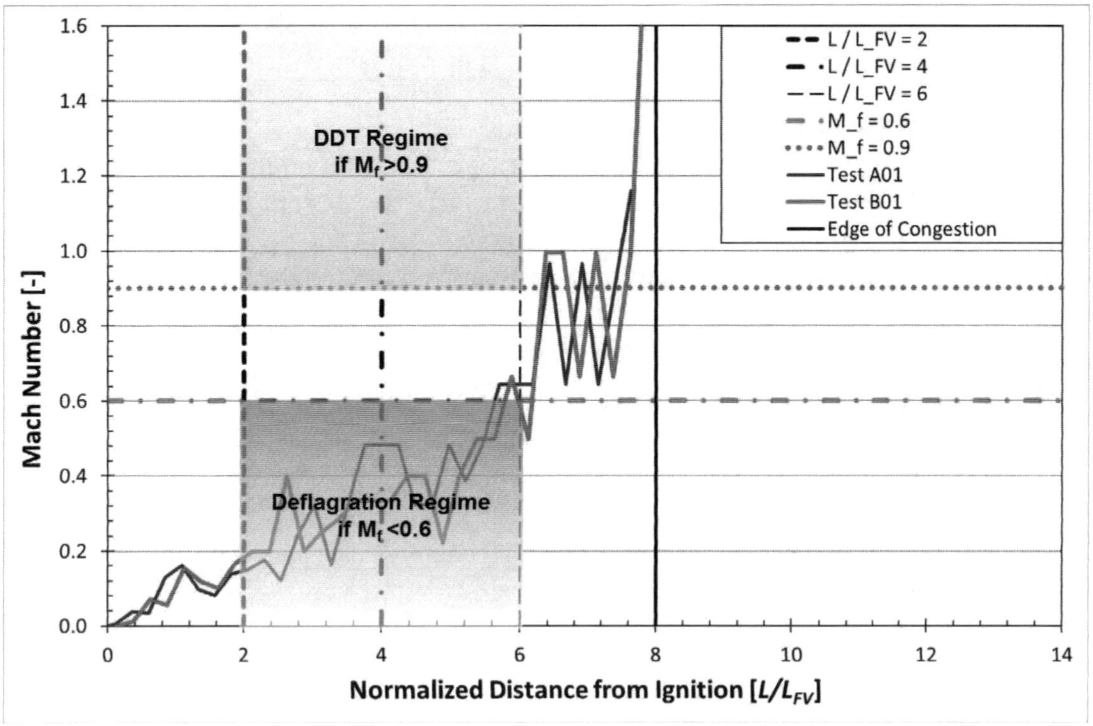

Figure 12. Observed Flame Speed for Tests that Result in a Detonation

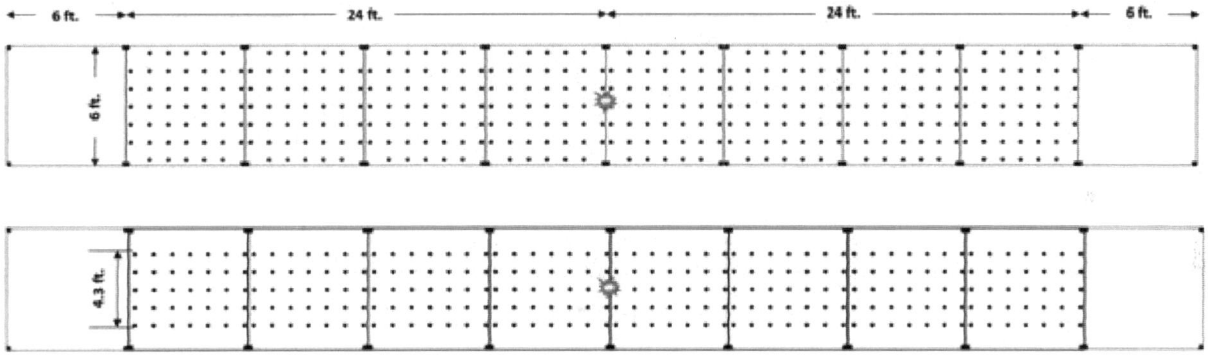

Figure 13. Original 7x7 Congestion Pattern (Top) and Modified 5x7 Array (Bottom)

Flame speeds for the final three tests that did not undergo a DDT are provided in Figure 14. This plot was generated by tracking the leading edge of the flame at an elevation of approximately 1 ft. (0.3 m) above grade. Note that the occasional oscillations may be due to the temporal resolution of the high speed video, "hiding" of the flame behind the congestion and/or test rig, or light reflecting off of the plastic. These oscillations do not necessarily indicate a sudden dramatic jump in flame speed. The presence of illumination at a given location within 2 frames of the high speed (HS) video vs. 3 frames after the previous data point results in the oscillations shown in the data. A near steady state flame speed was achieved after approximately 18 ft. (5.5m) of flame travel distance, equivalent to 9 D_{FVs}. Upon exiting the congested portion of the test rig, the flame rapidly decelerates from an exit velocity of 730 ft/s (Mach 0.65) to less than 200 ft./s (Mach 0.2) within 5 ft. (1.5 m) from the edge of congestion, equivalent to approximately 2 D_{FVs}.

The flame propagation along the mid-height of the test rig appears to be influenced by the 1-inch angle iron that was used to support the congestion pipes. The influence of these supports can be seen in the irregularly shaped flame front shown in the still images from the deflagration provided in Table 4. The angle iron supports were not accounted for in the preliminary FLACS model. Additional FLACS analysis was performed post-test which included the angle iron supports. The maximum normalized spatial pressure gradient predicted by FLACS with the refined geometry did not reach the 0.5 threshold, and hence a DDT would still be classified as "unlikely" (see Table 2).

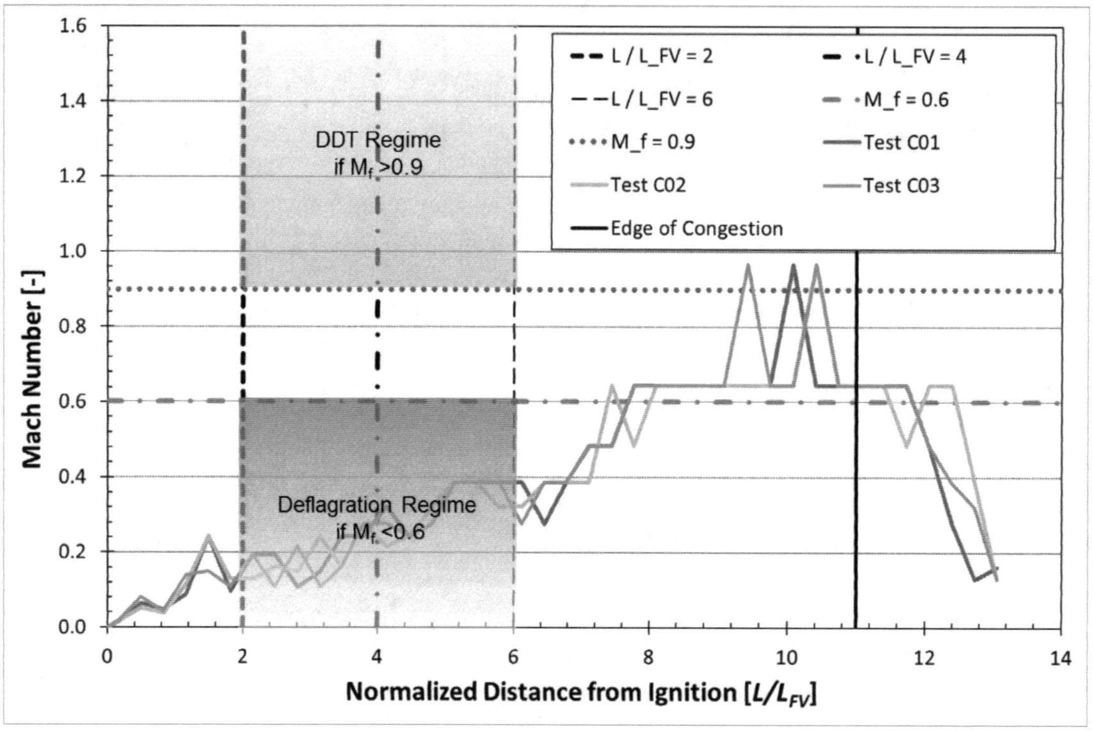

Figure 14. Observed Flame Speed for Tests that Result in a Deflagration

Table 3. Selected Still Frames Showing the Occurrence of a DDT

Frame 420 140.00 ms after ignition	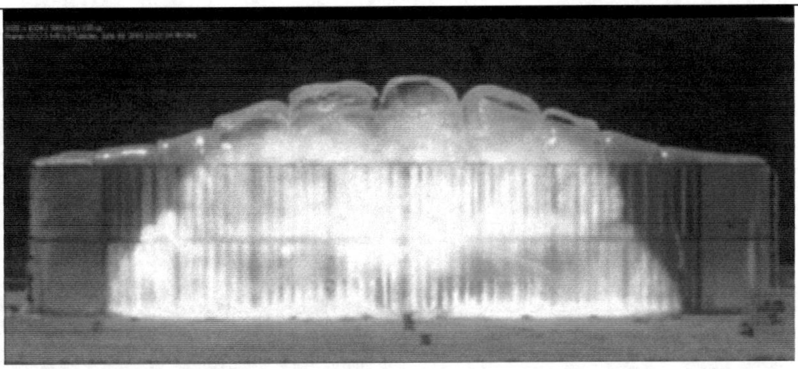
Frame 422 140.67 ms after ignition	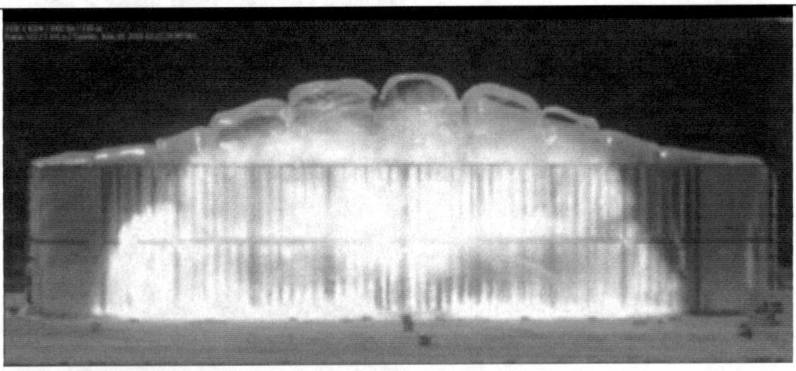
Frame 424 141.33 ms after ignition	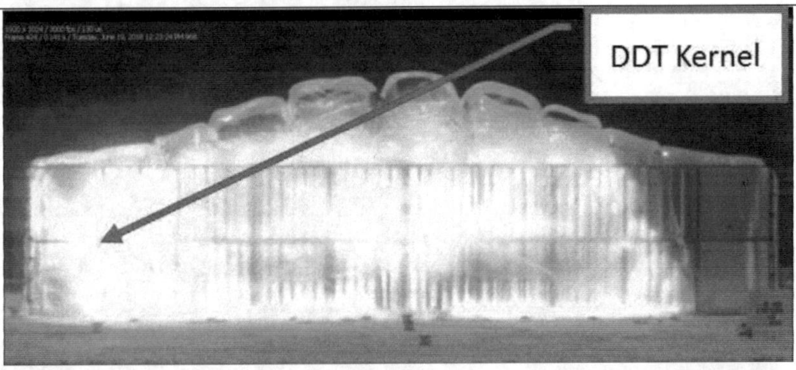
Frame 426 142 ms after ignition	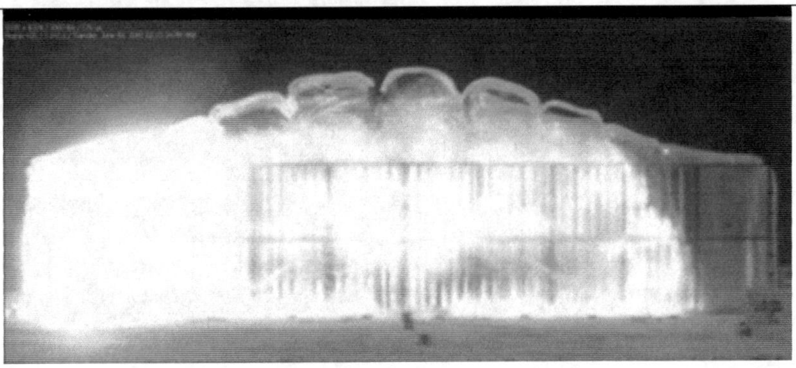

Table 4. Selected Still Frames Showing a Deflagration with Reduced D_{FV}

Test C03 100 ms after ignition	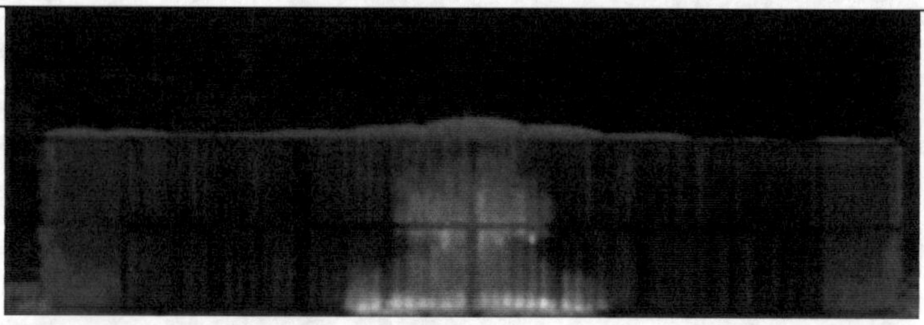
Test C03 106.67 ms after ignition	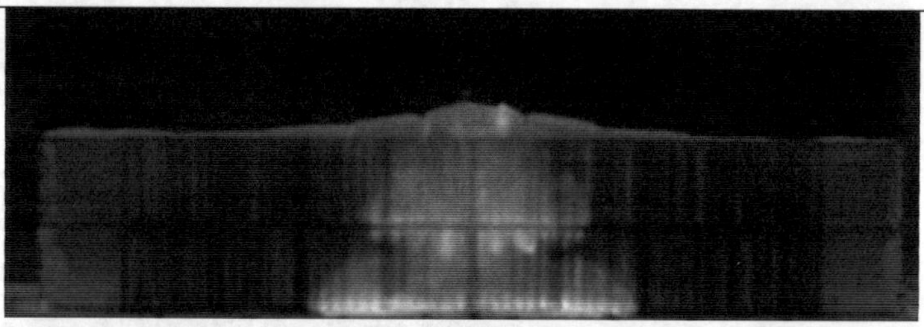
Test C03 113.33 ms after ignition	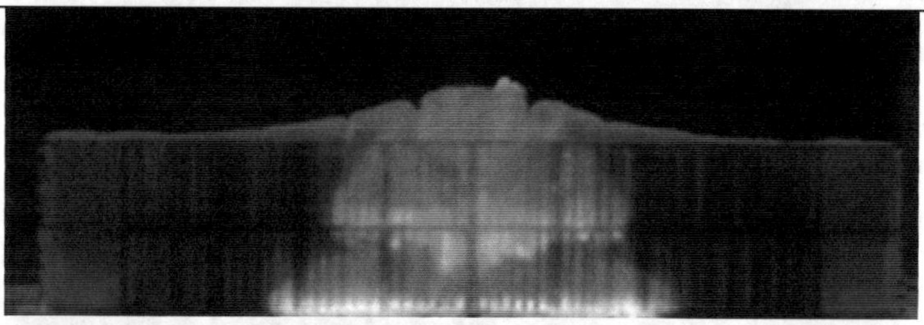
Test C03 120 ms after ignition	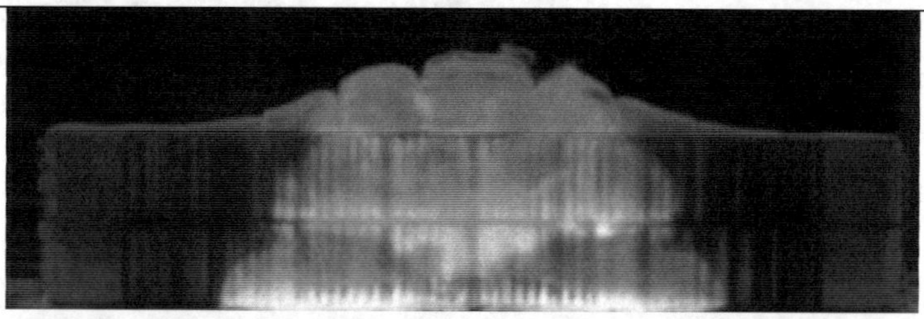
Test C03 126.67 ms after ignition	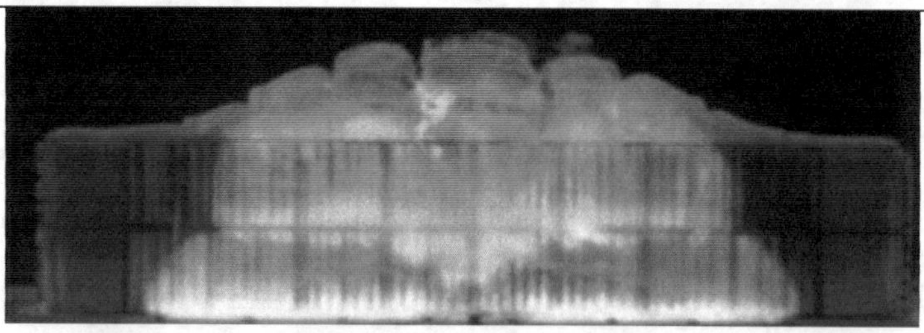
Test C03 130.00 ms after ignition	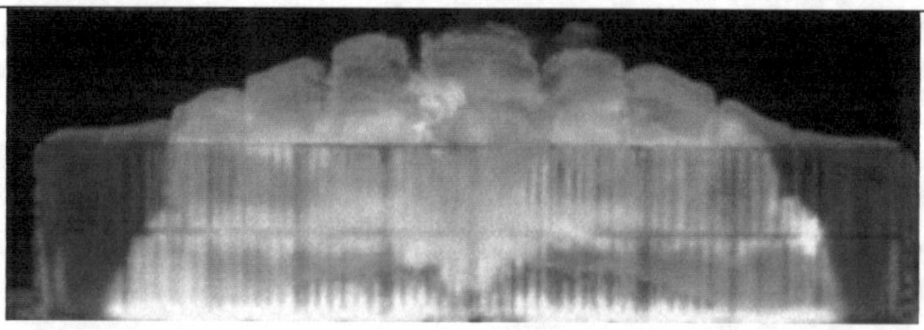

5.0 Conclusions & Recommendations for Future Work

The results from this test program indicated that it is possible to use D_{FV} as a design parameter to decrease the probability of a DDT. These tests were performed at a limited scale where the D_{FV} had to be held < 2.5 ft. in order to prevent a DDT; however, as shown in Figure 9 and Figure 10, the amount of congestion present in the test rig exceeds the levels seen in a normal process plant. Continued work in this area would help to support improvements in VCE modeling and could help influence future plant designs.

The current study represents only a small step towards understanding the impact of D_{FV} and Ln_f on VCE phenomena. Possible considerations for future related test programs are summarized below.

- For elongated congested geometry VCE tests, a congested volume with a length of at least 8-10 D_{FVs} should be employed (if possible) in order to confirm a steady-state flame front velocity.
- A parametric study should be conducted with congested volumes elevated above grade by 1 or more D_{FVs} to provide information on the elevation required so that the ground plane no longer contributes to flame acceleration in the congested volume.
    - In addition, elevating process equipment above grade would minimize the amount of congestion available for interaction with low lying clouds, such as those involved in the accidental explosions at Buncefield [8] and Caribbean Petroleum [9, 10].
- Additional D_{FV} testing should be conducted on a larger scale with a lower congestion level (i.e., one with a congestion surface area to volume ratio are less than 0.8/ft.) in order to further explore the application of this concept to realistic plant geometries.

6.0 References

[1] Mercx, W.P.M., A.C. van den Berg and D. van Leeuwen, "Application of Correlations to Quantify the Source Strength of Vapour Cloud Explosions in Realistic Situations, Final Report for the project: 'GAMES'," TNO Prins Maurtis Laboratory, Oct. 1998.

[2] Marx, J.D. and B.R. Ishii, "Revisions to the QMEFS Vapor Cloud Explosion Model," 13th Global Congress on Process Safety, AICHE, March 2017.

[3] Geng, J., T.J. Mander, J.K. Thomas and Q.A. Baker "Elongated VCE Blast Waves and Structural Damage," 13th Global Congress on Process Safety, AICHE, March 2017.

[4] B. A. Burgan, Dispersion and Explosion Characteristics of Large Vapour Clouds (Volume 2 – Appendices), SCI DOCUMENT ED024, 2014.

[5] FLACS user manual: Gexcon AS, "FLACS v 10.5 User's Manual," May 27, 2016.

[6] Pierorazio, A.J., J.K. Thomas, Q.A. Baker and D.E. Ketchum, "An Update to the Baker-Strehlow-Tang Vapor Cloud Explosion Prediction Methodology Flame Speed Table," Process Safety Progress, 24(1): 59-65, 2005.

[7] Baker, Q.A. et al. (2010) Guidelines for Vapor Cloud Explosion, Pressure Vessel Burst, BLEVE and Flash Fire Hazards, Second Edition, Center for Chemical Process Safety (CCPS) & John Wiley and Sons, New York, NY.

[8] Buncefield Major Incident Investigation Board (2008) The Buncefield Incident, 11 December 2005, Final Report of the Major Incident Investigation Board, ISBN 978 0 7176 6270 8.

[9] Baker, Q.A., J.K. Thomas, D.R. Malik and M.T. Edel, "Caribbean Petroleum Explosion Case Study", 15th Global Congress on Process Safety, New Orleans, LA, March 31 – April 3, 2019.

[10] CSB (2015) Caribbean Petroleum Tank Terminal Explosion and Multiple Tank Fires, U.S. Chemical Safety and Hazards Investigation Board, Final Investigation Report, CSB Report No. 2010.02.1.PR

Inherent Safety - How to Utilize Explosion Hazard Analyses as an Early Tool to Improve FPSO and FLNG design

Scott Davis
Gexcon US
4833 Rugby Avenue Suite 100
Bethesda, MD
sgdavis@gexcon.com

Drew Botwinick
Gexcon US
Houston, TX

Prepared for Presentation at
American Institute of Chemical Engineers
2019 Spring Meeting and 15th Global Congress on Process Safety
New Orleans, LA
March 31 – April 3, 2019

Inherent Safety - How to Utilize Explosion Hazard Analyses as an Early Tool to Improve FPSO and FLNG design

Scott Davis
Gexcon US
4833 Rugby Avenue Suite 100
Bethesda, MD
sgdavis@gexcon.com

Drew Botwinick
Gexcon US
Houston, TX

Keywords: inherent safety, FPSO, FLNG, early design phase risk assessments, explosion analysis, consequence modeling

Abstract

Process areas on floating production and storage (FPSO) and floating liquefied natural gas (FLNG) facilities are typically designed to be very compact due to the high costs of "floating real estate." Given the high congestion levels of FPSOs and FLNGs, explosion design accidental loads (DALs) can become very large and are very sensitive to the layout of modules, piping, equipment, and safety gaps. Safety gaps are regions of intentionally empty space between process modules to reduce the interaction between modules in the event of a hazardous event.

For these reasons, the analysis of explosion hazards on FPSOs and FLNGs is a critical step that should accompany the facility development from the very earliest design phase. This paper will discuss how quantitative analyses using computational fluid dynamics (CFD) modeling can be used to optimize the layout of a FPSO facility from a safety perspective. The methodology uses detailed 3D models of FPSO geometry (down to 1-2 inch pipe diameters by incorporating anticipated equipment blocks) within which ventilation, gas dispersion, and vapor cloud explosion scenarios are simulated using a CFD tool. The paper will present study results to compare the effectiveness of a number of design alternatives such as:

- Grated vs. solid decks;
- Safety gap size;
- Process module arrangement;

The paper will demonstrate how CFD-based methodologies can provide FPSO and FLNG designers with critical safety-related design implications previously not available in the early-design phase and help to ensure inherently safer designs. In addition, such studies can help enhance safety by evaluating the effectiveness of mitigation measures such as gas detection.

1 Introduction

Process areas on floating production and storage (FPSO) and floating liquefied natural gas (FLNG) facilities are typically designed to be very compact due to the high costs of "floating real estate." Given the high congestion levels of FPSOs and FLNGs, explosion design accidental loads (DALs) can become very large and are very sensitive to the layout of modules, piping, and equipment—including the size of safety gaps, gaps implemented between congested process modules in order to reduce the length of flame travel and acceleration to mitigate explosion consequences. In addition, the details associated with the facility layout play a major role in the design of mitigation measures, such as gas detection that activates water deluge. For these reasons, the analysis of explosion hazards on FPSOs and FLNGs is a critical step that should accompany the facility development from the very earliest design phase.

The layout and design decisions made during early design phases have a significant impact on flammable and toxic hazards that are often only assessed once the design is completed or in its later stages of development. The amount of re-engineering (and therefore cost) of mitigating the effects of these decisions increases as the project is developed, and eventually in construction or operation. One of the most effective strategies for limiting flammable or toxic hazards is to assess the impacts of design choices on the risk of these hazards at an early stage.

For example, one key question that commonly arises in early design phase is where to place the living quarters on an FPSO facility, especially relative to the more hazardous areas of the process (i.e., risers, separators, gas treatment, and compression). These are very complex issues and depend on the actual layout of the process and the mooring system. For example, (1) spread mooring systems that incorporate multi-point lines that moor to the vessel from the seabed and have a fixed heading relative to the seabed; and (2) turret mooring system that contains a bearing system that allows the vessel to rotate around the fixed geostatic part of the turret, which allows the FPSO to weathervane about the mooring system, in response to the environment. Weathervaning is the ability of the vessel to adapt its orientation with respect to the prevailing environmental direction that enables the FPSO to position the vessel favorably against the wind and sea currents so that it remains bow to wind and weather.

The upper image in Figure 1 shows an example of a FPSO facility that has the living quarters in the front of the vessel and a blast wall separating the turret and process areas from the living quarters. If weathervaning is used on the vessel, the living quarters will be well-ventilated because the wind and sea currents remain bow to ship; however, the process areas will not be as well-ventilated because the blast wall and living quarters will block most of the wind coming straight along the vessel. In addition, the living quarters and blast wall remain relatively close to the turret where the risers enter the vessel. In contrast, the lower image in Figure 1 demonstrates a spread mooring vessel where the living quarters are towards the rear of the vessel, with a blast wall separating the living quarters from the process areas.

To understand if putting the living quarters forward or aft for a given process layout or mooring system is inherently safer, quantitative risk assessments can be performed to assess the associated risks of each layout. Only after such an analysis can the relative risks be weighed against the cost of each design. This requires detailed representation of the physical process, statistical data associated with the weather and the frequencies of potential leaks, and advanced computational fluid dynamic (CFD) analyses to evaluate what scenarios drive the risk with the vessel. This paper will demonstrate how powerful analyses can be performed in the early design phase, which can help safety and design engineers make critical decisions that affect the inherent safety characteristics of the final design.

Figure 1: Upper image shows an FPSO design for the Petrojarl Knarr with the living quarters forward and the lower image showing the living quarters aft in the Cidade de Itaguai (Upper image: https://www.teekay.com/business/offshore/floating-production-storage-and-offloading/petrojarl-knarr/ and lower image: https://www.modec.com/fps/fpso_fso/projects/iracemanorte.html)

2 CFD Modeling

There are a variety of qualitative and quantitative techniques available to make decisions for an inherently safer design; however, the most powerful quantitative technique available is to make use of CFD modeling to be able to directly evaluate the effects of design decisions on hazards such as dispersion, fires and explosions. CFD modeling uses a representative 3D geometry model to be able to determine accurate ventilation, dispersion, fire, and explosion consequences.

The benefit of more accurate modeling comes with the drawback that a 3D model is required—and 3D CAD models of facilities are typically incomplete during the early stages of design, as illustrated in Figure 2. When congestion due to small components is not integrated into the explosion pressure analysis model, the design blast loads will be severely underestimated. Previous studies concluded that explosion pressures can differ by an order of magnitude between an early-phase CAD model and the final, detailed design. [1]

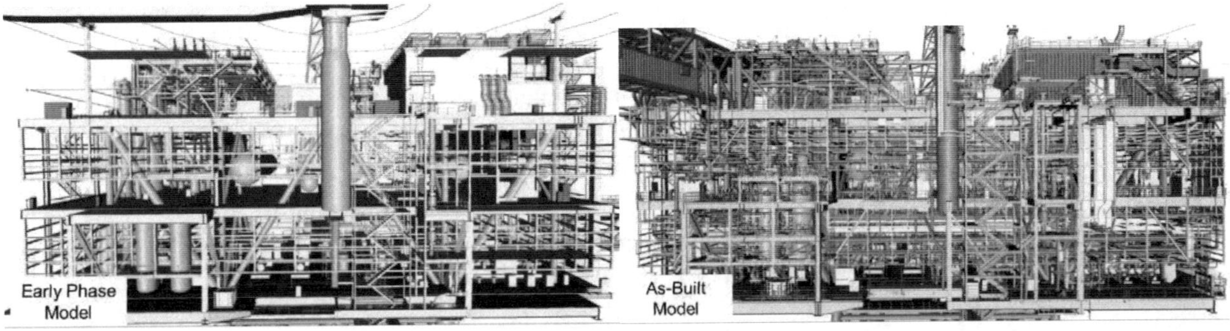

Figure 2: Illustration of difference in congestion levels between CAD models in the early design phase and in the final, as-built design [1]

There are two factors important for the development of explosion overpressure: congestion and confinement. Explosion overpressure will often be stated to be either confinement-driven or congestion-driven. In practical situations, explosion overpressures will be governed by a combination of these two factors.

Confinement is relatively straightforward to characterize. It is simply the presence of solid boundaries like plated decks or blast walls. When a gas burns, it will expand. The presence of solid boundaries will reduce the open volume available for said expansion and, the overpressure will be higher than that of an unconfined explosion. However, whether confinement is ultimately better or worse from an inherent safety perspective isn't always as obvious. Plated decks and walls can also act as barriers that affect ventilation and dispersion. Depending on factors such as layout, fuel composition, and wind conditions, solid boundaries could potentially reduce the maximum size of contiguous flammable clouds and reduce the overpressures of an explosion event. Various methods and documentation exist on whether plating or grating should be preferred; however, only CFD modeling can assess the finer points in the context of a particular FPSO or FLNG design.

The other key factor affecting explosion overpressures is congestion. The quantity, size, and placement of obstacles need to be taken into consideration in a three-dimensional domain. Ignition of a gas cloud in a congested area will result in a volume expansion and induce flow around all the surrounding obstacles ahead of the arriving flame. As turbulence is generated around each individual obstruction, it stretches and "wrinkles" the flame surface. As the flame surface increases, the area for chemical processes to occur increases, hence increasing the overall combustion rate and speed of the flame front. The increased combustion rate will then feed back into the above mechanism: increased combustion results in faster expansion of the unburned gases and increased flow around the surrounding obstacles. In turn, this increases turbulence ahead of the flame front, which increases the flame surface and combustion rate. Or more simply, the more obstacles in a volume, the higher the resulting explosion overpressure. Again, various methods exist to characterize these consequences; however, only CFD modeling can assess the finer points of congestion effects on turbulence and explosion overpressure.

This requires methods that will provide accurate overpressure predictions for a completed installation well before the installation is completed. Determining explosion overpressure risk is complex. The risk is affected by ventilation conditions, flammable vapor cloud formation, and geometry/layout configuration, in addition to the more directly related cloud position and ignition location. To this end, extensive research has been performed over the last 20 years into methods on how to supplement early-phase CAD models with anticipated congestion to achieve congestion levels of as-built facilities [1]. This methodology has recently been verified [1] and proven to yield reasonably accurate congestion levels to successfully evaluate and optimize concept-phase designs. An example of an early design phase process module is provided in the left image of Figure 3, while the right image includes the structural and process components anticipated to exist in final as-built design (i.e., structural support, railing, relief piping, blowdown piping, instrumentation, water deluge, access platforms, etc.).

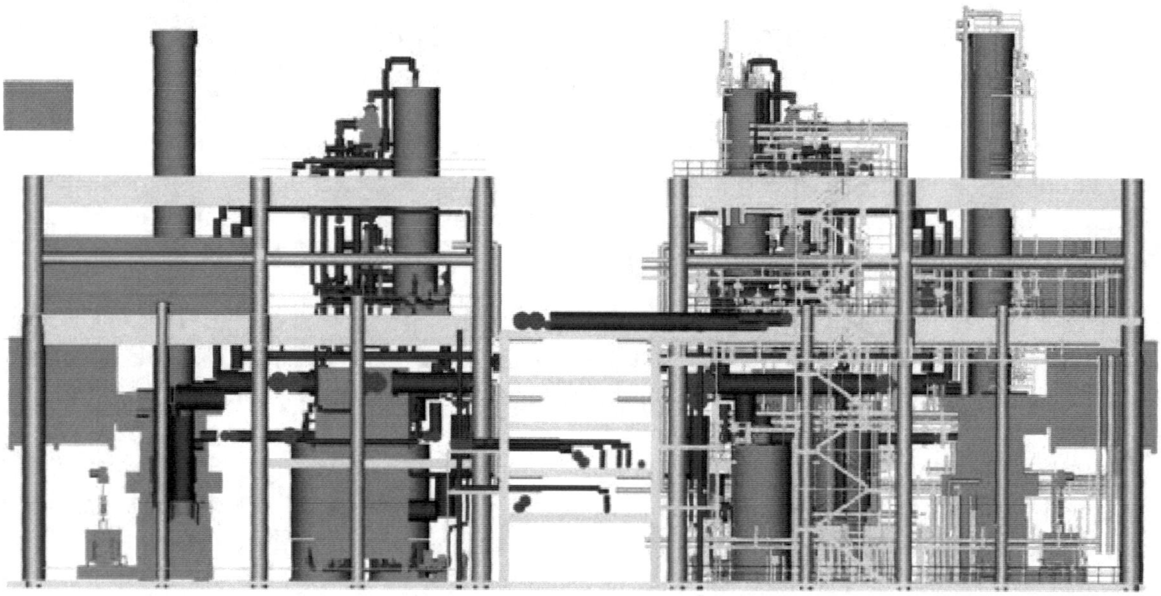

Figure 3: Two process modules in CFD. Left module early design phase, right module mirror image with anticipated congestion added

3 Explosion Risk and Design Accidental Loads

Once a detailed geometry layout has been determined, advanced explosion risk analyses can be performed to understand the inherent risks of a given facility and to help optimize features of the layout to further reduce the overall risk of process hazards onboard the vessel. For example, one of the major decisions that engineering teams need to evaluate is where to put the living quarters to minimize the risk to personnel in the event of an unintentional release, especially on congested process facilities such as FPSOs and FLNGs. As a specific example, for the Shell Prelude FLNG, it was decided to put the living quarters (LQ) aft of the vessel and to protect the LQ with a blast wall placed between the process area and LQ (see Figure 4). A key design parameter would be to determine the design accidental load (DAL) of the blast wall so that it will perform its function in the design accidental scenario. Determining the DAL requires detailed analyses that include the statistical weather conditions, likelihood of a wide array of leak scenarios, and the probability of certain explosion consequences. The first step is to determine the range of vapor cloud sizes that could reasonably contribute to explosion pressures, which is highly dependent on the vessel's ventilation.

Figure 4: Shell's Prelude FLNG with the LQ aft on the vessel (photo source: https://www.shell.com/about-us/major-projects/prelude-flng.html)

Weather data

As the size of a vapor cloud is typically inversely proportional to the wind speed, it if first important to obtain statistical weather data for the given facility. If we are considering turret-based FPSOs and FLNGs, the vessels utilize weathervaning, where the vessel rotates around the turret to orient itself with the wind and sea currents. This will lead to very distinct wind and ventilation patterns on the vessel. For example, the predominant wind direction will generally be along the vessel (0 degree ± 10 degrees wind heading) and the wind speeds will be that local to the region. Figure 5 shows the likelihood and wind speeds on a fictitious FPSO for different relative headings.

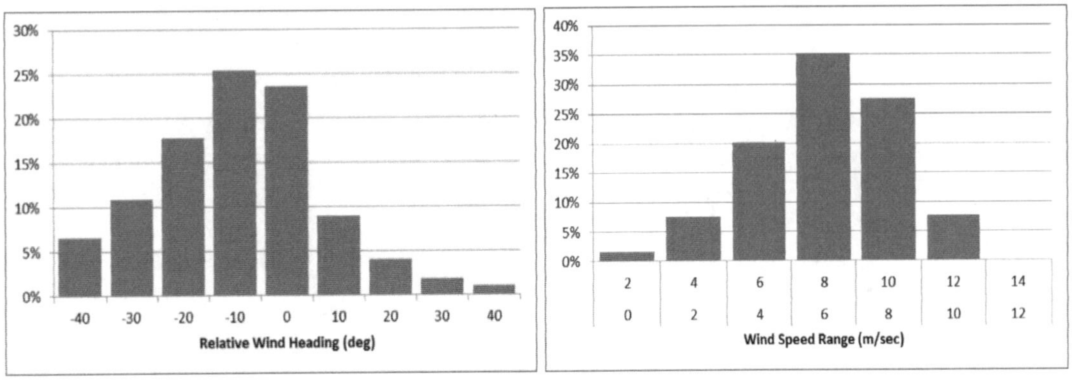

Figure 5: Relative wind heading and speeds for a fictitious FPSO

The angle in the plots is assumed to be the difference between the ship's heading and source wind direction, with negative angles associated with wind originating from the port side and positive angles from the starboard side. Simulations to determine ventilation yield results for selected volumes, generally measured as air changes per hour (ACH). If the wind originates from the port side, ventilation will generally be much higher for the portside modules and gaps between the modules; however, ventilation is much lower for the modules on the starboard side (see upper image in Figure 6). In addition, if the wind is oriented directly head-on towards the front of the vessel, the ventilation of the forward-most area is well ventilated, but the rest of the vessel is much less ventilated.

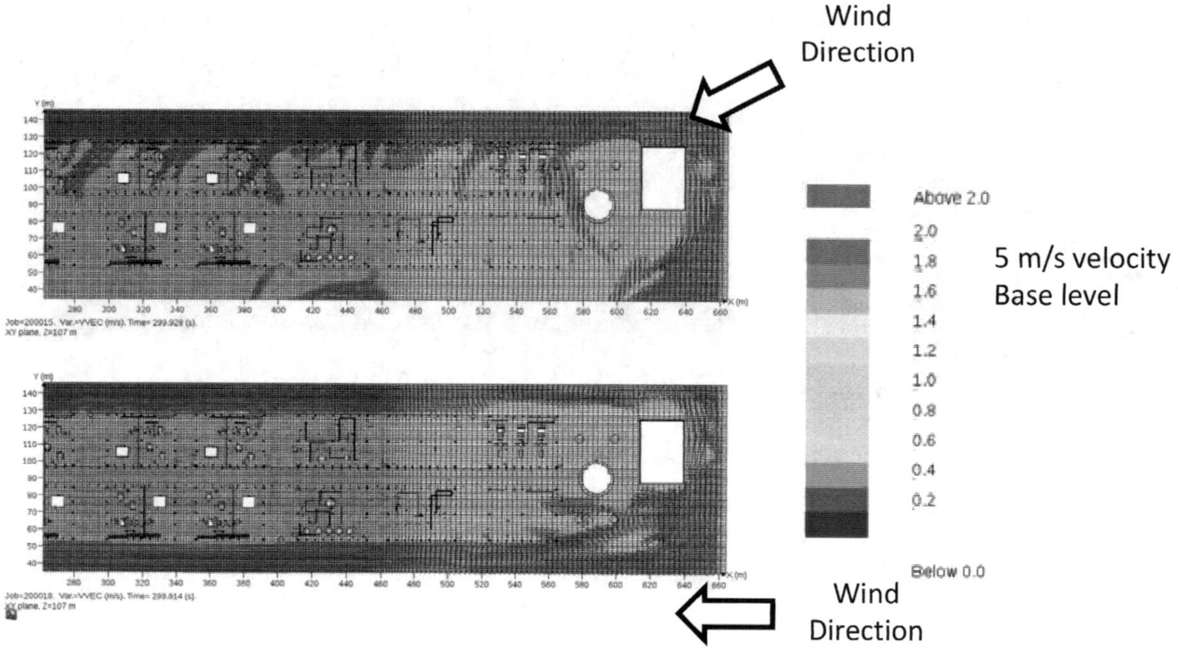

Figure 6: Ventilation patterns with wind coming from the port side (upper image) and head onto the vessel (lower image).

Leak scenarios and frequencies

Once the ventilation conditions are determined, a wide range of leak scenarios (ranging from 1 - 100 kg/s release rates) are performed at various locations and directions to establish temporal variation in vapor cloud formation in congested areas. An example of a release is shown in Figure 7.

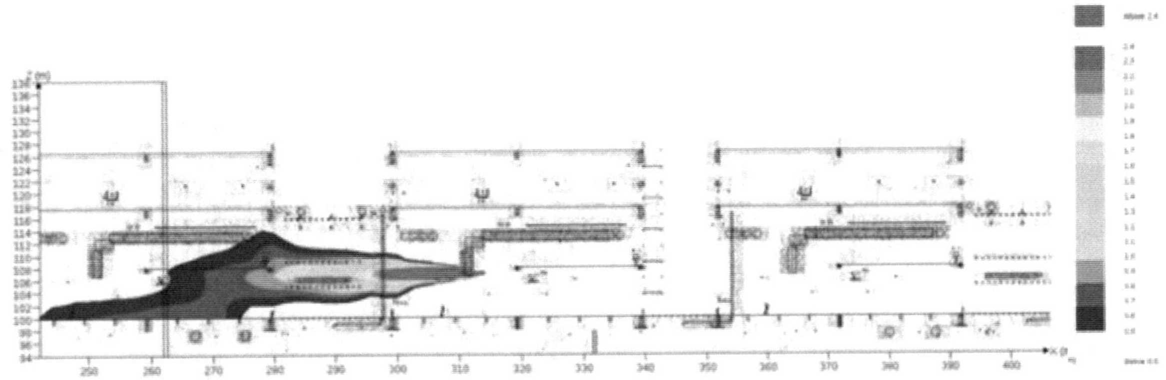

Figure 7: Illustrative example of flammable concentrations from dispersion simulations that account for congestion and confinement (scale is LFL to UFL)

A complete set of scenarios can be run that show the full impact of accounting for 3D congestion and confinement for the design. Figure 8 illustrates a large release for which the wind is oriented 15° from the ships heading, which causes the flammable vapor to cross process module boundaries.

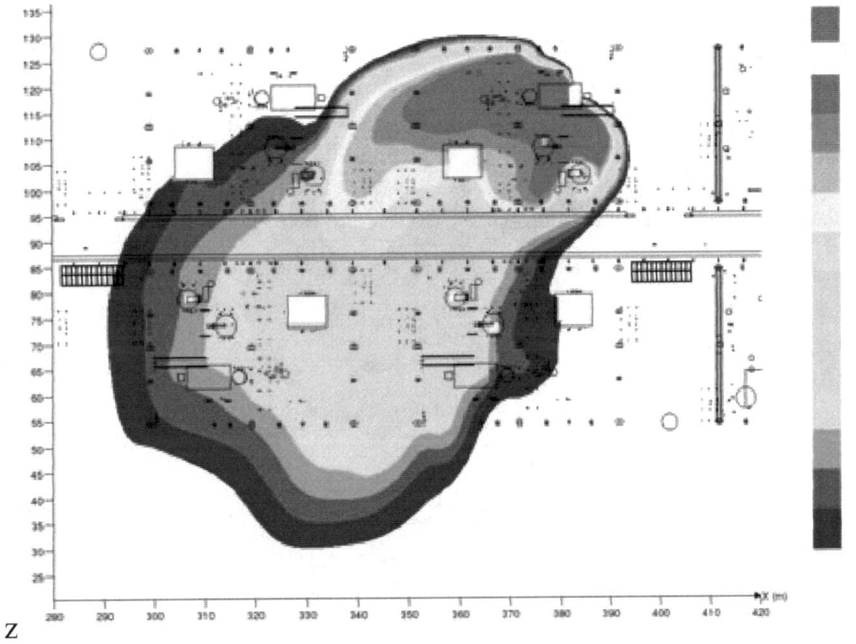

Figure 8: Illustrative example of large leak rate release dispersing into adjacent modules (scale is from LFL to UFL)

Many dispersion simulations can be combined to review the effects of design decisions on flammable cloud sizes. For example, perhaps the design team is debating between plated or grated deck regions to reduce explosion consequences at the blast wall. Figure 9 shows the maximum explosion cloud sizes for releases when comparing plated and partially grated design alternatives.

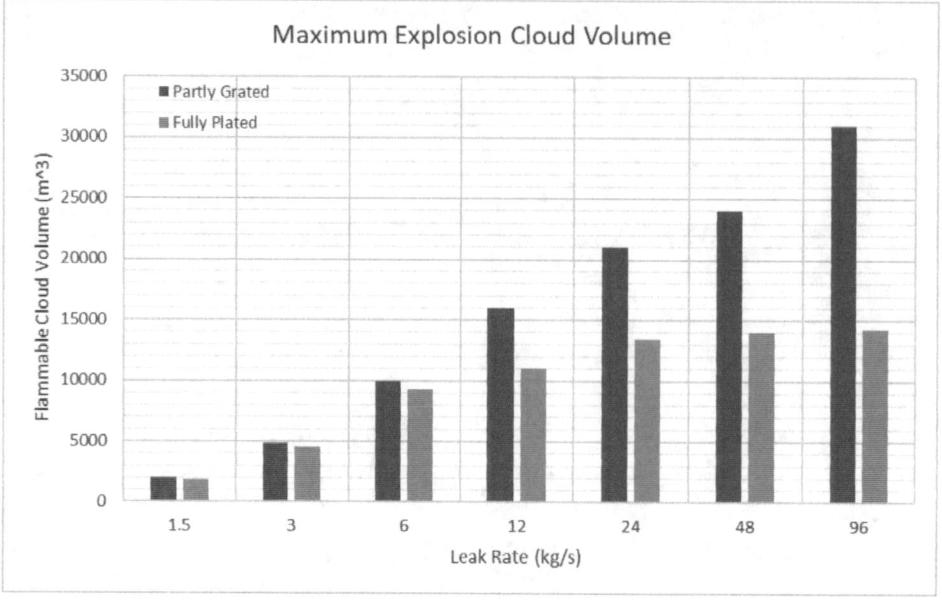

Figure 9: Flammable cloud volumes for dispersion scenarios comparing two design alternatives

Combined with frequency data based on the leak sources assumed in the design, it is possible to generate exceedance curves for cloud sizes such as the one shown in Figure 10.

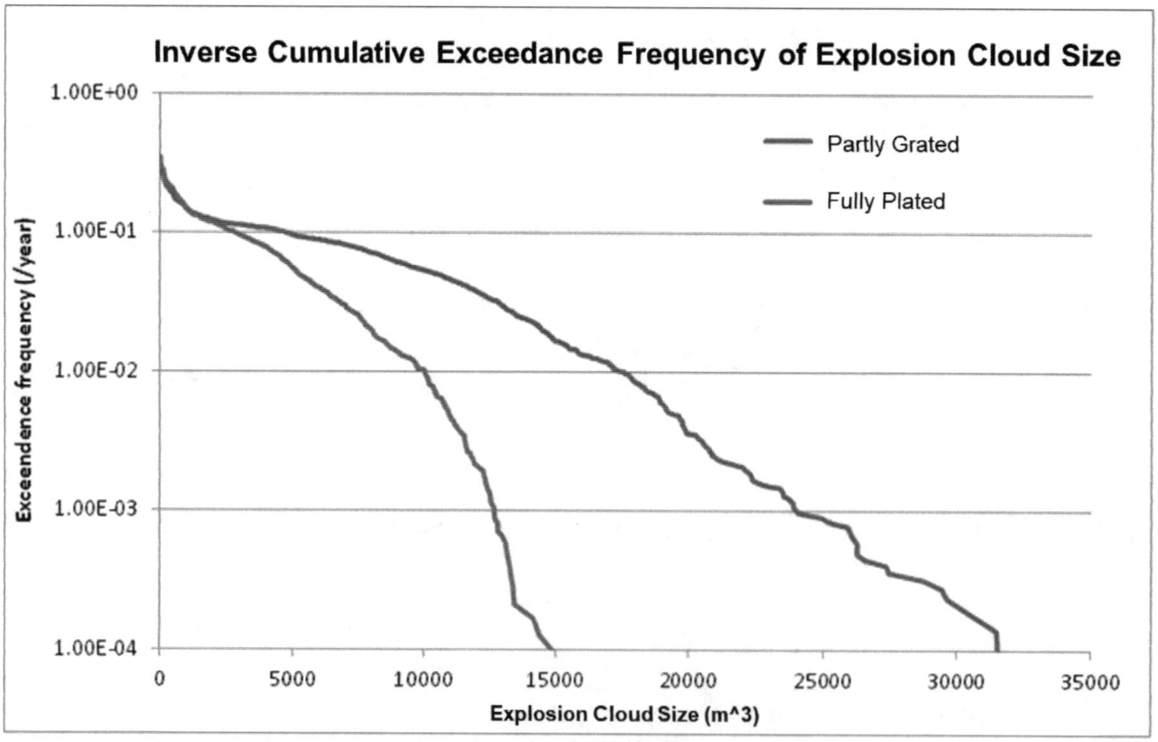

Figure 10: Exceedance curve for flammable cloud size

The results of the dispersion simulations can then be fed forward to a range of explosion simulations to evaluate the effects of grating versus plating from an explosion perspective. Combining the results of explosion simulations with the frequencies associated with the scenarios yields an exceedance curve for the blast wall as shown in Figure 11. The DALs are the overpressures associated with the project's risk frequency criteria.

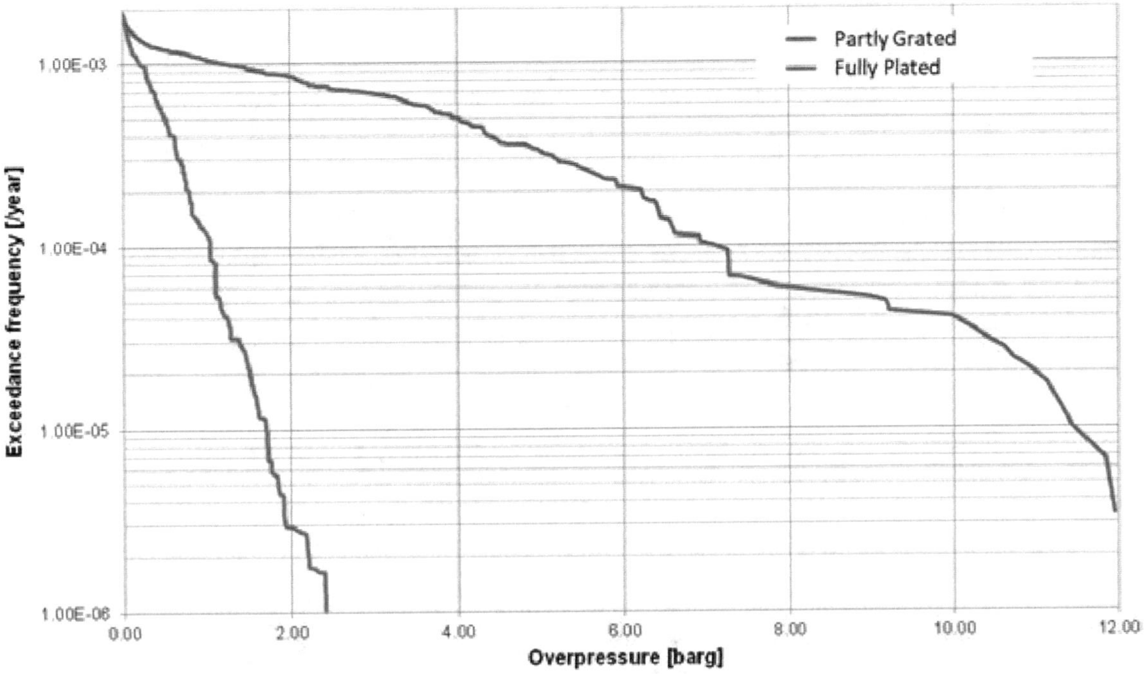

Figure 11: Exceedance curve for overpressure at the blast wall

This kind of approach ensures that design teams and project managers have good information to make decisions regarding the feasibility of various design alternatives as early as possible before significant engineering hours have been invested in a potentially unsafe design. For the above generic example, if grating were to be implemented, the blast wall would need to be designed to 6 barg for an exceedance frequency of $1x10^{-4}$ per year instead of 1 barg if plating were utilized.

4 Safety Gap Sizes

Safety gaps are regions of intentionally empty space between process modules to reduce the interaction between modules in the event of a hazardous event. It is typical to provide safety gaps between process modules that contain flammable material to reduce the length of flame travel and acceleration through congested regions to mitigate explosion consequences. Recently, safety gaps have been shown to affect both the dispersion and combustion in case of a loss of containment (and possibly ignition) event as will be shown in the next example.

These case studies evaluated the effectiveness of safety gaps on: (1) dispersion, where two different safety gaps (12.5 meters and 20 meters) between the liquefaction modules were evaluated (see Figure 12), and and (2) explosions, where three different safety gaps (10, 15 and 20 meters) between the liquefaction modules were evaluated in Figure 13 for similar cloud sizes.

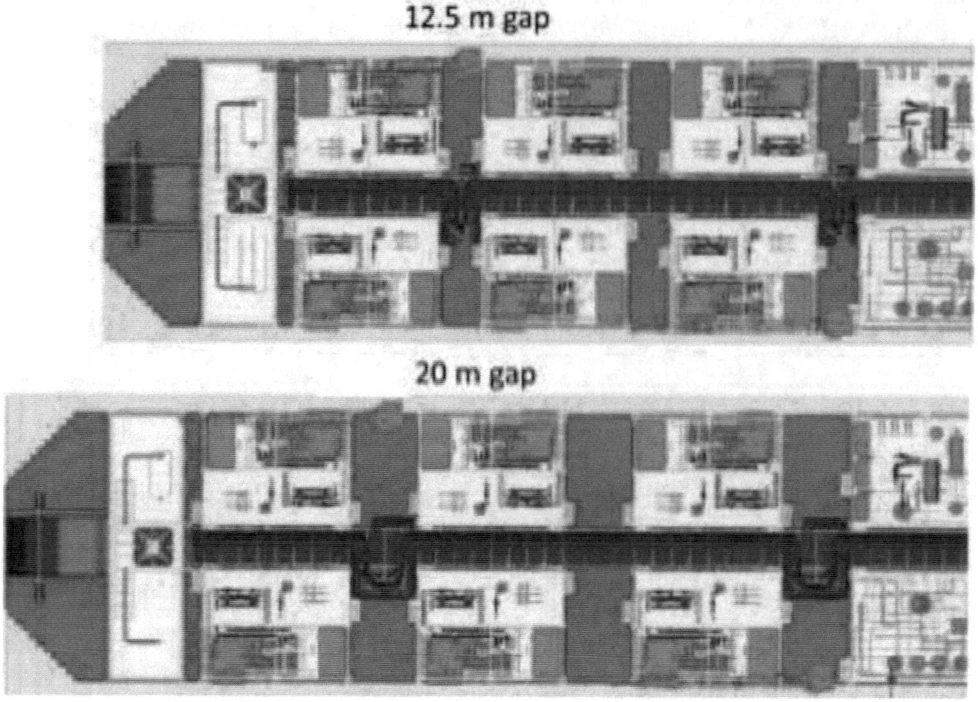

Figure 12: Safety gap configurations for ventilation and dispersion

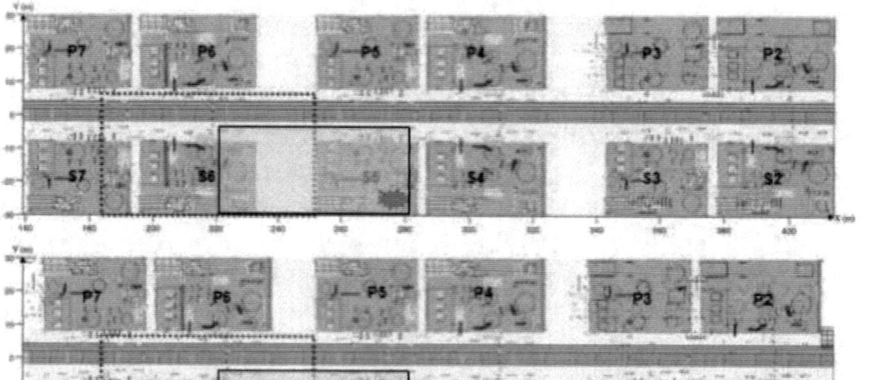

Figure 13: Safety gap configurations for explosion analysis

The results showed that the ventilation rate within the module (measured by air changes per hour or ACH) remained essentially unchanged as the safety gap was increased from 12.5 to 20 meters. Therefore, increasing the gap size from 12.5 to 20 meters did not significantly reduce the size of a gas cloud that remained inside the module where the leak originated. However, increasing the gap size had a sizeable effect on clouds that migrated out of the origin module and into neighboring modules, as shown in Figure 14. While the only variable that changed in the figure was the size of the safety gap, there was a difference of more than 2.5x the cloud size in the neighboring module (20 965 m^3 versus 7904 m^3). This was significant with respect to overall gas cloud size and ignition frequency.

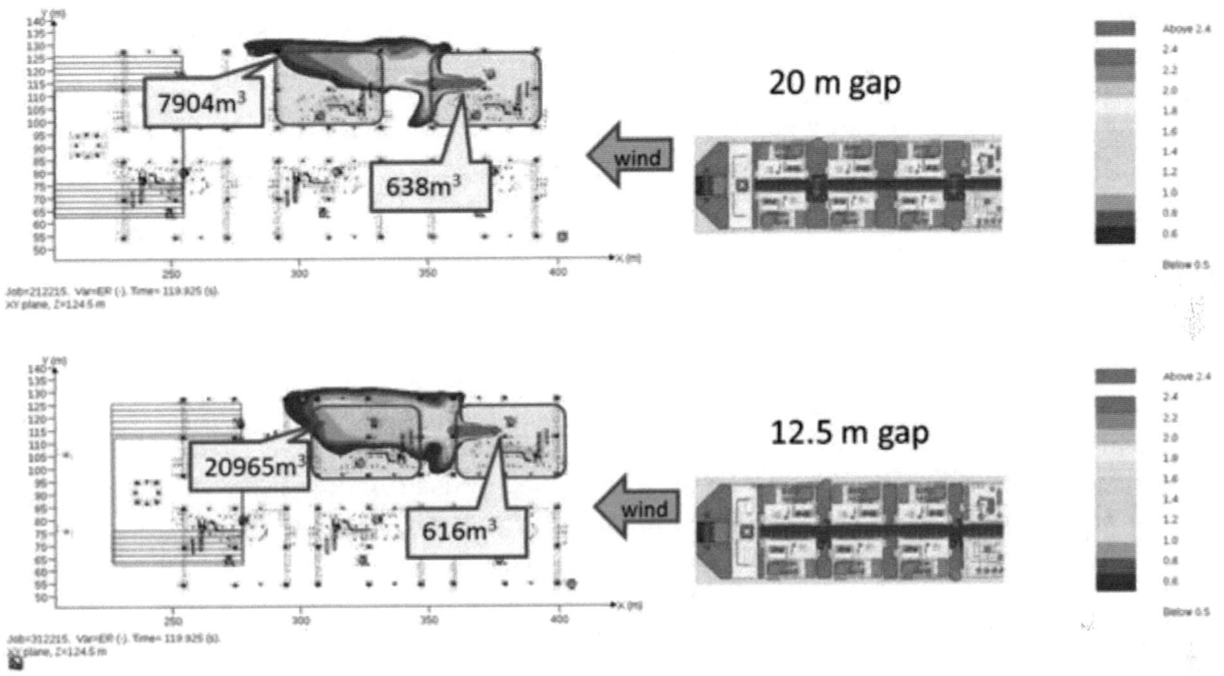

Figure 14: Dispersion example for the 20 meter gap and 12.5 meter gap

The results for the explosion case showed that the explosion consequences decreased significantly for the larger vapor clouds as the safety gap was increased from 10 meters to 20 meters, as shown in Figure 15. As the vapor cloud volume was held constant in the example, the reduction in local overpressures was due to the combined effect of the vapor cloud occupying less congested volume as well as the flame speed decreasing within the safety gap prior to arriving at the neighboring module.

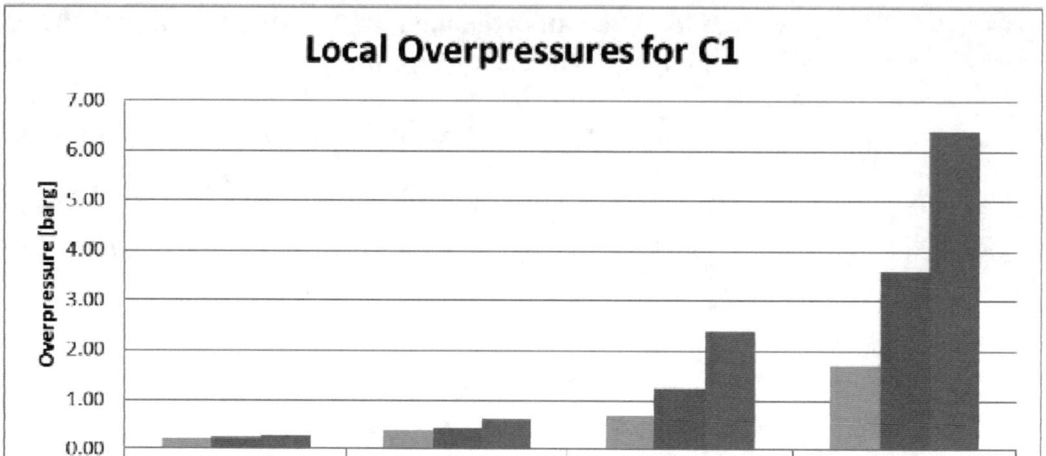

	8000 m3	12000 m3	19500 m3	24000 m3
20m	0.21	0.38	0.69	1.73
15m	0.25	0.44	1.25	3.59
10m	0.28	0.59	2.39	6.41

Figure 15: Local explosion pressures for various cloud sizes traversing a 20, 15, and 10 meter gap

5 Module Layout

CFD ventilation studies can also be used as part of selecting and/or optimizing layouts. Ventilation studies are often performed to validate that layouts meet industry criteria or as a step in more complex safety studies; however, their value in making critical decision decisions in early-design that can drastically alter the risk posed by toxic, fire, and explosion hazards is often overlooked.

Many ventilation studies provide results as a single measure, such as “adequate ventilation” measured by ACH for a given area. For example, certain hazardous area classifications require that the ventilation must be greater than 12 ACH at least 95% percent of the time (95th percentile ACH). However, the 95th percentile ACH only describes a limited range of the ventilation conditions at the facility and should not be used as the sole means of comparing alternative layouts or making design decisions.

A carefully constructed ventilation study that accounts for the nature of facility hazards can provide relatively fast and inexpensive insight into potential safety-oriented optimizations during early stages of facility design. This insight in the early design phase allows for facilities to be rigorously designed for safety—minimizing costly late-stage design changes. For example, it would not be practical to move around process modules after investing months in designing piping and cabling; however, changing the arrangement of process modules is more practical in early-design.

Figure 16 represents the base case layout of a FLNG that would be offloaded in a stationary fixed vessel position at port for offloading operations. In the figure, the blocks A, B, C, and D represent the liquefaction modules. For this example, the task was to select the layout that has the best ventilation of a fixed position offloading vessel by alternating the position and layout of the topside modules. Table 1 summarizes the differences between the various layouts.

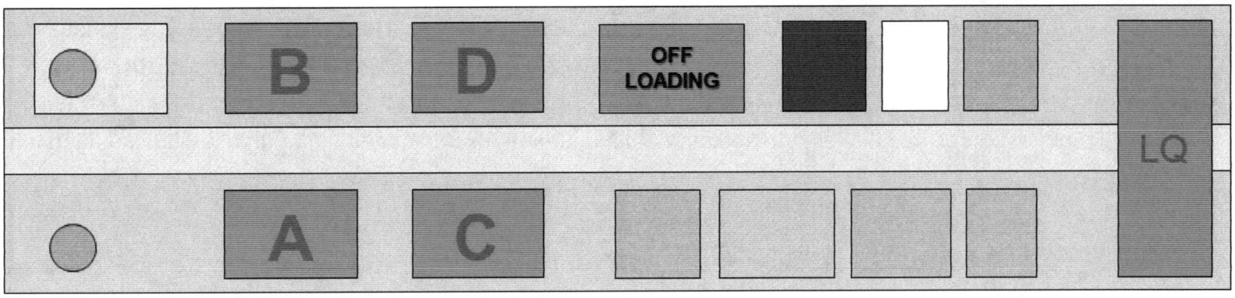

Figure 16:Sample FLNG layout

Table 1: Listing of alternative FLNG layouts

Layout #	Description of Case
1	Base Case
2	Larger Safety Gaps
3	Diagonal Liquefaction Modules Alternative 1
4	Diagonal Liquefaction Modules Alternative 2
5	Liquefaction on Non-Loading Side Alternative 1
6	Liquefaction on Non-Loading Side Alternative 2
7	Liquefaction on Loading Side with Alternate Offloading Position
8	Staggered Liquefaction on Non-Loading Side Alternative 1
9	Staggered Liquefaction on Non-Loading Side Alternative 2

Table 2 shows the 95th percentile for all of the layout alternatives, sorted in descending order. Note that all of the 95th percentile ACH values fall into a reasonably narrow range.

Table 2: Sorted 95th percentile ACH values for alternative FLNG layouts

Layout #	95th Percentile ACH
7	20.1
6	20.0
5	19.8
9	19.6
8	19.4
2	19.3
3	18.7
1	18.0
4	17.6

Figure 17 shows the ACH exceedance curves for the top five layouts. The curve for layout #7 (which has the highest 95th percentile ACH) is represented with a dashed line. Although layout #7 has the highest 95th percentile ACH, overall, it is not one of the better ventilated layouts. Especially considering the close range of 95th percentile values, layout #7 should not be selected as the optimal layout.

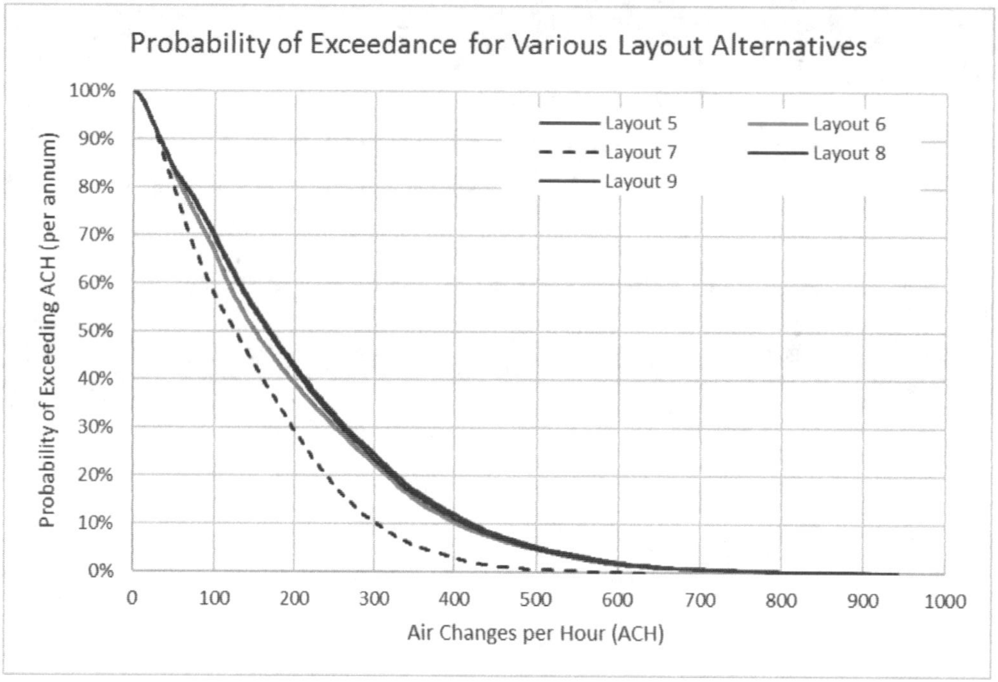

Figure 17: ACH Exceedance Curves for Top 5 FLNG Layout Alternatives by 95th Percentile ACH

Table 3 shows the 95th percentile ACH for each layout as well as two ranks for each layout—for the 95th percentile as well as for the area under the curve. Layouts #7 and #9 are both highlighted in the table because they are the top ranked layouts for the 95th percentile and area under the curve, respectively.

Table 3: 95th percentile ACH rankings and area under curve rankings for FLNG layout alternatives

Layout #	95th Percentile ACH	95th Percentile Rank	Area Under Curve Rank
1	18.0	8	9
2	19.3	6	8
3	18.7	7	4
4	17.6	9	6
5	19.8	3	3
6	20.0	2	5
7	20.1	1	7
8	19.4	5	2
9	19.6	4	1

Despite that layout #9 is the 4th highest ranked 95th percentile ACH, it has the best ventilation profile overall. Figure 18 shows the ACH exceedance curve for layouts #7 and #9 only. On this plot, it is very clear that layout #9 shows generally better ventilation performance.

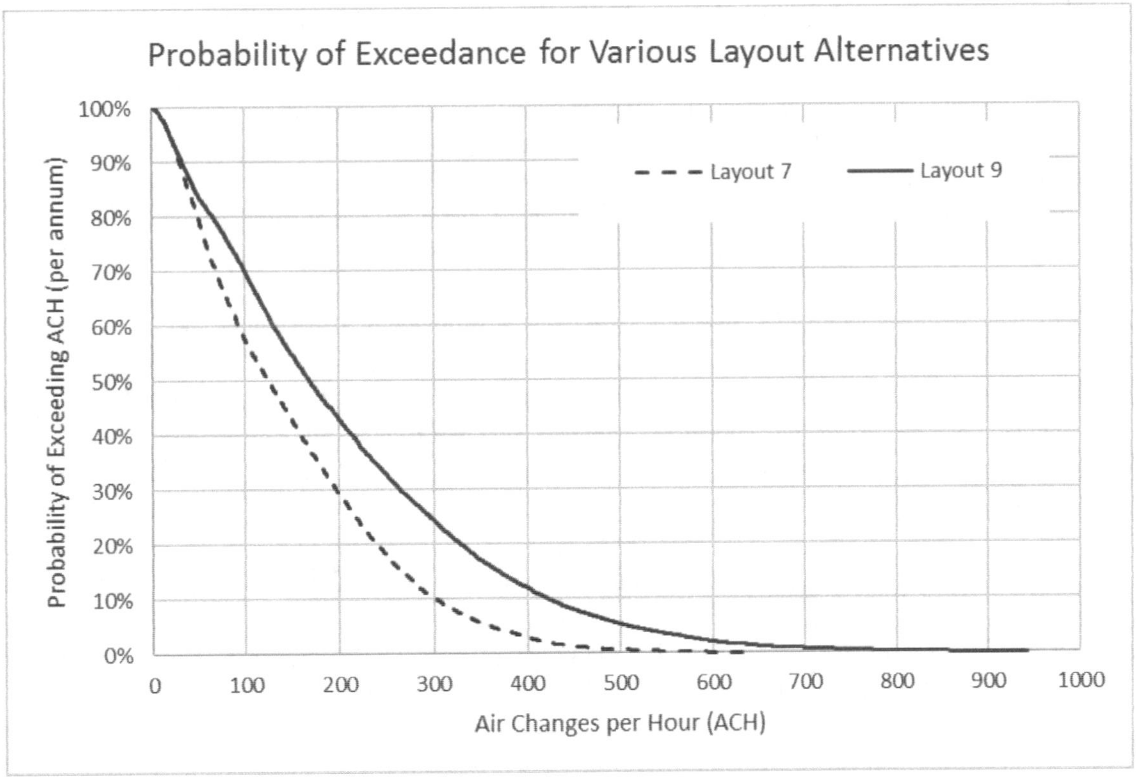

Figure 18: ACH Exceedance Curves for Top 95th Percentile ACH Layout and Top Area Under the ACH Curve Layout

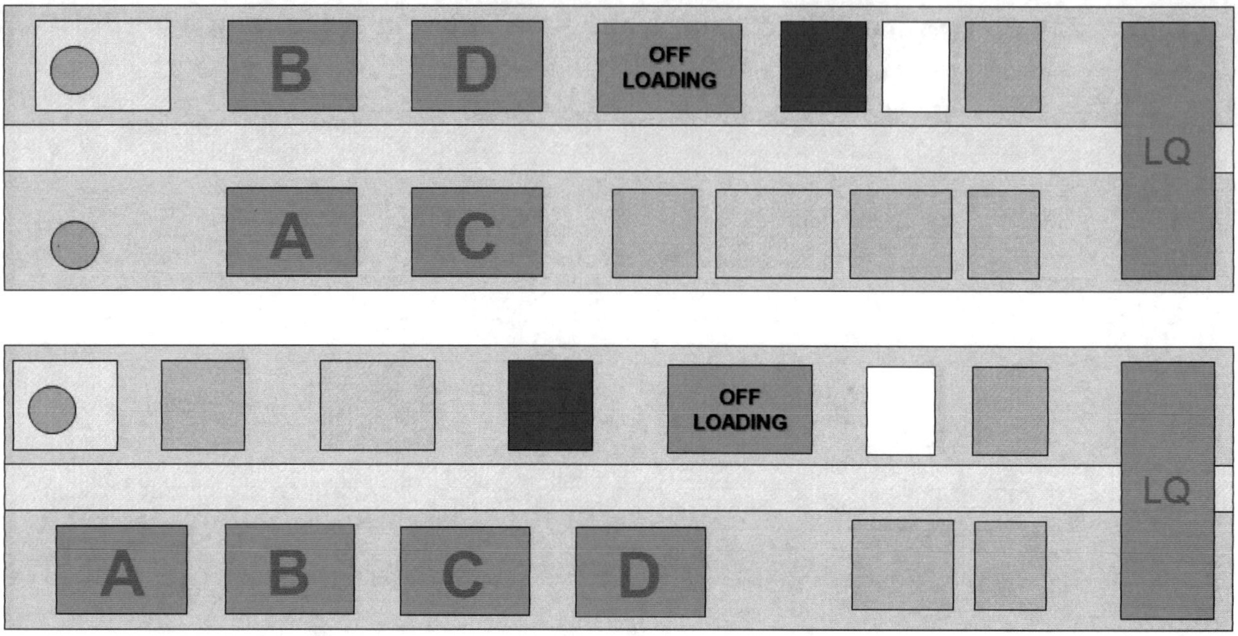

Figure 19: Comparative illustration between the base case layout (top, layout #1) and the best ventilated layout (bottom, layout #9).

As shown in Figure 19, there are significant differences in the layout of the modules between the base case (top of figure) and the best ventilated case (bottom of figure). The liquefaction modules are all located on the opposite side of the vessel from off-loading and the locations of the safety gaps on the offloading side of the vessel are aligned with the liquefaction modules. This combination yielded the overall best ventilation performance.

Ventilation studies can be used to provide tremendous insight into layouts early in design while less information about leak sources is available. When further dispersion and explosion studies are performed later in the design of this facility, the best results will likely come from the best ventilated case—which could have been determined extremely early in the project's life.

6 Conclusion

This paper examined the benefits of using advanced CFD simulations to evaluate for explosion risk in the early design phase of FPSO vessels. For example, the inherent safety of having the living quarters forward in the vessel for a FPSO with turret mooring versus the living quarters aft in the vessel for a fixed, spread mooring system can be quantitatively evaluated in concept or pre-FEED phase. There are a variety of qualitative and quantitative techniques available to make decisions for an inherently safer design; however, this paper provides examples to show the power of using CFD modeling in early phases of design.

Industry capability to represent as-built facilities using early CAD models is reasonably well-developed. This enables using CFD methods to facilitate answering complex questions regarding ventilation, dispersion, fire, and explosion hazards with a reasonable degree of certainty, and therefore, providing useful insight into safety decisions early in the design process when these decisions can have the largest impact for the smallest cost—resulting in a design that is inherently safer.

7 References

[1] D. Botwinick and S. Davis, "Getting It Right for Projects in Early Design Phase: Advanced Techniques to Ensure Accurate Explosion Assessments," Process Safety Progress, Dec 5, 2018, DOI: 10.1002/prs.11995.

Interpretation of Damage Caused by a Vapor Cloud Explosion

D.M. Johnson
DNV GL, Spadeadam Research and Testing
MOD R5, Brampton, Cumbria, CA8 7AU, UK
michael.johnson@dnvgl.com

A. Pekalski, Shell Research Ltd., London, UK. Andrzej.Pekalski@shell.com

V.H.Y. Tam, University of Warwick, University Road, Coventry, CV4 7AL, UK. V.Tam@warwick.ac.uk

B.A. Burgan, Steel Construction Institute, Silwood Park, Buckhurst Road, Ascot SL5 7QN, UK. B.Burgan@steel-sci.com

P. Hoorelbeke, Total, Paris la Défense cedex, France. pol.hoorelbeke@total.com

C. Savvides, BP Exploration, Chertsey Road, Sunbury-on-Thames TW16 7LN, UK. chris.savvides@uk.bp.com

D. Allason, Spadeadam Testing and Research Centre, MOD R5, Brampton, Cumbria, CA8 7AU, UK. daniel.allason@dnvgl.com

Prepared for Presentation at
American Institute of Chemical Engineers
2019 Spring Meeting and 15th Global Congress on Process Safety
New Orleans, LA
March 31 – April 3, 2019

Interpretation of Damage Caused by a Vapor Cloud Explosion

D.M. Johnson
DNV GL
Spadeadam Testing and Research Centre, MOD R5, Brampton, Cumbria, CA8 7AU, UK
michael.johnson@dnvgl.com

A. Pekalski
Shell Research Ltd., London, UK.
V.H.Y. Tam
University of Warwick, University Road, Coventry, CV4 7AL, UK.
B.A. Burgan
Steel Construction Institute, Silwood Park, Buckhurst Road, Ascot SL5 7QN, UK.
P. Hoorelbeke
Total, Paris la Défense cedex, France.
C. Savvides
BP Exploration, Chertsey Road, Sunbury-on-Thames TW16 7LN, UK.
D. Allason
Spadeadam Testing and Research Centre, MOD R5, Brampton, Cumbria, CA8 7AU, UK.

Keywords: Vapor Cloud Explosion, detonation

Abstract

The understanding of vapor cloud explosions (VCEs) has changed significantly over recent years. Whereas it had been previously considered that VCE incidents had involved only deflagrations restricted to well defined congested process regions, it is now widely accepted that Deflagration to Detonation Transition (DDT) has been involved in major VCE incidents. Once DDT occurs, high pressures will be generated throughout any remaining unburned cloud, not just the congested process region. The stimulus for this change in understanding was the research conducted following the Buncefield incident in the UK in 2005, however, the conclusions have been found relevant to other VCE incidents. In the process of both conducting the research and reviewing VCE incidents, a significant amount of data has been obtained on the effects of VCEs on items within and around the flammable cloud. This data will be of significant value in the interpretation of the forensic evidence produced in any future VCE. This paper reviews the data available from the research and shows how it can be used in the investigation of any future VCE incidents.

1 Introduction

The investigation of the VCE at Buncefield, UK, prompted research that provided unequivocal evidence that DDT was important in the explanation of major VCE incidents [1][2]. The research was conducted under the guidance of a Technical Group comprising experts in major hazards and explosions, referred to as the 'Buncefield Research Group'[1] in this paper, and considered a range of possible explanations; deflagration only, DDT and a 'new' explosion mechanism that could sustain a deflagration travelling at less than half the speed of sound across open uncongested ground. It was concluded that the Buncefield VCE could only be explained by a DDT that resulted in a detonation propagating through most of the vapor cloud. The Buncefield Research Group rejected the possibility of the proposed 'new' explosion mechanism (called episodic combustion) on the basis that large scale experimental studies failed to see any evidence of key elements of the process, even in very favorable conditions. Though some publications have continued to suggest that the episodic combustion is required to explain some VCEs [2], this has been shown to be based on a narrow interpretation of the very specific evidence rather than the totality of information available [4][5]. Also, as will be seen, any deflagration mechanism travelling well below the speed of sound would fail to explain some of the critical evidence within the vapor cloud.

Using the understanding gained from this research, evidence from other major VCE incidents has been reviewed and it has become evident that DDT and sustained detonation propagation has occurred in other VCEs [6].

The understanding gained through both the research and the review of incidents is used in future to aid the investigation of major VCE incidents. Illustrations of the types of evidence that might be observed following a major VCE are provided along with a description of what can be concluded from the evidence.

2 Important Research Findings

Before considering the interpretation of evidence in a VCE, it is worth providing a summary of some important aspects related to detonation and key findings from research.

2.1 Conditions for DDT

DDT of hydrocarbon-air mixtures is most likely to occur when the fuel concentration is close to stoichiometric as:

- The amount of 'explosive' energy required to initiate a detonation is at its minimum. The 'explosive' energy can be characterized by the amount of high explosive required to initiate a detonation [7].
- The flame acceleration in any congestion or confinement will be at its maximum, giving greater potential for the conditions required for DDT to occur [8].

[1] The members changed through the research and the personnel and organisations involved are detailed in [1] and [2]. The authors/organisations presenting this paper have all been involved throughout the research.

Large scale experimental studies with congested pipework have shown that DDT occurs only when the flame speed exceeds the speed of sound. The speed of the flame at the time of DDT seems to be dependent on the experimental arrangement, ranging from about Mach 1.2 [4] up to approximately Mach 2 [1][8][9] for alkanes from ethane and heavier. Given a stoichiometric mix of these fuels and sufficient congestion or partial confinement, DDT can occur within a relatively short flame propagation distance, with flame propagation distances of only 10-30m [1][4][8]. For example, in propane-air experiments studying flame acceleration in dense undergrowth at the DNV GL Spadeadam site [1], the flame accelerated from a spark ignition to DDT within only12m. Figure 1 illustrates this with the pressures measured at 6, 9 and 12m from the spark ignition in this test. At 3m the peak pressure is about 1.5bar, increasing to over 6bar at 9m and showing a detonation profile at 12m.

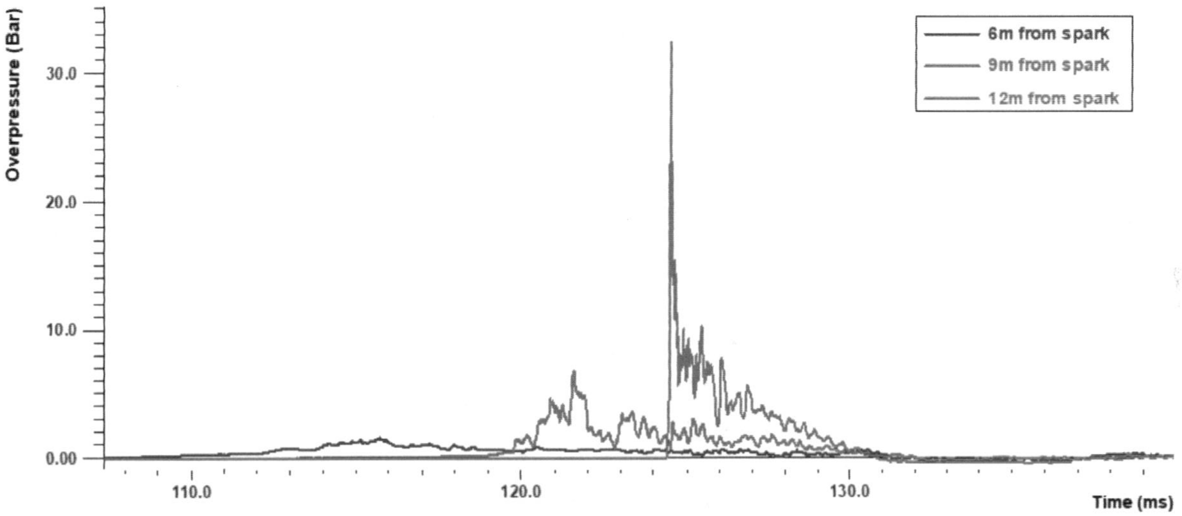

Figure 1. Pressures at 6, 9 and 12m from spark (time axis is time from ignition)

In several large scale propane-air tests performed by Gexcon and SRI [4], the flame accelerated from a low energy ignition source through relatively low congestion to DDT in a distance of about 30m.

Within a dense vapor cloud covering a large area, only part of the cloud needs to be close to stoichiometric and with enough congestion (and/or some degree of confinement) for DDT and sustained detonation to result. Once a detonation has been established, it will be far less sensitive to fuel concentration than a deflagration as it can be sustained by any fuel concentration within the detonable limits. The detonable limits are not as wide as the flammability limits (e.g. 3%-7% for propane air, compared to 2.1%-9.5% for the flammability limits [1]), but the detonation is likely to be able to propagate through substantial portions of the flammable cloud.

Large scale experiments have also shown that an established detonation was sustained as the depth of the cloud was reduced down to a depth of 200mm [1], thus even if there are localized pockets of the cloud with a concentration outside the detonable limits, the detonation is likely to be able to propagate around them even if the detonable part is relatively 'thin'.

2.2 *Pressures*

Once DDT has occurred, the propagating detonation will generate a shock wave of magnitude of up to 20bar and a duration of about 4ms [1]. The magnitude of the shock will be reduced in mixtures with a concentration significantly different from stoichiometric but will remain above 15bar.

In the Buncefield incident the severity of the pressure damage reduced rapidly beyond the cloud edge. Analysis of the results of simulations of large low-lying vapor clouds [1] showed that for these 'pancake' clouds, the magnitude of the overpressure outside the cloud (*P*) depended on only the height of the cloud (*H*) and the distance from the cloud edge (*D*), in the form of the equation below.

$$P = P_0 \frac{7.8H}{D}$$ where P_0 is the atmospheric pressure

In this formula, the magnitude of the pressures outside the cloud are independent of the cloud area. However, analysis of the results from detonation experiments showed that the external pressure field was relatively well predicted by the TNO Multi-Energy Method and TNT equivalence [1], which scale by cloud volume and therefore are dependent on the cloud height and area.

An explanation for this apparent conflict is that the simulations were for clouds significantly bigger (radius 100-200m) than those used in the experiments. The results suggest that there is a finite area of cloud in the horizontal direction that can contribute to the magnitude of the external pressure field [1]. Once the cloud area exceeds this maximum, the only way to change the volume contributing to the external pressures is by varying the cloud height, hence the dependence on only the cloud height. The experiments are below this critical size and the simulations above it.

It should be noted that this conclusion only related to the magnitude of the external shock wave. The impulse does not decay rapidly [1], indicating that the duration of the external shock wave must be increasing as it propagates further from the cloud to counteract the effect of the rapid decrease in the pressure magnitude. In addition, these conclusions relate to investigation of pancake clouds only.

2.3 *Net Loading and Directional Effects*

A very important finding from the Buncefield research related to the interpretation of 'directional indicators'[1]. These are items (illustrated later) where there is some asymmetry in the effects of the VCE such as physical translation or deformation in one direction or damage such as scorching or removal of paint on one side on an object. In the Buncefield investigation, the directional evidence within the vapor cloud was initially confusing, suggesting multiple explosions. The realization that the indicators within the cloud showed where the explosion was coming from, not where it was going, was a one of the most important findings of the post Buncefield research. This finding was confirmed by both simulations and large-scale experiments [1][2].

The simulations of both detonating clouds and fast deflagrations showed that for pancake clouds, the flow of the combustion products was in a direction opposite to the direction of propagation of the flame/detonation [2]. The impulse from the resulting drag forces exceeded those from the much shorter duration detonation shock loading and were in the opposite direction. With this realization, the directional indicators provided a consistent picture, pointing back towards the point of DDT in the cloud [2].

It is the directional indicator aspect that critically flaws the episodic deflagration mechanism that was proposed for Buncefield. The average flame speed was proposed to be 150m/s or less and therefore the average speed of the reverse flowing combustion products would be much less than for a detonation. The reverse drag force depends on the square of the flow speed and will therefore be between one and two orders of magnitude less in the proposed episodic propagation than in the detonation (where the reverse flow speed is approximately 400m/s) and would be too low to cause the observed translation of items. This is also confirmed by experiments with comparable flame speeds [1].

Review of previous incidents showed that this observation was not new. In report on a propane-air explosion in a rural area at Port Hudson, Missouri in 1970 [10], the authors concluded the vapor cloud had detonated and commented *"We think that it is significant that the wind direction was everywhere opposite to the postulated direction of the detonation"*. The Buncefield Research Group had the advantage that it was possible to understand how this occurred through a combination of explosion simulation and large scale experimental studies.

3 Interpretation of Forensic Evidence

To illustrate the types of evidence that might be observed in a major VCE and how this can be interpreted, reference is made to previous incidents, particularly the VCEs at Buncefield [1], Jaipur [11][12] and Puerto Rico [4].

3.1 Areas within the vapor cloud with relatively low levels of pressure damage

The ignition of a vapor cloud in a VCE may well not lead to immediate fast flame propagation and high pressures. There is a good chance that there will be some slower flame propagation generating little or no pressure. The Buncefield, Jaipur and Puerto Rico incidents provide good examples of how this early stage of flame propagation can vary significantly.

In the case of Jaipur, the cloud covered almost all the site, as shown in Figure 2. In the North-East corner, there was an area, also shown on Figure 2, where there was evidence of thermal damage from combustion of the vapor cloud but a low level of pressure damage. The directional indicators pointed to the initiation of the DDT in this corner of the site.

The interpretation of the evidence is that ignition occurred in the north-east corner, probably just off site and the flame propagated as a flash fire through at least the shaded area in Figure 2. The flame then underwent a DDT, possibly by entering a building and then venting out at high speed [12]. The small amount of less severe pressure damage in the shaded area was caused by the pressure field from the detonation in other parts of the cloud.

Figure 2. Full Jaipur site (left) showing approximate cloud boundary (yellow) and summary of directional indicators, and expanded view of north-east corner (right) showing area with little pressure damage (shaded yellow) and directional measurements made on specific objects (red arrows)

In the case of Buncefield, the investigation [2] identified that ignition occurred in a firewater pump house at the north end of the site, Figure 3. In this case, the DDT occurred following flame acceleration in the tree line to the north of the pump house. Examination of photographic evidence showed that there was a relatively low level of damage in a section of the tree line just adjacent to the pump house, as shown in Figure 3.

The interpretation of this evidence is that following ignition in the pump house, the flame entered the tree line and began to accelerate, causing a relatively low level of damage until DDT occurred. The sequence of events, based on closed circuit television (CCTV) evidence and witness statements, indicated that there was at least half a second between the initial explosion in the pump house and DDT [2], allowing time for the flame acceleration in the trees.

Figure 3. Firewater Pump House before and after Buncefield VCE showing area of relatively low damage to tree line from south (top right) and north (lower right)

In the Puerto Rico incident at the Caribbean Petroleum Company (CAPECO) facility, the flame propagation through the vapor cloud was captured on CCTV and it was evident that most of the vapor cloud burned at a relatively slow rate without the generation of significant pressures [4]. One explosion occurred within a drain, but a second more significant event occurred in the north east corner of the site where the flame was seen to travel at much higher speed, consistent with a DDT [4]. It was this second event that cause the offsite damage. As with Jaipur and Buncefield, the change between lower level of damage and low flame speeds to high flame speeds and severe pressure damage was sudden, being initiated in a localized area of the cloud.

Examples of directional indicators and 'severe' pressure damage are provided in the following sections, however, in relation to investigation of any future VCEs that may have involved DDT, the experience from the incidents described above indicates:

- Where there is evidence of thermal damage due to combustion of the cloud but little or significantly reduced pressure damage (compared to other parts of the vapor cloud), this is most likely to be associated with combustion of the cloud before high speed deflagration or DDT.
- In this case, the source of ignition of the cloud is likely to have been in, or on the edge of, this portion of the cloud.
- The transition from low levels of damage to severe pressure damage indicates that there has been a change to either rapid deflagration (flame speeds close to the speed of sound) or DDT.
- Some means of accelerating the flame would be expected near this area. This may be congestion provided by process pipework (or dense trees and bushes as was the case at Buncefield), or confinement, or a combination of both.
- The experimental evidence indicates that given a near stoichiometric concentration, the region where flame acceleration takes place up to DDT can be small (10-30m) in proportion to the scale of the vapor clouds present in most VCEs.

3.2 Directional Indicators

Some of the directional indicators observed in the vapor cloud have been widely published [1][11][12] and are illustrated in Figure 4. The effects include surface damage on one side, such as scouring of paint or scorching. Items can also be displaced, such as broken or fallen trees, tilted lampposts and collapsed weaker structures such as camera towers.

Figure 4. Directional indicators from Jaipur and Buncefield incidents

The damage to trees is very variable, ranging from uprooting or breaking of relatively substantial trees, but with less damage to other nearby trees, as shown in Figure 5. Other types of directional evidence may be present. For example, Figure 5 shows damage to a pipe guide on one side of a firewater pipe at Jaipur. This evidence was consistent with other directional indicators, though it only provides a more approximate indication of the direction of travel of the explosion.

Figure 5. Variability in damage to trees and directional damage to pipe guide

If this type of evidence is present in a VCE incident involving a pancake cloud, the following points are of note:

- Directional indicators can be generated by an intense deflagration in a process region but will only be in the 'reverse' direction within the congestion. High flame speeds will be required for any significant effect as the reverse load is a drag force proportion to the square of flow speed. As already indicated, the reverse flow for say a 100-200m/s flame will generally not be sufficient to produce the Buncefield and Jaipur indicators.

- If 'reverse' directional indicators are widespread though open areas of the cloud, then DDT must have occurred with a sustained detonation propagating though these areas of the cloud.
- As the effect is due to the reverse flow of the combustion products, the 'reverse' effect may disappear close to the cloud edge as there is insufficient volume of combustion products to give the net reverse impulse. This has been confirmed in experimental studies [1]. Outside the cloud, items will be displaced away from the cloud. Hence a sudden reversal in the directional indicators can be used to estimate the extent of the detonable cloud.
- If the detonation propagated around a significant obstacle such as a building or a tank, then the indicators may not point back to the source of the DDT but more indicate the local direction of the detonation wave.
- Care should be taken when there has been a significant confined explosion as this can spread debris in many directions. This evidence should not be confused with directional indicators produced by explosion propagation through open areas of the vapor cloud.

The 'reverse' directional indicators described above rely on the loading or flow produced by the high velocity expansion of the combustion products in the opposite direction to the flame/detonation propagation. The reverse directional indicators would not be generated by central ignition or detonation of a spherical or hemispherical cloud as the combustion products would be contained within the flame envelop and there would be no reverse flow.

A good practice in any investigation would be to conduct a site walk around and where directional indicators are observed, systematically record their location and bearing they indicate. This should be the bearing in which an item has fallen, broken or is tilting, or for items scorched or where paint has been scoured off, the bearing taken by standing behind the item looking at the face affected. The recording of this data allows a picture of the directional indicators to be produced.

3.3 Pressure Damage

The research conducted following the Buncefield incident provided a much greater understanding of the pressures required to cause different levels of damage to items such as cars, instrument boxes and oil drums [1]. This data has been compared to the evidence collected at both Buncefield and Jaipur [13] and records of other VCEs [6] and found to be consistent.

Figure 6 shows the damage caused to cars placed inside and near a detonated gas cloud at DNV GL Spadeadam. The incident pressures on the cars outside the cloud are also shown. Figure 7 shows damage to vehicles at Buncefield, Flixborough and Jaipur and shows a high degree of consistency with the experimental studies.

Figure 6. Levels of Damage to cars placed inside (bottom) and near (top) a detonated propane-air cloud

Figure 7. Damage to vehicles: top left - Buncefield inside cloud, top right - Buncefield about 20m outside cloud, bottom left - Flixborough UK, 1974, bottom right - tank torn off road transporter, Jaipur

The evidence from the Buncefield [1][2]and Jaipur [12] indicate that the damage to buildings exposed to a detonation can be severe, but that some building components, such as reinforced concrete framework, will remain standing. Examples of building damage within the detonated parts of the Buncefield and Jaipur clouds are shown in Figure 8.

Figure 8. Damage to buildings directly exposed to a detonating cloud (Jaipur top, Buncefield bottom)

The key aspect of any VCE involving a DDT and a sustained detonation through the remaining cloud is that the pressure damage will be severe and can occur anywhere in this part of the cloud, including open areas. The level of pressure damage will also be seen to reduce rapidly beyond this area. There is a degree of variation in the damage caused to items most likely due to factors such as differences in construction, the direction the detonation wave was incident on the item and, in the case of tanks and drums, the degree to which they are full.

4 Suggested Investigation Process

The recent change in the understanding of VCEs has come from research, including large scale experimental studies. These have shown that DDT and sustained detonation needs to be considered during the investigation of future major VCE incidents.

In investigating a VCE involving a low-lying dense vapor cloud, the flowing process is suggested:

1. Obtain evidence related to the extent of the vapor cloud from sources such as CCTV cameras, eye witness reports and thermal damage to susceptible materials.

2. Collate information on the pressure damage caused to items such as buildings, vehicles, oil drums, instrument boxes.
3. Systematically record the location and bearing of directional indicators.
4. Compare information on the estimated extent of the vapor cloud with the directional indicator data. If directional indicators outside the estimated cloud boundary comprise items blown over or displaced away from the cloud, this supports the proposed cloud extent.
5. If none of the pressure damage within the vapor cloud is comparable to that caused by direct exposure to a propagating detonation, as illustrated in this paper, then the VCE most probably involved a deflagration only.
6. If such damage is present, then the possibility of a DDT should be considered.
7. In this case, the distribution of severe pressure damage and the directional indicators can be used to determine if DDT has occurred.
8. If severe pressure damage is widespread in open, uncongested, areas of the cloud then a DDT has occurred. In this case, directional indicators pointing in the opposite direction to that of the explosion flame would also be expected in these areas of the cloud.
9. If such evidence is only present within congested regions such as process units, then it is likely that the incident involved a high speed (closed to the speed of sound) deflagration.
10. If DDT is indicated, the directional indicators in the areas of the cloud with severe pressure damage will tend to point back to the point at which DDT occurred.
11. If there are areas where there is evidence of thermal damage from the combustion of the vapor cloud, but little or reduced pressure damage, this part of the cloud most likely burned prior to the DDT. In this case the source of ignition for the VCE is likely to be in or on the boundary of this area.

This information combined with other data such as the sequence of events or witness statements can allow the nature of the VCE to be explained in some detail.

5 Facility Design

Though the VCE research discussed in this paper has been viewed from the point of view of incident investigation, it is important to provide some comment on the implications for facility design. Clearly the primary protection against a VCE of any type is the implementation of an effective process safety management system. Prevention of a release that could form a vapor cloud is the most effective way of avoiding DDT. However, there will always be some residual risk of a VCE and the realization that DDT needs to be considered has implications for the design of facilities.

This paper has shown that the damage to buildings and many other structures within a detonating vapor cloud is generally very severe, but it would not be practicable to design such structures to withstand detonation loadings on new facilities. Other than moving buildings well outside potential vapor cloud dispersion distances, an alternative is to address this problem by taking a risk-based approach to facility design. This involves assessing the many different outcomes that are possible following an accidental release, modelling both vapor dispersion and explosions.

This approach allows the residual risk to be estimated and compared against risk tolerability criteria. It also allows the practicability of any risk reduction measures to be assessed. Risk reduction measures could involve strengthening or moving buildings, but they could equally involve aspects such as:

- Increased detection and more effective isolation to reduce the amount of fluid that could be released.
- Reducing the size of individual process regions to reduce the potential for flame acceleration.

6 Summary

The evidence obtained from detailed investigation of the Buncefield and Jaipur incidents and the subsequent research has been reviewed in terms of the implications it has for the investigation of VCEs. This has allowed a process to be suggested for collating evidence in any future incident and drawing conclusions from this evidence.

7 References

[1] Burgan, B.A. (2014). "Dispersion & Explosion Characteristics of Large Vapour Clouds", Steel Construction Institute, Document ED024, 2015, available at http://www.fabig.com/video-publications/OtherPublications.

[2] Health and Safety Executive, "Buncefield Explosion Mechanism Phase 1", Research Report RR718, 2009.

[3] Atkinson, G., Hall, J., McGillivray, A., (2017). "Review of Vapour Cloud Explosion Incidents", Health and Safety Executive, Research Report RR1113.

[4] Davis, S., Pagliaro, J., Botwinick, D., DeBold, T., van Wingerden, C.J.M., Allason D. & Johnson, D.M., "Do not believe the hype: Using case studies and experimental evidence to show why the HSE is wrong about excluding deflagration-to-detonation transitions", Process Safety Progress, October 2018.

[5] Johnson, D.M., Pekalski, A, Tam, V.H.Y. & Burgan, B.A., "The Importance of Deflagration to Detonation Transition in Explaining Major Vapour Cloud Explosion Incidents", 12th International Symposium on Hazards, Prevention, and Mitigation of Industrial Explosions (ISHPMIE), Kansas City, 2018.

[6] Chamberlain G., Oran E., Pekalski, A., "Detonations in industrial vapour cloud explosions", 12th International Symposium on Hazards, Prevention, and Mitigation of Industrial Explosions (ISHPMIE), Kansas City, 2018.

[7] Bull D.C., Elsworth J.E., McLeod M.A., Hughes D., "Initiation of Unconfined Gas Detonations in Hydrocarbon-air Mixtures by a Sympathetic Mechanism". Progress in Astronautics and Aeronautics, Volume 75, Gas Dynamics of Detonations and Explosions, Pages 61-72, 1981.

[8] Harris R.J. and Wickens M.J., "Understanding Vapour Cloud Explosions – An Experimental Study", IGE comm. 1408, 1989.

[9] Pekalski A., Puttock J., Chynoweth S., "Deflagration to Detonation Transition in a Vapour Cloud Explosion in Open but Congested Space: Large Scale Test", Journal of Loss Prevention in the Process Industries, vol 36, p 365-370, 2015.

[10] Burgess, D.S. & Zabetakis, M.G., "Detonation of a flammable cloud following a propane pipeline break", US Bureau of Mines, Report of investigation 7752, US Department of the Interior, 1973.

[11] Johnson, D.M. "Characteristics of the Vapor Cloud Explosion Incident at the IOC Terminal in Jaipur, 29th October 2009". 8th Global Congress on Process Safety, Houston, TX, April 1-4, 2012.

[12] Johnson D.M, "Characteristics of the Vapour Cloud Explosion Incident at the IOC Terminal in Jaipur, 29th October 2009", DNV GL Report 11510, August 2011, available at http://www.fabig.com/video-publications/OtherPublications.

[13] Johnson, D.M. and Tam, V.H.Y., "Why DDT is the Only Way to Explain Some Vapor Cloud Explosions", Process Safety Progress, February 2017.

Automated Methods in Relief Vent Stack Sizing

Michael James
Principal Chemical Engineer
Eastman
P.O. Box 511, Building 18
Kingsport, TN 37662
mjames@eastman.com

Prepared for Presentation at
15th Global Congress on Process Safety
New Orleans, LA
March 31 – April 4, 2019

UNPUBLISHED

Automated Methods in Relief Vent Stack Sizing

Michael James
Senior Chemical Engineer
Eastman
P.O. Box 511, Building 215
Kingsport, TN 37662
mjames@eastman.com

Keywords: Dispersion modeling, consequence analysis, risk assessment, dosage, shelter-in-place, temporary haven, evacuation, toxicity, toxic chemical release, toxic load, downwind concentration, vent stack sizing, Visual Basic, scripting, coding, MS Excel

Abstract

Quantifying the impact of a relief event to personnel is critical to site safety and is mandated by Process Safety Management (PSM) [1]. Within sites and chemical parks containing several occupied structures, analyzing the impact to all affected personnel from a stack release is highly labor intensive. Performing comparisons between several potential locations can be both cost and time consuming.

This paper details a methodology for utilizing automated scripting to study the impact from a stack release, assessing concentration values inside of each building on-site and estimating the number of potential exposures to building occupants from a stack release. The total number of potential exposures is minimized through optimization of stack sizing parameters (i.e. height and diameter). The model is iterated in all potential wind directions, with the worst-case wind direction used in setting stack size.

The paper also discusses methods of assessing safe relief rates through existing vent stacks. In addition, if stack sizing and relief rate information are not available, the model can still be utilized to generate estimates all of parameters.

In this model, the number of potential exposures to a chemical release are calculated based on dose-response methodology [2]. The dose-response methodology returns a probabilistic score for potential impact. This is scaled by the count of personnel inside of an occupied structure to estimate the total count of exposures. The number of potential exposures is utilized as specifying consequence targets in Layers of Protection Analysis (LOPA).

This paper will discuss the required inputs to the model, detail on the procedural flow of the automated methods, how to interpret the results including some caveats noted around the model behavior during its development.

1. Introduction

This work was developed in support of a corporate standardization and simplification initiative to facilitate consequence targets within Layers of Protection of Analysis (LOPA) [3]. LOPA is a semi-quantitative method for estimating the risk of a single loss of containment event. Risk is defined here as the product of the number of exposed individuals and the frequency of the potential loss of containment event. Companies employing LOPA techniques may set targets, which are tolerable levels of frequency for a single loss of containment that would result in a set number of exposed individuals.

To estimate the number of exposed individuals, first the concentration of toxic vapor must be calculated at the distance from the release source to the group of individuals. This can be calculated using the Pasquill-Gifford dispersion modeling equations. As an example, the following equation calculates concentration downwind (x), crosswind (y), and at elevation (z) from a release at a rate of "Qm" grams per second [4]:

$$C(x,y,z) = \frac{Q_m}{\pi \sigma_y \sigma_z u} \exp\left[-\frac{1}{2}\left(\frac{y^2}{\sigma_y^2} + \frac{z^2}{\sigma_z^2}\right)\right]$$

Qm ~ chemical release rate, g/s

σ_y ~ dispersion coefficient representing standard deviation of concentration in the crosswind direction, m

σ_z ~ dispersion coefficient representing standard deviation of concentration in the vertical direction, m

u ~ wind speed, m/s

y ~ crosswind offset between release source and point of concern, m

z ~ vertical offset between release source and point of concern, m

The dispersion coefficients $\sigma_y$and σ_z are empirically derived functions of downwind distance and atmospheric conditions.

At a chemical facility, chemical releases may occur in either liquid phase or vapor phase. Releases can occur from systems that are under pressure from piping of any size through holes of any size. Releases can also originate from vent stacks designed to relieve systems that are experiencing a process upset. Due to the many potential modes of release, the rate term, Q_m, can be difficult to calculate. This rate term is commonly referred to as the "source term".

Several applications, including DNV PHAST (2018) [5] and SAFER TRACE (2012) [6], provide models for different types of releases, including stack releases, tank leaks, and piping cracks. These dispersion modeling applications will calculate the source term based on the type of model input. For example, when modeling a stack release, the stack sizing parameters and release rate are entered. The dispersion modeling application will calculate and plot concentrations of the chemical as it disperses downwind, crosswind and vertically.

If a plant is relieving vapor through a stack at a high rate, there is a potential for air to ingest into the releasing vapor cloud. This phenomenon is referred to as jetting. The dispersion model mostly employed for jetting conditions is the Ooms model [7]. The Ooms model solves for concentration at points downwind by integration of a set of differential equations. There are no known elementary solutions for the set of integrals, and most dispersion modeling applications will utilize a numerical method such as a Runge-Kutta method to approximate a solution for concentration.

With the values for concentration known at the target locations of personnel, an infiltration analysis may be needed if the personnel are inside of a building. Infiltration analysis will calculate the concentration inside of a building based on air turnover rates and outdoor concentration. Infiltration analysis assumes a perfectly mixing in the volume of air being studied [2]. Performing a detailed analysis without the assumption of perfect mixing requires use of Computational Fluid Dynamics (CFD), which is extremely costly and time-consuming and requires very specialized training.

When concentration is known either at the indoor location of personnel or at their outdoor position, a dose-response analysis can be performed to integrate concentration over time. This integrated value is transformed via probit analysis to a probability of exposure [2].

An estimation for total number of exposed individuals can be made by scaling the count of personnel in the studied area by the probability of exposure.

There are two generally accepted methods of approach for utilizing the number of potentially exposed individuals as a means for assessing safe stack sizing parameters, maximum tolerable relief rate and ensuring safe stack location:

1. Utilize a dispersion modeling application to run scenarios of varying stack height, stack diameter and release rate. Analyze the results of all scenario runs to determine concentrations at indoor and outdoor locations of personnel. Determine an optimal set of parameters that maintains total number of exposures to within the tolerable LOPA target. Dispersion modeling tools are reasonably inexpensive. However, the described analysis would require days, if not weeks, to complete.

2. Include the potential release scenarios in a Quantitative Risk Assessment (QRA) [8]. A QRA will analyze the aggregated risk of all potential losses of containment at a facility, including the stack release with thousands of other release scenarios. A QRA can be a great tool in facility siting. However, QRA's are mostly conducted by contracted companies, and determining the sensitivity of site risk to the size and relief rate of a single stack may not be practical. If a QRA has not been performed at a site, completing one may be a multi-year endeavor, depending on the size of the facility.

Conducting this analysis by a facility employee would likely be performed with a dispersion modeling application. Most dispersion modeling applications will model a single release event at one time. If utilizing the application as a tool for determining appropriate relief stacks sizing parameters, maximum safe relief rate, or the safest location of a vent stack on-site (i.e. siting), it is possible that dozens of dispersion models may need to be run. This is both cost and time-consuming. In addition, it is not valuable utilization of resources.

The process of siting and sizing relief stacks as well as providing information on maximum safe relief rate, a script can be automated by use of scripting. A scripted model was developed for this purpose based on first principles. The model will not give as accurate of results as a CFD study. However, the model will provide best estimates on optimal values for the parameters previously discussed.

Optimal stack sizing parameters are the most cost-effective parameters that result in no greater than the tolerable level of consequence for a specific LOPA target. Cost-effectiveness is assessed solely on the cost of steel used in building the relief stack.

The model currently utilizes correlations based on model runs from SAFER Trace (2012) to determine concentrations at distances to occupied buildings. However, future releases will incorporate the Ooms model methodology and numerical solution.

All cases discussed are hypothetical. Any resemblance to reality is coincidental.

2. Required Inputs

The following inputs are required for performing analysis with this tool:

- Mapping personnel locations
  - Identify the latitude and longitude of building locations on-site as well as occupancy of all buildings.
  - A potential extension can be inclusion of on-site outdoor occupied areas as well as off-site residential locations. Currently, the model only studies impact to building locations.
- Air Handling Unit Elevations
  - For occupied buildings, identify the elevation of air handling intakes.
  - If the height of an air handling unit is unknown, the model will assume that the air handler is at the centerline height of the relieving cloud. The centerline height is the point of greatest concentration of the cloud's vertical profile.
- Target Location of Vent Stack
  - Identify the latitude and longitude target for the vent stack.
  - Another potential extension could be to compare several potential siting locations of the stack and report the safest location.

- Source Term
    - Input chemical name or chemical mixture components
    - If stack sizing or relief rate are known, these values can be entered. The model will still run if neither of these values are given.
- Steam Injection
    - For heavier than air releases, the tool will calculate the required concentration of the relieving chemical in steam that maintains buoyancy while the relief is dispersing downwind.
    - The concentration of chemical in steam is calculated from the following equations, derived from the Ideal Gas Law [5] and Archimedes' Principle [6].

$$a = 1 - \frac{\rho RT}{P * MW_{Chem}}$$

$$b = \frac{\rho RT}{P * MW_{Water}} - 1$$

$$m = \frac{a}{a+b}$$

ρ ~ Density of releasing cloud at the exit of the vent stack
R ~ Ideal Gas Constant
T ~ Ambient temperature (Absolute)
P ~ Atmospheric pressure
MW_{Chem} ~ Molecular weight of releasing cloud at the exit of the vent stack
m ~ Mass fraction of chemical (or of chemical mixture) in steam
a & b ~ intermediate values

    - Steam injection is specifically for relief of heavier-than-air compounds. Neutrally buoyant compounds will be modeled without steam injection. Lighter-than-air compounds are not considered for modeling in this study due to their nature of lofting and rising without impacting nearby buildings.

3. Target Risk Guidelines

The model employs a semi-quantitative risk approach, utilizing a company's target LOPA guidelines. Each organization may employ different target values, but the intent is to quantify consequence (injuries, exposures, etc.) and frequency (number of events per year). Many/most companies define target frequency values for different consequence severity levels in thier LOPA gudielines (or risk matrix). The target frequency values are referred to here as "LOPA criteria".

4. Automated Scripting Methodology

The model is built in the Visual Basic environment of Microsoft Excel (2016). Excel provides a spreadsheet interface for input and output of data. The model finds optimal parameters for stack height, diameter, and/or relief rate by employing a simple iterative approach. The user may enter either stack sizing parameters (height and diameter) or relief rate. Alternately, the user can leave all three input fields empty. Due to model complexity, discussed later, the model will exhaustively iterate between the upper and lower bounds of all studied parameters.

The model boundaries are as follows:

- Stack Height – 50 ft to 300 ft
  - These values are applied by the author. Other values are acceptable.
- Stack Diameter – 2" to 7"
  - This limitation currently is imposed because the model utilizes data correlated from runs in SAFER Trace to estimate downwind concentration.
  - Future releases of the tool will utilize the Ooms model. This will allow for a much greater range of diameters.
- Relief Rate
  - Lower bound for jetting is based on the rule-of-thumb given in the SAFER Trace help file, 10 m/s [6]. The author utilizes this value as the lower bound based on absence of a more accurate source.
  - Upper bound (u_{max}) is restricted by sonic velocity in the relieving fluid (assuming fluid ideality) [9]:

$$u_{max} = \sqrt{\frac{\gamma RT}{Mw}}$$

u_{max} ~ Sonic Velocity [m/s]
γ ~ Heat Capacity Ratio [C_P / C_V]
R ~ Ideal Gas Constant [8,314.47 J / kmol / deg K]
T ~ Absolute Temperature [deg K]
M_W ~ Molecular Weight [kg / kmol]

The model is first run at a given set of input parameters (stack sizing and/or release rate) to study the impact to buildings that are within wind directions of 0□ (due north) and 15□ (north-northeast) from the release source. The concentration and number of potential exposures are modeled at each building within this range of wind directions. The model then iterates over ranges of wind directions 360□ around the release source. The range of wind directions with the highest number of potentially exposed personnel is maintained for comparison with the worst-case wind direction range modeled for all studied input parameters. The optimal parameters for each "LOPA criterion" are determined by the set of model inputs that result in no more than the company's target guidelines.

Within a given wind direction, the following steps detail the method to determine the number of potential exposures. For each building on-site, the code:

- Confirms if the building latitude/longitude is within currently studied wind direction range.
- Confirms that the building has not already been studied in an overlapping wind direction range.
- Calculates the centerline concentration of the dispersing cloud at the distance of the building.
- Calculates the concentration at the vertical offset distance between cloud centerline height and the height of the building air handling unit intake.
- Applies a dose-response methodology to determine the probability of impact to personnel sheltering inside the building.
- Estimates the number of impacts inside of the building by multiplying the probability of impact by the total building occupancy.
- Adds the number of impacted personnel in the building to the running total for the current wind direction range.

The model will then proceed with its exhaustive search method across the bounds of each model input and all buildings.

5. Model Limitations: The Problem of Multiple Solutions

As stated previously, the model searches for optimal conditions via iteration over all possible inputs. Attempts at implementing more efficient algorithms (e.g. Secant Method [10]) were met with difficulty. The problem with implementing "smarter" algorithms is the result of the existence of multiple solutions.

Consider the following hypothetical example where the vent stack sizing is fixed, and the model is searching for an optimal relief rate. Table 1 illustrates impact to a nearby occupied building to varying relief rates.

Table 1. Impact to occupied structure based on varying relief rate.

Relief Rate (lb/min)	Illustration	Description	Personnel Impacted inside Occupied Building
50	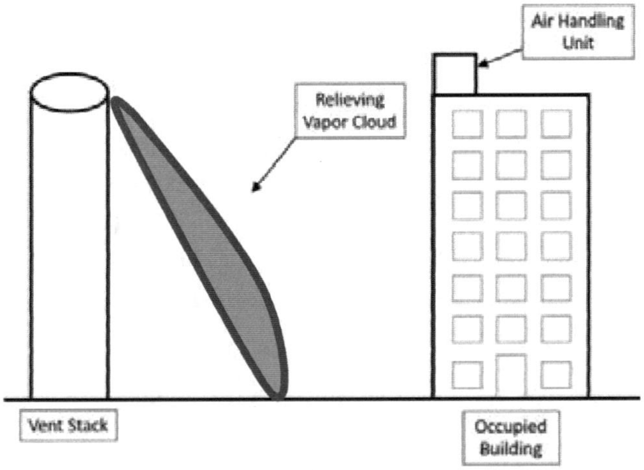	Potential for cloud to slump, impacting personnel outdoors, but no impact to building.	0
60	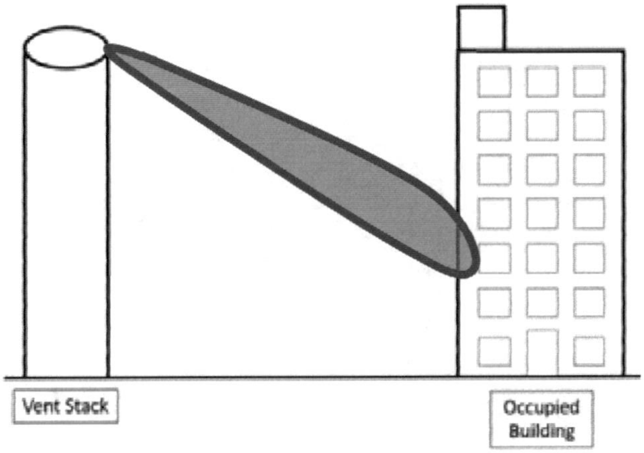	Relieving cloud starts to impact building, but not at air handler.	3
70	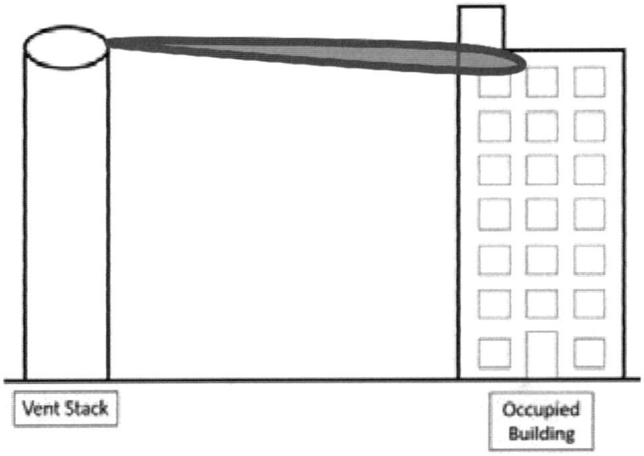	Minor impact at air handler.	7

Relief Rate (lb/min)	Illustration	Description	Personnel Impacted inside Occupied Building
80	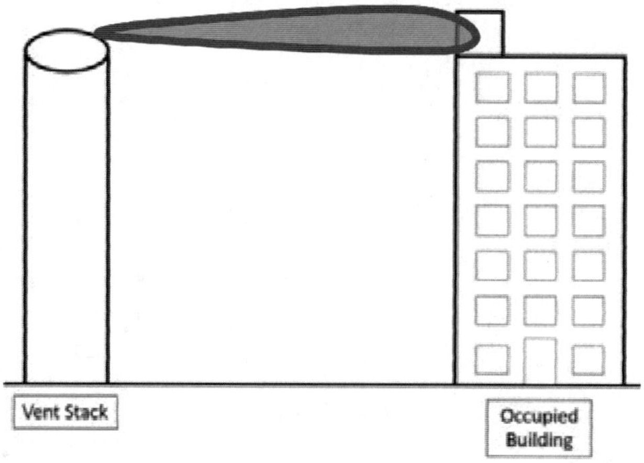	Relieving cloud impacts air handler at cloud centerline. This results in maximum possible consequence.	31
90	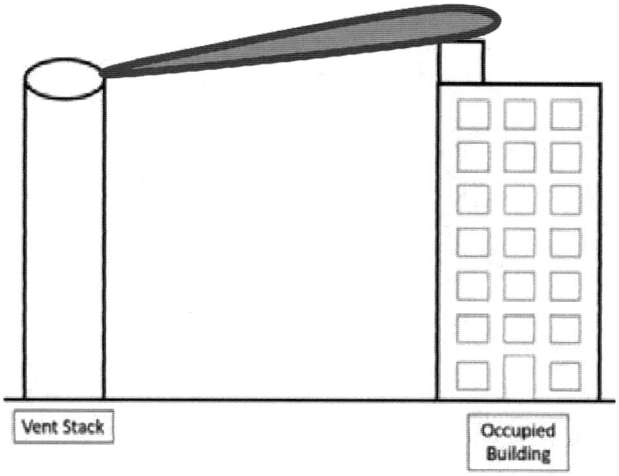	Cloud relieves to elevation above building. Minor impact at air handler.	7

Approved for External Use

Relief Rate (lb/min)	Illustration	Description	Personnel Impacted inside Occupied Building
100	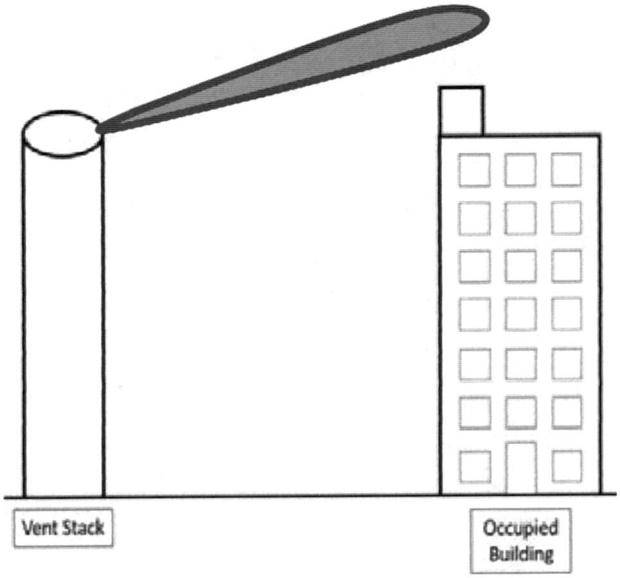	Cloud is entirely above building. Air handler is outside of cloud.	0
110	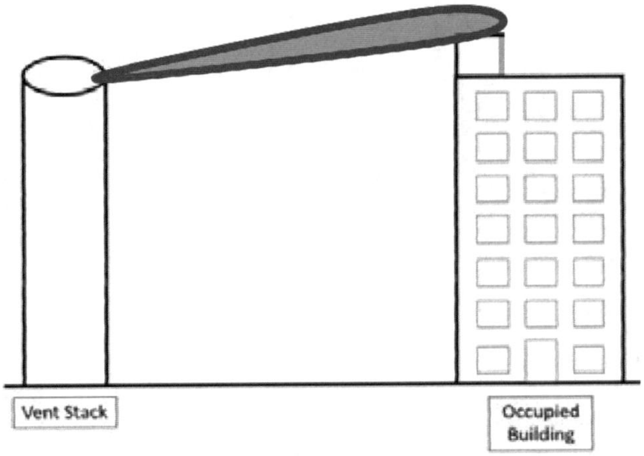	Mass of dispersing cloud is heavy enough to weigh cloud and start to again impact air handler.	11

Relief Rate (lb/min)	Illustration	Description	Personnel Impacted inside Occupied Building
120	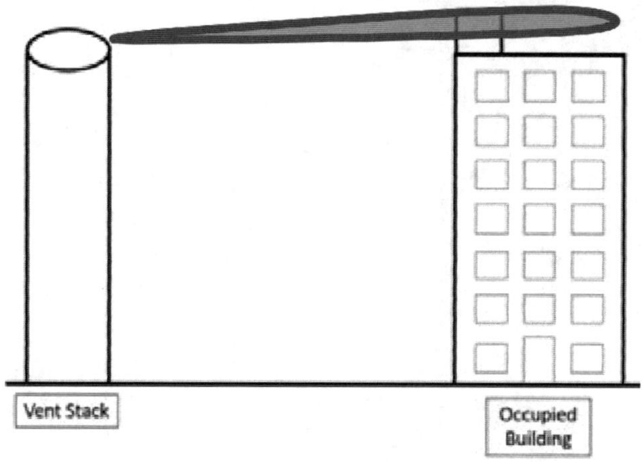	Building air handler is impacted by cloud at its centerline, resulting in maximum possible consequence.	40
130	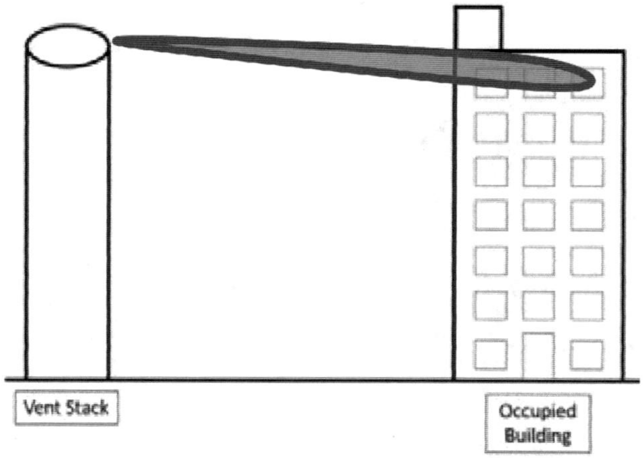	Minor impact at air handler.	10
140	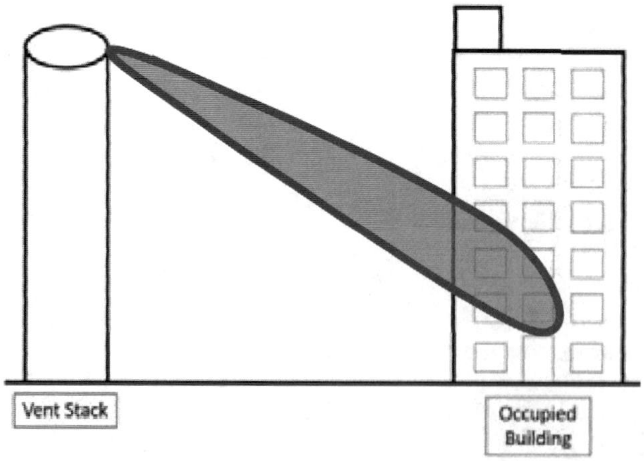	Building impacted, but not at air handler.	6

Relief Rate (lb/min)	Illustration	Description	Personnel Impacted inside Occupied Building
150	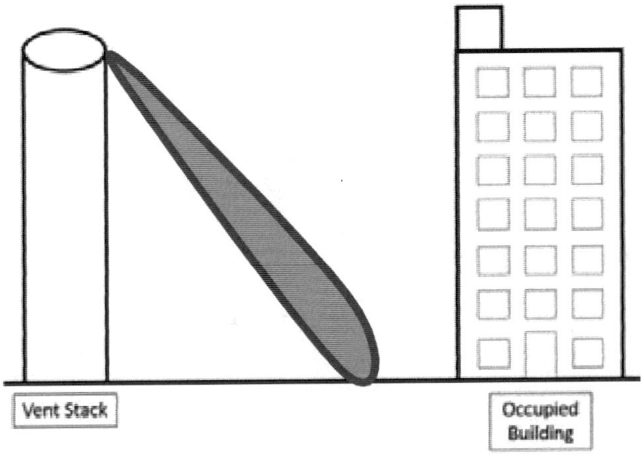	Larger potential impact to personnel at ground level, but no impact to building occupants.	0

Figure 1 shows the relationship between relief rate and impacted personnel. In this case, three relief rates (50 lb/min, 100 lb/min, and 150 lb/min) could be considered "optimal" as they result in zero consequence to building occupants.

Figure 1. Personnel Impacted inside Occupied Building vs Relief Rate

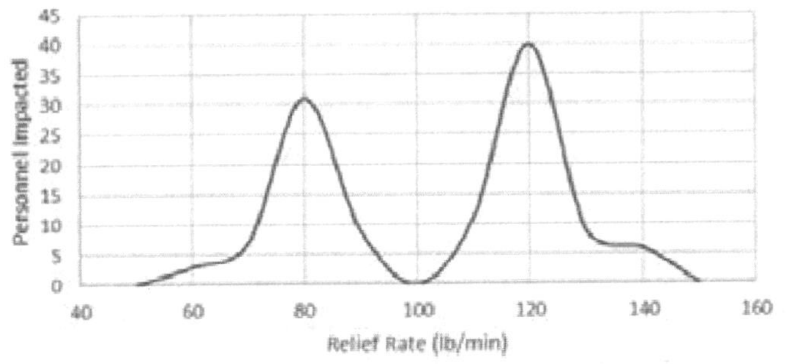

When the relief rate is low, the cloud may slump and not affect any buildings. As the relief rate increases, the cloud will stop slumping and will begin to moderately impact the downwind occupied building. As the relief rate continues to increase, the cloud will directly impact the air handler, and eventually it will have enough velocity to loft above the building air handler. As the rate continues to increase, it will become heavy enough to start to slump, again impacting the downwind building, and eventually impacting only ground level locations.

The exhaustive search method of iterating over all possible input values will ensure that all results are considered. More efficient algorithms may report only one solution, which is highly dependent on the model's initial guess on relief rate.

While the current model will output an optimal solution, the application will also produce a complete report to a separate tab in the Microsoft Excel v16 (2016) workbook. The report details the results of the models run on all model input configurations. The user can reference this complete report to further understand which solution may be most optimal for their site. An example of this report is given in Appendix A.

6. Model Interface

The interface for the model is on one sheet in a Microsoft Excel v16 (2016) workbook.

The interface is divided into sections, explained below.

As shown in Figure 2, Section A is used for the input of the chemical name and target stack location. If the relieving stream is a mixture, then a name for the mixture can be provided here.

SECTION A		
Chemical or CAS No	Bromine	
Cas No Confirmation	7726-95-6	
Stack Latitude	24.9144566	(Decimal Degrees or DMS)
Stack Longitude	17.7653585	

Figure 2. Section A close-up

As shown in Figure 3, Section B is for inputting any known release parameters. If the stack sizing is known, then both stack height and stack diameter must be entered. If no information is known, then the input parameter fields can be left blank.

For companies using this tool across multiple sites, the site building locations and occupancies can be organized separately. The name of the site currently being studied is selected from a drop-down list.

SECTION B		
If Stack Release Rate or Stack Height/Diam is known, input the values below. If no values are known, all values will be optimized. Optimizing all values will take roughly 15 min.		
Stack Release Rate		kg/min
Stack Height		m
Stack Diam		in
Site Name	Site XYZ	

Figure 3. Section B close-up.

As shown in Figure 4, Section C details the limits used for the LOPA targets. These can be in terms of a concentration of the dispersing chemical in air, a dosage, or a count of Severe Injuries (i.e. total number of personnel exposed).

SECTION C			
LOPA Target	Consequence	Consequence Unit	Type
A	xyz	ppm	Concentration
B	7.00		Severe Injury
C	16.00		Severe Injury
D	400.00		Severe Injury

Figure 4. Section C close-up.

As shown in Figure 5, Section D provides the user with information on the availability of chemical data for the studied chemical. If no data is available for this chemical, the user will be prompted to enter the minimum amount of information required to run the model. This includes threshold concentration of interest and molecular weight. For mixtures, the average molecular weight is entered here. The threshold concentration is a known toxic limit, ERPG [11] for example, of a chemical of interest. While this is the current minimum amount of information, additional input will be required when the tool is updated to use the Ooms model.

SECTION D			
Chemical is available for Dose-Response Methodology.			

Figure 5. Section D close-up.

As shown in Figure 6, Section E is the results section, detailing the output of the model run. The first column contains "Stack Diameter," "Stack Height," and "Stack Rate" repeated for each "LOPA criterion". Solutions are populated in the column labeled "Lowest Cost" when the relief rate is known but the stack sizing is not. When stack sizing is known but relief rate is not, solutions will be populated in the "Highest Stack Rate" column. If none of the input parameters are known, the solutions will be populated in the "Both Lowest Cost and Highest Stack Rate" column.

At the bottom of Section E, the model outputs the recommended concentration of the relieving cloud in steam. This value is calculated using the equation given in the "Required Inputs" section of this paper.

SECTION E				
LOPA Bucket		Lowest Cost	Highest Stack Rate	Both Lowest Cost and Highest Stack Rate
Personnel Impacted 1	Stack Diameter (in)			
	Stack Height (m)			
	Stack Rate (kg/min)			
Personnel Impacted 5	Stack Diameter (in)			
	Stack Height (m)			
	Stack Rate (kg/min)			
Personnel Impacted 10	Stack Diameter (in)			
	Stack Height (m)			
	Stack Rate (kg/min)			
Personnel Impacted 15	Stack Diameter (in)			
	Stack Height (m)			
	Stack Rate (kg/min)			
Recommended Concentration of Chemical in Steam			mass%	

Figure 6. Section E close-up.

7. Sample Run and Explanation of Output

The following hypothetical example illustrates the use of the model in siting a relief stack for a bromine storage vessel. This vessel is at ambient temperature and a pressure equal to bromine's saturation pressure. The stack is sited at 24.9144566N, 17.7653585E, at the center of a hypothetical plant site. The site consists of 100 occupied buildings, the closest is 50 ft from the stack, and the furthest is 2500 ft from the stack. The building air handling units are all between 50 and 100 ft above ground level.

The site uses 1, 5, 10, and 15 personnel impacted as its "LOPA criteria".

The model will be used to estimate risk in the following scenarios:

1. There is an existing stack 15 m tall, 4" diameter at the target location. Determine maximum safe relief rates for each "LOPA criterion".

   The model runtime for this condition is on the order of seconds. The relief rate is the only unknown in this case. The model will iterate across a range of valid relief rates. The lower bound is a rate equivalent to a linear velocity of 10 m/s at the exit of the vent stack. The upper bound is a rate equivalent to the sonic velocity of the fluid in air at the exit of the vent stack.

SECTION E				
LOPA Bucket		Lowest Cost	Highest Stack Rate	Both Lowest Cost and Highest Stack Rate
Personnel Impacted 1	Stack Diameter (in)		4.00	
	Stack Height (m)		15.00	
	Stack Rate (kg/min)		27.03	
Personnel Impacted 5	Stack Diameter (in)		4.00	
	Stack Height (m)		15.00	
	Stack Rate (kg/min)		24.42	
Personnel Impacted 10	Stack Diameter (in)		4.00	
	Stack Height (m)		15.00	
	Stack Rate (kg/min)		21.82	
Personnel Impacted 15	Stack Diameter (in)		4.00	
	Stack Height (m)		15.24	
	Stack Rate (kg/min)		16.62	
Recommended Concentration of Chemical in Steam		48.1%	mass%	

Figure 9. Stack Diameter, Stack Height, and Relief Rate combinations for each "LOPA criterion" based on scenario of existing stack.

The results are output in the "Highest Stack Rate" column. These results are the maximum relief rate that result in the consequence mapped to each "LOPA criterion".

The results appear counter-intuitive, as lower rates show higher consequence. A release rate of 27.03 kg/min results in an impacted personnel count of 1. A release rate of 16.62 kg/min results in an impacted personnel count of 15. This counter-intuitive result is described previously, and the case may warrant further specific consequence modeling outside of the boundary of reported values.

2. In this case, a relief analysis was performed on this system. The governing credible case required flow of 50 kg/min of bromine. Determine the most cost-effective sizing for this vent stack.

   For the case of known rate, model runtime is on the order of 1-2 minutes. The model iterates over diameter and height within a set range of values.

SECTION E				
LOPA Bucket		Lowest Cost	Highest Stack Rate	Both Lowest Cost and Highest Stack Rate
Personnel Impacted 1	Stack Diameter (in)	3.00		
	Stack Height (m)	15.24		
	Stack Rate (kg/min)	50.00		
Personnel Impacted 5	Stack Diameter (in)	4.00		
	Stack Height (m)	18.29		
	Stack Rate (kg/min)	50.00		
Personnel Impacted 10	Stack Diameter (in)	4.00		
	Stack Height (m)	15.24		
	Stack Rate (kg/min)	50.00		
Personnel Impacted 15	Stack Diameter (in)	7.00		
	Stack Height (m)	15.24		
	Stack Rate (kg/min)	50.00		
Recommended Concentration of Chemical in Steam		48.1%	mass%	

Figure 10. Stack Diameter, Stack Height, and Relief Rate combinations for each "LOPA criterion" based on scenario of known relief rate.

   In general, wider stacks result in higher consequences. The 7" diameter stack is modeled as causing 15 potential personnel exposures while the 3" stack is modeled as causing only 1 potential personal exposure. This is as expected because wider stacks have a lower linear velocity, providing lower initial momentum to the relieving cloud.

3. In this case, there is no available data on the relief of this bromine tank. The model will be used to provide estimates on stack sizing/relief rate combinations that fit within each "LOPA criterion".

   When no data is given, the number of required runs grows very large. The model runtime is approximately 9 minutes. This case suffers the most from lack of a more efficient algorithm. Creating a model with a solver that can manage multiple solutions is beyond the scope of this paper.

SECTION E				
LOPA Bucket		Lowest Cost	Highest Stack Rate	Both Lowest Cost and Highest Stack Rate
Personnel Impacted 1	Stack Diameter (in)			4.00
	Stack Height (m)			18.29
	Stack Rate (kg/min)			58.43
Personnel Impacted 5	Stack Diameter (in)			4.00
	Stack Height (m)			15.24
	Stack Rate (kg/min)			58.43
Personnel Impacted 10	Stack Diameter (in)			4.00
	Stack Height (m)			15.24
	Stack Rate (kg/min)			50.40
Personnel Impacted 15	Stack Diameter (in)			4.00
	Stack Height (m)			15.24
	Stack Rate (kg/min)			34.34
Recommended Concentration of Chemical in Steam		48.1%	mass%	

Figure 11. Stack Diameter, Stack Height, and Relief Rate combinations for each LOPA target based on scenario of no available data on stack size or rate.

Results show the counter-intuitive increase in consequence from decreasing relief rate. In addition, as is highly intuitive, increased vent stack height impacts fewer buildings, thus resulting in lower consequence.

In all above scenarios, the model runs were on the order of seconds to minutes. Compare this to the current generally accepted methodologies:

1. Utilizing a dispersion modeling tool that may only run one combination of model input parameters at a time. Studying the consequence from hundreds of input parameter combinations to hundreds of occupied structures would take weeks to complete.
2. Including the range of parameters in a QRA. This would be an accurate method and would be recommended if price was not an issue. In addition, it would only be recommended if there was an existing QRA study that had been completed. The cost to study the sensitivity of site risk to the range of input parameters may cost between 50,000 USD and 100,000 USD or more. Results from a QRA may be received on the order of a few months.

The automated stack sizing tool is an incredible time and resource savings. A study that would otherwise take weeks to complete can now be performed in 20 minutes or less. The model can both provide matching values to targets and more in-depth reports. A sample in-depth report is included in Attachment A.

8. Logistics in Implementing the Modeling Tool

This modeling tool has been developed for users with basic Excel (2016) experience to be able to input required data and launch the automated script. However, as currently implemented, the results can appear counter-intuitive. Until a more intelligent solving algorithm can be implemented, it is recommended that results be performed or, at a minimum, reviewed by a user with experience in consequence modeling.

This model is contained within an Excel workbook, so implementation is as simple as copying the file to the user's local computer or accessible network location.

9. Conclusion

The intent of this paper was to detail an automated method to use consequence modeling in sizing relief vent stacks. The automated tool will estimate the stack height and stack diameter as well as maximum safe relief rate at a specified vent stack location. The tool utilizes a map of occupied building locations and calculates distance to each from the target stack location. Based on the building distances and potential concentrations, it will estimate total number of exposed personnel, and will report the stack height, stack diameter and maximum relief rate that matches the site's specified LOPA criteria.

The automated tool can analyze venting of buoyant and heavier-than-air releases. The density of a heavier-than-air release is assumed to be made buoyant by injection with steam. The required amount of steam can be calculated using equations presented in this paper.

Due to model complexity, the automated modeling tool utilizes a simple exhaustive search algorithm. It utilizes a specified range of stack heights and diameters as well as a range of relief rates. The model iterates over all valid combinations of model parameters, and determines the optimal conditions matching each Layers of Protection Analysis (LOPA) criteria.

The automated tool provides results that may otherwise not be able to be determined due to practicality, timing or cost. The results of this tool would be very costly and time-consuming to study as part of a Quantitative Risk Assessment (QRA). In addition, analyzing all potential release cases over the range of potential inputs by means of a dispersion modeling tool would require several hundred modeling runs, each performed one at a time. This time-consuming method would not be practical.

A QRA may cost tens if not hundreds of thousands of dollars and may be a multi-year project. Studying the release conditions one-at-a-time in a dispersion modeling tool would take months to complete. In comparison, the longest runtime of the automated tool is 20 minutes.

This tool has been an enormous savings in time and expense. It can be employed with minimal knowledge of the release conditions and can provide estimates on the optimal stack sizing parameters that fit within each of the site's LOPA criteria.

References

[1] "Process safety management of highly hazardous chemicals," Feb. 8, 2013, https://www.osha.gov/pls/oshaweb/owadisp.show_document?p_table=STANDARDS&p_id=9760 (Accessed Mar 11, 2019).

[2] James, M., "Simplified Methods of Using Probit Analysis in Consequence Analysis," *Process Safety Progress*. First Published Online 3 JUN 2014. DOI: 10.1002/prs.11686

[3] Center for Chemical Process Safety (2001). *Layer of protection analysis: Simplified process risk assessment.* New York, NY: American Institute of Chemical Engineers.

[4] Crowl, Daniel A., & Louvar, Joseph F. (1990). *Chemical process safety: Fundamentals with applications.* Englewood Cliffs, NJ: Prentice-Hall, Inc.

[5] DNV GL AS (2018). PHAST [computer software]. Hovik, Norway

[6] SAFER Systems (2012). SAFER Trace [computer software]. Westlake Village, CA.

[7] Spicer, T., & Havens, J. (1989). EPA-450/4-89-029: User's guide for the Degadis 2.1 dense gas dispersion model [PDF File]. Retrieved from https://www3.epa.gov/ttn/scram/userg/other/degadis2.pdf.

[8] Center for Chemical Process Safety (2000). *Guidelines for chemical process quantitative risk assessment.* New York, NY: American Institute of Chemical Engineers.

[9] Melhem, G. A. (2006). Advanced consequence analysis [PDF File]. Salem, NH

[10] Kaw, A. K., Kalu, E. E., & Besterfield, G. (2009). *Numerical methods with applications.* Morrisville, NC: Lulu Enterprises.

[11] 2018 ERPG/WEEL Handbook (2018) [PDF File]. Retrieved from: https://www.aiha.org/get-involved/AIHAGuidelineFoundation/EmergencyResponsePlanningGuidelines/Documents/2018%20ERPG-WEEL%20Handbook%20(ERPG).pdf.

Appendix A – Review of In-Depth Report Generated as Supplementary Data

The model outputs a report of the worst-case result for each combination of model input parameters. The worst-case result is the total impact for the wind direction resulting in the greatest number of personnel impacts.

This report is provided because of the counter-intuitive results that may be output from the model. To select an optimal combination of release parameters, a user can reference this table.

The model inputs are on the left-hand side of the table. "Tot Dosing" refers to the total dose to all exposed individuals in the worst-case wind direction of the set of model input parameters for each row. "Tot Impacts" is the total count of potential personnel exposures. The critical angles are the wind directions that reported the greatest amount of personnel exposures. "Tot Impacts" is the total count of potential personnel exposures.

Table 2. Modeling Tool report (left-hand side)

Stack Diam (in)	Stack Height (ft)	Rel Rate (lb/min)	Tot Dosing (ppm^n-min)	Critical Angle for Dose	Time of day for Dose
2.00	50.00	1.22	354,838,472,204.47	195.00	nighttime
2.00	50.00	5.65	325,742,737.18	225.00	nighttime
2.00	50.00	10.07	1,167,083,471.73	225.00	nighttime
2.00	50.00	14.50	2,306,294,693.68	195.00	nighttime
2.00	50.00	18.93	19,484,615.40	-	nighttime
2.00	50.00	23.35	17,601,133.64	195.00	nighttime

Table 3. Modeling Tool report (right-hand side)

Tot Impacts	Critical Angle for Impacts	Time of day for Impacts	Peak concentration at any bldg	Peak Centerline Conc at any bldg	Cost Estimate (USD)
90.00	195.00	daytime	1,988.60	1,988.60	3,139.97
104.50	-	nighttime	192.25	192.25	3,139.97
147.33	-	nighttime	292.45	292.45	3,139.97
104.86	195.00	nighttime	364.73	364.73	3,139.97
37.89	-	nighttime	69.61	137.37	3,139.97
30.26	15.00	daytime	72.02	75.12	3,139.97

What Is My Flare Capacity, Really? Best Practices for Flare QRA Tools

Ben Pratt
Nancy Faulk
Siemens Energy, Inc
4615 Southwest Fwy, Suite 900
Houston, TX 77027
ben.pratt@siemens.com
nancy.faulk@siemens.com

Prepared for Presentation at
American Institute of Chemical Engineers
2019 Spring Meeting and 15th Global Congress on Process Safety
New Orleans, LA
March 31 – April 3, 2019

UNPUBLISHED

What Is My Flare Header Capacity, *Really?* Best Practices for Flare QRA Tools

Ben Pratt
Nancy Faulk
Siemens Energy, Inc
4615 Southwest Fwy, Suite 900
Houston, TX 77027
ben.pratt@siemens.com

Keywords: Flare, Flare QRA, Quantitative Risk Analysis, Flare Header

Abstract

Flare Quantitative Risk Analysis (QRA) is a systematic approach to determine the adequacy of a flare header and can be applied to provide a more realistic assessment of the risk associated with vessel accumulation due to common mode scenarios, in accordance with American Petroleum Institute (API) Standard 521 [1]. The QRA process takes the results of a traditional flare study and adds QRA inputs. These include the frequencies of the common mode scenarios, and the effects of instrumentation, operator intervention, and other layers of protection that tend to reduce the severity of the common mode scenario. The process generates a system risk profile, such as an accumulation versus frequency relationship for each vessel discharging to the relief header, which in turn can provide an estimate of the risk of catastrophic loss of containment associated with the relief header system. Flare QRA may reduce costs associated with the need to retrofit flare headers or construct new ones.

Experience has shown there are several challenges when performing QRA, including (a) how the system level risk is defined, (b) verification of reliability data used as QRA inputs, and (c) how considering only individual equipment may skew the view of resulting risk. This paper discusses these challenges, as well as what results can be expected from Flare QRA and what issues (e.g., flare tip) are *not* addressed through this analysis process.

1 Introduction to Flare Quantitative Risk Analysis (QRA)

1.1 Introduction

Most existing pressure relief header and flare systems were originally designed by taking little or no credit for any of the multitude of mitigating measures, commonly referred to as safeguards, which are present in a typical operating facility. However, as plants have increased throughput and added process units, the relief header and flare systems are usually no longer considered adequate using the same conservative methods that were used in the original design. As such, operating companies are faced with the decision either to install additional relief headers and flare capacity, to bring the system into compliance with the original design methods, or to perform a more detailed engineering analysis, consistent with recognized and generally accepted good engineering practices (RAGAGEP).

It is imperative to distinguish between design or capacity modifications with a real safety benefit and those with a theoretical benefit. Flare QRA (Quantitative Risk Analysis) is a systematic approach to determine the adequacy of a flare system, and this technique can be applied to better understand the risk profile associated with these systems. Flare QRA uses standard statistical analysis and, optionally, Monte Carlo methodology to provide an objective assessment of the risk associated with multiple releases to relief header and flare systems (i.e., "global scenarios"). A QRA can account for the effect of instrumentation, operator intervention, and other layers of protection, as well as the frequency of all relieving scenarios of interest. The Flare QRA method has been utilized by several major operating companies and this method is already included in both the 5th and 6th editions of API STD 521 [1].

1.2 Industry Guidance

American Society of Mechanical Engineers Boiler & Pressure Vessel Code (ASME BPVC) Section VIII governs the design of pressure vessels and associated pressure relief requirements but does not give detailed guidance on the design or analysis of relief header and flare systems [2]. For example, ASME Section VIII Appendix M (which is non-mandatory, however considered to be RAGAGEP) states, "The sizing of any section of common-discharge header downstream from each of the two or more pressure relieving devices that may reasonably be expected to discharge simultaneously shall be based on the total of their outlet areas, with due allowance for the pressure drop in all downstream sections." ASME Section VIII does not give details on what constitutes a reasonable scenario and, therefore, leaves this decision up to the judgment of the designer.

API STD 520 and API STD 521 are the most commonly followed industry practices for the design of relief header and flare systems in oil, gas and chemical facilities [1][3]. Of these two documents, API STD 521 provides the majority of the guidance on the evaluation of overpressure scenarios and the design of relief header and flare systems. API STD 521 provides clear guidance on the selection and analysis of overpressure scenarios for individual process equipment. A list of typical overpressure scenarios that should be considered is presented in API STD 521 along with clear guidance not to consider the positive response of instrumentation when evaluating relief protection for an individual piece of equipment. However, in the design of the relief header and flare systems the guidance is considerably less clear. Per API STD 521 §4.2.6 and §5.3.43, credit should typically not be taken for favorable instrumentation response when sizing relief devices and disposal system with respect to individual equipment overpressure scenarios. However, such credit may be taken for sizing disposal systems such as flare headers, etc. with respect to global overpressure scenarios. The type, design, reliability, and availability of each safeguard, as well as the effects of each safeguard on the associated flare load must be evaluated carefully. In addition, credit for operator intervention may also be taken to reduce flare loads from specific systems when certain criteria are met. Although the above guidance is certainly not prescriptive, it is apparent that consideration of existing safeguards in the design of the relief header and flare systems may be acceptable.

In addition, there are resources available from industry groups, such as the Center for Chemical Process Safety (CCPS), to explain how users can make better decisions about QRA [4] [5].

1.3 Reasons for Performing Flare QRA

Flare QRA is a detailed engineering analysis, and as such, takes considerable time and must be performed by a relief systems engineer with specialized knowledge. It is also important to understand when it is worthwhile to apply the detailed QRA technique to a system . Flare studies which are good candidates to consider evaluating further with this technique usually meet one of the below criteria:

- When baseline relief header and flare evaluation (completed per 29 § CFR 1910.119) identifies significant issues; typically, first pass flare hydraulics are performed without taking credit for any safeguards [6]. If there are backpressure concerns or in some cases, network equipment calculation concerns, Flare QRA can be performed before making any major changes to the existing flare header;
- To help design the flare header system when contemplating future changes which would increase the relief load to the flare system, such as tying atmospheric relief valves into the existing flare header; or,
- When interested in evaluating the impact of flare maintenance or debottlenecking projects on the flare header.

2 Establishing Risk Analysis Criteria and Parameters

2.1 Data Necessary to Perform Flare QRA

Flare QRA is typically performed after issues are identified in the baseline study, or after tying in additional relief valves into existing flare header; therefore, relief device loads, and flare header and relief device outlet piping isometrics are generally available. Piping & Instrumentation Diagrams (P&IDs) for the participating equipment and units, and vessel and relief device information are also typically available. In addition, the following information is required:

- Initiating event frequency of global scenarios
- Risk acceptance criteria
- Potential safeguards to mitigate the loads
- Probability of failure on demand (PFOD) for those safeguards

It is important to gain agreement on these parameters before beginning the study, to avoid challenges of the study results and rework. These parameters are discussed in more detail in the sections below.

2.2 Initiating Event Frequency

Assessment of the expected frequency of the initiating event (total power failure as an example) must first be established. For facilities that have been in operation for some time, operating history can be used to aid in estimating the expected frequency of various types of failures or initiating events. For example, a total power failure may be assumed to occur once every 10 years. If historical data is not available, published reliability data or existing reliability models (such as for the electrical power distribution) can be used to estimate a reasonable conservative frequency for each initiating event. The Center for Chemical Process Safety (CCPS) provides published reliability data for various equipment and instrumentation in the book, "Guidelines for Process Equipment Reliability Data", and the OREDA Participants have published similar data in "OREDA Handbook 2015" [8] [9].

2.3 Risk Acceptance Criteria (RAC)

Similar to a Risk Matrix, the Flare QRA Risk Acceptance Criteria (RAC) takes the form of a relationship between vessel accumulation and frequency of occurrence. The RAC defines what the acceptable risk is and how frequently a vessel can exceed a particular accumulation limit. (Note that this does not account for contents or for capacity of any vessel, in determining risk.) The vessel accumulation is the percentage over the maximum allowable working pressure (MAWP) that the pressure in the vessel reaches during a relieving event, while the frequency is typically reported as an allowable interval between occurrences, such as once per 100 years. The vessel accumulation is indicative of the potential severity of consequence of the event in terms of loss of containment. Table 1 lists accumulation levels of significance based on standard ASME Section VIII vessel design [2]:

Table 1. Vessel Accumulation vs Severity of Consequence

Accumulation	Significance	Potential Consequence
10%	ASME code allowable accumulation for process upset cases (non-fire) protected by a single relief device.	No expected consequence at this accumulation level. Lowest consequence from qualitative risk matrix.
16%	ASME code allowable accumulation for process upset cases (non-fire) protected by multiple relief devices.	No expected consequence at this accumulation level. Lowest consequence from qualitative risk matrix.
21%	ASME code allowable accumulation for external fire relief cases regardless of the number of relief devices	No expected consequence at this accumulation level. Lowest consequence from qualitative risk matrix.
30%	ASME standard hydrotest pressure for newer designs	Catastrophic vessel rupture not expected at this accumulation level. Possible leaks in associated instrumentation, etc. Medium consequence from qualitative risk matrix.
50%	ASME standard hydrotest pressure	Catastrophic vessel rupture not expected at this accumulation level. Possible leaks in associated instrumentation, etc. Medium consequence from qualitative risk matrix.
~90%	Minimum yield strength (dependent on the materials of construction)	Catastrophic vessel rupture remotely possible. Significant leaks probable. High consequence from qualititative risk matrix.
~300%	Ultimate tensile strength (dependent on the materials of construction)	Catastrophic vessel rupture predicted. Highest consequence from qualitative risk matrix.

Using a facility's Risk Matrix is a good way to develop RAC, as potential consequence in this table can be linked to a severity level in the risk matrix. For example, if the accumulation in a vessel exceeds the hydrotest pressure once every 10 years, and if the target interval for exceeding hydrotest pressure is less than 1 event in 10 years, then the vessel would not meet the RAC. Acceptable frequencies should be defined for various levels of the variable of interest. RAC must be finalized before running a QRA. Based on Table 1, RAC could be established by assigning the tolerable interval for each level of accumulation; example vessel RAC are presented in Table 2, along with definitions of the criteria.

Table 2. Example Individual Vessel Risk Acceptance Criteria

Accumulation	Tolerable Interval	Notes
0%	Once or more per year	Any number of occurrences is acceptable
21%	Once every 10 years	Consistent with "several occurrences in the facility lifetime"
50%	Once every 50 years	Consistent with "one occurrence in the facility lifetime"
90%	Once every 1000 years	Consistent with "not allowed to occur in the facility lifetime"
110%	No credible occurrences	Most conservative

2.3.1 Individual Vessel Accumulation Frequency Criteria

The purpose of these criteria is to assure that no single vessel is subjected to unacceptable accumulation too frequently. All vessels connected to the Flare Header are evaluated against these criteria regardless of whether relief from the vessel is expected. In order for the system to be deemed acceptable, all vessels in the system should meet these criteria.

For simplicity, these criteria are typically specified uniformly for all equipment, without consideration for contents or capacity. However, as actual consequence may vary, depending upon a number of factors, it may be appropriate to adjust the criteria for certain systems to align with corporate risk guidelines or applicable regulations.

2.3.2 Virtual System Accumulation Criteria

The purpose of these criteria is to assure that the overall system does not reach unacceptable accumulation too frequently. Unlike the Individual Vessel Accumulation Criteria, the Virtual Vessel Accumulation Criteria approach tracks only the single highest resulting vessel accumulation for each simulation run. It is called a "virtual" vessel approach because the highest accumulation tracked for each simulation run could occur on different vessels or equipment within the system. That is, the accumulation versus frequency performance is not for any single vessel but for the entire system – or for a "virtual" vessel in the system. This measure does not indicate how many other vessels could be overpressured. These criteria are approximately one order of magnitude less than those used for individual vessels, since the flare system could involve hundreds of vessels. The virtual vessel should meet all Accumulation Criteria in order for the system to be deemed acceptable.

2.3.3 Average System Accumulation

The purpose of these criteria is to assure that the average of all vessel accumulation is not excessive. This approach is different from the virtual vessel approach in that it keeps track of all vessel accumulations that result from each simulation. Because it keeps track of all vessel accumulations and frequencies, the results can be converted to an "average" vessel. For example, a 100% average accumulation once every 1000 years would mean that one vessel (any vessel) in the system would reach 100% accumulation every 1000 years. These criteria are identical to the virtual vessel criteria and approximately one order of magnitude less than those used for individual vessels. The "average" vessel should meet all Accumulation Criteria in order for the system to be deemed acceptable.

2.4 Safeguards and Probability of Failure on Demand (PFOD)

The QRA engineer identifies potential safeguards that currently exist that could mitigate an individual vessel's relief load to flare in the event of a global scenario. The safeguards which can reduce or eliminate relieving events can typically be identified by analyzing the P&IDs and reviewing each system with operating personnel. The list below highlights some typical safeguards that are present in most operating facilities and often aid in reducing or eliminating relieving loads:

- Operator intervention in field
- Operator intervention from control room
- Basic process control system (BPCS)
- Spare pump auto-starts
- Independent high pressure shutdowns
- Independent high or low liquid level shutdowns
- Independent high temperature shutdowns
- Specific pump or system operation time
- Safety instrumented system (SIS)

Each safeguard is assigned a probability of failure on demand which indicates the safeguard's reliability. Mitigated relief loads are calculated for each vessel that has the benefit of safeguards.

Examples of typical safeguards and associated PFODs are given in Table 3.

Table 3. Typical Safeguards and Associated PFODs

Safeguard	PFOD, percent
Conventional instrumentation	10 – 20%
High pressure / high temperature override	5%
Spare pump autostart	10%
Operator field intervention	30 – 80%
SIL-1	1 – 10%
SIL-2	0.1 – 1%
SIL-3	0.01 – 0.1%
Mode of operation	Varies based on usage

The presence of safeguards leads to high possible variance in the total global scenario relief load when that global scenario actually occurs. For example, consider the equipment in Figure 1, consisting of 10 vessels which relieve during a global scenario. Assuming each has 1 layers of protection, which either function or fail to function, there may be as many as two different possible relief rates for each vessel; therefore, each global scenario in this example could have 1,024 different global flare load permutations (2^{10}). Ignoring the safeguards would mean that the flare analysis would be based on the most conservative case, out of more than a million permutations.

Flare QRA allows for the analysis of the flare system to be based on thousands, if not all, of the possible permutations.

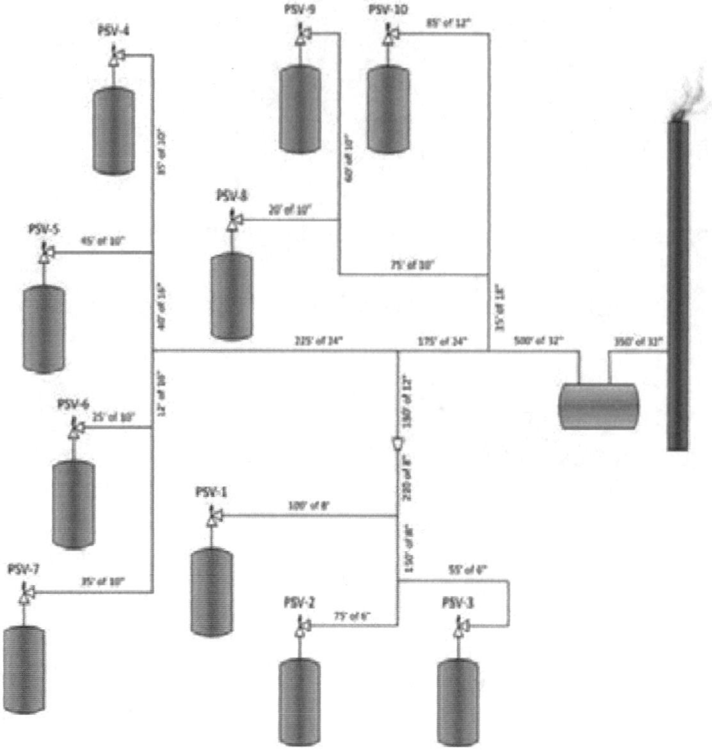

Figure 1. Example Relief Header Network

3 Performing the Flare QRA

3.1 Calculation of System Risk Profile

With the data collected in the above steps, the relationship between vessel accumulation and frequency of occurrence for the system can be calculated for comparison to the established RAC. This is accomplished by generating all possible permutations of safeguards, calculating the probability of each of these permutations, analyzing the relief header pressure profile for each permutation, and calculating the vessel accumulations from the back pressure obtained from the pressure profile. Table 4 shows an example of the type of data stored for each permutation, for the flare system in Figure 1. Note that the vessel accumulations at the locations which do not relieve are set equal to zero, as the pressures in the vessels are not expected to reach the MAWP due to operation of the relevant safeguards.

Table 4. Global Scenario Example – Sample Run Data

Relief Device	Vessel	Safeguard Operates?	Probability	Relief Rate (lb/hr)	Vessel Accumulation
PSV-001	V-001	No	10%	110,000	36%
PSV-002	V-002	No	10%	90,000	54%
PSV-003	V-003	Yes	90%	0	0%
PSV-004	V-004	No	10%	225,000	25%
PSV-005	V-005	No	10%	350,000	67%
PSV-006	V-006	Yes	90%	0	0%
PSV-007	V-007	Yes	90%	0	0%
PSV-008	V-008	No	10%	215,000	38%
PSV-009	V-009	No	10%	230,000	16%
PSV-010	V-010	Yes	90%	0	0%
Total				1,220,000	

3.2 Accumulation

The vessel accumulations in Table 4 were calculated based on flare header backpressure. The vessel accumulation analysis needs to consider the specific relief device type (e.g., conventional, bellow, pilot-operated). Two simplifying assumptions are made. First, the relief rate is treated as a function of safeguard activity only, and not vessel accumulation. While it is theoretically possible to eliminate this assumption and utilize a relief rate vs. accumulation matrix, doing so would exponentially increase the complexity of an already computationally intensive simulation. Second, the pressure in the relieving systems would be increased such that each relief device maintains full lift at the hydraulically determined backpressure.

Several complications and considerations follow from these assumptions.

- While it may not be practical to adjust individual relief rates based on accumulated pressures, it is a straightforward exercise to determine the maximum possible accumulated pressure (i.e., the internal pressure at which there would be no theoretical relief rate) for each system. This value can then be utilized as an upper bound in the evaluation.
- Below certain backpressure limits, balanced bellows relief valves may require less differential pressure to maintain full lift than conventional relief valves. Above these limits, however, the bellows can be expected fail. It should be determined prior to starting calculations how to account for backpressure exceeding bellows limits. Typical bellows rating limits can be obtained from API Standard 526, though manufacturers may have published values for their valves [7].
- If the disposal system includes pilot-operated relief devices which are not expected to relieve under the scenario under evaluation, it should be noted that excessive backpressure may result in the opening of these devices and lead to backflow as discussed in API STD 520 Part II [3]. Many modern pilot-operated valves include backflow preventers, but the potential for backflow, leading to potential system accumulation where none was expected, should be considered.

3.3 What Results Can Be Expected from a Flare QRA

Assuming the safeguards are independent, the overall probability for this particular permutation is equal to the product of the respective individual probabilities for each safeguard corresponding to the stated load from each vessel (in the above example from Table 4), or 6.56x10-7. The results can then be sorted per the accumulation ranges in the RAC; see Table 5.

Table 5. Global Scenario Example – Overall Probability per Single Safeguard Permutation

Relief Device	Vessel	Accumulation Exceeds 21%	Accumulation Exceeds 50%	Accumulation Exceeds 90%
PSV-001	V-001	$6.56x10^{-7}$	0	0
PSV-002	V-002	$6.56x10^{-7}$	$6.56x10^{-7}$	0
PSV-003	V-003	Yes	0	0
PSV-004	V-004	$6.56x10^{-7}$	0	0
PSV-005	V-005	$6.56x10^{-7}$	$6.56x10^{-7}$	0
PSV-006	V-006	Yes	0	0
PSV-007	V-007	Yes	0	0
PSV-008	V-008	$6.56x10^{-7}$	0	0
PSV-009	V-009	No	0	0
PSV-010	V-010	Yes	0	0

The probability of exceeding a given level of accumulation from the RAC can then be computed for all possible permutations by first summing the respective overall probability for each safeguard permutation. Then the frequency of occurrence for the initiating event (total power failure) should be factored, i.e. once per 10 years or 0.1 occurrences per year. As such, the frequency at which the vessel exceeds that level of accumulation is estimated. The reciprocal of the frequency yields the interval of occurrence in terms of years between occurrences. The overall results from the entire analysis are shown below in Table 6.

Table 6. Global Scenario Example – Calculated Risk Profile

Relief Device	Vessel	Accumulation Exceeds 21% (years between occurrences)	Accumulation Exceeds 50% (years between occurrences)	Accumulation Exceeds 90% (years between occurrences)
	RAC	**10**	**50**	**1000**
PSV-001	V-001	100	10,000	Never
PSV-002	V-002	100	526	Never
PSV-003	V-003	100	100	10,000
PSV-004	V-004	948	Never	Never
PSV-005	V-005	361	1,000	8,074,283
PSV-006	V-006	11,137	148,528	$1x10^{10}$
PSV-007	V-007	36,340	36,340	$3.6x10^{9}$
PSV-008	V-008	100	3,835	$1.886x10^{8}$
PSV-009	V-009	484	90,690	Never
PSV-010	V-010	100	Never	Never

3.4 Consequence Modeling

If desired, these permutations may now be analyzed with dispersion models to describe how the material is dispersed downwind. If the release involves flammable materials, fire and explosion models are used to convert the information determined from the source models into energy hazard potentials (*e.g.*, thermal radiation, explosion overpressure) [4] [10].

4 Challenges

4.1 Challenge 1 – Considering Only Individual Equipment May Skew the View of the Resulting Risk

Individual Vessel Accumulation Criteria are readily understood, whereas the Virtual System and Average System Criteria are comparably abstract concepts. This can lead to an inappropriate bias focused only on Individual Vessel Accumulation Criteria and occasionally the failure to adopt System Level Criteria. This would be a significant error. The Individual Vessel Accumulation Criteria evaluates the risk to each source in isolation, giving little explicit information on how stressed the disposal system is as a *system*. This data is provided by the System Level Criteria.

As an example of the Virtual System Criteria, see Table 7 below, which shows the maximum predicted accumulation levels for five equipment, with each occurring during a different global scenario:

Table 7 Virtual System Example

Relief Device	Vessel	Scenario	Accumulation Exceeds 50% (years between occurrences)
	RAC		50
PSV-011	V-011	Scenario 1	200
PSV-021	V-021	Scenario 2	250
PSV-031	V-031	Scenario 3	150
PSV-041	V-041	Scenario 4	225
PSV-051	V-051	Scenario 5	200
Virtual System			40

V-031 shows the highest risk of any individual vessel with a predicted accumulation exceeding 50% once every 150 years, or a 0.67% chance each year, a value which is only one-third the maximum accepted probability. Based on the Virtual Vessel Criteria, however, it is observed that the probability of *some* vessel exceeding 50% accumulation is once every 40 years, or a 2.5% chance each year. Clearly stated, the disposal system has 2.5% chance of reaching excessive pressure any given year, which provides a better data point for assessing the adequacy of the disposal system than would be obtained from the Individual Vessel Accumulation Criteria alone.

For Average System Criteria, using input from Table 6 above, the probabilities of accumulation can be summed at each level for all equipment, giving the results in Table 8:

Table 8. Average System Example

	Accumulation Exceeds 21% (years between occurrences)	Accumulation Exceeds 50% (years between occurrences)	Accumulation Exceeds 90% (years between occurrences)
Average System	18	75	9,987

Accordingly, on 'average', a vessel is predicted to exceed 21% accumulation once every 18 years, or a 5.6% chance each year. A vessel is also predicted to exceed 50% accumulation once every 75 years, or a 1.3% chance each year. Again, this Average System Criteria provides better measure of the overall disposal system adequacy than the Individual Vessel Criteria alone.

4.2 Challenge 2 – How System Level Risk is Defined

Recognizing the interpretive challenge posed by the System Level Criteria to non-relief specialists, additional methodologies can be utilized to assess overall performance. In particular, header pressures can be tracked at specific points of interest. As with probabilities of accumulation levels in the RAC, the probabilities of reaching certain pressure levels at these points can be reported and are seen in Table 9:

Table 9. Pressure Tracking

Pressure Tracking Node	Header Location	5 psig (years between occurrences)	10 psig (years between occurrences)	20 psig (years between occurrences)
001	KO Drum D001	10	10	50
002	KO Drum D002	10	10	50
003	Unit 001 Tie-in	15	60	240
004	Unit 002 Tie-in	10	100	10,000
005	Unit 003 Tie-in	35	1,000	10,000
006	Unit 004 Tie-in	15	30	150

Such approaches can be particularly useful in flare debottlenecking as well as in evaluating locations for new load sources, whether from atmospheric tie-in or expansion projects. Nevertheless, for an accurate representation of risk, the best practice should be to consider these methods as supplemental to, and not replacement for, the System Level Criteria.

4.3 Challenge 3 – Not Evaluating All Global Scenarios

It must be stressed that the RAC is defined in terms of total, or overall, risk –not just risk from any one initiating event. As such, it is important to evaluate all potential global scenarios, and not only the worst-case or controlling event(s). Consider Table 10 below, for which Scenario 1 is the known Design Case, resulting in the highest disposal system pressures. By comparison, all other global scenarios are minor and are qualitatively recognized to carry less risk.

Table 10. Cumulative Effect of Evaluating Multiple Scenarios

Vessel	Global Scenario	Accumulation Exceeds 50% (years between occurrences)
RAC		**50**
V-001	Scenario 1 (Design Case)	55
V-001	Scenario 2	750
V-001	Scenario 3	1000
V-001	Scenario 4	800
V-001	*Cumulative*	*46*

Due to the low risk for Scenarios 2, 3, and 4; it could be tempting to short-cut the analysis and only evaluate Scenario 1 – and indeed the qualitative risk assessment would be confirmed. Yet doing so would result in the wrong conclusion, as the calculated 1 in 55 year interval for Scenario 1 exceeds the minimum given 1 in 50 year interval required for risk acceptance. Accounting for the other scenarios, the cumulative interval is 1 in 46 years, which does not meet the criteria.

4.4 Challenge 4 – Verification of Reliability Data Used for Flare QRA Input

As with any risk-based analysis, Flare QRA is a probabilistic tool, and uncertainties in available reliability data introduces corresponding uncertainties in the assessment. This reliability data is needed in two areas: (1) the reliability of the systems; be they electrical, steam, cooling water, instrument air, etc. for which a failure can result in a global overpressure scenario; and (2) reliability of the protective systems, or safeguards, for which mitigative credit is being taken.

Typically, this information is taken from facility experience; from published generic reliability data; or from vendors. In some cases, such as SIL rated protective equipment, PFOD data may be well known. Electrical distribution systems can be rigorously evaluated based on each component to calculate overall dependability. In all cases, the utilized reliability data should be reasonably conservative. As long as specified initiating event frequencies are not less than actual, and specified safeguard reliabilities are not greater than actual, the Flare QRA methodology will give conservative results.

Nevertheless, variances on both sides can significantly alter the outcome. The frequency of any single event, or permutation, is the product of the frequency of the initiating event and cumulative probability resulting from the action of all safeguards:

$F_i = I_i \prod S_i$, where

F_i is the frequency, in years between occurrences, of a specific load permutation
I_i is the frequency, in years between occurrences, of the initiating event
S_i is the probability of action, or inaction, of each safeguard

For example, say a scenario expected once every 10 years results in relief from two equipment, both of which have one safeguard of 90% reliability. The result is four possible permutations with the probabilities given in Table 11:

Table 11. Load Permutations for Two Equipment with One Safeguard Each

	Safeguard Acts			
Equipment A	Yes	Yes	No	No
Equipment B	Yes	No	Yes	No
Probability	81%	9%	9%	1%
Frequency (Years between Occurrences)	12.35	111.11	111.11	1,000

However, if there is a 25% uncertainty in the frequency of the initiating event, the table would instead be (Table 12):

Table 12. Load Permutations with Initiating Event Uncertainty

	Safeguard Acts			
Equipment A	Yes	Yes	No	No
Equipment B	Yes	No	Yes	No
Probability	81%	9%	9%	1%
Frequency (Years between Occurrences)	9.3 – 15.4	83.3 – 139	83.3 – 139	750 – 1,250

If compounded by a 10% uncertainty in the reliability of the safeguard for Equipment A, the results become (Table 13):

Table 13. Load Permutations with Initiating Event Uncertainty and Safeguard Uncertainty

	Safeguard Acts			
Equipment A	Yes	Yes	No	No
Equipment B	Yes	No	Yes	No
Probability	73 – 89%	8 - 10%	1 - 17%	0.1 - 2%
Frequency (Years between Occurrences)	8.4 – 17	76 – 154	44 – 1,389	395 – 12,500

As demonstrated, as the probability for a specific permutation gets lower – which often means failure of safeguards, and greater consequence – the variance becomes every greater, and the interval between expected events becomes harder to define. No reliability datapoint will ever be known with 100% accuracy, but it is critical to eliminate as much uncertainty as possible, while remaining appropriately conservative.

5 Issues *Not Directly Addressed* by Flare QRA

With regards to consequence, Flare QRA focuses on the potential for loss of containment resulting from excessive overpressure of equipment during a global scenario. However, flare and disposal systems present other risks that are not explicitly accounted for by the RAC in a Flare QRA and meeting all RAC does not absolve the responsible party from adequately addressing these other issues using other approaches. While not directly addressing these issues, data resulting from Flare QRA can help in further evaluation.

5.1 Flare Tip

Emergency releases resulting from global overpressure scenarios often exceed the original design flow rates for installed flare tips, especially since facilities have steadily expanded and increased throughput over the years. This can lead to increased thermal radiation, and in extreme cases, issues such as flame-out.

Flare QRA data, in the form of expected flare load probabilities can assist in evaluating these risks. The level of granularity with regards to flare load interval size can be specified to suit the precision required. Further, as each load interval may represent hundreds, or even thousands, of different permutations, the probability of each permutation can be assessed to determine the most representative run, or set of runs, for each interval, as seen in Table 14:

Table 14. Expected Frequency of Flare Load

Flare Load (lb/hr)	Years Between Occurrences
> 200,000	5
> 300,000	10
> 400,000	30
> 500,000	100
> 600,000	500

Once representative permutations have been identified, further flare tip evaluations – including rigorous radiation and dispersion modeling – can be performed. These results would then be assessed against the applicable risk policies.

Note that further complications may arise if significant variability in relief load composition is possible even within the defined buckets. This challenge may be overcome through a layered approach, wherein the same bucketing methodology is applied to each discrete flare load set based on molecular weight or total load fraction of components of interest.

5.2 Knockout Drums

Knockout drums present additional disposal system equipment which may not perform as desired if actual conditions exceed the original design. API STD 521 §5.7.9.6 briefly discusses some considerations for potential knockout drum overfill [1]. Other risks may apply if vapor velocities do not permit adequate residence time to limit liquid drop size, or result in re-entrainment.

While perhaps more challenging, if appropriate consequence models are applied a similar approach as that suggested for Flare Tips can be utilized to determine a risk profile. At the simplest level, a yearly probability for overfilling a drum can readily defined.

5.3 Acoustic and Flow Induced Vibration

Significant pressure reduction at relief devices and any choke points in the disposal system can generate considerable acoustic energy. In addition to the noise hazard, this can induce vibration within the pipe wall that may lead to fatigue failure, from Acoustic Induced Vibration (AIV). Turbulence within the fluid flow may also lead to piping vibrations, called Flow Induced Vibration

(FIV). Methods for calculating the energies inducing these vibrations are available in the literature [11]. API Standard 521 §5.5.12 discusses acoustic fatigue in some detail and provides some mitigative strategies [1].

Global relief scenarios can present an increased AIV and FIV hazard, as the sound power levels (dB) from AIV sources, such as relief devices, are additive and high flow rates increase turbulence. Though consequences must be defined separately, Flare QRA can determine probabilities for significant AIV or FIV threats, aiding the evaluation of further mitigations.

5.4 Defining a Design Capacity

In many cases, it is helpful to have a defined design rate or capacity. Under traditional approaches, it will be at this rate that the flare tip, knockout drum, vibration, etc. criteria discussed above are evaluated. The design rate is also utilized when evaluating the disposal system impact of projects of various kinds.

Clearly, a full Flare QRA, in which all possible load permutations are assessed and weighted based on probabilities does not terminate in the production of a design rate. Further, rerunning the Flare QRA analysis may not be practical for every change or 'what-if' analysis. However, explicit information is given on load probabilities, and provided there is buy-in from all stakeholders good engineering judgment can be applied to identify an appropriate rate, or rates, to use for these purposes.

6 Conclusions

Quantitative risk analysis can be applied to evaluate complex relief header and flare systems. Consistent with current industry practices, credit can be taken for safeguards to mitigate relief loads to the header system. By considering availability for each safeguard and the initiating event frequency, a relationship between the predicted vessel accumulations and overall frequency of occurrence can be developed to represent the risk profile for each individual vessel and for the overall system. In a manner similar to the qualitative risk evaluation process used at many facilities, the calculated risk profile can be compared to corporate risk acceptance criteria.

There are challenges and limitations when performing QRA, including how the system level risk is defined, verification of reliability data used as QRA inputs, and identifying a flare system design capacity. A thorough understanding of QRA inputs, the scope of the QRA, the risk acceptance criteria, and the different accumulation criteria is necessary before performing the analysis.

The Flare QRA methodology can serve as a practical approach to resolve a previously intractable problem and may allow operating companies to better understand and manage the risk associated with relief header and flare systems.

7 Bibliography

[1] American Petroleum Institute, *API Standard 521 "Pressure-relieving and Depressuring Systems,"* 6th edition, 2014.

[2] American Society of Mechanical Engineers, *ASME Boiler and Pressure Vessel Code, Section VIII, Div.1 "Rules for Construction of Pressure Vessels,"* 2017 Edition.

[3] American Petroleum Institute, *API Standard 520 "Sizing, Selection, and Installation of Pressure-Relieving Devices,"* Part I, 9th Edition (2014) and Part II, 6th Edition (2015)

[4] Center for Chemical Process Safety, "*Guidelines for Chemical Process Quantitative Risk Analysis*," 2nd ed., John Wiley and Sons (2000).

[5] Health & Safety Executive, *Research Report RR25 "Application of QRA in Operational Safety Issues,"* (2002).

[6] U.S. Occupational Safety and Health Administration, *29 CFR § 1910.119 Process Safety Management of Highly Hazardous Chemicals*, 1994.

[7] American Petroleum Institute, *API Standard 526 "Flanged Steel Pressure-relief Devices,"* 7th edition, 2018.

[8] Center for Chemical Process Safety, "*Guidelines for Process Equipment Reliability Data, with Data Tables*," John Wiley and Sons (1989).

[9] SINTEF (*Stiftelsen for industriell og teknisk forskning)*, Offshore and Onshore Reliability Database, OREDA 2015 Handbook, 6th Edition, 2015.

[10] N. Prophet, "Understand Your Vulnerabilities with Quantitative Risk Analysis," Chemical Engineering Progress, pp. 49-52, July 2016.

[11] F. Eisinger, R. Sullivan, P. Feenstra, D. Weaver, *"Acoustic Vibration in a Stack Induced by Pipe Bends,"* Proceedings of the 5th International Symposium on FSI, AE, &FIV, at 2002 ASME Int'l Mechanical Engineering Congress and Exposition.

8 Acronyms

AIV	Acoustic Induced Vibration
BPCS	Basic Process Control System
FIV	Flow Induced Vibration
MAWP	Maximum Allowable Working Pressure
P&ID	Piping and Instrumentation Diagram
PFOD	Probability of Failure on Demand
RAC	Risk Acceptance Criteria
RAGAGEP	Recognized and Generally Accepted Good Engineering Practice
QRA	Quantitative Risk Analysis
SIS	Safety Instrumented System

551353 Using Process Simulators to Quantify Flashing Two Phase Release Conditions

Process simulators are a readily available tool for making complex thermodynamic calculations which are necessary for quantifying the conditions for flashing two-phase flow under sudden depressurization. Many common toxic industrial chemicals (TICs) such as chlorine and ammonia are stored as liquids at ambient temperature, and these materials will flash when depressurized to atmospheric pressure forming an aerosol. This paper considers how to use a process simulator to estimate the mass flow rate from a simple hole for flashing two-phase flow. Analysis shows that the typical approach for determining the flash fraction underestimates the amount of vapor formation for realistic release conditions. Previously published data are considered to demonstrate the effectiveness of this approach. The result of this work is the development of a general approach for quantifying the mass release rate of flashing two phase releases for TICs and other flashing liquids such as cryogenics.

The Effects of Oxygen Enrichment

on the

Explosibility Properties of Flammable Gases and Vapors

Richard W. Prugh
DEKRA Process Safety
113 Campus Drive
Princeton, NJ 08540
www.dekra-process-safety.com
prughrw@aol.com

Prepared for Presentation at
American Institute of Chemical Engineers
2019 Spring Meeting and 15th Global Congress on Process Safety
New Orleans, LA
March 31 – April 3, 2019

The Effects of Oxygen Enrichment on the Explosibility Properties of Flammable Gases and Vapors

Richard W. Prugh
DEKRA Process Safety
113 Campus Drive
Princeton, NJ 08540

Keywords: Flammables, Oxygen, Explosion

Abstract

In the chemical industry, oxygen-containing chemicals are used in the production of partial-oxidation products, including many in the pharmaceutical industry. The vapor spaces in oxidation reactors can contain high concentrations of oxygen gas, in combination with flammable vapors. Thus, it is important to determine the explosibility properties of organic-compound vapors in atmospheres containing high oxygen concentrations.

1. Introduction

In the chemical industry, "nascent" oxygen is involved in the production of partial-oxidation products, including many in pharmaceutical operations. The oxygen is initially in the form of peroxides, such as hydrogen and sodium peroxides, and also as chlorates and perchlorates, periodates, nitrates, and dichromates [1]. The vapor space in oxidation reactors can contain high concentrations of oxygen gas, in combination with the flammable vapors of organic compounds. To evaluate the possible explosion hazards of such mixtures of fuel and oxidizer, it is important to determine the explosibility properties of organic compounds in atmospheres containing high oxygen concentrations.

The basic flammability/explosibility data for this study are from the following references:

Table 1. References for Flammability Data

REFERENCE	FLAMMABILITY/EXPLOSIBILITY PROPERTY					
	Flammable Limits	Auto-Ignition Temperature	Minimum Ignition Energy	Maximum Explosion Pressure	Deflagration Index Kg	Oxygen Conc. Limit
NFPA 49; NFPA 325 [2]	√	√				
NFPA 68 [3]				√	√	
NFPA 69 [4]						√
NFPA 77 [5]	√ *		√			
NFPA 497 [6]	√	√	√			
USBM 503 [7]	√ *					√
USBM 627 [8]	√ *	√ *				
Babrauskas [9]	√	√ *	√ *			
Britton [10]	√ *		√ *			√
Crowl and Louvar [11]	√ *	√	√	√	√	√
Yaws [12]	√	√				

* Limited data for oxygen

Several methods are presented in this paper for estimating explosibility properties for preliminary process design. However, it is prudent to confirm the fuel/oxygen relationships with tests where process safety depends on the reliable control of oxygen concentration.

2. Flammability/Explosibility Properties

The explosibility properties of combustible materials can be divided into Ignition Sensitivity and Explosion Severity [13, 14]. The Lower Flammable Limits, the Autoignition Temperatures, and the Minimum Ignition Energies would be important in evaluating the Ignition Sensitivity. The Maximum Explosion Pressure and the Deflagration Index – a measure of burning rate – would be important in evaluating the Explosion Severity. The following sections describe the relationships of these properties to oxygen concentration.

2.1 Lower Flammable Limit

Data from many sources [15, 16, 17, 18] indicate that there is no significant effect of oxygen concentration on the Lower Flammable Limit [LFL] from the 21 volume-percent oxygen in air to 100% oxygen. At oxygen concentrations greater than the LFL in air, the specific heat of oxygen is about the same as the specific heat of nitrogen; that is, excess oxygen behaves as a diluent at the LFL.

For comparison with the Minimum Explosible Concentrations of combustible dusts and for evaluating the hazards of mixtures of flammable gases/vapors of combustible dusts, the values of LFL can be converted to mass densities by:

$$C = [\,LFL / (100 - LFL)\,]\,[\,MW / 0.024\,] \quad \text{grams per cubic meter}$$

where 0.024 is the number of cubic meters per gram-molecular weight [at 25°C and one atmosphere pressure (14.7 psia)].

For example, the LFL for propane in air is 2.1 volume-percent and, for the molecular weight of 44.1, the Minimum Explosible Concentration would be given by:

$$C = [\,2.1 / (100 - 2.1)\,]\,[\,44.1 / 0.024\,] = 39 \text{ grams per cubic meter}$$

Since the LFL is not significantly affected by oxygen concentration, oxygen enrichment would not affect the ventilation and dilution or inerting precautions that should normally be taken for a flammable vapor or gas.

2.2 Autoignition Temperature

The Autoignition Temperature [AIT] of a fuel is a kinetic phenomenon; that is, the AIT is low for very-long exposures to a given temperature, and it is quite high for short exposures to a given temperature. This relationship can be shown to follow the Arrhenius "law" [19], as:

$$r = A\,e^{-Ea/RT} \quad \text{fraction per second}$$

where A is the pre-exponential factor (in "per second");
Ea is the activation energy (in calories per gram-mole);
R is a constant (1.987 calories per gram-mole per °K); and
T is the temperature (in °K).

If equal reaction rates are assumed for combustion in air and oxygen, at two different temperatures, a value for the activation energy can be obtained, from:

$$Ea = 1.987\,[\,\ln(100/21)\,] / [\,(1/T_{AIR}) - (1/T_{O2})\,] = 3.1 / [\,(1/T_{AIR}) - (1/T_{O2})\,] \quad \text{cal./gmol}$$

The Autoignition Temperatures for 90 chemical compounds in air and in 100% oxygen are available [20]. The average AIT in air was calculated to be about 394°C [about 667°K], and the average AIT in oxygen was about 361°C [about 634°K]. Thus, the differences in AIT for air and oxygen is quite small, at about 5% lower.

The calculated activation energy Ea, as based on the referenced data, would be about 40,000 calories per gram mole, which is a “reasonable” value for an oxidation/ combustion reaction [21].

Thus, the AIT for a given oxygen concentration can be estimated from:

$$AIT_{O2} = AIT_{AIR} / [1 + (1.987\ AIT_{AIR} / 40{,}000) (\ln (C_{O2}/21)] - 273 \quad {}^{o}C$$

Oxygen enrichment would not affect the precautions that should normally be taken for a flammable or combustible vapor, such as avoiding exposure to hot surfaces in ovens, in dryers, and on electrical equipment such as motors.

2.3 Minimum Ignition Energy

Previous investigators had found that the Minimum Ignition Energy [MIE] was inversely related to oxygen concentration [22]. The Minimum Ignition Energies for 13 chemical compounds in air and in 100% oxygen are available [23]. The average MIE in air [excluding hydrogen (very low) and trichloroethylene (very high)] was calculated to be about 0.30 millijoule, and the average MIE in oxygen was about 0.0027 millijoule. Thus, oxygen concentration has a very strong effect on MIE, with about a factor of 100 reduction for 100% oxygen. The exponent for oxygen concentration for the 13 chemical compounds [including hydrogen and trichloroethylene] is about 2.9.

Thus, the MIE for a given oxygen concentration can be estimated [+/- 50%] from:

$$MIE_{O2} = MIE_{AIR} / [(C_{O2} / 21)^{2.9} = 6{,}200\ MIE_{AIR} / C_{O2}^{\ 2.9} \quad \text{millijoule}$$

As indicated above, oxygen enrichment would have a significant effect on the precautions [such as inerting] that should be taken for flammable gases and liquid against very weak electrostatic [and other] ignition sources, including corona discharges [24].

2.4 Maximum Explosion Pressure

The Maximum Explosion Pressure [Pmax] – for combustion in air or in an oxygen-enriched atmosphere – can be calculated rather accurately if the heat of formation of the fuel is known and the degree of dissociation of the carbon dioxide and steam combustion products can be determined. For combustion in an oxygen-enriched atmosphere, the degrees of dissociation are very high; thus, the Pmax in 100% oxygen could be expected to show a rather small increase in pressure as compared to combustion in air.

There is a direct relationship between flame temperature and explosion pressure, as governed by the Ideal Gas Law, as:

$$P_{Max} = 14.7\ (\ T_{FLAME} + 273\ )\ /\ (\ 25 + 273\ ) - 14.7 \qquad \text{pounds/sq.in. [psig]}$$

The available flame-temperature explosion-pressure data are as follows [25]:

Table 2. Flame Temperatures and Explosion Pressures for Fuel/Oxygen Mixtures

FUEL	FLAME TEMPERATURES (°C)			MAX. PRESSURES (barg)		RATIO (bara/ bara)	IDEAL-GAS-LAW RATIOS (°K/bara)	
	In Air	In Oxygen	Ratio [°K/°K]	In Air	In Oxygen		In Air	In Oxygen
Methane	1,950	2,705	1.37	7.1	10.4	1.41	274	261
Propane	1,988	2,822	1.37	7.9	10.2	1.26	254	276
Ethylene	2,088	2,902	1.34	8.0	10.7	1.30	262	271
Acetylene	2,262	3,069	1.32	10.6	11.2	1.05	219	274

Thus, approximate values for Pmax in mixtures of oxygen and in air would be given by:

$$P_{Max,O2} = [\ (\ T_{FLAME} + 273\ )\ /\ 261\ ] - 1 \qquad \text{barg}$$

Thus, if the flame temperature can be calculated, from heats of formation and specific heats – with consideration of dissociations – the maximum explosion pressure for a hydrocarbon can then be calculated. It is likely that this relationship would also apply to other hydrocarbons and carbohydrates. Also, as indicated in the preceding Table 2, the values of P_{MAX} in 100% oxygen are about 30% greater than the value in air [excluding acetylene].

The data in Table 2 above can be used to estimate the Maximum Explosion Pressure [P_{MAX}] for a given concentration of oxygen as:

$$P_{MAX,C} = 0.035\ C_{O2} + 0.92\ P_{MAX,AIR} \qquad \text{bara}$$

The following Table presents the results of calculations, based on the above relationship, for two typical values for P_{MAX} in air [26]:

Table 3. Maximum Explosion Pressures for Fuel/Oxygen Mixtures

$P_{MAX,AIR}$	MAXIMUM EXPLOSION PRESSURES IN FUEL/OXYGEN MIXTURES (barg)							
	30 %O_2	40 %O_2	50 %O_2	60 %O_2	70 %O_2	80 %O_2	90 %O_2	100 %O_2
7 barg	7.4	7.7	8.1	8.4	8.8	9.1	9.5	9.8
8 barg	8.3	8.6	9.0	9.3	9.7	10.0	10.4	10.7
9 barg	9.2	9.6	9.9	10.3	10.6	11.0	11.3	11.7
10 barg	10.1	10.5	10.8	11.2	11.5	11.9	12.2	12.6

Thus, oxygen enrichment would have a significant effect on the extent of damage that could result from ignition of confined flammable gases and vapors, and could affect the design of explosion-protection systems and devices, such as explosion vents.

2.5 Deflagration Index, Kg

The Deflagration Index [Kg for flammable gases and vapors] is a measure of the speed of combustion and, thus, is directly related to the speed of flame through a mixture of fuel and oxidant. In the absence of strong turbulence, the flame speed is related to the fundamental burning velocity [FBV] by the ratio of the densities of the unburned mixture and the density of the combustion gases, by [27]:

$$S_{FLAME} = FBV\,(d_{UNBURNED}/d_{BURNED}) \quad \text{centimeters per second}$$

Typically, the ratio of the flame speed to FBV is about 6.

Previous investigators had observed that there is a relationship between the FBV and the Minimum Ignition Energy [MIE], as given by [28]:

$$FBV = K/MIE^{0.8} \text{ centimeters per second}$$

Some data are available for both the FBV and the MIE, as indicated in the following Table [29] :

Table 4. The Relationship Between FBV And Oxygen Concentration

GAS/VAPOR	FBV, Air (cm./sec.)	Kg, Air (bar-m/s)	MIE, Air (millijoule)	FBV, O_2 (cm./sec.)	MIE, O_2 (millijoule)	Kg, O_2 (bar-m/s)	Kg Ratio O_2/Air
Hydrogen	312	638	0.016	1400	0.0012	3,000	4.5
Methane	45	55	0.21	450	0.0027	450	4.6
Ethane	47.5	103	0.24	375	0.0019	670	6.5
Propane	46	100	0.25	370	0.0021	720	7.2
Butane	-	92	0.25	-	0.009	-	-
Hexane	-	-	0.24	-	0.006	-	-
Heptane	42.5	-	0.24	-	-	-	-
Averages	-	-	-	-	-	-	5.7

As indicated in the above Table, the Deflagration Index – and the speed of combustion – in oxygen is almost six times greater in 100% oxygen than it is in air [at 21% oxygen]. The data in the preceding Table can be combined to yield two approximate expressions for the Deflagration Index in given concentrations in oxygen, as:

$$Kg_{,O2} = 0.04\,(K_{G,AIR})\,(C^{1.2}) \qquad \text{bar-meters/second}$$

$$Kg_{,O2} = 1.85\,C^{1.1} / MIE_{AIR}^{\,0.46} \qquad \text{bar-meters/second [excluding } H_2\text{]}$$

Thus, if the value of the Deflagration Index in air is known, or if the Minimum Ignition Energy in air is known, an approximate value for the Deflagration Index at a given oxygen concentration can be determined.

3. <u>Combined Relationships</u>

The explosibility/flammability properties of a given flammable gas or vapor can be compared with a standard or typical flammable gas, through the use of Ignition Sensitivity and Explosion Severity. These property combinations were first developed and utilized for the comparisons of the hazards of coal mines, starting in 1940, where combustible dusts presented an explosion hazard [30].

3.1 Ignition Sensitivity

The product of LFL [in terms of grams per cubic meter], AIT, and MIE form the basis for a calculation of Ignition Sensitivity. A high value of this product indicates a low likelihood of ignition by relatively-weak ignition sources, such as electrostatic discharges, friction, impact, and hot surfaces. For propane in a mixture with 40% oxygen, the product is only 39 x 445 x 0.035 = 600. For comparison, the similar product for typical combustible dusts is about 45,000.

To determine the Ignition Sensitivity of a combustible dust, the LFLxAITxMIE product would be divided into the corresponding product for Pittsburgh [bituminous] Coal Dust, at 2,000,000. Thus, the Ignition Sensitivity for typical combustible dusts is about 2,000,000/45,000 = 45. By comparison, the Ignition Sensitivity for propane in a 40% oxygen mixture would be about 2,000,000/600 = 3,300, indicating the much greater sensitivity of this propane mixture to ignition sources.

As another indication of the effects of oxygen enrichment, the Ignition Sensitivity for a stoichiometric [worst-case] mixture of propane and air would be about 450. Thus, the 40% oxygen mixture would be about 7 times more sensitive to ignition [3,300/450].

3.1 Explosion Severity

The product of Deflagration Index Kg and the Maximum Explosion Pressure Pmax [in terms of barg] forms the basis for a calculation of Explosion Severity. A low value of this product indicates a low likelihood of severe damage from explosion. For propane in a mixture with 40% oxygen, the product is about 275 x 9 = 2,500. For comparison, the similar product for typical combustible dusts is about 1,500.

For combustible dusts, the product is divided by the corresponding product for Pittsburgh [bituminous] Coal Dust, at 840. Thus, the Explosion Severity for typical combustible dusts is about 1,500/840 = 1.8. By comparison, the Ignition Sensitivity for propane in a 40% oxygen mixture would be about 2,500/840 = 3.0, indicating the greater likelihood of damage if a confined mixture of propane and 40% oxygen were to be ignited.

The effect of oxygen enrichment can be shown by the Explosion Severity for a stoichiometric [worst-case] mixture of propane and air, at about 0.9. Thus, the 40% oxygen mixture could be about 3 times more damaging than an air mixture [3/0.9].

3.1 Combination of Ignition Sensitivity and Explosion Severity

If the Ignition Sensitivities of fuel/oxidant mixtures are less than 0.2 AND the Explosion Severities are less than 0.5, then such mixtures “are generally not considered to be significant explosion hazards” [31].
For the mixture of propane and 40% oxygen – for which the properties are shown above – the combination of Ignition Sensitivity and Explosion Severity is about 5 orders of Magnitude [a factor of almost 100,000] above the “no significant explosion hazard” criterion. Thus, it is unlikely that a mixture of air with any flammable or combustible vapor could be considered to present “no significant explosion hazard”.

3.1 Combination of Ignition Sensitivity and Explosion Severity as Explosibility Index

The product of Ignition Sensitivity and Explosion Severity can be considered to be an Explosibility Index [32 to 37]. This Index has been found to be useful in comparing the hazards of combustible dusts, but had not yet been applied to flammable gases and vapors.

4. <u>Graphical Presentation of Flammability/Explosibility Properties</u>

Figure 1 presents the flammability/explosibility properties of a mixture of flammable fuel, oxygen, and nitrogen, in terms of Ignition Sensitivity [IS], Explosion Severity [ES], and Explosibility Index [EI] at atmospheric pressure and at an initial temperature of 25°C, as compared to the similar explosibility properties of propane at room temperature and atmospheric pressure. The ratios are based on the available data and the results of the calculations that are described above. As could be expected, the explosion hazards increase as the stoichiometric line – where all carbon is oxidized to carbon dioxide, and all hydrogen is oxidized to steam – is approached.

5. Conclusions.

To evaluate the explosion hazards of mixtures of fuels with oxygen-enriched atmospheres, the available data were utilized to obtain methods for estimating the values of pertinent explosibility properties. The relationships are as follows:

Lower Flammable Limit	Equal to that of fuel/air mixtures	
Autoignition Temperature	$AIT_{O2}=AIT_{AIR}/[1+(2.0AIT_{AIR}/40{,}000)(\ln(C_{O2}/21)]-273$	°C
Minimum Ignition Energy	$MIE_{O2} = 6{,}200\ MIE_{AIR} / C_{O2}^{\ 2.9}$	millijoule
Max. Explosion Pressure	$P_{Max,O2} = [[0.035\ C_{O2}) + (0.92\ (P_{Max,AIR} + 1))] - 1$	barg
Deflagration Index	$Kg_{,O2} = 0.04\ (K_{G,AIR})\ (C^{\ 1.2})$	bar-meters/second
	$Kg_{,O2} = 1.85\ C^{\ 1.1} / MIE_{AIR}^{\ 0.46}$	bar-meters/second

In addition, a method is presented for evaluating the Ignition Sensitivity and the Explosion Severity of fuel/oxygen mixtures, using the above relationships, together with a method for graphically showing the relative hazards of such mixtures.

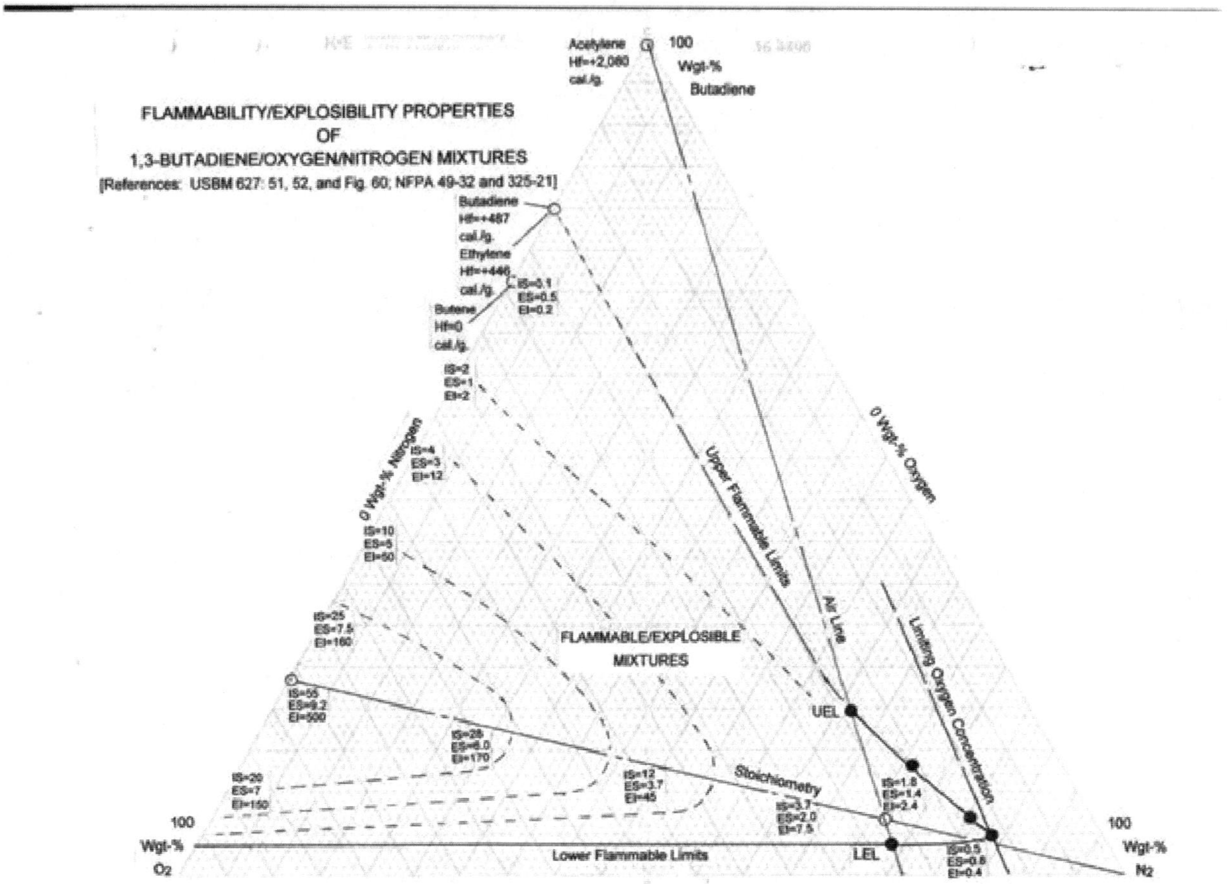

Figure 1 – Weight Concentrations of Gas Mixtures

with

Ignition Sensitivities, Explosion Severities, and Explosion Index

as compared to the properties of Propane

6. References.

1. Urben, P. G. [editor], "Bretherick's Handbook of Reactive Chemical Hazards", 6th Ed., Vol. 2, Page 287 (1999).

2. National Fire Protection Association, "Fire Protection Guide to Hazardous Materials", 13th Ed., including NFPA 49 and NFPA 325 (2002).

3. National Fire Protection Association, "Guide for Venting of Deflagrations", NFPA 68, Table D.1 (2002).

4. National Fire Protection Association, "Standard on Explosion Prevention Systems", NFPA 69, Table C.1(a) (2014).

5. National Fire Protection Association, "Recommended Practice on Static Electricity", NFPA 77, Table B.1 (2014).

6. National Fire Protection Association, "Recommended Practice for the Classification of Flammable Liquids, Gases, or Vapors and of Hazardous (Classified) Locations for Electrical Installations in Chemical Process Areas", NFPA 497, Table 4.4.2 (2017).

7. Coward, H. F., and G. W. Jones [U. S. Bureau of Mines], "Limits of Flammability of Gases and Vapors", Bulletin 503, Table 44 [and many "nose" Figures] (1952).

8. Zabetakis, M. G. [U. S. Bureau of Mines], "Flammability Characteristics of Combustible Gases and Vapors", Bulletin 627, Table A-1 [and many "nose" Figures] (1965).

9. Babrauskas, V., "Ignition Handbook", Table 1A (1992).

10. Britton, L. G., "Avoiding Static Ignition Hazards in Chemical Operations", Page 223 (1999).

11. Crowl, D. A., and J. F. Louvar, "Chemical Process Safety", 3rd Ed., Pages 251, 259, 261, 287, and 654 (2011).

12. Yaws, C. L., "Chemical Properties Handbook", Table 25-1 (1999).

13. National Fire Protection Association, "Recommended Practice for the Classification of Combustible Dusts and of Hazardous (Classified) Locations for Electrical Installations in Chemical Process Areas", NFPA 499, Paragraph A.5.2.3 (2017).

14. Frank, W. L., and S. A. Rodgers, "NFPA Guide to Combustible Dusts", Page 92 (2012).

15. Britton, L. G., "Avoiding Static Ignition Hazards in Chemical Operations", Page 223 (1999).

16. Babrauskas, V., "Ignition Handbook", Page 1023 (1992).

17. Zabetakis, M. G. [U. S. Bureau of Mines], "Flammability Characteristics of Combustible Gases and Vapors", Bulletin 627, Page 9 (1965).

18. Crowl, D. A., and J. F. Louvar, "Chemical Process Safety", 3rd Ed., Pages 263, 265 to 267, and 269 (2011).

19. Crowl, D. A., and J. F. Louvar, "Chemical Process Safety", 3rd Ed., Pages 398 and 400 (2011).

20. Babrauskas, V., "Ignition Handbook", Pages 1023 and 1024 to 1058 (1992).

21. Green, D. W., and M. Z. Southard, "Perry's Chemical Engineers' Handbook", 9th Ed., Page 19-24 (2019).

22. Babrauskas, V., "Ignition Handbook", Pages 75, 76, 203, and 625 (1992).

23. Babrauskas, V., "Ignition Handbook", Pages 1024 to 1055 (1992).

24. Babrauskas, V., "Ignition Handbook", Page 625 (1992).

25. Lewis, B., and G. von Elbe, "Combustion, Flames, and Explosions of Gases", 3rd Ed., Pages 717 and 719 (1987).

26. National Fire Protection Association, "Guide for Venting of Deflagrations", NFPA 68, Table D.1 (2002).

27. Babrauskas, V., “Ignition Handbook”, Page 35 (1992).

28. Babrauskas, V., “Ignition Handbook”, Page 74 (1992).

29. Zabetakis, M. G. [U. S. Bureau of Mines], “Flammability Characteristics of Combustible Gases and Vapors”, Bulletin 627, Page 43 and Figures 44 and 45 (1965).

30. Nagy, J., et al. [U. S. Bureau of Mines], “Explosibility of Carbonaceous Dusts”, Report of Investigations 6597, Page 2 (1965).

31. National Fire Protection Association, “Recommended Practice for the Classification of Combustible Dusts and of Hazardous (Classified) Locations for Electrical Installations in Chemical Process Areas”, NFPA 499, Paragraph A.5.2.3 (2017).

32. Dorsett, H. G., et al. [U. S. Bureau of Mines], Laboratory Equipment and Test Procedures for Evaluating Explosibility of Dusts”, Report of Investigations 5624, Page 2 (1960).

33. Jacobson, M., et al. [U. S. Bureau of Mines], “Explosibility of Dusts used in the Plastics Industry”, Report of Investigations 5971, Tables A-1, A-2, and A-3 (1962).

34. Jacobson, M., et al. [U. S. Bureau of Mines], “Explosibility of Metal Powders”, Report of Investigations 6516, Tables A-1, A-2, and A-3 (1964).

35. Nagy, J., et al. [U. S. Bureau of Mines], “Explosibility of Carbonaceous Dusts”, Report of Investigations 6597, Page 2 (1965).

36. Dorsett, H. G., and J. Nagy [U. S. Bureau of Mines], “Dust Explosibility of Chemicals, Drugs, Dyes, and Pesticides”, Report of Investigations 7132, Tables A-1, A-2, and A-3 (1968).

37. Grossel, S. S., “Safety Considerations in Conveying of Bulk Solids and Powders”, Page 62 in the “Journal of Loss Prevention in the Process Industries”, Butterworths, Vol. 1, No. 2 (April 1988).

7. Legal Disclaimer and Liability

(a) Limitation of Liability. The consulting services conducted by DEKRA Process Safety (the "Company") were performed using generally accepted guidelines, standards, and/or practices, which the Company considers reliable. Although the Company performed its consulting services pursuant to reliable and generally accepted practices in the industry, the Company does not guarantee or provide any representations or warranties with respect to Client's use, interpretation or application of the findings, conclusions, and/or suggestions of the consulting services provided by the Company. Moreover, the findings, conclusions, and the suggestions resulting from the consulting service are based upon certain assumptions, information, documents, and procedures provided by the Customer. AS SUCH, IN NO EVENT AND UNDER NO CIRCUMSTANCE SHALL THE COMPANY BE LIABLE FOR SPECIAL, INDIRECT, PUNITIVE OR CONSEQUENTIAL DAMAGES OF ANY NATURE WHATSOEVER, INCLUDING WITHOUT LIMITATION, ANY LOST REVENUE OR PROFITS OF THE CUSTOMER OR ITS CUSTOMERS, AGENTS AND DISTRIBUTORS, RESULTING FROM, ARISING OUT OF OR IN CONNECTION WITH, THE SERVICES PROVIDED BY THE COMPANY. The Customer agrees that the Company shall have no liability for damages, which may result from Client's use, interpretation or application of the consulting services provided by the Company.

(b) The Company's pricing of the consulting services provided does not contemplate that the Company shall have any liability resulting from its performance of the consulting services, except as otherwise set forth in the Quotation from the Company. Accordingly, the Customer shall indemnify and hold harmless the Company, its shareholders, directors, officers, employees and agents (the "Indemnified Parties") from and against any and all loss, cost, liability and expense, including reasonable attorney's fees and costs, which any of the Indemnified Parties may incur, sustain or be subject to, as a result of any claim, demand, action, investigation or proceeding arising out of or relating to either: (a) the consulting services provided by the Company; or (b) any material, equipment, specifications or safety information (or lack thereof) supplied to the Company (or which should have been supplied to the Company) by Customer and/or any failure of such materials, equipment, specifications and safety information to comply with any federal, state or local law or safety standard.

(c) For additional terms and conditions, which apply with respect to the provision of this report, see the Quotation provided by the Company and executed by Customer. If any terms set forth in the Quotation conflict with the terms set forth herein, the terms set forth herein shall apply.

19 AIChE Spring Meeting
15TH GLOBAL CONGRESS ON PROCESS SAFETY
NEW ORLEANS • 2019

Bridging Chemical Engineering (Risk Based Process Safety) to the Humanities – 7 Mega Elements' Approach (Protocol) of Accident Investigation in Light of Ethical Dimension, Architecture & Epistemology

Long(Steven) Zhang
No affiliation since Jul 2014 off-hiring from COP
11455 Bissonnet St Apt #514, Houston TX 77099
longz.houston@gmail.com

Prepared for Presentation at
American Institute of Chemical Engineers
2019 Spring Meeting and 15th Global Congress on Process Safety
New Orleans, LA
March 31 – April 3, 2019

Bridging Chemical Engineering (Risk Based Process safety) to the Humanities – 7 Mega Elements' Approach (Protocol) of Accident Investigation in Light of Ethical Dimension, Architecture & Epistemology

Long(Steven) Zhang
No affiliation since Jul 2014 off-hiring from COP
11455 Bissonnet St Apt #514, Houston TX 77099
longz.houston@gmail.com

Keywords: Investigation, Diagnosis, Humanities, *the Art of War, Narrative, Epistemology*

Abstract

In this paper, facing the challenges of Incident Investigation, actually whole CCPS 4th Pillar "Learn from Experience", chemical engineers' (process safety practitioners') brand new vocabulary (7 Mega Elements' narrative on failure diagnosis) would be presented in context of bridging chemical engineering (RBPS) to the Humanities, of which 3 core topics i.e. ethical dimension, architecture and epistemology would be explored and reflected upon so that a new era of RBPS diagnosis (Learn from Experience) predicted.

1 Introduction

In 2007 Center for Chemical Process Safety (CCPS) unprecedentedly presented to process industry Risk Based Process Safety (RBPS) 4 Pillars (and relevant 20 Elements), among which 4th Pillar "Learn from Experience" consists of 4 Elements: Incident Investigation, Measurement & Metrics, Auditing, and Management Review & Continuous Improvement[1].

In latest decades chemical engineers (process safety practitioners) have been facing some big challenges from accident/incident investigations, actually from whole CCPS 4th Pillar (Learn from Experience), for which there might have been many efforts, discussions, and even some 'solutions'. The author argues, however, the challenges comes not only from the needs of high quality and consistency of accident investigation, but, more fundamentally, from building up a brand new vocabulary (protocol) for chemical engineers to competently communicate of, and negotiate on, accident investigations or "why they occur and how ultimately to prevent".

Toward building a new vocabulary/protocol/narrative on accident investigation or, more broadly, on how Experience could be truly Learned (CCPS 4th Pillar), since 9th Global Congress, the author has had a series of efforts for exploration and reflections on the possible way forward, e.g. RBPS Core value (landlord for temple), Process Safety Cultural Communication (landlord's livelihood) & Egg model[2]; 4th Pillar as "espionage" not to enemies but ourselves[3][4]; 7 Mega Elements & 28 infra cultural activities (practices); RBPS Socio-Technical systems[5]; an ethical dimension of Vision 20/20[6] etc.. And in broader context of bridging chemical engineering (Risk Based Process Safety) to the Humanities, the author finally arrived at 7 Mega Elements' vocabulary/ protocol/ narrative/ "cosmology" on RBPS diagnosis for chemical engineers (process safety practitioners)[7].

2 7 Mega Elements' Protocol – Chemical Engineers' Vocabulary

Louis L. Bucciarelli claimed that he had taken seriously the "two cultures" challenge set by C. P. Snow and been working at bridging engineering to the Humanities for 35 years[8]; instrumental rationality does not suffice itself in engineering design, diagnosis and education[9]. And the author's experiences during the humble quest (through relevance of anthropology, intercultural communication, *the Art of War* etc.) for an ultimate solution to catastrophic (process safety) accidents (chemical engineering failure) seems not only resonating with Bucciarelli's bridging journey but a little bit unique, not only toward chemical engineering (Risk Based Process Safety) but also to engineering discipline at large, i.e., through benchmarking RBPS approach against Sun Tzu's *the Art of War,* 7 Mega Elements (vocabulary/ protocol/ narrative) not only were discovered for (chemical) engineers but function as 7 fundamental (chemical) engineering thoughts, which eventually sheds light on those entailed 28 practices of (chemical) engineering discipline.

Unlike Bucciarelli who tried to bridge philosophy and engineering, the author proceeded another way, i.e. first making relevance of anthropology & intercultural communication to chemical engineering (Risk Based Process Safety), then benchmarking Egg model against *the Art of War*, what the author calls 1st main ancient source for bridging (chemical) engineering to the Humanities, compared to 1st main modern one - Bucciarelli's *Engineering Philosophy.*

When guiding all visitors to some biggest lessons ("more original" two cultures) of ancient cultural traditions or civilizations of east and west, the author claimed that there should be an landlord (core value) to live within the temple building (CCPS 4 Pillars) and, furthermore, the landlord should have livelihood (Process Safety Cultural Communication (PSCC)) as well, and that should the landlord died the building seems still beautiful outwardly, however, with a purpose of tomb. In another words, when (chemical) engineers' vocabulary/ protocol/ narrative becomes deficient, or landlord and his livelihood are absent in the vocabulary, catastrophic (process safety) accidents are bound to happen, no matter how technically over competent relevant chemical engineers (process safety practitioners) might be, e.g., the lesson from the case of Bian Que–Duke Cai.

Through anthropology and intercultural communication, **RBPS Core Value** was defined in two ways, an unfolded one: *RBPS Core Value functions as the root of the tree of Process Safety*

Culture. Deficiencies are symptoms or signals for a dying or dead state of a core value, which means that the organization has strategically failed at the highest levels in managing process safety through a risk based approach – even if there has been no occurrence of process safety accidents till that moment. In another words, the RBPS temple is bound to collapse, sooner or later, due to the change or drift from the original design intent of the temple. And RBPS Core Value's simpler definition, through the concept of Power Distance in value studies: *there is no power, authority or governance on process safety except on behalf of process safety expertise (PSE).* (PSE means the divine essence of the highest strategic insights and vision of process safety management; the Tao for process safety culture; and the soul of *The Art of War* against catastrophic process safety accidents)

And process safety culture does not exist except through ***PSCC****, whose ultimate purpose is to convey the message of RBPS Core Value. Without nourishment from RBPS Core Value, the tree of Process Safety Culture or PSCC would get dry and dead.* So the temple building was eventually upgraded to Egg model, as showed within the scheme 1st. And these 2 new universes (or 2 Mega Elements) might be deemed what Bucciarelli called the rhetoric of engineering as the parcel of instrumental explanation[10], and probably 2 most fundamental (chemical) engineering thoughts as well.

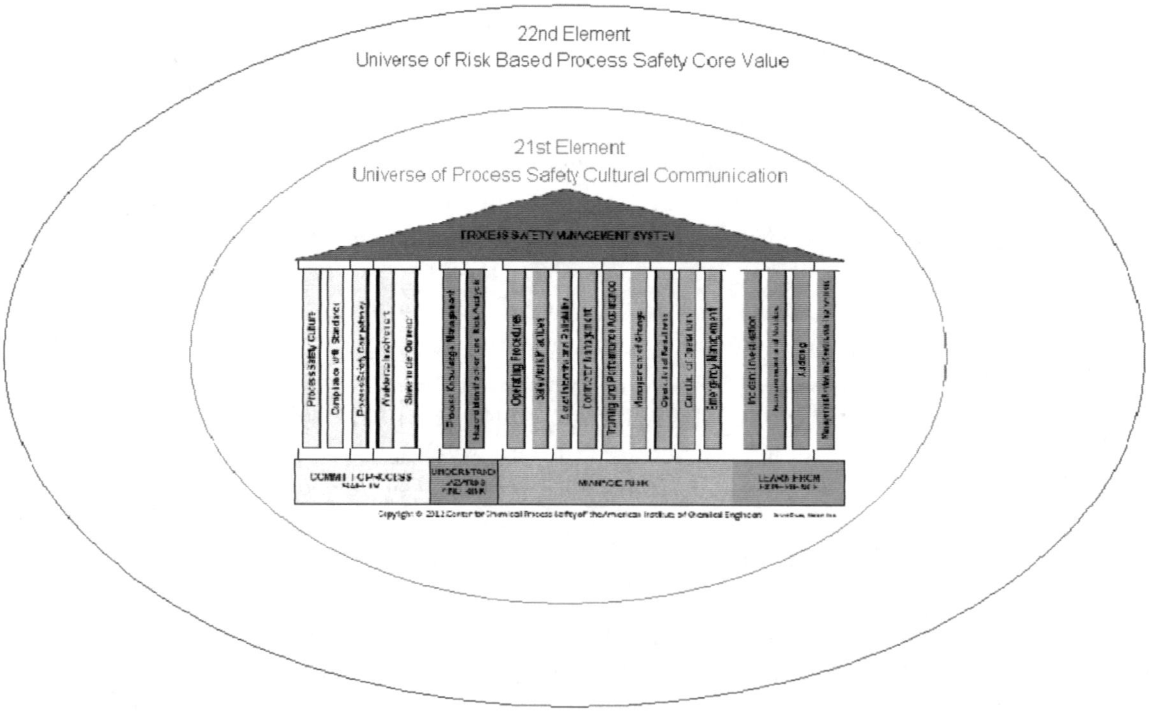

Egg Model (CCPS Temple + 2 universes) - Scheme 1st

The Temple (4 pillars) Figure Source: CCPS, Process Safety Boot Camp, contributed by David Guss, Nexen Inc. Copyright 2010 AIChE and used with permission

Then, in spite of the formidable strengths of most, actually 17, elements of CCPS 4 Pillars, except Process Safety Culture, Process Safety competency & Stakeholder Outreach, the outcome of benchmarking Egg model (4 Pillars + 2 Universes) against *The Art of War,* what the author calls a matured ancient vocabulary on engineering (military) diagnosis, not only shows that those 4 Pillars' (20 elements') narrative, even together with the new 2 Universes, would probably be not so architecturally organized at negotiation and communication on why process safety (chemical engineering) accidents (failure) occur & how to prevent, but makes the author further believe that Sun Tzu's vocabulary (6 Mega Elements, refer to Table 1st) might potentially become a narrative/ protocol paradigm for future (chemical) engineering diagnosis and *The Art of War* an main ancient source of bridging engineering to the Humanities.

For the fact that, however, we could not neglect the historical context of narrative/ protocol of *The Art of War*, the author, in light of modern globalization and global operations, fundamentally redefined Process safety Culture into Process Safety Cultural Communication (PSCC) as the 7th Mega Element, and upgraded another 3 ones i.e. Leadership, Weather/Climate & Espionage as below:

Leadership: a new chain of command, inherently from RBPS Core Value, rather than blindly from superiors;

Weather/Climate: RBPS Core Value enlightened outreach to the stakeholders & community, rather than only tactically maintaining seemingly good relationship with community; &

<table>
<tr><th colspan="2">RBPS Egg</th><th>The Art of War</th><th>Philosophy of Process Safety</th><th></th></tr>
<tr><td colspan="4">RBPS Core Value - Tao - RBPS Worldview</td><td>Worldview</td></tr>
<tr><td rowspan="4">Commit to Process Safety</td><td>Process Safety Competency</td><td rowspan="2">Leadership</td><td rowspan="2">People</td><td rowspan="7">Process Safety Cultural Communication (Process Safety Culture) RBPS Methodology</td></tr>
<tr><td>Workforce Involvement</td></tr>
<tr><td>Compliance with Standards</td><td rowspan="2">Weather/Climate</td><td rowspan="3">Plant</td></tr>
<tr><td>Stakeholder Outreach</td></tr>
<tr><td>Understand Hazards & Risk</td><td></td><td>Terrain</td></tr>
<tr><td>Manage Risk</td><td></td><td>Discipline</td><td>Procedure</td></tr>
<tr><td>Learn from Experience</td><td></td><td>Espionage</td><td>Engineering & Humanities</td></tr>
</table>

Table 1st

Espionage: 4th Pillar *Learn from Experience* (4 elements: *investigation, metrics, audit & review*) plays role "espionage", not to the enemies but to ourselves our organizational "national security"

And 7 Mega Elements might also be deemed as 7 (chemical) engineering thoughts.

3 Ethical Dimension (2 Universes) - What Chemical Engineers Count & How

While declared *there is no power, authority or governance on process safety except on behalf of Process Safety Expertise (PSE)* the author claimed that PSE shall eternally and exclusively chair at the highest strategic levels throughout all the processes of decision making and behavior determining, i.e. chemical engineers (process safety practitioners) need getting inspired from RBPS Core Value and persistently conveying the message of RBPS Core Value, both of which make up the ethical dimension (of 7 Mega elements' vocabulary/ protocol/ narrative), probably what Bucciarelli calls "best left sacred"[11], which ensures, or anthropologically designs, all relevant decision making and behavior determining correct. In this sense, at highest strategical levels, chemical engineers do not design at all but RBPS Core Value does.

From philosophical perspective, Bucciarelli claimed that the tensions at the interfaces between different design participants could not be solved by instrumental means alone [12]. And imitating American anthropologist Edward T. Hall's saying the author ever described process safety culture as a prison, unless we have RBPS Core Value as the key to unlock it and PSCC the brightness of the key. And Bucciarelli pointed out that *designing is a social process of negotiation, of iteration, of rectifying mis-steps, even misconceptions – a process rich in ambiguity and uncertainty* [13]. Similarly, the author in metaphor described PSCC as the photosynthesis of tree of process safety culture, the impulses from the heart (RBPS Core Value) and nourishment (highest designing principles) transportation from the root (RBPS Core Value). What does Bucciarelli mean by that 'social process'? What does he mean by "contribute to a sense of community among the citizenry" [14]? And what does the author mean by "photosynthesis" and "nourishment"? With hesitation & obligations, the author claims that the said social process, what the author puts PSCC, traditionally should be called Constitutional Democracy; and RBPS Core Value a nickname for Universal Value in chemical engineering.

As Universal Value & Constitutional Democracy in (chemical) engineering seem more subtle than those traditional ones the author here might call them Universal Value & Constitutional Democracy 2.0, what Bucciarelli described "old themes but in a new light" [15]. And this 2.0 might be acknowledged as two fundamental (chemical) engineering thoughts and the ethical dimension of (chemical) engineers' vocabulary/ protocol/ narrative on accident investigation and whole (chemical) engineering diagnosis (4th Pillars) as well. And probably only in this way could (chemical) engineers truly acquire what Bucciarelli called "new ways of perceiving and reading the world" [16], eventually leading the industries and communities into a new enlightened era of accident investigation and (chemical) engineering diagnosis. And the author might further states that, as the ethical dimension of (chemical) engineers' vocabulary (protocol), Universal Value & Constitutional Democracy 2.0 should function as the highest communication (designing) principles on accident investigation and (chemical) engineering diagnosis, and on accident prevention (foreknowledge) as well.

4 7 Mega Elements' Architecture – 28 Practices' Diagnosis Map

Inspired from Edward T. Hall's map of culture, the author developed a map of 28 infra cultural activities of Process Safety Culture (refer to 7 ones in the diagonal from upper left to lower right and 21 ones above), what Bucciarelli might like to call another, slightly different, "more complex vision of engineering thoughts and practices" [17]. And each infra cultural activity was defined as the critical statements & behaviors related to the interaction between Mega Element A and Mega Element B, e.g. infra activity 04 means: verbally & behaviorally communicate that the actual death of RBPS Core Value might be one of the root causes of accidents. And those 7 ones, extremely, are in fact of the critical verbal statements & behaviors related to the definition of the Mega Element (or engineering thought) concerned, e.g. infra activity 22: verbally and behaviorally embrace and reflect on Leadership - the engineering thought of new chain of command inherently from RBPS Core Value rather than blindly from superiors.

This matching between verbal statements & behaviors, or the inherent relationship between the message of RBPS Core Value and all relevant decision making & behavior determining, seems some resonance with Wittgenstein's recognition that rules and meanings (what the author calls ***tone***) in talk and texts (what the author calls vocabulary/ protocol/ narrative) were mixed with practice [18], or with what Bucciarelli put another way "the meaning of statements is embedded in practice" [19]. So these 28 infra cultural activities' map, an unfolded version of what & how chemical engineers (process safety practitioners) count, might accommodate as highest strategic encompassing diagnosis protocol (map).

	RBPS C.V. 0	P.S.C.C. 1	Leadership 2	Weather/C. 3	Espionage (Diagnosis) 4	Terrain 5	Discipline 6
RBPS Core Value 0	00	01	02	03	04	05	06
Process Safety Cultural Communication (PSCC) 1	10	11	12	13	14	15	16
Leadership 2	20	21	22 4 dimensions	23	24	25	26
Weather/ Climate 3	30	31	32	33	34	35	36
Espionage (Diagnosis) 4	40	41	42	43	44	45	46
Terrain 5	50	51	52	53	54	55	56
Discipline 6	60	61	62	63	64	65	66

28 Infra Cultural Activities' Diagnosis Map (Table 2nd)

As the Humanities e.g. philosophy, anthropology & intercultural communication, Sun Tzu's *The Art of War* etc. sheds more lights on the old themes (i.e. RBPS Core value, PSCC, Leadership and Weather/Climate), upon this Diagnosis Map, "another" complex vision of engineering thoughts and practices, the author claims that *the yellow grounded, or more exactly red number lettered area, are all about organizational "national security", and that the infra activity 44 (Espionage)'s 2nd rationality, next to the instrumental one, whose protocol area is painted green, SHALL be reactivated so that "espionage" to our organization would shift to a more proper name - "counter-intelligence", on accident investigation & prevention and on engineering diagnosis at large.*

Through maintaining the architecture/protocol of 7 Mega Elements could all these 22.5 infra activities of "national security" area be reliably visible, open and "problematic" over there, together with "traditional" green area (5.5 infra activities), to critique and diagnosis among board of directors, executive management, operation managers, strategy, legal, HR, chemical engineers (process safety practitioners), and even beyond. Pragmatically, from designing perspective, chemical engineers should pursue this more matured architecture/protocol/vocabulary, getting inspired from RBPS Core Value, persistently conveying its message and maintaining alive this complex vision of engineering thoughts & practices i.e. this 28 Practices (infra activities)' Diagnosis Map, otherwise the insightful topic "social construction of failure/ accident" [20] would turn out to be a truism. In another words, either RBPS Core Value highest strategically designs safety (and tranquility) through PSCC, or otherwise (chemical engineering/ process safety) failures/ accidents/ catastrophes are socially constructed.

5 Epistemology – Future (Chemical) Engineers' Knowledge Hierarchy

Walter Vincenti's understanding of *engineering knowledge,* as J. C. Pitt ever explained*, is task-specific and aims at the production of an artifact to serve a predetermined purpose* [21], of which, the author believes, CCPS 4 Pillars (20 Elements), as of (chemical) engineering/RBPS knowledge, might have been satisfactorily reflecting the main characters. Bucciarelli discovered, however, that engineering knowledge, of which he described *what engineers do, the reasons they do what they do, and their justification for believing the latter are good and true,* is *historically and culturally contingent, not fixed by the dictates of science and empirical fact alone.* [22]. The author might explain here, echoing the former statement "the landlord ... even to buy a new house with new style of design to live within", that the (chemical) engineering knowledge we have here with CCPS 4 Pillars might not be the single unique design/house for the landlord [23], but the landscape (landlord, livelihood & house) of the contingencies, what Bucciarelli calls "a picture of both possibilities and limitations of engineering thinking" [24], is unique. In another words, both the landlord (RBPS Core Value), deeply rooted in most cultural traditions and civilizations, and his livelihood (PSCC), what the author likes to call Constitutional Democracy, should preside the landscape of the contingencies and chair and guide at the highest strategic levels the whole engineering design process, for surety including on accident investigation and diagnosis, so that the objects of design, which "traditionally" have been confined to artifacts e.g. a mega petrochemical project/ operations, might also need to include all processes of decision making and behavior determining (refer to Table 2nd – 28 Infra Activities' Diagnosis Map).

The author might state that *a complete (chemical) engineering design process, including on accident investigation, diagnosis & prevention, shall consist of 3 components: RBPS Core Value (Universal Value), PSCC (Constitutional Democracy) and the application of (chemical) engineering knowledge (e.g. one with CCPS 4 Pillars).* And at same time, we recognize that *future (chemical) engineers need a new epistemology - a new hierarchy of knowledge which consists of Universal Value Knowledge, Constitutional Democratic Knowledge & the relevant (chemical) engineering knowledge.* Edward T. Hall had an earth-breaking saying: *culture is communication and communication is culture* [25], from which the author got inspired and developed the concept of PSCC. When (chemical) engineers verbally (of text) and behaviorally (of practice) abandon the ethical dimension of 7 Mega Elements' vocabulary/ protocol/ narrative, they lose the whole process safety culture or, we could say, both their knowledge hierarchy is deficient, akin to a genocide, and their engineering design process seriously flawed, like building concentration camps.

6 Conclusion

In context of bridging chemical engineering to the Humanities, on accident investigation, diagnosis and prevention, and on engineering design process at large, the author presents 7 Mega Elements' Vocabulary/ Protocol/ Narrative, 28 Infra Cultural Activities' Diagnosis Map and a newly galvanized old Hierarchy of Knowledge, so that some outlooks might be made: a) as one of the largest communities of engineering professions, AIChE would pursue leading roles at dialogues between engineering & the Humanities; b) some small groups of chemical philosopher/anthropologist-engineers would emerge; c) an enlightened era of (chemical/process safety) accident investigation, diagnosis & learning, and engineering design process at large, would arrive; & d) engineering education's future would get brighter than ever.

7 References

[1] Center for Chemical Process Safety, *Guidelines for Risk Based Process Safety*, John Wiley & Sons, Inc., 2007

[2] Zhang, L., An Upgraded Definition of Risk Based Process Safety Core Value and Its Impact in Light of Anthropology & Intercultural Communication on the Future Evolution of Process Safety Culture & Management, 2013 AIChE Spring Meeting and 9th Global Congress on Process Safety, San Antonio, April 2013

[3] Sun-Tzu and Ames, Roger T., Sun-tzu: *The Art of Warfare*, Ballatime Books, 1993

[4] Zhang, L., Benchmarking Risk Based Process Safety Egg (RBPS Temple+2 Universes) against The Art of War, 1st CCPS Asia-Pacific Conference on Process Safety, Qingdao, Sept 2013

[5] Zhang, L., People Means Leadership, NOT Simply mean People – 4 New Dimensions of Process Safety Competency, 2014 AIChE Spring Meeting and 10th Global Congress on Process Safety, New Orleans, April 2014

[6] Zhang, L., How Committed to Process Safety is Committed Enough? – Doing Process Safety Cultural Communication (PSCC) as the Inevitable Ethical Dimension for Vision 20/20, 2016 AIChE Spring Meeting and 12th Global Congress on Process Safety, Houston, April 2016

[7] Zhang, L. and Wasel, W., Corporate Memory Loss Ongoing – Except CCPS 4th Pillar (Learn from Experience) as Espionage, 2018 AIChE Spring Meeting and 14th Global Congress on Process Safety, Orlando, April 2018

[8] Bucciarelli, L. L., *Engineering Philosophy*, Delft: DUP Satellite, 2003, p.2

[9] Ibid. p.99

[10] Ibid. p.6

[11] Ibid. p.4

[12] Ibid. p.4-5

[13] Ibid. p.7

[14] Ibid. p.5

[15] Ibid. p.2

[16] Ibid. p.7

[17] Ibid. p.101

[18] Ibid. p.6

[19] Ibid. p.6

[20] Ibid. p. 101

[21] Pitt, J. C., "What Engineers Know", *Techne* 5:3, Spring, 2001, p.25

[22] Bucciarelli, *Engineering Philosophy*, p.99

[23] Pitt, "What Engineers Know", p.26

[24] Bucciarelli, *Engineering Philosophy*, p.99

[25] Hall, E. T., *The Silent Language*, Anchor Books, 1990, p.186

548354 Applying PSM in Vaca Muerta Shale Oil & Gas Operations: The Experience

How to develop, implement and sustain Process Safety in greenfield shale oil and gas projects? This has been a challenge to many new ventures.

Although conventional oil operations have been in place for decades, the exploration, drilling and fracking operations in unconventional shale fields are relatively new in the World. Most of the projects have kicked off less than eight years ago and there are some aspects of the operations that are unique to unconventional fields.

The objective of this paper is to share with the Process Safety global community, how a fit for purpose Process Safety program was developed and is being implemented in the Upstream Unconventional shale oil and gas YPF & Chevron joint venture in Vaca Muerta / Argentina. The project started in 2013, has currently 14 D&C rigs, 5 fracking sets, several separation and treatment production facilities, 3,500 employees and is expected to grow even more in the next 5 years. The paper describes how the project team developed the idea, planned the steps, engaged all stakeholders and selected the key elements of the Process Safety Management System that would add more value in each stage of the project, including drilling, fracking and oil and gas production. The paper also explains the key challenges, such as the difficulties faced to instill Operational Discipline and other key cultural and technical aspects related to Contractor field Operation crews. The journey to implement Process Safety in the Unconventional shale oil & gas operations is very dynamic and is in progress, however, there are already several positive aspects as well as pitfalls to take away as a lesson learned.

548596 A Survey of Process Safety Workers' Attitudes, Behaviors, and Outcomes: Moving Beyond Human Error and Toward a Systems Model

In an effort to move beyond the "human error" explanation for safety incidents, we explored issues with procedures from the worker's point of view by using an anonymous, holistic survey developed from interviews with then currently employed operators. The current sample (*N* = 174) included individuals employed in the process safety industry (MKO/Advanced Procedures Initiative) and were primarily from the Oil & Gas and Chemical industry. The survey was deployed over the course of approximately four weeks. Twelve distinct constructs emerged from the survey (e.g., perceptions of procedure quality, procedure deviation, attitudes toward the procedure change process, etc.). Results indicated that perceptions of procedure quality was the focal variable in all analyses including positive relationships with attitudes toward the procedure change process and negative relationships with procedure deviations, and safety incidents and near-misses. We conclude that these elements are important factors to consider when evaluating and developing procedure systems and provide additional information beyond what the industry considered paramount.

546680 A Method of the Evacuation Pathway Optimization Under Toxic Gas Releases in Chemical Plants

Toxic gases release in chemical plant is a key factor contributing to major accidents and may cause tremendous human life and economic losses. A scientific planning evacuation pathway is an important guarantee of reducing the losses of accidents. Many studies regarding evacuation planning in chemical plant have been carried out by simulating the accident scenarios. However, at present, a systemic method of evacuation pathway optimization has not been developed, and the effectiveness of emergency preparedness has been compromised. To address this problem, an approach based on the CFD modelling of hazardous gas release consequences to plan evacuation pathways and appropriate shelters is proposed in this paper. The methods stochastic programming and dose-response model are mainly employed in this paper. The toxic gas concentration can be converted into poisoning fatality probabilities by dose-response model. Then, stochastic programming method is used to select all the optimal pathways from the evacuation point to the safe area. Setting the minimal fatality probability as the optimization goal, the optimized evacuation pathways under all the leakage scenarios are finally determined through solving the stochastic programming model. As for places with limited evacuation opportunities, an approach of setting up a safe room is proposed in this paper and relevant design guidance are put forward in this paper considering the detailed situation of the chemical plant. A case study of the optimization of evacuation pathways and safe room setup in a salt chemical plant that contains 88 leakage scenarios of chlorine leakage is investigated. With the proposed methodology, the optimized evacuation pathway and the appropriate safe room can provide scientific references for decision makers and evacuees on reducing losses when toxic gases release in chemical plants.

Four Major Gaps That Are Preventing Most Companies Worldwide from Achieving Excellent Process Safety Performance

William G. Bridges
Process Improvement Institute Inc. (PII)
1321 Waterside Lane, Knoxville, TN 37922 USA
wbridges@piii.com

Jeff Thomas
Process Improvement Institute Inc. (PII)
jthomas@piii.com

Prepared for Presentation at
American Institute of Chemical Engineers
2019 Spring Meeting and 15th Global Congress on Process Safety
New Orleans, LA
March 31 – April 3, 2019

FOUR MAJOR GAPS THAT ARE PREVENTING MOST COMPANIES WORLDWIDE FROM ACHIEVING EXCELLENT PROCESS SAFETY PERFORMANCE

William G. Bridges
Process Improvement Institute Inc. (PII)
wbridges@piii.com

Jeff Thomas
Process Improvement Institute Inc. (PII)
jthomas@piii.com

Keywords: process safety, PSM, human factors, near misses, PHA, HAZOP, procedures

Abstract

Process safety requires implementing many management systems, specific engineered features, and operating and maintenance practices effectively. Most companies believe they have done just that, and yet major accidents continue to occur. Why is that? What is missing?

This paper looks at the statistics of major accidents, combined with results from audits and assessments from more than 50 chemical, petrochemical, oil/gas, and related processing companies world-wide. The paper illustrates the four major gaps that are common to the companies/sites that keep having major accidents, compared to those companies/sites that do not have such accidents:

- Accurate and clear operating and maintenance work instructions (procedures)
- PHA of all modes of operation, especially startup, shutdown, abnormal operations, and online maintenance modes
- Getting Near Misses reported and investigated
- Addressing human factors that are missing from most management systems

The paper also provides guidance for closing each of these gaps.

Introduction

"Process safety management (PSM) is the application of management principles to the identification, understanding, and control of process hazards to prevent process-related incidents"[1]. PSM entails development and implementation of programs or systems to ensure that the practices and equipment used in hazardous processes are adequate and are maintained appropriately. The primary categories of programs or systems have come to be called elements of PSM. However, the basic elements of PSM have been defined by many groups in a number of ways. Table 1 lists the elements of PSM systems from various industry and government groups. Many of the elements with different names have essentially the same meaning. For instance, "maintenance and inspection of facilities," together with some aspects of "personnel" practices, both under American Chemistry Council's (ACC's, formerly CMA's) Process Safety Code of Responsible Care™[2] are essentially the same as the single element, "mechanical integrity," under 29 CFR 1910.1190)[3].

Table 1. Comparison of PSM systems

OSHA 29 CFR 1910.119 EPA 40 CFR 68	AIChE/CCPS Risk-Based Process Safety (RBPS) Standard	Responsible Care© Process Safety Code
Management System	**Commitment to Process Safety**	**Management Leadership**
Employee Participation	Process Safety Culture*	Commitment*
Process Safety Information	Compliance with Standards	Accountability*
Process Hazard Analysis*	Process Safety Competency*	Performance Measurement
Operating Procedures*	Workforce Involvement*	Incident Investigation
Training*	Stakeholder Outreach	Information Sharing
Contractors	**Understand Hazards and Evaluate Risk**	CAER Integration
Pre-Startup Safety Review	Process Knowledge Management	**Technology**
Mechanical Integrity	Hazard Identification and Risk Analysis	Design Documentation
Hot Work Permit	**Manage Risk**	Process Hazard Information
Management of Change	Operating Procedures*	Process Hazard Analysis
Incident Investigation	Training and Performance*	Management of Change
Emergency Planning and Response	Safe Work Practices *	Facilities
Compliance Audits	Asset Integrity and Reliability	Siting
Trade Secrets	Contractor Management	**Codes and Standards**
	Management of Change	Safety Reviews
	Operational Readiness*	Maintenance and Inspection
	Conduct of Operations*	Multiple Safeguards*
	Emergency Management	Emergency Management
	Learn from Experience	**Personnel**
	Incident Investigation	Job Skills*
	Measures and Metrics	Safe Work Practices*
	Auditing	Initial Training*
	Management Review and Continuous Improvement	Employee Proficiency*
		Fitness for Duty*
		Contractors

* Contains some of the 6 elements of a Human Factors element, but all missing elements not covered completely. Mostly covered in CoC/OD element

Note that the newest definition of process safety is CCPS's *Risk Based Process Safety (RBPS)*[4] replaces their earlier process safety definition. In the older definition from CCPS [1], there was an element on Human Factors, which brought strong focus to this necessary element; in RBPS, the human factors subelements are now spread across 6 different elements, though some human factors were inadvertently weakened in the transition to *RBPS*.

Although nearly the entire industry agrees that implementing PSM is the right thing to do, interpreting and converting the PSM requirements into practices is unique to each company, and even unique to each plant site. Not only can the requirements be interpreted differently for each site based on local needs, but each site also starts from a different point when they begin to implement a system that is consistent with this regulation.

This paper looks at the statistics of major accidents, combined with results from audits and assessments from more than 50 chemical, petrochemical, oil/gas, and related processing companies world-wide. The paper illustrates the four major gaps that are common to the companies/sites that keep having major accidents, compared to those companies/sites that do not have such accidents:

- Accurate and clear operating and maintenance work instructions (procedures)
- PHA of all modes of operation, especially startup, shutdown, abnormal operations, and online maintenance modes
- Getting Near Misses reported and investigated
- Addressing human factors that are missing from most management systems

The paper also provides guidance for closing each of these gaps.

1. Procedures Need to be More Accurate and Clear

Operating and maintenance procedures have always been crucial to the safety, quality, and productivity of process systems. With the advent of new safety and quality standards such as OSHA's process safety management (PSM) regulation, OSHA's personal protective equipment (PPE) regulation, EPA's risk management program (RMP) regulation, and the ISO 9000 quality standard, many companies are facing the daunting task of developing or upgrading their procedures to satisfy varied and sometimes complex and conflicting requirements.

Although there are overlapping characteristics that the various regulations and standards share, each approach the procedure-writing process from a somewhat different perspective (e.g., quality or safety, protection of workers or protection of the public/environment), and the required level of detail for documenting the procedures differs greatly between these regulations and standards. For instance, the ISO 9000 quality standard advocates documenting procedures and work instructions that impact quality and suggests keeping the procedures as simple as possible, while the OSHA PSM regulation requires detailed procedures that address: all operating modes (startup, shutdown, etc.), operating limits, consequences of deviations, means to avoid hazards, safety and health considerations, and safety systems and their functions. How can companies develop and maintain procedures that ensure productivity and simultaneously satisfy the different regulations and standards? The key is to remember the ultimate goal of the regulations and standards: to reduce human errors that can impact quality, productivity, and/or safety.

Procedure-related errors are errors that occur because some characteristic of the procedure caused task performance to fail. **This is currently the most critical human factor at most sites since 90% of accidents have at least one root cause related to mistakes within procedures**. Reducing these procedure deficiencies can reduce human error rates by a factor of 2 to 20 [5].

1.1. Identified problems

Deficient Procedures are the most prevalent problem in process industries since procedures have not traditionally been developed from the perspective of optimizing human factors; instead procedures have been traditionally developed to meet a compliance requirement. Examples of procedure deficiencies (inaccuracies) include:

- Incorrect/incomplete/nonexistent (most procedures we have audited have been only 70-75% accurate – the inaccuracies include missing critical steps, steps as written are not what needs to be done, or the steps are out of sequence) [5]
- No/misplaced/incorrect information in warnings (for example, a warning should **never contain** the action to take; it should instead **emphasize** the action to take) and warnings should be located BEFORE the step applicable to the warning
- Poor format and presentation rules

1.2. Objective

The flowchart (Figure 1) nd the procedure quality checklist (Table 2) describes the basic sequence of steps for writing effective procedures. Note that some basic rules are contained in the flowchart, such as "operators write operating procedures," "maintenance craft-persons write maintenance procedures," and "Lab technicians write lab procedures." Procedures are NOT written by engineers or department superintendents. This is a critical to ensure procedures are written in the user's common language and are written at the right level of detail.

Figure 1. Best Approach to developing written procedures, once the unit has decided which procedures are needed (Extracted from PII Course, Writing Effective Operating and Maintenance Procedures © PII 2003-2017) [6]

The procedure content controls the human error related to "procedure-based" error – if content is wrong, the trainee is very likely to learn the step wrong (trainer uses the procedure as a basis for training or the trainee refers to the procedure step that is wrong when executing the procedure).

Even if the content is correct, not following page formatting best practices can still increase human error rates. If the content of the step is wrong it does not matter if the presentation is clear. If the content is perfect, you can further reduce error rates by 50% to 70% by presenting the information properly.

Rules for developing content[7]

- The procedures should be written at the level of someone who has just completed the basic training for that task. Do not write the procedure for someone just hired or for the 10+ veteran.
- Clearly identify ahead of time what activities need procedures to help ensure error rates are controlled low enough, and identify which activities are considered common "skills" of all of the staff.

> *Example: Starting a pump is typically considered a "skill." This means a procedure step can simply be "Start the benzene recycle pump (P-119)." The procedure does not need to explain how to start the pump. But, starting a pump is a skill that nonetheless must be learned; it requires doing several sub-steps related to positioning of intake and discharge valves, how the pump is throttled at startup, checking local pressure gauge, etc. So, a training module is needed for starting pumps of various types, and the operator needs to learn this necessary skill. Once the skill is learned, it will be applied to so many various cases and so often, that a procedure is no longer needed as a "refresher" on how to "start the pump."*

- The first draft of the procedure must be walked-down in the field by another user. Simply reading through the procedure does not catch enough of the mistakes.
- The revised draft should be reviewed in the field by a technical staff person, such as a process or operations engineer (for operating procedures).
- The procedure needs to be checked to ensure it follow the page format and step writing rules described in the next section. This ensures the accurate steps are clearly presented.

The target is to reach an accuracy of 95% or better[7] (so, no more than one wrong or missing step out of 20 steps). The following have been observed in the field by PII staff by direct data collection at more than 110 sites/plants around the world, based on walk-down review of several to dozens of procedures at each unit/plant:

- At 95% accuracy or better and when the same procedures follow 80% of the rules for procedure clarity (presented next), then most users will follow the written procedures and will try to keep the procedures up-to-date
- At about 85% accuracy or less, about half of the users stop using the procedures
- At about 75% accuracy or less, less than 10% of the users will refer to the procedure or will try to keep it up-to-date. **So, the written procedures are not very effective.**
- Unfortunately, the typical operating procedure walked down by PII staff (accompanied by senior operators, a process engineer, and a shift supervisor) is about 75% accurate (so one step in four is missing or wrong). Usually these inaccurate procedures also follow less than half of the best practices for procedure format, presented next [7].

Procedure Clarity

The importance of the accuracy of the procedure steps were discussed above. The clarity of the procedure steps (how they are written and how the page is formatted) is also important. Following best practices for step and page format reduces human errors by a multiplying factor of 3 to 5. [5], [7]

The best practice rules for writing and validating procedures have been published for many years *(see Bridges & Williams, 1997; Madden & Bridges, 2016, 2017)* [7],[8],[9]. These rules include best practices for formatting of the pages and steps. These have been gradually improved over the past decades and now are incorporated into PII's training materials.[6]

Table 2. Procedure Quality Checklist (courtesy PII, copyright 2003-2017)[6]

#	Issue	Response
	Procedure Content Checklist	
1	Is the procedure drafted by a future user of the written procedure? (Engineers should not author procedures to be used by operators or maintenance staff.)	
2	Is the procedure validated by a walk-down in the field by another future user of the procedures?	
3	Is the procedure reviewed and commented on by technical staff (engineers or vendors)?	
4	Is the procedure checked versus the Page and Step format rules below?	
5	Is a hazard review of step-by-step procedures performed to make sure there are sufficient safeguards (IPLs) against the errors that will occur eventually (when a step is skipped or performed wrong)?	
6	Is the content measured using "newly trained operators" to judge the % of steps that are missing, steps that are confusing or wrong, and steps that are out-of-sequence? (A score of 95% accuracy of content is good; 98% should be the targeted average.)	
	Page Format Checklist	
1	Is the title of the procedure the largest item on the page?	
2	Is the procedure title clear and consistent with other titles, and does it uniquely describes the topic?	
3	Are the document control features the smallest items on the page?	
4	Are temporary procedures clearly identified?	
5	Is white space used effectively? • Is there one blank line between each step? • Does indentation help the user keep their place? • Are the margins large enough to reduce page congestion?	
6	Is type size is 12 pt font or larger?	
7	Is mixed case used for words of steps, with ALL CAPS used only for special cases (such as IF, THEN, AUTO, and WARNING)?	
8	Is the step number very simple (single level of number used)? Only an integer?	
9	Have sections or information not necessary to performing the steps been moved to the back or to another part of the manual or training guide?	
10	Are section titles bold or larger than the text font? Do sections have clear endings?	
11	Is the decision on electronic presentation versus hard copy correct? Are electronic linkages to procedures clear and accurate and easy to use? If paper is chosen, is the color of the paper appropriate?	
12	Is the overall page format (such as Outline format or Two-Column [T-Bar] format) appropriate to the use of the procedure?	
13	Are play script features added for tasks that must be coordinated between two or more users? • Play script is normally used when there are two or more hand-offs of responsibility for steps.	
14	Are rules followed for formatting of Warnings, Cautions, and Notes? (See annotated rules, such as Warnings are for worker safety and Warnings must clearly standout from rest of page.)	
	Step Writing Checklist	
1	Is each step written as a command?	
2	Is the proper level of detail used throughout? This is judged based on: • *Who* will use the procedures • Same level of detail used in similar procedure steps	
3	On average, is there only one implied action per instruction? Best practice is to average 1.2.	
4	Does the procedure indicate when sequence is important? • If sequence matters, each step should be numbered (with an integer or letter)	

	• If sequence does not matter, bullet lists should be used	
5	Are only common words used? Apply "education" level test (5 grade reading level is best)	
6	Do all acronyms, abbreviations, and jargon aid understanding? • Develop a list of such terms for use in procedures *and* communication. • Use terms that users use (within reason)	
7	Is each step specific enough? No room left to guess/interpret: • The meaning of a word or phrase (Check vs. Make sure) • The intent of a step or series of steps • A desired quantity or value • To what equipment the step applies	
8	Is the procedure free of steps that require in-your-head calculations? • Values expressed as ranges rather than targets with error bands • Conversion tables, worksheets, or graphs provided where needed	
9	Are graphics_to the user's advantage? • No explanatory paragraphs or lengthy instructions that could be replaced by a picture • No impressive graphics that provide no real advantage	
10	Are references to the user's advantage? • No lengthy explanations or instructions that could be replaced by branching to a reference • No references to a procedure that references still another • No gaps or overlaps between this procedure and a referenced document • If branching, must branch to a procedure, not to a specific step in a procedure	
11	Are rules followed for writing warnings, cautions, and conditional steps? • No more than 2 per page • No actions within a warning or caution (actions must always be numbered steps) • Warnings and Cautions contain descriptions of potential consequences	

For a complete description of best practices, refer to PII's paper *"Best practices for writing operating procedures and trouble-shooting guides"*. [7]

2. PHAs (Process Hazards Analyses) Need to Close Two Major Gaps

Process Hazards is a broad term. The relationship between many components define process hazards. For a PHA team, Process Safety Information (information on the chemicals, technology and equipment) and operating procedures are necessary to identify specific components of and analyze a hazard scenario. These components include: process deviations (parametric and procedural), their causes, the process safety consequence and various safeguards for detecting deviations and causes, preventing causes, detecting and mitigating consequences and intervening safeguards. A thorough review of incident investigations also offers an opportunity to further analyze process hazards.

2.1. Identified problems

Over the last twenty years the CSB has brought needed attention, through its investigations of industry incidences, to two significant gaps in many current PHAs; where the teams did not identify and analyze:

- Process hazards unique to non-routine modes of operations resulting in catastrophic consequences, and
- Process hazards related to Damage Mechanisms.

Though the intent of the US OSHA PSM regulation has always been for PHA teams to analyze the hazards for all modes of operation, OSHA did not give specific requirements for addressing damage mechanisms and only recently have they increased their enforcement attention on PHA of non-routine modes of operation.

Industry has responded by modifying the *Guidelines for Hazard Evaluation Procedures*, 3rd Edition, 2008, CCPS/AIChE[10] to improve the coverage of these two main issues. API in turn issues guidance on addressing damage mechanisms[11]. Industry best practices now exist for addressing both of these weaknesses. Now, with proper approaches, a team with process safety competencies and process knowledge can thoroughly address the issues during a PHA.

Another less significant gap noted by regulators and by the US CSB has been an incomplete consideration, within the PHA, of previous incidents.

This section of the paper lays out the case from the US CSB and US OSHA on these PHA gaps:

- Poor or no coverage of hazards during non-routine modes of operations
- Poor or no coverage of damage mechanisms

2.2. Objectives

2.2.1. Hazard Evaluation of Non-Routine Modes of Operation

Non Routine modes of operation include start-up, shut-down, online maintenance and other abnormal operations. The hazard evaluation of non-routine modes of operation involves reviewing procedures using a HAZOP, simplified HAZOP, or What-if analysis to uncover potential accident scenarios associated with non-routine operations, for continuous or batch operations. Human error is more likely and more critical during non-routine operations. By analyzing procedural steps where human error is more likely, and where human error or component failure could lead to a consequence of interest, risk can be reduced. The hazard evaluation team's objective is to evaluate the risk associated with skipping steps and performing steps wrong.

The purpose of a hazard evaluation of non-routine modes of operation (governed by written procedures) is to make sure an organization has enough safeguards for the inevitable instance when a step is either performed wrong or skipped (inadvertently or due to shortcutting or other reasons).

Industry has found that a HAZOP or what-if analysis, structured to address procedures, can be used effectively for finding the great majority of accident scenarios that can occur during non-routine modes of operation[10],[12],[13],[14]. Experience shows that reviews of non-routine procedures have revealed many more hazards than merely trying to address these modes of operation during the P&ID driven hazard evaluations.

General guidelines for analyzing non-routine modes of operation or batch processes

- Define the assumptions about the system's initial status. "What is assumed to be the starting conditions when the user of the procedure begins with Step 1?"
- Define the complete design intention for each step. "Is the step actually 3 or 5 actions instead of one action? If so, what are the individual actions to accomplish this task?"
- Don't analyze safeguard steps that start with ensure, check, verify, inspect, etc., or where the consequence of skip is "loss of one level of safeguard/protection against…" There is no reason to analyze these steps since they will show up as safeguards of deviations of other steps. This approach is similar to not analyzing a PSV during a HAZOP of continuous mode (i.e., during a parametric deviation analysis); instead the PSV is shown as a safeguard against loss of containment.
- Together with an operator before the review, identify the sections of the procedures that warrant use of:
  - 7-8 Guide Words (extremely large consequences can happen if deviations occur)
  - 2 Guide Words (the system is complex, mistakes are costly, or several consequences could occur)

  - On others, use What-If (no guide words or guide phrases; for use on simpler or lower hazard systems)
- Decompose each written step into a sequence of actions (verbs)
- Apply guide words directly to the intentions of each action

The following preparation steps may also be needed prior to the procedure PHA:

- Walk through procedure in the plant with one or more operators to see the work situation and verify the accuracy of the written procedure.
- Determine if the procedure follows the best practices for "presentation" of the content; the best practices will limit the probability of human error
- Discuss generic issues related to operating procedures, such as:
  - staffing (normal and temporary)
  - human-machine interface
  - worker training, certification, etc.
  - management of change
  - policy enforcement
- Review other related procedures such as lock out/tag out and hot work
- **If the procedures are NOT >90% accurate, then redo procedures before the PHA.**

2.2.2. Damage mechanisms

Recently, more and more regulators, including the US CSB, are making analysis of Damage Mechanisms (DMs) a PHA requirement. This requires that potential process damage mechanisms and their potential consequences of interest be identified. Damage mechanisms can be broken down into three main categories. These categories and examples of each are shown in the following table (Table 3).

Table 3. Types of Damage Mechanisms [15]

Category	Examples
Mechanical	• Mechanical loading failure • Mechanical fatigue • Buckling • Cracking • Embrittlement • Ductile fracture • Brittle fracture

Chemical	• Corrosion
	  - Uniform
	  - Localized
	  - Pitting

Physical	• Thermal failures
	  - Creep
	  - Thermal fatigue
	  - Transformation

The team may use aids such as piping specs, established literature and standards, any applicable MOC documents, etc. to help identify DMs. The PHA leader has several options of how to address damage mechanisms, but PII preferred approach is to discuss during HAZOP under "Loss of Containment" deviation in each node/section.

Damage mechanisms can (and sometimes ***should***) be reviewed prior to the PHA. For example, all MOCs should consider damage mechanisms and their potential consequences of interest before the change is approved. The MOC should also require updates to process safety information which might be relevant to protect against DMs.

Coverage of Damage Mechanism (DM) for each major section [14]

As mentioned above, this is PII's preferred method for thoroughly covering all DM within a PHA/HAZOP of an entire unit. The benefits are that the team can more easily catch changes in DM from section to section and they can more easily identify when unique safeguards, such as Remote Isolation Valves, segregated containments (dike), and unique materials of construction are needed. To facilitate this, a list of generic causes of loss of containment and a generic list of safeguards against loss of containment are covered in the Loss of Containment (LOC) deviation of each section.

For each node (each line, each vessel, each column, etc.), the PHA team should discuss and document each damage mechanism listed in Table 3 (as a cause of loss of containment), consequence of the failure if the damage mechanism occurs, and the safeguards in place to prevent the damage mechanism, detect the mechanism before failure, prevent the release, detect and response to the release, and mitigate or contain the release.

The documentation style varies between PHAs. There are two different styles that have been acceptable in the past.

- One style (Example A, Table 4) uses a reference to a summary table of typical causes (this summary table is not shown but is similar to Table 5 of this paper) instead of listing each individual damage mechanism. In Example A, the same approach is used for safeguards. This approach saves redundant text and some time, but it requires the leader and scribe to be diligent to cover everything in the Typical tables that were referenced in the Cause and Safeguard columns. A modified approach is to relegate some generic causes and safeguards to Typical tables for reference and then to carefully list the specific cause and/or safeguards of interest for LOC for each specific node.

Table 4. Example A – Referencing Generic Tables for Typical Causes and Typical Safeguards [14]

Dev	Deviation	Causes	Consequences	Existing Safeguards	Recommendations
9.10	Loss of containment	Accelerate corrosion External fire High pressure (linked from 9.7) Typical causes of loss of containment (see Table A.1)	Release to atmosphere leading to potential injury of workers and/or community	Pressure Relief Valve's 231A, B, C on Reactor and no valves in line to reactor Most lines / connections are welded construction; only a few flanges Drills conducted each year on evacuation, rescue and isolation Emergency Response personnel are trained at SABIC FTC Generic safeguards protecting against or mitigating process material releases (see Table A.2)	Safety 7. Consider changing the ITPM schedule for managing most PSVs whose inspections are too infrequent based on industry standards and best practice. For instance, the current inspection frequency for PSV-8220 on the Ammonia Receiver (I-2005G) is 9 years, whereas, consensus codes typically recommend testing/inspection every 1-4 years for PSVs in highly toxic services.

- Example B (**Table 5**) does not use or reference a generic or Typical list of causes (damage mechanism) or Typical safeguards, but instead develops a specific listing for the LOC deviation of each node. This style has proven easier to justify to regulators and other outside reviewers but takes more documentation effort.

Table 5. Example B – Specific Listing of Causes and Safeguards [14]

Dev	Deviation	Causes	Consequences	Existing Safeguards	Recommendations
1.9	Loss of containment	Corrosion/erosion External fire and/or flame impingement Gasket, packing, or seal failure Improper maintenance Material defect Operator failing to close or inadvertently opening a valve to the atmosphere (e.g., a valve at a hose connection) Railcar inadvertently derailed Valve leaking to the atmosphere High pressure (linked from 1.5) Acid corrosion caused by high concentration of water (linked from 1.8) High ambient temperature External impact (such as from a mini-engine or another railcar)	Catastrophic release of chlorine from a ruptured railcar Steady release of chlorine from a ruptured connection Steady release of chlorine from a leaky connection High pressure caused by thermal expansion of liquid chlorine if railcar is also over-full	Chlorine repair kit Derailer and warning flag to prevent impact by a mini-engine or another railcar Limited vehicular access to area Maintenance/operator response as required, including isolation if needed Operator periodically monitoring the railcar valves while unloading Personal protective equipment in the area Plugs installed in all chlorine valves to the atmosphere when the valves are not in use Relief valve on each railcar for mitigating releases caused by overpressure Supplier maintenance of railcars (per strictly enforced US DOT requirements) Video monitoring of the unloading area Concrete crossties on rail spur Dike preventing any combustibles spilled nearby from reaching the unloading rack area Concrete railroad ties in chlorine unloading area to prevent fires near railcar	10. Consider installing a chlorine detection system in the unloading and vaporizing area to help detect chlorine releases (especially at likely release points) 11. Verify that periodic maintenance and inspections are being performed in accordance with Chlorine Institute recommendations 12. Review the drainage system for the unloading are, and identify the areas that may be affected by a large chlorine release 13. Consider prohibiting the use of heavy equipment (e.g., cranes) in the unloading and vaporizing area unless special precautions to prevent equipment damage are enacted 33. Consider providing a high pressure alarm for each vaporizer 40. Consider providing a water deluge system in the unloading area to help mitigate chlorine releases from the railcar

Regardless of method, the team leader and scribe must ensure the team rigorously discusses all of the damage mechanisms in Table 5 and adequately documents the results of the damage mechanism review in the PHA/HAZOP analysis tables.

For a complete description of PHA best practices, refer to PII's paper *"Recipe for a Complete Process Hazard Analysis"*. [14]

3. Near Misses Need to be Reported and Enough of Them Investigated

We must learn from accidents and near misses to prevent recurrence. The first step in the learning process is investigation to determine the causes and underlying reasons why accidents and near misses occur. A thorough investigation of root causes will identify the management system weaknesses. Learning which management system weaknesses are leading to near misses and accidents is one of the highest value activities in which a company can invest, and learning from near misses is much cheaper than learning from accidents. Many chemical companies have implemented process safety management systems, and now they are beginning to focus on getting near misses reported and on root cause analysis. This is a very exciting trend. Unfortunately, the chemical industry gets very few near misses reported (the chemical industry is certainly not the only industry with this problem).

To understand more about near misses and getting them reported, it is best to first review the basic definitions.

> *An incident is either an Accident or a Near Miss.*
>
> *An Accident is a sequence of unplanned events and conditions that result in harm to people, environment, process, product or image.*
>
> *A Near Miss is an unplanned sequence of events that could have caused harm if conditions were different or is allowed to progress, but did not in this instance.*

Given a consistent understanding of the definition of a near miss, it is possible to estimate how many near misses should be reported for every accident. Studies in several industries indicate that there are between 50 and 100 near misses for every accident. Also, data indicates that there are perhaps 100 erroneous acts or conditions for every near miss. This gives a total population of roughly 10,000 errors for every accident. **Figure 2** illustrates the relationships between accidents, near misses and non-incidents.

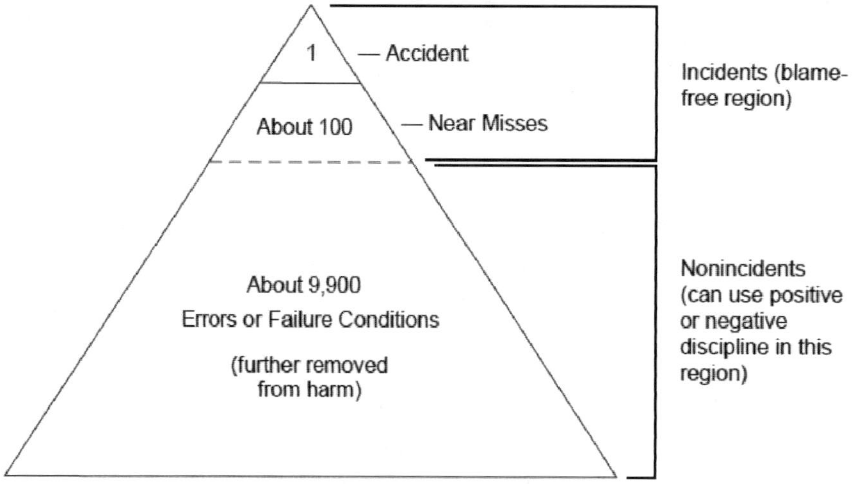

Figure 2. Relationship between Errors and Potential or Actual Impact [16]

Investigating near misses is critical to preventing accidents, because near misses share the causes and root causes of accidents; they are one or two barriers away from the loss/accident. We are very likely preventing many apparently unrelated accidents when we prevent the ones that are obviously related.

3.1. Identified problems

During the past 15 years, we have asked more than 5,000 students of our process safety management (PSM) courses and 3,500 students from our investigator leadership training courses how many near misses they get reported for every accident. The students represented about 400 companies, predominantly in chemical-related process industries (chemical, polymer/plastic, petrochemical, refining, oil and gas exploration, pharmaceutical, pulp and paper, etc.). The answers are quite disturbing. More than 95% said that their ratio of near misses reported to accidents reported falls in the range of 0 to 20. Less than 5% of the individuals said the ratio was greater than 5, and less than 2% said the ratio was higher than 10. Students noted that fear of disciplinary action, lack of management commitment, and lack of understanding of the difference between a near miss and a non-incident were the main reasons why near misses do not get reported. [16]

In conducting about 150 PSM audits and 7000 process hazard analyses (PHAs) during the past 22 years, we have found ratios from 0 to 105. For the first half of the 1990s, more than 90% of the facilities we talked to had ratios in the range of 0 to 0.5, and more than 95% had ratios in the range of 0 to 1. In the last half of the 1990s, more than 90% of the facilities had ratios in the 0 to1 range, and more than 95% had ratios in the 0 to 2 range. In addition, only a few (less than 2% of the facilities) had a ratio higher than 5. By 2005, many facilities had ratios in the 20 to 100 range and each of these had seen great gains (more on this in future articles), but the average across all companies we talked to increased only slightly. By 2011, the ratio of near misses to accidents (any loss event) was about 2 for companies we deal with. But by the end of 2011, many major companies had ratios well above 50! [16]

Our auditors and PHA leaders commented that the primary reasons for lack of reporting were about the same as those found in the survey. The barriers to getting near misses reported, discussed in detail later, are:

- Fear of disciplinary action.
- Fear of teasing by peers (embarrassment).
- Lack of understanding of what constitutes a near miss versus a non-incident.
- Lack of management commitment and lack of follow-through once a near miss is reported.
- An apparently high level of effort is required to report and to investigate near misses compared to low return on this investment.
- There is No Way to investigate the thousands of near misses per month or year!
- Disincentives for reporting near misses (e.g., reporting near misses hurts the department's safety performance).
- Not knowing which accident investigation system to use (or confusing reporting system).
- Company discourages near-miss reporting due to fear of legal liability if these are misused by outsiders.

The good news is that near-miss reporting appears to be improving, and the industry appears to recognize most of the barriers to near-miss reporting. The bad news is that the ratio is still very poor and improvement appears slow!

The number of companies that participated in the informal surveys suggests that the results are a statistically significant representation of the chemical industry.

A formal (written) survey was developed and e-mailed, faxed, and/or mailed to more than 100 companies. Of these, more than 12 replied originally in 1997. Since then, we have reviewed data in detail from more than a dozen major companies. These 25+ companies or affiliates provided data from more than 400 facilities, including more than 150,000 employees in manufacturing and operations. The data were from the chemical industry, polymer industry, refineries, drug/pharmaceutical companies, pulp and paper mills, petrochemical companies, and oil exploration/production.

Table 6 summarizes the results of both the informal and formal surveys.

Table 6. Summary of survey results [16]

Survey Items	Informal surveys		Formal Survey (or actual organization data)
	Classroom polls	Found during Audits and PHAs	
Current Near-miss Reporting Ratio Range (0-5 years prior)	1 to 105	2 to 20	0 to 105
Previous Near-miss Reporting Ratio Range (5-10 years prior)		1 to 5	NA
Current Near-miss Reporting Ratio Average (0-5 years prior)	2-3	3	5
Previous Near-miss Reporting Ratio Average (5-10 years prior)		1	NA
Goal for Near-miss Reporting Ratio (average)	20+	NA	20
Theoretical Upper Limit on Near-miss Reporting Ratio (average)	About 100	NA	100
Number of Participating Companies	400	300+	25+
Number of Participating Facilities	8500+	500+	400+

3.2. Objective

A company should strive to reach a ratio of 50-100 and investigate about 20 near misses per accident. This will provide a statistically significant sample of all incidents (and all important errors) and provide a company with sufficient feedback on which management system weaknesses are causing the errors and component failures. Various companies with different cultures have achieved high ratios with great return on investment.

Table 7 shows a brief description of possible solutions to overcome each one of the barriers identified in the survey.

Table 7. Summary of Survey Results[16]

Barriers	Solutions
Fear of disciplinary action	Implement a policy to NOT punish individuals when their errors lead to accidents and Near Misses.
Fear of teasing by peers (embarrassment)	Ensure that all employees understand the importance of near-miss reporting; demonstrate, through feedback of lessons learned.
Lack of understanding: Near miss vs Non-incident	Develop a list of "in-context" examples that illustrate what you consider to be Near Misses and what you consider to be non-incident
Lack of management commitment and lack of follow-through on reported near misses	Hold management accountable for achieving a Near-Miss reporting ratio
Apparently high level of effort is required to report/investigate Near Misses	Ensure that the data are entered in a database and queried regularly. Share the results with employees so they can see the value of the reported near misses
There is no way to investigate the thousands of Near Misses per month	Let front-line foremen or supervisors decide if a Near Miss or accident needs to be investigated
Disincentives for reporting Near Misses	Ensure that goals and incentives are not tied to lower incident rates (since this discourages reporting), but instead provide incentives for high Near-Miss reporting ratios
Not knowing which accident investigation system to use	Have ONE incident reporting system with ONE approach
Company discourages Near Miss reporting due to fear of legal liability	Involve legal on major Near Misses and accidents to ensure the results are protected as much as possible under attorney/client privilege.

For a complete description of the possible solutions for each barrier, refer to PII's paper *"Gains from Getting Near Misses Reported"*. [16]

4. Key Human Factors are Not being addressed

All accidents (or nearly all, if you consider that there are some natural phenomena that we either cannot guard against or choose not to guard against) **result from human error**. This is because humans govern and accomplish all of the activities necessary to control the risk of accidents. Humans influence other humans in the process – not only do humans cause accidents (unintentionally) by making errors directly related to the process itself, but they also cause errors by creating deficiencies in the design and the implementation of management systems (i.e., we make errors in authorities, accountabilities, procedures, feedback, proof documents and continual improvement provisions). Ultimately, these management systems govern the human error rate by directly causing or indirectly influencing the process. The process-related activities where errors have the most influence include:

- Designing a process
- Engineering a process
- Specifying components

- Receiving and installing equipment
- Commissioning
- Operating a process
- Maintaining, inspecting and repairing a process
- Troubleshooting and shutting down the process
- Managing process, procedure, materials, facility and personnel changes

The 10 human factors categories to be controlled by companies are:

- Available Time (includes staffing Issues) – for responses only
- Stress/Stressors (includes staffing issues)
- Complexity & task design
- Experience/Training
- Procedures
- Human-Machine Interface (includes tools)
- Fitness for Duty
- Work processes & supervision
- Work Environment
- Communication

4.1. Identified problems

Recent major accidents have highlighted the need for increased focus on human factors. The U.S. Chemical Safety Board (CSB) cited (US CSB Video, 2006) human factor deficiencies as one of the main contributors of the catastrophic accident at the BP Texas City Refinery in March 2005[17]. The human factor deficiencies included lack of control of worker fatigue, poor human-system-interface design, poor communication by radio/phone, out-of-date and inaccurate operating procedures, and poor (no) communication between workers at shift handover. The CSB cited similar issues from many other accidents[18] and has urged industry and the U.S. OSHA (the regulator) to pay much more attention to human factors. As a result, the recent U.S. OSHA National Emphasis Program for Refineries included human factors as one of the 12 core elements.

Implementing human factor engineering and policies to prevent accidents is not a new concept. Nearly all (or all, from a more complete perspective) of the causes and root causes of major accidents in the past 30 years have been the result of poor control of human factors. This has been cited in many root causes analysis reports and papers concerning these major accidents.

Process Safety Management (PSM) systems based on OSHA's PSM standard are likely lacking the fundamental human factor elements and implementation guides that, if applied across the applicable PSM elements, would work together to reduce human error. The *Risk Based Process Safety* guideline from the CCPS/AIChE (2007) does contain these human factor requirements.

Many site and company staff do not know there are standards for the control of human factors. However, at last count there were more than 300 non-governmental standards (such as from ANSI, ISO, IEEE, etc.) for the control of human factors, and there are more than 100 government regulations and standards and guidelines. Some of the better government regulations cover aviation, marine operations (shipping) and aerospace. An organization should devote resources to finding and implementing such best practices.

The following is a description of each of the human factors mentioned above and specific ways to control each. Our data from review of more than 2000 incidents reveal that:

- 90% of accidents had at least one root cause related to deficiencies in written work instructions.
- 70% of accidents had at least one root cause related to miscommunication between workers or between workers and supervisors.

- 40% of accidents had at least one root cause related to excessive fatigue of the worker. [5]

4.2. Objective

Not all errors can be prevented. Since the beginning of time, humans have tried to control error rate with more or less success. Human errors have been measured for hundreds of years. Psychologists have studied why humans make mistakes and have gradually put a science around human error probability. Today, the best models for control of human error in the workplace are generally agreed to be related to control of human factors; in turn, these have grown out of what was previously denoted performance shaping factors. What can be done to control each Human factor is discussed in greater detail next.

For a complete description of each human factor, its importance and how to control them; refer to PII's paper *"Human factors and their optimization"* [5].

Time available for the task

Task design is directly linked with this category. Process design, procedure development and worker scheduling need to take into consideration not only the time it takes to complete each task but also the time it takes to manage usual as well as potentially unusual or unexpected situations. True; we do not create norms for exceptions, but exceptions sometimes are more usual than realized.

Stress/Stressors

A well-designed, organized and managed work process helps maintain and promote individual health and well-being, but stresses will arise through no fault of the organization. Supervisors should be trained to recognize stress in workers and, in severe cases, reassign them and help them get assistance to deal with the stress and in other cases, reduce their job pressures for a day by shifting assignments. A wellness program may also be necessary to help cope with stress. Simple approaches, such as training workers to recognize and deal with stress, can help most individuals. Approaches to deal with stress include positive self-talk and practicing relaxation techniques (i.e., breathing control).

Complexity & task design

If roles and responsibilities for accomplishing tasks are not clearly defined, then there will be a risk of serious errors. It is also important that employees and engineers/designers understand the characteristics of the work elements involved, how each element passes information to the other, how each person involved communicates, and that each person has guidelines to adhere to during task design and procedure development.

For optimal performance, task designers often must integrate human and automated equipment in their task analysis.

It is crucial for task designers also to have an understanding of the appropriate allocation of roles and responsibilities to the various participants in a task. Understanding the characteristics and limitations of all of the humans and automated systems involved in the task (particularly critical tasks) will allow for establishing additional controls where needed.

The task analysis process includes *Task definition* which is the initial operational description of the operator tasks required for execution of a given system function; *Performance analysis* of all tasks with respect to human performance requirements; *Error and workload analysis* to be included in the design process, particularly when the likelihood of unacceptably operator error is high; and *Evaluation* since evaluating the design of a new or complex task is important to ensure that the task can be effectively performed as designed. For very high consequence scenarios involved complex human interaction, it is likely prudent to invest in a human reliability analysis (HRA).

Experience/Training

Training to master the Knowledge, Skills, and Abilities (KSAs) required for a job may be obtained from a variety of sources. Companies provide the training or it is obtained from other sources (e.g., trade schools, universities, contractor organizations, etc.) to reach the skills and knowledge established in job descriptions. There are several factors that may result in personnel not mastering the required KSAs for a job. These include course design and delivery methods, course completion, practical skills demonstrations or simulation, and training frequency. Course design begins when the learning objectives have been identified. The design process consists of determining the delivery methods (simulator, on-the-job training, etc.), number of hours required to cover the materials, instructor qualifications, etc. Although some methods, materials and instructors may be more effective or efficient than others, the important issue is that the course content is complete and addresses all of the relevant KSAs, so that the learning objectives are met and the KSAs mastered.

Another determinant of KSA mastery is course completion. Although this factor appears obvious, there are often competing demands on personnel that may pull them out of training at times. As a result, they may miss the instruction related to specific KSAs. Testing may not identify the KSA deficiency because it is impossible to test mastery of all KSAs. Sampling techniques are used to generate examinations. If attendance and participation are not controlled, some personnel may miss training on specific KSAs and testing may not identify the deficiencies before an error is committed.

Another factor affecting KSA mastery is forgetting. An individual's ability to perform a task will degrade over time unless the relevant KSAs are refreshed. Proficiency training (i.e., refresher) will be required for some tasks to maintain the level of mastery that was demonstrated following initial training. One function of training programs is to identify those tasks that require proficiency training.

If certain tasks are performed frequently, proficiency training may be unnecessary. By performing a task, personnel practice the task and obtain feedback on where they have weaknesses. Task performance refreshes the KSAs and successful task performance verifies that proficiency has been maintained. Furthermore, since humans learn by comparison to similar activities, as a worker practices one task, they are learning about similar tasks. Care must be taken here to avoid the practice drifting away from what the procedure requires. Also, if new operators are trained by experienced operators, you must ensure that the new employee is trained according to the procedure, not whatever practice is actually being used. (Of course, ideally, the practice and procedure will be identical.)

Operating and maintenance procedures

Many procedures do not follow best practices for controlling human error, and so the written procedure actually "contributes" to increased error rates. Additionally, many organizations do not have guides on how to troubleshoot (what to do when process deviations occur). The best practice rules for writing and validating procedures have been published for many years. [6], [7], [8], [9]

Procedures have not traditionally been developed from the perspective of optimizing human factors; instead, procedures have been traditionally developed to meet a compliance requirement to have written procedures. Examples of procedure deficiencies (inaccuracies) include:

- Incorrect/incomplete/nonexistent (most procedures we have audited have been only 70-85% accurate – the inaccuracies include missing critical steps, steps as written are not what needs to be done, or the steps are out of sequence)
- No/misplaced warnings or misuse of warnings (for example, a warning should never **state** the action to take; it should instead **emphasize** the action to take an the statement of the action should be in a numbered step)
- Poor format and presentation rules

Human-machine interface

Design for operability refers to designing the Human-System Interface (HSI) to be consistent with the abilities and limitations of the personnel who will be operating it. Weaknesses in the design processes can result in an HSI that is not well suited to the tasks that personnel must perform to ensure plant safety, resulting in increased workload, decreased performance by personnel, and an increased likelihood of errors.

Design for maintainability refers to designing the HSI and associated plant equipment to ensure that personnel are able to perform necessary maintenance activities efficiently. Weaknesses in the design process can result in systems that impose excessive demands on personnel for maintenance and, therefore, are prone to maintenance errors or problems with reliability and availability. *If it is hard to reach, workers will make more errors, including the error of deciding it is not worth it.*

Design for flexibility refers to the way that changes, such as upgrades to the HSI, are planned and put into use. A new HSI component may require the user to perform functions and tasks in new ways. Skills that the user developed for managing workload when using the former design, such as ways for scanning information or executing control actions may no longer be compatible with the new design. The new HSIs must also be compatible with the remaining HSIs so that operators can use them together with limited possibilities for human error. Also, HSI modifications may not be installed or put into service all at one time, causing the user to adapt to temporary configurations that are different from both the original and final configurations. Weaknesses in HSI implementation can increase operator workload and the likelihood of errors.

Fitness for duty

Implement a company fitness-for-duty program with a primary responsibility for detecting and preventing impaired personnel from performing tasks that may affect productivity and safety. Medical evaluations of personnel, behavioral observation programs, employee assistance programs, and drug and alcohol testing are used to detect impairment. Weaknesses in this program may allow impaired personnel to have access to vital areas in a plant where they could commit errors.

Overtime Policies and Practices – Most companies establish limits for work hours to reduce on-the-job fatigue and the potential consequences for poor task performance. Routine authorization for work hours in excess of those recommended may result in fatigued workers. Furthermore, a practice of excluding training or meetings that occur outside of an individual's normal work schedule from work-hour limitations will also contribute to fatigue. But in many cases these have weak enforcement, especially during shutdowns when the company is under stress to re-start soonest.

Shift Scheduling – Shift scheduling may also affect the likelihood that personnel will show performance decrements due to fatigue. A change in the assigned shift or a rotating shift schedule will disrupt circadian rhythms and may increase the likelihood of errors. So, a company must choose the proper shift rotation to allow adjustments to sleep patterns.

Work processes & Supervision

Written plant policies and procedures are meaningful only when they are enforced; otherwise, worker practices are the policy. Once discrepancies are tolerated, individual workers have to use their own judgment to decide what tasks are necessary and/or acceptable. Supervisors and managers should vigorously enforce plant policies/procedures and ensure that plant policies/procedures and practices are revised as necessary to be consistent. To enforce well, a supervisor must (1) lead by example, (2) watch workers often to ensure they are following company policies and procedures, and (3) determine the correct course of action (including recommending changes to policy and procedures) when workers are found deviating from procedures.

Work environment

Programmatic causes of task environment errors are typically found in the company's processes for designing human-system interfaces or in managing maintenance activities. Other programs may also be implicated. Common programmatic causes of task environment errors include:

Industrial Hygiene and Radiation Protection – These programs are responsible for ensuring that task environments have been evaluated to identify hazards and that needed controls are implemented to minimize exposures. Weaknesses in these programs may result in personnel working in task environments that are conducive to errors.

Work Planning and Control – Weaknesses in the work planning and control system may allow work to be planned without consideration of adverse environmental conditions and performed without the necessary compensatory measures. For example, communication devices may not be provided in noisy environments to support task performance. For tasks that involve unusual physical positions or cramped workspace, additional time to complete the task may not be scheduled. Rest breaks for hot and cold environments may not be planned into the work, or additional temporary lighting may not be provided if the work site is not adequately lighted.

Procedures – Weaknesses in the company's procedure development process may result in the design of procedures that are inappropriate for the conditions in which they will be used. For example, procedures that may be used at night, outside and in the rain should be laminated and the type size should be larger to ensure the procedure can be read. Procedures that will be used in vibration conditions may also require larger type size than procedures read in the stationary environment of the control room.

Human Factors Engineering – Weaknesses in the human factors engineering program may result in the installation of new equipment or systems without consideration of task environment characteristics. For example, the impact of control room lighting on the visibility of digital displays or effects of vibration on the legibility of dials or gauges at local control stations should be considered before installation.

Communication

Successful communication requires several steps. The sender first develops the intention to communicate either verbally or in writing. The sender then composes a message that presents the meaning as clearly as possible. The receiver must pay attention to the message and then interpret its meaning. If the communication is successful, the receiver interprets the message consistently with the sender's intended meaning. Table 8 summarizes the rules for successful verbal communication.

The similarity of the meanings given to the message by the sender and receiver can be verified through feedback. An example of feedback verification in verbal communication is when the receiver "repeats back" the message and the sender either agrees with the receiver's repeat back or corrects it. Verification feedback serves an important error-checking function in the communication process. It also allows supervisory oversight of communications to catch errors before they have consequences.

A sender and receiver must both be active for communication to be effective. The sender and receiver share responsibility for ensuring successful communication. However, when companies analyze the causes of events, errors in sending messages are more often identified than errors in receiving. The reasons for the difference are unclear. A company's investigation should consider sending and receiving errors and corrective actions should address both to be effective.

Shift handover is another critical aspect of communication. The status of equipment, repairs that occurred, problems faced, upcoming chores, etc., are all key information the oncoming worker will need to know. One good guide on shift turnover is U.S. DOE 1038-93. [19]

Table 8. Rules for effective verbal (face-to-face, radio, phone) Communication (© PII, 2012)

#	Communication rules
1	Senders MUST require repeat back by the receiver; ask questions to make sure of understanding.
2	Receivers of messages MUST repeat a message they receive.
3	Communicate one task at a time.
4	Sender & receiver must use the standard name and component number (and building name and number) and/or for related reference material (drawing, procedure name/number) in each message.
5	Sender must require receiver to give feedback of foreseeable conflicts. Sender (if message is to subordinate) should follow-up in the field if possible, to make sure message was understood.
6	Receiver must request clarification anytime they believe the message is confusing.
7	Do NOT give verbal instructions to workers who do not have the demonstrated skills to correctly understand and perform the task assignment (unless you plan to supervise/coach the task yourself).
8	Communicate to the most senior member on the work crew.
9	Know your audience & change message accordingly to compensate for lack of knowledge (when appropriate; see other related rules).
10	Use units (and use approved units, such as SI) for process parameters.
11	Always count from left to right when giving instructions related to multiple choices of equipment.
12	Use approved jargon only.
13	Use the words **check**, **make sure**, and **actions** consistently (see procedure writing rules).
14	When possible, write out the task (use a special procedure or work order) rather than communicating verbally. Use formalized templates for consistency and make sure the receiver is provided all information (such as work orders and drawings) necessary for the task.
15	Perform pre-job meeting with work crew (including with contractors or construction company).
16	Perform walkthrough at the location where the work will be completed.
17	Have a backup communication method. Don't rely on one mode of communication. In emergency situations have hand signals as backup to loss of verbal communication.
18	Talk to the operator or technician who is doing the work (no delegation of work); but see Rule 8.
19	Receiver (worker or group) must report back when work is complete.
20	If confusion exists in implementing a task, the shift supervisor (and higher, if supervisor is unavailable) must be contacted to make sure he/she understands the problem.
21	Do NOT use a PA (public address system) for process instructions, since this method of communication does not allow repeat-back.
22	Use Open questions and non-confrontational questioning methods when requesting clarification.
23	When communicating remotely (by phone or radio), if the message is not understood on the second attempt at clarification the Sender must find the Receiver and communicate face-to-face.
24	For large or complicated jobs, in addition to repeat back, the sender must ask the receiver for an assessment of the pre-job briefing to ensure the workers are not confused.
25	Workers and supervisors must keep a shift log to aid in turnover between shifts. Workers and supervisors and support staff (if necessary) must have a minimum or 15 minutes overlap with a relieving shift. *See shift turnover standard for more details.*

Conclusion

Closing the gaps in these 4 areas has a greater than 100:1 payback!

Acronyms and Abbreviations

ACC - American Chemistry Council
AIChE – American Institute of Chemical Engineers
API – American Petroleum Institute
CCPS – Center for Chemical Process Safety (of AIChE)
CMA - Chemical Manufacturers Association
CSB – Chemical Safety Board
DM – Damage Mechanisms
FMEA – Failure Mode and Effects Analysis
GCPS -- Global Congress of Process safety
HAZOP – Hazard and Operability; as in HAZOP Analysis or HAZOP Study
HRA – Human Reliability Analysis
HSI – Human-System Interface
KSA – Knowledge, skills and abilities
LOC – Loss of containment
LOPA – Layer of Protection Analysis
MOC – Management of Change
OSHA - US Occupational and Health Administration
PHA – Process Hazard Analysis
PSM – Process Safety Management
RBPS – Risk Based Process Safety
SOP – Standard Operating Procedure

References

[1] *Guidelines for Technical Management of Chemical Process Safety,* CCPS/AIChE, 1985.

[2] "The Cost and Benefits of Process Safety Management: Industry Survey Results," W. Bridges, *Process Safety Progress*, AIChE, January 1994.

[3] "Necessity of Performing Hazard Evaluations (PHAs) of Non-normal Modes of Operation (Startup, Shutdown, & Online Maintenance)", W. Bridges and Mike Marshall (US OSHA), *18th Annual International Symposium, Mary Kay-O-Connor Process Safety Center, College Station, TX, October 2015.*

[4] *Guidelines for Risk Based Process Safety*, CCPS/AIChE, 2007.

[5] "Human factors and their optimization" W. Bridges and G. Collazo-Ramos, *8th Global Congress on Process Safety*, Houston, AIChE, April 2012.

[6] PII Course 5, *Writing effective operating and maintenance procedures*, © 2003-current

[7] "Best practices for writing operating procedures and trouble-shooting guides", W. Bridges and R. Tew, *13th Global Congress on Process Safety*, San Antonio, AIChE, March 2017.

[8] "Effective Procedure Writing – A Two-Part Series (Part 1: Reducing Risk of Worker Errors; Part 2: Addressing Process Safety Management Requirements for Operating Procedures)," *Chemical Engineering Progress*, 1997, W. Bridges with T. R. Williams (both of JBFA).

[9] "Best Practices for Writing Operating Procedures and Trouble-Shooting Guides", L. Madden and W. Bridges, *12th Global Congress on Process Safety*, Houston, TX, AIChE, March 2016.

[10] *Guidelines for Hazard Evaluation Procedures*, 3rd Edition, 2008, CCPS/AIChE.

[11] "API RP 571 - Damage Mechanisms Affecting Fixed Equipment in the Refining Industry", API, 2nd edition, 2011.

[12] Bridges, W.G., et. al., "Addressing Human Error During Process Hazard Analyses," *Chemical Engineering Progress*, May 1994.

[13] "Necessity of Performing Hazard Evaluations (PHAs) of Non-normal Modes of Operation (Startup, Shutdown, & Online Maintenance)", W. Bridges and Mike Marshall (US OSHA), *18th Annual International Symposium, Mary Kay-O-Connor Process Safety Center, College Station, TX, October 2015 and 12th Global Congress on Process Safety, April, 2016.*

[14] "Recipe for a Complete Process Hazard Analysis – Especially Addressing the Key Demands from US CSB", R. Tew and W. Bridges, *13th Global Congress on Process Safety*, San Antonio, AIChE, March 2017.

[15] PII Course 8, *PHA/HAZOP Leadership training*, © 2003-current.

[16] "Gains from Getting Near Misses Reported", W. Bridges, *8th Global Congress on Process Safety*, Houston, AIChE, April 2012.

[17] US Chemical Safety and Hazard Investigation Board (CSB), *Anatomy of a Disaster: Explosion at BP Texas City Refinery,* published 2006.

[18] US Chemical Safety and Hazard Investigation Board (CSB), *"Pesticide Chemical Runaway Reaction Pressure Vessel Explosion – Bayer CropScience, LP",* published 2008.

[19] "DOE-STD-1038: Guide to Good Practices for Operations Turnover", US DOE, 1998.

546668 Recycle Line from NGL Spheres to Residual Gas Header

Abstract:

HNGL plant is received 1.6 BSCFD of Haradh gas flows into the DGA Gas treating units where CO2 is removed and the produced sweet gas is directed to the NGL recovery units .Hawiyah sales gas is pretreated at HGP for CO2 removal ; hence the produced 2.4 BSCFD flows directly to the NGL units where it is mixed with the Haradh gas. The mixed gas then is processed in the NGL units producing C2+ NGL and sales gas stream. The NGL is pumped to storage spheres and subsequently shipped to NGL piping network through Shedgum Gas Plant. The Residue Gas flows to the sales gas compression area which increase the pressure allowing routing back the gas to the sales gas grid.

The purpose of this project is to install vent line from the two NGL Spheres to Sales Gas (Residue Gas) common header at Hawiyah NGL Plant. The aim of this project is to minimize the continues flaring from the NGL spheres , safe plant resources and protect the environment

Sphere PVs were venting the NGL vapor to the cryogenic burn pit when the pressure of the sphere exceeds 350 psig. In order to lower sphere pressure without losing NGL production, newly installed PVs are set at lower pressure set point of 305 PSIG to recycle the spheres vapors to NGL trains residue gas line, downstream of the brake-compressors..

During summer months July and August (if associated with maximum plant production), spheres are venting about 4.0 MMSCFD of Methane. Based on Methane price of 4.4 $/MMBTU then total saving is 1.05 MM $/year.

547776 Practical Notes on Ragagep Compliance

Process Safety Management (PSM) and Risk Management Plan (RMP) regulations have long required facilities to ensure that they comply with Recognized and Generally Accepted Good Engineering Practice (RAGAGEP). In recent years, authorities having jurisdiction have placed increased emphasis on RAGAGEP compliance. Specifically, emphasis has been placed on documentation and compliance with the industry codes and standards to which a facility is designed and maintained. This poses a unique challenge to older facilities, which have been designed to codes and standards long out of date and especially to owners and operators that do not have access to engineering documents used in the original design of their facility. For facilities using outdated codes, OSHA requires that the operating company document that "equipment is designed, maintained, inspected, tested, and operating in a safe manner."

Over time, the gradual evolution of industry codes and standards may cause facilities to fall out of compliance with the RAGAGEP requirements of the regulations. This paper provides example cases based on authors' experience with various owners and operators of older facilities. Practical methods are discussed for achieving and maintaining RAGAGEP compliance. It includes examples of comparing systems designed to outdated codes and standards against the latest codes, to ensure that they are able to maintain compliance with RAGAGEP.

548482 Propane Safety: Investigation Findings and Lessons Learned in the 2014 Philidelphia FOOD Truck Explosion

On July 1, 2014, a propane cylinder catastrophically failed and exploded on the back of a food truck in Philadelphia, PA. The explosion led to a sudden release of pressurized propane vapor and superheated propane liquid, which rapidly evaporated resulting in a large white cloud that entered and engulfed the rear of the food truck. This cloud ultimately found an ignition source within the food truck and resulted in a significant fireball burning both inside and outside of the food truck. The explosion and ensuing fire fatally injured the food truck owner and her daughter, and also caused injuries to 11 other individuals. This paper will first present the results of our investigation into the cause and origin of the catastrophic failure and explosion event, which included: (1) analyzing video footage of the explosion and ensuing fire; (2) applying propane phase diagrams to determine the expansion and resulting pressures within the cylinder prior to failure; and (3) applying blast techniques to determine the post catastrophic blast and ignition during the event. These techniques included advanced computational fluid dynamic modeling.

The paper will also discuss safety critical design features and processes in the propane industry that must be in place to ensure the safe use and filling of propane cylinders, which include: cylinder requalification; having a fixed maximum liquid level gauge; having a pressure relief valve; and filling cylinders by weight or volume.

548391 Using the Risk Analysis Screening Tool (RAST) in Process Design to Manage Risk

The Risk Analysis Screening Tool (RAST), available through the Center for Chemical Process Safety (CCPS), courtesy of the Dow Chemical Company can enable users to identify hazards and risk scenarios to better manage process safety risks. RAST is a spreadsheet based tool that uses simplified algorithms to identify and evaluate risk scenarios of a unit operation. The risk scenario identification is based on generic HAZOP deviations applied to the unit operation. An engineer can use RAST to identify risks to proposed process designs or plant modifications in the design phase of a project, before P&IDs are developed and a full process hazard analysis study is done. This can save time in process design by weeding out higher risk process alternatives as the process flow diagram is being developed.

The utility of RAST is demonstrated by application to actual case studies. RAST is applied to the runaway reaction at T2 Laboratories in 2007 and to the chlorine release at DPC Enterprises in 2003 showing how the hazardous events and consequences could have been analyzed during the design phase of the projects.

19 AIChE Spring Meeting
15TH GLOBAL CONGRESS ON PROCESS SAFETY
NEW ORLEANS • 2019

Drug Product vs. Process Safety: Integrating Both into the Design Process from Concept Development to Start-up

Jeffrey Heil, P.E.
Lead Process Engineer
DPS Group Global
959 Concord Street, Suite 100
Framingham, MA 01701
Jeffrey.Heil@dpsgroupglobal.com

Prepared for Presentation at
American Institute of Chemical Engineers
2019 Spring Meeting and 15th Global Congress on Process Safety
New Orleans, LA
March 31 – April 3, 2019

Drug Product vs. Process Safety: Integrating Both into the Design Process from Concept Development to Start-up

Jeffrey Heil, P.E.
Lead Process Engineer
DPS Group Global
959 Concord Street, Suite 100
Framingham, MA 01701
Jeffrey.Heil@dpsgroupglobal.com

Keywords: Process Safety, Process Design, Biotechnology, Inherently Safe Design

Abstract

Process safety has been around for over 350 years but may easily be overlooked or treated as an afterthought in the design process, often leaving hazard reviews and safety valve considerations to the end of detailed design. In the chemical industry, emphasis on safety is generally in respect to process safety, but in the growing biotechnology industry the emphasis is often shifted to product and patient safety. Due to this subtle yet significant difference in focus, it is often challenging for the design team on these projects to secure the necessary project resources to ensure both robust product and process safety in the facility being designed, often on an accelerated schedule. Leveraging experience from working in the oil and petrochemical industry for over 10 years, an in-depth understanding of the project design cycle, and by asking some simple questions along the way; process safety in design can be assured for even the most challenging projects, without a significant impact on project resources.

1 Introduction

1.1 Brief History

The first safety valve, developed in 1681 by Frenchmen Denis Papin, protected the first ever pressure cooker, one of which was used to make meat jelly for England's King Charles II. [1] Papin first invented the steam pressure cooker in 1676 to make the jelly and other food products but then realized no one would be willing to operate the cooker given the risks. Having identified the hazard, he set out to invent a way to mitigate that hazard while ensuring the King got his jellied meat pies.

Following development of the first weighted lever style safety valve invented by Papin, little progress was made until the early 1800's when locomotive steam engines gained notoriety both as a superior means of transportation and a hazardous machine often operated beyond its limits in the name of more power and speed. Safety valves were revised throughout the early 1800's to make these locomotives safer and formally modernized in 1856 with the invention of the first spring actuated pressure relief device that could not be modified to a different pressure, while in service [2].

Modern process safety, as we can still see it in existence today, dates to the early 1900's with the development of modern oil refineries and all the hazards that they include. With these refineries and their complex chemical processes came a wealth of experimentation, safety system design, and industry supported safety initiatives. The goal was the same as it is today, ensuring safety to the employees, the environment, and the facility. The problem with many of these early safety initiatives is that different companies had different ideas of what safe meant and relied on historical data to predict the future. The notion that "it has been running like this for 20 years and will continue for another 20 years" lead to a raising number of deadly process safety incidents, particularly throughout the 1970's to 1980's (i.e. Bhopal, India Methyl Isocyanate release and Piper Alpha explosion in the North Sea.). Plants were operating to the status quo, not improving their practices new technology became available. In parallel, urban sprawl forced industrial plant operation closer to populated areas, increasing risk to the community. The OSHA process safety management (PSM) initiative was introduced into law in the United States to address this very problem [3]. The PSM standard was performance-based meaning that as new technology was developed and safer systems could be designed, those safer designs had to be implemented; i.e., the status quo was no longer good enough. The PSM standard was ratified in 1992: hence, even though the oil and petrochemical industries in the United States have been around for about 100 years, they have only been operating under this standard for 26 years.

Contrast this with today's biotechnology companies. Gentech, one of the earliest, received FDA approval for the insulin drug Humulun in 1982. Like any new technology, investment and growth in biotechnology was slow to start, not taking off until early 2000's. There are many metrics for tracking growth of the Biotechnology industry such as venture capitalist investment and the number of companies in operation, but the high number of mergers and acquisitions over the past 30 years makes such comparisons incomplete. The Nasdaq Biotechnology Index (NBI) was established in 1993 with a baseline index value of 200 and provides a good metric for tracking industry growth over time. After holding steady in the low 200-300s throughout the 1990's, the NBI surged to 1,300 in 2000 as the Biotechnology bubble was forming. The index has steadily risen to a peak of 4,003 on 01-July-2015 and is hovering currently around 3,500 (Q1 2019), exemplifying the continued growth of the industry over the past 20 years. [4]

Pharmaceutical companies (e.g., Aspirin, invented by Bayer in 1899) have been around almost as long the oil companies have and are sometimes confused with biotech. Biotechnology companies are adolescent at best, having typical company life of less than 20 years with little process safety oversight beyond building code requirements as they do not meet the hazard requirements of the OSHA PSM Standard. Not meeting the hazard requirements of the PSM standard does not exempt these companies from a responsibility to ensure a safe working environment. If anything, their proximity to high density population areas and minimal operating experience places them at a higher risk with respect to process safety, but in many cases the operators and designers have only worked in these types of facilities and may not fully appreciate the safety concerns present. Coming from a background where process safety is paramount and the product quality, while important, takes second place, the author has made it his personal mission to increase awareness throughout the Biotechnology industry one project at a time.

1.2 Process vs. Product Safety

In the chemical industry, emphasis on safety is always in respect to process safety; ensuring that the public, employees, environment, and facility (in that order) are sufficiently protected from inherent hazards of a process. "Safety" refers to containment of the process and protection of personnel from harm. If a hazardous material cannot escape the process; it cannot be breathed in, released to the environment, set on fire, or result in damage to equipment that may cause injury to nearby personnel. The previously described process safety management standard formalized the set of tools used by these industries to ensure current best practices are used when it comes to process safety in hazardous environments. The goal of the PSM standard and implications of it being "performance based" is summarized as follows:

> *"... Recognizing the dangers associated with highly hazardous chemicals, Congress mandated (by the Clean Air Act Amendments of 1990) that OSHA set requirements for the management of highly hazardous substances to prevent and mitigate hazards associated with catastrophic releases of flammable, explosive, reactive, and toxic chemicals that may endanger workers. As a result, OSHA developed the Process Safety Management (PSM) standard (issued in 1992), which covers the manufacturing of explosives and processes involving threshold quantities of flammable liquids and flammable gasses (10,000 lbs.), as well as 137 listed highly hazardous chemicals. The PSM standard (29 CFR 1910.119) requires employers to implement safety programs that identify, evaluate, and control these hazards. Unlike some of OSHA's standards, which prescribe precisely what employers must do to comply, PSM is a "performance-based" standard-it outlines key features of safety programs for controlling highly hazardous chemicals, and employers have the flexibility to tailor their safety programs to the unique conditions at their facilities..." [4]*

The biotechnology industry on the other hand often has a more focused emphasis on product safety. Product safety ensures that all current good manufacturing practices (cGMP) are being followed with the goal of protecting the end-user or patient from hazards of a contaminated biological drug product. Product safety was important for the pharmaceutical industry because the medicines being produced were intended for use in the human body and would chemically alter the bodies function. Product cross contamination could result in serious health side effects, especially with highly potent or specialized substances. Biotechnology processes work by growing viruses and manipulating this viral material to deliver highly specialized medicines to the patient. Poor manufacturing conditions could allow unintended bacterial growth alongside the desired product, leaving the future patient with a potentially deadly virus or bacterial infection instead of a cure. The US Food and Drug Administration (FDA) passed the first cGMP regulations in 1969 [6], making them older and with many respects better understood in the industry than the PSM standard is in the oil and petrochemical industry. The practiced engineer will note several similarities in the quoted statements describing the PSM standard and the cGMP standards.

> *"The CGMP requirements were established to be flexible in order to allow each manufacturer to decide individually how to best implement the necessary controls by using scientifically sound design, processing methods, and testing procedures. The flexibility in these regulations allows companies to use modern technologies and innovative approaches to achieve higher quality through continual improvement. Accordingly, the "C" in CGMP stands for "current," requiring companies to use technologies and systems that are up-to-date in order to comply with the regulations. Systems and equipment that may have been "top-of-the-line" to prevent contamination, mix-ups, and errors 10 or 20 years ago may be less than adequate by today's standards." [7]*

Product safety requires process containment too, but in many cases the consequences are less likely to cause bodily harm so long as product quality control procedures ensure quarantine of the affected drug product. Note that this statement applies to the biotechnology industry, the consequences of a loss of containment in the pharmaceutical industry can be as deadly as in a chemical plant, depending on the substance released. Fifty years ago, when the cGMP's were adopted, pharmaceutical operators already knew about process safety and adopted it as readily as oil refineries and industrial chemical plants did. Twenty years ago, biotechnology was in its infancy and there were few large-scale biological processes in operation. Ten years ago, most biological processes ran at atmospheric pressure using water-based chemistry. The risk associated with potential process safety events was minimal in comparison. Today, many biotech processes use advanced solvent-based chemistries needing large volumes of hazardous flammable solvents, a key element in the severity of an incident when a loss of containment occurs. Add to this unique process chemistries such as dichloroacetic acid, methyl chloride, hydrochloric acid, and sodium hydroxide; all of which can easily corrode stainless steel, a common metal alloy used in biological processing, and unique hazards become a reality even at small scale.

2 The Design Process

Designing processes to produce goods takes time, time that the future producer of those goods doesn't always like to part with because "time is money". Each month spent designing or building the facility to manufacturer a product is a month of revenue missed or often in biotech, a month of trials that could be running. A typical design process includes several independent stages or phases. An ideal project would separate each of these phases with well-defined checks creating structure, an opportunity for detailed reviews, and adding time to the project. Initial concept development, feasibility, and design of large-scale chemical or industrial facilities often involves acres of real estate, spans years, and costs millions of dollars. By contrast small-scale bioprocessing facilities, often less than 40,000 ft^2, are fully designed in terms of weeks or months and leverage parallel work paths where possible.

The impact of such a short design schedule is that time for staggered activities, separated by well-defined stage-gates, is often sacrificed in the name of the project schedule. It is possible to crash a project schedule, completing the design early, but doing so limits the amount of design work that can be completed based on previously completed design elements. If a project schedule allots for 6-8 weeks between establishing the basis of design for a facility and documents that are to be issued for vendor bid, scheduling one week for a hazard analysis and another week to implement the resulting design changes (25% of the schedule) is a challenge. By the time the hazard analysis is complete, many aspects of the design have been progressed and resulting changes will result in unbudgeted rework. The same is true for the design of key safety systems such as relief devices or process interlocks that require a complete system design to be finalized. The engineer can anticipate credible overpressure scenarios and safe operating limits for a process but cannot complete these design elements until piping to and from the relief device is complete and the hazard analysis is approved, accepting the level of protection assigned to each process step. It is for these reasons that it is not uncommon to see relief devices for a process being sized and planned for as one of the final project tasks. Explaining to an architect that a storage tank must be moved to meet a flammable liquid set-back identified during a hazard review or that another roof penetration is required for a relief device tail pipe two weeks before the completion of a design are just two examples of real-world challenges that project teams may be faced with.

Understanding the different design phases or stage-gates, the tasks that occur during them, and their impact on the ultimate facility design is important if one is to create a plan for addressing process safety on such a short schedule. Table 1 summarizes one example of a design process, key stage-gates, and design elements to be considered in each phase. Note that jumping a phase and for example combining schematic design and design development (the most common violation of the design process) into one phase can confuse the separation of tasks between these phases and eliminate checks important to the design process.

Table 1 – Key Aspects of the Design Process

Phase	Process Design	Process Safety
Feasibility Study	• Identify product to be manufactured and major unit operations • Establish block flow diagram of process • Identify facility requirements • Confirm business case for proposed facility	• Identify hazardous materials used in the process • Identify hazardous operating conditions (i.e. high pressure) • Establish acceptable risk criteria for operation of the facility

Phase	Process Design	Process Safety
Concept Design	• Breakdown individual processing steps to the equipment level • Develop process flow diagrams for each major process step • Develop preliminary heat and material balance • Develop equipment list and identify long lead equipment (to prioritize design development tasks) • Detailed layout of process equipment • Identify utility requirements and size utility systems • Size and layout major process headers	• Develop a comprehensive hazardous and flammable materials inventory / determine if additional design precautions are required • Identify potential material compatibility concerns and material specifications for each process step • Complete a code review in collaboration with the full design team to ensure all building, fire, and design codes are being followed • Ensure facility citing requirements are met in the layout
Design Development	• Develop initial P&IDs for the process • Identify and document key control loops and automation requirements • Finalize design of all process equipment (size, pressure rating, operating conditions) • Complete detailed sizing and specification of major process headers and utility lines • Finalize process heat and material balance • Prepare equipment and piping specifications and datasheets	• Confirm inherently safe design where possible (pressure profile versus equipment design) • Identify safe operating limits for each unit operation, add control points as required • Review P&IDs for relief devices and confirm their discharge locations • Complete scenario analysis and preliminary sizing for all relief devices. Based on preliminary sizing, define relief device inlet/outlet piping • Confirm piping specifications versus services • Confirm vessel sizing versus hazardous material inventory and update as required • Complete a preliminary hazard analysis on the full process at the end of the design development stage

Phase	Process Design	Process Safety
Detail Design	• Incorporate all recommendations from the design development preliminary hazard analysis • Finalize all equipment and line sizing • Fully specify all process components including equipment, lines, valves and instruments	• Ensure that all actions from the preliminary hazard analysis are incorporated into the design • Review and finalize relief device sizing • Complete sizing and design of effluent handling system for relief device discharges • Double check the set points of key process variables versus the operation's safe operating limits, previously defined. • Review final equipment specifications against material compatibility matrix and equipment specifications

3 Safety Throughout the Design Process

Process safety should be considered in the earliest stages of design. One principle of process safety design is that inherently safe design, creating a system that physically cannot fail, is the only way to prevent an incident from occurring. The easiest example of an inherently safe system is designing a pressure vessel to withstand maximum system pressure, even if all process controls were to fail. While an easy goal on paper, inherently safe design is not always possible and often never fully achievable. A vessel can be designed to the proper pressure, but an unforeseen corrosion issue could reduce the design pressure making the vessel susceptible to premature failure. In anticipation of this and to protect against it, engineers design-in multiple layers of protection to a process starting with inherently safe design at the concept level and ending with proper emergency response procedures developed during commissioning and simulated during plant operation. Continuing with the vessel example, minimizing the quantity of material stored and having an inventory management plan to support smaller on-hand inventories is another way to reduce hazards associated with the process. Additional design principles or layers of protection used are summarized in Figure 1. Different layers of protection against a process safety event are established during concept development and built-up throughout the design process.

Figure 1. Typical Layers of Protection Applied to a Design

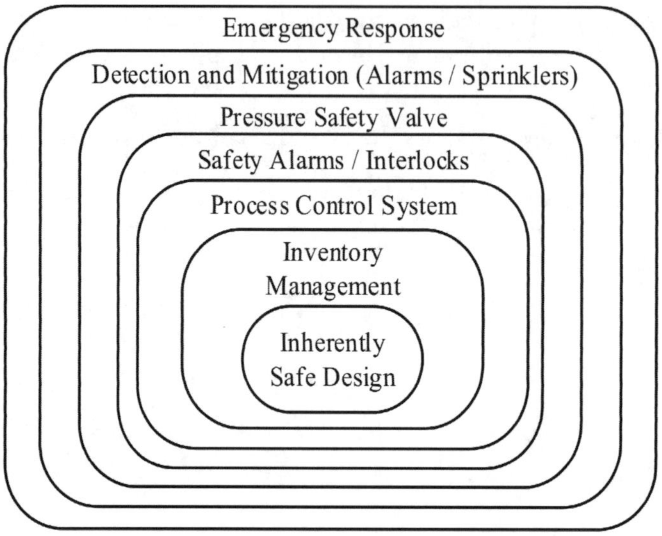

3.1 Concept Design

Concept design is one of the earliest stages of the design process. During concept design types of process equipment are first selected, their location mapped out, and consideration given to how materials will be heated up, cooled down, transferred, and stored when not in use. By asking a few questions as these fundamental design decisions are made, the design predisposes its self to being inherently safer.

- What is the worst-case operating condition, can the system be mechanically designed to handle those conditions?
- Where will hazardous materials be present and can that equipment be installed away from high concentrations of personnel?
- What hazards do the materials pose (toxic, flammable, corrosive)?
- How much of a potentially hazardous raw material must be on-hand? Weekly vs. monthly delivery may reduce stockpiles by 75%.
- Is there a safer way to produce the desired product?
- Is there a safer way to complete this unit operation?

3.2 Detail Design

Concept design develops enough detail to prove that the processing goal is obtainable and business case can be made. Concepts developed such as system location, connectivity, and required unit operations are further developed into a constructible design that can ultimately be operated to produce a finished product. In detailed design, detailed questions are needed as more layers of protection are added. These questions are to be asked informally throughout design and will be formalized in a process hazard analysis (PHA), completed during the detailed design phase.

- Where are the areas of high pressure, temperature, and hazardous material inventories; can they be reduced?
- Are there any specific code requirements or industry best practices for this system?
- What are the expected operating parameters, how can they be exceeded, identified ahead of time, and mitigated?
- Are all systems protected by a mechanical pressure relief device (if needed), do those devices discharge to safe locations?
- How and where could a loss of containment occur, how can it be quickly identified (leak or fire detection)?
- How will the consequences of an event be mitigated once started; how can the event be stopped (process shutdown or fire protection)?
- How can the public, employees, and environment be protected from the consequences of an event if it were to occur (emergency management)?

3.3 Start-Up

Following detailed design and construction, a process must be started, commissioned, qualified, and transitioned to manufacturing. Throughout this process, design assumptions must be confirmed, formal procedures verified, and changes to the design noted so that their impact on the process can be determined. Process safety considerations during start-up must include the full team from engineering to operations and maintenance.

- How does the as-built design compare to the final design, what are the impact of any changes?
- Were changes made to the operating envelope for any of the systems (pressure, temperature, level, concentration)?
- Has a pre-start-up safety review (PSSR) and as-built PHA been completed?
- Do operating procedures match the actual operation of the process?
- Have emergency drills been completed, are there lessons learned from these?

Addressing process safety throughout the design process, versus at the end of design when final design reviews are completed; minimizes late stage design changes, results in a better designed facility, and ensures safe facility operation for years to come.

4 Conclusion

Process safety is a well-established element in the design of oil and petrochemical facilities, taking a front row seat in the design process and being an area of focus once operations begin. Product safety holds a similar role and has been common place for a similar period in pharmaceutical and recently bioprocessing facilities. While the gap between these two elements of design has been closing in recent years as more pharmaceutical companies enter the bioprocessing arena and bioprocessing operations grow, a considerable gap remains. Based on the authors experience designing processes and facilities in both of areas, a few causes for the gaps are easily identifiable:

- Historically, bioprocessing facilities have not contained hazardous processes. Areas of pressure were limited to gas delivery systems (i.e. nitrogen and compressed air) and storage tanks contained aqueous based chemistries (i.e. bioreactors) minimizing the hazards with a loss of containment of these products.
- The shortened design timeline for bioprocess facilities challenges the ability to incorporate process safety into the design at distinct stage gates. The schedule does not allow for a discrete process safety design effort while other design work is on hold.
- There is a general lack of knowledge regarding safety system design in the biotechnology industry. This goes back to the opening statement that in many cases engineers responsible for these systems don't know what they don't know.

Product and patient safety are well defined requirements for biomanufacturing companies and are considered in the earliest stages of the drug discovery process. Applying similar principles to process safety throughout the design lifecycle, and not at distinct stage gates, will ensure a smooth transition from process development to design, validation, and ultimately commercial manufacturing.

5 References

[1] F.M. Editors of Encyclopedia Britannica. "Denis Papin." *Encyclopedia Britannica*, Encyclopedia Britannica, Inc., 18 Aug. 2018, www.britannica.com/biography/Denis-Papin.

[2] Ramsbottom, J. "On an Improved Safety-Valve." *Process Institute of Mechanical Engineers*, vol. 7, 1856, pp. 37–47.

[3] US Department of Labor Final Rule, "Final Rule on Process Safety Management of Highly Hazardous Chemicals; Explosives and Blasting Agents", 1910.109, *Fed Register #: 57.6356*, 24 Fen. 1992.

[4] NASDAQ Biotechnology Index, NBI, https://finance.yahoo.com/quote/%5Enbi/.

[5] *Process Safety Management Introduction*. Occupational Safety and Health Administration, www.osha.gov/dsg/psm/index.html.

[6] US Food and Drug Administration, *Good Manufacturing Practices (GMPs) for the 21st Century - Food Processing*, "GMPs Section One: Current Food Good Manufacturing Practices", 09 Aug. 2004, www.fda.gov/food/guidanceregulation/cgmp/ucm110907.htm.

[7] US Food and Drug Administration, *"Facts About the Current Good Manufacturing Practices (CGMPs)"*, 25 Jun. 2018, www.fda.gov/food/guidanceregulation/cgmp/ucm110907.htm.

546706 LOPA and Sil Risk Analysis As Critical Tools for PSM Culture and Accidents Prevention

Industrial facilities that operate with dangerous products, whether chronic or acute, are increasingly common to be present and part of society at large, often allowing an aggregate of increasingly rich and economically beneficial professions and specialties to the community, be it directly or indirectly. Accidents in these types of units, especially those operating with hazardous products of acute consequence (those that have the capacity to cause fatalities when released to the environment), such as flammable, toxic, explosive products, due to the possibility of occurrence of undesirable events and the behavior of the effects resulting from releases of these types of product, could cause serious damage to the people and / or the installations that are in its vicinity. The various types of accident studies are known as EAR (Risk Analysis Studies), some of which are used more widely and others are specific to specific response needs.

According to BERNSTEIN (1997), the word "risk" derives from the ancient Italian risicare, which means "to dare". While DUARTE (2002) describes that the meaning of the word risk is an expression of thought that has a lot to do with the social, scientific and technological evolution of man. As BILLINTON and ALLAN (1992) note, society has great difficulty distinguishing between a hazard, which can be prioritized in terms of its severity, but does not take into account its probability, and risk, which considers not only the dangerous event, but also their probability of occurrence. KAPLAN (1997) explains that when a person asks "What is the risk?", 3 (three) questions are actually being questioned: What can happen? How can this happen? And if this does happen, what are the consequences? The answers define the scenario, frequency and consequence.

Within the agencies that tend to manage corporate risk, it is noticeable that the risk associated with a particular accident scenario is calculated according to its frequency of occurrence and its severity. Frequency is related to the temporal chance of occurrence of the initiating event, which leads to the release of dangerous product, such as LPP (Liquefied Petroleum Gas) emanation by ball rupture through overpressure of this equipment or, as a second example, the intrinsic rupture of a stretch of line. The consequence, also called severity, considering personal safety, is the capacity that the physical effects, due to the release of dangerous product, associated to the analyzed initiating event, can cause fatality (s), related to the potential of damage related to each effect analysis. The risk, qualitatively or quantitatively observed, and defined in a simplified way, is evaluated as the multiplication of frequency of occurrence and severity. The total social risk related to a particular facility, which can be defined as the risk that a community outside the facility is subject to by its presence in its vicinity, can be calculated as the sum of the risks of all the initiating events arising of the same. In order to do this, for each accident scenario, not only the generating causes, known as Initiating Events, but also the likelihood of ignition, the probability of affecting people, and several other parameters are considered, being one of them very important and permissible of analyzes, which are the failures

safeguards. These, if they act effectively when demanded, can prevent the accident from occurring, since the Initiating Event happens.

Hazardous product risk studies, such as qualitative (Hazard Operational Preliminary Analysis (APPP), Hazard Operability (HAZOP, among others) or those known as quantitative analyzes (Vulnerability Analysis and Quantitative Risk Analysis per se), in addition to being increasingly requested by regulatory agencies, have often been seen as an internal requirement of companies, representing one of the pillars of industrial safety management, in order to keep their risks at low or acceptable levels. However, these analyzes are usually performed in stages

until the final definition of the risk, comparison with acceptability criteria, and later mitigation, if necessary, each of these stages having a minimum time required for its execution, which may make the analysis complete rather time-consuming. Situations similar to these are detrimental to Risk Management when, in fact, what is desired is not a complete risk analysis, but only the achievement of a specific response, such as if my identified risk scenario is effective and properly protected.

The extent of possible damage related to each identified accident is delimited by the intensity of the physical effect causing the damage, and the relationship between the intensity of the physical effect and the corresponding damage is established through the vulnerability models.

On the other hand, semiquantitative risk analyzes such The extent of possible damage related to each identified accident is delimited by the intensity of the physical effect causing the damage, and the relationship between the intensity of the physical effect and the corresponding damage is established through the vulnerability models.

On the other hand, semiquantitative risk analyzes such as LOPA (Layer of Protection Analysis) and SIL (Safety Integrity Level), even though they are less demanded by chemical and petrochemical companies than qualitative or quantitative steps can have beneficial and effective results in terms of knowledge of all the characteristics of the analysis scenarios and of their protection barriers and, in a practical way, much faster to be developed. In addition, they can provide results that do not make up the analysis range of qualitative and / or quantitative studies in depth, such as knowledge and prior review of safeguards, many of which are known as Independent Protection Layers (CIP or IPL - Independent Protection Layers).

Therefore, with the anticipation of an increase in the number of future installations operating with hazardous products and the current installations that already do so, the semi-quantitative studies of LOPA represent an enormous value for the observance of the existing protections in each scenario of process accidents, also serving as the basis for the determination of new Independent Protection Layers or even the reduction of their number, tending to simplify the steps of determining the frequency of the scenario., even though they are less demanded by chemical and petrochemical companies than qualitative or quantitative steps can have beneficial and effective results in terms of knowledge of all the characteristics of the analysis scenarios and of their protection barriers and, in a practical way, much faster to be developed. In addition, they can provide results that do not make up the analysis range of qualitative and / or quantitative studies in depth, such as knowledge and prior review of safeguards, many of which are known as Independent Protection Layers (CIP or IPL - Independent Protection Layers).

Therefore, with the anticipation of an increase in the number of future installations operating with hazardous products and the current installations that already do so, the semi-quantitative studies of LOPA represent an enormous value for the observance of the existing protections in each scenario of process accidents, also serving as the basis for the determination of new Independent Protection Layers or even the reduction of their number, tending to simplify the steps of determining the frequency of the scenario.

546375 Controlling Static Discharge Using Antistatic Additives

Controlling Static Discharge using Antistatic Additives.

Abstract

The presentation will discuss controlling static discharge in non-fuel, non-conductive/low-flash point fluids. The scope of the presentation will concentrate on industrial producers, distributors, consumers of these fluids. Furthermore,furthermore, the presentation covers several industries where these fluids are most commonly found. Common Static-Accumulating Flammable Liquids That May Form Ignitable Vapor-Air Mixtures:

- VM&P naphtha
- Cyclohexane
- n-Heptane
- Benzene
- Toluene
- n-Hexane
- Xylene
- Ethyl benzene
- Styrene

Static discharge has caused catastrophic explosions and fires (U.S. Chemical Safety Board, Static Spark Ignites Explosion Inside Flammable Liquid Storage Tank No. 2007-06-I-KS).

Issues

- Nonconductive flammable liquids can accumulate static electricity during transfer and storage.
- Static sparks can readily ignite flammable vapor-air mixtures inside storage tanks.
- Material Safety Data Sheets (MSDSs) often do not adequately communicate hazard data and precautions.

Key findings:

- Safety publications omit antistat options. (see Process Safety Beacon, August, 2016)
- US affiliates of multinational corporations have different guidelines for antistatic use than their parent companies.
    - NFPA®77 and the American Coatings Association's, *Generation and Control of Static Electricity in Coatings Operation* has guidelines that are different than EU regulations ATEX- Appareils destinés à être utilisés en ATmosphères EXplosibles and ESIG's *Flamability: Safe Working With Industrial Solvents.*
    - The presentation will highlight these differences and how they contribute to the disparity in antistatic additive applications.
- Normal Bonding and Grounding May Not Be Enough! Companies that handle, transfer, and store flammable liquids should contact manufacturers to determine if these liquids can accumulate dangerous levels of static electricity, and if they can form explosive vapor-air mixtures inside storage tanks. If so, *extra precautions*—beyond normal bonding and grounding—may be necessary. Antistatic Additive is an extra precaution.

Recommendations: Knowledge members can use

- Who adds the Antistat?
- How much Antistat is added?
- How is the Antistat added?

546677 Led Your Process Safety Culture, Lead Beyond Hierarchies

The Oil & Gas industry is trying to advance at the next crossroads: "How to achieve that leaders without skills and without training in the management of culture, transform their own culture?"

Our research presents three key ideas that leaders from all over the world

can apply immediately to lead risk management with effectiveness:

frequency, consistency, meaning.

When specific leadership competencies to manage risk have not yet been developed,

Leaders can use this three concepts of easy adoption.

Increasing frequency of interactions with people will allow us to generate safer operations. In full attention economy and contexts of increasing complexity, for risk management, more communication is better communication. But it is not enough to improve the frequency of interactions. That frequency must be accompanied by consistency. The challenge is to achieve possible consistency between what we say and what We do. Leading by example is one of the central keys of risk management. The lack of consistency provides symbolic permissions to normalize deviations. We must maintain a high frequency of face-to-face interactions, but in addition, we must be consistent at every opportunity. And finally, to give meaning to our collaborators. It is necessary to ensure the understanding of risks. Each collaborator need to build his or her own sense of vulnerability and Operational Discipline.

We need workers who operate according to procedures and not adjusted to what the bosses say. This is what meaning allows.

548192 The Importance of Using Color Scale in Process Safety

Everyone knows the importance of color. From physics to fashion going through the automotive industry she will always be present. There is no world without color. Often, they come up on the scale form or not, but the variation in the scenarios are always important. It is so on the traffic sign and the ph tape. From the Diamond of Hommel to across the street. Colors, colors and colors. In addition to not necessarily needing ididomas, such as color of the sky and molten metal have a variety of shades. The main idea of the work is to choose a suitable color to be used in communication within a work area, choosing which and how many elements can evaluated suggest a color scale from the chosen one and present its contribution in the process safety.

550372 Flammable Gas Release Modeling for Real-Time Analysis Using Adversarial Variational Bayes

In this study, we have suggested a method to calculate high-accuracy damage range from flammable material in petrochemical complex. There are many flammable and toxic substances in chemical processes. Computational Fluid Dynamics(CFD) is one of the best way to calculate and simulate scenarios of release range of various fluids or gases. However, it is not used to calculate damage range, because of a lot of time to simulate one scenario. Thus, lots of researchers have enthusiastically studied the surrogate model or reduced order model to get the results of release range in real-time [1-3]. Recently, deep learning based generative model has receive attention because of its superior performance of non-linear manifold estimation [4]. Herein, we propose surrogate model based on adversarial varational Bayes (AVB) [5] for real-time analysis of probability of death. The proposed surrogate model can extract the latent space of high-dimensional flammable and toxic substances dispersion contours. Furthermore, it can generate the estimated contour image from variables of interest much faster and more precise than CFD data memorization with linear interpolation because it does not memorize the data but train the non-linear manifold regarding exact posterior distribution on given observation which basis is predefined in standard variational autoencoder (VAE) [4, 6].

The acrylonitrile was studied about risk assessment for health damage like cancer hazard.[7] But the component has flammable and toxicity, so we should calculate range of damage. The main scenario in this study is the acrylonitrile released from underground model, and the result is shown different damage following hole size, mass flow rate, wind velocity, and wind direction. The superior performance of the proposed model was exemplified by comparing with other surrogate models.

[1] Chen, T., Hadinoto, K., Yan, W., & Ma, Y. (2011). Efficient meta-modelling of complex process simulations with time–space-dependent outputs. *Computers & chemical engineering*, *35*(3), 502-509.

[2] Wang, K., Chen, T., Kwa, S. T., Ma, Y., & Lau, R. (2014). Meta-modelling for fast analysis of CFD-simulated vapour cloud dispersion processes. *Computers & Chemical Engineering*, *69*, 89-97.

[3] Long, K. J., Zajaczkowski, F. J., Haupt, S. E., & Peltier, L. J. (2009). Modeling a Hypothetical Chlorine Release on a College Campus. *JCP*, *4*(9), 881-890.

[4] Na, J., Jeon, K., & Lee, W. B. (2018). Toxic gas release modeling for real-time analysis using variational autoencoder with convolutional neural networks. *Chemical Engineering Science*, *181*, 68-78.

[5] Mescheder, L., Nowozin, S., & Geiger, A. (2017, August). Adversarial Variational Bayes: Unifying Variational Autoencoders and Generative Adversarial Networks. In *International Conference on Machine Learning (ICML)* (pp. 2391-2400). PMLR.

[6] Kingma, D. P., & Welling, M. (2013). Auto-encoding variational bayes. *arXiv preprint arXiv:1312.6114*.

[7] Haber, L. T., & Patterson, J. (2005). Report of an independent peer review of an acrylonitrile risk assessment. Hunam & Experimental Toxicology, 24, 487-527.

546615 Modeling and Analysis of Flammable Gas Dispersion and Deflagration from Offshore Platform Blowout

The operational safety of offshore platform is challenged by a series of adverse factors including harsh environmental conditions and process accidents. Blowout is a type of major accident that may result in catastrophic consequences. This paper uses a computational fluid dynamics (CFD) model to simulate the blowout accident on an offshore drilling platform. The simulation includes the dispersion of flammable gas and the resulting deflagration based on dispersion results. Through the modeling, the blowout gas dispersion processes under different wind speeds and directions are analyzed, and a deflagration scenario due to the ignition of flammable gas is predicted in which the impacts of jet fire, overpressure and heat radiation are assessed. The simulations indicate that wind speed and direction have a great impact on the dispersion behavior of blowout gas. During the deflagration, the maximum overpressure is 280 KPa, the jet fire mainly distributes in the derrick, and the maximum radiation flux of thermal radiation field is 2300 W/m^2. Essentially, the outcomes from simulations could help to conduct a prior risk assessment and make an emergency response plan. Additionally, it could give a reference for the inherent safety design of offshore platform to mitigate the unexpected consequences due to blowout accidents.

548682 Application of Inherently Safer Design Tool (*i*-SDT) for Hazards Assessment and Safer Operation

Inherently ***S***afer ***D***esign ***T***ool (*i*-SDT) is a newly developed semi-quantitative metrics for hazards identification and characterization[1]. It is built on property platform, probabilistic analysis of historical accidents information and characteristic equations of safety parameters (i.e., flammability, toxicity, explosiveness, reactivity, etc.) which are developed in-house to assess potential hazards under various operating conditions. Designers can identify dangerous components and vulnerable equipment of the investigated industrial process and can take necessary corrective actions by applying the inherent safety principles[2] (e.g., minimization, substitution, moderation and simplification). The final output of this *i*-SDT tool is a cluster safety parameter score (C_{SP}) which provides insights regarding the development of safe design and operational parameters to manage the hazards using a very limited amount of process information and safety incidents information.

In this work, the application of the proposed *i*-SDT tool is explained for assessing different alternative routes of an industrial process supply chain. The proposed *i*-SDT safety metric identifies the chemical components with highest risk and the major accident-prone operational units. The detail analysis using *i*-SDT tools shows how different operating conditions can affect the safety performance of the given process. The ultimate object is to find the safer route while considering the techno-economic and environmental performance. It provides Pareto curve to compare alternative routes and gives the freedom to the end-users choosing their safer operating route/technology and process conditions. Therefore, safety can be considered as an important selection criteria for choosing best alternative from the beginning.

Final cluster safety parameter scores (C_{SP}) obtained from the analysis enables designers to compare between available technologies and provides information on various hazardous situations. It can also provide in depth knowledge regarding the cause behind the hazardous behavior. The proposed *i*-SDT metrics allows designer to do the safety analysis simultaneously while doing the initial process synthesis which will reduce the future modification cost of any add-in safety features. This work shows how the unique features of *i*-SDT tool can be applied in a real life scenario while doing inherent safer design during early stage of design.

1. Kazi, M.–K, Eljack F. T., & AlNouss, A. Application of a property-based inherent safety quantification framework for integrating risk assessment into process safety life cycle. *AIChE Spring Meeting, Orlando, Florida, USA, 2018.* (156c), p. 2260.

2. Kletz, T. A., & Amyotte, P. (2010). Process Plants: A Handbook for Inherently Safer Design, Second Edition: Taylor & Francis.

546631 An Approach to Understanding Worker Interactions with Procedures Under Normal Working Conditions

Incidents in high-risk industrial settings are often attributed to the way in which procedures are used. Despite the identification of procedures misuse, little research has been conducted on how workers have interacted with procedures in normal conditions. To understand these interactions a behavior assessment technique has been developed which can describe how a worker interacts with a procedure to complete tasks in everyday work situations. interactions under normal conditions. Using the concept of work-as-imagined/work-as-done, the techniques describes the interactions at a macro whole-of-task level as well as an individual step level. At the macros level, assessment considers if the task was completed correctly and safely (Tier 0). At the micro step-level, assessment considers if the step was completed as it was imagined (Tier 1) or if the step was done in an alternate way (Tier 2). For the alternate approaches, assessment considers if the worker struggled or required assistance, completed the steps out of order or with iterations or if the step was not completed as intended.

Two studies, one involving industrial tasks Second Life® (Study 1), the other at Shells Robert, LA training facility (Study 2), were used to test and develop the behavior assessment technique. In Study 1, the behavior of 13 participants from offshore drilling conducting tasks in either day or night conditions was assessed. In Study 2 the behavior assessment technique was used to test if experience accounted for behavior difference in 25 participants from high-risk industries. Results demonstrate the success of applying the assessment techniques through identifying how the behavior of the participants changes according to their experience and familiarity with the tasks the were performing. Further analysis identified recommendations for design and work flow changes of both written procedures and specific equipment. In conclusion, the initial results suggest the assessment technique is an important contribution to the methods use for understanding early interventions to improve safety in the process industry.

PROCESS HAZARD ANALYSIS (PHA) PREPARATION: A ROLE BASED OPTIMIZATION

Susan S. Frizzell, PE
Adley Services, LLC
Lafayette, LA
Susan@AdleyServices.com

Biography:

Susan Frizzell, PE is the President of Adley Services. Shortly after high school, Susan began her career alongside her father on a drilling rig. Upon graduating from the University of Louisiana with a BS in Mechanical Engineering, she obtained her MBA and began working for a firm performing engineering and process safety reviews on refineries and other land facilities in the New Orleans area. After Susan relocated to Lafayette, LA, she began engineering offshore facilities, complimented with process hazard analysis facilitation. During her almost 20-year career, she has worked upstream, midstream, downstream, and all points in between. She is a licensed professional engineer and an executive member of various organizations.

Abstract

A PHA meeting consumes large resources and can place a strain on departmental manpower. Teams are often faced with the want to be more efficient in PHA conflicting with the need to be thorough and accurate. The key to solving this paradox is efficient preparation. By examining the lessons learned from past PHA's, a role by role analysis of best practices with regards to preparation will allow a PHA team to reduce the costly time spent in the meeting and yet still stay true to the methods and deliver a great, safe result.

When it comes to an optimal PHA, preparation is everything. For example, in a 6-member team PHA, 60 minutes of misused time equates to 6 hours of misused manhours. Whether it is managerial, operations, engineering, or even facilitation, there are lessons learned from previous PHA's that when shared, will not only make the PHA more efficient and the facility safer, but the industry as whole better. The goal being that those reading this be able to take back to their facility these practical preparation tips to share with their present and future PHA team members so that the resources of the facility can be better spent.

1. PHA Preparation Optimized

To fully optimize the Process Hazard Analysis (PHA) preparation, the team needs a clear vision of the end product: "To identify, evaluate, and control the hazards" [1, 2] of an operation or process to minimize the likelihood of occurrence and the severity of consequences of release of a dangerous substance [3]. With that in mind, the team can focus time and energy preparing for the PHA on factors within their control to meet that goal. An efficient, cost-effective PHA begins with including the different roles in the preparation. The PHA planning team should be comprised of management, engineering, operations, the scribe, and the facilitator.

An active commitment to reduce risk begins by assigning tasks to each role prior to the PHA. From the start, management should provide support and direction for the PHA team. They set the tone for obligation and importance. Second, engineering involvement in the preparation ensures accurate and complete process safety information (PSI) is available and accessible. Without correct information, the hazards may not be fully identified. Next, operations should walkdown the facility and identify areas of concern. They have the most intimate knowledge of the workings of the process, which is extremely valuable to the PHA when properly communicated to the team for hazard recognition. The scribe should have the technical skills to ensure equipment is functional and to properly and efficiently document the team's findings. Lastly, the facilitator needs to coach the team before and during the PHA to ensure the methodology is followed.

Successful PHA preparation can improve risk identification and limit cost through appropriate management direction, current and detailed documentation, and clear expectations of the timeline leading up to the PHA. We identify ways that companies can assist in preparation in the months leading up to a PHA and the benefits that preparation has on hazard recognition, team morale, and, most importantly, the bottom line.

2. Management – "*Set the Tone*"

It is important for management to set the tone of the PHA. More specifically, management needs to set the tone before, during, and after. In a PHA, the attitude, the posture, and the dedication of the team is usually a direct reflection of management's attitude, posture, and dedication. In our experience of both facilitating and participating in PHAs, we've seen attitudes of management in a variety of flavors. The following are observations we've made in the past regarding a successful PHA and one not as successful.

The corporate culture and support infrastructure set by management begins well before the first day of the PHA. A typical PHA is done every 5 years [1]. Given that schedule, it is imperative that management starts out on the right foot and begin planning well in advance. A management component that is very serious and very organized will typically induce an attitude of seriousness and organization to the team members. For example, a management team that waits until the last moment and rushes preparation will typically have PHA members that delay and do little preparation, if any at all, for the PHA. PHAs are very expensive and very time consuming. Lack of adequate and efficient preparation will be very damaging to an efficient use of PHA budget and resources. When management in advance sets the tone that a PHA needs to be done correctly and efficiently, the PHA team members tend to take preparation seriously. Just as it is well documented that an effective safety culture of a company starts at the top with management, effective preparation for a PHA also starts at the top with management [4]. The management teams we've seen in the past that do this most effectively set the tone of preparing and doing a PHA as part of your job, not as additional work on top of your daily duties. In other words, management sets the tone that PHAs are in your job description and that we are all PHA team members. For example, a PHA is not additional work given to engineers, it is in an engineer's job description.

The other things we have observed that effective management does by setting the tone is setting up a support team for the PHA team members. These PHA team members will have to transition out of their daily duties, be present at the PHA, and transition back into their regular duties. Often, their project managers and/or functional managers may give some resistance to that. It is imperative for upper management to set the tone with the managers of the team members that they need to have a plan for them to transition into the PHA and transition out of it with regards to their regular duties. Most of these facilities run around the clock all year long and the work never stops. Therefore, it is imperative for management to have a plan for these employees to be present at the PHA while it is going on. A PHA which is constantly stopped for employees dealing with issues outside of the PHA is an unacceptable waste of resources. For example, if a PHA consists of ten team members and one team member must stop the PHA to deal with outside problems unrelated to the PHA, the nine other team members are wasting company resources. When management sets this tone early on that we are dealing with real human lives and significant company resources, the employees and their managers tend to put greater weight on a PHA. And sometimes, the employees just need the help of upper management to persuade lower and mid-level management that their presence, uninterrupted, is necessary to conduct a PHA.

During the PHA, management's role is typically only at the very beginning. It is very effective during the opening remarks of a PHA for a member of management to also make an opening remark. During this address, management will set the tone on several things. It is not uncommon for management to begin opening remarks with a thank you for the team members for their hard work in preparing and committing of their time. Some members of management stress employees' safety; others may stress the importance of safety to the community or environment. Opening remarks is a great time for management to state known worst-case consequences initiated within the limits of the PHA scope or define the assumptions regarding severity. This guides the team towards consistent company-wide risk rankings and avoids the team paranoia over risk ranking too high or too low. It's not management's position to force upon the team a subjective belief; it is management's position to reaffirm the team uses best judgement and not be hesitant of ranking the risks as they believe they should be. Finally, management can give the PHA team members feedback from previous PHAs as to the wording, substance, and ranking of the recommendations to assist with review and completion post-PHA.

Once the PHA begins, management's role is very limited. We've seen effective managers from time to time pop-in, review recommendations, and discuss with team members their thoughts on such. Effective managers will pick a recommendation, look at the supporting data, and ask specific questions of how the team arrived at its conclusion. This serves the purpose of correcting any doubts that the PHA team may be having and serves the purpose of reaffirming to the PHA team what they are doing is correct. Management can also review recommendations with only the facilitator at specific progress intervals of the PHA. The facilitator can then bring suggestions back to the team without the direct watchful eye of management. This minimizes the intimidation factor but gives the team the opportunity to clarify wording or details.

We find in PHAs that the PHA team members want to be very cooperative with management. That is why it's imperative for management to not only set the expectations and tone, but also provide a supportive means and network for a high-quality PHA to be conducted. By management providing a support team and stressing the importance of a PHA during the preparation phase, the PHA team members are motivated to attend the PHA prepared with their full attention. It goes back to the old saying, garbage in, garbage out. Anyone who has sat in a PHA will attest to the opinion that PHAs are grueling and a marathon of being attentive. The PHA team members tend to be very technically sound and understand the PHA process and goal. The biggest hurdle is usually morale and minimizing outside distractions. A great opening statement by management and periodic check-ins to keep morale and encourage the team members goes a long way. Nothing lifts the moral of battle wary troops like a visit to the battlefield by the general. After the PHA is over, feedback and encouragement go a long way in instilling confidence for the PHA team members that will carry over to their next PHA. Although hardly any of this is required by the statutes, these are observations we have noticed with regards to effective management participation in PHAs.

3. Engineering – "The Devil is in the Details"

As an engineer, it is understood that your job is to ask the questions that have not been answered, review the material that has not been presented, and observe that which is not obvious. PHAs are no different. One of the duties of the engineer during a PHA is to gather all the process safety information (PSI). However, we've observed the more effective engineers will take this list, check how current it is, and review the design pressures and temperatures, such as pump discharge pressures. This review helps determine if the PSI is accurate. It also serves the dual purpose of observing any abnormalities that will need to be discussed during the PHA. Looking at this before a PHA is very effective when the engineer's eyes are fresh, and his mind is clear. Once the PHA is ongoing, this task becomes a little more complicated with lower morale and PHA fatigue. This holds true to the majority of the engineer's tasks.

Drawings, drawings, drawings. One of the largest time and money wasters we observe in PHAs is showing up to a PHA with either outdated drawings or needing drawings that are missing. It is imperative for the effective engineer to not only review the drawings to make sure they are up to date and accurately reflect what is present in the field, but it is equally important to review which drawings have been gathered and submitted to ensure that there are no surprises during the PHA. Anyone who has sat in a PHA and heard the words "these are not the most up to date drawings" or "where are all the other drawings at", will attest that it is going to be a long day. The PHA facilitator, the scribe, and pretty much everyone else in the room relies on the drawings being both accurate and complete in their preparation. If the drawings submitted for the PHA are either outdated, incomplete, or just wrong; it will have a ripple effect for the rest of the team both before and during the PHA. If the reader of this paper takes nothing away but one thing, please note that accurate and complete drawings are imperative for an effective PHA team.

PHA team members include those who are present in the room full time and other team members who may be called upon as needed. Effective engineers anticipate ahead of time what possible other resources will be needed during the PHA. For example, engineers for adjacent facilities need to be made aware of an upcoming PHA and given a heads-up as to be ready to answer questions regarding upset conditions and their effects. Likewise, engineers for skid packages or other equipment requiring expertise of vendor engineers need to also be made aware of an upcoming PHA and request their availability if called upon. Similarly, engineers of other disciplines may have better knowledge of failure mechanisms or a specific design review already conducted which is beneficial to the PHA team. It is not uncommon in a PHA to have a question that needs to be answered by a vendor engineer, subject matter experts, or an adjacent facility engineer. Examples of information needed include possible flow rates or pressure to/from receiving/feeding units, adequacy of safe guards in packages provided by vendors, or electrical area classifications. Oftentimes, a simple phone call to these engineers asking if they would be available by phone during the PHA is enough. Other times, effective engineers will require that other personnel be available to come in as requested. PHAs involve a lot of moving parts with each part having its own calendar or schedule. Giving the other possibly necessary personnel a heads up will ensure that should their presence become necessary, the PHA team will not have to recess until they are available.

Effective engineers will typically review MOCs and make phone calls or e-mails to update their current status. The adage that change is constant is very applicable to MOCs. MOCs may have been implemented, postponed, in progress, or completed. It is important to update the status of an MOC. One that has been canceled may require different attention than an MOC that is in progress or completed.

Effective engineers provide the technical expertise of the specific process under review [1]. Before a PHA, effective engineers will take a closer look to evaluate operating trends and upset conditions looking for a pattern that is not readably apparent. Engineers should discuss the facility and how it runs day-to-day with operations prior to the PHA. They should review procedures to ensure they are current and understood. It can be an awkward moment during a PHA when the engineer says, "we've never had a problem with that", and the operator rebuts with "we have a problem with that every day". Engineers know that engineered controls are the strongest defense against hazards. They look for ways to solve problems without relying on administrative controls.

Finally, effective engineers are present during a PHA. This does not mean they are sitting in a chair, it means they are focused, engaged, and completely dedicated to the PHA while they are in that room. Engineering is a life of dealing with problems. To be an effective PHA team member, it is important to make sure the engineer has support to deal with problems while he/she is in the PHA. Effective engineers can have a clear mind because they know that if normal, everyday problems occur, someone else is taking care of them. An engineer that is constantly worrying about what great surprises are awaiting them when they check their email at the end of the day, are not present during a PHA.

4. Operations – "The Ambassador of the field"

The operator is the ambassador of the conditions in the field. The operator, not engineering, management, or the facilitator, is the eyes and ears of the field. As an ambassador, it is important that the operator speak to his constituents (i.e. opposite crews), learn how to effectively communicate, and be present for the meeting of the PHA team. Effective operators speak to their coworkers and get a history of problems, concerns, or other issues they may be having that might not be well documented. Effective operators also walk around and observe conditions of the facility and document such. Documentation being key. Effective operators also review previous incidents not only in their facility, but other similar facilities or neighboring facilities to bring up any concerns that may not be readily apparent when reviewing documentation during a PHA. A great ambassador becomes the eyes and ears of his constituents by listening to what the other operators have to say and physically observing the condition of the facility.

Other things effective operators will do is review operating procedures and note deficiencies. Sometimes operating procedures are written in a way that no one understands them making them ineffective as a safeguard. Other times, there is no written procedure, forcing operations to react to conditions "the way we always do it." If operators are not following the procedures because they are vague, ambiguous, confusing, or out of date, it is up to the operator during the PHA to bring that to the attention of the team members. Frankly, effective operators understand that management and engineers assume that everything they write is perfect, clear, and easy to follow. It is up to the operator to explain why that may not always be the case. Some of the most effective operators we have encountered in a PHA not only describe what the problems are but are very specific about why these are problems.

The next characteristic of the operator as an ambassador for the field conditions is communication. For an operator to communicate with a PHA team, they must speak the same language. Effective operators know that to communicate with a PHA team, they must become a master at reading P&IDs. Being able to understand how the elements of a facility and its attributes are represented on a P&ID are extremely important for an operator. Effective operators can look at a P&ID with little or no guidance and interpret it to see if it is an accurate reflection of what is going on in the field. When the engineer asks a question regarding the P&ID, effective operators are on the same page. If the reader wanted to take away one important thing from this section on operators, please take away that an operator being a master of a P&ID is extremely important to be an effective operator in a PHA.

5. Scribe – "The Show Must Go On"

In American football, a great offensive lineman only receives attention when things do not go as planned. The same can be said about a scribe in a PHA. An effective scribe will give the outward appearance of very little responsibility or having very few complicated tasks. Nothing can be further from the truth. With the goal of quickly and accurately recording information [4], effective scribes understand that a lot of thought and preparation must go into making sure that nothing goes wrong, and if something does go wrong, the PHA suffers minimal interruption. Also, effective scribes are great at being very efficient in the tasks they complete.

At the end of the day, a scribe's responsibility is to ensure that the audio/video equipment works as planned. Although some facilities may be tasked with providing projectors and so forth, the scribe cannot scribe unless all this equipment works. Therefore, an effective scribe will oversee the equipment being used to make sure the parties who are responsible for providing it are communicating and that all the items are compatible. The scribe will ensure that both the scribe's computer and the display have the proper connectors and cables needed. An effective scribe not only brings the connectors and cables needed, but also bring the connectors and cables that were not anticipated as being necessary. An example is the conference room being used for the PHA has HDMI and ethernet and Wi-Fi…or at least that what was told to the PHA facilitator. In the past, we have shown up to conference rooms to find that the HDMI port is in fact a VGA port, or that the HDMI cable is three feet long and the scribe's position is ten feet away. Very effective scribes keep with them a toolbox of supplies such as VGA to HDMI converters, extra batteries for mice and Bluetooth keyboards, and extra cables. Sometimes, the right connectors and cables are present but are malfunctioning. The truly great, effective scribes seamlessly work through these problems and the show continues without interruption. Along those lines, a great practice for scribes to have is to not only back up the PHA file regularly, but also have a written procedure that has been reviewed regarding the backup. We always joke that we PHA the PHA. In other words, what are the risks for file corruption or data loss. Which risk are we willing to accept, and which risk are we going to mitigate? Knock on wood, so far, it's been very effective.

A lesser problem, but a problem none the less, is having to wait on a scribe to catch up. Although not as catastrophic as the PHA having to be stopped completely while technical issues are worked out, a scribe who is not able to keep up with the PHA team stifles the free flow of thought and can leave the door open for other interruptions such as taking phone calls and checking email. Very effective scribes will have experience and practice with the software, know the shortcut keys, and have a vocabulary list prepared for technical terms and names. A typical vocabulary list will be something as simple as a word document open where the scribe has already checked the spelling of names and technical equipment and as the scribe goes along, they can copy and paste the names and equipment labels very quickly without having to correct for spelling. A truly exceptional scribe will have already created templates with standard verbiage for safeguards, recommendations, and comments. For example, when the team decides to make a recommendation, the scribe can copy and paste it into the software and the team can edit the few words that need to be changed to modify the recommendation for the specific application. What this does is it allows the team to make a recommendation and move on to the next node without getting bogged down in the weeds of language choice. We find it is always easier to give a group of people a starting point for a recommendation as opposed to generating the language of a recommendation by consensus of the group. Along those lines, effective scribes review the drawings, MOCs, and other information to become familiar with the vocabulary in the drawings and have an inventory of what PSI they have and what node it corresponds to. Therefore, if a team member erroneously refers to a drawing that does not exist or refers to a drawing or MOC by incorrect name, the scribe will be quick to point out that the item does not exist in the inventory.

Effective scribes will also begin opening vital documents for a specific node as soon as the review of that node begins. Therefore, when the team would like to reference a specific document such as a drawing, MOC, or incident report, the document will be readily available. A common, effective means of doing this is having a second monitor only seen by the scribe, in addition to the screen shared with the projector. The scribe can work on his/her monitor opening documents or correcting language. As the team is ready to view specific information, he/she can drag the window over to the extended screen that is being displayed overhead. This allows the scribe to work without being micromanaged by the team and minimizes the distraction of the team from the scribe flipping through windows or searching for files. Again, the overall theme for an effective scribe is to keep the free flow of thoughts and the team moving. Anytime the team must pause or take a break because of the scribe performing its duties, the team's momentum and productivity can be lost.

Finally, truly effective scribes are great at covering the intangibles. This can include filling all the supplies or doing things to the environment in the room to lift morale. For example, effective scribes will typically have a box of random office supplies, whether it be pens, hi-liters, or post-it notes. At the beginning of each day and after each break, he/she should refill the table or the workspace of the PHA team with extras. So, if a team member would require a specific color pen, they would simply grab it and not have to pause the PHA. This gets back to the overall theme that an effective scribe keeps the evaluation moving forward without interruption. Another intangible that we often see effective scribes will do is they will place motivational posters around the room, such as large check-off list of nodes to visually see progress, to hang inspirational quotes, and so forth. PHAs are really an endurance test of morale and motivation and these little intangibles that the scribes do are of very limited cost, but of great benefit. In all honesty, we often do not realize how great and effective our scribe is until we do a PHA with an ineffective scribe. When you do a PHA with an ineffective scribe, you feel like you're stuck in the mud and going nowhere fast. Although realistically, the delay may add just a few minutes a day, with regards to morale and motivation, it feels like an eternity.

6. Facilitator – "What's Your Plan, Coach?"

Like a good coach, an effective facilitator enables great people to do great things. Effective facilitators setup the team members to be successful and, during the PHA, guide them to produce a high-quality product. Although a facilitator's duty is to ensure the methods are adhered to and the regulations concerning a PHA are complied with [1], a great PHA facilitator is so much more.

A great PHA facilitator, like a coach, ensures that the team is prepared come game day. Effective PHA facilitators will do such things as set deadlines for collection and development of information and suggest further actions to better prepare for game day. Not only do they set the deadlines, an effective facilitator will check in with the team members prior to those deadlines to ensure that progress is proceeding as planned and that the deadline will likely be met. Effective facilitators are great at helping team members overcome obstacles and hurdles in collecting this information. A team being well prepared on game day when they take their chairs and listen to management's opening remarks is no accident. Also, during the preparation, an effective PHA facilitator will ask the hard questions of the team members, such as: Do you have a transition plan for someone to fulfill your role during the PHA? Or, are there any allergies, physical, or dietary restrictions a team member may have so that these can be addressed well ahead of the PHA? To be an effective PHA facilitator, it really comes down to having the team prepare for the work they anticipate doing and addressing as many possible obstacles as one can prior to the PHA commencing. The litmus test for a PHA facilitator is simple - once the PHA commences, no surprises.

In addition to ensuring the rest of the team is preparing for the PHA, the effective facilitator puts in just as much effort as the rest of the team into preparing the PHA file, determining the node segregation, and sharpening his skills as a technical guide. On the day of the PHA, the PHA file should be prepopulated with node descriptions, drawings, common safeguards, etc. The facilitator should have reviewed the node list and noded drawings with other member(s) of the team to ensure the segregations make sense and all parts of the process are included. The effective facilitator should also review recent changes to regulations or process conditions that might affect the hazard identification. He should gain a basic understanding of the process under review along with the potential hazards specific to the facility. Effective facilitators have a list of questions to pose to the team during the appropriate node to ensure all hazards have been addressed.

Again, following the theme of the PHA facilitator being a coach, once the PHA commences, it is game day. It is the PHA facilitator's job to adhere to the methods and regulations regarding PHA facilitation. However, great, effective PHA facilitators understand that the true challenge in a PHA is morale, motivation, and focus. Effective PHA facilitators will typically start off the PHA with a "rules of the road" presentation. They do not assume that everyone in a room regularly participates in PHAs and is familiar with the process. They review the PHA process, the end goal of risk mitigation, the timeline, the daily schedule, acceptable and unacceptable behaviors, and expectations of participation. An effective PHA facilitator will go over these rules of the road and get each team member to commit to their role, to fully participate, and to strive towards the goal. It creates a sense of responsibility and obligation among the members. Once the PHA commences, the effective PHA facilitator will have a sense of when progress is slowing down and when to step in to move the team along. The effective PHA facilitator knows that the team must be given opportunity to discuss causes and consequences, but also has goals and objectives that need to be met within the time allotted.

As the team leader, the facilitator should plan on how to react to differences among team members. The great PHA facilitators have a way of getting conflicts resolved without upsetting or demeaning anyone in the room. Typically, in an argument, somebody must be wrong and somebody must be right. However, the great PHA facilitators have a way of allowing both parties to express their beliefs, guiding them to understand the beliefs of others, and concluding with a common solution. A great way to kill the free flow of ideas in a PHA and stifle productivity is to tell someone, "you're wrong, please be quiet". Great PHA facilitators know how to do that in a way that allows the person to maintain dignity and continue to perform in a PHA.

Another attribute of great PHA facilitators is that they are systematic and thorough. They have a plan of attack for each node. This usually includes the node description and limits review, a review of the appropriate drawings, written list of deviations to encourage the team, and review of MOCs, previous recommendations, and incidents related to the node. The effective facilitator also has a plan in place to limit idle time when critical information necessary to properly review a node is found to be missing. The action is documented as a follow-up for a specific team member to complete prior to the next day allowing the PHA to move forward. Any outstanding to-dos are documented and presented to the team members at the end of each day prior to dismissal. The PHA facilitator is responsible for meticulously documenting the team's findings, which cannot be done without adequate preparation of a game plan.

Finally, great PHA facilitators understand they have an obligation to future PHA facilitator to send their team off with a sense of accomplishment and pride in PHA participation. If the team members of the PHA leave with a bad taste in their mouth or feel discouraged, their participation in the next PHA may not be to the level it could be. Great PHA facilitators compliment their team members and make sure they leave the PHA on a positive note. Therefore, when the team members approach their next PHA, the new facilitator will not be dealing with baggage leftover from the previous PHA. It is a courtesy that great facilitators give to and expect from other facilitators. PHA facilitators are often asked by team members during a PHA, "how can you do this every day?" The response we often use is, "what can we do better to make you volunteer for the next PHA?" Soliciting feedback from team members besides a multiple-choice survey is imperative to be an effective facilitator. The feedback needs to be of substance that can be acted upon and improved upon. We truly love doing PHAs and at the end of every PHA, we express that to the team and solicit their feedback to what we can do better to make them truly love doing it too.

7. Summary

In conclusion, when it comes to an optimal PHA, preparation is everything. Each team member plays an important and specific role. Management sets the tone with their level of commitment to preparation, resources, and action. Engineering ensures the quality of information available to the team and contributes a deep understanding of the process design. Operations represents the field conditions, the procedures, and the most intimate knowledge of the process. The scribe provides accurate documentation through technical skill, proficiency with hardware and software, and team morale support. The facilitator leads the team to identify, evaluate, and control the hazards utilizing a systematic approach to preparation. It has been proven that a small amount of time spent preparing for each role will result in improved hazard identification, higher team morale, and reduced PHA cost.

8. References

1. Occupational Safety and Health Standards, Toxic and Hazardous Substances, Hazard Communication, Available at https://www.osha.gov/laws-regs/regulations/standardnumber/1910/1910.1200 Accessed on December 12, 2018.
2. Bureau of Safety and Environmental Enforcement, Department of the Interior, Offshore, Oil and Gas and Sulphur Operations in the Outer Continental Shelf, Available at https://www.ecfr.gov/cgi-bin/text-idx?tpl=/ecfrbrowse/Title30/30cfr250_main_02.tpl Accessed on December 12, 2018.
3. American Petroleum Institute Recommended Practice 750: Management of Process Hazards, First Edition, American Petroleum Institute, Washington D.C., 1995.
4. CCPS. *Guidelines for Hazard Evaluation Procedures*. Center for Chemical Process Safety, American Institute of Chemical Engineers, New York, NY, 2008.

551204 Building Process Safety Intelligence: What Should Every Graduating Chemical Engineer Know?

Process safety intelligence is derived from derived from a vast corpus of sources - including textbooks, handbooks, journals, conferences, incident reports, safety datasheets, and available materials and thermophysical property data. Dedicated decision-support tools enabling practitioners to harness this intelligence in their daily work, such as Knovel are an important component of chemical engineering practice today. However, many students don't realize the breadth and depth of technical knowledge needed to support process design and safety – they tend to overrely on non-specialist search, and miss pedigreed information available on specialized knowledge platforms, such as Knovel.

The undergraduate engineering space offers an excellent opportunity to plant the foundational information seeking behaviors in future process safety professionals and leaders. Knovel focuses on applied a preview of the types of insight they will be forced to leverage in industry careers, without a direct cost to students and faculty.

Common Process Safety Management Gaps Based on Occupational Safety and Health Administration Process Safety Management of Highly Hazardous Chemicals Violation Trends

Janet Persechino
Engineering Planning and Management, Inc.
959 Concord Street
Framingham, MA 01701
jap@epm-inc.com

Ryan Welch
Engineering Planning and Management, Inc.
959 Concord Street
Framingham, MA 01701
rdw@epm-inc.com

Prepared for Presentation at
American Institute of Chemical Engineers
2019 Spring Meeting and 15th Global Congress on Process Safety
New Orleans, LA
March 31-April 3, 2019

Janet Persechino
Engineering Planning and Management, Inc.
959 Concord Street
Framingham, MA 01701
jap@epm-inc.com
Keywords: OSHA, violations, regulations, process safety

Abstract
On July 17, 1990 the U.S. Occupational Safety and Health Administration (OSHA) published in the Federal Register the proposed standard *Process Safety Management (PSM) of Highly Hazardous Chemicals (HHCs)*. Approximately four months later it was enacted into law via Clean Air Act Amendments (CAAA). These amendments require that the U.S. Secretary of Labor, in coordination with the U.S. Administrator of the Environmental Protection Agency, promulgate, pursuant to the Occupational Safety and Health Act of 1970, a chemical process safety standard to prevent accidental releases of chemicals that could pose a threat to employees [1]. The full text of the final standard was issued in the Federal Register on February 24, 1992.

The OSHA standard is listed under Title 29 (Department of Labor) of the Code of Federal Regulations (CFR) as 1910.119. This standard leans towards a performance-based guidance document that outlines methods and programs to decrease the likelihood of flammable, toxic, reactive and explosive chemical releases. The intent is to reduce the risk to employees, visitors and contractors that may enter or work at chemical processing facilities.

Facilities are subject to the PSM regulation if they hold a threshold quantity of any HHC that is listed in Appendix A of the standard, with some exceptions. The violations that OSHA issues for the 1910.119 standard will cite the specific element and its relevant sub-section. Published data allows for review of the violations, and based on that information an indication of common PSM program gaps can be assessed.

A ten-year span of OSHA PSM violation data was examined (2007-2017). The data shows the top four OSHA cited PSM sections (often referred to as elements) are (highest to lowest):

1) Process Safety Information
2) Mechanical Integrity
3) Operating Procedures
4) Process Hazard Analysis

Each of these elements have known solutions, and specific corrective actions can be easily implemented given the lessons learned, the current level of expertise that exists in the process safety industry and the cost-effective software tools that are now available.

Regulatory compliance is not the only goal of adequate PSM programs, nor is it the goal of this review. The purpose of reviewing the OSHA violation information is not meant to help facilities avoid OSHA citations, but to uncover gaps found by OSHA and bring awareness and attention to some common issues, hopefully leading to focused solutions for safer, more reliable facilities.

1.0 Introduction

There are approximately 24,000 sites that are subject to the OSHA PSM standard nationwide [2]. Facilities are subject to the standard if they process any of the 130+ chemicals listed in Appendix A (of the standard) in excess of the associated threshold quantities, also listed in Appendix A, on-site in one location. They are also subject to the standard if they process flammable liquids and/or gases in quantities equal to or greater than 10,000 pounds. OSHA defines "on site in one location" as areas under the control of an employer, or affiliated employers, that have control over any group of vessels that are interconnected, or in separate vessels that are located in close enough proximity that a HHC could be involved in a catastrophic release.

Exceptions to the applicability include [3]:

- ✓ Retail facilities
- ✓ Oil/gas well drilling or servicing operations
- ✓ Normally unoccupied remote facilities; the standard defines "remote" as a facility that is operated, maintained or serviced by employees who visit the facility periodically to check its operation, no permanent employees are stationed at the facility, and is geographically remote from all other buildings and persons.
- ✓ Hydrocarbon fuel used solely for workplace consumption as a fuel if such fuels are not part of a process containing another HHC covered by the standard
- ✓ Flammable liquids with a flashpoint below 100°F stored in atmospheric storage tanks or transferred, which are kept below their normal boiling point without benefit of chilling or refrigeration

The above information outlines specific threshold quantities that trigger, and exclude, the PSM regulatory requirement. It is strongly recommended that some form of PSM be practiced at plants that may not be subject to PSM, especially those that fall into the following categories:

- Plants that store quantities of HHCs close to the threshold quantities listed in Appendix A of the standard
- Plants that utilize hazardous chemicals not listed in Appendix A of the standard
- Processes exempt from the standard, for example, those listed as exceptions in the standard, plants that have runaway reaction potential, etc.

An overview of "PSM Lite", and related fire code requirements, can be found in these conference proceedings in the paper titled *Fire Code Chemical Process Safety Requirements*. Implementation of an abbreviated form of PSM should be evaluated on a case-by-case basis which will take into account various factors that would clarify the actual risk and the residual risk that would remain after programs are incorporated into the operations.

2.0 Violation and Corrective Action Evaluation

The PSM elements are listed in the 1910.119 standard in the following order/sections:

(c) Employee Participation
(d) Process Safety Information
(e) Process Hazard Analysis
(f) Operating Procedures
(g) Training
(h) Contractors
(i) Pre-startup Safety Review
(j) Mechanical Integrity
(k) Hot Work Permit
(l) Management of Change
(m) Incident Investigation
(n) Emergency Planning and Response
(o) Compliance Audits
(p) Trade Secrets

OSHA has been administering violations for the 1910.119 standard since 1992. For purposes of relevancy, the last 10 years of violation data was examined (2007-2017). These violations were issued when OSHA completed an inspection and/or accident investigation visit to a particular facility.

Each one of the elements listed above has subsections pertaining to the requirements for that element. For example, when OSHA administers a PSM violation during a visit it is listed as: 1910119 E03 I. The nomenclature correlates to a violation for section (e) Process Hazard Analysis (PHA), part 3(i), or (3) the process hazard analysis shall address (i) the hazards of the process. This particular example indicates that the PHA did not properly address any or all of the hazards of the process. The OSHA inspection information also includes other data such as penalty dollar amounts, dates and type of inspection. The information provided in this summary reviews a few segments of this data from years 2007-2017 [4].

3.0 Violation Data Interpretation

All OSHA PSM violations were tallied over the 10-year period (2007-2017), this included:

- ✓ All citation classifications types (serious, repeat, willful etc.)
- ✓ All types of inspections/visits (accident, planned etc.)
- ✓ All types of facilities

This review does not distinguish whether violations were given during the same visit or during separate visits.

The data shows that there are 4 elements, or 4 deficiencies, that clearly stand out as the most prevalent (see Figure 1):

Most Commonly Cited PSM Elements (highest to lowest):

1) Process Safety Information (PSI)
2) Mechanical Integrity (MI)
3) Operating Procedures (OP)
4) Process Hazard Analysis (PHA)

Total Monetary Amount (highest to lowest):

1) Mechanical Integrity
2) Process Safety Information
3) Operating Procedures
4) Process Hazard Analysis

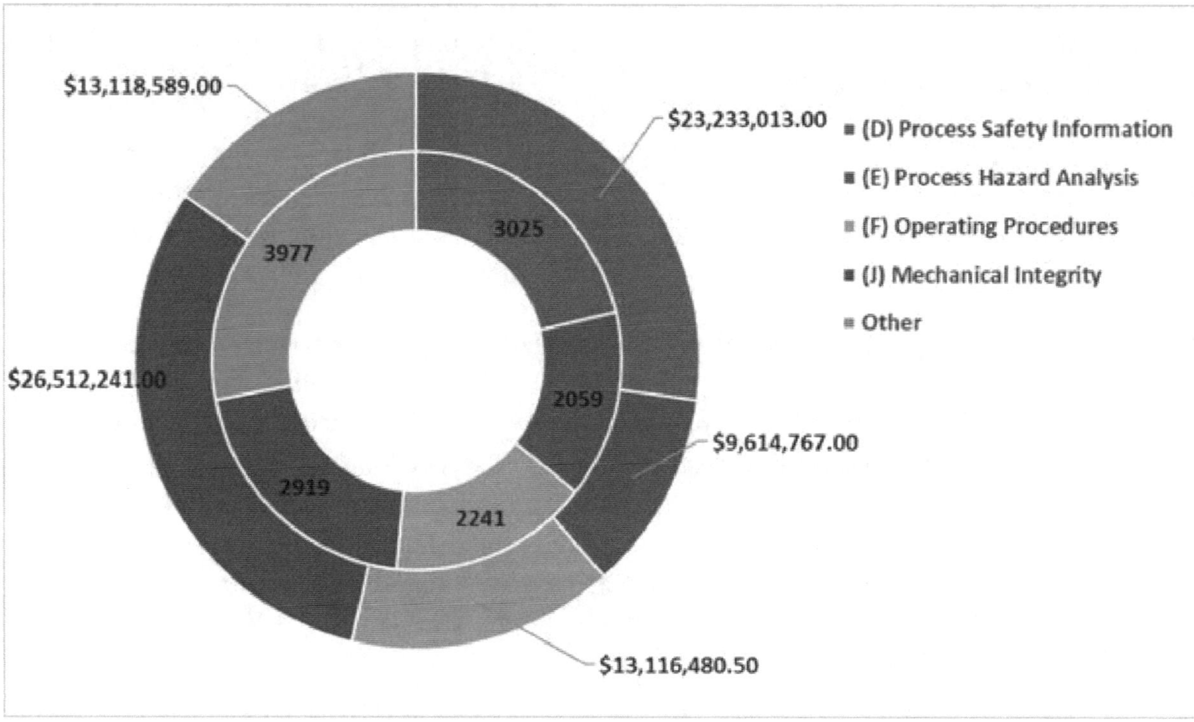

Figure 1: Total Number of OSHA PSM Violations and Total Citation Dollar Amounts by Element: 2007-2017

The top two elements (PSI and MI) have close to the same number of violations and the same goes for the third and fourth most common violations, giving the impression that the struggle to execute PSI and MI is about equal and OP and PHA both come in as close second.

It is alarming that PSI is the most cited element. Having correct and adequate PSI is the foundation for implementing, not only all PSM elements correctly, but any and all chemical safety precautions. PSI is crucial in completing an adequate PHA, the analysis that examines the hazards of the process and ensures that the proper safeguards are in place. The lack of PSI could lead to process hazards being missed, or incorrectly evaluated, during a PHA. Incomplete PSI can also lead to having inappropriate MI programs since some of these programs can be based on the hazards of the process, for example, more frequent inspections than the manufacturer's recommended practices based on the characteristics of the material being processed.

MI is the second most cited element but has the largest monetary amount, this could point to more instances of MI being noted within a single violation. OSHA will increase a fine for a single violation based on the number of "instances", or how many times they see a violation for the same element during the same visit. This review did not take into account the number of instances per violation.

MI can be difficult to implement, but new software solutions and training programs exist to help with proper execution. There are MI software products on the market that include capabilities to capture PSI, PHA and OPs, addressing all of the top-cited elements within one program.

It is helpful to utilize PSM experts for implementation and execution of PSM programs, as they are aware of the depth of information that is needed to meet the intent of the standard and for overall safety. They can often recommend products depending on your particular operation. Individuals dedicated to PSM will have access to resources that might not be available to others who are not embedded in the PSM industry. For instance, access to fee-based databases with previous incident data, one item that must be documented as part of a PHA. The expenditure for bringing in this type of expertise can provide a beneficial return on the investment when examining the average cost per violation and average number of violations per visit (Figures 2 and 3).

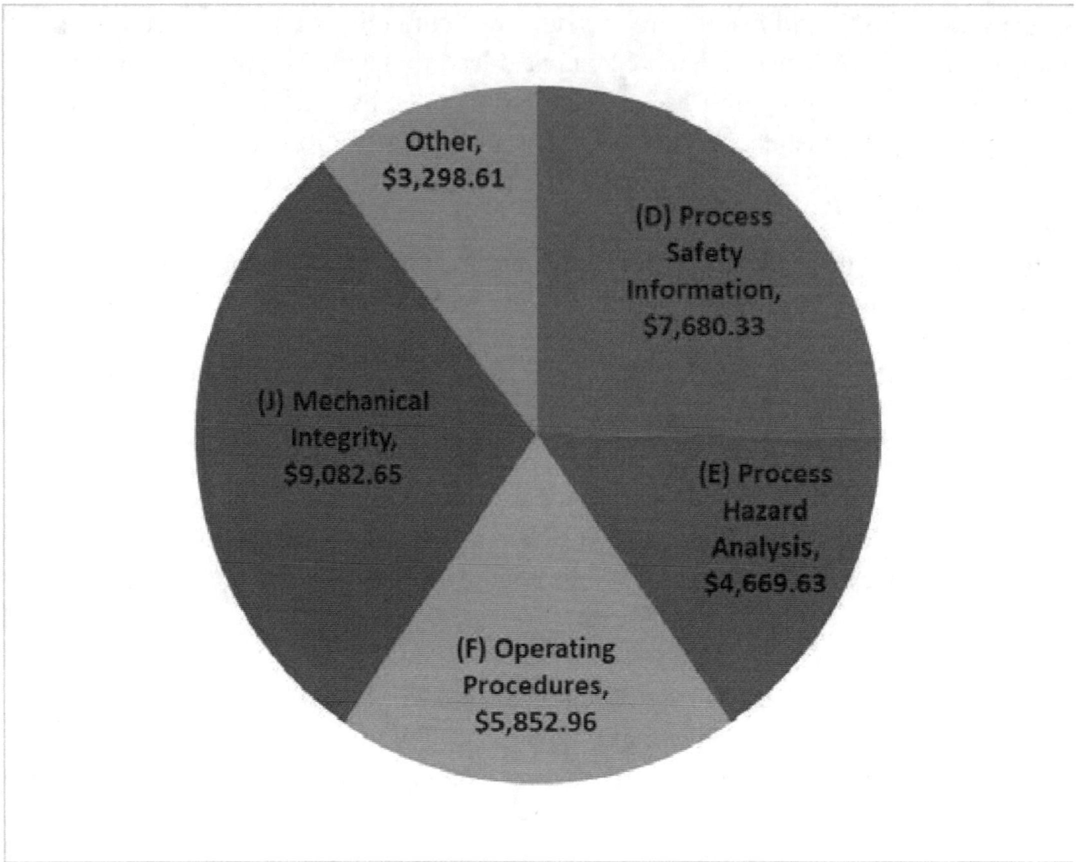

Figure 2: Average Cost per Violation by Element 2007-2017

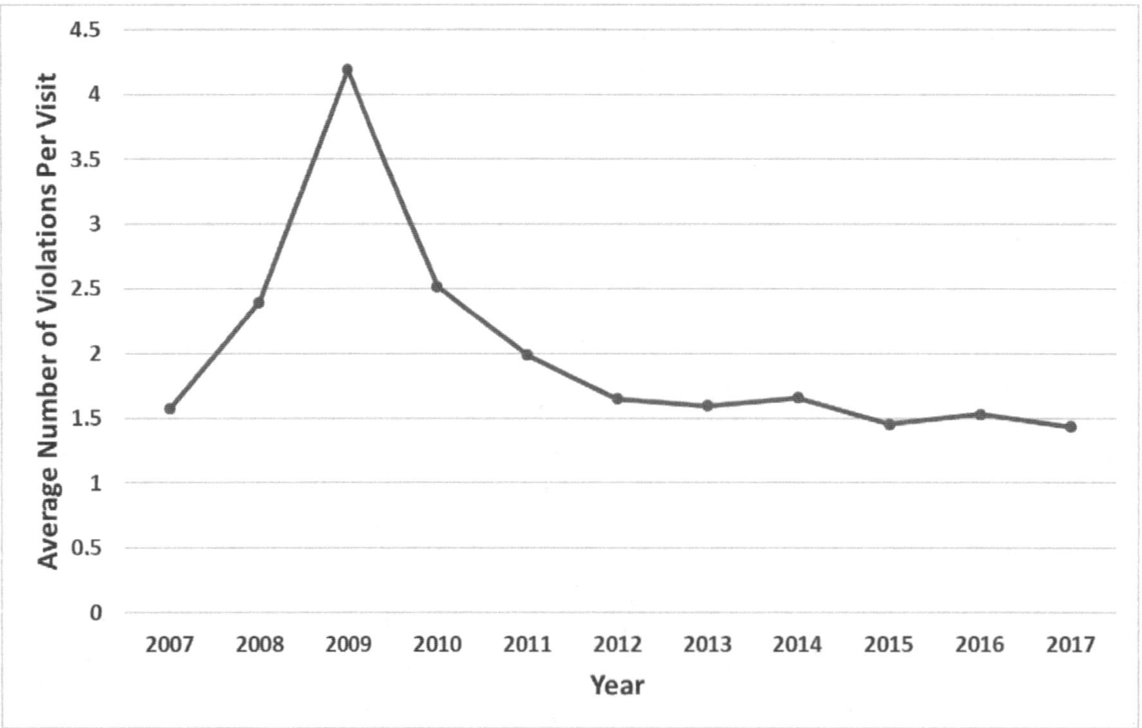

Figure 3: Average Number of Violations per Visit 2007-2017

(To note about Figure 3: the spike in the number of violations in 2009 could be due to the National Emphasis Program (NEP) in which OSHA extended to include a pilot on PSM-covered facilities from the 2007 NEP which only included refineries. The NEP went on from 2011 to only include chemical facilities and not petroleum refineries)

The citation costs are only one aspect to consider when reviewing the return on the investment. Other benefits that can be quickly realized once the programs have been established include but may not be limited to:

- ✓ Improvement in operating efficiencies
- ✓ Increased employee morale
- ✓ Lower turnover rates
- ✓ Safer working conditions
- ✓ Uptime gains
- ✓ Insurance benefits
- ✓ More consistent product quality
- ✓ Decreased likelihood of an incident

A deeper dive into the subsections of the most cited elements yielded the results in Figure 4.

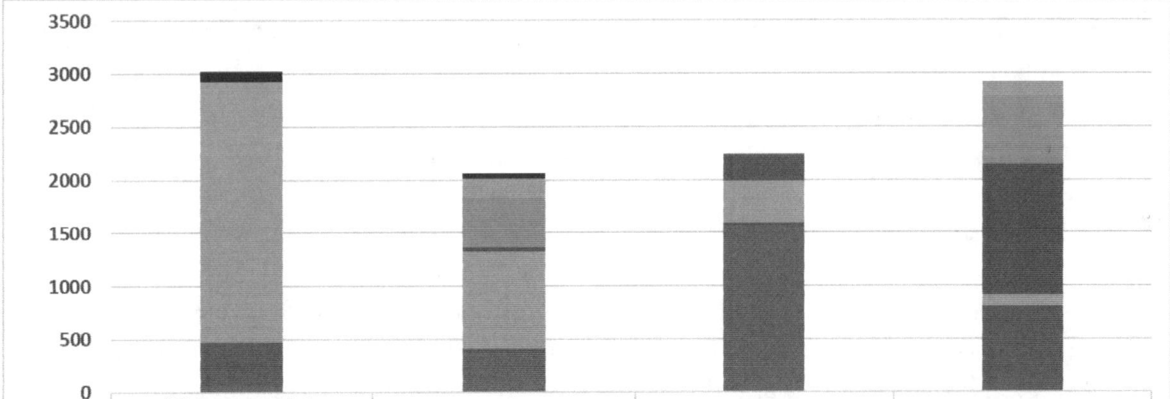

	(D) Process Safety Information	(E) Process Hazard Analysis	(F) Operating Procedures	(J) Mechanical Integrity
■ Blank	101			
■ 07		48		
■ 06		186		135
■ 05		463		640
■ 04		37	255	1235
■ 03	2452	918	391	105
■ 02	418	26	23	798
■ 01	54	381	1572	6

<u>Figure 4:</u> Sub-section Cited in the Top 4 OSHA Most Cited PSM Elements 2007-2017

Figure 4 shows that the biggest gaps over the last 10 years include the following sub-sections of these elements:

1. Process Safety Information (d)
    - Equipment Information (section 3)
2. Mechanical Integrity (j)
    - Written Maintenance Procedures (section 4)
    - Inspection and Testing (section 2)
    - Correcting Equipment Deficiencies (section 5)
3. Process Hazard Analysis (e)
    - Inadequate coverage of the hazards of the process (section 3)
4. Operating Procedures (f)
    - Develop written operating procedures (section 1)

There is a common theme among these four (4) subsection violations. While PSI was the most cited element, the most cited PSI sub-section is Equipment Information, which ties into the MI portion of PSM. Based on this, deficiencies in equipment related information and MI ranks very high on the list. MI related programs help to ensure that chemicals remain inside the vessels, pipes, etc. A review of the OSHA data shows that the industry may be struggling with proper execution of this critical piece of process safety.

As noted above, MI programs are complex and can be difficult to establish, but there are current solutions on the market to help correct these deficiencies, such as economical software solutions that improve compliance by organizing maintenance tasks. The right solution can alert staff to overdue maintenance, and store the maintenance history for each particular piece of equipment. Some tools can also track inventory of spare parts and link Management of Change documents related to the process and/or the equipment. Consultants versed in the regulations and MI techniques can help establish appropriate programs and find solutions useful to your specific operation.

Development of operating procedures for both process and maintenance activities has a high number of citations in the MI and OP elements. It is essential that these procedures are comprehensive, accessible, updated and easy to interpret. Mistakes in operating steps, or failure to recognize abnormal conditions, can quickly lead to catastrophic releases.

There are various writing and auditing techniques that can be used to ensure that these procedures include the correct information and address all stages of the process (start-up, normal conditions, abnormal conditions and shut down). Software tools currently on the market can also assist in organizing this critical information. This approach, while time consuming up-front, is time well spent in providing a safer and more efficient workplace.

Once authored, it is easier to audit the documents for accuracy and audit the operators using a behavior-based system to determine that procedures are being followed correctly.

Vast benefits can be gained using this approach, and many PSM consultants offer services to help author these documents and provide suggestions for document control and review processes.

4.0 Resources

Despite the intent for OSHA PSM to decrease the likelihood and severity of accidents, the OSHA data is showing that PSM gaps still exist in the most fundamental PSM elements. These gaps present an unacceptable risk. This is evidenced by the accidents that continue to occur. The various incident investigations performed by the U.S. Chemical Safety and Hazard Investigation Board often attribute inadequate PSM programs as the root cause.

There are many resources available for assistance with PSM implementation and improvement for plants of all types and sizes. There has been an increase in consulting organizations providing guidance and software solutions for PSM. The list of resources readily available to even the smallest of plants are much more prevalent than they were 10 years ago.

OSHA has published documents with guidelines for PSM for storage facilities and for small businesses. They also have a PSM resource page: https://www.osha.gov/SLTC/processsafetymanagement/. The Center for Chemical Process Safety has many publications, classes and tools to help with all aspects of PSM. Some of these resources are free, or provided at reasonable cost, and CCPS/AIChE (American Institute of Chemical Engineers) membership increases those benefits.

Small businesses often find it difficult to execute PSM. In 2016-2017, in almost the same number of inspections, OSHA administered 57 PSM related violations to companies with under 100 employees and only 18 violations to companies with over 100 employees. OSHA also points to a study that estimates "*employers with 1-25 employees are 47 times more likely to have a release and 17 times more likely to suffer an injury, per employee, than facilities with 1500 or more employees.*" [5]

There are a few PSM consulting firms that target small to medium companies, the companies that cannot dedicate resources to PSM, nor have the expertise in-house to meet the intent of the standard. Many of these consulting firms not only focus on OSHA compliance but becoming a partner to help embed PSM concepts into the plant's overall safety culture. Bringing in these third parties is worth the cost because they are embedded in the PSM realm and can bring along expertise from other industries, the latest information and years of experience.

There are cost effective software solutions, some aimed at small businesses from a price point and feature standpoint. These solutions can improve PSM while improving the bottom line through improved efficiency, production uptime gains, insurance benefits and violation and incident avoidance. The smaller chemical processing companies are yet to realize that these solutions exist at very reasonable prices, and the small upfront investments have very beneficial returns that easily justify the expenditure.

It is ideal to use the same consulting firm for various locations to standardize PSM solutions and audits across multiple sites that may have different risk. This type of PSM implementation helps to normalize the oversight and tracking to give a better picture of risk at the corporate level.

5.0 Conclusion

The research objective of this paper was to spot the potential for common gaps and issues in implementing PSM across the United States at all types of chemical processing industries using OSHA violation data, and to use that information to identify opportunities for solutions and preventative/corrective actions. The OSHA violation picture is only one piece of the gap analysis puzzle, but it is one place that collects enough data to increase awareness of PSM weaknesses nationwide regardless of the type of operation. This information can aid in creating education/awareness programs and campaigns aimed at improving PSM execution.

The top four cited violations are PSI, MI, PHA and OPs. Solutions to the deficiencies revealed by the OSHA data are well-known and PSM experts are aware of the many practical and feasible methods that exist to close these deficiency gaps. The return on investment on items outside of the safety aspect are usually realized soon after the solutions are put in place, including: more efficient and consistent work procedures (which can lead to improved product quality), increased up-time and insurance benefits to name only a few.

A review of the OSHA data can help quantify the costs that can be expected from OSHA PSM violations and citations. However, other impacts of violations and/or PSM incidents that are not so easily quantified include the business, legal, liability, company reputation and the impact that the injuries have on families. Sometimes plants are not able to rebuild in the same location or regain the customer share they may have lost due to an incident. Additionally, the OSHA violations are public record and can be accessed by lawyers, local authorities having jurisdiction, insurance companies, activists, etc. Companies may also find it difficult to obtain insurance due to OSHA violation or incident history. While these impacts are difficult to quantify, they must be considered in the overall cost to the business.

PSM implementation is a continuous improvement journey with new lessons learned, new technologies and new resources being created every year. Using the plethora of tools available can help close the deficiency gap.

References

[1] U.S Department of Labor OSHA 3132 Publication *Process Safety Management* 2000
[2] Washington State Department of Labor & Industries July 2010
[3] U.S Department of Labor OSHA 3133 *Publication Process Safety Management Guidelines for Compliance*, 1994
[4] OSHA Violation Data 2007-2017
[5] OSHA *Process Safety Management for Small Businesses* 2017

546725 Safety Performance Evaluation: A Benchmark of Practices across Various High-Risk Systems

The concept of performance has invaded all management spheres and imposed itself as a strong driver of decision making. Following the funding work of Kaplan and Norton (1996), performance appears as a multidimensional concept which evaluation and use require adapted tools and well calibrated decision processes.

Safety performance does not escape these requirements as it englobes various dimensions (workplace safety, process safety, environmental safety), can be approached through various evaluation paradigms (auto evaluation, external assessment through audits and KPIs) and may be used by several stakeholders (risk owner, regional and national authorities, public perceptions). Indeed, different combinations of these three dimensions allow for a large variety of performance governance paradigms carrying various understandings and practices with regard to the way performance should be assessed and used.

By comparing safety performance management in three different risk systems (process, railway and health) in Europe, this works aims at providing safety management actors with a description of the richness and variety of safety performance understandings and practices. Insights gained from this work should enhance safety managers' abilities to grasp the strengths and weaknesses of their approaches and explore the possibilities of a beneficial benchmarking.

19 AIChE Spring Meeting
15TH GLOBAL CONGRESS ON PROCESS SAFETY
NEW ORLEANS • 2019

Lessons learned from analyzing an explosion accident at Shanghai Secco chemical plant

Hanhai Dong
China University of Petroleum (East China)
No. 66, West Changjiang Road, Huangdao District, Qingdao, China
donghanhaiupc@126.com

Yi Liu
China University of Petroleum (East China)
No. 66, West Changjiang Road, Huangdao District, Qingdao, China
liuyi@upc.edu.cn

Dongfeng Zhao
China University of Petroleum (East China)
No. 66, West Changjiang Road, Huangdao District, Qingdao, China
zhaodf@vip.sina.com

Jian Chen
China University of Petroleum (East China)
No. 66, West Changjiang Road, Huangdao District, Qingdao, China
836771528@qq.com

Yanchun Wang
China University of Petroleum (East China)
No. 66, West Changjiang Road, Huangdao District, Qingdao, China
Wychun25@163.com

Wei Wang
China University of Petroleum (East China)
No. 66, West Changjiang Road, Huangdao District, Qingdao, China
15621097585@163.com

Prepared for Presentation at
American Institute of Chemical Engineers
2019 Spring Meeting and 15th Global Congress on Process Safety
New Orleans, LA
March 31 – April 3, 2019

Lessons learned from analyzing an explosion accident at Shanghai Secco chemical plant

Hanhai Dong
China University of Petroleum (East China)
No. 66, West Changjiang Road, Huangdao District, Qingdao, China
donghanhaiupc@126.com

Yi Liu
China University of Petroleum (East China)

Dongfeng Zhao
China University of Petroleum (East China)

Jian Chen
China University of Petroleum (East China)

Yanchun Wang
China University of Petroleum (East China)

Wei Wang
China University of Petroleum (East China)

Keywords: event sequence diagram; Multilayer Bow-tie Analysis; FLACS

Abstract

On May 12, 2018, six workers from Shanghai Aijinke Engineering Construction Services Co., Ltd. died on the spot of an explosion when they were maintaining a benzene tank in the tank farm at Shanghai Secco Petrochemical Co., Ltd. The accident was characterized as a disastrous accident caused by hot work in a confined space, which resulted in massive casualties and a direct economic loss of 11.66 million yuan. This paper aims to analyze the accident occurrence process in a chronological order firstly by using the event sequence diagram (ESD). Multilayer Bow-tie Analysis was adopted to explore and classify the causes of the explosion of the benzene tank, along with the corresponding consequences, and the preventive measures. The analyses show that the primary inducements were the unsafe behavior of on-site workers, hazardous conditions of in-tank substances, risky factors in the construction environment, and deficiencies in management. Casualties, fire and explosion, equipment damage, and property losses were all included as consequences. The FLACS software was used to simulate the variation of explosion pressure and explosion temperature varying with time when benzene vapor was ignited at the tank bottom during the accident, and a flow field diagram for the explosion was achieved correspondingly. The results obtained are of direct significance for investigation of similar accidents in the future and are instructive to equipment safety evaluation as well. Based on the qualitative and quantitative analyses, we emphasize three considerable reasons leading to the accident, which includes inapposite flammable gases detection points before the hot work, failing to test the flammable gases during the hot work, and deficiency of the Management of Change (MOC). Finally, the paper concludes with a summary of lessons learned from the accident as well as suggestions for accident prevention and safety improvement.

1 Introduction

The term confined space refers to closed or semi-closed facilities and places with limited entrance and exit, and poor ventilation, which may lack oxygen or contain flammable, explosive, toxic, or harmful substances. The main causes of accidents in a confined space include fire, explosion, poisoning, asphyxiation, and physical injury [1]. According to the statistics of the Ministry of Emergency Management of the People's Republic of China, from January 2012 to April 2018, a total of 64 accidents involving three or more deaths in a confined space occurred in China, resulting in 225 deaths and 82 injuries [2]. Among them, on April 8, 2014, three people were killed and two injured by an explosion during the repair and welding of the desulfurization tank at a factory in Wuhai, China. On March 18, 2015, the operators of a chemical company in Binzhou, China, suffered an explosion accident during an operation in a confined space, resulting in four deaths and two injuries. Hot work in a confined space is the common cause of the two accidents above.

Due to the poor ventilation, a large amount of flammable gas may be accumulated in a confined space. When construction workers perform hot work inside, they will face potential safety risks and problems, such as difficulty in escaping, delayed personnel rescue, and inconvenient communication. Hot work in a confined space can easily lead to fire, explosion, and personal injury accidents, which threaten not only the personal health of operators but also the property of enterprises.

Researchers have done a lot of risk analysis and exploration for the particular operating environment in a confined space. Selman et al. [3] analyzed the mortality rate of confined space operations in similarly industrialized countries from 0.05 to 0.08 per 100,000 workers and pointed out that the main risks are toxic atmospheric hazards and physical hazards. Selman et al. [4] divided the emergency rescue of confined space accidents into three levels: individual self-rescue, non-entry rescue, and progressive rescue. Botti et al. [5] applied the Internet of Things (IoT) technology and proposed a holistic approach to analyzing the procedure of confined space operations. Meanwhile, many studies adopted experimental and numerical simulation methods to observe the evolution of characteristic parameters during the diffusion and detonation of flammable gas in a confined space. The results obtained are helpful to the quantitative analysis of confined space accidents. Bang et al. [6] established a theoretical model to predict the pressure rise caused by the deflagration in a confined space. He et al. [7] presented an LES turbulence model to simulate the diffusion process of LPG in the underground confined space.

This paper mainly analyzes a disastrous accident caused by hot work in a confined space at Shanghai Secco Petrochemical Co., Ltd (Secco). On May 12, 2018, the workers from Shanghai Aijinke Engineering Construction Services Co., Ltd. (Aijinke) triggered the explosion of flammable gas when they were performing inspection and maintenance inside a benzene tank. The serious accident resulted in the immediate death of six on-site workers and direct economic loss of 11.66 million yuan. Through a series of qualitative and quantitative analysis methods, we reveal the root causes of the accident, simulate the development process of the explosion, and propose corresponding rectification suggestions. Here we want to emphasize that the reason why we write this paper is not to pursue responsibility and punishment but to reflect on how to avoid the recurrence of similar accidents.

2 The occurrence of the explosion accident

On the morning of May 12, 2018, Aijinke workers arrived at the utility tank farm of Secco to prepare for maintaining TK-75-0201 benzene tank (0201 benzene tank). Before the start of the operation, a Secco field operator used a hand-held gas detector to measure oxygen and flammable gas at the manhole outside the tank. The on-site supervisor of Aijinke, the on-site supervisor of Secco, and the on-duty monitor of Secco utility tank farm signed the work ticket successively without verifying the measurement data and checking the personal protective equipment (PPE) and explosion-proof tools of the on-site workers according to the requirements specified in the operating permit.

At 1:15 p.m., six workers of Aijinke continued to dismantle the floating pontoon in the tank and two workers worked outside for monitoring. At the scene, another Secco field operator was in charge of regular oxygen and flammable gas measurement. At 3:25 p.m., a sudden explosion occurred in the 0201 benzene tank.

After the explosion, Secco immediately launched their emergency response plan and reported the accident information to the response center of Shanghai Chemical Industry Zone Management Committee (SCIZMC), the fire brigade, the medical center, and the production operations monitoring center of Sinopec. At 3:50 p.m., the fire brigade put out the fire by spraying water. At 5:50 p.m., the on-site rescue work ended and a total of six Aijinke workers in the tank died on the spot. Figure 1a shows the scene of the 0201 benzene tank after the explosion.

Figure 1. The 0201 benzene tank and floating pontoons

2.1 Description of the timeline

The occurrence of accidents is a complex system problem, and it gives useful information on how accidents arise in practice [8]. In March 2018, Secco found that the breather valve of the 0201 benzene tank leaked slightly, which caused the VOC concentration to exceed the standard. After inspection, it was judged that the floating pontoon leaked. In order to find the root causes, we describe in detail the fundamental behaviors of the primary responsible party and the contractor prior to the explosion.

- On April 16, Secco established a maintenance order and a work list in the SAP system.
- On April 19, Secco began to empty the 0201 benzene tank, performing four days of steam dilution and eight days of nitrogen replacement.
- On April 24, Secco informed Aijinke to formulate maintenance plan.
- On April 26, Secco carried out the Job Hazard Analysis (JHA) around the maintenance work.
- From May 1 to 7, Secco opened the manhole of the tank for natural ventilation and inspection, and found that more than half of the floating pontoons had effusion.
- On May 8, the relevant departments of Secco held a special meeting and decided to change the original plan to replace all the floating pontoons.
- On May 9, Secco asked Aijinke workers to enter the tank and destructively dismantle all floating pontoons at the regular OSBL equipment maintenance meeting.
- From May 9 to 11, Aijinke workers entered the tank and dismantled the floating pontoons. A total of 38 floating pontoons were dismantled, as shown in Figure 1b.

2.2 Analysis by event sequence diagram

Event Sequence Diagram (ESD) is an accident scenario-oriented simulation method, which treats events as the basic research object, and takes into account the domino effect and risk transfer effect [9]. ESD is also an intuitive reflection of the accident scenario. It provides a more literal representation of the system state, indicating the behavior of crucial process variables, and even operator and hardware state changes [10].

In order to further illustrate and point out the critical links of the accident, this paper makes an intuitive description of events by using ESD. As shown in Figure 2, the basic events of accidents are divided into initial events, judge events, action state events, and termination events. At the same time, the path indicated by the dashed line represents the potential events that may occur in this accident, and helps to make a comprehensive and logical analysis of the whole process. The graphic symbols in the ESD are explained in the Appendix.

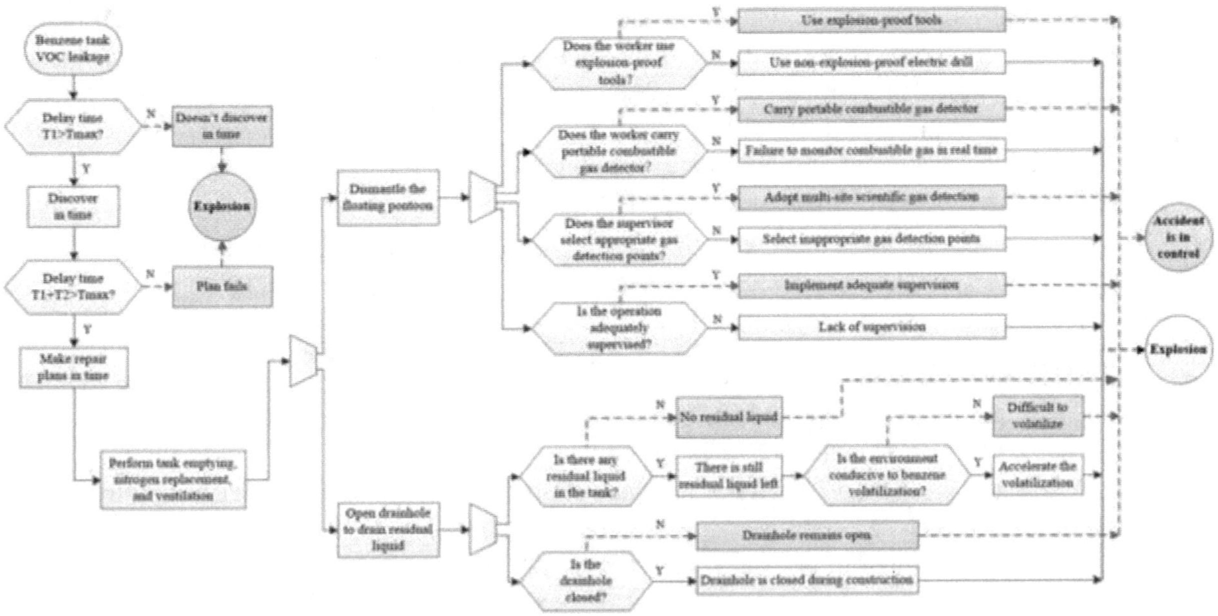

Figure 2. The ESD of the accident

As can be seen from the ESD, there are nine pivotal judgment nodes on the occurrence of the event chain, which can determine the direction of the incident. Among them, whether to discover leakage in time and whether to make a repair plan in time are the nodes at the time aspect. When the detection instrument and worker cannot find leakage and formulate the response plan in the maximum allowable delay time, they will directly lead to the explosion. Enterprises need to enhance equipment reliability and inspection frequency, so as to reduce emergency response time. Whether there are residues in the tank and whether environmental conditions are conducive to benzene volatilization are two nodes at the environmental aspect, which are also the leading causes of accidents. They require the workers to make a detailed hazard identification of the construction environment, and improve scientific detection. The critical nodes regarding human factors include whether the drainhole is closed, whether the on-site workers use explosion-proof tools, whether they carry portable combustible gas detector, whether they select appropriate detection points, and whether the supervisors take adequate monitoring [11]. Strengthening safety training before the operation, improving individual risk identification ability, perfecting risk control measures, and enhancing safety supervision level are essential means to avoid accidents caused by human factors.

All of these factors, organizational and operational, contributed to the accident, with a hierarchy of casualties among them. Through the comprehensive description, ESD can not only find out the nodes where the problem lies, but also guide the prevention of similar accidents. Therefore, enterprises should pay attention to the basic events expressed by nine judgment nodes and take appropriate interventions to block the development path of dangerous accidents.

3 Multilayer bow-tie analysis

3.1 The process of MBTA

Among many techniques available, bow-tie analysis is prevalent because it organically combines the two sophisticated risk analysis methods of Event Tree Analysis (ETA) and Fault Tree Analysis (FTA) [12], systematically summarizes the causes and consequences of accidents, and further integrates the prevention and mitigation of accidents or risks [13]. Bow-tie analysis has been proven to be a reliable and efficient tool, mainly used for process safety analysis, risk management, and accident risk analysis. It helps to understand which significant event combinations will lead to the top events, and which security barrier failures would escalate the top event to a certain consequence.

Traditional bow-tie analysis mainly includes top event, causes, consequences, preventive barriers, and mitigation barriers, which shows a lot of information. However, the single-level structure lacks logic and is not conducive to people's deep understanding. Thus, this paper establishes a Multilayer Bow-tie Analysis (MBTA), which resolves the top-event-related ETA and FTA individually into hierarchical structures, to deeply analyze the root causes, specify the severe consequences, and propose the corresponding safety barriers and scientific disposal methods. MBTA uses graphics to visually display the relationship between different levels and avoid repeated analysis effectively. Figure 3 shows the MBTA model, in which the code explanations are listed in Table 1.

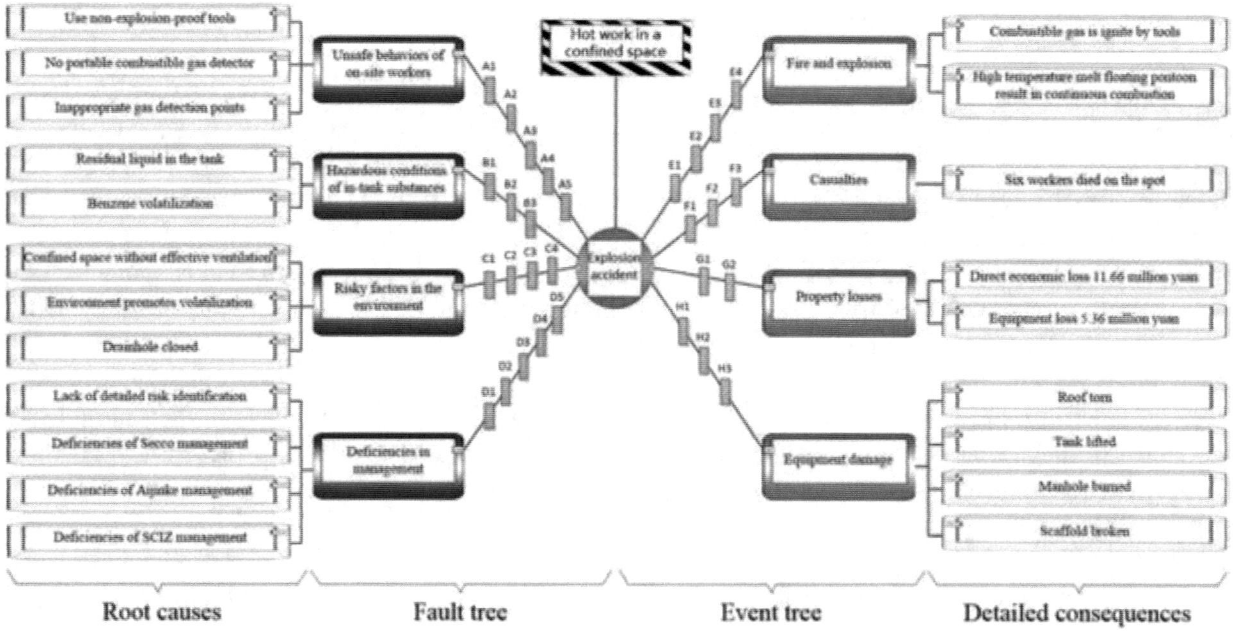

Figure 3. The MBTA model

Table 1. The explanations of MBTA

Code	Safety barrier	Code	Safety barrier
A1	Use explosion-proof tools	D4	Implement the main responsibility
A2	Carry portable combustible gas detector	D5	Improve contractor management
A3	Strengthen operation training	E1	Call the fire brigade in time

A4	Enhance supervision and assessment	E2	Spray water to extinguish fire
A5	Set up warning signs	E3	Fire isolation
B1	Remove the residual liquid	E4	Launch the emergency plan
B2	Hazardous material isolation	F1	Call for medical treatment
B3	Combustible gas detection in the tank	F2	Search and rescue the wounded
C1	Forced ventilation	F3	Launch the emergency plan
C2	Open drainhole	G1	Insurance claim
C3	Continuous detection during operation	G2	Launch the emergency plan
C4	Take environmental risk assessment	H1	Dangerous equipment isolation
D1	Compliance with rules and regulations	H2	Equipment repair
D2	Increase the intensity of punishment	H3	Equipment replacement
D3	Strengthen process safety management		

In the above MBTA model, we adopt the analytic hierarchy process (AHP) to extract the common characteristics of 12 causes of accidents and classify the causes into four categories: unsafe behaviors of on-site workers, hazardous conditions of in-tank substances, risk factors in the construction environment, and deficiencies in management. Moreover, we could conclude the direct and indirect reasons for the accident.

3.1.1 Direct reasons

Benzene is a colorless and transparent liquid with a strong aromatic odor. It has a boiling point of 80.1 ℃, a saturated vapor pressure of 10 kPa at 20 ℃, a flash point of -11 ℃, an explosion limit range of 1.2%-8.0%, and minimum ignition energy of only 0.20 mJ. Benzene is a highly flammable substance. The explosive mixture of benzene vapor and air can burn and explode when exposed to ignition source or high temperature.

At the construction site, the temperature is 24.7 ℃, and the pressure is 100.68 kPa. Under these conditions, benzene is easy to volatilize and form a gas explosion environment in a confined space. During the process of dismantling, residual benzene in the floating pontoons leaked onto the bottom of the tank and was not cleaned up in time. Because there was no effective ventilation in the tank, benzene vapor mixed with air to form an explosive environment and benzene concentration reached the explosion limit. The non-explosion-proof tools used by the operators produced ignition energy, which brought about the deflagration of the flammable gas. The high temperatures generated by the combustion melt the other aluminum alloy pontoons, resulting in sustained combustion for a short period.

3.1.2 Indirect reasons

Aijinke failed to abide by the relevant safety production rules and regulations. Managers did not fulfill their duties of safety management and supervision. Aijinke failed to give safety technical information to workers following the operating rules before the construction. Without implementing and supervising PPE, carrying portable gas detector, and using explosion-proof tools, operators were arranged to work in a confined space. When significant changes occurred in the job conditions, Aijinke did not promptly request workers to stop working and implement the Management of Change (MOC).

Secco had some omissions in safety risk management, professional management, and special operation management. In addition, Secco did not follow the safety production rules and regulations. On-site inspectors failed to carry out gas detection in accordance with specifications. After the change of operation, the manager did not ask the contractor to stop the process in time and revise the construction plan according to the regulations. Secco failed to report the maintenance plan to SCIZMC as required.

SCIZMC failed to implement the safety management system fully and did not discover that Secco performed maintenance operations without permission.

3.2 Lessons learned from MBTA

This paper puts forward a series of preventive measures, such as strengthening operation training, removing residual benzene in the tank, improving risk assessment of the environment, and enhancing process safety management. It is noticed that the environment in a confined space is harsh and there are many potential unsafe factors. Furthermore, the contractor's maintenance work cycle spans the statutory weekend. Based on the two reasons above, the enterprise should upgrade the level of hot work and adopt a broader system of operation approval and management. On-site workers need to stop working immediately when safety problems are found in the construction process and report to their superiors by relevant regulations.

Through refining the consequences of accidents, we summarized that detailed safety emergency plans and energy isolation methods are universally applicable in various measures to control and mitigate the impacts of accidents. They require Secco and Aijinke to formulate a comprehensive emergency response plan before, including fire extinguishing, personnel search and rescue, equipment maintenance, property damage remediation and so on. When an accident occurs, workers have to stay away from dangerous substances and places immediately.

As with many disasters, what ended up in a sequence of equipment damage and human errors started mostly as a management problem. Many unsafe practices were tolerated and even encouraged. The risk management of confined space operations should be divided into ten steps according to relevant regulations, namely training, pre-operation planning, job approval, hazardous energy isolation, ventilation, gas detection, personal protection and tool safety, warning signs, monitoring, and emergency response, so as to strengthen process safety management comprehensively. It is necessary to enhance the issuance of management of work tickets for hazardous operations such as hot work and confined space operations. Enterprises must strictly inspect illegal operations and supervise safety precaution measures to ensure that the operation process is safe and controllable. For changes occurring in the construction process, enterprises ought to strictly implement the MOC, analyze the potential risk and possible accidents, formulate corresponding safety measures, and carry out safety education for all operators.

MBTA provided a detailed analysis of the accident and established the corresponding safety barriers. However, there are still some certain subjective problems, and its content should also be flexible. Therefore, we encourage employees, accident investigators, case learners, and other relevant personnel to improve and develop the content of MBTA continually, thereby effectively expanding the application scope and effectiveness of MBTA.

4 Quantitative simulation of the explosion

The analysis of the accident should include not only qualitative analysis but also quantitative analysis [14]. We use FLACS software to simulate and analyze the occurrence of the benzene explosion. According to the relevant information, the 0201 benzene tank is an internal floating roof tank with a volume of 10,000 m^3. Its diameter is 30 m and height is 19 m. The design temperature range and pressure range are -10 – 65 ℃ and -0.49 –1.96 kPa, respectively. The tank adopts box-type aluminum alloy internal floating pontoons, and a manhole with a diameter of 0.6 m is set at the position of 0.75 m above the ground. Figure 4 shows the 0201 benzene tank model.

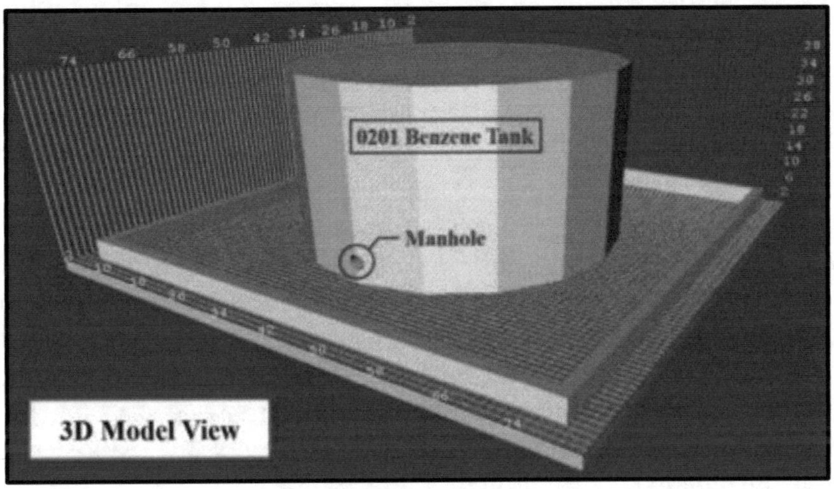

Figure 4. The 0201 benzene tank model

The simulated environmental conditions are fixed at the temperature of 24.7 ℃ and the pressure of 100.68 kPa, which are the same as the construction environment. We make reasonable assumptions: (a) The residual benzene in the tank evaporates continuously into the confined space and forms a uniform mixture of flammable gas. When the volatilization concentration of benzene reaches the lower explosion limit (LEL), the ignition source that persists during the construction operation detonates the flammable gas and results in an explosion. (b) Six on-site workers dismantled the float pontoons at the bottom of the tank, so the ignition source is set at the height of 0.75 meters from the geometric plane center, as shown in Figure 5 point I. There are several detection points of pressure and other characteristic parameters in the model of benzene tank, as shown in Figure 5, points M1-M5. (c) The benzene tank is windless, and the flammable gas accords with the real gas state equation. (d) The gas explosion process satisfies a series of fundamental equations, such as the continuity equation, momentum equation, and energy equation.

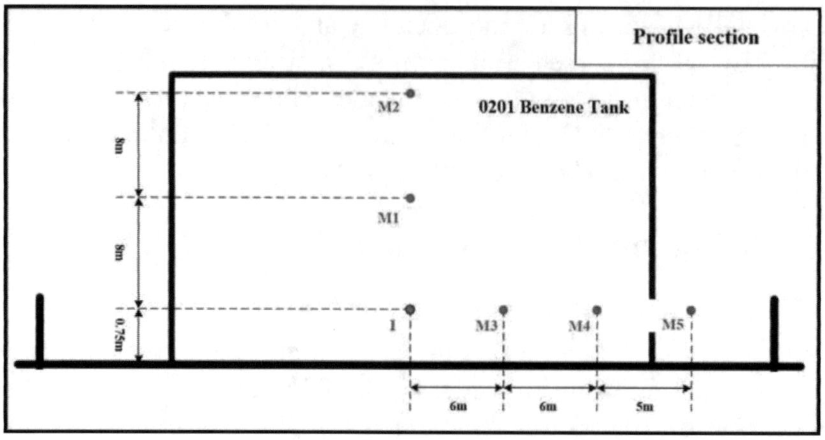

Figure 5. The profile section of the model

We selected the maximum profile section of the tank as the research object and intercepted the temperature field and pressure field at four different moments in the explosion process, as shown in Figure 6 and 7. The simulation results intuitively describe the development process of the explosion. When the flammable gas in the tank was ignited, the flame diffused around in the form of a spherical wave and reached the bottom of the tank at 0.35 s, as shown in Figure 6a and 7a. Since the radial distance of the tank is less than the vertical axial distance, the flame propagated horizontally to the wall, as shown in Figure 6b and 7b. At 1.52 s, the flame reached the manhole and sprayed out to the outside of the tank in the form of jet flow. During this period, the pressure and temperature increased continuously, which are the leading causes of the manhole being burned, as shown in Figure 8a. When the flame propagated to the top of the tank at 2.11 s, the uncombusted gas was recirculated by the viscous effect and blocking effect of the wall, which further caused the recirculation vortex. The recirculating vortex enlarged the contact area of the flame front, resulting in flame acceleration and pressure increase, as shown in Figure 6c and 7c. Figure 6d and 7d illustrate the flammable gas burned out at 3.24 s. Simultaneously, the explosion temperature and pressure reached the peak value contributing to the tearing of the roof and the lifting of the tank, as shown in Figure 8b and 7c.

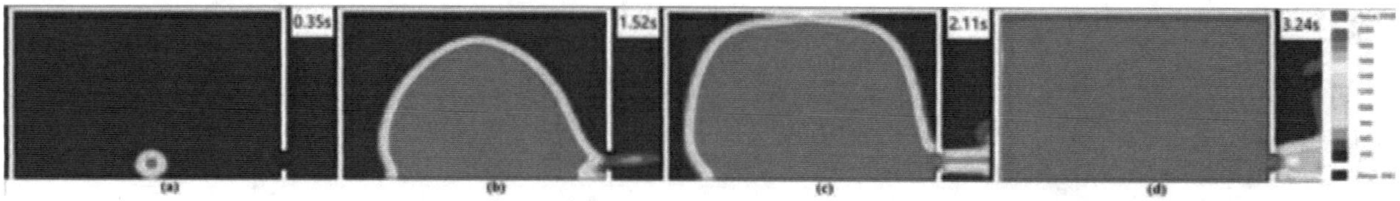

Figure 6. Temperature simulation of the explosion

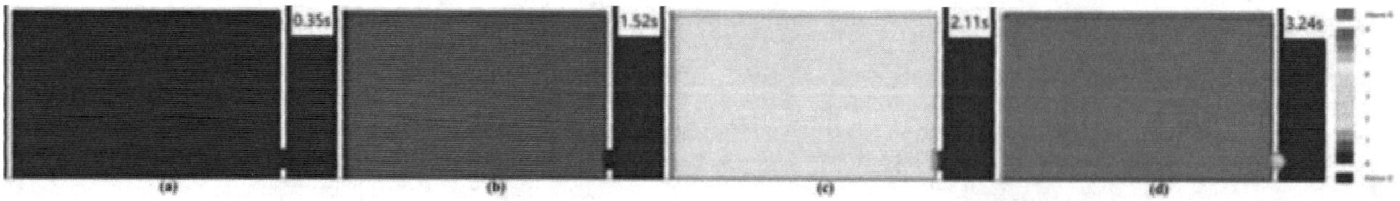

Figure 7. Pressure simulation of the explosion

Figure 8. The results of the accident

Figure 9 reveals the change of temperature and pressure recorded by the detection points M1-M5 during the explosion. Table 2 describes the data of the detection points, including maximum and minimum explosion pressures, maximum and minimum explosion temperatures, peak data arrival times and so on. Combining the data of detection points and the pressure field diagram in Figure 7, it is shown that the pressure in a confined space will be uniform immediately for a certain time and further proves the accuracy of the analysis of the flame propagation process. In particular, the external detection point M5 was affected by the propagation of explosion flame in the tank at 1.6 s, resulting in a peak explosion overpressure of 0.031 bar and a peak temperature of 1895 ℃. Subsequently, due to the pressure rise in the tank, the pressure at the detection point M5 dropped to negative pressure until the flammable gas burned out.

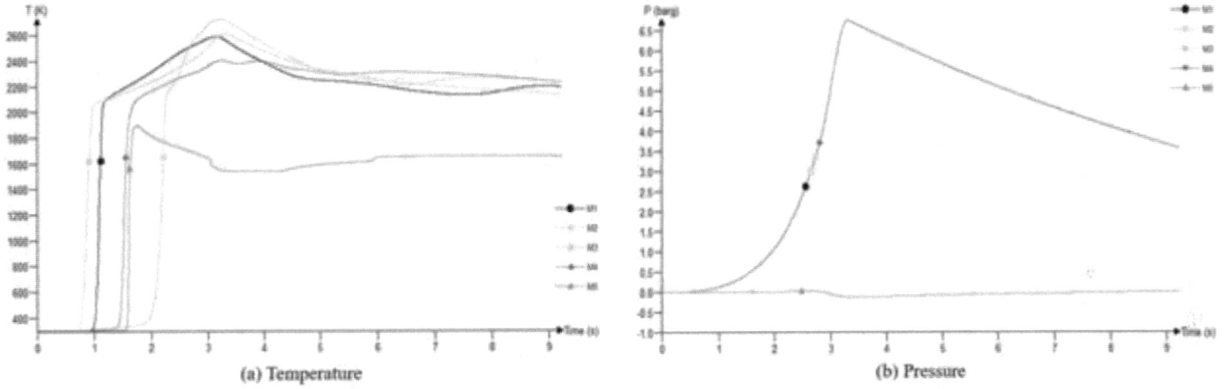

Figure 9. Parameters change of the detection points

Table 2. Data from the detection points

No.	Temperature				Pressure			
	Tmax/K	Time/s	Tmin/K	Time/s	Pmax/bar	Time/s	Pmin/bar	Time/s
M1	2589.4	3.11	298.1	0.04	6.75	3.24	0.00	0.04
M2	2732.2	3.24	298.1	0.04	6.76	3.24	0.00	0,04
M3	2609.3	3.28	298.1	0.00	6.76	3.24	0.00	0.00
M4	2306.2	3.25	298.1	0.00	6.75	3.24	0.00	0.00

M5	1895.5	1.68	298.0	0.69	0.05	1.65	-0.15	3.29

Although there is only a small amount of residual benzene, the benzene volatilization forms a flammable gas with the air inside the tank, which generates a substantial destructive power when ignited. The simulation results show that the flame propagation velocity of the flammable gas increases constantly, with the peak value reaching up to 41 m/s. Since the flame is propagated in the form of subsonic velocity, it can be judged that there are mainly the phenomena of deflagration occurred in the process of explosion [15]. Soloviova-Sokolova et al. [16] calculated the flame burning rate of the benzene vapor was 0.6 m/s at 298 K, 0.1 MPa and 0.6 eq., which clearly illustrated that the peak propagation velocity of the benzene vapor in a confined space would be several times higher. Meanwhile, the explosion pressure in a confined space is usually greater than that in an open area due to lack of heat loss and volume expansion.

When workers are engaged in hot work in a confined space, they should carry out comprehensive risk identification, select several representative flammable gas detection points, and detect the flammable gas in real time. Workers ought to stop work immediately and withdraw in an orderly manner once the flammable gas exceeds the standard. Through the analysis of the explosion process, the flame propagates to the wall in a very short time. Thermal radiation and explosive shock wave are the primary causes of casualties in the tank. Owing to the special working environment in a confined space, the phenomenon of mass casualties and injuries often occurs under the effect of high temperature and high pressure, which needs people's sufficient attention.

5 The subsequent event

The explosion accident of Secco has aroused widespread concern, and 22 responsible persons have also been punished. But unfortunately, lessons from the accident did not prevent a subsequent similar event. On November 26, 2018, three contractor workers of the Ningbo Engineering Company (NEC) and one worker of Secco inspected the discharge tank of cracking gas compressor in the olefin separation unit. Without ascertaining the concentration of oxygen in the tank, a contractor worker of NEC entered the tank for construction and fainted unexpectedly. A Secco worker immediately went into the tank to rescue him, but he also fell into the tank. The accident resulted in the death of two workers who had been rescued after asphyxia.

It is the second confined space operation accident in the same company within six months. The main causes of the accident are still the deficiency of contractor management and risk identification in a confined space. While expressing sadness, the public should profoundly reflect on why accidents continue to occur, what is the role of punishment and accountability, and how relevant regulations and reform measures can be truly implemented. Drawing lessons from the previous accidents is necessary. Enterprises ought to raise the safety awareness of their employees and use strict supervision to ensure safe production of chemical plants.

6 Conclusions

This paper provides an introduction of an accident caused by hot work in a confined space, together with the detailed ESD-based description of nine pivotal judgment nodes leading to this disastrous accident. It concludes that enterprises should take suitable intervention measures to block the pathway of the accident. The MBTA, covering the causes and consequences of the accident, offers proactive and reactive measures based on an intuitive analysis. Meanwhile, it emphasizes the importance in selection of appropriate detection points, real-time detection of in-tank operation, and implementation of MOC. Also, we apply the FLACS software into quantitative simulation of the explosion process in the tank. The results contribute to the guidance of accident investigation and safety measures selection. Finally, it suggests that enterprises adopt technical remedies and better safety management practices, since they are the essential basics for effective implementation of risk control measures.

Appendix

Symbol	Annotation
	Initial event: Beginning event of the ESD.
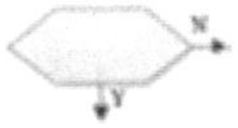	Judge event: The event has two possible outcomes that determine whether subsequent events occur.
	Action state event: Describe the action state of the event.
	Termination event: Ending event of the ESD.
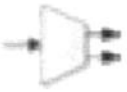	Output AND gate: Input event lead to multiple output events occurring.

7 References

[1] D.Burlet-Vienney, Y.Chinniah, A.Bahloul and B.Roberge, Design and application of a 5 step risk assessment tool for confined space entries, Safety science 80 (2015), 144-155.

[2] The Office of the State Council Security Committee Notified 8 Cases of Blindly Rescuing in a Comfined Space Leading to Casualties Enlargement, Available at http://www.chinasafety.gov.cn/default.htm Accessed on June 8, 2018.

[3] J.Selman, J.Spickett, J.Jansz and B.Mullins, Safety science 109 (2018), 333-343.

[4] J.Selman, J.Spickett, J.Jansz and B.Mullins, Safety science 113 (2019), 78-90.

[5] L.Botti, V.Duraccio,M.G.Gnoni and C.Mora, A framework for preventing and managing risks in confined spaces through IOT technologies, ESREL (2015), 3209-3217.

[6] B.Bang, C.Ahn, J.Lee, Y.Kim, M.Lee, B.Horn, D.Malik, K.Thomas, S.James, A.Yarin and S.Yoon, Theoretical, numerical, and experimental investigation of pressure rise due to deflagration in confined spaces, International Journal of Thermal Sciences 120 (2017), 469-480.

[7] B.He, X.Jiang, G.Yang and J.Xu, A numerical simulation study on the formation and dispersion of flammable vapor cloud in underground confined space. Process Safety and Environmental Protection 107 (2017), 1-11.

[8] I.M.Shaluf, F.Ahmadun and A.M.Said, Fire incident at a refinery in West Malaysia: the causes and lessons learned, Journal of Loss prevention in the Process Industries 16.4 (2003), 297-303.

[9] S.Swaminathan and C.Smidts, The event sequence diagram framework for dynamic probabilistic risk assessment, Reliability Engineering & System Safety 63.1 (1999), 73-90.

[10] J.Zhou, G.Reniers and N.Khakzad, Application of event sequence diagram to evaluate emergency response actions during fire-induced domino effects, Reliability Engineering & System Safety 150 (2016), 202-209.

[11] E.Zarei, M.Yazdi, R.Abbassi and F.Khan, A hybrid model for human factor analysis in process accidents: FBN-HFACS, Journal of Loss Prevention in the Process Industries 57 (2019), 142-155.

[12] N.Khakzad, F.Khan, P.Amyotte, Dynamic risk analysis using bow-tie approach, Reliability Engineering & System Safety 104 (2012), 36-44.

[13] M.B.Mulcahy, C.Boylan, S.Sigmann and R.Stuart, Using bowtie methodology to support laboratory hazard identification, risk management, and incident analysis, Journal of Chemical Health and Safety 24.3 (2017), 14-20.

[14] C.Zhu, J.Zhu, L.Wang and M.S.Mannan, Lessons learned from analyzing a VCE accident at a chemical plant, Journal of Loss Prevention in the Process Industries 50 (2017), 397-402.

[15] I.C.Tolias and A.G.Venetsanos, An improved CFD model for vented deflagration simulations–Analysis of a medium-scale hydrogen experiment, International Journal of Hydrogen Energy 43.52 (2018), 23568-23584.

[16] J.V.Soloviov-Sokolova, V.A.Alekseev, S.S.Matveev, I.V.Chechet, S.G.Matveev and A.A.Konnov, Fuel 175 (2016), 302-309.

Strengthening Process Safety Culture throughout Learning from Each Other´s Vulnerabilities

Luciana Poli SIlva
PETROBRAS
Av. Republica do Chile, 65, 4th floor, aisle 402.
lucianapoli@petrobras.com.br

Prepared for Presentation at
American Institute of Chemical Engineers
2019 Spring Meeting and 15th Global Congress on Process Safety
New Orleans, LA
March 31 – April 3, 2019

Strengthening Process Safety Culture throughout Learning from Each Other´s Vulnerabilities

Luciana Poli Silva
PETROBRAS
Av. República do Chile, 65, 4th floor, aisle 402.
lucianapoli@petrobras.com.br

Keywords: andragogy, dialogue, leadership, safety leadership culture, trust, vulnerabilities.

Abstract

A fundamental approach to any program that targets the implementation of a process safety culture is to work within reality, going beyond procedures and compliance. To facilitate the emergence of the unspoken is essential to stimulate the dialogue among leaders that brings about their vulnerabilities and opens new perspectives to address them effectively. This paper presents a pilot program designed to increase the Process Safety Leadership Culture in a Brazilian Energy Company. Based on a benchmark made within the industry and in andragogic principles, the program created a trust ambiance among peers and inspired leaders to recognize and share their vulnerabilities, as well as to take actions towards the strengthen the role of being a safety culture leader.

1 Introduction

A lot has been said and written about Leadership Process Safety Culture, but there is a gap between what the compliance procedures dictate and what managers really do. When it comes to the Brazilian Energy Industry, the pressure leaders are under in order to achieve the results they are expected to can be considered significant. How to balance the production challenges with a safety culture that must be among the core values of the operation? Are these factors compatible?

The pilot program entitled '*Strengthening Process Safety Culture'* dealt with reality, using andragogic approaches to bring about the complex factors forty managers face in their routine, such as the existence of several levels separating top managers from the operational team, privileging safety over production goals, and being an effective representative of the Company´s policies. All these factors can jeopardize the safety culture from becoming stronger every day.

This paper describes the implementation of a methodology designed to create a safe environment, in order to build trust among participants and at the same time, make them reflect and dialogue about how they can evolve from compliance to commitment.

In the coming sections, we will present the central concepts used to develop the program – Culture, Safety Culture and Leadership - exploring its implication on an actual work environment.

In the sequence, we will talk about what it takes to make the unspoken, unwritten rules emerge. From there, how we stimulated the increase in the self-conscious level as well as in the collective one, building a matrix entitled 'Personnel Influence on Safety' to diagnose how one really sees himself concerning his leadership in safety culture.

We will then present the method used to address the issues managers pointed out, in order to transform the discussions into actions to strengthen the safety process leadership culture.

In the conclusion, we will present our findings and discuss possible future approaches to amplify the safety leadership culture further.

2 Safety Culture and Leadership: definitions and implications

There are many possible definitions for culture. Taylor (2015) says culture has to do with the behaviour patterns that are encouraged, discouraged or tolerated over time in an organization. Schein (2015) says that culture is "the total sum of everything an organization has learned in its history in dealing with the external problems – which would be goals, strategy, how we do things – and how it organizes itself internally."

Since we humans are gregarious social beings, we are naturally wired to fit in with our community. We use our perceptions to read what the tribe accepts and we adapt by behaving accordingly, because we want to fit in and be rewarded. Therefore, we reinforce the existing norms. This process is perpetuated over time; established behavior influences the behavior of new employees and so on.

Most cultural sources are non-verbal. Taylor (2015) says it is a myth to believe that culture reflects the values statements that are declared in the company's annual reports. Most commonly, the statement will be the organization´s intents, not the culture as it actually is.

If we want to change culture, it is necessary to have a clear direction and a focused, effort-driven leadership to lead people towards the goal culture. Culture has to do with what is encouraged and allowed. For Taylor (2015), three major sources send cultural messages: behaviours - of other people, especially important, recognized ones; symbols - artefacts, events and decisions that are meaningful to the community; and systems - mechanisms that manage people and tasks.

In this sense, culture has to do with the manifestation of what is valued in the organization.

According to Lafraia (2014), culture shapes reality. Cultural lenses structure the perception of the people involved in the organization.

When it comes to safety culture, its meaning may vary from one organization to another, or even in different departments of the same company. It is perfectly possible to have diverse lenses that translate the meaning and the importance of safety practices in diverse ways.

OGP, in its 452 Report (2013), says that a positive Safety Culture has to do with a culture in which safety is a core value for everyone who works in the organization, and it is pursued in an everyday basis. In order to build a Safety Culture, there are factors that must be considered, such as "leadership commitment to safety; employee involvement and motivation; responsibilities and accountability; production and bottom line pressures versus quality issues; and actions, or lack of action, to correct unsafe behaviours and unsafe conditions"; among others.

In order to build a consistent safety culture, leaders must educate their teams by demonstrating, by their practices, that some policies are fundamental. Safety culture should use this lens, rather than a command and control opaque one. Leaders must educate people to understand why safety is a core value that concerns the most precious gift: one´s life.

Safety culture is not about only following the rules, but making the team understand that the rules serve a greater principle. This approach can influence values and beliefs and change the individual and team´s behavior. The challenge leaders face is to conciliate the educational role with the policeman role, guaranteeing that compliance is required and everyone in the organization, starting by himself, values that as an unnegotiable principle.

In order to develop a strong safety culture, some elements must be considered (OGP, 2013):

Elements of a Strong Safety Culture

- an informed culture: the company has quality data concerning safety and uses it well;
- a reporting culture: people feel they can report safety concerns without worrying about being punished;
- a learning culture: the organisation actually learns from its unsafe conditions and changes them, banning any equipment, practice or whatever takes place that hazards safety;
- a flexible culture: the company changes its command chain if necessary to adapt;
- a just culture: people understand what is acceptable and unacceptable. What is not tolerated is dealt fairly.

For Peter Drucker (Cohen, 2018), leadership can be defined as the lifting of a man's vision to higher sights. Throughout leadership, people perform in higher standards, they are inspired and evolve in multiple levels.

When it comes to safety leadership, we can quote Ram Charan (2007) when he says that "a leader who does not produce other leaders is not a great leader." A safety leader should act in a way to enhance in everybody´s minds and actions a safety leadership commitment.

In order to make it possible, it is crucial that the leader himself is aware of his beliefs and constantly monitoring his behaviours. It takes time and a lot of effort to build a strong safety culture. Despite that, a safety culture is fragile and easily threatened by wrong messages. (OGP, 2013)

In the next section we explore the method used to address self-consciousness about safety.

3 Creating a trust environment that favors Safety Process Leadership Culture

According to Malcon Knowles, who develop the paradigm of andragogy, adults have a learning style characterized, among other factors, by changing one´s concepts according to new experiences one lives; the pursuit for autonomy; increasing one´s own motivational level to match the social role assumed; and acquiring knowledge to immediately solve practical issues (Muresan, 2013).

Considering this practical approach, we began the program to strengthen safety leadership culture by making each participant reflect on its daily practices concerning management, as well as the safety implications involved in routine decisions.

To start this reflection, we did a benchmark and decided to invite a senior top manager from another company in the Energy Industry to give a testimony. He made an honest statement about his team perception of his management practices. Concerning safety process, he mentioned that he realized that, in order to influence people, he must first change himself. He then shared with an amazed group of participants some practical examples of safety interventions he changed in his auditing procedures, for example, in order to be coherent with the message he wanted to pass forward.

The testimony had an impressive impact on the managers. A remarkable phrase that was repeated by managers was: "*Simplify. Be the Powepoint. Less Presentations, More You.*"

After the testimony, there was a presentation on the invited company's safety philosophy, about its journey into safety leadership. The practical steps mentioned were followed by videos with managers from different parts of the world talking about the way they see safety culture benefits.

With that background environment set, we forwarded to a self-perception diagnosis. The goal was to stimulate the increase in the individual conscious level as well as in the collective one. We used a matrix entitled 'Personnel Influence on Safety' that was obtained in the benchmark process with the above mentioned company.

It aimed to diagnose how one really sees himself concerning its leadership in safety culture. It involved two dimensions: the personal influence towards oneself; and the personal influence concerning other people´s safety.

The figure below illustrates the matrix:

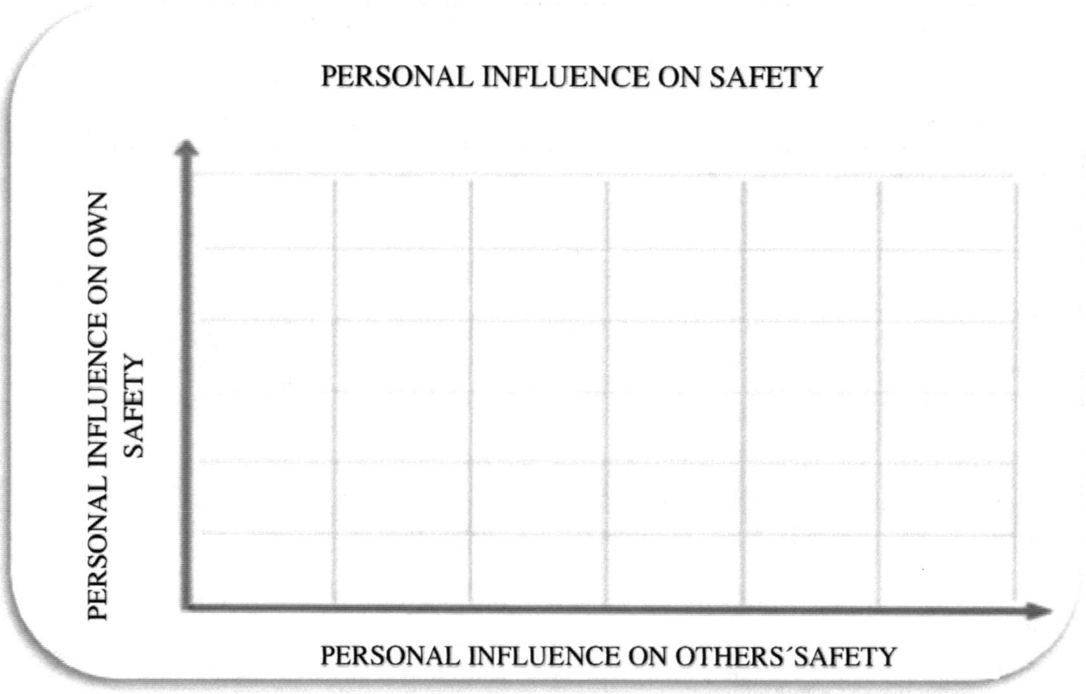

Figure 1. Personal Influence on Safety Matrix

We used the Self-perception as a tool to promote reflection and consciousness. We asked the participants to fill in the matrix in two different papers: one in his scrapbook (explained below) and another one on a plain sheet to hand over to the program facilitator. All the matrices were compiled and the result is shown in figure 2. As indicated, most participants initially evaluated themselves as influencing more themselves than others.

The coordination staff evaluated the matrix result as requiring attention from the participants. The staff observed the fact that some managers attending the program were obese, others worked extra hours, and most of them had reported having trouble balancing personal and professional life. Therefore, to promote deeper self-reflection, the consultant who was facilitating the program posed some questions to the participants regarding personal care. The remarks made were an invitation to honestly look at oneself and feel if the level of personal influence corresponds to what one would recommend to a loved person. Quoting Brené Brown (2015), the consultant said "Treat yourself like someone you love. Do you do it on a regular basis? Do you put yourself first?" Those remarks made some participants change their self-evaluation.

The intention was to call leaders ´attention to the fact that genuine influence on safety starts with their own safety, and goes beyond process or operational safety. To be legit and truly be a safety leader, one should take good care of oneself.

Figure 2. Compiled Personal Influence on Safety Matrix with the results from all the participants

We also developed a scrapbook that was given to each participant as a job-aid material to be used both during the program and after. The content of the scrapbook followed the program plan and was an invitation to self-analysis. The participants had special pages to do some journaling about what they had just listened to or dialogued about. There was also a special page dedicated to register insights. Figure 3 illustrates the scrapbook.

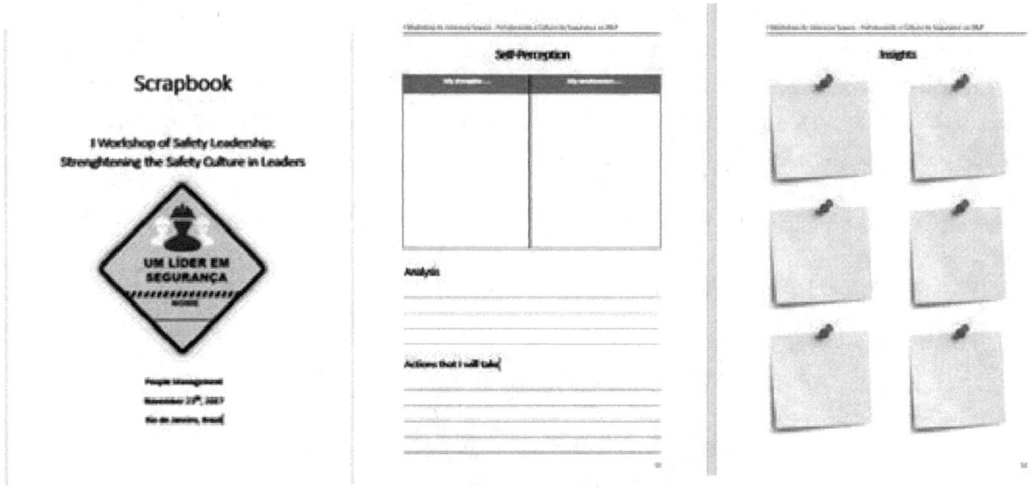

Figure 3. Pages from the Scrapbook used to stimulate self-consciousness and to plan actions

4 Exposing and Addressing Vulnerabilities

4.1 Control Room

We will now present the method used to address the issues managers pointed out, in order to transform the discussions into actions to strengthen the process safety leadership culture.

We prepared a special control room, where we designed a pathway decorated with images of accidents from the company and from the industry, and the number of fatalities. We also put an alarm sound playing during the route and videos showing accidents. These resources were intended to create an emotional ambiance to take the analysis to another level.

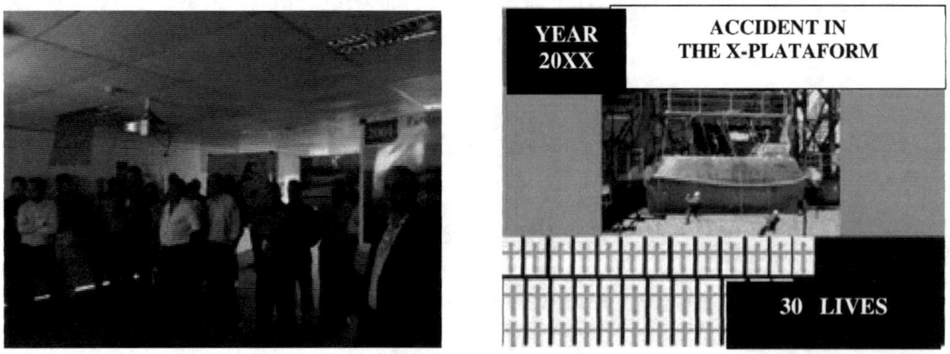

Figure 4. Picture of the Control Room and example of poster from accident used to decorate it.

Besides the planned journey passing several industry accidents, the Control Room was decorated with red material, to create an atmosphere of urgency. After going through the posters and watching the videos, the participants were invited to give their testimony of accidents they´ve been through.

The group tone got emotional as some of them shared unpleasant stories they had lived. The goal of inducing the reflection to an in-depth level was successfully achieved.

After that point, the participants went back to the regular room and dialogued about some questions that were in their scrapbook. The questions were related to the *Elements of a Strong Safety Culture* preconized by OGP (2013), mentioned in section 2 of this paper.

The methodology privileged social interaction among participants and the active listening of each other's considerations. Below is the list of the questions addressed to the participants.

Questions to Reflect and Dialogue:

1. Am I able to influence my operational team?
2. How many hierarchical levels are there between the operational team and me? To what extent does my speech actually makes sense to them?
3. How much of what is said by the operational team do I believe genuinely applies to me? Do I believe I receive good quality feedback?
4. Of the security actions implemented in my management, do I receive direct feedback on the effectiveness or otherwise of its implementation?
5. What degree of tolerance do I accept as a leader, regarding violations of safety standards?
6. Do I, as a leader, make it mandatory that the urgent inspection recommendations are executed immediately?
7. Do I, as a leader, take every opportunity to motivate, mentor and develop my team to enhance individual safety awareness and safety leadership?
8. Do I praise when I see a safe behavior?
9. Do I perform Behaviour Audits in a protocolled way, to meet the Scorecard, or do I do it because I believe in the effectiveness of what can arise from it?
10. Is Safety a value or is it a Priority for me as a leader?
11. Do I have the courage to step in and stop an activity when I see someone at risk? Do I always intervene?
12. Do my behaviours as a leader have safety as a foundation?
13. When I find that the environment is unsafe, do I privilege safety over other goals?
14. Safety comes first in my management?
15. How do I, as a leader, ensure that safety in my management is dynamic enough to ensure continuous improvement in operational procedures?
16. Thinking Safety as a chain, how can I, as a leader, be an effective link between operations and the company policy?

4.2 Results

In order to capture the precious discussions and share insights among the participants, in the final part of the workshop there was a collective presentation based on what each small group discussed about the questions above. Each group had to have at least one delegate, and others could participate if they wanted to.

Some of the findings reported were, for example: "*I have to go to the platform more often*"; "*Behaviour audits can be very useful, I should profit more from them*"; "*There are too many levels between my operational team and I. I have to get together with them on a regular basis.*"

After sharing insights, all managers filled in the Personal Influence on Safety Matrix again. The group reported that, this time, the self-perception was lower, since many managers realized they need to invest more efforts and more quality time in Safety, in both dimensions – with oneself and towards others.

After that, each participant revised and complemented his individual action plan (figure 3). They also wrote a personal commitment as a Safety Process Culture Leader.

At the very end of the program, we asked everyone to do a check-out by sharing with the entire group one action of the action plan. This intended to create a deeper commitment, as well as possible insights to others. Also, it was a practice that could be used with the managers team.

According to the group evaluation, presented in the word cloud below, the program succeeded in strengthening the safety culture in leaders, who mentioned as the top benefits from the program: (*Safety as a*) Value; Communication; Life; Commitment; Example; Integration; Consciousness; and Synergy.

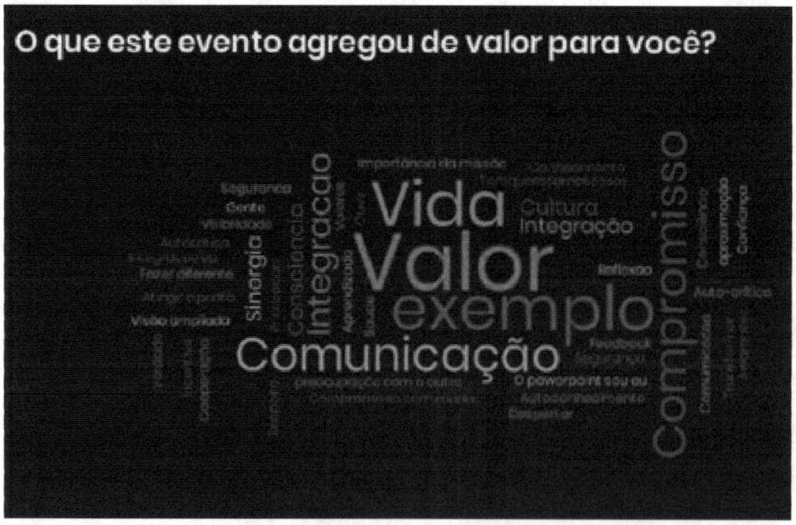

Figure 5. Evaluation: Word cloud containing the managers´ perception about the contribution brought by the program to participants.

5 Conclusions

The feedback given by the program´s participants was encouraging. It indicated that testimonies from partners, combined with relevant questions based on the daily problems managers face concerning safety, created a trust ambiance among managers.

The activities planned based on andragogic approaches favored dialogue and in-depth self-analysis. The use of the same instrument (Personal Influence on Safety Matrix) twice, as well as the individual scrapbook, facilitated the self-consciousness process and lead managers to review their initial assessment and to elaborate an action plan.

Leaders play a decisive role in strengthening the safety culture in the organizational world. Bringing about their vulnerabilities and having honest conversations with peers definitely helps managers to find ways to evolve from compliance to commitment.

Strengthening the safety leadership culture is a process that does not come to an end. It is an on-going process that must be continuously taken care of by leaders who value life and put safety first. Future projects that aim to implement effective actions in Safety Culture should take into consideration the company´s culture, as well as daily operational problems and safety as a core value to preserve life and business.

6 References

B. Brown, "*Daring Greatly: How the Courage to Be Vulnerable Transforms the Way We Live, Love, Parent, and Lead*", Penguin Random House, NY, 2015.

C. Migueles, J.R.B. Lafraia, G. C. Souza, "Criando o Hábito da Excelência: Compreendendo a Força da Cultura na Formação da Excelência em SMS", Quality Mark, Rio de Janeiro, 2014.

C. Taylor "*Walking the Talk: Building a Culture for Success"* Random House Business Books, London, 2015.

M. Muresan, "*Using Cybergogy and Andragogy Paradigms in Lifelong Learning*", 5th World Conference on Educational Sciences – WCES. Romania, 2013

OGP, *Shaping Safety Culture through Safety Leadership*. Report N° 452. International Association of Oil & gas Producers. October, 2013.

R. Charan, *Ram Charan on Leadership*,

Available at https://hbr.org/ideacast/2007/01/harvard-business-ideacast-27-r.html. Jan, 2007 (accessed Dec 4, 2018)

T. Kuppler, *Edgar Schein on Culture*,
Available at https://www.leadershipandchangemagazine.com/edgar-schein-on-culture/ Accessed on January 2, 2019.

W. Cohen, *Drucker's Top Tips to Successful Leadership*
Available at https://www.managementmattersnetwork.com/strategic-leadership/columns/druckers-top-tips-to-successful-leadership. Jun, 2018 (accessed Dec. 4, 2018)

7 Acknowledgments

I would like to thank Petrobras who fully supported this project, as well as Shell Brasil, for a very profitable partnership. A special thanks to my Human Resources, HSE and E&P work colleagues, who had a decisive role in making this project a reality.

548424 Development of Human Reliability Analysis Method for Oil and Gas Operations

The oil industry has grown in recent decades in terms of quantity of facilities and process complexity. However, human and material losses still occur due to major accidents at the facility. The analysis of these accidents reveals that many involve human failures that, if prevented, could avoid such accidents. These failures, in turn, can be identified, modeled and quantified through Human Reliability Analysis (HRA), which forms a basis for prioritization and development of safeguards for preventing or reducing the frequency of accidents. The most advanced and reliable HRA methods have been developed and applied in nuclear power plant operations, while the petroleum industry has usually applied Quantitative Risk Analysis (QRA) focusing on process safety in terms of technical aspects of the operation and equipment. A joint project from the B. John Garrick Institute for the Risk Sciences and a big company from the oil and gas industry is focusing on developing a HRA method focused on this industry operations, to bridge the existing gap between HRA and the Petroleum industry. The Phoenix HRA methodology was used as a basis, which has three layers i) a crew response tree (CRT), which models the interaction between the crew and the plant; ii) a human response model, modeled through fault trees, that identifies the possible crew failures modes (CFMs); and (iii) "contextual factors" known as performance influencing factors (PIFs), modeled through Bayesian networks. In addition to building on such a structure, the method development will rely on interviews with key personnel, visitations to installations, and analysis of past incidents data.

547327 Studying the Dependence of Operator Performance on Step Level Complexity

In the chemical, oil, and power industry, procedures are used by operators to perform their work correctly without compromising safety. Unfortunately, many procedures fail in adequately conveying information to their users. Heuristics tell us that technical jargon, vague language, and operator training all play a role in determining the effectiveness of procedures. To what extent these variables play a role is still not fully understood. Knowing this information will allow plants to more effectively write procedures. To counter this, researchers at Texas A&M's School of Public Health and Mary Kay O'Connor Process Safety Center are investigating the underlying factors that cause operators to deviate from procedures.

One possible factor may be that complex steps hinder an operator's ability to perform a task. This work aims to understand why certain steps are more complex than others. To do this, a set of procedures on how to operate a gas chromatography mass spectrometry (GCMS) machine is written. A study is performed in which participants use a set of procedures to analyze samples through gas chromatography. Using a scoring rubric, the complexity of each step is determined. The procedures have been designed to obtain an even distribution of complexity. Participants for this study are chemical engineering students with little to no prior experience in handling a GCMS machine. Thus, the experience, education, expertise, and training of the participants is controlled. The performance of each participant is reviewed and is grouped into one of four quadrants. For this study, a decrease in operator performance is expected as the complexity of each step increases.

548400 Prevent Them All or Stop Them Small: Tripod Best Practices in Reducing Chemical Accidents and Enhancing Response Preparedness

Prevent Them All or Stop Them Small: Tripod Best Practices in Reducing Chemical Accidents and Enhancing Response Preparedness

The connection between industry, government, and public safety, which we call the Tripod, is vital when implementing the 3-P's - Prevention, Protection, and Preparation. As identified by AICHE's Vision 2020 program, there are significant opportunities for organizations that address industrial process safety to work together. Responsible Collaboration between industry, regulatory and investigatory authorities, labor organizations, communities, and technical organizations promotes mutual understanding of risks, effects process safety and can reduce chemical accident risk in both small and large communities. Community based safety partnerships and leadership are fundamental to the success of such efforts.

For the past five years, a series of geographically focused Chemical and Ammonia Safety Days have cultivated and showcased chemical emergency prevention and planning efforts between Industry, NGOs such as the Ammonia Safety Training Institute (ASTI), and a myriad of public agencies including fire departments, health departments, LEPCs, SERCs, local regulators, and state and federal agencies such as EPA, CSB, DHS and OSHA. Within the context of Chemical Accidents and Releases, the focus of these efforts has been to Prevent Them All or Stop Them Small.

These Safety Day programs are designed for and with local community stakeholders from industry and first response agencies with a goal of seeding a permanent, engaged geographic collaboration for accident prevention and response readiness ownership and partnership. Agenda and curriculum elements include technical training focused on local risk, first responder field practicums and case studies, accident prevention program prioritization, practice and accountability, bowtie analyses of event propagation and management, and emergency response planning centered on the first 30 minutes and one-plan approach. The programs often include field demonstrations and exercises for fire departments and facility responders.

As a best practices demonstration, the presenters for this session will share the developed programs, recruitment methodologies and the demonstrated results of these efforts. In addition to Responsible Collaboration, other Vision 2020 principles embraced and promoted by the Safety Day Approach include Disciplined Adherence to Standards, Intentional Competency Development, Enhanced Application & Sharing of Lessons Learned, and Enhanced Stakeholder Knowledge. As a bonus, the societal theme of Harmonization of Standards is enhanced by the community of practice element of engaging federal, state and local regulators as well as industry stakeholders and process safety professionals.

548394 Risk Analysis Screening Tool – CCPS Project 280

A spreadsheet based Risk Analysis Screening Tool (RAST) is available through the Center for Chemical Process Safety (CCPS), courtesy of the Dow Chemical Company. The tool allows users to identify hazards and understand risks including causes, consequences and probabilities. RAST can be used for existing or proposed installations to facilitate MOC reviews, assess compliance with risk criteria, or evaluate proposed designs.

The users can input the type of equipment being used along with the equipment characteristics, chemical data, process conditions, and the plant layout. Using these inputs, the tool is used to help identify hazard scenarios and evaluate the consequences of representative loss of containment scenarios. RAST also allows the users to specify their own hazard scenarios based on their own knowledge of the process. Users can also input their own risk ranking matrix. With the hazard scenarios and the risk matrix, the user can select those scenarios of most concern and use the tool to create a Layer of Protection Analysis (LOPA) spreadsheet for the process.

As an example a comparison of RAST calculated scenario identification and consequence results to actual case studies from the CSB will be presented. Courses on how to use RAST, and how to adjust the pre-loaded default parameters are available.

550863 Impacts of Climate Change on Process Safety Management in Oil and Gas Operations

Impacts of Climate Change on Process Safety Management in Oil and Gas Operations

Chizaram Nwankwo

Coventry University

Project objectives and goals

The aim of this study is to analyse the impacts of climate change on process safety management in oil and gas operations. The objectives for this study include the following: -

- To evaluate past, present and future trends of climate change around various global regions
- To establish the relationship between climate change and process safety management in oil and gas operations
- To proffer adaptation and mitigation measures for climate change impacts on process safety management in oil and gas operations

In the last two decades, climate change has emerged as one of the most controversial topics that has been argued on several fronts (Hulme 2009). The idea of global warming and the negative impact of human-driven activity on the global weather has been fiercely debated (Cook *et al.* 2016). However, Field *et al.* (2012) argue that the facts and figures tend to reiterate the continuous occurrence and proliferation of climate change, as well as its increasing impacts in different industrial sectors. Indeed, the warming of the climate is unequivocal (Panteli and Mancarella 2015). According to IIGCC *et al.* (2017), 70% of global greenhouse gas (GHG) emissions are attributed to energy-related activities. The oil and gas industry contributes to 60% of these energy-related emissions due to production of fossil fuels as their main operational activity (Puig *et al.* 2015). Evidence from NASA (2018) suggests that there has been 1^0C increase in temperature which is close to the set mark in accordance to the Paris Agreement.

There are several case studies such as the flood in San Jacinto River, Houston Texas in 1984 and Coffeyville Refinery, Kansas in 2007. In the case of the San Jacinto flooding event in 1984, 8 pipelines were ruptured and 29 were undermined, 35,000 barrels of oil were spilled and ignited, while 547 people were injured and $23 million was incurred in damage costs (Matek 2017).

Hence, adaptive measures need to be applied to manage the severe consequences of climate change on oil and gas operations (Smith 2013). Such adaptive measures include more robust design specifications for oil and gas assets exposed to harsh climatic conditions (Cruz 2010). Ramasamy and Yusof (2015) suggest that this could be vital in offsetting the impacts of climate change on assets such as offshore rig platforms and equipment, transport pipelines, storage facilities etc. However, there is still little attention accorded to design and adaptation of oil and gas assets to offset the effects of climate change on them (Skjaerseth and Skodvin 2010). Therefore, this study seeks to analyse the impacts of climate change on deterioration of oil and gas assets.

Description of methods and results

Firstly, this study adopts a pragmatic paradigm since it explores the consequences of actions, is problem-centred, pluralistic, real-world practice oriented and allows for use of mixed models (Creswell and Creswell 2018). The area of investigation for this study is focused on impacts of climate change on oil and gas operations. The predominant approach taken for this study is a correlational research approach which tries to establish a relationship between climate change and oil and gas operations. The selected approach is due to the continuous occurrence of accidents in the oil and gas industry due to asset deterioration.

After thorough literature review of climate change trends and asset integrity issues in the oil and gas industry, this study identified a gap in knowledge that the oil and gas industry has not duly considered climate change during design of assets; hence posing a risk to oil and gas operations conducted under harsh climatic conditions. Therefore, the data collection for this study employed "concurrent triangulation" method of quantitative and qualitative data to ensure robust analysis of key variables such as climate change and oil and gas operations. Consequently, data collection tools for this study include questionnaires, interviews, archive data and documentary analysis. The data collection for this study was conducted over a period of 3 months.

Firstly, it involved piloting the questionnaire among academic and industry professionals to identify any errors, as well as testing and refining the data collection instrument. Next, the study was approved as a medium-risk project by Coventry University Ethics Committee, as primary and secondary data were collected from archive data, documents, questionnaires and interviews. The population sample for the study includes oil and gas industry professionals such as Mechanical Engineers, Facility Managers, Production and Well Engineers, Process Engineers, Health and Safety Officers, as well as Safety and Reliability Engineers. The questionnaires used for this study were designed using Bristol Online Survey (BOS) tool and covered questions around climate change and oil and gas operations. These data were predominantly collated and analysed using various software such as Zotero, Microsoft Excel and SPSS software. Statistical analysis conducted in this study include trend analysis, descriptive statistics, Pearson's correlation, Cronbach's alpha reliability analysis and Relative Importance Index (RII). Lastly, the findings of this study were benchmarked and justified with peer-reviewed academic literature.

Potential for application of results

This research could aid in curbing several accidents caused by asset deterioration due to increasing number of climate change events. It will help the oil and gas industry to avoid the enormous financial, legal and moral costs associated with accidents. Also, this study will provide a deeper insight on current climate change statistics and proffer adaptation and mitigation measures on how to deal with these problems. Similarly, it could broaden the scope of researchers to start incorporating impacts of climate change on various industrial sectors and subsectors asides the oil and gas industry.

References

Cook, J., Oreskes, N., Doran, P.T., Anderegg, W.R., Verheggen, B., Maibach, E.W., Carlton, J.S., Lewandowsky, S., Skuce, A.G., and Green, S.A. (2016) 'Consensus on Consensus: A Synthesis of Consensus Estimates on Human-Caused Global Warming'. *Environmental Research Letters* 11 (4), 048002

Creswell, J.W. and Creswell, D.J. (2018) *Research Design: Qualitative, Quantitative, and Mixed Methods Approaches*. 5th Edition (International Student Edition). Sage Publications

Cruz, A.M. (2010) *Vulnerability of Oil and Gas Infrastructure to Climate Change and Extreme Weather Events* [online] available from <http://indico.ictp.it/event/a09141/session/35/contribution/25/material/0/0.pdf> [5 March 2018]

Field, C.B., Barros, V., Stocker, T.F., and Dahe, Q. (2012) *Managing the Risks of Extreme Events and Disasters to Advance Climate Change Adaptation: Special Report of the Intergovernmental Panel on Climate Change*. Cambridge University Press

Hulme, M. (2009) *Why We Disagree about Climate Change: Understanding Controversy, Inaction and Opportunity*. Cambridge University Press

IIGCC, CERES, and IGCC (2017) *Global Climate Disclosure Framework for Oil & Gas Companies* [online] available from <http://www.iigcc.org/files/publication-files/Global_Climate_OG_DisclosureFramework.pdf> [14 November 2017]

Matek, B. (2017) *CHARACTERISTICS OF OIL AND GAS PIPELINE ACCIDENTS CAUSED BY CLIMATE CHANGE INTENSIFIED HURRICANES* [online] Baltimore, Maryland: Johns Hopkins University. available from <https://jscholarship.library.jhu.edu/bitstream/handle/1774.2/46088/Benjamin%20Matek.pdf?sequence=1&isAllowed=y> [26 June 2018]

NASA (2018) *Global Climate Change* [online] available from <https://climate.nasa.gov/> [9 May 2018]

Panteli, M. and Mancarella, P. (2015) 'Influence of Extreme Weather and Climate Change on the Resilience of Power Systems: Impacts and Possible Mitigation Strategies'. *Electric Power Systems Research* 127, 259–270

Puig, D., Haselip, J., and Naswa, P. (2015) *Adaptation to Climate Change in Colombia's Oil and Gas Industry: Recommendations to Promote Risk Management* [online] UDP-CRD-OLADE-01(EN). Denmark: UNEP DTU Partnership. available from <www.unepdtu.org> [19 September 2018]

Ramasamy, J. and Yusof, S.M. (2015) 'A Literature Review of Subsea Asset Integrity Framework for Project Execution Phase'. *Procedia Manufacturing* [online] 4, 79–88. available from <http://linkinghub.elsevier.com/retrieve/pii/S2351978915011336> [14 May 2017]

Skjaerseth, J. and Skodvin, T. (2010) *Climate Change and the Oil Industry: Common Problem, Different Strategies*. Manchester University Press

Smith, M.H. (2013) *Assessing Climate Change Risks and Opportunities for Investors: Oil and Gas Sector* [online] Analysis of Climate Change Risks and Opportunities. Technical Report. ANU and Investor Group on Climate Change. available here. [5 April 2018]

Further Information

Chizaram Nwankwo (E: nwankwo4@uni.coventry.ac.uk)

546246 Study of the Effects of Wind Speed and Surface Roughness Under Stable Atmospheric Condition Using Aloha Model and CFD Model

Study of the effects of wind speed and surface roughness under stable atmospheric condition using ALOHA model and CFD model

Many substances released in chemical processes are toxic, flammable, or even both. When these substances accidentally release to the atmosphere, human health and the environment could be threatened. As results, a reliable quantitative risk analysis tool is crucial to provide decision makers useful information on downwind concentration and hazards. ALOHA consequence model and Computational Fluid Dynamics (CFD) numerical method are widely used in chemical industries. Validation of these models under different scenarios is a necessary preliminary step before applying them to risk analysis. CFD models generally have acceptable concentration predictions on dense gas dispersion. ALOHA model is verified with DEGADIS model, but there is lack of study on effects of wind speed and surface roughness on dense gas dispersion. This paper compared field data from the Kit Fox experiments, which were CO_2 gas releases under stable atmospheric conditions with increased roughness, with CFD and ALOHA models' predictions. Also, statistical performance indicators were utilized to evaluate the accuracy of ALOHA model. Statistical performance indicators showed that ALOHA had systematic errors for dense gas dispersion with increased roughness under stable atmospheric conditions. Comparing effects of wind velocity profile and surface roughness values in CFD model and ALOHA model, this paper provided a better understanding of effects of key parameters in both ALOHA and CFD models. Furthermore, engineers would have better quantitative risk assessments under stable atmospheric conditions when they are using ALOHA consequence model and CFD model.

548222 Consequence Assessments and Collaboration to Streamline Project Siting

Collaboration is often sought after across disciplines, but all too often disregarded in favour of linear project development. For siting and consequence, as with other areas where design interacts with environment, health and safety constraints, this often leads to more costly design due to increased project iterations.

Stantec recently completed a consequence assessment for a compressor station to assess flammable extents during an emergency blowdown. At the same time, the client was experiencing siting issues due to sound emissions from the same process. The facility was grappling with competing factors: blowdown the facility in time to avoid catastrophic failure of their infrastructure, avoid excessive flammability extents from the blowdown stack location, and avoid extraneous noise exposure to nearby receptors. Using transient consequence modelling tools to model the blowdown sequence, Stantec was able to simultaneously and efficiently assess the flammable extents and sound emissions from the blowdown operations.

Consequence modelling for flammability was completed using a transient pipeline blowdown modelling tool to simulate the mass, energy, and momentum throughout the pipeline system. The model first simulated normal operating conditions, and then simulated the blowdown to estimate mass flow rate, liquid mass fraction, and release temperature from the blowdown stack. Several sizes of reducers upstream of the blowdown stack were introduced to estimate the pressure drops in the piping that would be required to appropriately reduce noise levels but also achieve full depressurization of the facility in time to meet safety criteria.

Consequence modelling of flammability was completed using SLAB, while the predicted change in sound pressure level from the blowdown stack was completed using Cadna acoustic modelling software. By combining these two analyses, the assessment costs streamlined for both required studies. In addition, the acoustics study benefited from additional insight provided by the consequence modelling tool.

548348 CFD Simulation of the Effect of Wind Speed and Release Elevation on Dispersion of Flammable Fluids at LNG Export Facilities

Facility siting studies for LNG export terminals are required by federal regulations to assess the potential hazard from accidental releases of flammable fluids. Federal regulations require under 49 CFR 193.2059 the evaluation of flammable vapor-gas dispersion distances from releases from flammable fluids in the facility for a range of representative wind conditions based on historical weather data expected to produce the maximum dispersion distances. A parametric CFD study will be conducted to evaluate not only the effect of wind speed but also the effect of the release elevation on the dispersion distance from flashing and jetting fluids. Hazard distances will be compared between the 2D software PHAST and the 3D software FLACS for a set of exemplar fluids found at LNG export facilities. The ability of the FLACS software to handle detailed 3D geometry provides for the evaluation of the effect of obstructions, congestion, confinement and mitigation measures on the dispersion distances. Release scenarios typical of LNG export facilities will be presented, with a focus on jetting and flashing releases of LNG, ethylene, propane, mixed refrigerants and heavy hydrocarbons.

546616 A Comparative Study of the Optimization Models for the Gas Detector Placement in Process Facilities

Gas detector network is an important layer of protection in process facilities for prevention gas release accidents of turning into fire and explosion disasters. But traditional standards just provide basic principles of the installation of gas detectors. Recently mathematical programming gradually became an important method to solve the problem. When using the mathematical programming method to optimize the detector layout, different optimization strategies will result in different optimization target models and detector layouts. In this study, three models have been developed and contrasted, namely minimal detection time P-Median model (MITP), minimal leakage concentration P-Median model (MICP), and minimal individual risk P-median model (MIRP), for the optimization of the gas detector placement based on stochastic programming. And as a basis, possible release scenarios were identified. In order to achieve the goal of using finite discrete scenarios to represent all scenarios effectively and quantitatively, clustering analysis was used to filter similar scenarios and select representative leak scenarios. The input data of these models was obtained from computational fluid dynamics. Furthermore, an adaptive particle swarm optimization (APSO) algorithm was applied to solve these models. The gas detector network optimization in a diesel hydrogenation unit was carried out as a case study. Results demonstrate that the MICP model is superior to the MITP model and MIRP model in describing the integrity of hazardous gas leakage and diffusion distribution. But the cumulative detection time of its optimized layout is slightly longer than that of MITP. Moreover, the control effect of MIRP model on personal risk caused by hazardous gas leakage was increased by 24.2%, although the minimum number of detectors was slightly higher than that of MICP model.

549411 Reducing Industrial Risk in a Converging It/OT World: 20 Experts Share Their Strategies

Rapid adoption of Industrial IoT requires tight integration between IT and OT. CISOs and corporate OT leaders are now scrambling to develop strategies, secure budget, and implement solutions to secure their mission-critical industrial control systems against burgeoning cyber threats.

Effective integration of IT and OT cybersecurity strategies is essential to address risks successfully. To secure industrial facilities and ensure safe, reliable production, IT and OT security teams – traditionally two separate disciplines with different priorities – must collaborate and share cybersecurity and risk management best practices.

In a series of interviews in late 2017 and early 2018, more than 20 different experts on the front lines of OT cybersecurity risk mitigation shared practical advice for making control systems more secure. These experts represented a diverse range of industries including oil and gas, chemicals and refining, and power generation.

In their interviews, they shared their insights into how to leverage IT/OT convergence and develop an effective OT cybersecurity program that meets challenges in real-world ICS environments. Technology plays an important role, but as these experts also point out, more technology isn't always the solution. Just as important is the need to establish effective communication and trust between cybersecurity professionals and OT engineers.

Security project teams must make a concerted effort to create alignment between business management, the cybersecurity community, and OT professionals. Without alignment, security initiatives will fail to attain the support required to ensure success.

In this presentation, attendees will learn:

- Practical, real-world recommendations from OT cybersecurity experts
- Best practices for breaking down barriers and building trust between cybersecurity professionals and OT engineers
- How to align business management and plant operations on cybersecurity strategies

549417 The Intersection of OT Security and Process Safety

The rise of cyberattacks on critical infrastructure creates troubling concerns for security, economic stability, and personnel and public safety. However, many companies are still relatively unprepared to successfully defend their facilities from cyberattacks launched by increasingly sophisticated threat actors. Safe production is at risk.

Safety focuses on incident prevention and measures risks to personnel. Cybersecurity focuses on protecting systems from threat actors and measures risk based on the threat landscape. The increased network connectivity that exists within industrial facilities today means that cybersecurity is now closely linked to the safe operation of industrial facilities and their processes. Successful defense of cyber-physical systems from nation-state attacks requires close coordination and collaboration between cybersecurity and process safety. Every safety initiative must now consider cyber risk. Every cybersecurity initiative must now consider safety.

Considerable overlap exists between risk analysis in safety and cybersecurity. Safety and cybersecurity disciplines can learn much from each other. However, safety and cybersecurity standards and corporate organizational structures remain siloed in far too many companies. This session focuses on the importance of executing a cohesive strategy for both safety and cybersecurity to ensure safe and reliable operations.

In this session, we will:

- Examine the existing cyber threat landscape
- Discuss how established process safety best practices can be applied to OT cybersecurity
- Provide practical guidance for improving cyber defenses and reducing safety and environmental cybersecurity risks

551385 Process Safety at Offshore Aerodromes (Helidecks)New Application for Process Safety

Offshore air transport is an importance activate of the oil & gas industry. In this way, the management of air operations is essential for maintaining the safety and reliability. The present work has the objectives of divulging the methods used by
Petrobras to audit maritime units with helideck and to demonstrate the contribution of such methods to the safety of process in the air operations. To achieve these objectives, the paper was based on the Center for Chemical Process Safety (CCPS) publication "Guidelines for the Safety of Risk-Based Processes", and aims to share the Bow tie methodology for the qualitative analyzes of risks inherent in Air Operations, focusing on the barriers typically faced by oil and gas industry, in order to contribute to the process by providing a graphical and structured approach to the required barriers, both to avoid catastrophic consequences and to prevent events recognized as precursors.

Thermodynamics and Computational Methods for Relief Valves

By Umesh Mathur, P.E.
umesh.mathur@aggreko.com , mathuru@gmail.com
Aggreko Process Services
3732 Magnolia Street, Pearland, Texas 77584, USA

15th Global Congress on Process Safety (AIChE, March, 2019, New Orleans)

1. INTRODUCTION

The computation of flow through a relief valve is one of the most important aspects of safety system design. This paper discusses the fundamental question of choosing the proper thermodynamic basis for such calculations. We also discuss the discrepancies we have observed in the literature on this crucial question and attempt to dispel some of the misconceptions we have found.

Our concern with this issue arose from a study of the relevant literature on this subject, which shows that the flow of a compressed gas (or gas-liquid mixture) through a rigidly held valve or hole in a pipeline must be treated as an irreversible, isenthalpic process. This reasoning flows simply from the First Law of Thermodynamics as applied to a flow process.

For example, as early as 1947, Hougen and Watson[1] discussed "Free Expansion" of gases (p. 551) and noted as follows:

"The unrestrained expansion of a gas is known as free expansion. Under conditions of no restraint no work is done, and under adiabatic conditions no heat is added. Free expansion under flow conditions is commonly known as throttling or as the Joule-Thompson effect referred to in Chapter XII.

The Joule-Thompson effect is measured experimentally by expanding the gas slowly and in steady flow through a well insulated porous plug; in this way, potential work is lost, and no heat is allowed to enter or leave the system through the walls. The fluid flows reversibly into and out of the process, but the expansion step is completely irreversible. Since no heat is added or lost, the process takes place under conditions of constant enthalpy, according to Equation (VII-11), page 207. However a change in internal energy results if any change in flow energy pV occurs."

Similarly, in a discussion of valves and throttles, Elliott and Lira[2] observe as follows:

*"A throttling device is used to reduce the pressure of a flowing fluid without extracting any shaft work and with negligible fluid acceleration. Throttling is also known as **Joule-Thompson expansion** in honor of the scientists who originally studied the thermodynamics. An example of a throttle is the kitchen faucet. Industrial valves are modeled as throttles....Changes in kinetic and potential energy are small relative to changes in enthalpy as we just discussed. When in doubt, the impacts of changes in velocity can be evaluated as described in Example 2.9. The amount of heat transfer is negligible in a throttle. The boundaries are not expanding, and there is also no mechanical device for transfer of work, so the work terms vanish. Therefore, a throttle is isenthalpic."*

Finally, in a discussion of nozzles, Klein and Nellis[3] observe that:

"...The potential energy terms of the inlet and outlet flows are not usually significant. Nozzles neither require nor produce power ($W_{in} = W_{out} = 0$) and are very nearly adiabatic. Note that the kinetic energy terms associated with the flows, particularly the outlet flow, cannot be neglected as the purpose of the nozzle is to increase the kinetic energy of the fluid. With these simplifications, an energy balance on the nozzle becomes:

$$H_{in} + ½V_{in}^2 = H_{out} + ½V_{out}^2 \qquad (4\text{-}54)"$$

In the light of these assertions by these noted authorities, and others[6][10], it is hard to understand how the flow through a relief valve, or a hole in a vessel that is open to the atmosphere, can be treated as isentropic (***reversible*** adiabatic) in nature, as opposed to the isenthalpic (***irreversible*** adiabatic) basis that is consistent with free expansion. This is because an isentropic expansion process would necessarily show an outlet enthalpy for the gas from the nozzle or valve (or an orifice in a punctured vessel) to have a lower enthalpy than the inlet, the difference in such enthalpies being equal to the work done, for example on a rocket or airplane. However, a relief valve or vessel remains stationary and absolutely no work is done during a free expansion by the gas flowing through.

Nevertheless, several standard references have used isentropic formulas for free expansion[4],[5] in stationary valves, or nozzles, from the inlet to the throat. We believe that this methodology is flawed as it violates the steady flow energy equation (i.e., the first law of thermodynamics written for an open system). This procedure would result in an erroneous prediction of outlet gas temperature when entropy is held constant. This, in turn, would predict an erroneous outlet gas density and thus, for a desired gas relief rate, the required cross-sectional area for the valve throat and also any conduit or pipe downstream of the valve.

The practical consequence of this anomaly in predicting outlet gas properties (especially the gas density at the throat) would be that the relief valve sizing could be incorrect. Therefore, it is important that the proper thermodynamic basis (isenthalpic expansion) be used in all such computations.

Conversely, for a given relief valve, errors in throat density based on an isentropic path would also affect the relief rate. Both types of errors could lead to potentially serious process safety concerns.

For these reasons, when calculating the release rate for gas at a high pressure through a relief valve, it is important to be able to use the proper basis for gas enthalpy (and therefore accurate computations of fluid thermodynamic properties). The same considerations apply to the gas release rate calculations resulting from a rupture in a pipeline or vessel containing a gaseous mixture under pressure.

To help clarify the proper methodology for a relief valve - or a hole in a pressurized vessel - the computations that should be performed are described below. We also show a few simple examples. Note that the possibility of sonic velocity at the throat, or choking, of the valve (or at the orifice in a ruptured vessel) is explicitly considered. In both such cases, the procedure described below is subject to the following assumptions:

- The opening can be represented by a convergent nozzle of exit area equal to the area of the throat of the relief valve, or hole in the case of pipeline rupture

- The compressed gas pipeline, or leaking container, is a large reservoir in which upstream pressure and temperature are maintained constant
- The compressed gas undergoes an irreversible, adiabatic process of expansion through the nozzle.

The flow through such a device from either (a) the inlet to the throat, or (b) from the throat to the outlet follows the steady flow energy balance equation (Equation 9.21, Himmelblau & Riggs[6]):

$$\Delta E = Q + W - \Delta(H + KE + PE) \qquad (1)$$

Where:
ΔE = energy accumulation (zero at steady state)
Q = net heat transfer into the system (very close to zero for a relief valve)
W = net work done by the system (zero for a relief valve)
ΔH = change in enthalpy of the system
ΔKE = change in kinetic energy of the system
ΔPE = change in potential energy of the system (zero for a relief valve)

Clearly, for steady flow through an orifice, valve (or relief valve), ΔE, Q, W, and ΔPE are all zero. Therefore, we are left with:

$\Delta(H + KE) = 0$, or

$\Delta H = - \Delta KE$, or

$$H_{out} = H_{in} - \Delta KE \qquad (2)$$

In other words, the outlet enthalpy is equal to the inlet enthalpy minus the change in kinetic energy in going through the throat or orifice. (It should be noted that the same logic would apply for discharge through an ordinary valve, a relief valve, or an orifice plate).

If there is no change in velocity, the outlet enthalpy will be the same as that of the inlet gas. This phenomenon is generally referred to as a Joule-Thompson (J-T) expansion. For an ideal gas, enthalpy is a function of temperature only; therefore, a J-T expansion would result in no change in temperature. However, for most real gases, a J-T expansion results in a decrease in temperature. There are a few notable exceptions, however, such as hydrogen or helium gas for which the J-T coefficient has the opposite sign[10]. On the other hand, if the gas speeds up, its kinetic energy will increase and, therefore, its enthalpy will decrease in accordance with equation (2) above. This kinetic energy increase would result in a decrease in outlet gas enthalpy, and therefore temperature, even for an ideal gas.

For ideal gases, the computation of enthalpy for ideal gases is straightforward as it is a function of temperature only. For real gases, however, enthalpy is also a function of pressure, temperature, and composition and use of an equation of state (EOS) becomes necessary[2] to calculate the enthalpy departure from the ideal, as discussed below.

COMPRESSIBILITY FACTOR AND ENTHALPY FOR REAL GASES

For real gases, enthalpy is found using a suitable equation of state (EOS), such as the Soave-Redlich-Kwong (SRK), Peng-Robinson (PR), or Lee-Kesler (LK). This procedure may be summarized as follows:

At the given upstream pressure (P), temperature (T), and composition, solve the EOS for the compressibility factor (z). This requires solving a cubic polynomial in z when using any cubic EOS, such as the SRK or PR, using either an explicit analytical or numerical trial and error method. For the LK, however, a trial and error procedure is always required as the LK EOS requires solving the 8-constant Benedict-Webb-Rubin[2] (BWR) equation, which is not a simple cubic in z, once for the simple fluid and again for the reference fluid.

Using the appropriate thermodynamic expressions for the EOS, use this value of z to calculate ΔH_d, the enthalpy departure from ideal.

Calculate the ideal gas enthalpy, H_0 (function of temperature T_0 only).

Find the real gas enthalpy using:

$$H_r = H_i - \Delta H_d \qquad (3)$$

Where:

H_r = real gas enthalpy
H_i = ideal gas enthalpy
ΔH_d = enthalpy departure from ideal

This procedure must be followed for calculating enthalpy at both upstream and downstream conditions.

The gas velocity at the throat may be sonic or sub-sonic. Also, as the gas speeds up in passing through the valve throat or hole (since the cross-sectional area is lower than that of the pipe), its kinetic energy increases and this results in a reduction of the enthalpy (Elliott & Lira[2], Eq. 2.53):

$$H_t = H_0 - \Delta KE \qquad (4)$$

Where:
H_0 = upstream gas enthalpy[J/kg]
H_t = exit gas enthalpy at the throat or hole[J/kg], and
ΔKE = change in kinetic energy[J/kg]

ΔKE is found as follows (Elliott & Lira's Eq. 2.53):

$$\Delta KE = (u_t^2 - u_0^2)/2 \qquad (5)$$

Where:

ut = throat (or hole) velocity[m/s], and

u_0 = upstream velocity[m/s].

Clearly, the outlet and inlet enthalpies are equal only if the change in velocity is negligible.

RELEASE RATE CALCULATION

The computation of the release rate from a pipe rupture should be done subject to the following equations:

Umesh Mathur, P.E., Aggreko Process Services, Houston, TX

Bernoulli's Equation

This equation relates upstream and downstream conditions for an arbitrary flow geometry, as follows:

$P_0 + \rho_0.u_0^2/2 + \rho_0.g.z_0 = P_e + \rho_e.u_t^2/2 + \rho_t.g.z_t + \Delta P_{friction}$ (6)

Where the subscripts o and e refer to the inlet and exit planes, respectively, and

P	= pressure[Pa]
ρ	= density[kg/m^3]
u	= velocity[m/s]
g	= gravitational constant[m/s^2]
z	= elevation above datum[m]
$\Delta P_{friction}$	= friction losses[Pa]

The friction losses have been ignored for simplicity in this presentation.

Obviously, for a hole, or throttle (or relief) valve, the elevations z_0 and z_t are the same.

Sonic Velocity

The sonic velocity in any medium[2] may be expressed as:

$u_s = (\gamma . z . R . T / M)^{0.5}$ (7)

Where:

u_s	= sonic velocity[m/s]
γ	= gas C_p/C_v ratio[]
z	= gas compressibility factor[]
R	= gas constant = 8314.46[J/(kgmol.K)]
T	= temperature[K]
M	= gas molecular weight[kg/kgmol]

In Equation (7), properties at the throat must be used for choked flow calculations. For ideal gases, the compressibility factor z is equal to 1. For real gases, however, the compressibility factor z is generally not equal to 1 and can be found only using an appropriate EOS.

C_p, C_v, and $\gamma=C_p/C_v$ ratio

Similarly, for ideal gases, the value of γ is found from the value of C_p, which is a function of temperature only, and the value of C_v is found from the following:

$C_v = C_p - R$ (8)

For real gases, however, the value of γ again requires use of an equation of state to calculate the ΔC_{pd} and ΔC_{vd} departures from ideality. The real gas C_p and C_v values are found using the ideal values and these calculated departures from ideality, by subtraction, in a manner similar to that described above for enthalpy:

$C_{p1} = C_{p0} - \Delta C_{pd}$ (9)

Where

C_{p1} = real gas C_p[J/(kg.K)]

C_{p0} = ideal gas C_p[J/(kg.K)]

ΔC_{pd} = departure in C_p from ideal gas value[J/(kg.K)]

Similarly, $C_{v1} = C_{v0} - \Delta C_{vd}$ (10)

In Equation (10),

C_{v1} = real gas C_v[J/(kg.K)]

C_{v0} = ideal gas C_v[J/(kg.K)]

ΔC_{vd} = departure in C_v from ideal gas value[J/(kg.K)]

In a conduit, orifice, or valve, the actual gas velocity, u_e, can never exceed the sonic limit, u_s. Note that u_s, in general, is different at the upstream conditions compared to the value at throat conditions. The upstream and throat temperatures and pressures of the gas may be different owing to the kinetic energy effect, Equation (4).

At a sufficiently low upstream pipeline or vessel pressure, the hole velocity would be lower than the sonic velocity limit. Above some critical limit for the upstream pressure, however, the maximum discharge flow is limited by the sonic velocity criterion at the hole.

Density

Gas density is computed using:

$\rho = P \cdot M / (z \cdot R \cdot T)$ (11)

Where:

ρ = density[kg/m^3]

P = pressure[Pa]

M = gas molecular weight[kg/kgmol]

z = gas compressibility factor[]

R = gas constant = 8314.46[J/(kgmol.K]

T = temperature[K]

The gas compressibility factor is 1 for ideal gases; it is calculated from the EOS for real gases.

1.2 Velocity

The velocity in a pipe or hole (or valve throat) is found using:

$u = W / (\rho \cdot A)$ (12)

Where:

u = velocity[m/s]

W = mass flow rate[kg/s]

ρ = gas density[kg/m^3]

A = cross sectional area[m^2]

For a circular pipe, the cross-sectional area is defined as:

$A = \pi . D^2/4,$ (13)

Where:
D = inner diameter[m]

For non-circular cross sections, the equivalent diameter D_e is defined using the concept of the hydraulic radius which is defined as:

D_e = 4 (Cross-sectional-area/Wetted-perimeter) (14)

It can readily be seen that this definition holds for a circular pipe also: $4 . \pi . D^2 / 4 / (\pi . D) = D$

As noted earlier, for ideal gases, the compressibility factor z is equal to 1. For real gases, however, finding the gas density ρ requires first calculating z from the EOS. This is done using the procedure as described above.

It should be noted that conditions at the throat of a valve (or at the hole in a pipe) are not known. Therefore, the computation of the gas flow rate must be done using an iterative procedure. When using Microsoft Excel, this is best performed using the Solver Add-in. Solver is a non-linear optimization technique that can simultaneously adjust the unknown quantities and solve the problem by enforcing the constraints imposed by the Bernoulli Equation (6) for throat velocity, and the Steady-Flow Energy Equation (1) for hole enthalpy. Also, the calculated throat velocity from the Bernoulli equation can never exceed the sonic velocity limit, Equation (7), and this constraint must be added explicitly when the upstream pressure exceeds a threshold value.

When solving for an unknown gas flow rate, if the upstream pressure is sufficiently low, hole velocity could remain sub-sonic. However, if the upstream pressure is sufficiently high, the pressure at the hole and also the gas mass flow rate would vary in a manner that obeys the sonic velocity limit at the hole.

Example

CASE (A): Methane is flowing through a 300[mm] diameter, above-ground pipeline at an absolute pressure of 46[bar] and a temperature of 300[K]. The pipeline develops a hole with an equivalent diameter of 50[mm]. Calculate the mass rate of leakage to atmosphere and also the velocity of the gas jet emerging from the hole. The pressure inside the pipeline may be assumed to be constant owing to continuous supply from the source.

CASE (B): Rework Case (A) assuming that the pressure in the pipeline is 1.5[bar] instead of 46[bar].

Case (A) P_o = 46[bar]

PROBLEM: Solve for release rate from a hole in a pipeline - sonic velocity at throat.

WE SHOW THE FOLLOWING DETAILED SOLUTION FOR THE IDEAL GAS CASE ONLY. Note that the solution of the real gas case requires use of complex computer programs to solve the chosen equation of state (e.g., Lee-Kesler) iteratively for the required thermodynamic quantities. Since these routines are imbedded in an outer loop for the unknowns (T and P at the throat, and the mass flow rate), the computational burden is too great for hand calculations.

The POLYMATH software[7] was used for solving this problem. All variables are defined in Table 1. The problem statement for Case (A) for the ideal gas case is as follows:

 # Initial guesses

$T_t(0) = 300$ # Initial guess for T_t[K]
$W(0) = 20$ # Initial guess for W[Kg/s]
$P_t(0) = 46$ # Initial guess for P_t[bar]

 # Aly-Lee ideal gas enthalpy constants:
A = 33298 # Ideal gas enthalpy constant A
B = 79933 # Ideal gas enthalpy constant B
C = 2086.9 # Ideal gas enthalpy constant C
D = 41602 # Ideal gas enthalpy constant D
E = 991.96 # Ideal gas enthalpy constant E
F = -120280000 # Ideal gas enthalpy constant F

 # The following three functions must be made equal to zero by adjusting the three unknown variables W[kg/s], T_t[K], and P_t[bar]:
$f(T_t) = h_t - h_{t2}$
$f(W) = u_t - u_{t2}$
$f(P_t) = u_t - u_s$

 # Initial guess ranges for non-linear equations
W(min) = 1
W(max) = 50
$T_t(\min) = 200$
$T_t(\max) = 400$
$P_t(\min) = 10$
$P_t(\max) = 46$

This problem requires solving the three simultaneous, non-linear algebraic equations (defined above) by varying T_t (the hole temperature), W (the mass flow rate), and P_t (the hole pressure):
$f(T_t) = 0$
$f(W) = 0$
$f(P_t) = 0$

In this case, the upstream pressure is high enough to cause the velocity at the hole to reach the sonic limit. This problem requires solving three simultaneous non-linear algebraic equations. The POLYMATH solution shown below for Example A was obtained using a software package called POLYMATH[7]. Note that any other similar package can also be used. We verified this numerical solution using the MATHCAD[8] and EXCEL[9] software packages also.

NOTE: The numerical solutions for Cases (A) and (B) are shown in Table 1, following the problem statement for Case (B).

Case (B) $P_0 = 1.5$[bar]

PROBLEM: Solve for release rate from a hole in a pipeline. Assume that the throat velocity is sub-sonic. This means that the outlet pressure is fixed at 1[atm] = 1.01325[bar].

When the outlet pressure is fixed, we are left with two non-linear equations. Also, for consistency with the methodology in (A) above, we use a very large upstream diameter to make the inlet velocity very small.

NOTE: If the calculated outlet velocity exceeds the sonic limit, we would revert to the problem formulation under (A) above. (This would add an equation requiring outlet velocity to equal sonic velocity, and thus again require solving three simultaneous non-linear equations.)

As noted earlier, the solution of the real gas case requires use of complex computer programs to solve the chosen equation of state (Lee-Kesler, in these examples) for the required thermodynamic quantities.

In this case, the upstream pressure is low enough so that the velocity at the orifice does not reach the sonic limit. Accordingly, this problem requires solving only two simultaneous non-linear algebraic equations. The solution shown below was obtained using a software package called POLYMATH. Note that any other similar package can also be used. As with Case (A), we verified this numerical solution using the MATHCAD and EXCEL software packages also.

The POLYMATH problem statement for Case (B) is as follows:

$f(T_t) = h_t - h_{t2}$

$f(W) = u_t - u_{t2}$

 # Initial guesses

$T_t(0) = 300$ # Initial guess for T_t[K]

$W(0) = 20$ # Initial guess for W[Kg/s]

Using POLYMATH, the solution for Case (A) shows that the gas release rate, W, for an ideal gas is 17.16[kg/s]. As noted above, identical results were obtained using the Mathcad and Excel software packages.

TABLE 1: SONIC OR SUB-SONIC FLOW THROUGH A VALVE OR HOLE

VARIABLES AND EQUATIONS

Variable	Description & Units	Explicit equations
A	Ideal gas enthalpy constant A	A = 33298.
B	Ideal gas enthalpy constant B	B = 79933.
C	Ideal gas enthalpy constant C	C = 2086.9
D	Ideal gas enthalpy constant D	D = 41602.
E	Ideal gas enthalpy constant E	E = 991.96
F	Ideal gas enthalpy constant F	F = -120280000.
P_t	Throat pressure [bar]	$P_t = 1.01325$
kcal_j	Conversion: 1[kcal] = 4184[J]	kcal_j = 4184
M	Gas molecular weight[kg/kgmol]	M = 16.0
T_o	Inlet gas temperature[K]	$T_o = 300$
P_o	Inlet gas pressure[bar]	$P_o = 1.5$
z_o	Inlet gas compressibility factor[]	$z_o = f(T_o,P_o,x)$
$(c_p/c_v)_o$	Inlet gas C_p/C_v ratio[]	$(c_p/c_v)_o = 1.32$
R	Gas constant[J/(kgmol.K)]	R = 8314.46
D_o	Diameter of pipe[m]	$D_o = 0.3$
D_t	Diameter of Throat[m]	$D_t = 0.05$
h_o	Inlet enthalpy[kcal/kg]	$h_o = (A*T_o + B*C/\tanh(C/T_o) - D*E/\tanh(E/T_o) + F)/M/kcal\_j$
rho_o	Inlet density[kg/m^3]	$rho_o = P_o*M*1.e5/R/T_o$
A_o	Pipeline cross-sectional area[m^2]	$A_o = 3.14159/4*D_o^2$
A_t	Throat flow area[m^2]	$A_t = 3.14159/4*D_t^2$
h_t	Throat enthalpy[kcal/kg]	$h_t = (A*T_t + B*C/\tanh(C/T_t) - D*E/\tanh(E/T_t) + F)/M/kcal\_j$
u_o	Upstream velocity[m/s]	$u_o = W/rho_o/A_o$
rho_t	Throat density[kg/m^3]	$rho_t = P_t*M*1.e5/R/T_t$
u_t	Throat velocity[m/s]	$u_t = W/rho_t/A_t$
u_s	Sonic velocity at throat[m/s]	$u_s = (z_o*(c_p/c_v)_o*R*T_t/M)^{0.5}$
u_{t2}	Throat velocity from Bernoulli Eq.[m/s]	$u_{t2} = (2/rho_t*(P_o*1.e5 + rho_o*u_o^2/2 - P_t*1.e5))^{0.5}$
dh_{KE}	ΔKinetic Energy = $\Delta(u^2)/2$[kcal/kg]	$dh_{KE} = (u_{t2}^2 - u_o^2)/2/kcal\_j$
h_{t2}	Calc. enthalpy at throat = inlet enthalpy - $\Delta(u^2)/2$[kcal/kg]	$h_{t2} = h_o - dh_{KE}$
T_t	Temperature at throat[K]	$f(T_t) = 0$
W	Mass flow rate[kg/s]	$f(W) = 0$
P_t	Pressure at throat[bar]	$f(P_t) = 0$

TABLE 1: SONIC OR SUB-SONIC FLOW THROUGH A VALVE OR HOLE (CONTINUED)

CASE (A) AND CASE (B) SOLUTIONS (USING POLYMATH)

Variable	CASE (A) Value (SONIC)	CASE (B) Value (SUB-SONIC)
A	33300.0	33300.0
B	79930.0	79930.0
C	2086.9	2086.9
D	41600.0	41600.0
E	991.96	991.96
F	-120300000.0	-120300000.0
P_t	27.71686	1.01325
kcal_j	4184.0	4184.0
M	16.0	16.0
T_0	300.0	300.0
P_0	46.00	1.50
z_0	1.00	1.00
$(c_p/c_v)_0$	1.32	1.32
R	8314.46	8314.46
D_0	0.30	0.30
D_t	0.05	0.05
h_0	226.2098	226.2098
rho_0	29.50683	0.96218
A_0	0.07069	0.07069
A_t	0.00196	0.00196
h_t	205.2641	210.3263
u_0	8.22527	7.71172
rho_t	20.86568	0.73254
u_t	418.7383	364.6543
u_s	418.7383	427.2981
u_{t2}	418.7383	364.6543
dh_{KE}	20.94576	15.88352
h_{t2}	205.2641	210.3263
T_t	255.6215	266.1833
W	17.15555	0.52433
P_t	27.71686	1.01325

CASE (A) RESULTS

This problem was also solved using the Lee-Kesler equation of state to calculate non-ideal gas properties, using a suitable computer program. This solution is provided below, with the ideal gas results quoted alongside for comparison:

	CASE (A)	
	IDEAL GAS	REAL GAS
Inlet pressure[bar]	46.00	46.00
Inlet temperature[K]	300.00	300.00
Calculated mass flow[kg/s]	17.28	18.25
Inlet velocity[m/s]	8.27	8.09
Throat velocity[m/s]	424.65	421.48
Sonic velocity at throat[m/s]	424.65	421.48
Inlet Enthalpy[kcal/kg]	225.61	214.87
Delta Kin. Energy[kcal/kg]	21.54	21.22
Outlet Enthalpy[kcal/kg]	204.07	193.64
Throat pressure[bar]	27.32	25.73
Throat temperature[K]	254.27	250.83

It should be noted that these results are quite similar for this case because the behavior of the gas (methane) is close to ideal at the conditions specified. However, these results would very likely differ significantly if, for example, the inlet temperature was considerably lower and the pressure much higher in which case the gas behavior would be markedly non-ideal.

CASE (B) RESUTS

Case (B) was also solved using the Lee-Kesler equation of state to calculate non-ideal gas properties. This solution is provided below, with the ideal gas results quoted alongside for comparison:

	CASE (B)	
	IDEAL GAS	REAL GAS
Inlet pressure[bar]	1.5	1.5
Inlet temperature[K]	300.0	300.0
Calculated mass flow[kg/s]	0.524	0.525
Inlet velocity[m/s]	7.71	7.70
Throat velocity[m/s]	364.7	364.2
Sonic velocity at throat[m/s]	435.06	435.3
Inlet enthalpy[kcal/kg]	226.21	226.21
Delta Kin. Energy[kcal/kg]	15.88	15.85
Outlet Enthalpy[kcal/kg]	210.33	210.36
Throat pressure[bar]	1.01325	1.01325
Throat temperature[K]	266.18	266.25

As might be expected, the gas mass flow rate is far lower for Case (B) compared to Case (A). Also, for Case (B), it was found that the conditions at the throat are virtually identical for the ideal versus the non-ideal gas.

2. ADDITIONAL EXAMPLES

To illustrate the importance of the issues discussed in this paper, we show a few additional examples for the differences in our numerical results compared to others.

2.1 Gas Release From A Hole In A Vessel

A 0.1[in] hole forms in a tank containing nitrogen at 200[psig] and 80[°F]. Determine the mass flow rate through this leak (Crowl[4], Example 4-4, p. 143).

The solution for this problem using the Lee-Kesler EOS was 4.26e-2[lb/s], compared to 3.86e-2[lb/s] in the reference (gas velocity at the throat was sonic). This difference is over 10% and not negligible.

2.2 Flow Through A Nozzle

Air at p_0=7.0e5[Pa] and T_0=293[K] discharges to the atmosphere. Compute the discharge mass flux, G, the pressure, Mach number, and velocity at the exit[5].

Using the Lee-Kesler EOS, the results are as follows:

Quantity	Lee-Kesler (this article)	Ideal gas (isentropic)[5]	Difference
Mass flux[kg/(m^2.s)]	1589	1650	-3.68%
Exit temperature[K]	238.5	244	-5.5 [K]
Exit pressure[Pa]	212041	370000	-42.7%
Mach number	1.0	1.0	-
Sonic velocity[m/s]	338	313	+8%
Velocity[m/s]	338	313	+8%

Note that the nozzle is not predicted to be choked when the downstream pressure is above 2.12[bar] (kinetic energy effects must be included), as opposed to 3.7[bar] in the original reference.

If a relief valve is to be designed for this scenario, the size would be under-predicted somewhat using the isentropic formulas used in the original reference. However, this discrepancy could be significantly larger in cases where an isentropic expansion incorrectly predicts partial condensation of the gas at the throat. In that situation, the density of the two-phase mixture at the throat could be far lower than that of the uncondensed gas (for the isenthalpic case) and the valve could be under-sized significantly.

3. CONCLUDING REMARKS

It has been shown that calculation of the release rate from a relief valve (or a hole in a vessel) should be done using an isenthalpic (i.e., irreversible adiabatic) path for the flow process. Using an isentropic thermodynamic path for this problem violates the First Law of Thermodynamics, as expressed for steady flow in a stationary relief valve or a hole in a pipeline. This could lead to an incorrect design both when sizing relief valves and estimating release rate from an existing valve or hole in a vessel.

In addition, it is important to use the proper equation of state to estimate real gas properties reliably when compared to the ideal gas assumption, especially when gas conditions are far from ideal. The Bernoulli equation (for throat pressure), and the sonic velocity limit (when flow is choked at the throat) must also be included for a complete definition of the problem. If the upstream pressure is low enough and the throat is un-choked, the isenthalpic methodology - as described in this paper - should again be followed to ensure a theoretically sound and consistent basis. Iterative computations for non-ideal cases (real gases) generally are too intensive for hand calculations and require use of computerized chemical engineering software.

REFERENCES

1. Hougen, O.A., K. M. Watson: "Chemical Process Principles" (John Wiley & Sons, New York, and Chapman and Hall, London, 1st Edition, Part II, p. 551, 1947, and 2nd Edition with R.A. Ragatz, Vol. II, p. 653, 1959)
2. Elliott, J.R., and C. T. Lira: "Introductory Chemical Engineering Thermodynamics" (Pearson Education, Inc. publishing as Prentice Hall, pp. 68-69, 2nd Edition, 2012)
3. Klein, S. and G. Nellis: "Thermodynamics" (Cambridge Univ. Press, 1st Edition, p. 168, 2011)
4. Crowl, D. A., and Louvar, J.F.: "Chemical Process Safety" (Pearson Education, 3rd Edition, pp. 130-131, 2011)
5. Green, D. W., and R.H. Perry: "Perry's Chemical Engineers' Handbook" (McGraw-Hill, 8th Edition, Section 6, pp. 23-24, 2008)
6. Himmelblau, D.M., and J.B. Riggs: "Basic Principles and Calculations in Chemical Engineering" (Pearson Education, 8th Edition, Fig. 9.18, Item 3, p. 525, 2012)
7. Cutlip, M.B., and M. Shacham: "Problem Solving in Chemical and Biochemical Engineering with POLYMATH, Excel, and MATLAB" (Prentice Hall, 2nd Edition, 2007)
8. Mathcad 2001 Professional: "Mathcad User's Guide With Reference manual", (Mathsoft Engineering and Education, Inc., 2001)
9. Freund, S.M., M. Jones, and J.L. Starks: "Microsoft Excel 2013: Complete (Shelly Cashman Series)", (Course Technology, 1st Edition, 2013)
10. Lewis, G.N., and M. Randall: "Thermodynamics", (McGraw-Hill, 2nd Edition, Revised by K.E. Pitzer and L. Brewer, p. 48, 1961)

548434 Engineered Solutions for Effective Flare System Analysis

Relief and flare system documentation is an important part of Occupational Safety and Health Administration (OSHA) Process Safety Management (PSM) 29 CFR 1910.119. Flare system analysis (sizing or rating) can be performed using commercially available software. Effective flare system analysis requires reliable Process Safety Information (PSI) such as process and instrumentation diagrams (P&IDs), relief system evaluation, Heat and Material Balance (H&MB), relief valve data sheets, isometrics, and disposal system loads. Good engineering methodologies can be implemented to effectively design header systems and rate existing header for any change. Some methodologies include use of dynamic simulation to predict realistic relief loads, reboiler pinch analysis for applicable systems, and selection of global relief scenarios. This paper will describes some methodologies and a combination of instrumented systems for effective flare system analysis. Effective flare analysis helps save capital expenditure spent on flare headers and possible flare load reduction yields reduced radiation footprint on a facility.

548364 Flares Are the End of the Line; How to Permit Correctly

INTRODUCTION

-

Flare is one of the key equipment in any refinery or LNG facility. Many relief valves relieve to flare, and if the flare is not permitted corrected – and functional - the plant is down. You cannot operate the plant without an active flare or flare field. The permitting of a flare is one of the prime activities in the permitting process. But the permit is dependent on the type of flare. This presentation covers various Federal Regulations involved in the permitting process, whether general or specific to each type of flare.

FLARE TYPES AND THEIR DESIGN FUCTURES

-

There are mainly three types of flares. The most common and most economical one is an elevated flare. There are elevated flares as high as 660 ft. The can be classified many ways; self-supported, guy wire supported, derrick supported, demountable type etc. One of the most important is the type of burner, whether a pipe tip or "furnace" burners. The second type is a ground flare with rows of furnace burners arranged in arrays at grade and surrounded with a wind fence. The last type is an enclosed flare with furnace burner enclosed with a tall stack. This type can be large as 70 ft in diameter and 150 ft tall. The flare type on a given project is selected based on the plot availability, location of the plant, sparing philosophy etc.

PERMITTING REGULATIONS

This presentation will discuss cover various Federal regulations that are applicable to all types of flares and the various parameter that the regulators are needing when apply for the permit. The presentation will cover the required destruction efficiency of the flaring gases, exit velocity for an elevated flare, alternate means of emission limitation for ground flares. In particular, the tip velocity, burner type, backpressure, emission testing and smokeless capacity are specific to each type of flare.

WHAT YOU WILL LEARN

-

· This paper will provide sufficient information to ensure the correct type of flare is selected and permitted

· This paper will cover various flares types employed in refinery and in LNG plants

· The limitations of these flare

· Importance of flare in any facility

· Application Federal Regulations

· Parameter required in the permitting applications

· Ground flare testing to establish minimum destruction efficiency

· Alternate means of emission limitation requirements in lieu of testing

548232 A Case Study on Application of HAZOP for a Brewery

Abstract

Beer is the most consumed alcoholic beverage in the world, and most popular beverage after water and tea. It's an alcoholic beverage produced by the saccharification of starch, often derived from the malted grains such as barley and wheat, and the fermentation of the resulting sugar. The brewing process can be broken into four main constituents before the bottling line, Mashing, Lautering, Wort boiling and Fermenting. Breweries commonly rely on ammonia refrigeration to keep their supplies cold. However, some brewers fail to realize that ammonia is one of the dangerous chemicals covered under 29 CFR 1910.119, OSHA's regulation for preventing or minimizing the release of toxic, reactive, flammable or explosive chemicals. "Safety matters at all levels", even for companies that may not be PSM level facilities.

A HAZOP study identifies hazards and operability problems. The study involves investigating how the plant might deviate from the design intent. It is based on the principle that several experts with different backgrounds can interact and identify more problems and failure scenarios working together and combining their results. This paper presents a review of consequences identified in brewery HAZOP. It also outlines some of the recommendations and action items identified from the case study of a brewery.

Keywords: Process Safety Management (PSM), PHA, HAZOP, Brewery, Ammonia, OSHA.

548410 A Basic Review of Fundamental of Thermodynamics and Its Application in the Analysis of Consequences and Accident Investigation

Once the stage of hazard identification has been concluded in a risk analysis, the next stage corresponds to select and specify the risk scenarios that will be modeled their effects or damages associated. This stage correspond to the consequences analysis and usually takes place in three basic steps: a) source term definition, b) dispersion model and d) effect or damage model.

Depending on the risk scenario to be modeled, several thermodynamic fundamentals or principles are applied in each of these steps supporting or sustaining the behavior and expected results of the consequences analysis model. In this work we propose to review the main thermodynamic fundamentals that support each of the models applied in each of the three steps that define a consequence analysis. This review of basic thermodynamic principles defines a set of different thermodynamic basic principles to apply to representative and different cases of consequence analysis. These basic thermodynamic principles are applied in the consequence modeling of a pair of cases of accident inventions to corroborate their validity.

548331 Evaluating Safe Location Criteria for Vents and Relief Devices Discharging to Atmosphere

In some process technologies, it may not be feasible to dispose the effluents from safety valves and vents to a closed system and discharging directly to the atmosphere can be the only alternative. Some examples include High Pressure Polyethylene Processes, Olefins and Storage units. The discharge can be directed to the atmosphere if:

- Local regulations permits it;
- Does not harm the environment;
- Does not harm the neighborhood;
- And is directed to a Safe Location.

The criteria for defining what a safe location is can vary, but it must include at least a consequence assessment of the flammable, toxic and thermal radiation effects. Some guidance can be found on industry standards and engineering references such as API 521 and the CCPS Guidelines for Pressure Relief and Effluent Handling Systems

One aspect to discharge hydrocarbons to a Safe Location is to evaluate the dispersion of the discharged hydrocarbons and ensure that it is diluted bellow the lower flammable before reaching the ground level or elevated structures. These areas are typically with people circulation and ignition sources. Another aspect that must also be evaluated is the thermal radiation generated due to an ignition of the discharge affecting personnel or damaging equipment or other assets. Lastly, if there is a possibility of the flammable mixture reaching confined spaces, the explosion possibility must also be assessed.

API 521 gives some guidance on chapter 5.8 for estimating the dispersion and concentration profile from a typical relief device discharge. It also describes that if the vapor relief velocity is high (> 30 m/s), the air entrance rate is also high and the vapors typically would be diluted below lower flammable limit in a very short distance.

This phenomenon explains the standard recommendation of locating the discharge of the relief device at least 3 m above any adjacent equipment, building or structure located within 15 m horizontally from the discharge point.

Experience has shown, however, that for some special cases these typical recommendations does not apply, such as:

- Heavy gas releases;
- Cryogenic releases;
- Control valve releases;
- Discharge piping with unconventional design (rain cap, horizontal, etc.);
- Discharge location within a building wake.

Adding to the complexity of the analysis, the relief devices can also be fitted with steam or other purge gas injection that could positively affect the dispersion.

After some case studies, which included the analysis of High Pressure Polyethylene plants Vent Stacks and Olefins unit relief and venting devices, Braskem developed an internal engineering standard with guidance and criteria for evaluating existing projects and new designs in order to ensure they comply with safe location. The guidance includes:

- Relief cases selection;
- Simplified and rigorous dispersion analysis;
- Consequence assessment endpoints;
- Thermal radiation effects to assets;
- Liquid, vapor and two-phase releases;
- Process Hazard Analysis;
- Diluting/snuffing steam;
- New systems and retrofitting criteria;

Applying the criteria in the developed standard allows the evaluation if the relief is in a safe location or not. If the discharge is not in a safe location, a risk ranking and recommendations to change the system must be made. One kind of a typical recommendation is to install the relief in a higher location. However, for some cases, achieving Safe Location is not possible and changes in the system are needed, such as relocating to a closed system or design changes to eliminate the relief scenarios associated with that relief device.

The development of an internal engineering guideline ensures a standard approach throughout the company for evaluating the Safe Location of hydrocarbon discharges and improves the quality of the risk evaluation during the PHA cycles.

548407 Experimental Study of Electrostatic Hazard inside Scrubber Column Using Response Surface Methodology

Experimental Study of Electrostatic Hazard inside Scrubber Column using Response Surface Methodology

Jingyao Wang[a,b], Yue Sun[a,b], Xiaodan Gao[a,b], Benjamin Wilhite[a,b], M. Sam Mannan [a,b]

[a]Mary Kay O'Connor Process Safety Center, Artie McFerrin Department of Chemical Engineering, Texas A&M University, College Station, Texas 77843-3122, USA

[b] Artie McFerrin Department of Chemical Engineering, Texas A&M University, 3122 TAMU, College Station, Texas 77843, United States

Abstract

Electrostatic ignition has triggered significant fire/explosion incidents in liquid processing industries including chemical, oil/gas, food, pharmaceutical and transportation. In this work, the impact of operating flowrate and temperature on the electrostatic energy accumulation in scrubber columns was studied. An *in-situ* charge measurement apparatus combined with a particle scattering laser function was installed inside a pilot-scale scrubber column, which can measure the aerosol particle size distribution, concentration and charge simultaneously. Four electrostatic ignition scenarios were considered including discharge from a single aerosol droplet, aerosol clouds, ungrounded conductor charged by the aerosol and ungrounded conductor charged by the liquid. Response surface methodology was applied to construct a quadratic empirical model correlating the electrostatic energy response with the flowrate and temperature. The obtained electrostatic response shows that corona from a charged single droplet was too small for safety concern. However, the ungrounded conductor polarized by charged aerosol or liquid can accumulate significant energy that is higher than the minimum ignition energy of some flammable chemicals. Temperature has two-sided effect where high temperature generates more concentrated charged aerosol but also facilitates the charge relaxation in the liquid phase. In all scenarios, increasing gas/liquid flowrate can cause significant electrostatic energy accumulation and can increase the hazards of many incidents. To reduce the electrostatic hazards, it is recommended to reduce the gas/liquid interaction flowrate and eliminate the presence of ungrounded conductors in liquid process operations.

Keywords: Scrubber column design, Electrostatic hazards, Response surface methodology, Process safety, Hazard identification, Aerosol hazards.

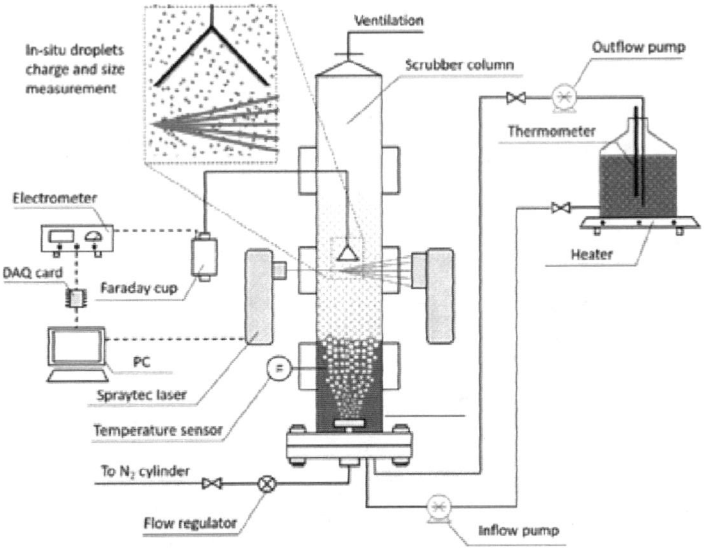

550852 Process Safety in Oil Refineries: Vacuum Distillation Column Hazop Study

Errors that can occur in the identification studies for hazards and risks which is one of the elements of a process safety management cause resulting major accident having undesirable long - term effects on human and environment. For preventing major accidents, identification and analysis of process hazards due to equipment failures, design or montage defects, improper maintenance, misuse, material defects, metal fatigue, corrosion, welding defects, human failure etc. are the main principles of process safety management. The aim of present study is to investigate possible accidental risks occurring in Vacuum Distillation Column of oil refining within contributing to the process safety frame.

In this cross - sectional study, semi quantitative HAZOP analysis was used for identifying accidental risks. By means of HAZOP studies 66 deviations, 238 deviation causes, 287 consequences, 651 barriers and 113 recommendation were identified. In the determination of accidental risk in the vacuum distillation column by applying semi quantitative HAZOP analysis method, effective results were obtained. The most common causes of deviations in the column have been found as control loop failure, leakage and mechanical failure. It is stated that the most common accidental risks are as a portion of 66.9 % operational deterioration, fire, explosion, release known as major accident risk. It is identified that the accidental risks mostly cause existence damage and the highest accidental risk occurs in the "vacuum column overhead system" and "bottom of the vacuum column system" 87.7 % of 113 recommendations are modification in the design, 5.3% is about modifications in the process safety management system.

551179 API Process Safety Site Assessment Program - Trends and Good Practices in Process Safety

The Process Safety Site Assessment Program is a collaborative industry effort managed by API and is focused on improving process safety in petroleum refining and petrochemical facilities. The program has conducted over 65 process safety assessments and 38 HF RP-751 assessments since its inception in 2012. The assessments are conducted by independent and credible third-party teams of industry-qualified process safety expert assessors using industry-developed protocols.

The presentation will review trends, both positive and low scoring, seen in the extensive benchmarking data collected. Good practices observed in the assessments will be identified and discussed. The emphasis will be to have the attendees leave with an understanding that typically some site(s) has developed a way to handle challenging areas of process safety.

545445 Emerging Trends in Process Safety Management Relating to Contractor Workforce Compliance

Following a Six Sigma Black Belt process improvement session conducted 2016 to 2018 for petrochemical complexes along the Gulf Coast the presentation identifies several emerging trends for Best Practices in industrial workforce compliance.

Safety Councils work for petrochemical plants and their contractors to facilitate innovative ways to reduce incident rates resulting in a 95% reduction in recordables. 2017 marked an evolution in Safety Council deliverables, slowly migrating away from site specific training to develop new workforce compliance and risk management processes. Working within the industry we developed several internationally recognized Best Practices for contractor compliance, risk management, labor cost management. ISTC collaborated with our partners bringing years of quantitative data to identify problems that are developing, have occurred in the workplace and formulating solutions, cost structures and most importantly compliance management to combat workforce issues. The presentation, best practices in contractor compliance highlights the journey from identifying the problems, and the associated issues with hiring, compliance, auditing, accident response and even Department of Homeland Security, CFATS inspections. Beyond the logistics of building solutions for a particular problem, the presentation reflects on how companies were able to share their issues among peers without violating confidentiality or company procedures to solve what was to be discovered as an industry trend. Using Six Sigma tools, the units were able to identify root problems, collaborate on a solution, model cost recovery and implement the solutions to the contractor workforce with a minimum of disruption to the business at hand.

Análisis Cuantitativo De Riesgo (QRA) Como Herramienta Para La Toma De Decisión Sobre La Infraestructura En Sistemas De Transporte De Fluidos Peligrosos

Daniel Mauricio Moncada Saavedra
CENIT -Transporte y Logística de Hidrocarburos
Cll 113 #7-80 Piso 13 Bogotá – Colombia
daniel.moncada@cenit-transporte.com

Maria Isabel Montoya Rodriguez
CENIT -Transporte y Logística de Hidrocarburos
Cll 113 #7-80 Piso 13 Bogotá – Colombia
maria.montoya@cenit-transporte.com

Prepared for Presentation at
American Institute of Chemical Engineers
2019 Spring Meeting and 15th Global Congress on Process Safety
New Orleans, LA
March 31 – April 3, 2019

Análisis Cuantitativo De Riesgo (QRA) Como Herramienta Para La Toma De Decisión Sobre La Infraestructura En Sistemas De Transporte De Fluidos Peligrosos"

Daniel Mauricio Moncada Saavedra
CENIT -Transporte y Logística de Hidrocarburos
Cll 113 #7-80 Piso 13 Bogotá – Colombia
daniel.moncada@cenit-transporte.com

Maria Isabel Montoya Rodriguez
CENIT -Transporte y Logística de Hidrocarburos
Cll 113 #7-80 Piso 13 Bogotá – Colombia
maria.montoya@cenit-transporte.com

Palabras clave:

Abstract

La pérdida de contención en activos utilizados para el transporte de hidrocarburos puede ocasionar impactos considerables a personas, medio ambiente, económicos y de reputación. Con el fin de mitigar la ocurrencia este tipo de eventos, se implementan estrategias sobre los activos, siendo la identificación y análisis de riesgo un elemento clave durante el proceso.

El presente análisis utilizó un enfoque cuantitativo, tanto para probabilidad como para consecuencia, con el fin de poder estimar el riesgo que genera la operación de una línea de transporte en un área de alta consecuencia clasificado como sitio identificado, planteando posibles alternativas de control y mitigación del riesgo.

Como resultado del estudio realizado se pudo establecer el nivel de riesgo actual asociado a la operación de la línea en el sitio de interés. Como actividades de mitigación para reducir el riesgo asociado se identificó: la reparación de los defectos, así como el realineamiento y reubicación de la tubería.

1 Introducción

Para el presente análisis se implementó un cálculo de la probabilidad de falla a partir del análisis de integridad estructural del sector de interés. Adicionalmente, se consideró el análisis de la susceptibilidad que se presenta a partir de las amenazas presentes, la cuales son identificadas a partir de las actividades de monitoreo e inspección desarrolladas sobre la infraestructura. El cálculo de la interacción entre las amenazas a la integridad; así como el resultado final de probabilidad de falla, se puede realizar mediante diferentes enfoques, el presente estudio utilizo cálculo de probabilidad de excedencia y aplicación de redes bayesianas para la estimación.

Para el cálculo de la consecuencia se buscó verificar la posible afectación a personas, producto de la ignición del producto transportado. Dentro de los sucesos finales se consideró: incendio de piscina, incendio por chorro de fuego, llamarada y la sobrepresión debida a la explosión. El cálculo de estos efectos se realizó en el sitio donde se presenta la perdida de contención, así como también para el caso en el que el hidrocarburo es transportado por un cuerpo de agua o se desplaza en tierra.

Una vez determinada tanto la probabilidad de que se desarrollen los sucesos finales identificados como la consecuencia que cada uno de ellos puede generar, se procedió a realizar el cálculo del riesgo expresando el resultado en términos de riesgo social e individual, comparando el resultado contra criterios de tolerabilidad referentes en la industria. Para la condición actual establecida se identificaron alternativas las cuales buscaban disminuir el nivel de riesgo, implementando actividades de mitigación del nivel de riesgo.

2 Análisis de Riesgo Cuantitativo

3 Frecuencia Final de Ocurrencia de los Sucesos Finales

En el sector bajo interés se tienen que la línea es susceptible a las siguientes amenazas:

3.1 Análisis de Condición.

En este aparte se relacionan los resultados de análisis de probabilidad de falla, obtenidos en el sector de interés, PK 43+000 al PK 45+000 del poliducto, para las amenazas de corrosión externa, corrosión interna, geotecnia y daños por terceros involuntarios:

3.1.1 Corrosión Externa (CE).

A continuación, se describe el estado de salud de las barreras de CE del poliducto en el sector solicitado (ver Ilustración 1), así como el análisis de probabilidad de excedencia de las anomalías ILI identificadas con la corrida del 2014. En la Ilustración 2 se observa la disponibilidad de la URPC y se reporta que la tubería cuenta con recubrimiento en buen estado. Bajo las condiciones y premisas consideradas para el cálculo y proyección de la POE (Probabilidad de excedencia; entre ellas una rata de corrosión de 2-3 mpy aproximadamente; no hay riesgo de excedencia de los criterios de profundidad y ERF (Factor Estimado de Reparación) de las anomalías ILI hasta el año 2021 en el sector 43+590 y 42+746 (Anomalías ILI con DR(Distancia de registro): 2646.76, 3490.08, 3492.38, 3492.71) (Ilustración 2).

Ilustración 1. Perfil de riesgo y salud de barreras CE.

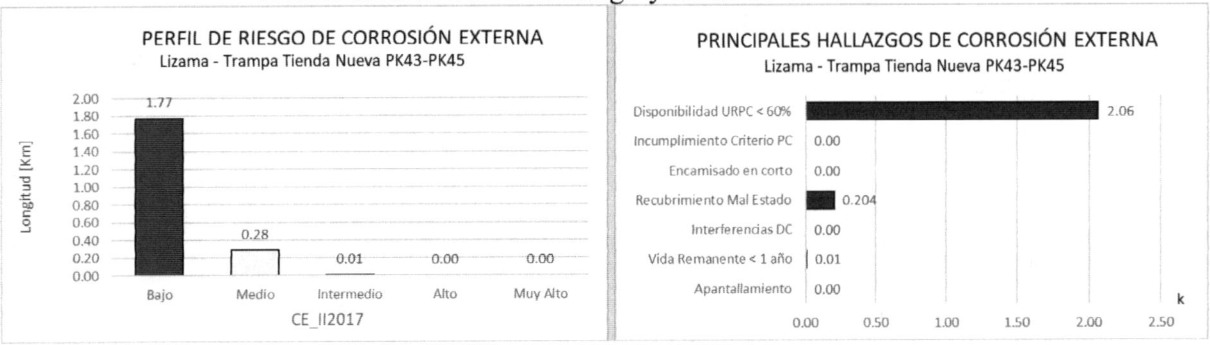

Ilustración 2. Análisis de POE de anomalías de CE en el tiempo.

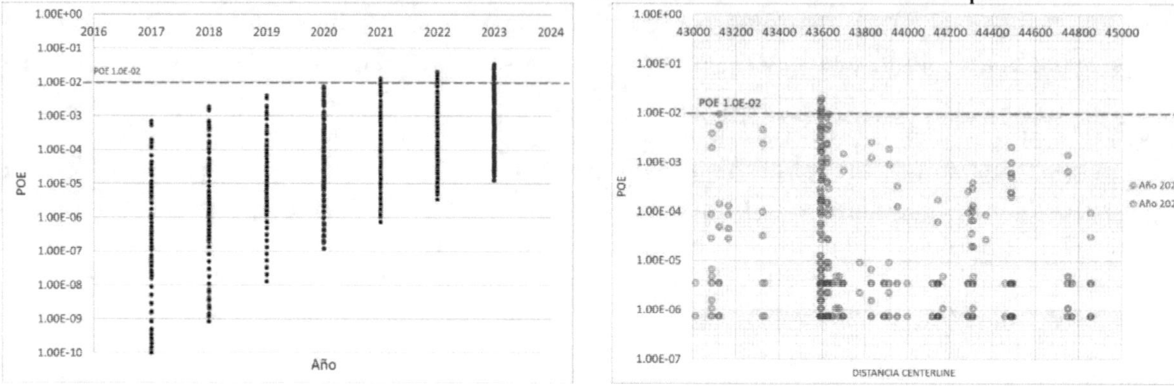

3.1.2 *Corrosión interna (CI).*

A continuación, se describe el estado de salud de las barreras de CI de la tubería en el sector solicitado (ver Ilustración 3), así como el análisis de probabilidad de excedencia de las anomalías ILI identificadas en la corrida del 2014. De acuerdo con los resultados de valoración de riesgo, se presenta un correcto estado de salud de las barreras de corrosión interna. Adicionalmente, de acuerdo con las condiciones y premisas consideradas para el cálculo y proyección de la POE, no hay riesgo de excedencia de los criterios de profundidad y ERF de las anomalías ILI en el mediano plazo (Ver Ilustración 4).

Ilustración 3. Perfil de riesgo y salud de barreras CI.

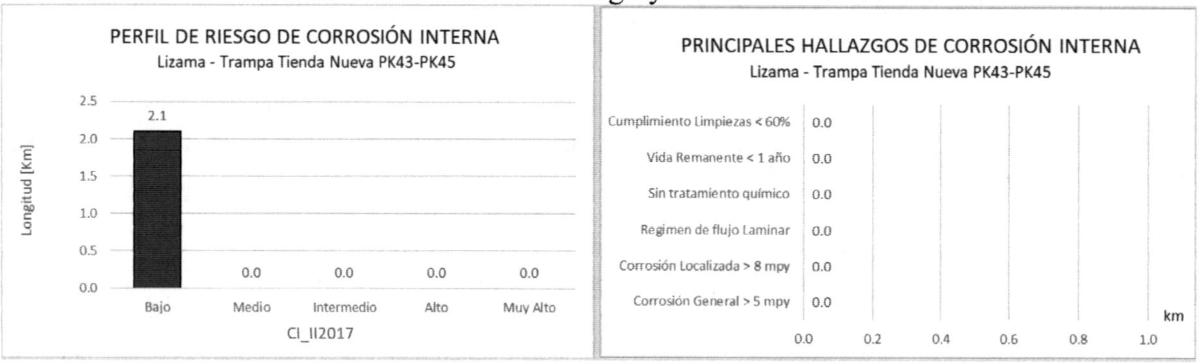

Ilustración 4. Análisis de POE de anomalías de CI en el tiempo.

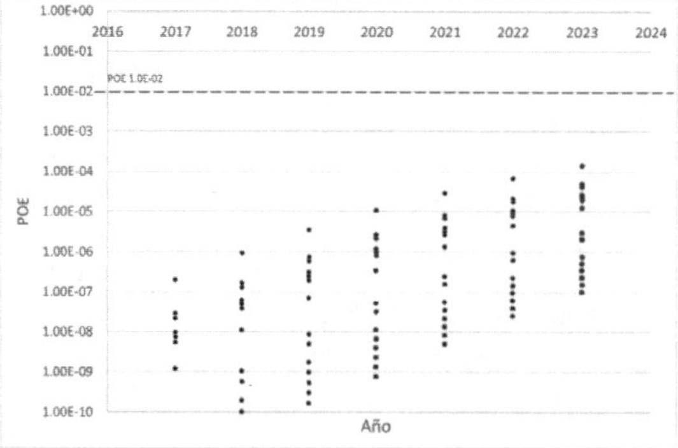

3.1.3 Geotecnia.

En el sector de interés se reporta hallazgo de geotecnia en la distancia centerline 44643 – 44703 (PK 44+400), donde se presenta proceso erosivo incipiente. Actualmente, no se identifica avance de la condición, razón por la cual, se continuará seguimiento a través de la inspección visual del derecho de vía (IVDDV).

Ilustración 5. Perfil de riesgo y salud de barreras de Geotecnia.

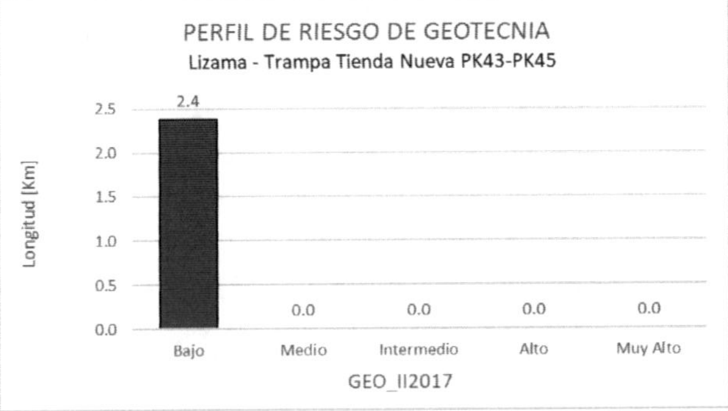

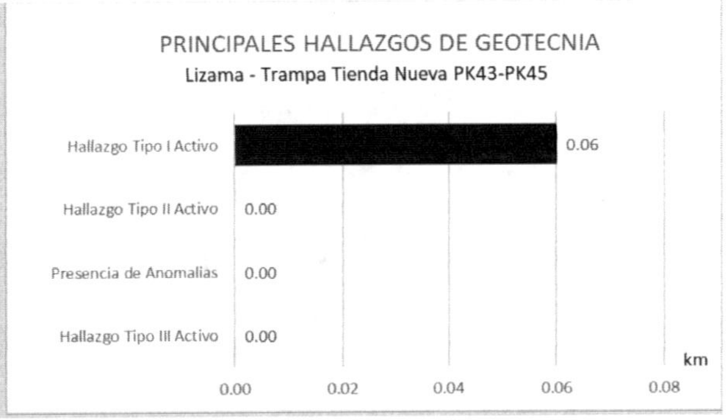

3.1.4 *Daños por Terceros Involuntarios (DPT).*

Debido a la condición de la tubería y a la gestión realizada por el operador para el control de la amenaza, ésta presenta una susceptibilidad baja de ser golpeada por un tercero sobre el derecho de vía. A continuación, se relaciona el perfil de riesgo de la amenaza de DPT en el sector analizado:

Ilustración 6. Perfil de riesgo y salud de barreras de Daños por Terceros Involuntarios.

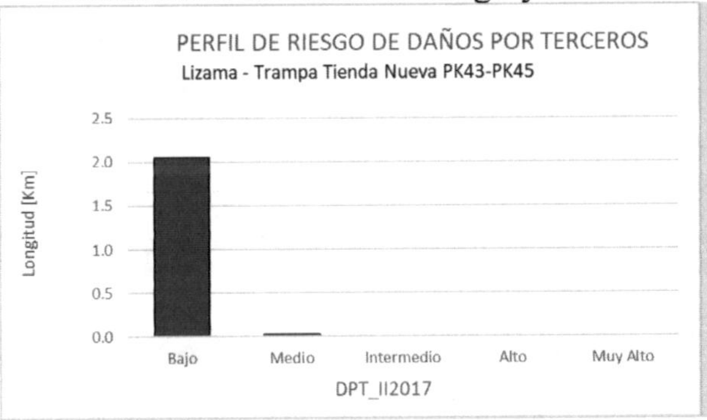

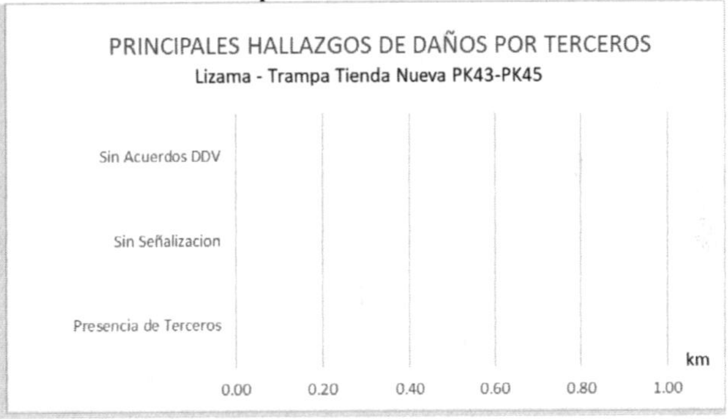

De acuerdo con el análisis ejecutado, el evento más probable obedece a una pérdida de contención generada por la amenaza de corrosión externa; donde el análisis estructural identifica que el modo de falla se daría por fuga. Sin embargo, las anomalías encontradas no presentan riesgo de excedencia de los criterios de profundidad y ERF hasta el año 2021.

Con el fin de poder estimar la probabilidad de falla de la línea en el sector de interés, se realizó el análisis estructural considerando principalmente defectos por perdida de espesor, bajo estas premisas se obtuvo un valor de probabilidad de excedencia de los criterios de profundidad y ERF de 7.09E-4. Del análisis de condición se concluye que el iniciador más probable en este sector es la falla de la tubería por fuga asociado a corrosión externa, para el cual a continuación se presenta el análisis de riesgo.

4 Análisis de consecuencia

A continuación, se presenta el análisis de consecuencia para los sitios en los cuales se ha identificado que se puede llegar a presentar una pérdida de contención.

4.1 Identificación de escenarios

4.1.1 Selección de sustancias.

En el poliducto se transportan las siguientes sustancias: Gasolina, GLP, Jet A1 y B2

4.1.2 Identificación de iniciadores.

Los iniciadores representan las situaciones iniciales de pérdidas de contención o exposición que se pueden presentar en los das diferentes equipos. Los iniciadores pueden ser de dos tipos, genéricos o específicos. Los iniciadores genéricos, involucran una pérdida de contención debida a causas no específicas del proceso u operación, como pueden ser corrosión, errores de construcción, fallas de la soldadura, etc. Los iniciadores específicos se caracterizan por tener causas concretas debido a las condiciones de proceso, el diseño, los materiales utilizados, etc.

A continuación, se describen los iniciadores tenidos en cuenta para cada uno de los sitios en los cuales se identificó la inestabilidad de taludes.

- **Sector PK 43+590.** Este sitio tiene la menor distancia desde la infraestructura a la escuela bajo análisis, 30 metros aproximadamente, sumado a esto en las cercanías de este se ha identificado la presencia de anomalías las cuales, bajo las condiciones actuales de operación y deterioro hacia el año 2021, pueden exceder los límites de profundidad critica a la falla.

- **Sector PK 43+805 – PK 44+110.** En este sitio se tiene que ante la posible pérdida de contención del producto se puede llegar a presentar la formación de una ruta de derrame en tierra, por la cual se desplazaría el producto hacia el interior de la Escuela.

4.1.3 Identificación de sucesos finales.

La identificación de los sucesos finales que pueden tener lugar después de que se da el iniciador, se debe realizar utilizando los árboles de eventos o sucesos. Dichos árboles permiten ver el comportamiento de la sustancia tras la descarga y estimar la probabilidad de ocurrencia de cada suceso, teniendo en cuenta los sucesos intermedios que condicionan la evolución del suceso final.

Esta identificación de escenarios se lista en Tabla 11 para los escenarios que involucran sustancias inflamables tales como Gasolina o Nafta. Los posibles sucesos finales identificados en el poliducto son: derrame continental, incendio de piscina, incendio de chorro y llamarada.

Tabla 1. Identificación de escenarios.

Escenarios	Descripción	Iniciadores	Sustancia	Tipo de pérdida
Afectación directa				
E1	Poliducto	L3	Gasolina	Incendio de piscina
				Incendio de chorro
				Llamarada
E2	Poliducto	L3	GLP	Incendio de piscina
				Incendio de chorro
				Llamarada

Escenarios	Descripción	Iniciadores	Sustancia	Tipo de pérdida
Afectación directa				
Afectación indirecta				
E3	Poliducto	L3	Gasolina	Incendio de piscina
				Llamarada

4.2 Estimación del área de afectación por los efectos que involucran producto

4.2.1 Condiciones para la simulación.

Para determinar las consecuencias, es necesario definir las condiciones propias del entorno de la infraestructura que servirán como base para la simulación de los efectos. Estos parámetros son presentados en la Tabla 2. Las estabilidades atmosféricas fueron determinadas a partir de la rosa de vientos basada en datos estadísticos y modelos meteorológicos, la cual es presentada en la **¡Error! No se encuentra el origen de la referencia.**.

Tabla 2. Condiciones atmosféricas

Parámetro	Valor	Estación de Referencia
Temperatura promedio (°C)	31	Colosal
Humedad relativa promedio (%)	83	
Rugosidad el terreno (m)	0,08	
Estabilidad atmosférica más probable y velocidad del viento (m/s)	2F	
Estabilidad atmosférica F y velocidad del viento (m/s)	2F, 5D	

4.2.2 Distancias de afectación directa

Las distancias de afectación representan las longitudes dentro de las cuales se pueden manifestar los efectos correspondientes a los niveles de afectación, para cada suceso final (incendios de piscina o llamaradas). Se denominan distancias de afectación directas, ya que los sucesos finales de que dependen se originan y desarrollan su efecto directamente en el sitio en el que se produce la pérdida de contención del producto

4.2.3 Estimación de Descargas.

Para la simulación de los posibles efectos producto de una pérdida de contención, es necesario conocer cuánto producto podría derramarse en cada punto de estudio sobre la infraestructura. El volumen derramado corresponde al estimado teniendo en cuenta el llenado de la tubería, las condiciones de operación y el tiempo de respuesta operativo, es decir, el tiempo de detección más el tiempo de actuación (generalmente el seccionamiento del sistema o parada de bombeo).

Tabla 3. Tiempos de actuación

Equipo	Tiempo de Detección (min)	Tiempo de Actuación (min)
	Fuga	**Fuga**
Poliducto	30	180

Tabla 4. Volumen y Tasa de descarga

Volumen (US bls)	Tasa de descarga (bls/hr)
59	22

4.2.4 *Simulación de Efectos de los Sucesos Finales.*

En este numeral se presentan los resultados de las distancias de afectación directa, obtenidas para cada uno de los niveles de afectación asociados a los sucesos finales. En la **¡Error! No se encuentra el origen de la referencia.** se presentan las distancias de afectación obtenidas para el evento incendio de piscina, es importante resaltar que este depende en gran medida del área de empozamiento que se forma una vez se produce la descarga del producto.

En la **¡Error! No se encuentra el origen de la referencia.** se presentan las distancias de afectación para incendio de chorro obtenidas durante las simulaciones, es importante resaltar que la distancia obtenida depende de las propiedades químicas de la sustancia y de las condiciones en las cuales se produce la descarga.

Tabla 5. Distancia de afectación incendio de piscina.

Equipo / Abscisado	Iniciador	Sustancia	Volumen Derrame	Niveles de Afectación por Incendio						
				Incendio de Piscina						
				1.6 Kw/m²	5.0 Kw/m²	7.3 Kw/m²	9.5 Kw/m²	14.5 Kw/m²	20.9 Kw/m²	37.5 Kw/m²
			(Bbl)	Distancia (m)						
K43+590	Rotura mínima	GLP	59	46,94	28,63	24,58	21,98	18,01	14,55	8,38
K43+590	Rotura mínima	Gasolina	59	94,63	49,69	35,17	27,20	20,98	20,98	-

Tabla 6. Distancia de afectación incendio JET FIRE.

Equipo / Abscisado	Iniciador	Sustancia	Volumen Derrame	Niveles de Afectación por Incendio						
				Chorro de fuego						
				1.6 Kw/m²	5.0 Kw/m²	7.3 Kw/m²	9.5 Kw/m²	14.5 Kw/m²	20.9 Kw/m²	37.5 Kw/m²
			(Bbl)	Distancia (m)						
K43+590	Rotura mínima	GLP	59	45,9	33,61	30,93	29,33	27,13	25,23	23,42
K43+590	Rotura mínima	Gasolina	59	44,43	31,20	28,46	26,82	24,58	22,96	20,84

En la **¡Error! No se encuentra el origen de la referencia.** se presentan las distancias de afectación por llamarada; las cuales están relacionadas con la volatilidad del producto junto con las condiciones climáticas de las zonas del evento.

Tabla 7. Distancia de afectación Llamarada.

Equipo / Abscisado	Iniciador	Sustancia	Volumen de Derrame	Niveles de Afectación por Llamarada más probable		Niveles de Afectación por Llamarada más estable	
				LII	LII/2	LII	LII/2
			(Bbl)	Distancia (m)		Distancia (m)	
K43+590	Rotura mínima	GLP	59	24,12	58,53	26,48	66,7
K43+590	Rotura mínima	Gasolina	59	-	-	-	-

4.2.5 *Distancias de Afectación Indirecta*

Las distancias de afectación indirecta representan las longitudes de potencial afectación por sucesos finales que se originan en una trayectoria o ruta de derrame (en tierra o en agua) o, a partir de estancamientos o acumulaciones de producto. En este numeral se presentan los resultados de las distancias de afectación obtenidas para cada uno de los niveles de afectación asociados a los sucesos finales que se pueden presentar en la ruta de derrame. Estos resultados corresponden a la radiación generada por piscina de fuego, evento que en los cuerpos de agua depende de la sección del cuerpo de agua, así como por el frente de llama generado por el volumen derramado.

En la Tabla **88 y ¡Error! No se encuentra el origen de la referencia.**9 se referencian los corredores de afectación generados por incendio de piscina sobre las rutas de derrame en tierra identificadas en el poliducto, en el sector comprendido entre los PK 43+000 - PK 45+000.

Tabla 8. Distancias de afectación indirectas POOL FIRE.

Abscisado	Medio de Transporte	Sustancia	Incendio						
			Incendio en piscina						
			1.6 Kw/m²	5.0 Kw/m²	7.3 Kw/m²	9 Kw/m²	14.5 Kw/m²	20.9 Kw/m²	37.5 Kw/m²
			Distancia (m)						
Km 43 - 45	Ruta en tierra	Gasolina	82,02	54,68	27,34	24,61	19,14	13,67	8,20

Tabla 9. Distancias de afectación indirecta FLASH FIRE

Abscisado	Medio de Transporte	Sustancia	Niveles de afectación por llamarada más probable		Niveles de afectación por llamarada más estable	
			LII	LII/2	LII	LII/2
			Distancia (m)		Distancia (m)	
Km 43- 45	Ruta en tierra	Gasolina	2,14	4,74	8,64	32,24

4.2.6 Áreas de Afectación

En sistemas de ductos, el área de afectación directa se define por un corredor medido desde el eje central del ducto. Este corredor está dado por la mayor distancia probable de afectación obtenida en la simulación de efectos para todas las sustancias analizadas. Por otro lado, el área de afectación indirecta está dada por varios corredores a lo largo de las posibles rutas de derrame tanto en suelo como en agua. En la Ilustración **77** se presenta el área de afectación identificada.

Ilustración 7. Área de afectación identificada.

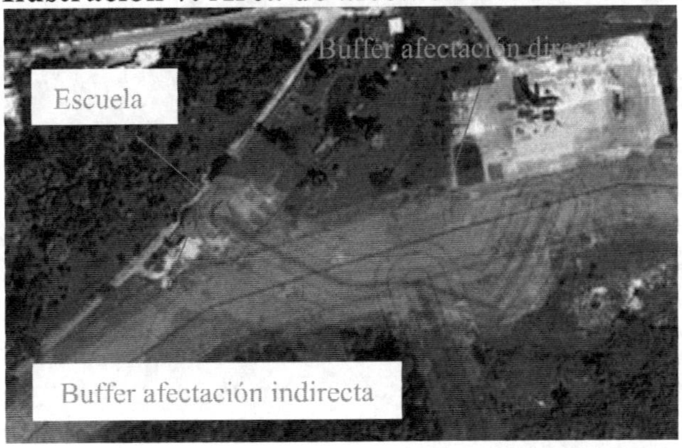

4.2.7 Identificación de Elementos Vulnerables

Para calcular el riesgo de la población respecto de la posible afectación producida por la pérdida de contención descrita, se requiere el número y localización de las personas, en la Tabla 100 se presenta la población identificada al interior de la escuela.

Tabla 10. Presencia de población Escuela

Presencia de Población							
Escuela - Dia				Escuela - Noche			
Niños	63	Profesores	6	Niños	0	Profesores	2

5 Evaluación del Riesgo.

A continuación, se presentan los resultados del análisis de riesgo el cual tuvo en cuenta la información presentada en los numerales anteriores asociada a las posibles causas de falla, los modos en que puede darse cada falla identificada, la frecuencia de ocurrencia de la falla, la probabilidad de ocurrencia de los sucesos finales, y finalmente, la estimación de la magnitud de las consecuencias de tales sucesos finales en el entorno en términos cuantitativos.

5.1 Riesgo Individual.

Los resultados de riesgo individual expresados en número de fatalidades por año (o año-1), sí permiten establecer zonas de aceptabilidad del riesgo según los diferentes criterios que se utilizan en la industria. En la Ilustración 8 se presentan los resultados de riesgo individual en el sector comprendido entre los PK 43+000 y 45+000.

Ilustración 8. Riesgo Individual.

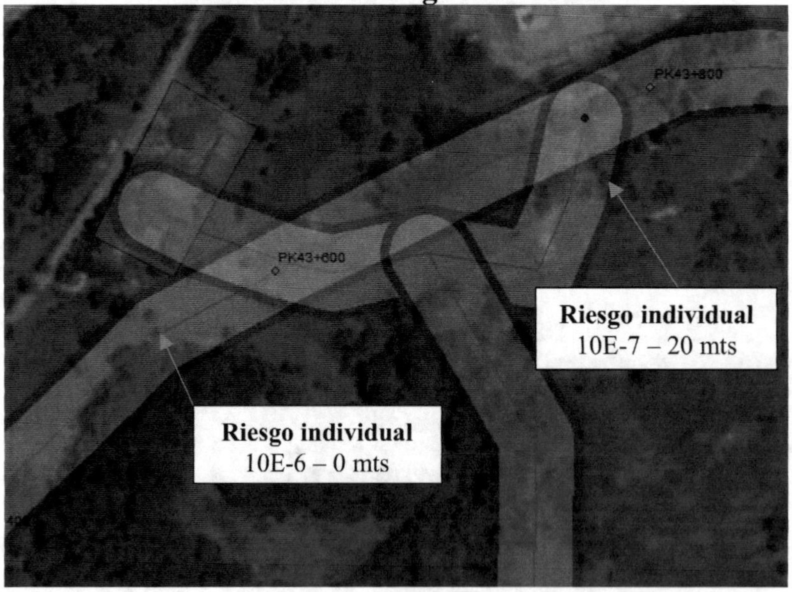

5.2 Riesgo Social

El riesgo social representa el número de fatalidades que pueden llegar a ocurrir en los sitios evaluados en el poliducto en el PK 43+000 – PK 45+000, al revisar la información contenida en la Ilustración 9 se puede apreciar como el riesgo se mantiene en la región tolerable, muy cercana al límite inferior ALARP. El evento dimensionante corresponde a una llamarada en la ruta de derrame, con el ducto transportando gasolina, la cual puede llegar a afectar a ocho (8) personas.

Ilustración 9. Riesgo social Escuela PK 43+000 – PK 45+000.

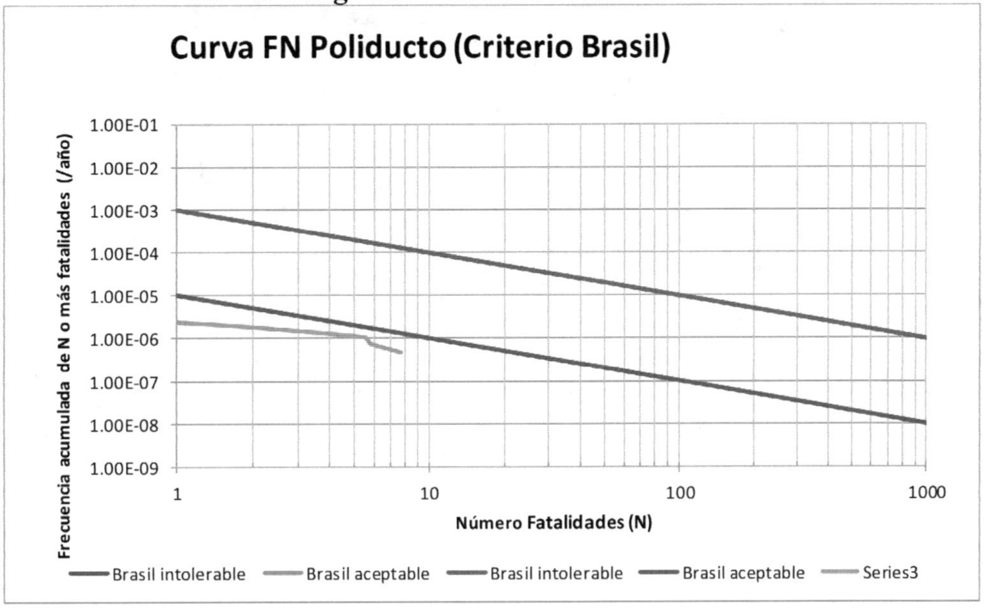

6 Alternativas de mitigación de Riesgo

Considerando los resultados del análisis de riesgo cuantitativo sobre el Poliducto para el sector PK 43+000 - PK 45+000, se tiene que ante una rotura total el evento dimensionaste corresponde a una llamarada con una distancia de afectación de 135 metros, para el caso de una fuga por corrosión el evento dimensionaste corresponde a una llamarada con una distancia de afectación de 32,24 metros, finalmente como resultado de una perdida de contención se puede llegar presentar la formación de una ruta de derrame en tierra la cual ingresaría a la escuela, con una distancia de afectación de 8.6 metros generada por una llamarada, tal como se presenta en la Ilustración 10.

Ilustración 70. Distancias de afectación

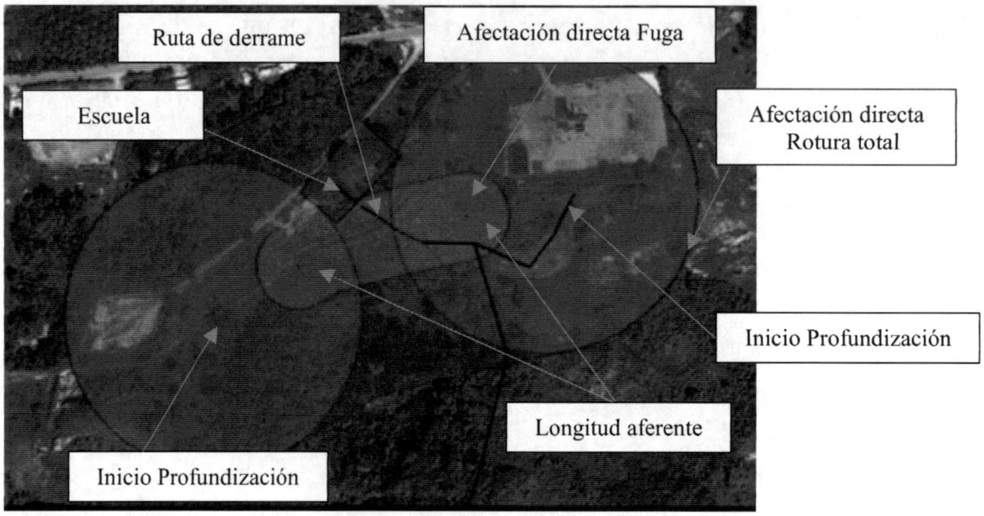

Teniendo en cuenta que el evento corresponde al presentado en la ruta de derrame, se hace necesario considerar la evaluación de alternativas de mitigación de riesgo. Con base en lo anterior, se consideró que la actividad de realineación y profundización cubra la longitud aferente de esta ruta, la cual es de 155 metros e inicia en la coordenada 74°33'35.466"O - 6°9'8.213"N y finaliza la coordenada 74°38'15.025"O - 6°9'8.737"N. Lo anterior, tiene como objeto de eliminar la ruta de derrame actual que conduce a la escuela en caso de una falla y perdida de contención, dado que el vertimiento se genere en sentido contrario al actual.

Sumado a lo anterior es necesario considerar una distancia de protección adicional para el dado caso en el que se llegue a presentar la rotura total de la línea, esta distancia se establece a partir del radio de impacto del evento llamarada, localizando los puntos sobre la línea que no presentan afectación a la escuela. A partir del análisis se estableció que esta distancia es de 316 metros e inicia en la coordenada 74°31'33,57146"O - 6°9'8,22442"N y finaliza en la coordenada 74°32'28,98241"O - 6°9'9,70756"N, esta distancia cubre la longitud aferente descrita anteriormente.

Dado que la alternativa de mitigación tiene como fin eliminar la afectación indirecta y el impacto directo sobre la escuela a continuación se presentan dos alternativas de trazados propuestos, que aseguran que las rutas de derrame en caso de una perdida de contención tendrán dirección opuesta a la ubicación de la escuela.

Ilustración 80. Opción 1 realineamiento entre PK 43 + 695 a PK 44 + 205.

Ilustración 91. Opción 2 realineamiento entre PK 43 + 625 a PK 44 + 460 de 752 metros.

En la ilustración 12 se presenta las dos alternativas propuestas y la distancia de separación del trazado actual de tubería, así como de la escuela.

Ilustración 102. Alternativas propuestas para mitigación de riesgo

7 Conclusiones

Del análisis realizado se tienen las siguientes conclusiones.

- El máximo número de fatalidades a generar es de ocho (8) y es generado por una llamarada en la ruta de derrame en tierra la cual, en el dado caso de que se presente el evento, ingresaría a las instalaciones de la escuela.
- La mayor susceptibilidad de la línea es por la amenaza de corrosión externa, especialmente en algunos sectores en los cuales el recubrimiento se encuentra en mal estado, esta condición junto con las anomalías encontradas en la línea, impactaron la frecuencia de falla establecida para el sector.
- En términos generales el nivel de riesgo se encuentra en la zona tolerable, muy cercana al límite inferior ALARP.

8 Recomendaciones.

- Es recomendable realizar actividades tendientes a disminuir el volumen disponible a desplazar en la ruta de derrame, para ello se puede realizar la profundización de la línea lo cual además de disminuir el efecto de los eventos por JET FIRE y FLASH FIRE disminuye la cantidad de producto disponible a desplazar al favorecer la evaporación de este.

- A partir del análisis realizado se tiene que la distancia a realinear y profundizar es:

  Opción 1 realineamiento entre PK 43 + 695 a PK 44 + 205 con una longitud de 516 metros
  Opción 2 realineamiento entre PK 43 + 625 a PK 44 + 460 con una longitud de 752 metros.

- Con ambas alternativas se elimina la afectación indirecta y la afectación directa, sobre la escuela, estas distancias cubren todos los posibles eventos que se pueden presentar y tienen afectación sobre la escuela en mención.

- Se pueden construir cortacorrientes los cuales eviten que la ruta de derrame ingrese o haga su tránsito en las cercanías de la escuela, junto con la construcción de estructuras de retención las cuales detengan el avance del producto mientras se realizan actividades de recuperación de este.

- Es recomendable rehabilitar el recubrimiento en aquellos sectores en los cuales se encuentra en mal estado, junto con la atención de aquellas anomalías que sean críticas.

9 Referencias

[1] Fingas, M. (2011). Oil Spill Science and Technology. Kidlington, UK: Gulf Professional Publishing Elsevier.

[2] D.O.T. Cornell Law School. Obtenido de 49 CFR 192.905 - How does an operator identify a high consequence area?: https://www.law.cornell.edu/cfr/text/49/192.905 Accesado 15 Enero, 2019

[3] D.O.T. Cornell Law School. Obtenido de 49 CFR 192.5 - Class locations: https://www.law.cornell.edu/cfr/text/49/192.5 Accesado 15 Enero, 2019

[4] Yellow Book. TNO. Methods for the calculation of Physical Effects Due to releases of hazardous materials (liquids and gases) – Third edition Second revised print 2005

[5] Purple Book. TNO. Guidelines for quantitative risk assessment – First edition 1999/2005

Referencias no citadas:

[6] ASME.ASME B31.8S. Managing System Integrity of Gas.New York, USA.2005.

[7] ASME.PIPELINE TRANSPORTATION SYSTEMS FOR LIQUID HYDROCARBONS AND OTHER LIQUIDS ASME B31.4-2002. New York, USA.2002.

[8] API.API Std 1160. Managing System Integrity for Hazardous Liquid Pipelines. Washington, USA.2008.132 Pags.

548036 Best Practice in MOC and Pssr Automation for Opex

Management of Change (MOC) and Pre Start-up Safety Review (PSSR) are still the most challenging elements of OSHA's Process Safety Management (PSM) standard and the EPA's Risk Management Program (RMP) rule. Now MOC and PSSR are requirements of the US Department of Interior's Bureau of Safety and Environmental Enforcement's Safety and Environmental Management Systems (SEMS). Even though the PSM standard has been around since 1992 and the industry has been managing change for several decades, we still can get it wrong, sometimes with disastrous results. The mindful or diligent efforts of many working in concert are necessary to ensure that change is identified, analyzed and executed in a quality way. But just when we get the process on paper and get the workforce up-to-speed, we have employee turnover, complacency sets in and sometimes there is regulatory change.

Many high-profile companies have addressed this challenge by creating and automating standard processes across their manufacturing or production facilities. MOC and PSSR are two of these major processes that have been successfully automated or **digitized**. Technology has been used to **automate** each step of the MOC workflow, enabling the collection of change ideas, formalizing the analysis of the ideas, and documenting the approval and execution of the change. Reports and dashboards can provide visibility of the health or current state of your MOC and PSSR processes. The use of **software** can enable the standardization of best practices.

On a well-meaning quest to standardize on a MOC/PSSR software solution and process, many companies are stifling creativity and innovation. Operational Excellence cannot be achieved in such ways. This paper discusses how the chemical process industry has defined the MOC and PSSR best management practices and how they should be automated for operational excellence.

Chemical Process Safety Requirements in the Fire Code

Ronald Melucci
Engineering Planning and Management, Inc.
959 Concord Street
Framingham, MA 01701
rbm@epm-inc.com

Prepared for Presentation at
American Institute of Chemical Engineers
2019 Spring Meeting and 15th Global Congress on Process Safety
New Orleans, LA
March 31-April 3, 2019

Ronald Melucci
Engineering Planning and Management, Inc.
959 Concord Street
Framingham, MA 01701
rbm@epm-inc.com
Keywords: NFPA Fire Code, International Fire Code, Process Safety

Chemical Process Safety Requirements of the Fire Code

Abstract

When most Chemical Process Safety professionals think about regulation and enforcement, they think about the applicable Occupational Safety and Health Administration (OSHA) and Environmental Protection Agency (EPA) regulations. However, the majority of US states adopt NFPA 1 or the International Fire Code (IFC) as their state fire code which contain requirements for chemical process safety. These requirements are applicable for processes of all sizes and levels of complexity below PSM and RMP thresholds and they address both fire and non-fire related chemical hazards. These requirements are enforced by municipal or state fire officials/marshals, who are becoming more aware of these requirements and chemical process safety in general. This paper will review some of the code requirements with a focus on requirements that may be commonly overlooked in the industry.

Introduction

The OSHA Process Safety Management (PSM) and EPA Risk Management Program (RMP) regulations, enforced by OSHA and EPA, respectively, are familiar to most safety professionals who work in industries that include processing of large quantities of chemicals. However, the applicable fire code requirements related to chemical process safety are often more restrictive than the federal regulations and are generally lesser known to such safety professionals.

The two model fire codes adopted by states in the US, are: *NFPA 1, Fire Code,* published by The National Fire Protection Association (NFPA), and the *International Fire Code* (IFC) published by the International Code Council (ICC). The states that do not adopt the IFC will typically use NFPA 1 and each state may also institute their own NFPA 1 amendments. Local amendments are almost always more restrictive than the model code verbiage, and many states have legislation that prohibits fire code amendments that reduce the level of safety/standard of care intended by the model fire code.

Both of these model fire codes contain many requirements that are similar to PSM and RMP regulations and contain references to them for additional guidance. Generally speaking, the model fire codes do not regulate all hazardous materials, but focus on hazards that are acutely dangerous to people or property. The PSM and RMP regulations address these hazards as well but additionally include materials that present a risk of chronic health or environmental hazards.

However, many states and municipalities have local fire code amendments that often include lower chemical quantity thresholds, such as those of the Commonwealth of Massachusetts as described in the following sections. The enforcement of these requirements is therefore not limited to OSHA and the EPA, as is commonly believed. The requirements are also enforceable by the local (municipal) fire authorities through the fire code references to these regulations.

Fire Code Requirements

The Commonwealth of Massachusetts currently adopts NFPA 1 – 2015 edition, with amendments, as 527 CMR 1.00: *Massachusetts Comprehensive Fire Safety Code*. Since 2006, 527 CMR has included chemical process safety requirements that are more restrictive than the federal regulations. The impetus for this was several accidents that occurred involving chemical process quantities below the OSHA and EPA thresholds. These Massachusetts (MA) amendments have been added to Chapter 60 of NFPA 1 – 2015 edition and include specific reference to OSHA PSM and EPA RMP regulations. Chemical processes with quantities above the PSM/RMP thresholds are defined in MA as "Category 5" processes. However, the 527 CMR amendments have also defined processes with lower quantities, i.e. Category 1, 2, 3 and 4, each with specific and unique requirements as described herein.

The NFPA 1 committee is currently reviewing proposals to adopt requirements with a "tiered" quantity threshold approach similar to MA. This is in recognition of the fact that severe hazards often exist, and severe accidents have occurred, in chemical processes with quantities below the PSM/RMP thresholds. It is likely that such a "tiered" approach will be incorporated into NFPA 1 and subsequently adopted by many US states. A preview of the applicable NFPA 1 and 527 CMR requirements are summarized herein.

Massachusetts Amendments to NFPA 1, Section 1.12.8.34.2 introduces the requirements as: *Process or Processing of Any Hazardous Material at any Facility* and it states the following:

(1) "A permit shall be required for the process or processing of any hazardous material at any facility identified in this *Code* as Category 2 through Category 5, as described in Chapter 60.
(2) The Authority Having Jurisdiction (AHJ) may require technical assistance in accordance with *1.15* to evaluate the adequacy of a Category 3 or Category 4 facility process safety conditions, programs, procedures, and practices undertaken at the facility, but only after a notice of denial has been properly served upon the person making application."

As part of the MA amendments to NFPA 1 – 2015 edition, Section 60.8, ***Hazardous Material Process or Processing,*** was added to 527 CMR. Section 60.8 applies to both new and existing facilities that process hazardous materials, with a list of exceptions such as:

- Motor vehicle service stations
- Refrigeration systems using a refrigerant other than ammonia or LPG,
- Certain wastewater treatment operations
- Storage of hazardous materials in atmospheric vessels.

Definitions:

NFPA 1 defines "**Hazardous Materials**" as "A chemical or substance that is classified as a physical hazard material or a health hazard material, whether the chemical or substance is in usable or waste condition."

OSHA and EPA define "**Process**" as "Any use, storage, manufacturing, handling, or the on-site movement of hazardous chemicals, or combination of these activities." The MA state fire code defines "Process" as "A sequence of operations in which the sequence can be inclusive of physical operations such as heating, cooling, mixing, distilling, compressing, and pressurizing, and chemical operations, such as polymerization, oxidation, reduction, and other chemical reaction processes. The sequence can involve but is not limited to: preparation, separation, combination, purification, or any actions that cause a change in state, energy content, or chemical composition."

MA Amendments to NFPA 1 Section 60.8 categorizes hazardous material processes into five (5) categories, as follows:

Table 1: Hazardous Material Threshold Quantities in the MA Amendments to NFPA 1

Category	Quantity (gallons)
Category 1	< 2.5
Category 2	2.5 – 60
Category 3	60 – 300 (or Group H)
Category 4	300 and not Category 5
Category 5	≥ the threshold quantities stated in 29 CFR 1910.119

The MA Amendments also add Section 60.8.5.1 to NFPA 1, which requires that facilities operating hazardous material processes must maintain the required documents and procedures for periodic inspection and review by the head of the Fire Department to remain in compliance with the state fire code. Requirements for each process category are as follows:

Table 2: Outline of Regulatory Requirements in the MA Amendments to NFPA 1

Category	Requirements
Category 1 (Vessel Capacity ≤ 2.5 gal)	• OSHA 1910.1450 (Occupational exposure to hazardous chemicals) • OSHA 1910.1200 (Hazard communication to employees) • Policy per NFPA 1, Ch. 66 (Flammable liquids) and Ch. 67 (Flammable solids) • Requirements of NFPA 1, Section 60.1 – 60.5
Category 2 (Vessel Capacity ≤ 60 gal)	• All requirements of Category 1 • Permit required per NFPA 1, Sections 1.12.8.34.2 and 60.8.1.1
Category 3 (Vessel Capacity ≤ 300 gal OR Group H Occupancy)	• All requirements of Category 1 & 2 • Category 3 Hazard Evaluation • Requirements of NFPA 1, Section 60.6
Category 4 (Vessel Capacity ≥ 300 gals AND not Category 5)	• All requirements of Category 1-3 • Limited Safety Program (PSM "lite")
Category 5 (Vessel Capacity ≥ the threshold quantities stated in 29 CFR 1910.119)	• All requirements of Category 1-4 • Process Safety Management (PSM) Program 29 CFR 1910.119 (OSHA) • Risk Management Program (RMP) 40 CFR Part 68 (EPA)

Category 1 – Requirement Details

Category 1 processes, as stated in Table 2 above, must possess documentation of compliance with OSHA 1910.1450, *Occupational Exposure to Hazardous Chemicals in Laboratories*, OSHA 1910.1200, *Hazard Communication,* maintain and implement a policy in compliance with NFPA 1, Ch. 66, *Flammable and Combustible Liquids*, and Ch. 67, *Flammable Solids*, demonstrate compliance with NFPA 1, Section 60.1.5.1, *Emergency Action Plan*, and Sections 60.5.1.4.3.2(5) through 60.5.1.4.3.2(9) and comply with permitting requirements in Sections 1.12 and 60.8.4 of the MA Amendments to NFPA 1.

OSHA 1910.1450 and 1910.1200

Compliance with OSHA 1910.1450 requires facilities to set and adhere to exposure limits, monitoring plans, hygiene plans, precautions, training, medical plans, and other plans, as found to be acceptable by the AHJ, to react to and minimize potential exposure to hazardous chemicals. OSHA 1910.1200 requires hazard communication methods for employees such as inventory documentation, hazard programs, labels, safety data sheets, and training.

NFPA Chapter 66 and 67

In addition to demonstrating compliance with OSHA standards, facilities meeting Category 1 requirements must implement and maintain a written policy in compliance NFPA 1, Ch. 66 and Ch. 67.Chapter 66 is divided into 29 sections, most of which contain multiple requirements. Section 66.6 contains the prescriptive requirements for fire and explosion prevention and risk control, and Section 66.6.4 goes on to list requirements related to a facility's hazards analysis, which is required to be reviewed to ensure that fire and explosion hazards are addressed by fire prevention, fire control, and emergency action plans. The AHJ should be consulted with for the development of the hazards analysis to ensure that the extent of the fire prevention and control is deemed acceptable or an engineering evaluation of the operation and application of fire protection and process engineering principles is soundly applied. The hazards analysis must include, but not be limited to, analysis of the following:

(1) Fire and explosion hazards
(2) Emergency relief from process vessels
(3) Applicable facility design requirements in Chapters 17, 18, 19, 28, and 29 of NFPA 30
(4) Applicable requirements for liquid handlings, transfer, and use, as covered in Chapters 17, 18, 19, 28, and 29 of NFPA 30
(5) Local conditions/exposures
(6) Emergency response capabilities of the local emergency services

Additional guidance for conducting a fire risk assessment can be found in NFPA 551, *Guide for the Evaluation of Fire Risk Assessments.* Facility management must repeat the hazards analysis when any of the following occur:

(1) Changes occur in the materials in process
(2) Changes occur in process equipment
(3) Changes occur in process control
(4) Changes occur in operating procedures or assignments

This is consistent with the requirements of OSHA 1910.119 Sections (e) and (l). Not all sections of Chapter 66 may be related to processes; however, Section 66.17 is for processing facilities. Section 66.17 includes prescriptive requirements for processing facilities such as location of process vessels and equipment (Section 66.17.4), containment, drainage, and spill control (Section 66.17.10), and management of operations hazards (Section 66.17.15). Later sections of Chapter 66 include additional requirements for dispensing, handling, transfer, and use of liquids (processes) (Section 66.18) and specific operations such as recirculating heat transfer system, vapor recovery and vapor processing system, solvent distillation units, and cooking oil storage tank systems in commercial kitchens (Section 66.19).

NFPA 1 Chapter 60

In addition to Chapters 66 and 67 of NFPA 1, Category 1 processes must also adhere to the requirements of 60.1 – 60.5 of NFPA 1 and the MA Amendments to NFPA 1. Depending on whether the quantities of materials at the facility exceed the MAQ or not, determines which requirements of Section 60.1 are applicable (Section 60.1.3).

Section 60.1.5 of NFPA 1 requires that an emergency action plan, consistent with the available equipment and personnel, shall be established to respond to fire and other emergencies in accordance with requirements set forth in NFPA 1. Section 10.8 of NFPA 1 includes additional requirements and guidance on generating emergency action plans, such as what must be included in the plans, purpose, responsibilities, etc. Section 60.4 of NFPA 1 contains requirements related to permissible storage and use locations and sets prescriptive requirements on the maximum allowable quantities for each individual control areas based on the area's occupancy type.

Section 60.5 contains the fundamental requirements for hazardous materials, including requirements for personnel training, ignition sources, systems, equipment, and processes (Section 60.5.1.4, 60.5.1.5, and 60.5.1.6 respectively). In Massachusetts, Section 60.5 includes multiple requirements that come from both NFPA 1 and the MA Amendments to NFPA 1. Despite Category 1 processes having a vessel capacity much less than Category 2-5 processes (Category 1 processes have a vessel capacity ≤ 2.5 gallons), there are numerous NFPA and state fire code requirements that must be followed. Therefore, the importance of fire protection engineers and their understanding of the NFPA and state fire codes cannot be understated for processing facilities. Each subsequent category process contains additional fire code fire protection/chemical process safety requirements of increasing complexity.

Category 2: Requirement Details

Category 2 processes must comply with all Category 1 requirements and additionally submit for a permit to the AHJ, as required per 1.12.8.34.2 and 60.8.1.1 of MA Amendments to NFPA 1. Permits submitted to the AHJ must comply with the requirements of Section 1.12 of NFPA 1 – 2015 edition. The nature of the operations and materials at the facility determine the permit requirements. Depending on the nature of the process, the AHJ may require technical assistance to evaluate the adequacy of facility process safety conditions, programs, procedures, and practices undertaken at the facility, but only after a notice of denial has been properly served upon the person making application.

Category 3: Requirement Details

While meeting all the requirements of Category 1 and 2 processes, Category 3 processes must also provide the following documentation and evaluations:

- Compliance with the requirements of Category 1 and 2 processes;
- Category 3 hazard evaluation; - per Section 60.8.2
- Implementation of process safety controls to mitigate hazards of normal and abnormal conditions identified in the hazard evaluation;
- Maintain documentation for review for a minimum of two (2) years;
- Third-party review (if required by the fire department/AHJ);
- Post-Incident Analysis per 60.8.6 (to be initiated within 48 hours of incident);

MA Amendment to NFPA 1, Section 60.8.2.2 provides requirements and details for the hazard evaluation that is required for Category 3 processes. **Hazard evaluations** are meant to identify hazards, including adjacent vessels that contain hazardous materials, and determine the required preventative, protective, and safety control measures in conformance with **r**ecognized **a**nd **g**enerally **a**ccepted **g**ood **e**ngineering (and safe work) **p**ractices (RAGAGEP) associated with a particular process or condition. Section 60.8.2.5 requires that a person with *specialized knowledge* beyond that of an average person, about risk assessment, process hazard analysis, and/or process safety management principles, for the process or processes being evaluated *conduct the necessary evaluations*. Section 60.8.2.5 further emphasizes the importance of both chemical process safety specialists and fire protection engineers for analysis of hazardous processing facilities.

In addition to generating a hazard evaluation, Category 3 processing facilities (vessel capacity ≤ 300 gal OR Group H Occupancy) must also meet the requirements of NFPA 1, Section 60.6, *Emergency Action Planning, Fire Risk Control and Chemical Hazard Requirements for Industrial Processes.* Where quantities of materials in use require compliance with Protection Level 1, Protection Level 2, Protection Level 3, or Protection Level 4 based on materials exceeding the Maximum Allowable Quantity (MAQ) in the following categories shall comply with the requirements of Chapter 7 of NFPA 400, *Hazardous Materials Code*:

(1) Unpackaged organic peroxide formulations that are capable of explosive decomposition in their unpackaged state
(2) Oxidizer Class 3 and Class 4: solids and liquids
(3) Pyrophoric solids, liquids, and gases
(4) Unstable reactive Class 3 and Class 4: solids, liquids, and gases
(5) Highly toxic solids, liquids, and gases
(6) Water-reactive liquids, Class 3

NFPA 400, Chapter 7, *Emergency Planning, Fire Risk Control, and Chemical Hazard Requirements for Industrial Processes,* contains requirements and guidance on topics such as:

- Process review and plan preparation
- Operating and maintenance procedures
- Safety reviews
- Incident investigation plans

Section 7.2.3 contains specific requirements for Emergency Action Planning that facilities must comply with. Some requirements of Section 7.2.3 include:

- Procedures for notifying and coordinating with the fire department and other applicable agencies
- Procedures and schedules for conducting drills
- Alternative measures for occupant safety
- Maintenance of fire protection equipment

Lastly, in lieu of complying with Chapter 60 in its entirety, occupancies containing high hazard Level 1 to high hazard Level 5 contents shall be permitted to comply with Chapter 10 of NFPA 400, as stated in NFPA 1, Section 60.7. MA Amendments to NFPA 1 further states that complying with NFPA 1, Section 60.7, is subject to an independent review in accordance with Section 1.15 and a copy, including its recommendations, shall be submitted to the Building Official.

Category 4: Requirement Details

Category 4 processing facilities (vessel capacity ≥ 300 gal) must meet:
All the requirements of Category 1, 2, and 3 processes
And also create and maintain a **Limited Safety Program (PSM "lite")**. The Limited Safety Program documents evaluation, policy and/or required procedures to ensure compliance with the following requirements of the MA Amendment to NFPA 1:

(1) Process information including, but not limited to, MSDS for the chemicals and products being processed, process chemistry, piping and instrumentation diagram, safety relief design, process control safety alarms and interlocks;
(2) Facility suitability including, but not limited to, the *Building Code* compliance, electrical hazard classification, ventilation design, fire alarm and fire protection, spill containment and control;
(3) A process hazard safety analysis including, but not limited to, effects in the event of failure, suitable administrative and engineering controls to minimize failure and to control unanticipated releases, and emergency responses to safeguard life and property;
(4) Written procedures, including routine operating and maintenance, as well as precautionary, shut-down and emergency response measures;
(5) A written training program for operating and maintenance personnel and outside contractors whose work or activity may affect process safety;
(6) A written records management protocol, which tracks any changes, including but not limited to changes to chemicals, equipment, operating procedures training program. Such records shall include the date of such change and the name of the manager responsible for such change; and an internal review at a maximum every three years.

These requirements are similar to those found in OSHA 29 CFR 1910.119, *Process Safety Management of Highly Hazardous Chemicals*.

Category 5: Requirement Details

Category 5 processing facilities must provide evaluation and documentation per 29 CFR 1910.119 OSHA, *Process Safety Management of Highly Hazardous Chemicals*, and 40 CFR Part 68 EPA, *Risk Management Program Rule*, all while meeting the requirements for Categories 1-4. OSHA's *Process Safety Management* (PSM) *of Highly Hazardous Chemicals* was issued in February 1992 and contains 14 elements that focus mainly on worker safety and catastrophic loss of containment. The Process Hazard Analysis (PHA) is the "heart" of the standard and is the most time consuming. The PHA must be completed every five (5) years and *is required to be completed by someone knowledgeable with the PHA methodology*. Additionally, compliance audits for the standard must be completed every three (3) years and may be completed in-house or by a third party (preferred). EPA's *Risk Management Program* (RMP) *Rule* consists of 12 elements and is focused on public safety and the environment. The program contains Chemical Accident Prevention Provisions (CAPP) and Accidental Release Prevention/Risk Management Plan Rule. The RMP must be submitted to the EPA every fire (5) years or if there are any changes.

Conclusion

Some fire code requirements for chemical process safety apply to processes with quantities much lower than the PSM/RPM thresholds. Failure to comply with these requirements can result in fire department citations, fines, remedial work and in extreme cases, shutdown of operations. Between 2013 and 2018, 214 OSHA PSM violations were recorded in Massachusetts alone, which resulted in over $485,000 in penalties. There are far more non-PSM facilities than PSM facilities in MA, so the magnitude of the problem can be seen by simple extrapolation. Failing to comply with the fire code and the federal requirements can also result in significant property loss, injuries and fatalities with much higher costs than the violation penalties. Compliance with the local fire code and federal regulations should be determined by experienced chemical process safety specialists and fire protection engineers to minimize the risk of significant loss.

Fire protection engineering and science is its own discipline and chemical process safety is its own discipline usually practiced by the chemical engineering field. Finding ways to create awareness and bridge the gap between the two disciplines, and sharing chemical process safety management concepts with those in the fire protection arena, will likely improve the oversight, understanding and enforcement of process safety resulting in better overall safety. There may be a way of joining these disciplines into one overarching safety category and therefore giving a better sense of how interrelated they really are.

546769 Poly-Butadiene Formation in Olefins Debutanizer Condenser

Abstract Title : Poly-butadiene formation in Olefins Debutanizer condenser

Discipline : Process Safety

Topic : Poly-butadiene formation in Debutanizer condenser

Author Name : Ali Bazroun

Participating team: Fahad AlShammari, and Eldridge Monette, Bertrand, Da, Cruz, Fahad AlSaad, Abdullah Al-Ismail

ABSTRACT

In one of Sabic olefins plant Plant, the condenser of the debutanizer tower was fouling. This was evident through the pressure increase in the tower and the cooling water control valve reaching the maximum opening (69%) limited by the mechanical stopper. The condenser process service is butadiene, which has the tendency of polymerizing and formation of popcorn, which could plug relief piping or accumulate in the condenser until the vessel shell is over stressed causing loss of containment and leading to fire or explosion.

The risk assessment team evaluated that the risk as significant, meaning that it could result into fatality and multiple lost time injury with a likelihood of 10E-3. With the propagating shell deformation, the likelihood could increase to L3 (10E-2) over a longer period. The risk could become C2-L3 and remain Significant. To ensure effective risk mitigation, a major plant modification was executed by constructing a new process line that bypasses the debutanizer tower and sends the material to the flare. The project execution was remarkable as it was completed in about a week. 15 tons of C4 is being flared per. This, however, allowed the olefins unit to mitigate the risk effectively without any interruption of the production. The change was not done before in the local industry, and it reflected how effective measures can be taken to mitigate risk through strong leadership and excellent engineering. Interim mitigations, moreover, were in place until the modification was completed: for example, inhibitor injection in the exchanger to inhibit the popcorn formation as soon as possible, and an emergency procedure in case loss of containment. Without the effective mitigation, Sabic operation team would have had to shutdown the olefins unit for over 2 months. This olefins unit feeds two poly ethylene that would shutdown consequently, causing significant losses over the mentioned period.

After isolating the exchanger, borescope test proved the formation of polymer. A root cause analysis was initiated, along with cleaning the tower and fabricating a new exchanger. The details of the RCA will be shared in the full paper.

19 AIChE Spring Meeting
15TH GLOBAL CONGRESS ON PROCESS SAFETY
NEW ORLEANS • 2019

Industrial process plant safety by collaboration of a Safety Instrumented System and a Dynamic Simulator

Tadaaki Ando
Yokogawa Corporation of America
12530 West Airport Blvd, Sugar Land, TX 77478, USA
Tadaaki.Ando@us.yokogawa.com
(Tadaaki.Andou@jp.yokogawa.com)

Yasuhiko Yamashiro
Yokogawa Electric Corporation
2-9-32 Nakacho, Musashino-shi, Tokyo, 180-8750 Japan
Yasuhiko.Yamashiro@jp.yokogawa.com

Prepared for Presentation at
American Institute of Chemical Engineers
2019 Spring Meeting and 15th Global Congress on Process Safety
New Orleans, LA
March 31 – April 3, 2019

Industrial process plant safety by collaboration of a Safety Instrumented System and a Dynamic Simulator

Tadaaki Ando
Yokogawa Corporation of America
12530 West Airport Blvd, Sugar Land, TX 77478, USA
Tadaaki.Ando@us.yokogawa.com
(Tadaaki.Andou@jp.yokogawa.com)

Yasuhiko Yamashiro
Yokogawa Electric Corporation

Keywords: Industrial process plant, SIS (Safety Instrumented System), Dynamic Simulator

Abstract

Safer HMI design for operators' quick and appropriate response is very important. Allowed maximum delay for operator's action from the alarm is not same all the cases because it is depends on many factors, i.e. ambient temperature, density of material, O2, and etc. So design of SIS includes HMI should be based "worst case". However, the 'worst case' is difficult to be found.

A "Dynamic Plant Simulator" can precisely simulate behaviors of the real process by visualizing how a hazard is diffused at various stages of a chemical reaction. And combination of DCS/SIS/HMI simulator with "Dynamic Plant Simulator", safety logic design is optimized about trip point and alarm threshold, and HMI design is modified safer manner.

The effective result of application of "cyber model by Dynamic Process Simulator" to a vinyl acetate monomer (VAM) plant and its value for operator training will be discussed.

1 Introduction

Industrial process plants in Japan have been seeking for the highest productivity and optimization of their processes, same with any process plant all over the world. These plants were renovated and expanded as time passed for the purpose, the plants themselves became complicated and troublesome as a result. Few years ago, severe process plant incidents occurred in Japan. The most probable causes of these severe incidents are out of "incorrect judgement and operation" and "shortage of prediction for technical aspects" during exceptional manual work that has been done by incidents. Understanding the hazardous situation by risk analysis and designing safety measures utilizing a SIS (Safety Instrumented System) are important and effective for resolving "incorrect judgement and operation" and "shortage of prediction of technical aspect". Safety measure includes human intervention via SIS man-machine interface which showing alarm, event, and process data.

The safety measure and its function are developed on the desk by brainstorming or so-called "hazard and operability study (HAZOP)." The verification of the safety function hardware could be established by reliability based analysis, SIL calculation in other word. However, the verification of safety logic is the issue because it has only been verified "on the desk". Especially the determination of the criteria for the emergency shutdown is the most important, but it is not possible to be confirmed with "try and error" on a running plant.

2 Safe HMI Design

Safer HMI design with considering different conditions and situations in the real process for operators' quick and appropriate response is very important. To design it, knowing the behavior of the process itself and related factors which affect the process is necessary.

For example, a reaction speed varies on the ambient temperature, density of material, O_2, catalyst, and etc., so allowed maximum delay for operator's action from the alarm is not same all the cases. So design of SIS includes HMI should consider "worst case". However, finding the worst-case scenario with how much is the allowed operator reaction is difficult because of many factors.

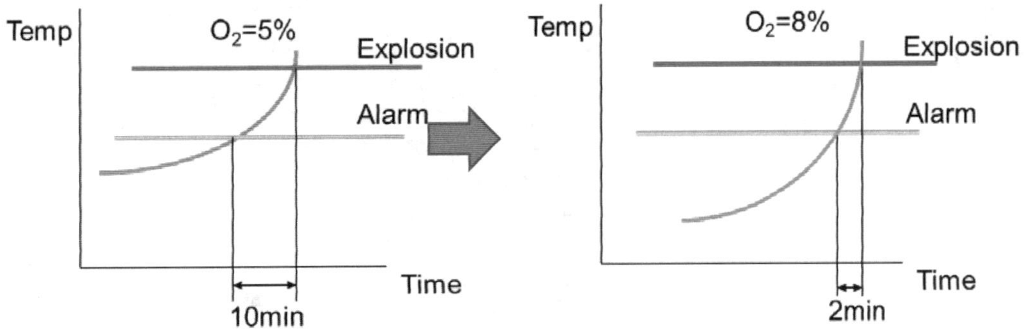

Fig.1 Example: Difference of O_2 density vs. allowed reaction time

Another example is the 'thermal runaway'. Even if feeding of material, heat, and catalyst were stopped by ESD because of a high temperature of a reactor, reaction continues due to a feed-forward loop of heating, means reaction generates heat to continue the reaction itself.

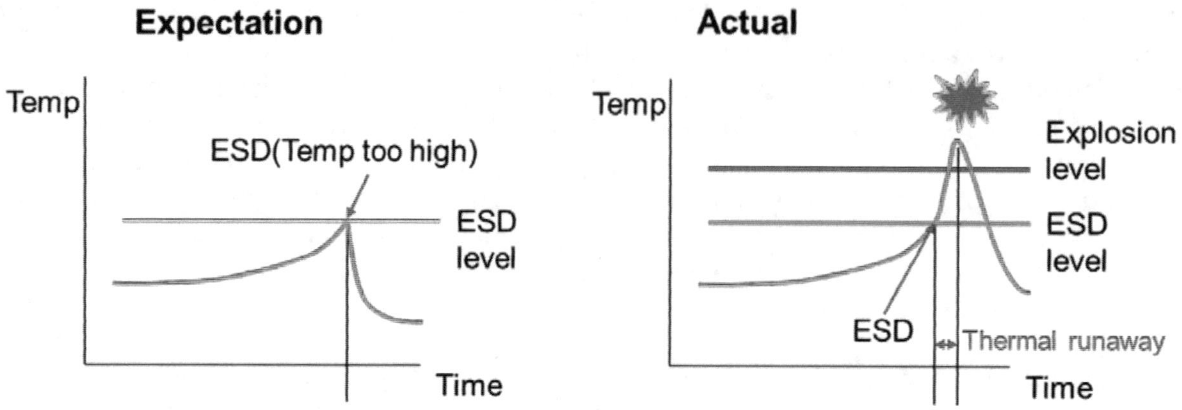

Fig.2 Even after ESD, thermal runaway leads the process to explosion

Thus, considered 'worst case' could not be a 'worst', or could be 'too much pessimistic'.

A "Dynamic Plant Simulator" can precisely simulate behaviors of the real process by visualizing how a hazard is diffused at various stages of a chemical reaction. And combination of DCS/SIS/HMI simulator with "Dynamic Plant Simulator", safety logic design is optimized about trip point and alarm threshold, and HMI design is modified safer manner.

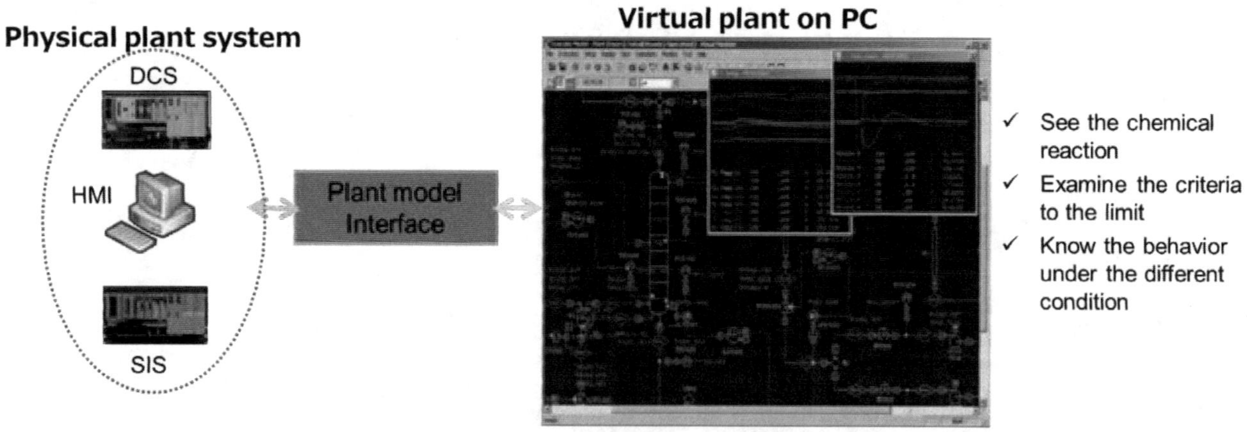

Fig.3 Dynamic Plant Simulator

3 "Dynamic Plant Simulator" at VAM (Vinyl Acetate Monomer) process

"Dynamic Plant Simulator" with "DCS/SIS/HMI simulator" was applied for several shutdown conditions of VAM plant. "Dynamic Plant Simulator" can simulate a variation of factors like ambient temperature and thermal runaway, because is based on a theory of chemical reaction. And "DCS/SIS/HMI simulator" is copying a real function and operation of DCS/SIS/HMI.

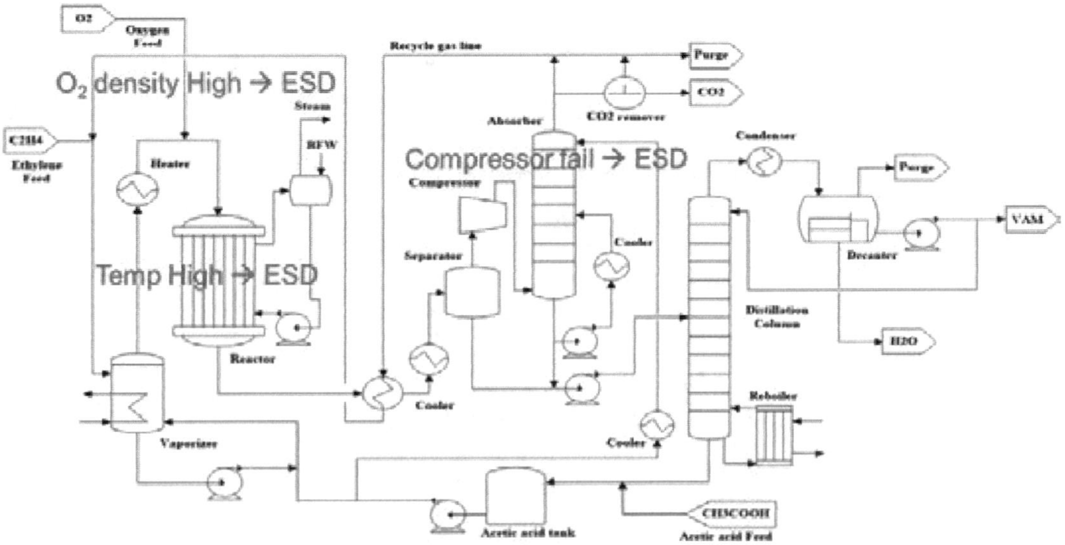

(Main Reaction) $C_2H_4 + CH_3COOH + 1/2O_2 \rightarrow CH_2{=}CHOCOCH_3 + H_2O$

(Side Reaction) $C_2H_4 + 3O_2 \rightarrow 2CO_2 + 2H_2O$

Fig.4 VAM process model

3.1 Shutdown by high temperature to avoid a damage to reactor

Reactor temperature must be monitored and controlled lower than 200 Degrees C (392 Degrees Fahrenheit) to avoid a damage of the reactor. If the temperature reached to a set point for emergency shutdown, emergency shutdown action stops feeding of Ethylene and O_2.

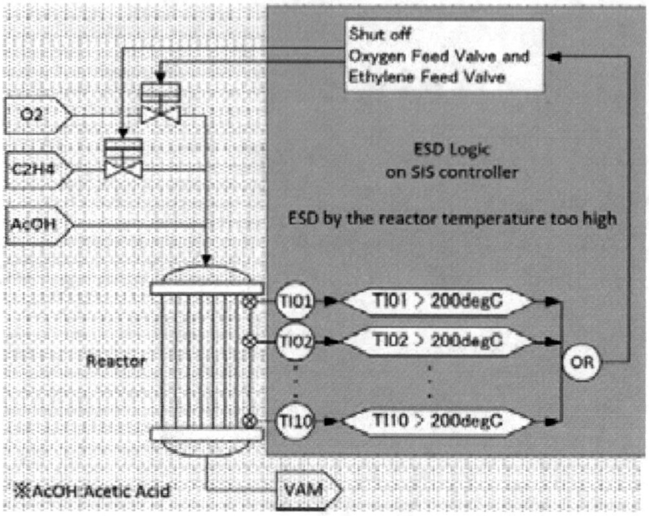

Fig.5 ESD logic for high temperature shutdown

When 190 Degrees C (374 Degrees Fahrenheit) is selected as the set point to shutdown, the result of simulation is, the reactor temperature must become over 200 Degrees C due to a runaway reaction in the reactor. It means reactor must be damaged. When 180 Degrees C (356 Degrees Fahrenheit) is selected, the simulation shows the reactor temperature can be kept lower than 200 Degrees C, so no damage to the reactor.

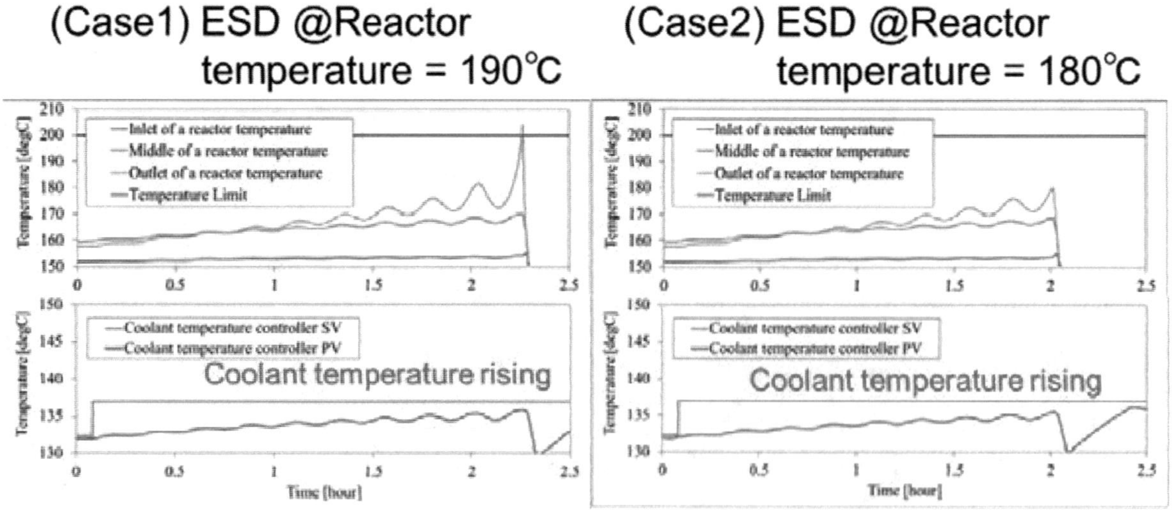

Fig.6 Simulation results

And simulation suggested the reactor temperature must be monitored in multipoint, middle of the reactor and top of the reactor. When the reaction temperature is lower than 160 Degrees C (320 Degrees Fahrenheit), temperature of a middle of the reactor is higher than other points, but when the reaction temperature is higher than 160 Degrees C, the temperature of top of the reactor is higher than other points.

3.2 Shutdown by high O_2 density to avoid explosion

If the density of O_2 in the reactor exceeds 8%, there is a risk of explosion. However, the O_2 density varies depends on other factors, i.e. feed pressure of ethylene to the reactor. Simulation has been done about O_2 density when the feed pressure of ethylene is suddenly changed. Simulation of some cases of dumping the feed pressure to lower side which increase the density of O_2 in the reactor were done.

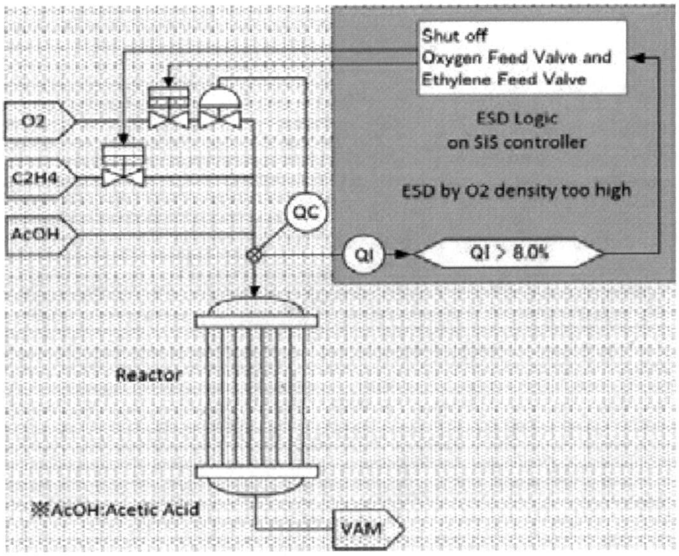

Fig.7 ESD logic for high O_2 density shutdown

When the feed pressure of ethylene dumped 10%, density of O_2 in the reactor doesn't reach to 8%. So there is no risk of explosion. However, the dumping of ethylene feed pressure is 20%, O_2 density will be higher than 8%, then an emergency shutdown must be activated. Shutdown does affect to business.

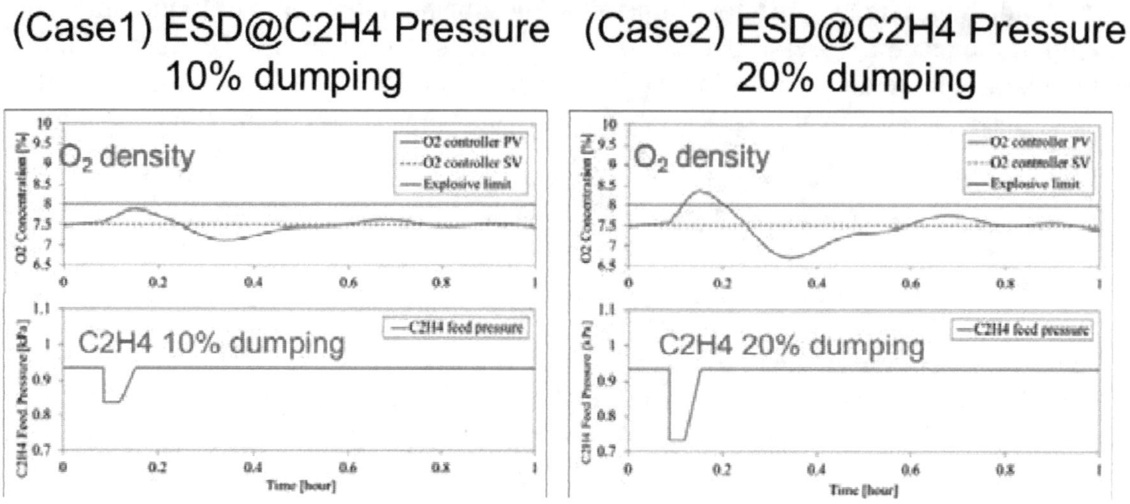

Fig.8 Simulation results

If there is a possibility of dumping ethylene feeding pressure, feeding O_2 in normal operation should be lower to avoid the risk of reactor explosion. On the other hand, lower O_2 density leads lower reaction speed, less production. Finding a best balance with safety and productivity is the key.

Using dynamic process simulator, many factors of process can be manipulated not only within normal operation range, but also close to abnormal, emergency situation or more un-realistic range, and results can be confirmed. This type of “try-and-error” approach is not possible on the real process, but possible in the “cyber” world to find a best balance of setting for maximize production with keeping or improving safety.

4 “Dynamic Plant Simulator” for Operator training

The training of operators is important in many reasons. Especially training for response to critical / emergency situation is indispensable, because operator rarely or never face such a situation and consequent secondary action to DCS/SIS or response to alarm flooding in a real plant. Therefore operator are impossible to familiarize with their required response in right procedure to critical / emergency situation. If operators are not capable or not knowing how to response such a situation, risk must propagate real hazard or accident. Furthermore, operators need to re-start the process after confirming the cause of emergency doesn’t exist anymore and checking all the conditions of materials are steady, status of equipment are ready to re-start. This procedures are also not normal for operators.

"Dynamic Plant Simulator" realizes operator training with "equal to real" environment. Operators will be trained for correct and timely manner response until proficient. And operators will have more understanding for process itself by playing on the simulator.

5 Conclusion

The effectiveness of "Dynamic Process Simulator" for safer design of DCS/SIS/HMI is confirmed on VAM plant. The best balance of safety and productivity can be found and optimization of safety action is created by simulation. The created new design of DCS/SIS/HMI is possible to use an actual system directly. And operator training with "Dynamic Process Simulator" will increase a total safety of the plant.

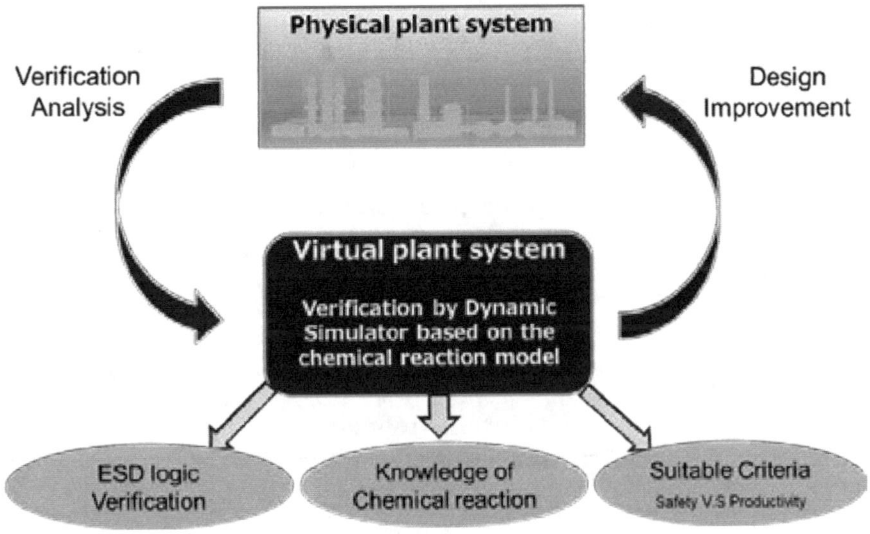

Fig.9:Virtual plant system realizes well balanced plant operation

Analysis of Chemical Effluent Discharge through the Emergency Relief System using Air-Water Properties

Basar Ozar
Jensen Hughes
600 W Fulton St. Suite 500
Chicago, IL 60661
bozar@jensenhughes.com

Jens Conzen
Jensen Hughes
600 W Fulton St. Suite 500
Chicago, IL 60661
jconzen@jensenhughes.com

Damian D. Stefanczyk
Jensen Hughes
3610 Commerce Dr. Suite 817
Baltimore, MD 21227
dstefanczyk@jensenhughes.com

Prepared for Presentation at
American Institute of Chemical Engineers
2019 Spring Meeting and 15th Global Congress on Process Safety
New Orleans, LA
March 31 – April 3, 2019

Analysis of Chemical Effluent Discharge through the Emergency Relief System using Air-Water Properties

Basar Ozar
Jensen Hughes
600 W Fulton St. Suite 500
Chicago, IL 60661
bozar@jensenhughes.com

Jens Conzen
Jensen Hughes
600 W Fulton St. Suite 500
Chicago, IL 60661
jconzen@jensenhughes.com

Damian D. Stefanczyk
Jensen Hughes
3610 Commerce Dr. Suite 817
Baltimore, MD 21227
dstefanczyk@jensenhughes.com

Keywords: two-phase flow, hydrodynamic load, emergency relief system, chemical effluent

Abstract

A sudden opening of a pressure relief device (typically a pressure relief valve or a rupture disk) during an overpressure event will cause the chemical effluent to accelerate through the emergency relief system (ERS) piping. The rapid change of momentum of the effluent creates transient hydrodynamic loads on the piping. ASME B31 in the United States and European Directive 2014/68/EU requires these transient loads to be considered along with other applicable loads (i.e., dead weight, thermal) in the system design. Additionally, the pressure distribution along the piping may be of importance, particularly when other vessels are connected to the same ERS. The back pressures generated on these vessels should be evaluated and compared with the vessels' MAWP (maximum allowable working pressure) as well as the set pressure of any isolation devices. As such, a detailed time history of the effluent pressure and hydrodynamic loads are required to perform a proper evaluation. The effluent may be composed of proprietary chemical mixtures containing various materials. Therefore, the exact material properties of the effluent during the transient might not always be known. In this study, a previously published methodology is used for calculating the dynamics of the effluent and the resultant hydrodynamic loads for a typical ERS by using the RELAP5 (Reactor Excursion and Leak Analysis Program) software. Air-water properties are assumed for the effluent. Key output parameters, such as pressure drop and hydrodynamic loads, are then compared with established hand calculation methods using the actual effluent properties. Given the mechanical nature of the hydrodynamic load event, it was found that the use of air-water properties provides a reasonable representation of the important characteristics of the actual effluent discharge in mechanical design space.

1 Introduction

Pressure build up in a system may occur as a result of a process upset, instrument failure or fire. These upset scenarios can result in either, or both, vaporization due to exothermic runaway reactions or generation of gases due to the decomposition process, which are of particular interest in terms of system and public safety. The extent of the potential hazards caused by such incidents are generally determined by the rate and amount of the energy being generated, the chemical substance makeup, and the quantity of the effluent.

The vessels or systems, which contain such chemicals (e.g. chemical reactors or storage tanks), are typically protected by an ERS that prevents from over-pressurization of the equipment as a consequence of the vaporization of chemicals and/or generation of gases. The process for designing of an ERS was previously discussed by Conzen and Wood (Ref. [1]) and Ozar (Ref. [2]). The design process consists of a process hazard analysis (PHA), defining credible and limiting upset scenarios, adiabatic calorimetry testing, vent sizing calculations, fluid transient analysis, pipe stress analysis, and a dispersion analysis. The central task in the design process is the fluid transient analysis (Refs. [1] and [2]), which is the main focus of this study.

The ERS is typically equipped with either a rupture disk or a pressure relief valve (PRV). Some systems may have both pieces of equipment. The sudden opening of the relief device during an overpressure event will typically result in a two-phase discharge and cause the chemical effluent to accelerate through the ERS piping. The rapid change of momentum of the effluent creates transient hydrodynamic loads on the piping. ASME B31 (Ref. [3]) in the United States and European Directive 2014/68/EU (Ref. [4]) requires these transient loads to be considered along with other applicable loads (i.e., dead weight, thermal) in the system design. Additionally, the pressure distribution along the piping may be of importance, particularly when other vessels are connected to the same ERS. The back pressures generated on these vessels should be evaluated and compared with the vessels' MAWP as well as the set pressure of any isolation devices. As such, a detailed time history of the effluent pressure and hydrodynamic loads are required to perform a proper evaluation.

The two-phase behavior of the chemical effluent has been studied extensively. Following earlier developments by Epstein et al. (Ref. [5]) and Grolmes and Leung (Ref. [6]), a simplified correlation was suggested by Leung (Ref. [7]) to predict flashing choked flow in a frictionless duct or nozzle. The correlating parameter, ω, was based on an isenthalpic (constant enthalpy) assumption.

Leung and Grolmes (Ref. [8]) integrated the momentum equation for homogenous two-phase flow in a horizontal duct using "ω" as a closure relationship for two-phase mixture specific density. Leung and Epstein (Ref. [9]) extended the derivation to an inclined duct configuration. Leung (Ref. [10]) provided the derivation of the reaction forces during a two-phase discharge from ERS piping.

Computational methods have also been used by various industries to investigate two-phase flow behavior in pipes and the hydrodynamic loads due to a sudden acceleration or deceleration of the fluid and the propagation of pressure waves. RELAP5 is a tool for analyzing small-break loss of coolant accidents and system transients in nuclear power plants. The program has the capability to model one dimensional single and two-phase fluid flow and heat transfer problems. The software has undergone extensive review and validation by the US Nuclear Regulatory Commission (Ref. [11]). It has also been benchmarked (Refs. [12], [13] and [14]) for predicting pressure wave propagation and hydrodynamic loads in piping systems. Refs. [15] and [16] provide a discussion about some of the nuclear power plant applications. However, RELAP5 is limited to non-condensable gas-water and steam-water two-phase flow applications.

This study provides a discussion about using the analytical methods developed by Leung and Grolmes (Ref. [8]) and Leung (Ref. [10]) to demonstrate the similarity between the two-phase mixture of common industrial chemical effluents and air/steam-water two-phase flow. Computational and analytical calculations for three different discharge scenarios are also provided. The comparisons between the computational and analytical results are used to demonstrate the applicability of using air/steam-water properties to simulate chemical effluent discharge to reasonably predict the hydrodynamic loads and pressure distribution in the ERS following the actuation of a pressure relief device.

2 Theoretical Background

The governing steady-state conservation equations for one-dimensional homogenous two-phase flow in a horizontal duct with a constant flow area are (Ref. [8]):

Mass

$$G = \text{constant} \tag{1}$$

Momentum

$$vdP + G^2\left(vdv + \frac{4fv^2dL}{2D}\right) = 0 \tag{2}$$

Energy

$$h_0 = h + \frac{G^2v^2}{2} \tag{3}$$

G, v, P, f, L, D and h are the mass flux, specific volume, pressure, two-phase friction factor, length of the duct, hydraulic diameter of the duct, and specific enthalpy, respectively. The specific enthalpy and specific volume are expressed in terms of quality (x) and phasic properties, i.e., $h_0 = h_{f0} + x_o h_{fg0}$, $h = h_f + xh_{fg}$, and $v = v_f + xv_{fg}$. Subscript 0 denotes stagnation conditions and fg the difference between vapor and liquid properties.

Epstein et al. (Ref. [5]) and Grolmes and Leung (Ref. [6]) suggested a simple relation between the two-phase specific volume and pressure as given by:

$$\frac{v}{v_0} - 1 = \omega\left(\frac{P_0}{P} - 1\right), \tag{4}$$

where ω is a non-dimensional parameter. Leung (Ref. [7]) correlated ω for flashing choked flow in a frictionless duct or nozzle. The correlating parameter was shown to be in the following simplified form:

$$\omega = \frac{x_0 v_{fg0}}{v_0} + \frac{C_{f0} T_0 P_0}{v_0}\left(\frac{v_{fg0}}{h_{fg0}}\right)^2 \tag{5}$$

C_{f0} is the heat capacity of the liquid. The last term on the right-hand side is zero for non-flashing gas-liquid two-phase flow and the non-dimensional parameter, ω, is provided by:

$$\omega = \frac{x_0 v_{fg0}}{v_0} \tag{6}$$

Leung and Grolmes (Ref. [8]) considered a two-phase horizontal discharge from a large reservoir (Figure 1). They integrated Equation 2 over the length, L, by substituting Equations 4 and 5 and using the isenthalpic (adiabatic and reversible) flow approximation. The momentum equations after integration yields to:

$$N \equiv 4f\frac{L}{D} = \frac{2}{G^{*2}}\left[\frac{\eta_1-\eta_2}{1-\omega} + \frac{\omega}{(1-\omega)^2} ln\left[\frac{(1-\omega)\eta_2+\omega}{(1-\omega)\eta_1+\omega}\right]\right] - 2ln\left[\frac{(1-\omega)\eta_2+\omega}{(1-\omega)\eta_1+\omega}\left(\frac{\eta_1}{\eta_2}\right)\right] \tag{7}$$

The non-dimensional parameters are defined by:

$$\eta_1 \equiv \frac{P_1}{P_0}, \eta_2 \equiv \frac{P_2}{P_0}, \eta_3 \equiv \frac{P_3}{P_0},\ G^* \equiv G/\sqrt{\frac{P_0}{v_0}}, N \equiv 4f\frac{L}{D} \tag{8}$$

η, $G^*$and f are the dimensionless pressure, dimensionless mass flux and two-phase friction factor, respectively. Leung and Grolmes (Ref. [8]) recommended a range between 0.003 and 0.005 for the friction factor.

G^* and η_1 are related via Bernoulli's equation for compressible fluid (Ref. [8]):

$$G^* = \frac{\sqrt{-2[\omega ln\eta_1 + (1-\eta_1)\omega - (1-\eta_1)]}}{\omega\left(\frac{1-\eta_1}{\eta_1}\right)+1} \tag{9}$$

The exit condition can be either subcritical, in which case $\eta_2 = \eta_3$, or critical where the local choking statement would require (Leung and Grolmes, Ref. [8]):

$$G^* = \frac{\eta_2}{\sqrt{\omega}} \tag{10}$$

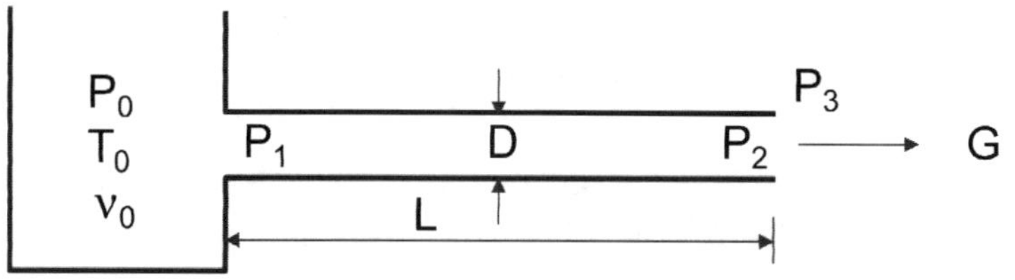

Figure 1. Two-phase horizontal discharge from a large reservoir.

With $\eta_1 = \eta_2 = \eta$, the above two equations satisfy the following transcendental equation (Leung and Grolmes, Ref. [8]):

$$\eta^2 + (\omega^2 - 2\omega)(1-\eta)^2 + 2\omega^2 ln\eta + 2\omega^2(1-\eta) = 0 \quad (11)$$

Similarly, the generalized thrust expression (T_R) is provided by Leung (Ref. [10]):

$$\frac{T_R}{P_0 A_e} = \frac{G^2}{P_0 \rho_0}\left(\frac{v_2}{v_0}\right) + \left(\frac{P_2}{P_0} - \frac{P_3}{P_0}\right) \quad (12)$$

The first term represents the fluid momentum expulsion rate and the second term represents the exit plane pressure (imbalance) force. For the case of unchoked exit, the exit pressure imbalance term vanishes, and the trust is only dependent on the exit momentum rate (Ref. [10]):

$$\frac{T_R}{P_0 A_e} = \frac{2[-\omega ln\eta_2 - (\omega-1)(1-\eta_2)]}{\omega\left(\frac{1}{\eta_2}-1\right)+1} \quad (13)$$

This approximate analysis shows that for two-phase flow in ducts (e.g. piping systems) ω is the sole independent variable and hence is the scaling parameter.

3 Transient Fluid Analysis using RELAP5

3.1 Description of the benchmark scenarios

A chemical reactor with a volume of 2.7 m^3 is considered for the benchmark scenarios. The reactor vessel is rated for 3 bars-g. The reactor is equipped with an emergency relief nozzle and a rupture disk of 41.1 mm inner diameter (DN40, 3.6 mm thickness). The nozzle is located near the top of the vessel. The piping downstream of the rupture disk has an inner diameter of 80.9 mm (4 mm thickness) and is 20 m long. It is a straight and horizontal section of piping.

Three different relief scenarios are used for the benchmarks. A summary of the postulated scenarios is provided below:

- Scenario 1: Polyurethane (PUR) and toluene diisocyanate mixture at 60 °C.
- Scenario 2: N-methyl pyrrolidone, formic acid, trimellitic anhydride and methylene diphenyl diisocyanate (MDI) at 90 °C.
- Scenario 3: Acetone under-fire scenario with vessel initially filled to 20% of its volume.

The chemical mixtures for Scenarios 1 and 2 are gas (CO_2, carbon dioxide) generating reactions. Alternatively, the scenarios are also called gassy reactions. A foamy mixture is expected inside the reactor vessel for these scenarios. Therefore, the chemical effluent discharging through the ERS is a two-phase gas-liquid mixture.

Scenario 3 is the boiling of a solvent due to a fire load applied on the vessel. The level of the boiling mixture is low, and liquid acetone is not entrained into the ERS. Only acetone vapor discharges through the ERS for this scenario.

3.2 *Description of the computer model*

The calculation was performed using the RELAP5/MOD3.3 (Patch 05) computer code, which is a light water reactor transient analysis code developed for the United States Nuclear Regulatory Commission (US NRC). It is suitable for the simulation of a wide variety of hydraulic and thermal transients in both nuclear and nonnuclear systems involving mixtures of steam, water, non-condensable gases and solutes under single phase and two-phase conditions (Ref. [17]).

A RELAP5 model was developed for the simulation of the venting scenarios from the reactor vessel into the ambient. The model was used to determine the hydrodynamic loads resulting from the sudden opening of the rupture disk. The mass flow rate of the two-phase effluent and the pressure distribution along the ERS piping were also calculated. Figure 2 provides an overview of the nodalization diagram of the model.

Time dependent volumes (TDVs) were used for the pressure boundary conditions and the thermo-physical condition of the fluid at those boundaries. Pipe components (P) were used to model the piping. Single junctions (SJs) connected the components such as pipes with pipes or pipes with TDVs. The valve junctions (V) was similar to the SJs but with the additional ability of the junction to open and close.

The length-to-diameter ratios (L/D) were set close to 1.0 for each pipe node and the time step was selected to be below the acoustic Courant limit. This is recommended for simulations involving propagation of acoustic waves and more generally for models where rapid fluid momentum changes are anticipated. The rupture disc was assumed to open linearly over a period of 10 msec, which is consistent with typical data (Ref. [18]). A steady-state calculation was performed for the first 5 seconds of each simulation to initialize the model. The transient portion of the simulation was initiated at 5 seconds, when each rupture disc started to open and was fully open at 5.01 sec. No heat transfer through the pipe walls was considered for the hydrodynamic fluid analysis. This is often a reasonable assumption for industrial processes due to the short duration of the relief event.

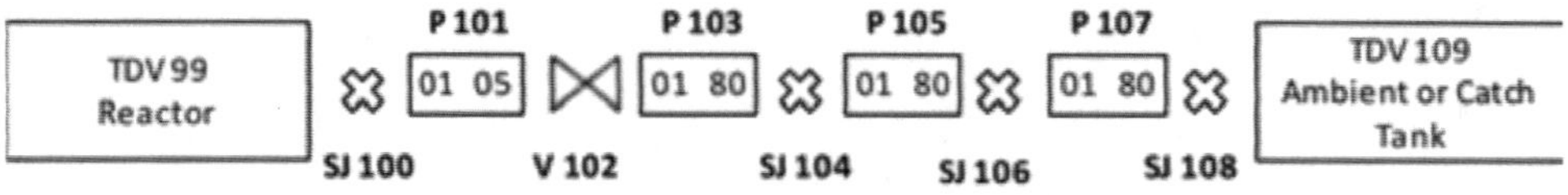

Figure 2. Nodalization diagram for the RELAP5 model.

The hydrodynamic loads (also commonly referred to as fluid force) on the modeled piping were computed through command files by using the RELAP5 output data. The generalized force equation in one-dimensional form can be resolved by a pipe segment bounded by two elbows as shown in Equation (14) (Ref. [19]). Note that the sign convention for the force is considered positive in the nominal direction of the flow as defined within the physical system.

$$F_W = -\frac{d}{dt}\int \rho A V dx \qquad (14)$$

F_W is the hydrodynamic load, ρ is the fluid density, A if the inner flow area of the pipe, V is the fluid velocity, and x is the distance along the pipe. For pipe segments with an open end an additional force component needs to be considered, which is regarded as the thrust load. Examples of open ended pipe segments are pipes that connect to pressure vessels, large manifolds, or a catch tank. The total thrust load is calculated as shown in Equation 15, which is made up of the momentum flux (due to fluid momentum expulsion) and pressure terms:

$$F_T = -(P_E - P_A)A - \dot{m}_E V_E, \quad (15)$$

where P_E, P_A, $\dot{m}_E$ and V_E are the pressure at the pipe exit, pressure of the reservoir the open piping is connected to, and the mass flow rate and fluid velocity through the pipe exit, respectively.

3.3 Boundary conditions for the RELAP5 model

A summary of the upstream boundary (reactor vessel) conditions for the RELAP5 models of the three scenarios are provided in Table 1. 10% over-pressure (10% greater than the set pressure) was used for Scenarios 1 and 2, which is typical for relief system sizing of reactive chemicals. 21% over-pressure was used for Scenario 3 per API 520's (Ref. [20]) recommendation for under-fire scenarios. Two-phase gas liquid discharge was assumed for Scenario 1 and 2, where foamy mixtures inside the reactor vessel were established. Single-phase gas discharge was assumed for Scenario 3. The gas-liquid mixture of chemical effluent was simulated by using the air-water properties. The quality values of the air-water mixtures were adjusted so that the ω value of the gas-liquid mixture of the chemical and that of the air-water mixture would match. The quality and the ω value of chemical mixtures were determined based on the initial fill ratio and the homogenous two-phase mixture assumptions. A single-phase acetone vapor release scenario (Scenario 3) was simulated by using air. The determination of the choking condition at the rupture disk was made based on the analytical model, which was described in Section 2. The downstream boundary conditions were set to be at typical atmospheric conditions (air at 20 °C and 1 bar-a).

Table 1. Upstream Boundary Conditions for the RELAP5 Model

Upstream Boundary Conditions	Scenario 1	Scenario 2	Scenario 3
Pressure	4.4 bar-a	4.4 bar-a	4.8 bar-a
Flow regime	Two-phase gas-liquid	Two-phase gas-liquid	Single phase vapor
Simulant	Air-water	Air-water	Air
ω	0.085	0.336	Not calculated
Choking at the burst disk	No	Yes	Yes

4 Results and discussion

The time history of the mass flow rate at the rupture disk calculated using RELAP5 for Scenario 1 is provided in Figure 3. The flow quickly accelerates following the opening of the rupture disk. The effluent experiences the frictional and form losses only for the sections of the piping, which it has already occupied at a given time. Therefore, the transient mass flow rate is slightly greater than the steady-state mass flow rate while the effluent is filling up the ERS piping. Once the entire ERS piping is filled, the mass flow rate achieves a steady-state value.

Figure 4 shows the pressure-time histories downstream of the rupture disk and at the end of the ERS piping for Scenario 1. The pressure downstream of the rupture disk gradually increases to its steady-state value as the effluent fills the piping. The pressure at the exit of the piping remains near atmospheric pressure (downstream boundary condition).

Figure 5 shows the dynamic load calculated on the ERS piping. The total load and its constituents are provided. There is a spike for the wave load as the rupture disk opens. This spike is caused by the depressurization wave traveling from the rupture disk to the reactor vessel. Typically, this section of the piping is filled with gas or vapor for which the speed of sound is well known. Therefore, the initial spike is captured accurately and the speed of sound of the two-phase mixture (which is typically unknown) does not have a significant role on the results. The wave load continues to be the dominant mechanism of the load as the effluent continues to fill up the ERS piping. Once the piping is filled up, the flow momentum term replaces the wave load. The flow momentum is due to the discharge of the effluent from the ERS piping into the ambient or a catch tank and it is the first term on the right-hand side of Equation 15. The pressure force is zero since the exit of the ERS piping is at the same pressure as the ambient; there is not choking at the exit of the piping.

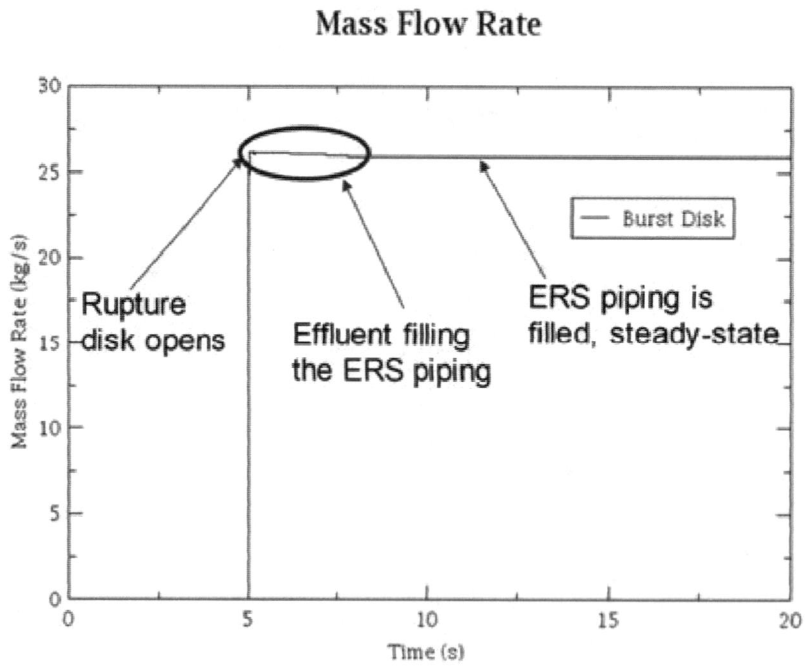

Figure 3. Mass flow rate for Scenario 1.

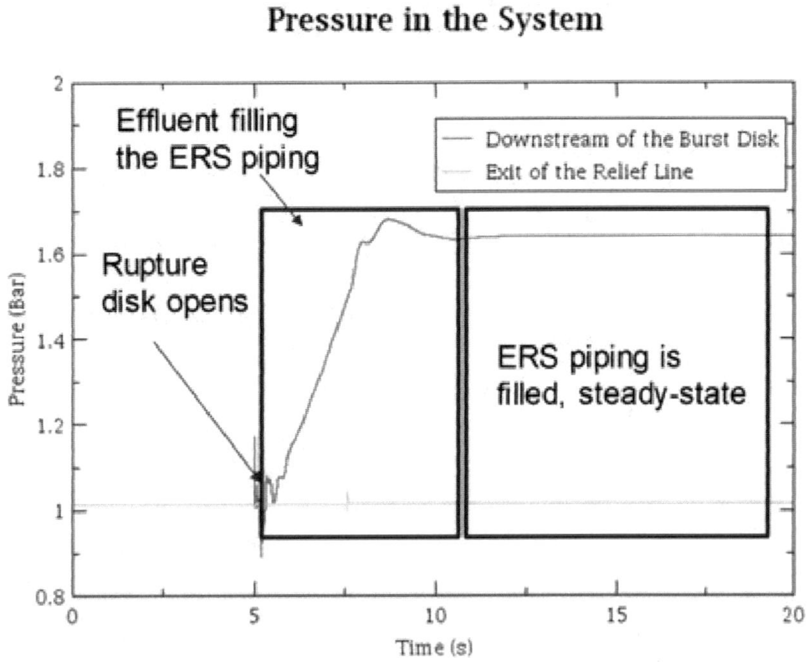

Figure 4. Pressure in the ERS piping for Scenario 1.

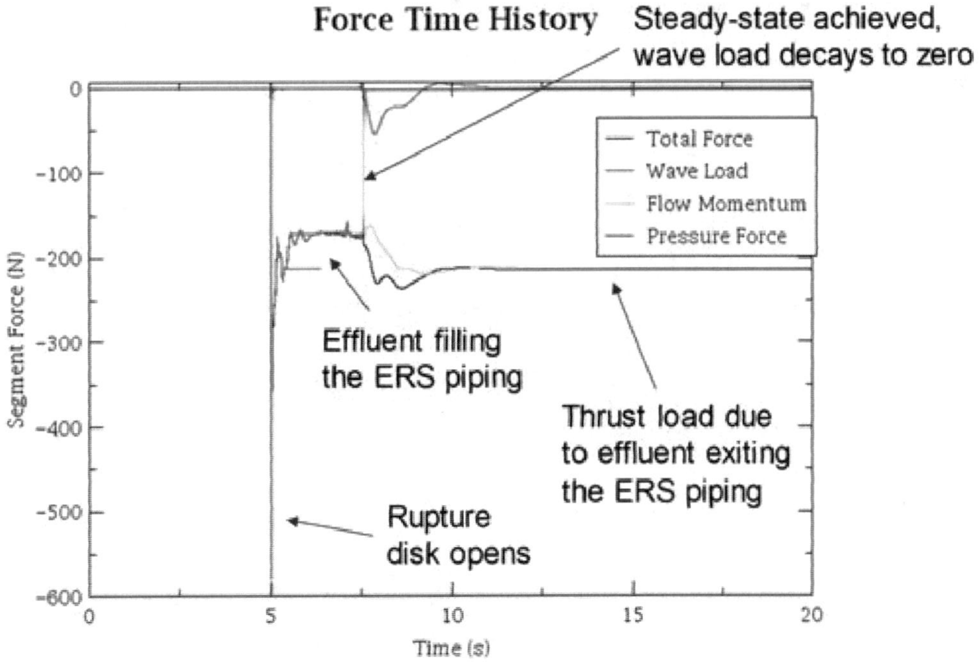

Figure 5. Dynamic loads for Scenario 1.

It is interesting to note that the wave load (while the ERS piping is filling up) and the load due to flow momentum (at steady-state) are almost similar. This is explained using the mathematical relations for the wave and thrust loads. The wave load for an effluent while filling up an empty piping (using Equation 14) is mathematically same as the flow momentum term in Equation 15.

The steady-state pressure, mass flow and thrust load values were compared between RELAP5 calculations and the analytical model, which was provided in Section 2. Actual chemical properties were used for the theoretical model. The comparisons between the two methods are provided in Table 2. The results demonstrate that the two-phase chemical effluent can be simulated with air/steam-water properties as long as the ω values are matched for these mixtures.

Table 2. Comparison between calculations using air-water (RELAP5) and actual two-phase chemical properties

Parameter	Scenario 1	Scenario 2	Scenario 3
Pressure downstream of the burst disk (bar-a)	1.6\1.7	1.6\1.7	1.9\1.6
Mass flow rate (kg/s)	25.9\26.1	16.3\16.4	1.2\1.2
Steady-state thrust load (N)	212.5\191.4	176.3\199.9	219.3\235.0

Black: RELAP5 results Red: Theoretical model

In the chemical manufacturing industries, the effluent can be composed of proprietary chemical mixtures containing various materials. Therefore, it can be challenging to know the exact material properties at any given time during an emergency relief event. The method presented in this study requires only the stagnation properties of the chemicals at the venting conditions for predicting key parameters, such as the pressure distribution and wave loads during the transient. These properties are typically established by adiabatic calorimetric testing or similar characterization of the chemicals. The method presented in this study may also be used with other commercially available software, which can predict transient fluid dynamics and wave loads.

5 Conclusions

In this study, a previously published methodology is used for calculating the dynamics of the effluent and the resultant hydrodynamic loads for a typical ERS by using the RELAP5 software and assuming air-water properties for the effluent. Key output parameters, such as pressure drop and hydrodynamic loads, are then compared with established analytical methods using the actual effluent properties. The following conclusions were observed:

- The fluid transient analysis is a key aspect of the design of the emergency relief systems. It provides the wave loads required for the dynamic pipe stress analysis. It also provides key parameters such as the pressure distribution, mass and volumetric flow rates through the system, and wave loads. These key parameters are often used to determine back pressure effects on other vessels, for chemical dispersion analysis and dynamic pipe stress analysis.

- The analytical approach shows that there is a similarity between the air/steam-water and the gas/vapor-liquid chemical mixtures if the "ω" values of the two mixtures are matched. This allows for an emergency relief event to be simulated with air/steam-water properties.

- Similar pressure distribution in the piping system and loads can be achieved through the similitude of "ω". Therefore, "ω" of the air/steam-water mixture should be matched with the "ω" of the chemical effluent at the stagnation conditions. The quality of the air/steam-water mixture can be adjusted until a similar "ω" is achieved.

- The sonic velocity of the effluent is typically not known. Also, it is practically difficult to match the sonic velocity of the air/steam-water mixture with the sonic velocity of the chemical effluent, which is a major requirement for tracking pressure waves and wave loads. This study demonstrates that the emergency relief scenarios are pipe filling events, which only require the momentum flux to be calculated for a reasonable prediction of the initial wave load. Therefore, the sonic velocity of the effluent is not required to capture the initial wave load.

- The effluent can be composed of proprietary chemical mixtures containing various materials. Therefore, it can be challenging to know the exact material properties at any given time during the transient. The method presented in this study may be used with most commercially available software and air/steam-water properties for a reasonable representation of the important characteristics of the actual effluent discharge in mechanical design space.

- The study needs to be extended further to identify applicability and limiting conditions for this method.

6 References

[1] Conzen, J., Wood. R.G., (2015), "Modeling and Computation of Reaction Forces on Relief Piping During Depressurization", AIChE 11th Global Congress of Process Safety.

[2] Ozar, B., (2018), "Analysis of Chemical Effluent Discharge through the Emergency Relief System using Stream-Water Properties," DIERS Fall Meeting, 15-17 October, Houston, USA.

[3] ASME. *ASME Code for Pressure Piping, B31.* The American Society of Mechanical Engineers, New York, NY, 2016.

[4] "Directive 2014/68/EU of the European Parliament and of the Council," Official Journal of the European Communities, May 2014.

[5] Epstein, M., Henry, R.E., Midvidy, W., Pauls, R., (1983), "One-dimensional Modeling of Two-phase Jet Expansion and Impingement," Thermal-hydraulics of Nuclear Reactors, 2nd International Topical Meeting Nuclear Reactor Thermal-hydraulics, Santa Barbara, CA.

[6] Grolmes, M.A., Leung, J.C., (1984), "Scaling Considerations for Two-phase Critical Flow," Multi-phase Flow Heat Transfer III, Part A: Fundamentals, T.N. Veziroglu and A.E. Bergles, eds., 549, Elsevier, Amsterdam.

[7] Leung, J.C., (1986), "A Generalized Correlation for One-Component Homogenous Equilibrium Flashing Choked Flow," American Institute of Chemical Engineers, Vol. 23, pp. 1743-1746, October.

[8] Leung, J.C., Grolmes, M.A., (1987), "The Discharge of Two-phase Flashing Flow in a Horizontal Duct," American Institute of Chemical Engineers Journal, Vol. (33), p. 524.

[9] Leung J.C., Epstein, M., (1990) "The Discharge of Two-phase Flashing Flow from an Inclined Duct," Transactions of the ASME, Vol. 112, pp. 524-527, May.

[10] Leung, J.C., (1992), "Reaction Forces During Two-phase Discharges," American Society of Mechanical Engineers Journal of Fluids Engineering, Vol. (114), p. 689-692.

[11] Information Systems Laboratories, (2016), "RELAP5/MOD3.3 (Patch03) code manual volume I to VIII, NUREG/CR-5535/Rev P5-Vol I to VIII", prepared for USNRC.

[12] International Agreement Report, (2007), "Simulation of the Propagation of Pressure Waves in Piping Systems with RELAP5/MOD 3.2.2: Comparison of Computed and Measured Results", TÜV SÜD - NUREG/IA-0206, June.

[13] Sokolowski, L., Koszela, Z., (2012), "RELAP5 Capability to Predict Pressure Wave Propagation Phenomena in Single- and Two-Phase Flow Conditions", Journal of Power Technologies 92 (3), page 150 – 165.

[14] Miller, J.S., Ramsden, K.B., (2005), "Validation of a Thermal Hydraulic Computer Code to Perform Two-Phase Multi-Component Force Calculations for Structural Evaluations", Proceedings of 13th International Conference on Nuclear Engineering (ICONE13), 16-20 May 2005, Beijing, China.

[15] Ozar, B., Henry, C.E., Harvill, R., Norton, D.A., (2012), "Analysis for Low Pressure Cooling Injection System Suction Hydrodynamics for a Boiling Water Reactor," Proceedings of 20th International Conference on Nuclear Engineering (ICONE20), July 30-August 3, 2012, Anaheim, CA.

[16] Henry, C.E., Lim, J., Ozar, B., (2012), "Investigation of Hydrodynamic Loads Associated with Pyrotechnic Valve Actuation," Proceedings of 20th International Conference on Nuclear Engineering (ICONE20), July 30-August 3, 2012, Anaheim, CA.

[17] Riemke, R.A., Davis, C.B., Schultz, R.R., (2009), "Recent Hydrodynamics Improvements to the RELAP5-3D Code", Proceedings of the 17th International Conference on Nuclear Engineering, ICONE17-75380, July 12-16.

[18] Health and Safety Executive, "Offshore Technology Report 2002/023: Testing and Analysis of Relief Device Opening Times," Pipeline Simulation and Integrity Ltd., Yorkshire/UK.

[19] Moody, F.J., (1990), "Introduction to Unsteady Thermofluid Mechanics", John Wiley and Sons, New York, Section 7.4, p. 365-366.

[20] API Standard 520, (2008), "Sizing, Selection, and Installation of Pressure-relieving Devices in Refineries," 8th edition.

547437 Evaluation of the Thermal Stability of Styrene in Contact with Impurities

Safe handling and storage of reactive monomers have long been an issue for polymer manufactures. The unsaturated double bond may self-react to initiate unintended polymerization reactions under many scenarios, for example, completely depletion of inhibitors, exposure to fire or external heat, and in contact with impurities. Usually, a trace amount of incompatible species is able to trigger unintended reactions and lead to catastrophic consequences. Thus, understanding how the impurities promote self-reaction is expected to provide valuable guidance to safe monomer-handling practices. Our goal in this study is to quantify the contamination effects of some most common contamination chemicals on styrene self-polymerization. Thermal analysis tools including DSC and ARSST are applied to examine the exothermic characteristics of styrene mixed with contaminating substances at different concentration levels and mixing ratios. From the preliminary results, we observed the onset of self-polymerization shifted to a lower temperature when in contacted with water and also alkaline. A trace amount of concentrated strong acid initiated unwanted exothermic activity even at room temperature. Self-heating rate was also elevated when diluted acid was present. Additional research is needed to understand what is the impact in of other types and concentrations of contaminants such as rust, metal ions. Our study demonstrates significant implications in the prevention of runaway incidents during transportation and storage of styrene.

549038 A Mixed-Integer Programming Framework for Placement of Fire Detectors in Chemical Processing Facilities

In chemical processing plants, gas and fire detectors provide an important layer of protection for personnel. However, given the large variety of chemical and physical sensors and number of factors influencing their capabilities to successfully detect hazards, the decision of where to place sensors is challenging. Gas detectors (e.g. ultrasonic, catalytic bead) must be in contact with gas leaks to be effective, while fire detectors (e.g. point infrared, open path infrared, optical) require a visual path to the fire, free of obstructions, for successful detection. Numerical optimization techniques provide an opportunity to overcome these challenges and determine effective sensor placement.

The optimal gas detector placement problem was formulated by [1] as a stochastic program based on the p-median facility location problem and relied on scenarios generated from CFD simulations within a petrochemical facility. Their results showed that coupling coverage with expected detection time using a scenario-based method outperformed the existing coverage-only approaches and has inspired formulations based on MINLP approaches for non-uniform failure probabilities [2].

But while gas detectors must be placed in the path of gas leaks for effective detection, fire detectors must be placed within unobstructed visual distance of flames. The optimal placement of visual fire detectors in processing plants can be particularly difficult due to the large number of physical obstructions (e.g. piping, valves, tanks).

This work focuses on an optimization formulation for placement of fire (visual) detectors. The scenario-based, p-median approach demonstrated by [1] for gas detectors is translated to a maximum coverage model for fire detectors in a 3D space with consideration to failure probabilities. We show that this approach allows for the placement of fire detectors for a fixed sensor budget while under uncertainty. We demonstrate the effectiveness of this formulation on several case studies. We further discuss how the framework is readily scalable for larger problem sizes and amenable to various objectives.

We would like to thank the Purdue Process Safety and Assurance Center for their support in this work.

References:

[1] Legg, S. W., Benavides-Serrano, A. J., Siirola, J. D., Watson, J. P., Davis, S. G., Bratteteig, A., and Laird, C. D. (2012). A stochastic programming approach for gas detector placement using CFD-based dispersion simulations. Computers and Chemical Engineering, 47:194–201.

[2] Liu, J., and Laird, C.D., A Global Stochastic Programming Approach for the Optimal Placement of Gas Detectors with Nonuniform Unavailabilities, to appear in Journal of Loss Prevention in the Process Industries, 2017.

546855 Advance Alarm Management Applied to DCS Improvement in a Petrochemical Plant: A Case of Study

Alarms use in process industry is fundamental for avoiding process accidents in which people, the environment or process equipment could be damage. Currently, Carbon Black production plants are equipped with distributed control systems (DCS) for setting alarms values and to provide indication of abnormal events to the control rooms operators. However, the basic control system could not meet the needs against the different production conditions and the presence of improper alarms could send a bad signal to control rooms and it could be the start of a process safety event.

In the present study, an alarm system assessment was performed in a petrochemical plant, located at Mamonal industrial park, in Cartagena – Colombia; based on data analysis, it was observed that when process conditions changes occured, the DCS was not be able to adjust alarms settings automatically, according to the load fluctuation and/or process conditions. In order to solve this problem, in first place the DCS performance was studied and it was reported an alarms list, which provided information regarding to the alarm flooding phenomena, alarm chattering and others alarm system issues.

On the other hand, the information earned and alarm rationalization methodology were used for establishing an action plan in order to improve DCS performance. Results showed the utility of these tools and it was identified the need to perform betters in the alarm system logic to avoid flooding alarm phenomena and to make operations safer.

548361 Overcoming Motivational Biases in Hazops

Hazard and Operability Studies (HAZOPs) are one method of assessing the risk in a facility. Based on the risk ranking of a scenario, recommendations may be required to reduce the risk. Often, these recommendations require additional capital expenditure and/or operational resources. In aging plants where resources are not readily available or in greenfield projects where there is a tight timeline to start up, there is sometimes a bias to risk rank scenarios such that there is no need to add additional safeguards. Motivational biases can outweigh common sense such that scenarios are not ranked according to their real risk. We have developed a method to overcome some of these motivational biases with the use of a polling system to engage all participants in the HAZOP. Participants are asked to vote anonymously on the severity and likelihood for a given scenario, then the results are discussed. During this discussion time, arguments for all the different risk rankings are discussed, including those voted on by the minority in the room. This facilitates an open discourse about the real risks for a given scenario rather than shifting the attention to the most vocal person in the room, who may be strongly influenced by a desire to reduce spending or minimize the effort required for maintenance. Following this discussion a second poll is performed to reach a consensus. We have found this to be an effective method for reducing motivational biases and ensuring that the right decisions are made based on the most logical argument instead of based on budget or time constraints. Tracking polling results increases the accountability of participants and makes them less likely to skew the risk ranking. The requirement for consensus ensures that all opinions are heard and considered, even if they are not popular. Together these make the polling system an effective method for overcoming motivational biases in HAZOPs and ensuring that safe decisions are made.

546515 Integrating Culture into Process Safety Assurance

Purpose: PETRONAS Chemicals Group Berhad (PCG) has embarked on a group-wide culture transformation, introducing the concept of the Results Pyramid® and a Culture of Accountability® i.e. a workplace culture where people take accountability to think and act in the manner necessary to achieve results.

Methodology: This paper will discuss two such activities anchoring on the Results Pyramid® that have been deployed across several plants and units within PCG.

The first activity is an enhancement of existing assurance programmes, where culture subject matter experts are assigned to observe plant leadership teams and front-liners during interviews and site verifications conducted by the technical assessors. The observations are synthesised into belief statements, which are presented together with the technical findings.

The second activity is Power Up, a session designed to incorporate powerful and specific feedback exchange between various teams, culminating in a specific list of actions to improve delivery of each team's activities in order to achieve the shared key results. This workshop can be done either as a follow-up to an assurance activity, in preparation for a project/turnaround activity or as a part of a periodic performance review.

Results: Examples of observations, feedback exchanges and specific actions from these activities will be shared, together with the impact on process safety leading and lagging indicators. Recommended practices and lessons learnt for continuous improvement will also be included.

Conclusion: Applying the Results Pyramid® concept to the design and implementation of Process Safety-related programmes has helped to identify the underlying leadership, mindset and behaviour issues impacting their success. By doing so, more specific interventions can be put in place to ensure full compliance to safety requirements and maximise effectiveness of these programmes.

548442 Emergency Plans Integration By Diferent Companies

In Brazil, different companies with pollution potential must have an emergency plan. However, each company analyzes its risks and contingency measures separately, based on the Conama 398 law. With the establishment of the Area Plan, companies from the same territory with pollution potential caused by oil in waters began to analyze how to combat a major accident quickly and effective together. This work shows how the integration of the individual contingency plans of each company and the cooperative result of the area plan in territory of significant scope and considerable environmental sensitivity.

No Brasil, diferentes empresas com potencial de poluição devem ter um plano de emergência. No entanto cada empresa analisa seus riscos e medidas de contingência separadamente, com base na lei Conama 398. Com a instituição do Plano de Área, empresas do mesmo território com potencial de poluição causadas por óleo em águas passaram a analisar como atender conjuntamente de forma rápida e eficaz a um grande acidente em seu território de abrangência. Este trabalho mostra como foi estruturada e organizada a integração dos planos de emergência individuais de cada empresa e o resultado cooperativo do plano de área em território de abrangência significativa e considerável sensibilidade ambiental.

548111 Moving Back to Flammable Refrigerants - Facility Siting and Personnel Safety Considerations

"Everything old is new again – Paul Allen"

Not just a catchy tune from the 1970's, this jingle summarizes a current technological shift in the world of refrigeration. Historically, refrigeration cycles used toxic and/or flammable materials as the working fluid. As the technology matured and made its way into transportation and consumer applications, the working fluid also evolved. Halogenated hydrocarbons, such as chlorofluorocarbons (CFCs), were developed as replacements to conventional refrigerants such as ammonia and propane. CFCs, which may have toxic effects, are desired for their tunable working range and for being non-flammable. However, the environmental drawbacks to the use of CFCs has been well researched and documented. As we embark into the future, where climate change and environmental quality are at the forefront of concern to consumers and industry alike, we are seeing a move back to flammable refrigerants. Everything old is, indeed, new again.

New construction and expansion projects can certainly consider and plan for the use of flammable refrigerants in commercial and industry facility design and siting, but what about retrofit applications? What are the risks and hazards associated with converting a CFC refrigeration system, designed and sited for the use of non-flammable refrigerants, to a system now utilizing flammable refrigerants? Beyond the need for additional fire protection systems, what risks are posed to the facility and personnel? In this paper, we seek to answer these questions by comparing the hazards and risks of mid-scale refrigeration systems using either non-flammable or flammable refrigerants. We explore how facility siting constraints may change with respect to the use of flammable refrigerants, and perform a quantitative risk assessment (QRA) for an example retrofit project in order to assess the potential for increased risk that should be considered in such projects.

548210 Making Sense of Risk Results

Risk based studies are very useful, but at the same time, the results are extremely complicated. While we have a basic set of results that can be used to illustrate the risks to people, these are relatively simple for risk practitioners and sometimes too complicated for site managers.

FN curves show the risk to all people from all sources, and individual risk contours illustrate the risk from all scenarios to people across specific locations, but how do we translate these results into actions. Generally, the action is to relocate people to lower risk areas, or upgrade buildings to mitigate the impacts identified. However, this is not always feasible. What if the site footprint is limited; there are public receptors who cannot be relocated; or building upgrades are cost-prohibitive or infeasible?

We must better quantify and evaluate the **sources** of risk rather than focusing on mitigating the impacts from the events. We need to evaluate the type of risks, the potential initiating events, and then we can assess the potential methods to effectively mitigate the risks. Existing risk result presentations are often very poor at illustrating the sources of risk. To effectively identify the sources of risk requires that we carefully dissect the risk results, while maintaining their validity and intent.

This paper will focus on how we find these risk drivers and make enough sense of them to create actionable recommendations that protect people.

551205 Ammonia Dispersion Modeling for Ammonia Storage with Multi-Scenario Failure System

PT Pupuk Kalimantan Timur is one of the largest fertilizer companies in Southeast Asia where located in Bontang City, East Kalimantan Province, Indonesia. The total production capacity of PT Pupuk Kalimantan Timur are 2.7 million Tons / year Ammonia, 3.43 million Tons / year Urea and 350 thousand tons / year NPK.

To accommodate ammonia production, PT Pupuk Kalimantan Timur currently operates 3 ammonia tanks with a capacity of 2 x 25,000 tons and 1 x 50,000 tons. In order to manage the hazard risk contained in the ammonia tank, it is necessary to do a modeling to map the potential and ammonia leakage potential areas and the impact of their exposure to the workforce and the community around the factory area.

The results of this modeling are expected to be used by Pupuk Kalimantan Timur's management in carrying out corrective actions for occupational health and safety performance, especially in controlling the potential hazard of ammonia, especially in ammonia tanks.

In this paper modeling is done using the AERMOD® Modeling System software. AERMOD Modeling System was chosen because it can be used on:

1. U.S. EPA regulatory Gaussian plume air dispersion model.
2. For near-field impacts (50 km max).
3. Prefered model for complex terrain.
4. Improved treatment of urban effects.

For modeling, we choose the AS-T-101 ammonia tank which has the largest capacity, which is 50,000 tons. Ammonia distribution models will be simulated based on weather conditions and ammonia leak source scenarios that can occur in the AS-T-101 ammonia storage tank system. Ammonia leakage sources will be displayed in 5 scenarios, i.e:

1. 1 Vent opens
2. 1 PSV poping
3. 2 PSV poping
4. 3 PSV poping
5. 3 PSV and 1 vent poping

The AERMOD Modeling System program is used to make ammonia distribution models for each event scenario assuming an exposure radius of up to 10 km. Based on the results of the calculation and modeling of the potential distribution of ammonia from AS-T-101 using the AERMOD Modeling System, the highest exposure of ammonia is in the 5th scenario, which is up to 36611 µg / M3.

546560 De-Mystifying the Concepts of Hazardous Area Classification

Purpose

The purpose of this topic is to address and clarify major items of confusion associated with hazardous area classification assessments and compliance auditing in classified locations. The ultimate objective is developing documentation that provides a framework for the control of electrical energy ignition sources in environments made hazardous by the potential presence of flammable vapors.

Background

In February of 1992 Occupational Safety and Health Administration (OSHA) 29 CFR 1910.119 Process Safety Management of Highly Hazardous Chemicals became law in the United States and confusion reigned supreme. Over the next several years industry professionals began the process of sorting through the responsibilities, regulations and impending deadlines that had been placed squarely in their laps. Acronyms such as PSM (Process Safety Management), PSI (Process Safety Information), RAGAGEP (Recognized and Generally Accepted Good Engineering Practices), MOC (Management of Change), and PHA (Process Hazards Analysis) became commonplace in the vocabulary of process safety professionals nationwide.

Employers now were required to meet stringent deadlines regarding the creation and management of a plethora of process information associated with equipment in the process covered by this law, i.e. any covered process. This information would be used to support the exhaustive efforts needed to create a baseline PHA for each covered process by May of 1997. In addition OSHA required all PHA's be placed on a five (5) year life cycle requiring their review and revalidation. Along with the revalidation of PHA's came the task of also revalidating all process safety information on the same five (5) year cycle.

As a result, every willing and able bodied employee was thrust into the quagmire of supporting, participating in, facilitating, or conducting PHA's. The required information used to support PHA's included the materials of construction, Piping and Instrument Diagrams (P&ID's), Relief System design and design basis, ventilation system design, material and energy balances, safety shutdown systems, and last but not least **electrical classification**. Yes, "electrical classification" was and still is legally considered an integral part of required process safety information.

This abstract addresses the following and provides an interactive discussion regarding the three (3) major myths surrounding electrical classification:

- assigning a classification to a location
- proper use of the classification information
- keeping the documentation evergreen

  The following topics will be addressed and clarified:

- What Is Hazardous / Electrical Classification?
- The Process of a Hazardous Area Classification Assessment
- The Concept of Electrical vs Hazardous Area Classification
- The Compliance Assessment (where does it fit in the process of controlling electrical energy ignition sources)
- Common Compliance Assessment Issues

546108 Maintaining Safety Excellence By the Implementation of Process Safety Inspection in Badak LNG

Badak LNG as one of the most experienced LNG Producers in the world, has undergone safety culture change throughout its operation. The goal of this change is to achieve excellent safety performances, indicated by the lower rate of Lost Time Injury Frequency (LTIF), lower number of Loss of Primary Containment (LOPC) Tier-1, High Potential Incident and multiple Train trip. This change was triggered by the occurrence of a fatal incident in 2005 as well as lost time injury incident in 2006. Since then, Badak LNG has integrated Process Safety Management (PSM) into its Company Management System, strengthened the roles and responsibilities of Managements, and pursued for interdependent safety behaviour of each personnel. One of PSM implementations is through process safety inspection.

Process safety inspection is an activity to assess the current equipment integrity based on visual observations compared with equipment's standard condition. The standard condition checklist for process safety inspection in Badak LNG is based on Physical Barrier Assessment tools, owned by Det Norske Veritas (DNV). The assessed items are divided into 5 categories: Inherent Safety, Hazard Prevention Measure, Hazard Control, Mitigations, and Support System. The overall score is determined by total number of standard items in all categories compared to the total number of verified items in all categories. The minimum acceptable score is at 80%.

During the year of 2012-2017, the average score of process safety inspections in Badak LNG is 90%. As a result of these inspections, Badak LNG had also identified the need of improvement in some items such as blinds, pipework and flanges, passive fire protection, fire & gas detection system and lock-open/lock-close block valves. In overall, by the implementation of PSM, Badak LNG has aligned on the right path to maintain its safety excellence achievement in LNG business, such as zero LTIF since 2007, zero multiple Train trip since 2012, zero LOPC Tier-1 since 2016, and zero High Potential Incident in 2017.

548411 Using HAZOP/LOPA to Create an Effective Mechanical Integrity Program

Many people view the conduct of a HAZOP/LOPA to address regulatory requirements as a chore, and stop there. However, the implementation of a quality HAZOP/LOPA has the potential to provide a framework for addressing numerous safety and operational optimization issues at plants, including the formulation/refinement of the Mechanical Integrity Program. The purpose of this paper is to focus on the mechanical integrity program, illustrate how a quality HAZOP/LOPA can support the effective implementation of some of the new requirements for California Refineries, and optimize key elements of an effective Mechanical Integrity Program, e.g.:

- Inspection/testing methods

- Testing intervals

- Maintenance outage periods

- Repair prioritization and allowable outage

- Identification of low priority equipment

544678 *"Identifying Critical Process Parameter to Ensure Asset Integrity" or "Pulse of Mechanical Integrity Management System"*

The management of equipment begins with an awareness that ageing is not about how old the equipment is, but is about what is known about its condition and the factors that influence the onset, evolution and mitigation of its degradation. The presentation provides an overview of Integrity Operating Window (IOW) parameters of any process that are key to understand the asset degradation and prevent Process Safety Incidents.

IOW are the situations in the process industries where a process parameters (temp., pressure, dosing rate, conc. etc...) when gets out of operating windows could result in significant loss of primary containment leading to potential severe consequences.

Arrazi - SABIC developed an IOW system that covers IOW identification, associated risk, mitigation (interim – long term), stewardship/management of IOWs and elimination.

The presentation shall cover Insight concept of the Topic, major Process Safety accidents related to breeching IOWs, actual example, IOW Management Process, its influence on PM plans etc.

three key bullet points that we can include in the agenda:

- Knowing operating limits with associated mechanical integrity or process safety risk to ensuring proper actions to be in operating envelope
- Developing the engineering leadership to systematically guide the operators, superintendents and engineers to be in safe zones
- Dynamic inspection frequency based on plant conditions and sharing some industrial major incidents for this critical item

19 AIChE Spring Meeting
15TH GLOBAL CONGRESS ON PROCESS SAFETY
NEW ORLEANS • 2019

The Practice of Process Safety Analysis and Management System Based on Safety Reliability
(THE NEW TITLE: The PHAMS system in SINOPEC)

Jiang Xue
SINOPEC Research Institute of Safety Engineering,
QingDao,China
Jiangxue.qday@sinopec.com

The PHAMS system in SINOPEC

Jiang Xue
SINOPEC Research Institute of Safety Engineering,
QingDao,China
Jiangxue.qday@sinopec.com

Keywords: Safety Barriers Reliability, PHAMS, HALOPA, SIL, Safety Management Online.

Abstract

The big data technique has been in trend for the process safety assessment. In China, it's the very need to transfer the qualitative safety management to quantitative safety management, which should rely on the professional risk analysis technique, as well as the standard safety reliability database. Based on the above, it's very important to integrate several professional risk analysis technique, as HAZOP, LOPA, SIL etc. to realize the integral risk identification and evaluation, grading and management plan making, supporting by the proprietary SINOPEC risk metrics and grading standards, as well as safety equipment reliability database,which is developed as the Process Safety Analysis and Management System(PHAMS) for SINOPEC, to support all SINOPEC branches to make dynamic risk analysis and management online

1 Introduction

With the smart factory's development, the static, periodic risk assessment and management can not satisfy the identification and real-time safety control of plant's risk events. The PHAMS which establishes the plant-wide dynamic risk analysis, can effectively solve the risk rating and monitoring of the unit and realize the plant-wide or corporate-wide management optimization.

The key data for HALOPA, SIL and Management module in PHAMS is the dependent protection barriers, which we rely on RBI correction, Personnel exposure probability and ignition probability to optimize the six mainly safety barriers' failure probability, simultaneously, do use of the Bayesian Correction Modelling to verify the instant failure probability on site, as well as the SIL identification, which can validly save the grading cost on time and personnel, and all of these are towards to the deeper research on dynamic risk analysis for PHAMS.

2 HALOPA Analysis System

2.1 SINOPEC RISK MARTIX

According to international common matrix criteria, as well as statistics from the SINOPEC accident data and the maintenance data, the frequency of risk occurrence and severity classes are coupled to the 8×7 matrix by a linear exponential model, and be assigned risk index to each risk level.

SINOPEC RISK MARTIX		Occurrence Frequency							
		1	2	3	4	5	6	7	8
		≤10⁻⁶/y	10⁻⁶-10⁻⁵/y	10⁻⁵-10⁻⁴/y	10⁻⁴-10⁻³/y	10⁻³-10⁻²/y	10⁻²-10⁻¹/y	10⁻¹-1/y	>1/y
Severity Classes	A	1	1	2	3	5	7	10	15
	B	2	2	3	5	7	10	15	23
	C	2	3	5	7	11	16	23	35
	D	5	8	12	17	25	37	55	81
	E	7	10	15	22	32	46	68	100
	F	10	15	20	30	43	64	94	138
	G	15	20	29	43	63	93	136	200

Figure 1 SINOPEC Risk Matrixp[1]

Based on this matrix, SINOPEC has established four-level risk list for establishing targeted management of branch network. For example, the executive board of the group is in charge of the red-level risk list, the branch directors is in charge of the orange-level risk list, etc.

2.2 HALOPA Analysis Modelling

2.2.1 System risk analysis based on equipment operation cycle

The HALOPA analysis model can be used to identify the hazard of the device or process unit, clarifying the initial event, process deviation, independent protective layer and consequence events, combined with the event tree analysis method to analysis the reduction influence of the occurrence frequency of initial events. The protection layer performs a failure frequency evaluation role to determine the final frequency of the on-site device risk events during each operating cycle under the coupling reduction of the existing protective layers at each level [2].

An important assumption of the model is that each protection layer is independent, that is, the failure of a single protection layer will not affect the risk reduction impact of other existing protection layers [3]. Each time a protective layer takes effect, the frequency of accidents is reduced by the risk reduction of the protective layer.

When the frequency of the initial event does not exceed 1 time/year or is less than twice the test frequency of the first protective layer, the evaluation of the consequence frequency can be referred to the following formula [4]:

$$f_I^C = f_I \times P_{enable} \times \prod_{j=1}^{J} PFD_{Ij} \quad (1)$$

where

f_I=Occurrence frequency of initial event I annually;

Penable=Moderation factors by enabling conditions and other conditional probabilities.

PFDIj=Failure probability of dependent protection layer j of initial event I.

When the frequency of the initial event exceeds 1 time/year or exceeds the test frequency of the first protective layer twice, the first protection layer failure shall be used as the initial event, and the calculation formula of the consequence frequency is as follows:

$$f_I^C = f_{\text{IPL\_first}} \times P_{enable} \times \prod_{j=2}^{J} PFD_{Ij} \qquad (2)$$

where

$\mathbf{f_{IPL\_first}}$= The failure probability of 1st dependent protection layer

The HALOPA analysis model in PHAMS designed a modified calculation model for equipment failure frequency and personnel exposure probability, and realized BPCS, SIS, physical protection, leakage storage and collection, fire and explosion suppression for equipment operation cycle and device in service. The system's six major types of protection layer failure frequency have the correction factors of risk reduction, and realize system risk assessment based on the device production cycle. For example, according to the system risk assessment of a diesel hydrogenation unit based on the PHAMS system, on-site data collection of various protective layers and key correction conditions under one cycle of the driving operation to date, online preliminary analysis of the risk overall value and grading of the feeding unit The results are as follows:

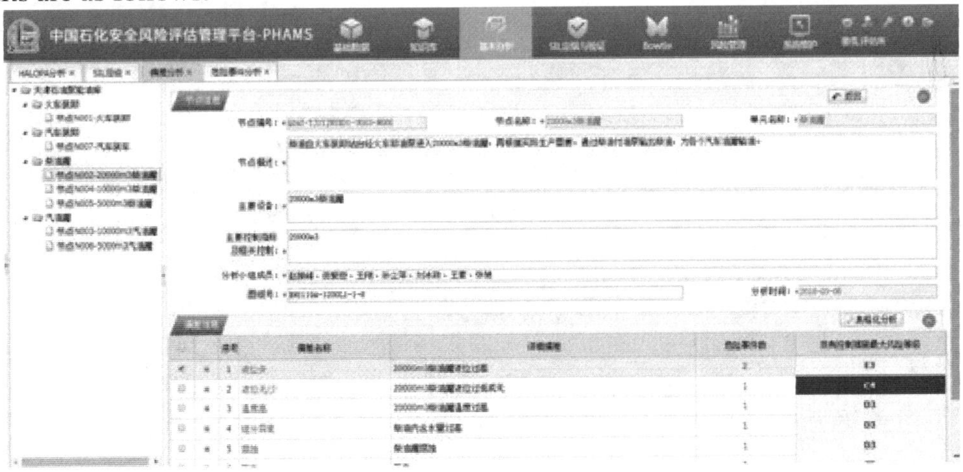

Figure 2 HALOPA Analytical Function Online

2.2.2 Dynamic risk assessment based on real-time data of safety control systems
According to the protection layer frequency analysis model, the failure probability of the protection layer lowering the frequency of a certain initial event can be regarded as θ , and the a priori risk probability distribution function is:

$$f_0(\theta) \propto \theta^a(1-\theta)^b \quad (3)$$

Its likelihood function $L(E|\theta)$ is

$$L(E|\theta) \propto \theta^m(1-\theta)^{n-m} \quad (4)$$

where

a, b=the prior distribution parameter of the failure probability of the protective layer;
m=The number of the posterior cumulative event occurred when the protective layer is valid;
n=The number of the posterior cumulative event occurred when the protective layer is failed.

With the posterior probability distribution function corrected by dynamic risk analysis as$f_1(\theta|E)$, the posterior probability distribution can be obtained by:

$$f_1(\theta|E) = \frac{f_0(\theta)L(E|\theta)}{\int f_0(\theta)L(E|\theta)d\theta} \propto \theta^a(1-\theta)^b\theta^m(1-\theta)^{n-m} \propto \theta^{a+m}(1-\theta)^{b+n-m} \quad (5)$$

Since m and n are parameters of the age change and the posterior probability distribution function seems to be continuous, the dynamic failure frequency of the protective layer at a certain moment can be obtained by the following formula[4-6].

$$P_i(\theta) = \int_{i-1}^{i} f(\theta)d\theta = (1-F_i(\theta|E)) - (1-F_{i-1}(\theta|E)) = F_{i-1}(\theta|E) - F_i(\theta|E) \quad (6)$$

In where $F_i(\theta|E)$ is $f_i(\theta|E)$ the cumulative failure frequency function as

$$F_i(\theta|E) = \int_0^i f(\theta|E)d\theta \quad (7)$$

And then bring in the following formula to get the dynamic risk frequency of the device or process unit:

$$f_I^C = f_I \times P_{enable} \times \prod_{j=1}^{J} P(\theta)_{Ij} \quad (8)$$

and

$$f_I^C = f_{IPL\_first} \times P_{enable} \times \prod_{j=2}^{J} P(\theta)_{Ij} \quad (9)$$

Where

$P(\theta)_{Ij}$=Posterior failure probability value of the jth independent protection layer

Based on this, the PHAMS system realizes the real-time update of the protection layer risk reduction frequency, and intelligently analysis the overall risk value and grading of the unit to plant.

3 SIL Intelligence Identification Online System

3.1 SIL Identification Modelling

The analysis data of the SIS protection layer of the HALOPA evaluation system in PHAM is updated and interconnected by means of the safety reliability database. After the HALOPA dangerous event using the same safety instrument loop is automatically brought into the SIL identification, the PHAMS system takes the same kind of consequences. C, to calculate without the cumulative frequency F^C before the SIF

$$F^C = \sum_{i=1}^{n} f_i^C \qquad (10)$$

where

$$f_I^C = f_I \times P_{enable} \times \beta \times \prod_{j=1}^{J} PFD_{IJ} \times \prod P_{condition} \qquad (11)$$

$$\beta = \frac{2^N - 1}{N} \qquad (12)$$

where

f_I^C= The occurrence frequency of the dangerous event I with consequence C

β= the common factors caused by several SIFs

N=the number of SIFs for the dangerous event I

Based on the modelling, the PHAMS system can intelligently group common risk events and assess the risk reduction factor RRF of SIF in each group [7] :

$$\text{RRF} \geq \frac{F^C}{TMEL} \qquad (13)$$

and

$$1 \times R_{\text{Factor}} < RRF_{SIL1} \leq 10 \times R_{\text{Factor}} \qquad (14)$$

$$10 \times R_{\text{Factor}} < RRF_{SIL2} \leq 100 \times R_{\text{Factor}} \qquad (15)$$

$$100 \times R_{\text{Factor}} < RRF_{SIL3} \leq 1000 \times R_{\text{Factor}} \qquad (16)$$

where

R_{Factor}= risk reduction factor between1 to 10.

Based on the SIL identification results of each group, the optimal solution is intelligently selected as the RRF of this safety instrument loop.

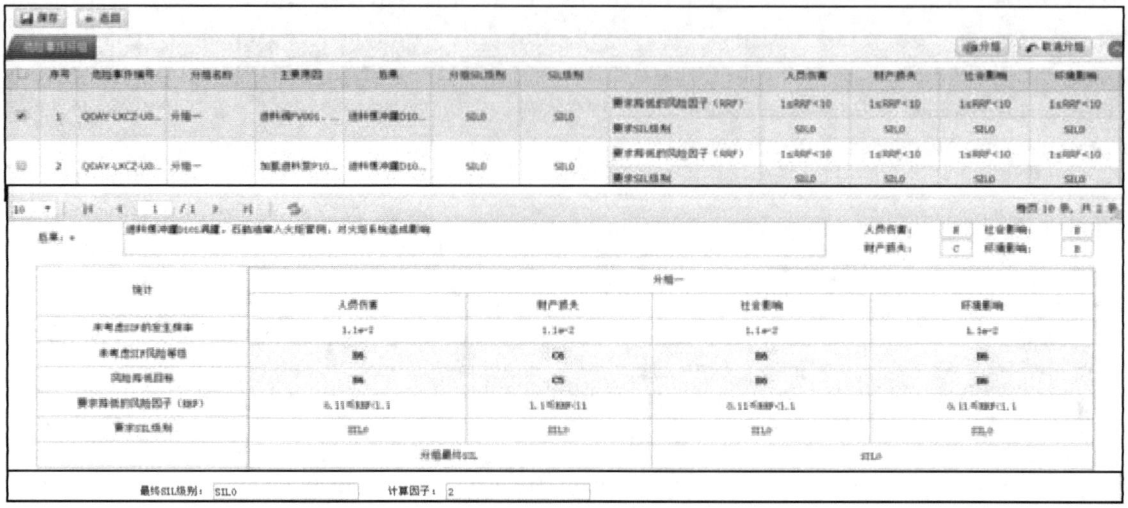

Figure 3 SIL Intelligence Identification Online

3.2 The SIL Identification Based on SINOPEC Risk Matrix in PHAMS

With the aid of data warehouse technology, PHAMS system can compare HALOPA assessment data which involved with SIS protection layers of the initial situation, the consequences classes and non SIS protective layer of data, and automatically take the residual risk of the events into the risk matrix. Then based on SINOPEC risk matrix, the SIL identification module can intelligently couple and assess the residual risk and RRF.

according to Sinopec's safety management focusing on high severity impact, PHAMS sets an independent TMEL requirements (as shown in the following table), and can automatically selects the most serious impact event as the final SIS identification data of this event. .

Table 1 TMEL of SIL Intelligence Identification[8-10]

Severity Classes	*Health & Safety (r/y)*	*Property Damage (r/y)*	*Social Impact (r/y)*
A	$\leq 1\times 10^{-1}$	≤ 1	≤ 1
B	$\leq 1\times 10^{-2}$	$\leq 1\times 10^{-1}$	$\leq 1\times 10^{-1}$
C	$\leq 1\times 10^{-3}$	$\leq 1\times 10^{-2}$	$\leq 1\times 10^{-2}$
D	$\leq 1\times 10^{-5}$	$\leq 1\times 10^{-4}$	$\leq 1\times 10^{-4}$
E	$\leq 1\times 10^{-6}$	$\leq 1\times 10^{-5}$	$\leq 1\times 10^{-5}$
F	$\leq 1\times 10^{-7}$	$\leq 1\times 10^{-6}$	$\leq 1\times 10^{-6}$
G	$\leq 1\times 10^{-7}$	$\leq 1\times 10^{-6}$	$\leq 1\times 10^{-6}$

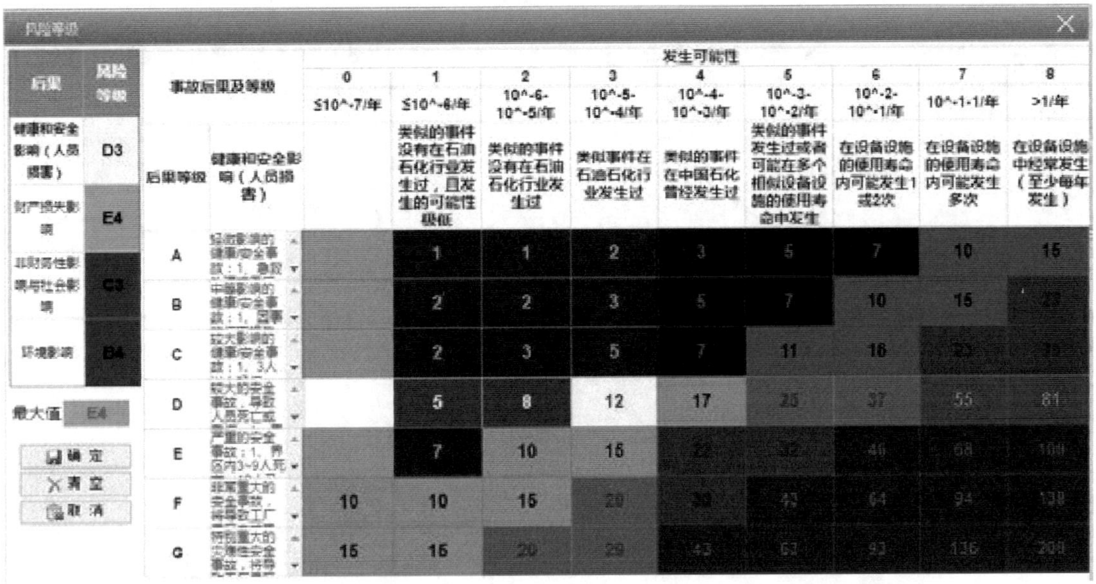

Figure 4 SIL Intelligence Identification on SINOPEC Risk Matrix

On the basis of SIL identification, the PHAMS system can also implement SIL verification based on multi-factor model, and evaluate the structural constraints, PFDavg and MTTF respectively. The verification result data is consistent with the accuracy requirements of IEC61508 and IEC61511.

4 Risk Management and Decision Support System

With the powerful safety and reliability database and expert experience database of PHAMS system, as well as the data mining of the result data of HALOPA and SIL system, the Risk Management and Decision Support System can implement the cumulative risk frequency calculation of independent devices and plants in construction or in service, to assess the overall risk based on different impact severities through the linear comparison [11-12].

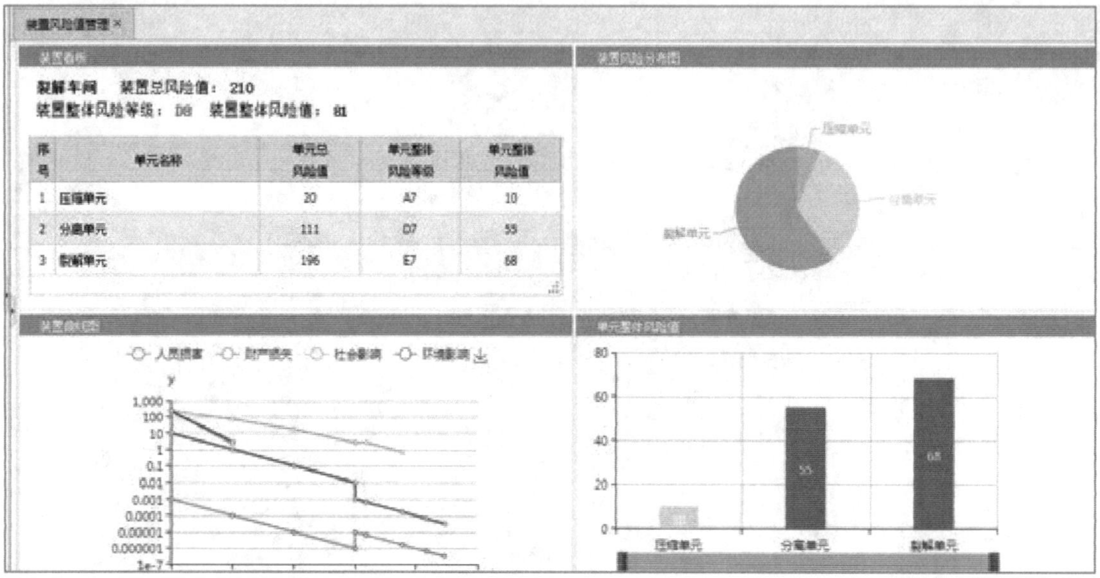

Figure 5 Overall Risk Rating and Risk Event Distribution Analysis of Ethylene Cracking Facility

5 Conclusion

1. PHAMS system sets up 9 categories of safety reliability database, expert experience data and Sinopec risk matrix, etc, which can support unified SINOPEC risk assessment and management standards, risk analysis template, process data storage and data mining for PHAMS system , as well as the interconnection and real-time update of data of each system.
2. HALOPA system implements plant overall risk analysis based on protection layer data, and performs data coupling evaluation from static analysis based on equipment cycle and dynamic risk correction to realize real-time evaluation of risk assessment technology and device operating conditions. The dynamic risk analysis research will be further strengthened to realize the system risk prediction function.
3. The SIL identificaiton and verification system can realize the evaluation of safety instrument function of in service and in construction. In the long run, it helps to improve the safety management level of the plant's overall safety instrument function
4. The risk management and decision support system can realize the overall risk assessment and risk distribution based on various evaluation results, through in-depth mining and statistical analysis, the PHAMS can realize the big data analysis management of Sinopec branches' risk management.

6 References

[1] The Bureau of Safe Production Supervision and Administration of SINOPEC. Guide for Safety Risk Identification and Assessment of SINOPEC[S].2018

[2] AQ/T 3054-2015. Guide for Application of Protective Layer Analysis Methods. Chapter 8：Scenery Frequency Calculation [S].

[3] Khan,Seyed Javad Hashemi,Nicola Paltrinieri,Paul Amyotte,Valerio Cozzani,Genserik Reniers.

Dynamic risk management: a contemporary approach to process safety management[J]. Faisal Current Opinion in Chemical Engineering.

[4] CCPS. Risk-based Process Safety [M]. 2013.

[5] Maryam Kalantarnia,Faisal Khan,Kelly Hawboldt.Dynamic risk assessment using failure assessment and Bayesian theory[J].Journal of Loss Prevention in the Process Industries . 2009 (5)

[6] Rakesh Ranjan,Sonam Singh,Satyanshu K. Upadhyay. A Bayes analysis of a competing risk model based on gamma and exponential failures[J] . Reliability Engineering and System Safety . 2015

[7]Zugravu-Soilita N. How does Foreign Direct Investment Affect Pollution? Toward a Better Understanding of the Direct and Conditional Effects[J]. Environmental & Resource Economics, 2017:1-46.

[8] Khakzad N, Yu H, Paltrinieri N, et al. Chapter 5 – Reactive Approaches of Probability Update Based on Bayesian Methods[J]. Dynamic Risk Analysis in the Chemical & Petroleum Industry, 2016:51-61.

[9] Villa V, Paltrinieri N, Khan F, et al. Towards dynamic risk analysis: A review of the risk assessment approach and its limitations in the chemical process industry[J]. Safety Science, 2016, 89:77-93.

[10] Mark Adrian van Staalduinen,Faisal Khan,Veeresh Gadag. SVAPP methodology: A predictive security vulnerability assessment modeling method[J]. Journal of Loss Prevention in the Process Industries.

[11] Podofillini L, Dang V N. Conventional and dynamic safety analysis: Comparison on a chemical batch reactor[J]. Reliability Engineering & System Safety, 2012, 106(4):146-159.

[12] R A F. Problems with risk analysis in the chemical industry. A detailed examination of the theoretical and practical problems faced by the risk analyst in the study of a chemical plant[J]. Plant/operations Progress, 1983, 2(3):185–190.

549550 Process Hazard Analysis

Process Safety Management System starts with managing the process related risks and from the first step of establishing a new process unit that is needed to evaluate the risks and define the necessary measurements. Process Hazard Analysis methodologies are the systematic tools which help to evaluate the risks in a process. For different needs there are several methodologies like DOW FETI, HAZOP, What if, HAZID, Fault Tree and Event Tree Analysis, Scenario Modelling, Facility Siting etc. Some of these methodologies are semi quantitative some are quantitative methodologies. In order to define and prevent major industrial accidents in our refinery we use DOW FETI, HAZOP, FTA and ETA and Scenario modelling. Moreover for some kind of projects we may also apply What if and HAZID analysis instead of HAZOP depending on the complexity. Facility Siting is another tool for risk related studies which is used to evaluate the risks for buildings in order to protect people inside the buildings. The workflow for a full unit PHA study starts with DOW FETI calculation which is used for risk ranking of al static equipment inside the unit. DOW FETI calculation is done in accordance with the equipment temperature, pressure, hold-up amounts, components and also reaction types if there is any. Then risk assessment process continues with HAZOP studies which is for scenario identification. In HAZOP, first step is chopping the unit into small parts to be evaluated which is called nodes. After that all nodes are evaluated using P&IDs and PFDs by keywords like temperature, pressure, flow, level etc. HAZOP helps to understand what kind of scenarios are possible inside the unit and these critical scenarios are taken through FTA and ETA studies. HAZOP is a semi quantitative risk assessment method where FTA and ETA are quantitative methods. FTA is applied for critical scenarios in order to find out the probability of top event then ETA is then applied in order to find out the probability of scenarios like fire, explosion, toxic dispersion etc. caused by this top event. FTA and ETA analysis all use the statistical PFD and failure rate values of all initiating events and barriers defined in HAZOP for the related scenario. Scenario Modelling is the following step which is used for understanding the effect of the probable scenario. The most probable scenarios coming from ETA is modelled to see the effect zones. These effect zones are used to plan Emergency Response Studies for all these potential major accidents.

548517 Site-Wide Asset Integrity Analytics & Benchmarking

All too often, reliability engineers focus on the equipment that is deemed as most critical or covered by some regulatory reference, while sometimes forgetting about those areas where something as simple as water can cause major issues. Non-destructive testing technicians and API inspectors alike can help facilities gather and compile data to perform predictive analytics and even benchmark sites with similar assets. This data can be used to identify assets where others may be seeing issues. This also allows facilities to compare reliability engineering effectiveness. This presentation will share some sample results of non-destructive testing and inspections of both regulatory required and non-required assets in the fuel ethanol production sector as well as an example of how that data can be assembled and shared across any sector to educate and learn from each other. Afterall, there are no secrets in safety!

548330 Reliability, Availability, and Maintainability (RAM). What Does It Provide?What Should You be Aware of before You Conduct a RAM Study?

Reliability, Availability and Maintainability (RAM) studies can significantly add value to your Process Safety Management (PSM) activities, especially with regards to Mechanical Integrity (MI). RAM studies are an excellent way to determine which equipment limits you from achieving optimum availability and throughput. Using industry failure rate data or even site-specific data can help you identify areas for improvement and determine how best to maintain your equipment. But RAM studies can go beyond purely theoretical studies by using the experience and knowledge gained from operating and maintaining similar facilities to develop a pragmatic approach to solving mechanical integrity-related issues. This paper will cover the pros and cons of site versus industry data and will present a case example of utilizing experience to better define the proposed problems and solutions. If done correctly, RAM studies can be more than statistics; they can be roadmaps to process reliability improvements.

546707 Driving Asset Integrity and Process Safety Consistency across a Multinational Organization

Mexichem is in the process of implementing world-class process safety management (PSM) and Asset Integrity Management (AIM) programs globally for their facilities. This is a major undertaking as Mexichem, to a large extent, has grown through acquisitions. Therefore, each plant has a safety culture and management system unique to that asset. As we all know, it's difficult to drive global consistency when dealing with different types of operations in various regions throughout the world under differing regulatory regimes. This is especially difficult when there is no common understanding of process safety. One of the first steps Mexichem has taken is to develop a "fit-for-purpose" management system with an emphasis on developing a common understanding of process safety and ensuring the culture is robust enough to support the system when it is implemented.

To aid Mexichem in this overwhelming task, Risktec Solutions were engaged to develop and deliver a bespoke set of interactive training workshops on a variety of PSM / AIM topics. The first of which was related to the identification and management of safety critical equipment (SCE). This presentation will outline the plan and methodology Mexichem and Risktec have used to transfer such knowledge and discuss some of the successes and challenges along with way.

548216 Alarm Floods and Their Connection to Plant Incidents and Near Misses

Investigations of many catastrophic plant events have shown alarm flooding to have been significant distractions for the operators just prior to the incident. Alarm floods have also been determined as a cause or contributing factor to smaller incidents and near misses.

In fact, many of the more recent findings by U.S. Chemical Safety and Hazard Investigative Board (CSB) include alarm management as one of the key issues in the investigative reports.

Even after several years of trying, many plants still struggle with controlling alarm floods. Many also have problems generating consistent and accurate alarm metric reports for the entire plant. Now that the ISA 18.2 Alarm Performance Metrics are considered "generally accepted good engineering practices," we should make haste to demonstrate where each of our plants fall (particularly the PSM plants) versus these metrics before the regulating authority moves in this area. If we are not able to prove we meet the metrics (particularly for alarm flooding) under all operating conditions, we should remediate.

Static rationalization can reduce your average number of alarms, but without controlling the alarm floods, there is no help for the operator when he needs it the most. For most, we must begin the process by recognizing that alarm floods occur most of all during a change of process state.

The means of controlling alarm floods has been available to us for many years. Unfortunately, only a select few have implemented the solution and are achieving the desired results today. These plants have achieved equal to or better than ISA 18.2 since before it was published.

This session will cover the justification for true alarm management from the safety and economic perspective with several practical steps that everyone should take if they are not currently meeting the ISA 18.2 specifications during all operating conditions.

548318 Do Incident Data & Lessons Learned Really Facilitate Process Improvement?

Regulations are a set of guidelines for the industry for protecting the people and environment, when it comes to process safety systems. While some companies tend to find that complying with regulatory requirements are always a challenge, on the other hand some tend to go overboard in their bid to stay compliant. In spite of the route taken, there are incidents – major, minor and near misses.

Other than the accumulation of this data, the investigations also result in lessons learned. Over the past four decades, lot of incident data and lessons learned has accumulated, been assimilated, and also analyzed. Complementing that are the audits, system reviews, hazard identification exercises, changes to regulations, and knowledge sharing forums.

It is a very efficient and effective mechanism to leverage these vast sources of current information developed, and implement its learnings, in the operations of existing processes. While the core concepts of safety are universal across the various industrial sectors, its application may differ. But once again, the lessons learned databases are effective sources of making this leveraging process more efficient.

This paper will discuss the some case studies of how the concept of lessons learned and the knowledge of the incident data has been leveraged by the authors in various manners ensuring success from one industry to another – example (1) Learnings from an incident due to gaps in the Emergency Response Plan, bringing about success at another chemical manufacturing site; (2) Incident data from a steel plant, when analyzed and applied to a chemical plant resulted in improved of effectiveness.

546673 Structured & Scrutinized Approach Towards HSE Excellence; A Success Story

Author: Usman Ali, Process Safety Engineer

Co-Author: M Ahsan Sarfraz, Unit Manager HSE

This paper studies the practices & programs that support reporting of Near misses, minor unsafe conditions, incidents and quantitative risk assessment strategies in our organization for analyzing and learning from past events of operational failures, injuries, safety and environmental incidents.

A near miss is a leading indicator to an incident that, if scrutinized appropriately, can prevent injuries and accidents. Some persons may be tempted to write off near misses as "no harm, no damage" situations, but organizations who track near misses, determine how and why they occurred, and take corrective action can prevent similar or more serious incidents from occurring in the future. Fatima Fertilizers Limited believes that, "We don't look at what happened; we look at what could have happened".

Keeping in mind the importance of near miss reporting, FFL initiates drive and implement systems for

1. Capturing and increasing near miss reporting at whole site.
2. Logging data for analysis and using this valuable data for identifying gaps in existing process safety system.
3. Implementing sustainable improvement plans and safety campaigns (Safe hands, Safe Routines, Job Cycle Checks, Safe Drive etc) in our process safety systems for making our site a safer workplace.

These actions has improved the site performance which is reflected as Site has now over 45 Safe Million Man Hours, Total Record-able Injury Rate (TRIR) is below 0.2 since last 03 years and No Loss time injury since last 06 years.

548162 Events Report Database Analysis for SHE/PSM Programs Improvements

Incident databases contain records of events that have occurred in a petrochemical facility during recent years. The most of chemical industries report databases are filled with accident valuable data (frequency and severity). Harnessing data mining to explore incident databases, it was used data mining to analyze the incidents database of the studied plant. Inferent approaches are ineffective for low frequency, high consequence events because of their rarity. Given this information limitation, this work uses database analyses to forecast incident frequencies, their relevant causes, equipment involved, and their possible consequences, in specific petrochemical plants.

Databases information analysis, also help to avoid future accidents, thereby reducing the risk. More specifically, this study presents dynamic analyses of incidents in a petrochemical plant ubicated in Mamonal industrial park, in Cartagena- Colombia. Differences in the chemical-specific predictions with the overall predictions are demonstrated. Furthermore, value theory is used for consequence of rare events by formulating distributions for events over a threshold value. Finally, the accidents number is used to estimate the capital at risk within an industry utilizing the frequency and loss-severity distributions.

In conclusion, it was recommended to update de maintenance program (predictive, preventive and corrective) in cases of abnormal changes in production rates, comparing the current production levels, with the analyzed plant production forecast. Also, it was suggested to improve the report culture, due to gaps founded regarding to events classification.

549483 The Process Safety Analysis and Management System in Sinopec

The big data technique has been in trend for the process safety assessment. In china, it's the very need to transfer the qualitative safety management to quantitative safety management, which should rely on the professional risk analysis technique, as well as the standard safety reliability database. Based on the above, it's very important to integrate several professional risk analysis technique, as HAZOP, LOPA, SIL,BOWTIE etc. to realize the integral risk identification and evaluation, grading and management plan making, supporting by the proprietary SINOPEC risk metrics and grading standards, as well as safety equipment reliability database,which is developed as the Process Safety Analysis and Management System for SINOPEC, to support all SINOPEC branches to make dynamic risk analysis and management online

The key data for HAZOP, SIL and Management module in PHAMS is the dependent protection barriers, which we do focusing on the failure probability, at the same time, we do use of the Bayesian Correction Modelling to verify the failure probability of safety equipent and protection barriers on site, which can validly save the grading cost on time and personnel, all of this is towards to the deeper research on dynamic risk analysis for PHAMS.

Thanks to the PHAMS, we has set up and implemented several standards on HAZOP, LOPA and SIL for SINOPEC.

552170 Analysis and Improvement of Safety Integrity Level Considering Human Factors

Safety Instrumented System (SIS) is an irreplaceable active protection layer that prevents the propagation of process accidents and plays an important role in reducing the risk of accidents. In addition to the sensor response triggering the SIS, a human emergency button is often provided in practice to ensure that the SIS is triggered by personnel in the event of the sensor part failure. Due to human factors added, the emergency button will improve the reliability of the SIS. However, there is no research to study the human reliability of the SIS upon sensor part failure in a continuous operation chemical plant, and how human factors can improve and affect the overall reliability of the SIS still remain undiscovered. In this paper, a verification methodology was proposed which integrated with Cognitive Reliability and Error Analysis Method (CREAM) and Reliability Block Diagram (RBD) to emphasize human reliability analysis of the SIS verification. The operator behavior is subdivided into several steps and the Human Error Possibility (HEP) is calculated by a modified CREAM. Then the RBD method is used to verify the SIS reliability considering human factors. An SIS system of one reacting furnace was taken as a case study using this method to show how human factors can improve the reliability of the SIS. The proposed methodology allows SIS verification considering human factors to improve safety integrity level in real practice thus help stakeholders saving costs and making more decisions on SIS implementation.

19 AIChE Spring Meeting
15TH GLOBAL CONGRESS ON PROCESS SAFETY
NEW ORLEANS • 2019

Caribbean Petroleum Explosion Case Study

Q.A. Baker
3330 Oakwell Court, Suite 100
San Antonio, TX 78218
QBaker@BakerRisk.com

J.K. Thomas, D.R. Malik and M.T. Edel
Baker Engineering and Risk Consultants, Inc.
3330 Oakwell Court, Suite 100
San Antonio, TX 78218
KThomas@BakerRisk.com, DMalik@BakerRisk.com, MEdel@Bakerrisk.com

Prepared for Presentation at
American Institute of Chemical Engineers
2019 Spring Meeting and 15th Global Congress on Process Safety
New Orleans, LA
March 31 – April 3, 2019

Caribbean Petroleum Explosion Case Study

Q.A. Baker, J.K. Thomas, D.R. Malik and M.T. Edel
Baker Engineering and Risk Consultants, Inc.
3330 Oakwell Court, Suite 100
San Antonio, TX 78218-3024
(210) 824-5960
QBaker@BakerRisk.com

Keywords: Accidental explosion DDT, VCE, deflagration, detonation, vapor cloud explosion

Abstract

A vapor cloud explosion (VCE) occurred at the Caribbean Petroleum Corporation (CAPECO) site in Bayamón, Puerto Rico on October 23, 2009. An above-ground gasoline storage tank was being loaded with gasoline from a cargo ship shortly before the incident. Overfilling of the tank resulted in gasoline discharge through elevated tank vents and formation of a very large vapor-mist cloud, which extended off the plant site into surrounding areas covered with heavy vegetation. The resulting VCE caused blast damage to both on-site and off-site structures. Fires were ignited in 20 storage tanks within the tank farm, which burned for several days. There were several minor injuries and no fatalities from this accident.

1 Background

The accident described in this paper occurred at the CAPECO facility located in Bayamón, Puerto Rico. The authors participated in the on-scene investigation, evidence collection, and data analysis. The information presented in this paper is the result of the authors' investigation.

The overall facility layout is shown in Figure 1. The facility included a refinery, tank farm, dock, and truck loading rack. The facility was near the north shore of the island, and the dock was about 2 miles from the north boundary of the facility. The CAPECO facility was being operated as a terminal. Fuel that was offloaded at the dock was pumped to the tank farm through a 20-inch diameter import pipeline. CAPECO delivered fuel to industrial and retail fuel distribution facilities on the island by pipeline and fuel trucks.

The refinery portion of the facility had been out of operation for a number of years. All the process equipment was depressurized and deinventoried. No operations or maintenance was taking place in the refinery area at the time of the incident.

The tank farm occupied a large percentage of the facility footprint, as can be seen in Figure 1. The north side of the tank farm is shown in Figure 2, with tanks of interest labeled in the figure. The tanks being filled by ship were Tanks 405, 504, 411 and 409. Tank 107 was also scheduled to receive gasoline, but the incident occurred before flow was started to Tank 107.

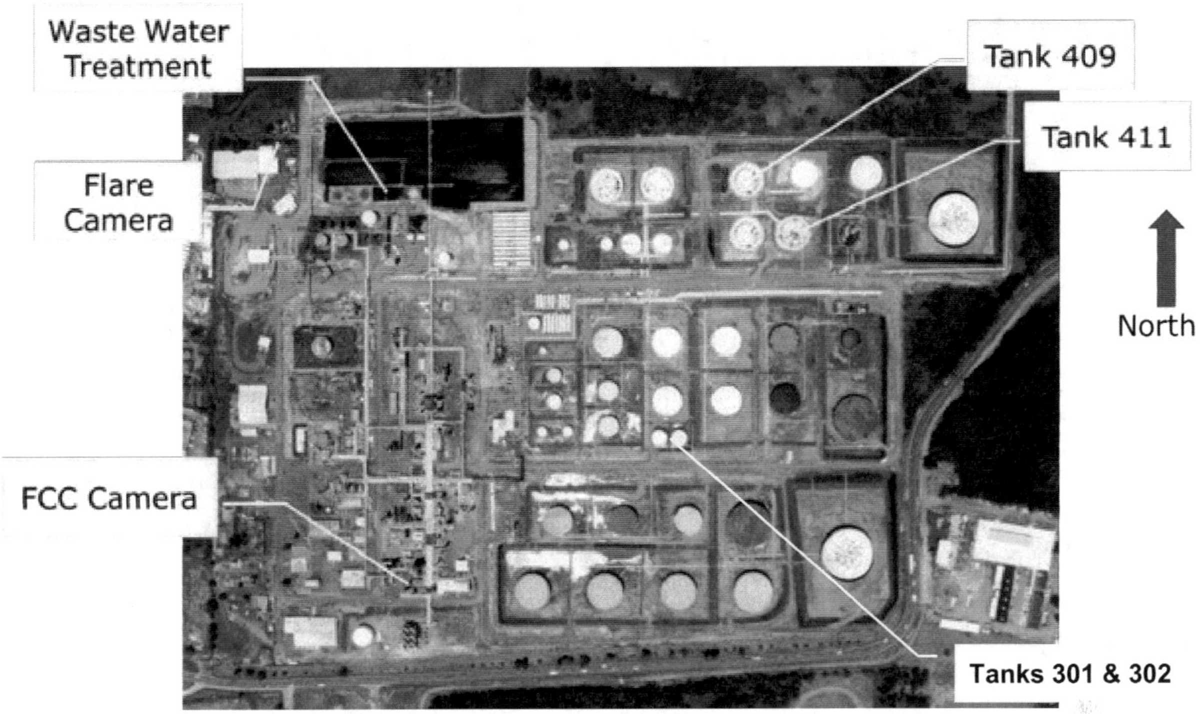

Figure 1. Satellite View of the CAPECO Facility

Figure 2. CAPECO North Tank Farm Layout

2 Sequence of Events and Witness Accounts

The sequence of events was developed from witness interviews, process data, inspector logs, ship logs, and surveillance video.

The VCE occurred during offloading of the tanker ship Cape Bruny on October 23, 2009 shortly after midnight. The ship arrived at the dock at 7:00 AM on October 21, 2009. Two independent inspectors arrived and measured the levels in the ship cargo tanks. The inspectors then went ashore and measured the levels in the tanks scheduled to receive gasoline. Inspectors also measured tank levels once they were filled and placed seals on the tank valves to "close" the tanks. The inspectors used their own tape measure to measure levels in the ship cargo holds and onshore tanks, not relying on ship or tank gauging systems. The inspectors' notes therefore provide an independent data set for the tank levels and fuel volumes transferred.

The planned sequence for filling tanks was Tank 405, 504, 411, 409 and 107. Switching of flow from one tank to another was done without stopping or slowing flow from the ship. The switching process was done by opening the inlet valve on the next tank in the sequence while continuing to pump to the tank which was being filled. The inlet valve on the tank that was just filled was then closed.

Unloading gasoline from the Cape Bruny began at 8:47 PM on October 21. Tanks 405, 504 and 411 were filled and "closed" before the accident occurred, with Tank 411 filling operations being complete around 10:00 PM on October 22. Tank 409 was the only tank receiving fuel for the remainder of the sequence until unloading was stopped at approximately 12:12 AM on October 23.

The facility had a general practice of "cracking" the inlet valve for the next tank in sequence when filling a tank, such that there was a secondary flow path to protect against dead-heading a pump. However, this practice was not consistently followed. Most importantly, when filling Tank 409 (i.e., the tank which overflowed), the inlet valve to Tank 107 (i.e., the next tank to be filled) had not been cracked open. All flow from the ship was therefore being directed into Tank 409 when it was being filled.

The tank monitoring system in the Control Room was not communicating with the Tank 409 level transmitter, as discussed in the next section. Operators were therefore making hourly rounds and manually checking the level using the index readings on the level housing on the side of the tank. Based on these hourly readings, operators estimated that Tank 409 would be filled at 1:00 AM on October 23, at which time they planned to divert flow to Tank 107. The last time the tank level was read was 11:00 PM on October 22.

The last persons in the tank farm were the inspectors, who arrived at Tank 411 at 10:54 PM to close Tank 411. They sealed the valve and climbed to the top of Tank 411 to gauge it. Tank 411 was located diagonally across from Tank 409, as can be seen in Figure 2, and the inspectors had a clear view of Tank 409. The inspectors reported that they observed no problems in the tank farm and that Tank 409 was not overflowing. They also reported that it was a clear night with no wind. The inspectors left the tank farm at 11:23 PM.

Overflowing of Tank 409 began sometime between 11:23 PM and midnight on October 22. Three operators arrived at the tank farm shortly before midnight to make their rounds. They started walking from the Wastewater Treatment plant to the east into the tank farm along the roadway south of Tank 409 (see Figure 1). As they approached the tank farm, they observed a light "fog" illuminated by a street light. The fog was likely higher than the street light, but they could not tell how much higher due to lack of lighting. As they walked closer, they observed a dense layer of fog about waist deep and felt the fog on their bare hands. They had reached the propane bullets shown on the left (west) side of Figure 2. At that point they decided to back out. The operators did not see or hear a tank overflowing or identify the source of the fog.

The operators radioed the ship to stop pumping, and then radioed other personnel on site to initiate an evacuation. Ship pumps are estimated to have stopped around 12:12 AM on October 23.

Other operators attempted to approach the tank farm in a car from the roadway on the east side of the tank farm. They observed a vapor cloud around Tanks 301 and 302 (see Figure 1). The cloud ignited before they could drive further north into the tank farm, and hence they did not identify the source of vapors. No witnesses directly observed the overflowing tank.

CAPECO had been transferring other fuels by pipeline to an export dock on October 22nd. Jet fuel, diesel fuel, and fuel oil were transferred to a different dock than the one at which the Cape Bruny was docked. The export dock as about ¼ mile away by water and 1 mile by road from the dock at which the Cape Bruny was unloading. All fuel transfers were completed by 9:00 PM. The investigation revealed that there was no path through the terminal pipe system for the fuel transfers to the export dock to have affected the fuel unloading from the Cape Bruny and filling of Tanks 405, 504, 411 and 409. Unloading of the Cape Bruny was therefore found to be the only feasible means to overflow Tank 409.

The VCE occurred at 12:22 AM on October 23. There were several explosions over a period of about 9 seconds following ignition, the last of which was most severe.

Two vehicles were present on the public roadway on the east side of the plant at the time of the explosions. A sedan was headed north and had passed the northeast corner of the plant, and a dump truck was headed south and was southeast of Tank 107. The sedan is not believed to have been in the flammable cloud, but it was determined that the dump truck was in the flammable cloud. Both vehicles were damaged, but operable, and the drivers drove away with minor injuries.

3 Tank 409 and Gauging System

Tank 409 was one of the newer tanks at the CAPECO facility. It was built in 2006, three years before the accident. The tank had a 120 ft diameter and was 63 ft high with a nominal capacity of 120,000 bbls. The tank had an internal floating roof. Figure 3 is a photograph of the tank taken at completion of construction. The photograph shows three of the six overflow openings; these openings were 3 ft wide by 9 inches high, and located at about 60 feet above grade.

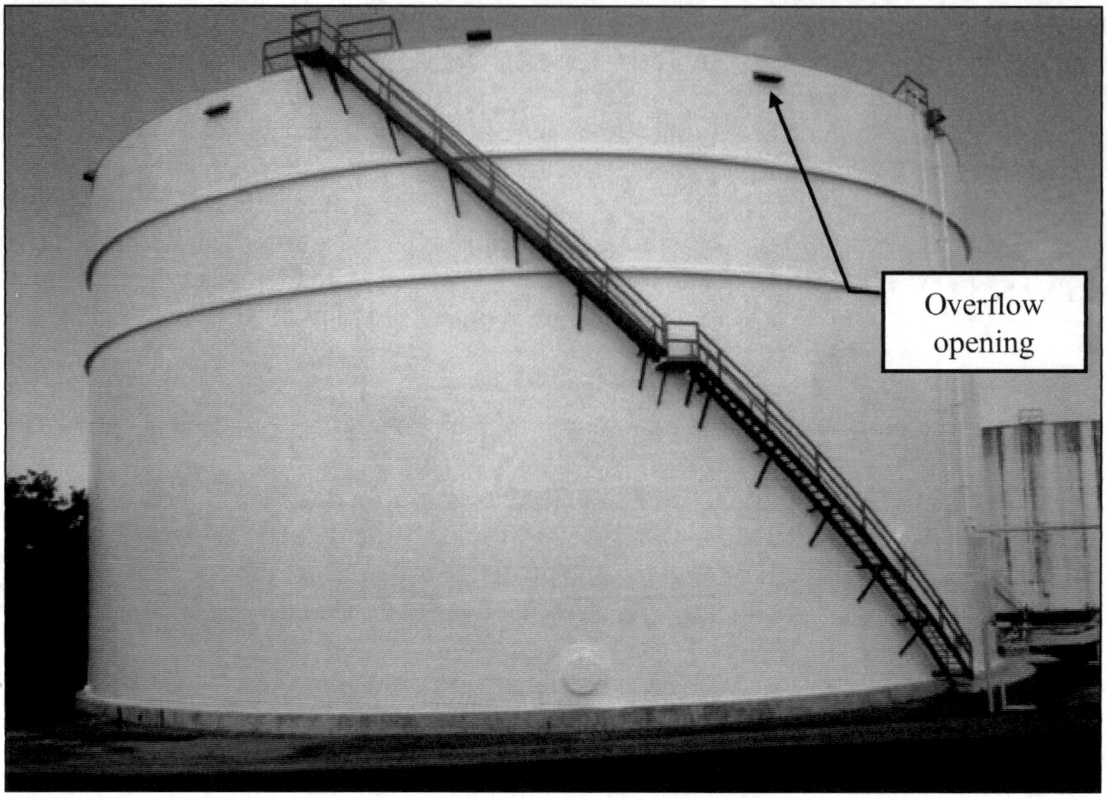

Figure 3. Tank 409 Photograph

Tank level was measured with a mechanical tank level system. This system used a float that ran on guide rods in a well penetrating through the floating roof of the tank. A metal tape connected to the float was routed through the roof of the tank through guide tubes to the gauge head. A spring actuated take-up reel wound the metal tape inside the gauge head. The gauge head was affixed to the side of the tank about 4 ft above ground level. It had an index head that operators could read at the tank. It also had a transmitter to communicate with a computer that monitored tank level in the tank farm control room. Unfortunately, CAPECO had a network communication system problem that prevented some tanks, including Tank 409, from being monitored in the control room.

The authors believe that the tank gauging system malfunctioned during filling of Tank 409, leading operators to incorrectly estimate the time required to fill the tank. There was no redundant high level alarm. More information on the level gauge system can be found in the Chemical Safety Board (CSB) report on this incident [1].

4 Tank 409 Filling and Tank Overflow

The initial level of Tank 409 was 3 ft 2.6 inches, which was below the level required to float the internal roof. The floating roof was therefore initially resting on its legs. The volume of fuel and water in the tank before filling was 6,737 bbls. The available volume was 117,562 bbls of gasoline at ambient temperature.

Cape Bruny had 12 centrifugal pumps available and was using six pumps during the unloading process. The ship could vary the pumping rate using combinations of pumps and pressure controls. The ship pumping rate varied during unloading from about 10,000 up to 11,400 bbls/hr. At the time of Tank 409 overflow, the pumping rate was estimated to be 11,400 bbls/hr at ambient temperature. Plant personnel reported that ships typically pump at the fastest rate that the receiving facility can accept to minimize the time the ship is docked and the associated expense. A hypothesis considered was that the ship increased the pumping rate in the final hours before the incident. However, despite investigator requests, the Cape Bruny never produced pump rate data, so this hypothesis could not be verified.

Based on a detailed material balance, accounting for the material off-loaded from the ship and received at the facility, approximately 4,600 barrels (193,000 gallons, 731,000 liters) of gasoline overflowed from Tank 409. The tank was overflowing for about 24 minutes before flow was stopped, and approximately 10 minutes later the VCE occurred. The combined time for gasoline vapor/mist to spread before ignition was therefore 34 minutes.

5 Weather Conditions

The weather was clear on October 22 and 23 with no precipitation. Plant personnel reported that the weather was mild. The weather reported in San Juan at the time of the incident was 82 °F with calm winds. The two inspectors who were on top of Tank 411 about an hour before the accident stated that the weather was clear with no wind.

6 Gasoline Dispersion

Gasoline overflowed from the six overflow openings arranged around the shell of Tank 409 about 60 ft above grade (see Figure 3). The tank also had two wind girders that protruded 8 inches (20 cm) from the shell, fabricated from 8 inch × 6 inch (20 cm × 15 cm) angle steel, with the open side of the angle facing down.

Gasoline spilled from Tank 409 collected in the dike. The dike had a drain through the north dike wall that connected to a sewer that discharged into the wastewater pond. The plant kept dike drain valves closed for environmental reasons. Rain water was collected in the dikes, sampled and tested, and then drained from the dikes. The dike drain valves were found closed, chain and locked after the accident. Plant personnel were ordered by government regulators to open the dike drain valve for Tank 409 after the accident to check the valve. Weeks later in a subsequent check with investigators using water, the valve was found to not be tightly closed, and some water passed the valve. The authors believe that the dike valve was closed and not leaking during the accident. However, if the dike valve was leaking, gasoline would have flowed through the sewer and discharged into the wastewater pond. This sewer did not connect to any other plant sewers.

The flammable cloud formed by a combination of vaporization of gasoline and mist formation from falling gasoline atomizing in air, impacting the tank wind girders, and impacting the ground. The momentum of the vapors and mist carried the cloud outside of the dike. In the calm wind condition, the denser-than-air cloud followed the topography downhill, spreading in all directions from the tank. A release which filled the tank dike, without the "waterfall" of gasoline issuing from the tank overflow openings (i.e., with only an evaporation source term), would have generated a much smaller flammable cloud.

The topography in and around the CAPECO facility is shown in Figure 4. The northeast portion of the tank farm was a high point in the topography. The ground sloped away outside of the dikes for the north group of tanks, which includes all tanks that were being filled. The arrows in Figure 4 show the downhill direction from the north dikes of the tank farm.

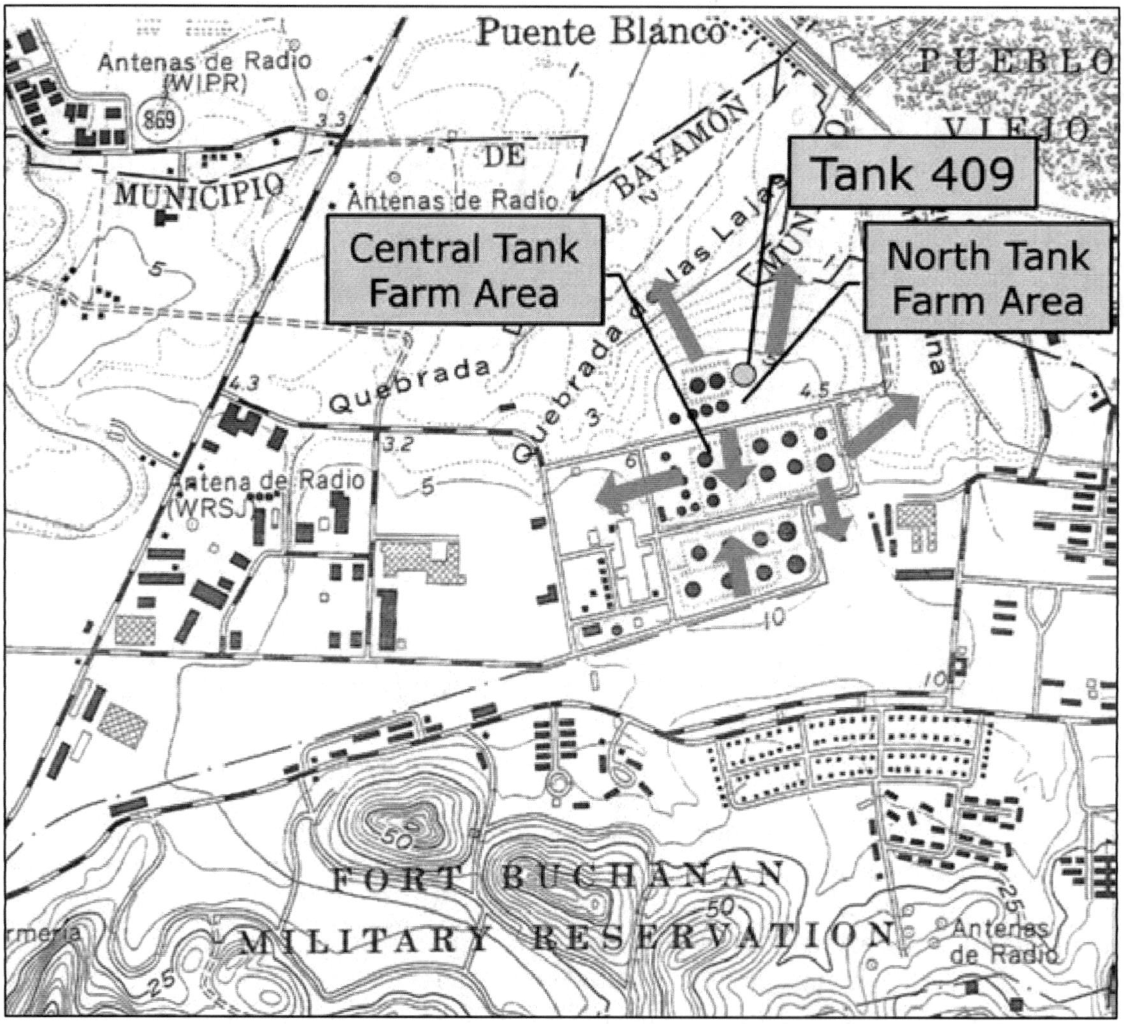

Figure 4. Topography at the CAPECO Facility [2]

A wetlands area was to the north of the facility and adjacent to it. The wetlands were downhill from the facility and had heavy vegetation. Only one onsite unpaved perimeter road separated the north dike wall from the wetlands. A public roadway ran along the east side of the facility, across which was the Fort Buchanan Military Reservation; both sides of this road also had heavy vegetation. A post-incident photo of an exemplar vegetated area just to the north of the tank farm is shown in Figure 5. Figure 6 shows an area within an adjacent wooded area. These areas are typical, and some areas were more heavily congested. Some heavily vegetated areas had much of the vegetation displaced to grade level and/or burned as a result of the event.

To the west of the north tank farm area was the Wastewater Treatment plant. The elevation of the Wastewater Treatment plant was below that of the north dikes. The lowest area of the facility was the central tank farm area, which was separated from the north tank farm dike area by a roadway.

Figure 5. Vegetation Adjacent to North Side of Tank Farm

Figure 6. Vegetation in Area Adjacent to Burned Zone

The area covered by the flammable cloud formed due to the release is shown in Figure 7. In general terms, the flammable cloud was about 1,000 ft (330 m) in diameter and centered on Tank 409. Based on observed damage to vegetation and analysis of surveillance videos, the flammable cloud area was roughly 3×10^6 ft^2 (70 acres, 0.3 km^2).

Figure 7. Flammable Cloud Overlaid on Pre-Incident Photograph and Compared to Post-Incident Photograph

7 Ignition Source

Ignition occurred about 34 minutes after Tank 409 began overflowing. The initial ignition source was on the western fringe of the flammable cloud as depicted in Figure 7. CCTV cameras were located at northwest corner of the site (rotating) and on a column in the refinery (fixed). The ignition of the vapor cloud was not captured directly by either CCTV camera. However, the general location of the ignition source was deduced by examining CCTV footage and noting the object (e.g., tanks, vessels, etc.) surfaces that were first illuminated; the first frame of this video after ignition is shown in Figure 8. This allowed the general area of the ignition source to be isolated to the dotted rectangle shown in Figure 9; the edges of the site lines for each CCTV camera are denoted as yellow lines in this figure. The most likely ignition area based on the reflections is shown as a red circle in Figure 9. The only identified potential ignition source within this circled area is an electrical equipment enclosure that housed pump switch gear, which is shown Figure 10. The white arrows in Figure 9 show the direction of illumination of the tanks from ignition at the electrical equipment enclosure. The metal panel siding for the enclosure was blown off the enclosure frame, consistent with flammable vapors entering the enclosure and being ignited within.

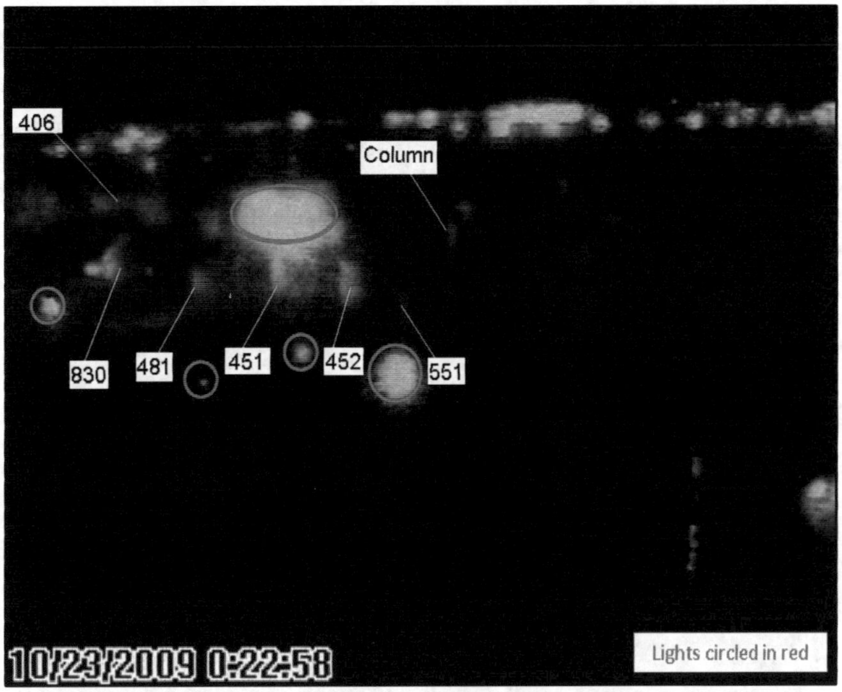

Figure 8. First Frame from CCTV Footage after Ignition

Figure 9. Location of Ignition Source

Figure 10. Ignition Source (Electrical Equipment Cabinet)

A secondary ignition was evident in the CCTV video roughly to the southwest of Tank 107 (i.e., flame appears in this area ahead of the main flame front), as shown in Figure 11. This secondary ignition source was about 1,380 ft (421 m) to the east of the initial ignition source. The hypothesis that this secondary ignition was a flame propagating through a sewer and jetting from a sewer opening southwest of Tank 107 was investigated in cooperation with the CSB. However, the only nearby storm sewer opening was a sump that always had standing water controlled by a level switch to keep a submersible pump submerged. In order to have reached this opening, the flame had to have been forced to propagate in the upstream direction through a check valve and flooded pump. However, inspection of this equipment did not indicate damage consistent with internal flame propagation. Process sewer pipes were also investigated and found to be liquid filled with no indications of internal blast pressure loading. The hypothesis of a flame jet originating from a sewer was therefore not borne out. Also, careful triangulation of the video placed this secondary ignition in line with an electrical equipment enclosure, not a sewer opening. The electrical equipment enclosure is shown in Figure 12. The panels forming the enclosure were all displaced in a manner consistent with internal pressurization; the circled items in Figure 12 are sections of this enclosure.

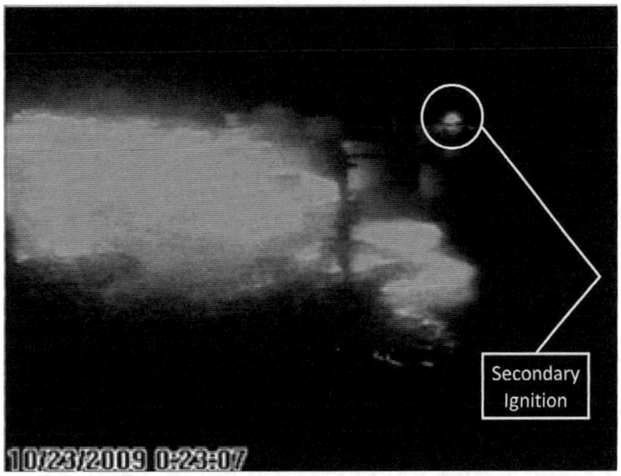

Figure 11. Frame from CCTV Showing Secondary Ignition

Figure 12. Secondary Ignition Location (Electrical Equipment Enclosure)

8 Vapor Cloud Explosion

VCE blast loads were evaluated using two approaches: (1) as a deflagration using the Baker-Strehlow-Tang (BST) VCE blast load prediction method [3, 4] and a flammable cloud consistent with the cloud outline shown in Figure 7, and (2) as a detonation based on the cloud predicted using the HSL methodology for a gasoline overflow [5, 6], which was developed following the Buncefield incident [7]. The BST VCE blast load predictions were developing assuming a 10-foot cloud height, which is a reasonably consistent average based on observations at the site and witness accounts. Confinement and congestion were mapped by BakerRisk investigators while on-scene both in the plant and the vegetated areas. The HSL cloud was modeled as being 1-meter high, consistent with the HSL guidance. The radius of the HSL cloud was 1,510 ft (461 m), about 50% greater than the observed damage radius. Coincidently, the flammable cloud volume in both cases was 23×10^6 ft^3; that is, the only significant differences between the two flammable clouds considered was the height and lateral dimensions. The predicted blast loads for these cases were compared to the observed structural damage.

The response of structural components (i.e., damage indicators) was compared to the predicted blast loads for the VCE scenarios considered. Damage indicator information was gathered for buildings and structures both on- and off-site. Examples of structural damage indicators are corrugated metal panels, cold-formed metal girts, open-web steel joists, wide-flanged steel roof girders, steel plate covers of firefighting equipment, and glass windows. Figure 13 through Figure 15 show typical damage indicators used for this purpose. The damage indicators were analyzed to determine the blast load that would result in the observed damage using the single degree of freedom (SDOF) and empirical pressure-impulse diagram methods [8]. Indicators that did not experience any observable damage served as upper bounds (i.e., an upper blast load limit, above which damage would have occurred). Indicators that had observable damage were used to determine blast load estimates. Pressure asymptotes were used to compare to blast predictions due to the long blast load duration.

Figure 13. Roof Joist Damage Indicator

Figure 14. Wall Girt Damage Indicator

Figure 15. Window Damage Indicator

The structural damage analysis indicated that a vapor cloud detonation using the cloud size indicated by the HSL guidance significantly over-predicts the VCE blast loads. On average, the detonation scenario over-predicted blast overpressures as compared to the blast damage indicator pressure asymptote values by a factor of approximately 40 (i.e., more than an order of magnitude). The detonation scenario gave blast overpressure predictions that were 5 to 300 times the damage indicator pressure asymptote values.

Blast load predictions for the deflagration of the same cloud volume using the BST method with a medium reactivity fuel, which is representative of gasoline vapor/mist mixtures, matched the structural damage indicators reasonably well. Neglecting the congestion present in the heavily vegetated areas significantly under-predicts the required blast loads, while accounting for only the vegetated areas still yielded reasonable agreement with the observed damage; this is expected since roughly 90% of the flammable cloud was located in the vegetated areas.

These analyses indicate that the flammable cloud did not undergo detonation, and that a deflagration of the postulated cloud using the BST parameters yields reasonable agreement with the observed damage. The vegetated areas were found to drive most of the VCE blast load predicted using the BST method. It is noted that, unlike the Buncefield incident [7], buildings were not present within or near the edge of the vapor cloud, and hence the observed building damage was not as dramatic as at Buncefield.

Damage observed at the facility was consistent with a deflagration and not a detonation. There were no locations of high airblast damage as would occur in a detonation. Extensive fire damage caused considerable secondary damage, some of which has been mistakenly used by others to conclude detonation occurred, such as a bent hand rail and a buckled tank shell. The authors investigated these damage indicators on-site and found that fire sources caused the damage. In addition, other nearby indicators did not suffer damage attributable to a detonation (e.g., a hand rail on the other side of the stairs from the one that was significantly bent).

9 Secondary Fires

Secondary fires were immediate and covered an extensive footprint. The surveillance videos showed that vegetation throughout the flammable cloud footprint was promptly ignited. This is an unusual situation in that the vegetation was very moist in the tropical environment and hence not particularly susceptible to prompt ignition. Nearby tanks and dikes were also quickly involved in the fire. It is believed that mist from the overflowing tank was a significant contributor to fire spread. Wetting of vegetation, tanks and other equipment by gasoline mist that ignited is suspected to be the reason for the abundance of fires throughout the area covered by the flammable cloud. Most of the damage in the facility was due to secondary fires, fed by a loss of containment from other tanks and vessels during the fires.

10 Conclusions

The following conclusions were drawn from the investigation.

- The VCE at CAPECO was the result of overfilling of Tank 409 during gasoline unloading from a ship.
- The volume of fuel estimated to have spilled was 4,600 barrels.
- The vapor-mist cloud dispersed in all directions in the calm wind conditions due to momentum and density of the vapors and mist, with the cloud spread generally following the topography. The cloud width was approximately 2,000 ft (610 m).
- The VCE was a deflagration. Thorough site inspection by the authors revealed no evidence of a DDT, even on a small, localized basis.
- Blast loads were consistent with a deflagration. By contrast, blast loads for the detonation of a flammable cloud size per HSL guidance for an overflowing gasoline tank were over-predicted by at least 5 times and by as much as two orders of magnitude. The over-prediction of the blast pressure was, on average, a factor of 40.
- Using a cloud volume consistent with the HSL methodology and applying the BST VCE blast load methodology would be expected to yield reasonable consequence predictions for similar events.
- Extensive secondary fires are suspected to have been rapidly ignited due to gasoline mist wetting vegetation, tanks and equipment. Such rapid fire spread is uncharacteristic of VCEs involving flammable gases.

11 References

[1] CSB (2015) Caribbean Petroleum Tank Terminal Explosion and Multiple Tank Fires, U.S. Chemical Safety and Hazards Investigation Board, Final Investigation Report, CSB Report No. 2010.02.1.PR.

[2] Anon, "Bayamon Quadrangle, Puerto Rico," Topographical map, US Department of the Interior, Geological Survey, no date.

[3] Baker, Q.A. et al. (2010) Guidelines for Vapor Cloud Explosion, Pressure Vessel Burst, BLEVE and Flash Fire Hazards, Second Edition, Center for Chemical Process Safety (CCPS) & John Wiley and Sons, New York, NY

[4] Pierorazio, A.J., J.K. Thomas, Q.A. Baker and D.E. Ketchum (2005) "An Update to the Baker-Strehlow-Tang Vapor Cloud Explosion Prediction Methodology Flame Speed Table," Process Safety Progress, 24(1): 59-65

[5] Atkinson, G. and S. Coldrick (2012) Vapour Cloud Formation, Experiments and Modeling, HSE Research Report RR908, UK Health and Safety Executive.

[6] SCI (2013) Vapour Cloud Development in Over-Filling Incidents, FABIG Technical Note 12, Steel Construction Institute, Berkshire, UK, April 2013

[7] Buncefield Major Incident Investigation Board (2008) The Buncefield Incident, 11 December 2005, Final Report of the Major Incident Investigation Board, ISBN 978 0 7176 6270 8

[8] Biggs, J. (1964) Introduction to Structural Dynamics, McGraw-Hill, New York.

Case Study - Characterizing and averting risk of catastrophic vessel failure for a reaction in dimethyl sulfoxide (DMSO)

Megan Roth
Merck & Co., Inc.
Rahway, NJ 07065 USA
megan.roth@merck.com

Prepared for Presentation at
American Institute of Chemical Engineers
2019 Spring Meeting and 15th Global Congress on Process Safety
New Orleans, LA
March 31 – April 3, 2019

Case Study - Characterizing and averting risk of catastrophic vessel failure for a reaction in dimethyl sulfoxide (DMSO)

Megan Roth
Merck & Co., Inc.
Rahway, NJ 07065 USA
megan.roth@merck.com

Keywords: catastrophic vessel failure, dimethyl sulfoxide, DMSO, process deviation, runaway reaction, reactive solvents, risk mitigation, undesired chemistry, vent sizing

Abstract

Process safety testing performed on a batch process for making a pharmaceutical intermediate revealed that one process stream, a reaction in dimethyl sulfoxide (DMSO), had a significant undesired reaction that could lead to catastrophic vessel failure as a result of a single process deviation. Comprehensive testing identified the worst-case process stream with the lowest exotherm initiation temperature (EIT) and most severe pressure consequences, for which corrective actions must be taken. With this information in hand, hazard assessments were performed to support processing; to determine the appropriate safeguards needed to mitigate the consequences and reduce the likelihood of reaching this activity to an acceptable level.

1 Introduction

The Environmental & Process Safety Engineering (EPSE) group at Merck & Co., Inc., Kenilworth, NJ, USA has the responsibility to characterize the desired and undesired process chemistry for synthesis of pharmaceutical candidates in order to mitigate low/moderate risks and avoid high risk processes.

Process safety testing performed on a batch process for making a pharmaceutical intermediate revealed that one process stream, a reaction in DMSO, had a significant undesired reaction that could lead to catastrophic vessel failure as a result of a single process deviation. This article will focus on the characterization and risk mitigation of the process step containing this undesired decomposition reaction. More specifically, this article highlights stream composition impact on the initiation and extent of exotherms, autocatalytic decomposition, the benefit of a solvent barrier to dissipate an exotherm, and the consequences of using a reactive solvent.

Identification and characterization of the thermal and pressure consequences of both desired and undesired reactions are critical components of assessing the safety of a process. Although the desired reactions are typically studied in depth, undesired reactions are frequently neither identified nor characterized. An undesired reaction is the reaction that occurs when as few as one process deviation such as a single human or equipment error occurs, causing the process to not run as intended. It is important to identify these hazards and understand the associated consequences in order to take the appropriate corrective actions to either mitigate or avoid the risks. This data may influence the design of the process or equipment as it feeds into associated hazard assessments (e.g. process hazard analysis (PHA), hazard and operability study (HAZOP), failure mode and effects analysis (FMEA))[i]. The end goal is to protect people, protect equipment/facilities, and prevent disruption to the supply of medicines. Even though it is uncommon to reach an undesired reaction that has severe consequences associated with it, when these reactions are present in a process, it is important that inherent protection is provided. A process may run for years without an incident before one day an error leads to a runaway reaction and vessel explosion. One relevant incident in industry that highlights the importance of characterizing undesired reactions is the event at Morton International, Inc.[ii]. Several other relevant incident investigations can also be found on the Chemical Safety Board (CSB) website (www.csb.gov).

2 Background

2.1 Process Overview

A batch process for making a pharmaceutical intermediate involved the addition of base to a pre-reaction solution composed of a brominated starting material, DMSO, and acid. The base addition was exothermic with an adiabatic temperature rise of greater than 40 K. The batch was then heated and aged to drive the reaction forward. Gas generation was not expected, based on the stoichiometry of the desired reactions.

2.2 Initial Assessment

Review of the chemistry raised the following process safety concerns:

- The use of DMSO; a reactive solvent with a high boiling point and large exothermic decomposition at elevated temperatures.[iii]
- The use of DMSO in the presence of a halogen (bromine) and base.[iv]

3 Process Safety Testing

3.1 Initial Thermal & Pressure Screening of the Desired & Undesired Chemistry

Thermal screening of all process streams was carried out by DSC, while initial pressure screening of select streams[v] (starting with the end of reaction) was carried out in scanning mode using the multiple module calorimeter (MMC). The screen showed that there were no major process safety risks identified with the desired reactions. There was no gas generation and no exothermic decompositions of concern at the normal operating temperatures, and the exothermic base addition could be controlled by addition rate. However, consistent with the known high temperature decomposition of DMSO, DSC thermal scanning of all process streams prior to the end of reaction showed a significant exothermic decomposition (undesired reaction). See Figure 1. The decomposition occurs at temperatures above the normal operating temperature but is potentially reachable by a reactor jacket loss of control (overheating) process deviation, which is a single point of failure that is considered a credible deviation. In this case, the reactor targeted for use has a silicone-based heat transfer fluid-filled jacket with a maximum jacket temperature of 170 °C versus the exothermic decompositions of approximately 400-700 J/g with a DSC onset temperature of 175-225 °C. Note that the general guidance is that the actual EIT can be up to 50 K lower than the DSC onset temperature. [1, vi]

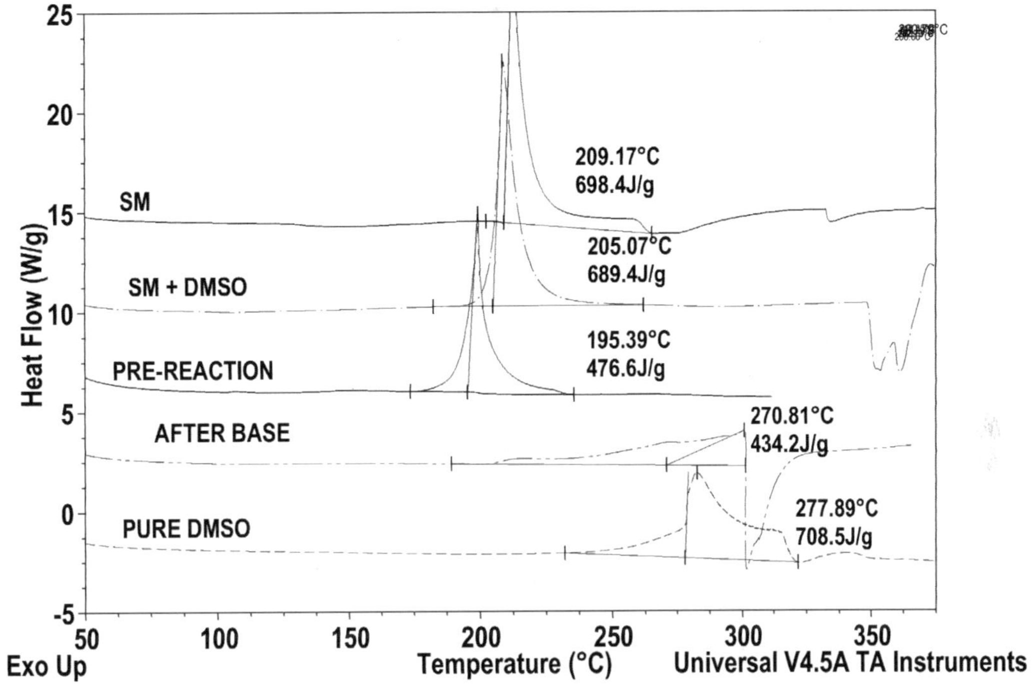

FIGURE 1. DSC scanning overlay plot of process streams prior to the end of reaction. SM, starting material. The temperature was increased linearly from room temperature to 350 °C (exotherm is shown upward). Tantalum-lined DSC cells with high pressure seals were used.

As shown by the shape of the exotherm peaks in Figure 1, prior to base addition the rate of the exotherm was rapid, while after the base addition the rate was much slower. Once base is charged, the presence of water in the base solution creates tempering and slows the decomposition reaction. Tempering occurs when the heat of evaporation of the solvent removes some or all of the heat generation from the decomposition.

3.2 Additional Screening Performed - Freezing Point & Material of Construction (MOC) Impact

---

[1] Onset temperature is used in discussing DSC results, for which typically about a 10 mW/g heat generation rate can be detected. Exotherm initiation temperature (EIT) is defined as a temperature rise rate of 0.02 K/min at a phi factor of 1.0, typically measured in an adiabatic calorimeter, such as an ARC, and corresponds to a heat generation rate of 0.7mW/g for a typical organic sample, compared with the typical 10mW/g threshold for DSC onset. The EIT is normally considered to be the temperature at which the reaction becomes hazardous.

Due to the presence of DMSO, with a freezing point of approximately 16-19 °C, DSC testing was also used to measure the potential freezing point of the stream prior to base addition (pre-reaction stream) in order to determine if a potential deviation involves the batch freezing. Batch freezing could lead to agitation issues and lack of thermocouple reliability. A DSC scan starting at sub-ambient temperatures confirmed that under normal operating temperatures the batch would not freeze, as the DSC freezing point of approximately 13 °C was lower than the minimum allowable batch temperature. See Figure 2.

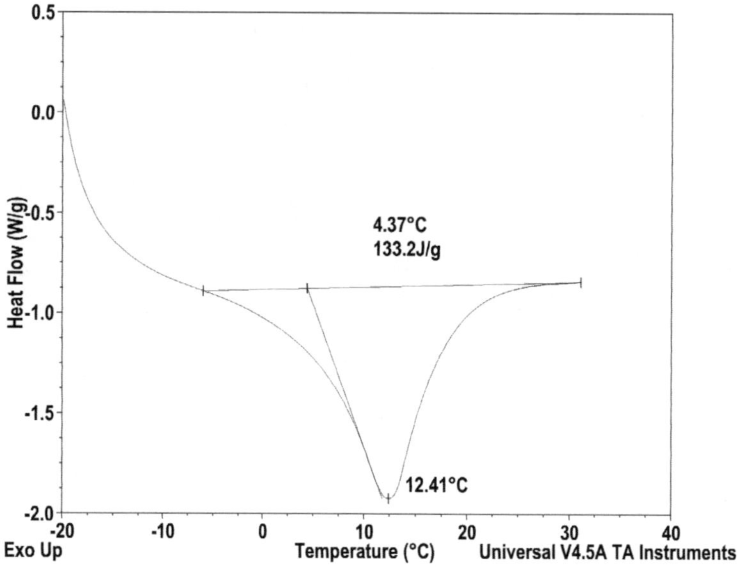

FIGURE 2. Pre-reaction stream freezing point determination using sub-ambient DSC scanning.

In addition, due to the presence of acid, bromine species, and DMSO, a DSC was used to determine if there was a MOC impact on the undesired decomposition reaction onset, size, or rate. DSC scans of the various process streams up through base addition were performed in tantalum, Hastelloy C, and stainless steel test cells. These results were used both to assist with reactor MOC evaluation and to confirm that test results across instruments were not impacted by different MOC test cells. The DSC results using tantalum and Hastelloy C were comparable, while results for streams in stainless steel showed comparable exotherm size and onset temperature, but had a slower exotherm rate. See Figure 3. Therefore, data generated in tantalum and Hastelloy C test cells was considered conservative. All of the MOCs assessed are considered acceptable, but based on the slower degradation rate, stainless steel would be the preferred reactor MOC.

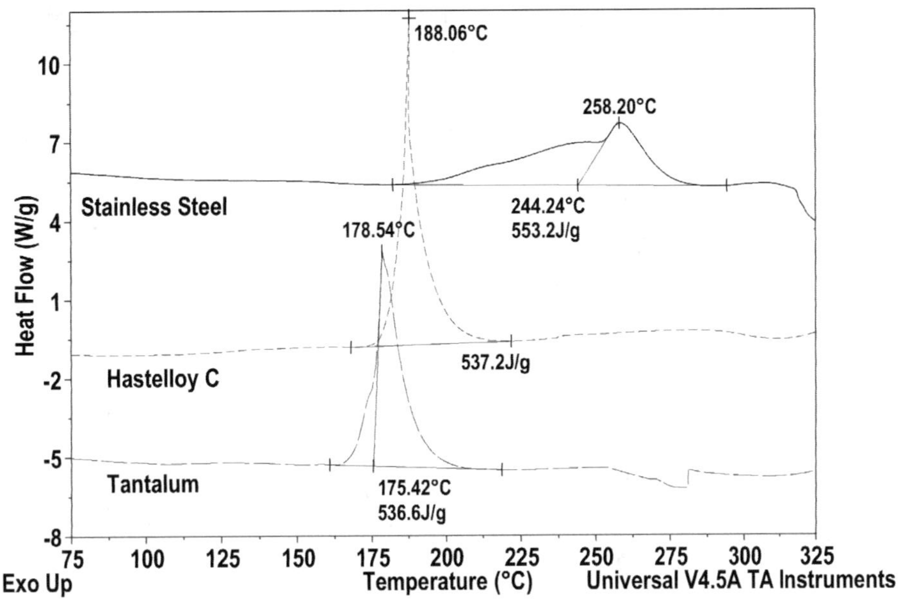

FIGURE 3. Material of construction evaluation of pre-reaction stream using DSC.

3.3 Advanced Pressure Testing & Consequence Assessment

With thermal screening complete, the next step was to determine the pressure consequences associated with the exothermic decompositions. It was known from testing of the prior step that the exothermic decomposition in the brominated starting material had significant and rapid gas generation associated with it. Accelerating rate calorimeter (ARC) testing of the brominated starting material showed an EIT of 158 °C. However, the undesired activity in the brominated starting material is considered a low risk since a double upset (two deviations) would need to occur in order for this activity to be reached: a mischarge of the starting material first to the reactor with a jacket overheating deviation at that time, prior to solvent addition.

Due to the onset temperature, large size, and rapid rate of the exothermic decomposition seen in the process streams prior to base addition, the screening step for pressure consequences of these streams was skipped, and the decomposition evaluation was performed using a more advanced instrument (ARC). The ARC gives the most accurate determination of the EIT. ARC testing was first performed on the stream prior to base addition (pre-reaction stream), as it was likely to be the worst-case stream for this part of the process. This was suspected to be the pre-reaction stream based on DSC results showing that it had the lowest exotherm onset temperature, with a rapid exotherm rate. The worst-case stream must be identified as this contains the undesired reaction for which corrective actions must be taken to either mitigate against or avoid the risks for this part of the process.

Figure 4 shows the plots from an ARC heat-wait-search conducted on the pre-reaction stream. The heat-wait-search was conducted from 110 °C, at which point a series of exotherms were immediately observed, which continued to approximately 350 °C before the test was ended. Once the exotherm initiated, gas generation occurred and the pressure immediately increased. The ARC test showed that within minutes the pressure would exceed the 100 psig maximum allowable working pressure (MAWP) of the reactor, and the rupture disc would open (by 1.16x MAWP)[vii]. If the rupture disc does not open or the venting capacity is inadequate then at approximately 170 °C, the maximum jacket temperature, the instruments would start failing (at 1.5x MAWP). Ten minutes later, the pressure is such that vessel failure would occur (at 3.5x MAWP). The adiabatic temperature rise associated with the first exotherm was 420 °C and the total residual pressure (permanent gas) from the ARC test was 145 psi (at an 8 vol. % test cell fill versus approximate 40 vol. % vessel fill upon scale-up). Note that DMSO boils at 189 °C, so most of the activity would be over before the temperature reaches the boiling point of the solvent.

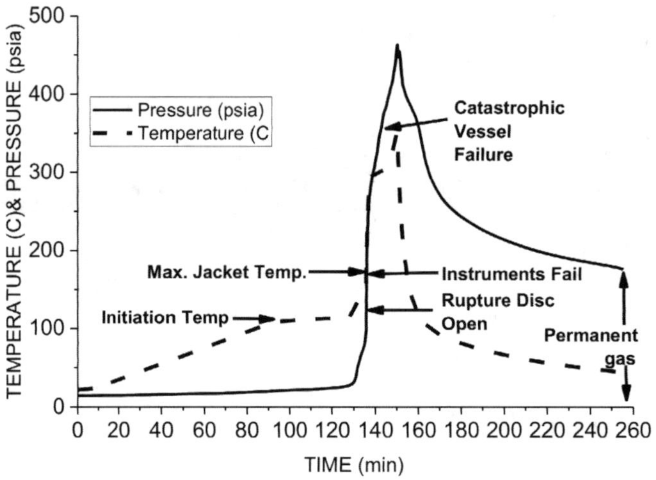

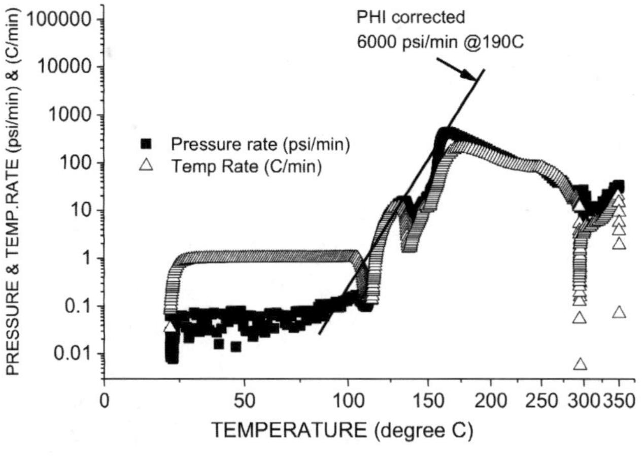

FIGURE 4. Plots of temperature & pressure vs. time and pressure & temperature rate vs. temperature from an ARC heat-wait-search on the pre-reaction stream conducted from 110 °C with 10 °C steps and an onset threshold of 0.04 °C. A Hastelloy C ARC cell was used with a phi = 2.2. This figure is not corrected for phi; therefore, the actual situation would likely be worse with a shorter timeframe.[viii]

This first ARC heat-wait-search was started at 110 °C based on the DSC onset of approximately 175 °C and the general guidance that the ARC EIT can be up to 50 K lower than the DSC onset.

Public

The fact that the exotherm initiated immediately at 110 °C indicates that the actual EIT may be lower; therefore, a second ARC was run starting at a lower temperature to determine the EIT of 84 °C. See Figure 5. In this case, the ARC EIT was greater than 50 °C from the DSC onset and did not follow the general rule of thumb; this is likely due to the fact that DMSO is a reactive solvent and the decomposition is suspected to be autocatalytic. The ARC data confirms that this exotherm is reachable by a reactor jacket loss of control (overheating) process deviation.

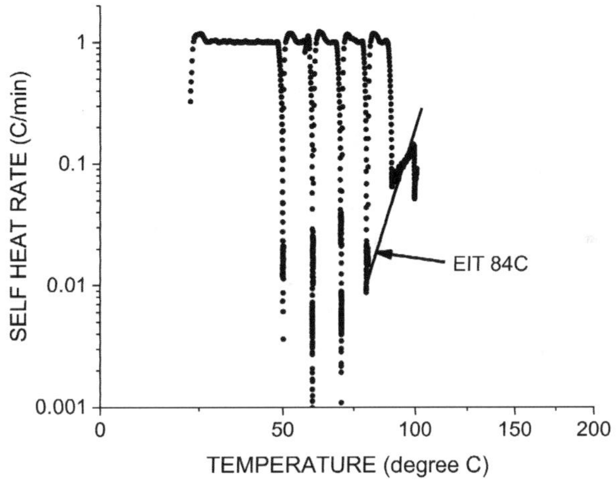

FIGURE 5. ARC heat-wait-search repeat run on pre-reaction stream starting at a lower temperature to capture the EIT. A Hastelloy C ARC test cell was used with a phi = 1.4. The EIT was obtained from extrapolating to a self-heat rate of 0.02 °C/min.

3.3.1 Identification and Final Testing of the Worst-Case Process Stream

ARC testing of additional process streams, before and after the pre-reaction stream, was completed and confirmed that the pre-reaction stream was the worst-case stream with the lowest EIT (see Table 1) and most severe pressure consequences. The higher EIT seen after the base addition is likely attributed to tempering from the presence of water.

TABLE 1. Exotherm Initiation Temperature (EIT) of Process Streams Prior to the End of Reaction.

Process Stream	EIT(°C)
DMSO + starting material	105
Pre-reaction	84
After base addition	144

Because this testing was for a multi-kilo scale process, in addition to evaluating the thermal and pressure consequences from a jacket runaway deviation, the consequences of several other credible single point of failure process deviations, identified in a hazard assessment, were investigated. None of the process deviations resulted in an exothermic decomposition worse than overheating of the pre-reaction stream. Examples of process deviations investigated included the following:

- Missed charge of the brominated starting material.
- Undercharge of DMSO/overcharge of starting material.
- Pre-reaction stream temperature lower than the normal operating temperature.
- Pre-reaction stream aged for several days.
- Rapid addition of base.
- Mischarge of a more concentrated base solution.

Now that the worst-case process stream, the pre-reaction stream, has been identified, further testing on this stream was needed to fully characterize the exothermic decomposition. First, an advanced reactive systems screening tool (ARSST) experiment was run to accurately identify vent sizing parameters, as the ARSST is an open system and most closely represents the deviation at scale. Vent sizing was needed since the ARC pressure results rapidly exceed the MAWP of the vessel and indicate that the decomposition has a high likelihood of over pressurizing the vessel. From the ARSST experiment it was concluded that the system is gassy (no tempering). The ARSST also provided the temperature at which the peak pressure rate occurs, 190 °C. As shown in Figure 4, this was then used to predict the peak pressure rise rate of 6000 psi/min from the ARC data. In order to predict (extrapolate) the peak pressure rise rate under adiabatic conditions (phi = 1.0) a straight line was drawn to describe the effect of temperature on the pressure rise rate.

3.4 Confirmation of the Exotherm Initiation Temperature (EIT)

It was suspected that the undesired reaction was autocatalytic due to the sharp peak observed in the DSC and the low EIT seen in the ARC. Due to this suspicion, several 24+ hour isothermal ages were performed in the DSC in order to gain further confidence in the ARC EIT. As shown in Figure 6, the isothermal-age at 85 °C showed an exothermic event approximately 440 minutes into the age, in alignment with the ARC EIT of 84 °C. The 65 °C and 75 °C iso-ages showed no decomposition after 30 and 72 hours, respectively. This confirmed the temperature of concern as 84°C.

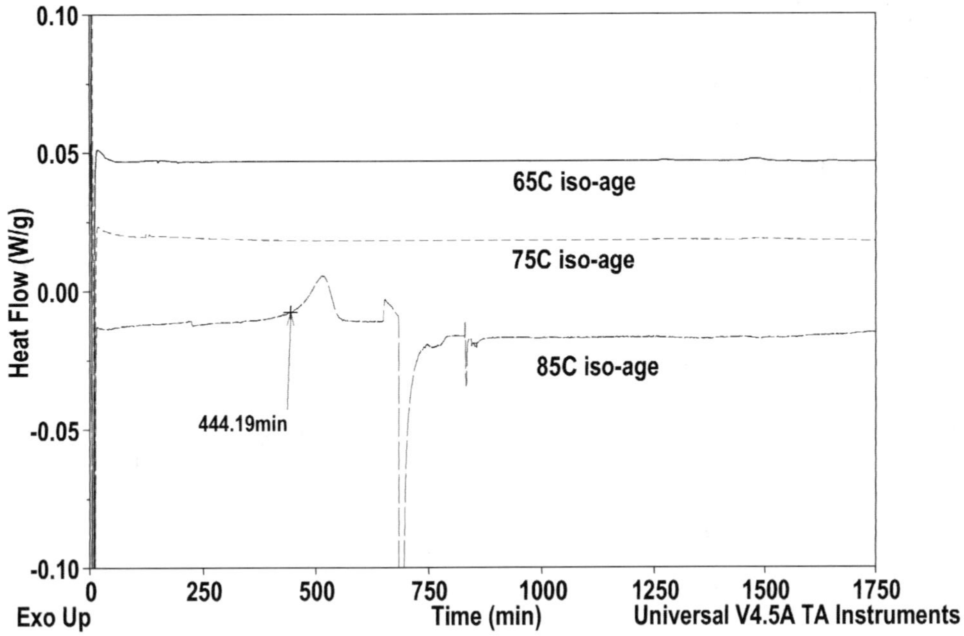

FIGURE 6. DSC isothermal age overlay plot of the pre-reaction stream aged at 65, 75, and 85 °C to gain further confidence in the ARC EIT.

3.5 Time to Maximum Rate (TMR)

To assist with the hazard assessments and determining appropriate process control set-points, the time to maximum rate (TMR)[2] for the pre-reaction stream was estimated using ARC data and Netzsch thermal analysis software. The TMR from 70 °C was estimated to be 84 hours.

3.6 Vent Sizing

With the pre-reaction stream fully characterized, vent sizing was performed to show the consequences of reaching the undesired activity. Using the ARC/ARSST results and Design Institute for Emergency Relief Systems (DIERS) [ix] vent sizing two-phase model for a gassy system, it was determined that a vessel vent diameter of 133-cm would be needed to prevent overpressurizing the vessel, an 11,300-liter reactor. This determination was based on an *L/D* [3] of 200 versus the actual *L/D* of greater than 900; therefore, the actual vent requirement would be even larger. As a final confirmation of the worst-case pre-reaction stream, vent sizing was also performed on the "DMSO+starting material" and "after base addition" streams, in addition to a rough vent sizing analysis on select process deviations; in all cases the vent sizing showed a smaller vent requirement. Since the 11,300-liter reactor had a significantly smaller nozzle, installing a 133-cm vent presented a major challenge. As this size vent is unreasonably large, a secondary vent could not be provided to protect against a vessel failure.

[2] TMR is the time taken for a material to self-heat to the maximum rate of decomposition from a specific temperature, under adiabatic conditions.

[3] *L/D* is the ratio of pipe length to pipe diameter, and correlates to pressure drop during flow; the higher the *L/D* the lower the flow at a given pressure, or the more pressure is required for a given flow.

3.7 Process Safety Testing Summary

In summary, the assessment performed by the EPSE group identified that the worst-case process stream, the pre-reaction stream, had a significant undesired reaction with an EIT of 84 °C versus the maximum jacket temperature of 170 °C, therefore the exotherm could have been reached as a result of a single process deviation, reactor jacket loss of control (overheating). Note that there was no heat sink or solvent barrier present due to the use of DMSO, a reactive solvent with a high boiling point (above the EIT) and large exothermic decomposition at elevated temperatures.

If reached, the undesired reaction could have led to a catastrophic vessel failure; the significantly smaller existing vent on the reactor would not have been sufficient, as a vent of at least 133-cm was required to prevent overpressurizing the vessel. This unreasonable vent size means that a secondary vent could not be provided to protect against vessel failure.

4 Risk Mitigation

With all this information in hand, the EPSE group worked with the site, safety & environmental, and engineering groups to manage this risk. A hazard and operability study (HAZOP) and layers of protection analysis (LOPA) were used to determine the appropriate processing safeguards needed to mitigate the consequences and reduce the likelihood of reaching this activity to an acceptable level. The risk mitigation strategy developed included:

- Additional above baseline controls (e.g. hard-wired interlocks) put in place, requiring equipment modification and capital expense.
- Operator training on hazard and response.
- Additional procedural controls and process oversight.

In order to completely eliminate the risk, either a change in chemistry or how the process was run would be necessary.

5 Conclusion

This case study demonstrates the importance of fully characterizing undesired reactions, including the consideration of process deviations. The process safety testing revealed a significant undesired reaction that could lead to vessel failure as a result of a single process deviation, reactor jacket loss of control (overheating). One must first be aware of the hazards and understand the associated consequences in order to take the appropriate corrective actions to either mitigate or avoid the risks. In this case, the data influenced the design of the equipment with additional controls put in place. It may also influence the design of future processes. The end goal is to protect people, protect equipment/facilities, and prevent disruption to the supply of medicines. A process may run for years without an incident before one day an error leads to a runaway reaction and vessel explosion.

This case study also validates why reactive solvent (e.g., DMSO) use is discouraged, especially high boiling point solvents, as the solvent participates in the decomposition activity rather than helping to absorb heat from the decomposition. In addition, the case study shows that small changes in stream composition can greatly impact the EIT, size, and rate of decomposition; for example, the pre-reaction stream containing acid had an EIT approximately 20 °C lower than the prior stream without the acid. The case study also illustrates the value of performing additional characterization on suspected autocatalytic decompositions, especially in the presence of a reactive solvent, as the ARC EIT may not follow the general guidance and may be more than 50 K lower than the DSC onset. This lower initiation temperature can greatly influence the safety assessment.

6 Acknowledgement

The author would like to thank Don Bachert, Dan Muzzio, Chuck Orella, and Tom Vickery from Merck & Co., Inc., Rahway, NJ 07065 USA, for their contributions to this work.

[i] (a) Center for Chemical Process Safety (2008) Selection of hazard evaluation techniques, in *Guidelines for Hazard Evaluation Procedures*, 3rd Edition, Wiley, Hoboken, NJ. doi: https://doi.org/10.1002/9780470924891.ch6;
(b) Center for Chemical Process Safety (1995) References, in *Guidelines for Chemical Reactivity Evaluation and Application to Process Design*, Wiley, Hoboken, NJ. doi: https://doi.org/10.1002/9780470938058.refs;
(c) Center for Chemical Process Safety (2010) Overview of LOPA, in *Layer of Protection Analysis: Simplified Process Risk Assessment*, Wiley, Hoboken, NJ. doi:https://doi.org/10.1002/9780470935590.ch2;
(d) Stoessel, F. (2008) Introduction to risk analysis of fine chemical processes, in *Thermal Safety of Chemical Processes: Risk Assessment and Process Design*, Wiley- VCH Verlag GmbH & Co. KGaA, Weinheim, Germany. doi:https://doi.org/10.1002/9783527621606.ch1.

[ii] Poje, G.V., Taylor, A.K., and Rosenthal, I. (2000). *US Chemical Safety and Hazard Investigation*. Report No. 1998-06-I-NJ, Chemical Manufacturing Incident at Morton International, Inc., 16 August 2000. https://www.csb.gov/assets/1/20/morton_report.pdf?13798 (accessed 16 October 2018).

[iii] (a) Lam, T.T., Vickery, T., and Tuma, L. (2006). Thermal hazards and safe scale-up of reactions containing dimethyl sulfoxide. *Journal of Thermal Analysis and Calorimetry* 85 (1): 25–30.
(b) Wang, Z., Richter, S.M., Gates, B.D., and Grieme, T.A. (2012). Safety concerns in a pharmaceutical manufacturing process using dimethyl sulfoxide (DMSO) as a solvent. *Organic Process Research & Development* 16 (12): 1994– 2000. https://doi.org/10.1021/op300016m.
(c)Brandes, B.T. and Smith, D.K. (2016). Calorimetric study of the exothermic decomposition of dimethyl sulfoxide. *Process Safety Progress* 35: 374–391. https://doi.org/10.1002/prs.11802.

[iv] Bassan, E., Ruck, R.T., Dienemann, E. et al. (2013). Merck's reaction review policy: an exercise in process safety. *Organic Process Research & Development* 17 (12): 1611–1616. https://doi.org/10.1021/op4002033.

[v] Frurip, D.J. (2008). Selection of the proper calorimetric test strategy in reactive chemicals hazard evaluation. *Organic Process Research & Development* 12 (6): 1287–1292. https://doi.org/10.1021/op800121x.

[vi] ASTM E1981-98(2012)e2 *Standard Guide for Assessing Thermal Stability of Materials by Methods of Accelerating Rate Calorimetry*, ASTM International, West Conshohocken, PA, 2012. www.astm.org (accessed 17 November 2017).
[vii] (a) ASME *ASME Boiler and Pressure Vessel Code, Section VIII Pressure Vessels, Division 1.* New York, NY: American Society of Mechanical Engineers http://www.asme.org/shop/standards/new-releases/boiler-pressure-vessel-code-2013.
(b) Crowl, D.A. and Tippler, S.A. (2013). Sizing pressure-relief devices. *Chemical Engineering Progress* 109: 68–76.
(c) Sandler, H.J. and Luckiewicz, E.T. (1993). *Practical Process Engineering*. New York: McGrew-Hill.
[viii] Townsend, D.I. and Tou, J.C. (1980). Thermal hazard evaluation by an accelerating rate calorimeter. *Thermochimica Acta* 37 (1): 1–30. https://doi.org/10.1016/0040-6031(80)85001-5.
[ix] Fisher, H. G., Forrest, H. S., Grossel, S. S., Huff, J. E., Muller, A. R., Noronha, J. A., Shaw, D. A. and Tilley, B. J. (1993) Front matter, in *Emergency Relief System Design Using DIERS Technology: The Design Institute for Emergency Relief Systems (DIERS) Project Manual*, Wiley, Hoboken, NJ. doi:https://doi.org/10.1002/9780470938317.fmatter

19 AIChE Spring Meeting
15TH GLOBAL CONGRESS ON PROCESS SAFETY
NEW ORLEANS • 2019

Investigation of a Polymeric Resin Deflagration Incident Using Advanced Calorimetric Techniques

Katie A. Mulligan[1]
The Dow Chemical Company
[1]Freeport, TX USA
KAMulligan@dow.com

Robert J. Bellair[1]
Steven E. Horsch[1]
Donald M. Huether[1]
Kefu Sun[1]
Kevin L. Simmons[1]
Amy L. Koziol[1]
Praveenkumar Boopalachandran[1]

RJBellair@dow.com
SEHorsch@dow.com
Huethedm@dow.com
KSun@dow.com
KLSimmons@dow.com
AKoziol@dow.com
PBoopalachandran@dow.com

The Dow Chemical Company
[1]Freeport, TX USA

Prepared for Presentation at
American Institute of Chemical Engineers
2019 Spring Meeting and 15th Global Congress on Process Safety
New Orleans, LA
March 31- April 3, 2019

Investigation of a Polymeric Resin Deflagration Incident Using Advanced Calorimetric Techniques

Katie A. Mulligan[1]
The Dow Chemical Company
[1]Freeport, TX USA
KAMulligan@dow.com

Robert J. Bellair[1]
Steven E. Horsch[1]
Donald M. Huether[1]
Kefu Sun[1]
Kevin L. Simmons[1]
Amy L. Koziol[1]
Praveenkumar Boopalachandran[1]

RJBellair@dow.com
SEHorsch@dow.com
Huethedm@dow.com
KSun@dow.com
KLSimmons@dow.com
AKoziol@dow.com
PBoopalachandran@dow.com

The Dow Chemical Company
[1]Freeport, TX USA

Abstract

In 2018, a Process Safety incident occurred during loading operations for a polymeric resin, resulting in a deflagration inside the shipping container. An investigation was conducted to determine the cause of the event. During the extrusion operation, an organic solid additive is added to base polymer resin at ppm levels to protect resin quality at elevated temperatures, and some additive degradation is expected. After the event, it was learned that a zone of the extruder had reached a higher-than-normal temperature shortly before the event, although it was unclear at the time whether this observation was connected to the event. Through the application of laboratory-scale thermoanalytical testing, it was subsequently determined that the higher operating temperature resulted in a higher percentage of additive decomposition than had been expected, creating the conditions that led to the event. Despite the maturity of this technology, the potential to accumulate a sufficient quantity of flammable additive degradation components during a non-routine operating mode had not been previously identified as a potential hazard scenario.

This paper will illustrate how reactive chemicals knowledge was combined with analytical expertise to help determine the cause of the event. This investigative work included the use of Thermogravimetric Analysis (TGA) as well as an innovative adiabatic Accelerating Rate Calorimetry (ARC) unit coupled to an online gas chromatograph/mass spectrometer (GC/MS). As a result of this reactive chemicals evaluation and design of key experiments, the root cause of the event was identified, enabling stakeholders to define mitigations that would permit safe return to operations and prevent a future reoccurrence.

Keywords: Reactive chemicals, adiabatic calorimetry, Gas Chromatography/Mass Spectrometry (GC/MS), Thermogravimetric Analysis, evolved gas analysis, deflagration, polymeric resin

1. Introduction

In 2018, a deflagration occurred in center railcar compartment during loading operations of polymer resin at a Dow facility. The deflagration ejected the loading spout from the center compartment and damaged the railcar and other miscellaneous equipment. There were no injuries.

The manufacturing process for the resin involves the polymerization of olefins to polymeric resin; a simplified process flow diagram is shown in Figure 1. Polymer resin that exits the reactor goes through a resin purge (N_2) operation to remove unreacted flammable olefin gas before moving to a pelleting system, where solid additives and resin are fed to an extruder in order to impart desired properties to the final resin product. Post-extrusion, pelletized resin is typically sent to short-term storage before being loaded into railcars for transport.

Thermal analysis/calorimetric techniques are routinely utilized to help determine the cause of industrial incidents, including unexpected exothermic activity, runaway reactions and flammability events [1, 2]. This report details how reactive chemicals and analytical expertise, including thermal analysis, flammability calculations, and spectroscopy was applied to explain the generation of a flammable atmosphere inside the railcar due to localized temperature excursion in the extruder barrel.

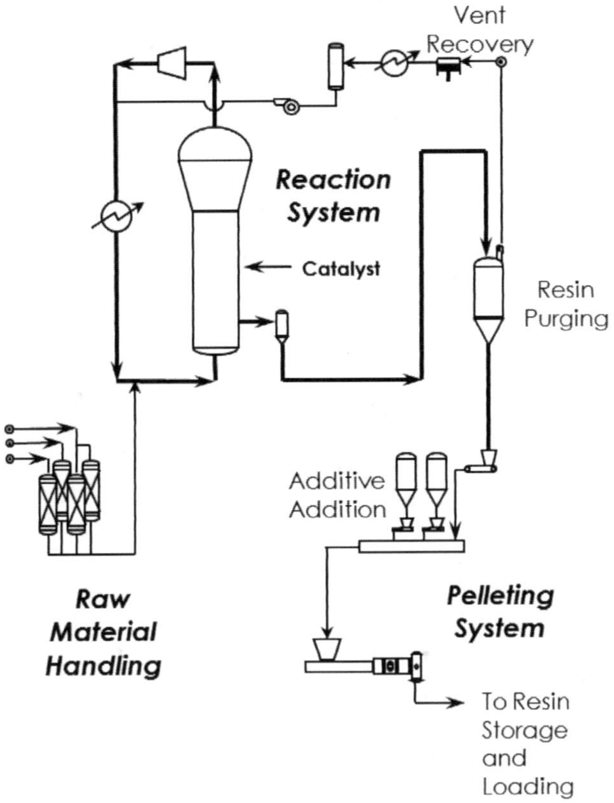

Figure 1. Simplified Process Flow Diagram for Polyolefin Manufacturing

2. Experimental

This report addresses the application of an analyzer, ARC coupled to Gas Chromatography/Mass Spectrometry (termed ARCxGC/MS), and Thermogravimetric Analysis (TGA) to support a Root Cause Investigation (RCI) for the deflagration event. Both techniques were utilized to help the RCI team assess decomposition properties of various samples. A brief comparison of these thermal methods is found in Table 1. TGA analysis was used to screen sample decomposition properties and provided fast data turnaround (high-throughput). Accelerating Rate Calorimetry (ARC) coupled to online Gas Chromatography/Mass Spectroscopy (GC/MS), which is a more time-consuming test than TGA (both data collection and analysis). However, ARCxGC/MS provides an exceptionally useful data set in a single test: temperature rate and pressure vs. temperature data *and* specific analytical identification of decomposition gases.

Table 1. Comparison of Thermal Techniques

	ARCxGC/MS	TGA
System	Closed	Open
Sample size (approximate)	≤ 5 gram	≤ 100 mg
Heating Method(s)	Heat-Wait-Search (HWS); Isothermal or Isoage	Isothermal; Continuous ramp
Temperature Range (°C)	Ambient-450 °C	0-800 °C
Data obtained	Modelable kinetics for exothermic reactions; gas generation	Temperature and mass change for decomposition reactions; Decomposition kinetics [3]
Headspace	Air, N_2, vacuum, specialty gas	Air, N_2, vacuum, specialty gas
Test time (typical)	Fast; 1-2 hours	Time intensive; >24 hours
Other	Screening; high-throughput	Quantitative & specific

2.1 ARCxGC/MS

The ARC is a thermal hazard evaluation calorimeter that was developed at The Dow Chemical Company to determine the time, temperature and pressure relationship of exothermic, adiabatic reactions [4]. Information from ARC experiments includes thermokinetic aspects of temperature and pressure associated with chemical reactions (heat and pressure generation rates, reaction rate constants) and is routinely utilized in thermal hazard evaluations [5, 6]. The thermodynamic and kinetic data from ARC experiments can be applied to runaway reaction scenarios to determine: adiabatic temperature rise, heat of reaction, and system characterization and emergency relief system design [7][8].

Reaction effluent can be comprised of desired reaction products, decomposition products, or volatile starting materials. The high-temperature thermal degradation of organic substances can lead to the evolution of flammable, gaseous reaction products. ARC coupled to Gas Chromatography/Mass Spectrometry (ARCxGC/MS) simultaneously allows for the assessment of reactive chemicals hazards and identification of reaction effluent(s) at a given set of conditions. A sampling system was designed for the safe sampling and transport of gas-phase effluent from the ARC sample sphere (which can contain pressures in excess of 3000 psi) to the GC. Details of the system design and function can be found in a prior report [9]. Figure 2 is a high-level schematic of the ARCxGC/MS setup. There are three main constituents: (1) ARC calorimeter, (2) Sampling System, and (3) an Agilent 6890 GC coupled with HP 5973 Mass Selective Detector (MSD) and PolyARC detectors. A Netzsch ARC 254 was utilized for the results presented herein. Agilent ChemStation Software was utilized to control the GC/MSD. Ultra-high purity helium and hydrogen carrier gases were supplied to the GC. Appendix A contains additional experimental details.

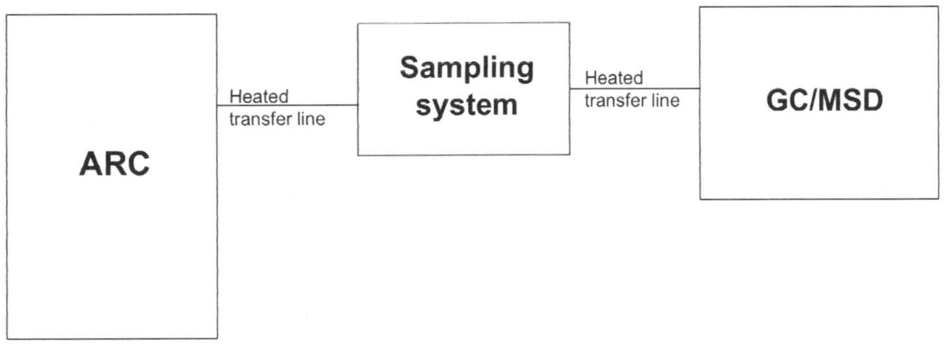

Figure 2. High-level Schematic of ARCxGC/MS analyzer. Reproduced from Reference 9.

To prepare samples for ARCxGC/MS testing, solid samples were loaded into 316 stainless steel ARC spheres and purged with N_2. After the ARC test was completed, the sample headspace was transferred to sampling system while still at the maximum test temperature; the sampling system transfer line was heat-traced with the temperature set to 150 °C (to prevent condensation of less volatile species). An injection was then made onto the GC for analytical analysis. Appendix B contains representative ARC test experimental details.

2.2 Thermogravimetric Analysis (TGA)

In a TGA experiment, the mass of a sample is measured using a microbalance as a function of increasing temperature (with constant heating rate) or as a function of time (with constant temperature and/or constant mass loss). Mass resolution is approximately ±0.1 µg or better (depending on TGA model). A TGA curve, which is a plot of mass change (%) vs. temperature (or time) is the common result of TGA analysis. The shape of the TGA curve can be utilized to understand sample properties, including thermal stability, oxidative stability and changes in physical and chemical properties of materials. Large mass losses in TGA experiments typically indicate thermal decomposition.

The TGA utilized for experiments detailed in this report is a TA Instruments Q500. Units which couple TGA to GC/MS and infrared analytical detection are commercially available [10-12], but were not utilized to support this RCI.

2.3 Fourier Transform Infrared (FTIR) Spectroscopy

FTIR-ATR spectrum of solid antioxidant additive was acquired on a Nicolet 6700 FTIR spectrometer equipped with a single bounce Czitek MicromATR Vision accessory. Appendix C contains specific details on FTIR-ATR experimental parameters.

3. Results and Discussion

3.1 Root Cause Investigation

A solid antioxidant additive was being fed to the extruder at a low concentration. The resin being produced was off-grade and spent minimal time in storage silos before loading operations commenced. Loading of the partially filled railcar compartment was paused to investigate an odor emanating from the pellets. Upon resuming the loading of resin pellets into the railcar compartment, the deflagration occurred. Photographs of damage caused by incident are given in Figure 3.

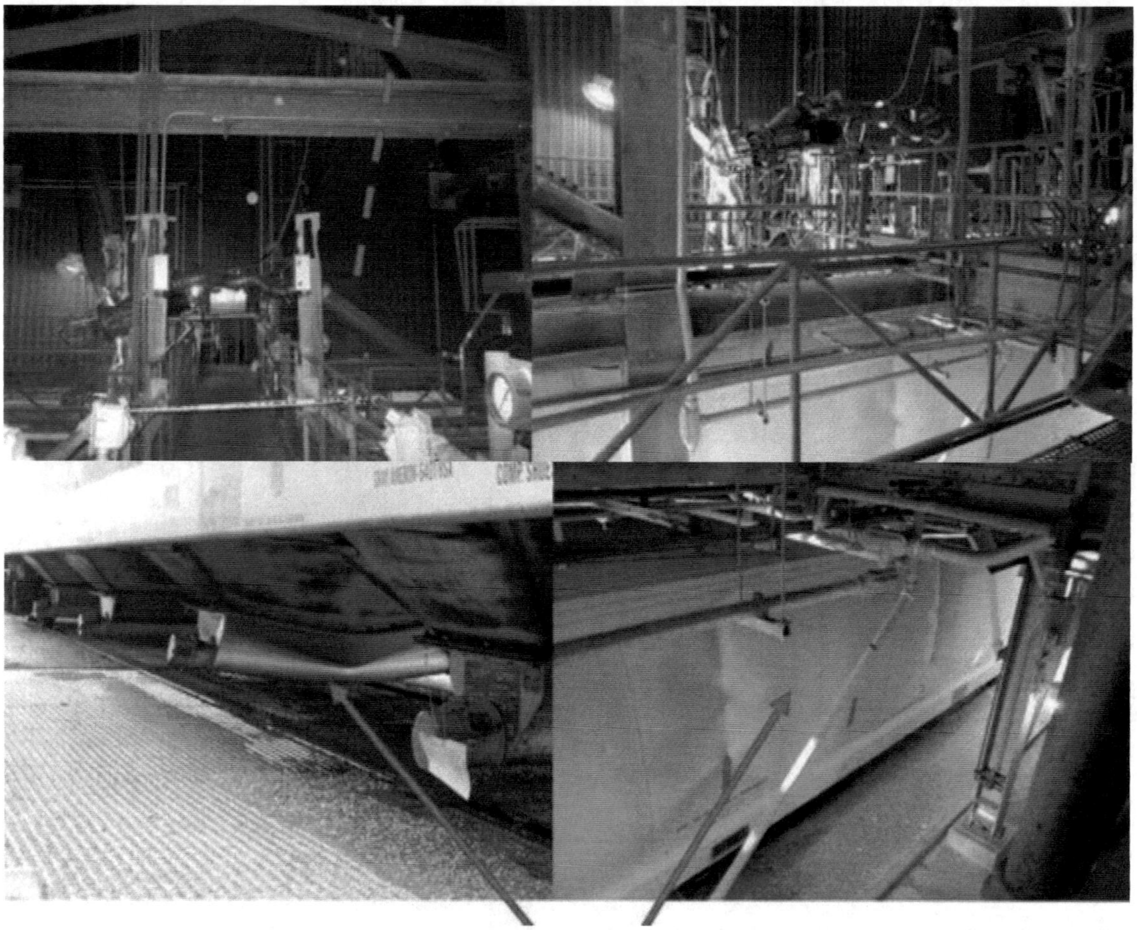

Figure 3. Top: Loading Spout damage. Red dashed line indicates original position of loading spout, which was ejected as a consequence of the deflagration. Bottom: Rail Car center compartment damage

A multidisciplinary Root Cause Investigation (RCI) team was assembled to fully investigate the deflagration incident. In-process flammability measurement data was used to rule out a hypothesis that residual, unreacted olefins from the reactor was the source of flammable gases which ignited in the railcar compartment. Additionally, it should be noted that the resin involved in the incident contained negligible fines concentration; as such, a dust explosion was ruled out.

This remainder of this report details how reactive chemicals and analytical expertise, including adiabatic calorimetry coupled to online GC/MS and flammability calculations, was applied to explain the generation of a flammable atmosphere inside the railcar due to a localized temperature excursion in the extruder barrel. This testing was performed to understand whether the RCI team's hypothesis of decomposed solid antioxidant additive being the primary source for flammable gas was credible.

Questions to Answer Using Thermoanalytical Techniques:

1. What is/are credible source(s) of flammable gas in railcar?
2. Does data support the hypothesis that a high temperature excursion at ~380 °C in the extruder degraded the additive and/or polymer resin at sufficient rate to produce flammable atmosphere inside the hopper car?

Samples to Test:

1. Solid organic additive (antioxidant additive at time of event)
2. Polyolefin resin, compounded (post-pelleting, compounded with antioxidant additive), from event (off-grade)
3. Polyolefin resin, granular (before pelleting, no antioxidant additive)

3.2 Application of Thermoanalytical Methods to Support Post-Incident RCI

TGA Results

TGA results (mass vs. time) for the three samples are overlaid in Figure 4. The samples were each subjected to an isothermal hold at 350 °C; N_2 gas was flowing during the experiment. The corresponding sample temperature is also plotted for reference (red curve). TGA results reveal that the antioxidant additive undergoes an apparent single step decomposition; it is completely decomposed (no sample residue remaining, 98% mass loss) after < 20 minutes at 350 °C. The additive is significantly decomposed before even reaching the isothermal test temperature. Derivative curves, which give mass loss % per minute as a function of both time and sample temperature, are shown in Figure 5. Extrapolation of this data to the measured event T of 380°C indicates that ~60 wt. % of the solid antioxidant additive is decomposed after just 1 minute. In contrast, both polymer resin samples (granular and compounded pellets) show excellent thermal stability (minor mass loss) under the exact same test conditions. There is no significant difference in TGA degradation profile between the pre- and post-pelleting polymeric resins.

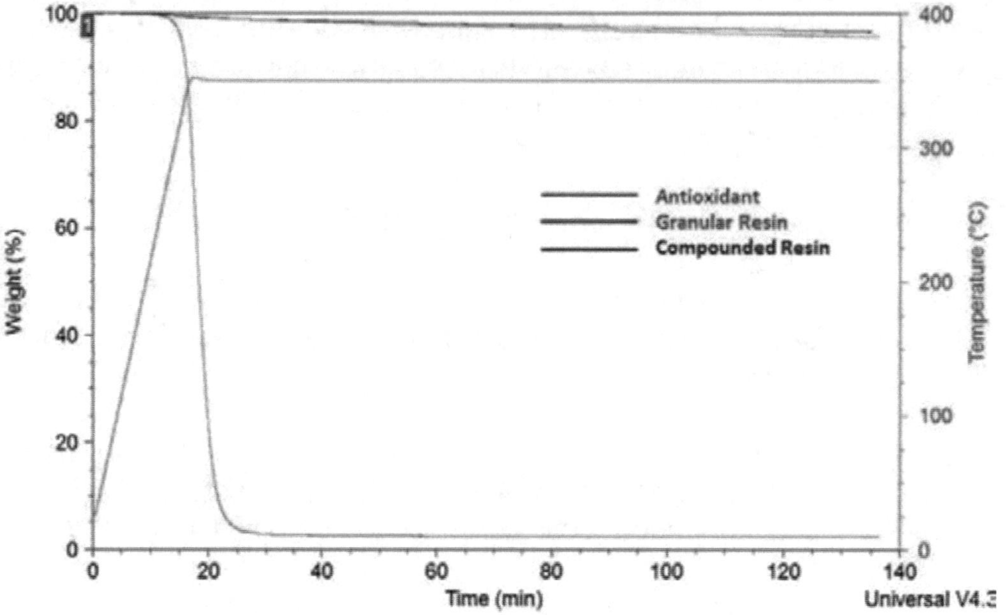

Figure 4. Isothermal (350°C) TGA curves for Granular resin (pink), Compounded resin (blue) and solid antioxidant additive (green). Red curve gives Temperature vs. Time.

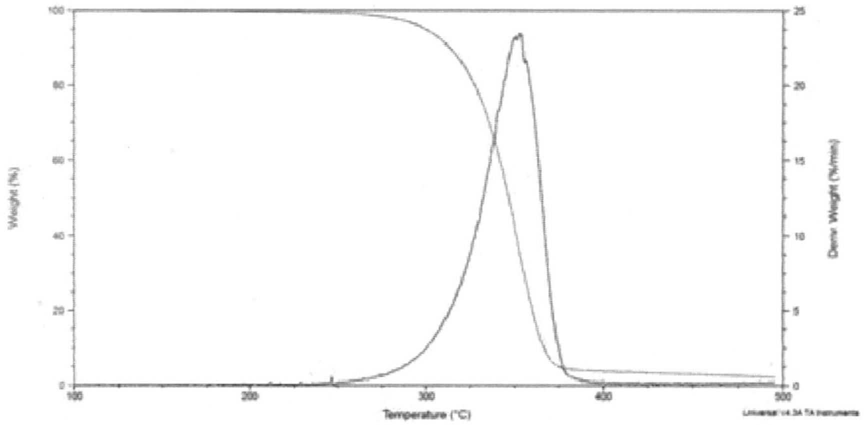

Figure 5. Overlaid TGA curves (weight % vs. T and derivative weight % vs. T) for solid antioxidant additive.

ARCxGC/MS Results

Samples Tested Using ARCxGC/MS:

1. Pure Antioxidant Additive (from event)
2. Compounded Resin (post-pelleting, contains additive in low final concentration)
3. Granular Resin (pre-pelleting, contains no additive)

ARC results for three tested samples are overlaid in Figure 6. All ARCxGC/MS testing was performed using a N_2 headspace, which prevents oxidative effects from atmospheric O_2 influencing the test results. An ARC test for the solid organic antioxidant additive (Heat-Wait-

Search (HWS) method, green curve) revealed exothermic activity detected at T> 300 °C, while significant gassy decomposition was observed at much lower T (T≈ 240 °C). The pressure drop at ~ 365 °C was due to a minor leak of the sample from the ARC sphere during the experiment. An Isothermal ARC test for the pure additive (results not shown) at 320 °C reveals >1.5 mol gas/mol solid additive (based on an ideal gas law calculation). No exotherm was detected during this isothermal test.

The two polyolefin resins were also tested using the ARCxGC/MS. The starting temperature for these two HWS ARC tests was 300°C (based on TGA thermal stability results). ARC test results for the compounded resin reveal that no exotherm was detected up to the maximum test temperature of 450 °C. Results for the granular resin sample reveal an unsustained, minor exotherm (phi-corrected temperature rise < 5 °C) at ~ 340 °C and a second, incomplete exotherm with a detected onset temperature of 440 °C. For both resins, significant gassy decomposition is observed before any exotherm is detected, at T ≈ 390 °C. It should be noted that there is significant time at temperature associated with the HWS ARC testing. Based on TGA and ARC data, it can be concluded that a very short time at T≥ 390 °C would not lead to significant resin decomposition.

Overall, the ARC data agrees well with TGA screening results for the additive and the two tested resins. Both techniques show that the additive is much more susceptible to significant decomposition at event conditions (temperature, short residence time) than the polymer resins.

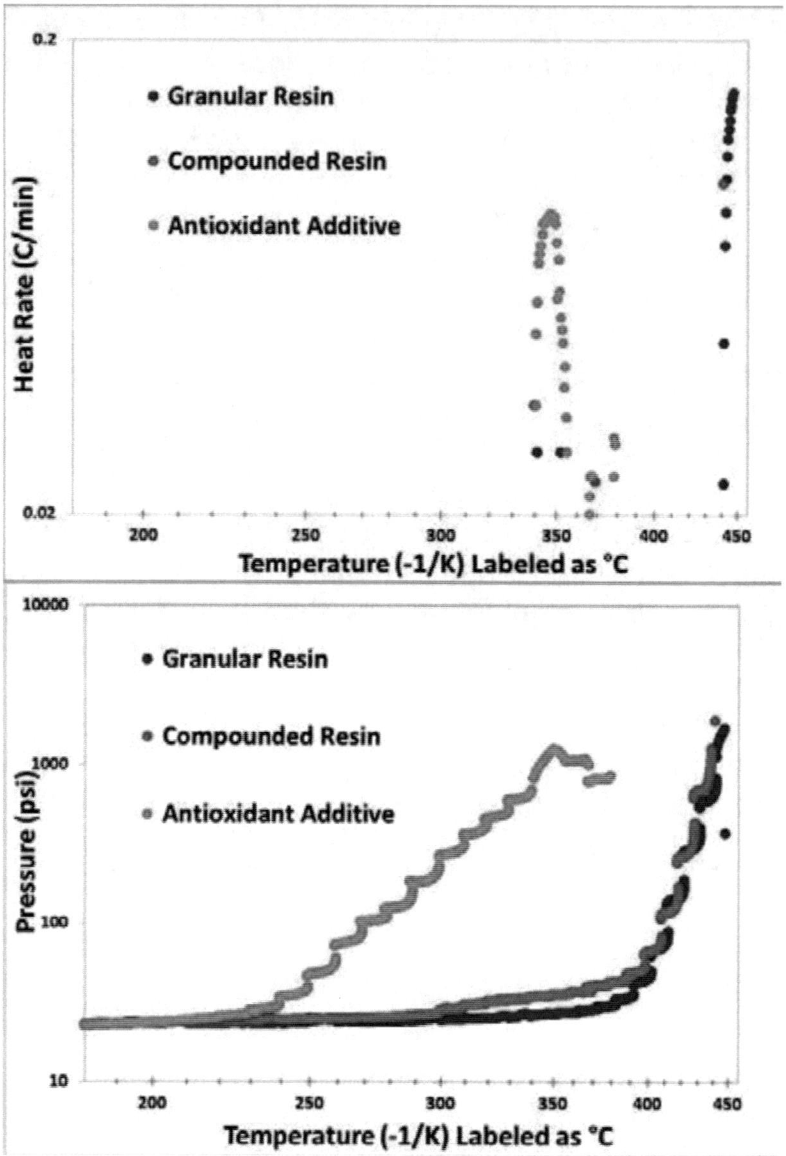

Figure 6. (top) Self-Heat-Rate vs. Temperature and (bottom) Pressure vs. Temperature curves obtained during HWS ARC experiments for Granular resin (blue), Compounded resin (red) and antioxidant additive (green). Note the zoomed-in scale for the Self-Heat-Rate plot.

For each sample, the sample headspace was transferred to the sampling system while still at the maximum test temperature; the sampling system heat tracing was set to 150 °C, and an injection was made onto the GC for analytical analysis.

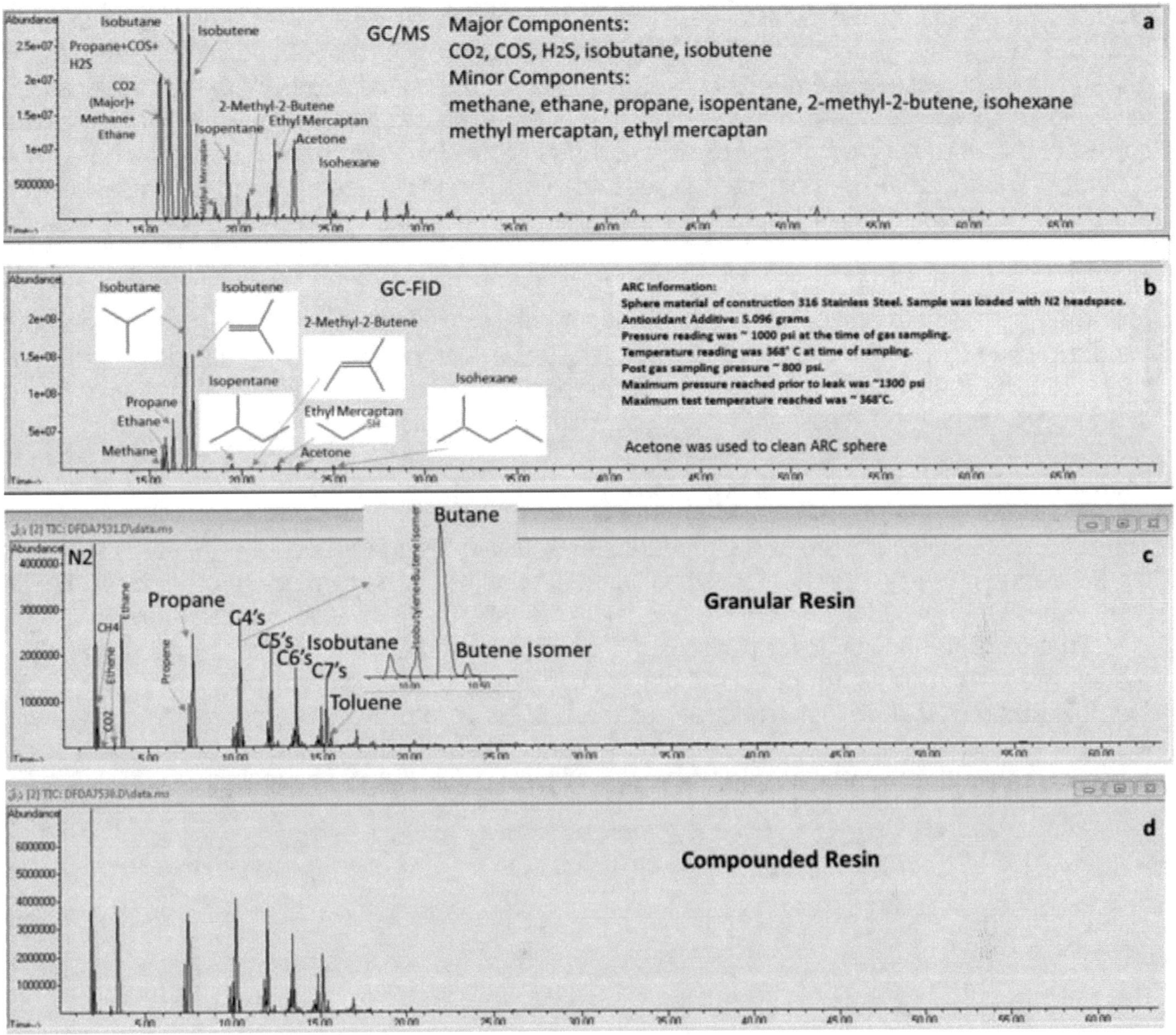

Figure 7. Total Ion Chromatograms of headspace generated in ARC experiment of (a-MSD and b-FID) organic antioxidant additive, (c) granular polyolefin resin and (d) compounded polyolefin resin.

Figure 7 shows the gas effluent analytical analysis results for the 3 samples tested using the ARCxGC/MS setup. Figures 7a and 7b show the Total Ion Chromatogram (Mass Selective Detector, MSD) and PolyARC Flame Ionization Detector (FID) signals for the decomposed antioxidant additive. Major components detected from antioxidant additive decomposition were carbon dioxide, carbonyl sulfide, hydrogen sulfide, isobutane, and isobutene. Minor components in the effluent stream included methane, ethane, propane, isopentane, 2-methyl-2-butene, isohexane, methyl mercaptan, and ethyl mercaptan. Most of these gases are flammable and the results aligned with observations of a "natural gas" odor while loading resin before the deflagration occurred. Methyl mercaptan is added to natural gas to act as a warning property; very small concentrations are easily detected.

ARCxGC/MS results for the granular resin and compounded resin samples, shown in Figures 7c and 7d, are effectively identical in both gas identity and relative ratio of evolved decomposition gas species. The gases were primarily hydrocarbons (C1-C8 chain length; note that the heavier, less volatile species are detected because the system is heat traced). The gas effluent results correspond well to evolved gas pyrolysis experiments for polyolefin resin (comparison data not shown). While acetone was detected in all 3 tests as a minor species, its trace presence was attributed to the fact that ARC spheres are cleaned with acetone before ARC testing.

Overall, thermoanalytical data (TGA and ARCxGC/MS) supported the hypothesis that, at the abnormal extrusion conditions (T≥ 380°C, approximate residence time in extrusion zone of 1 minute), can lead to significant antioxidant additive decomposition. In addition, the data supported that only minor polymer resin decomposition is credible, since long residence times at T>400 °C would be necessary to detect any significant decomposition.

Fourier Transform Infrared (FTIR) Spectroscopy

The solid additive is accepted at the manufacturing facility using Certificate of Analysis supplied by the manufacturer; no positive analytical identification test is performed. In order to eliminate a hypothesis of the addition of incorrect chemical into the extrusion process, spectroscopic investigation of the additive was performed. To support the RCI team, a high priority analytical request was made to verify that a sample of solid additive suspected of decomposing was, in fact, the desired additive. The spectrum of the additive (from event) compared to the known library spectrum exhibit perfect agreement, as seen in Figure 8. This result allowed the RCI team to quickly reject the hypothesis that the wrong chemical was fed to the extruder.

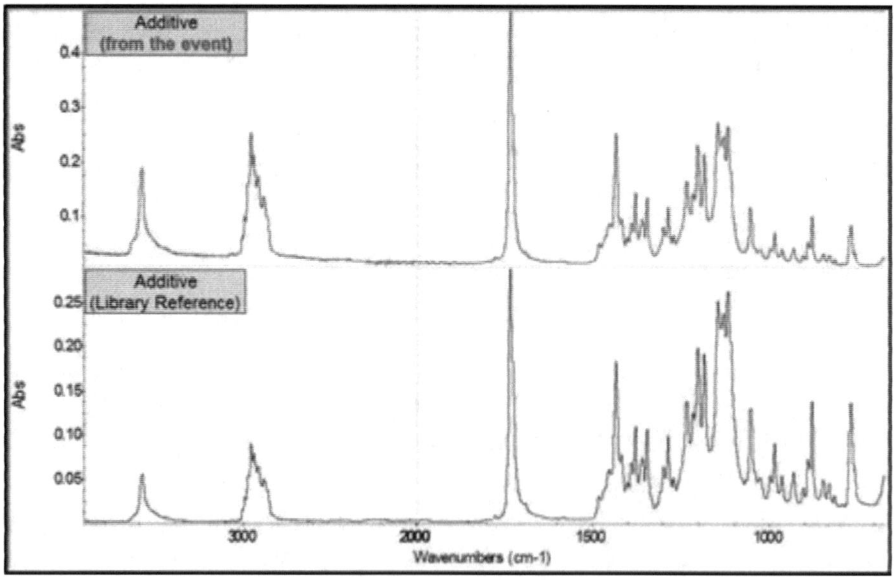

Figure 8. FTIR Spectral Overlay of Additive (from event) and Additive (known from Library Reference).

3.3 *Flammability Considerations*

Using the above data, an assessment was performed to determine if additive degradation could account for the amount of flammable gas necessary to generate a flammable atmosphere in the railcar.

Calculation assumptions:

- The gas distributes uniformly through the railcar compartment
- The amount of gas released from resin pellets before reaching the railcar compartment was nominal (short residence time in silo)
- 2 mol of flammable gas per mole of solid additive are generated
- The LFL of the gas mixture ≈ LFL of isobutene (1.4 volume %)
- No gas is generated from polymeric resin decomposition (conservative assumption)
- $T_{atmosphere}$= 35°C (pellets were still warm)

In the incident, off-grade resin pellets were being loaded directly into a railcar; loading was paused to investigate odor, and the resin was allowed to sit before loading was resumed. Based on gas generation data from ARCxGC/MS tests (quantity and % composition of decomposed gas-phase effluent that is flammable), a minimum of 3000 kg of polymeric resin would necessary to reach a flammable atmosphere in the railcar compartment. The incident investigation revealed > 3000 kg of resin had been loaded prior to the deflagration.

Static discharge from loading of resistive resin pellets was most likely the ignition source (~ 0.2 mJ MIE for hydrocarbon vapors) [13, 14]. It is difficult to ground/bond the polymeric resin itself (discontinuous phase), and therefore electrostatic discharges are likely as the charge builds during railcar loading [15]. Based on additive addition rates, ARC pressure results and GC/MS analysis, the relatively low antioxidant additive concentration was deemed sufficient to generate enough gas to create a flammable atmosphere in the railcar.

3.4 *Root Cause Summary*

- Extruder barrel section reached ≥380 °C
- Lack of understanding that a minimum residence time in silos (post-extrusion) is needed to protect against an increase in level of flammable gases in hopper cars
- Lack of knowledge regarding the quantity of flammable gas that can be generated by additive decomposition

4. Conclusions

A localized temperature excursion (outside of normal operation range) in the extruder barrel led to increased solid additive decomposition kinetics. Flammable decomposition gases were trapped in polyolefin resin pellets and slowly diffused in post-extrusion operations. The deflagration was the result of an ignition of hydrocarbon/air mixture inside the compartment, with flowing polyethylene material providing the static ignition source. The hydrocarbons were generated in the extruder system as a result of degradation of antioxidant additive at high temperatures. The minimal residence time provided in the silos for the off-grade resin pellets in the post-extrusion area contributed to event occurrence.

ARCxGC/MS and TGA thermoanalytical methods were utilized to support the deflagration RCI. TGA and ARC testing revealed that the antioxidant additive is unstable under the extruder conditions at the time of the event and decomposed considerably. ARCXGC/MS and basic flammability calculations revealed that the quantity and composition of evolved gas from significant additive decomposition is sufficient to have created a flammable atmosphere inside of the partially filled railcar compartment.

With new techniques to generate data, a previously unrecognized process safety scenario was identified. Reactive chemicals support helped to define and support both the temporary mitigative measures to allow for safe extruder startup as well as the implementation of the permanent modifications to prevent reoccurrence.

Leverageable takeaways from the event are summarized below:

- Relatively small concentrations of additive can degrade at high temperature in extruders and produce a flammable atmosphere in downstream equipment.
- Abnormal situations (in this case, odor) should be escalated with the intent of understanding the cause versus looking for confirmation that everything is OK.
- Two sides of the fire triangle are typically present in the railcars. The headspace is air and the flowing pellets are a discontinuous phase; grounding and bonding does not prevent static accumulation. Therefore, the fuel source must be managed.

Corrective actions include updates to logic control systems, operating discipline and new hardware installation. Flammable gas detectors were placed at strategic locations in the post-extrusion operations area, and high Lower Explosive Limit (LEL) readings prevent loading into railcar. In addition, the operating discipline for loading off-grade resin was adjusted to allow for sufficient residence time (for off-gassing from pellets) before railcar loading. High temperature alarms and automatic extruder shutdown limits were added to prevent extreme additive degradation and, if abnormal conditions are reached (such as measured high temperature in the extruder), an analysis will be performed to determine adequate post-extrusion purge/residence time. Finally, RCI learnings were communicated broadly, with the goal to prevent future incidents of this type.

5. Recommendations and future work

- The hazard presented in this paper is not specific to this specific polyolefin technology. Evaluation and mitigation of this hazard should be performed anywhere that there is the potential for additive and/or polyolefin degradation due to uncontrolled extruder temperature conditions.

- Additives are added to polymeric resin in extrusion operations across the chemical industry to impart desired properties (structural, chemical, mechanical, etc.). The characterization of a mosaic of additives using thermal analysis tools is recommended to understand the decomposition kinetics for specific additives.

6. Acknowledgments

The authors acknowledge the full RCI team, and Christopher Siegler and Tim Bourgeois for GC/MS technical support and expertise. In 2016-2017, Dow's Core R&D Analytical Sciences Department sponsored an internal effort to couple adiabatic Accelerating Rate Calorimetry with on-line GC/MS analysis to provide an innovative technical solution for the identification and quantitative speciation of effluent from adiabatic ARC calorimetric experiments that simulate process conditions. The authors want to acknowledge all the project team members and leadership which led to successful development of this capability.

7. References

1. Snee, T.J., *Incident investigation and hazard evaluation using differential scanning calorimetry and accelerating rate calorimetry.* Journal of Occupational Accidents, 1987. **8**(4): p. 261-271
2. Frurip, D., et al., *Reactive chemicals emergency response and post-event calorimetric testing.* Process Safety Progress, 2010. **29**(1): p. 2-10
3. *ASTM E1641-18 Standard Test Method for Decomposition Kinetics by Thermogravimetry Using the Ozawa/Flynn/Wall Method*, A. International, Editor. 2018, ASTM International: West Conshohocken, PA.
4. Townsend, D.I. and J.C. Tou, *Thermal hazard evaluation by an accelerating rate calorimeter.* Thermochimica Acta, 1980. **37**(1): p. 1-30
5. Kossoy, A. and E. Koludarova, *Specific features of kinetics evaluation in calorimetric studies of runaway reactions.* Journal of Loss Prevention in the Process Industries, 1995. **8**(4): p. 229-235
6. Frurip, D.J., *Selection of the Proper Calorimetric Test Strategy in Reactive Chemicals Hazard Evaluation.* Organic Process Research & Development, 2008. **12**(6): p. 1287-1292
7. Group, H.W., *HarsBook: A Technical Guide for the Assessment of Highly Reactive Chemical Systems.* 2002: DECHEMA.
8. Fisher, H.G., H.S. Forrest, and S.S. Grossel, *Emergency Relief System Design Using DIERS Technology - The Design Institute for Emergency Relief Systems (DIERS) Project Manual.* Design Institute for Physical Property Research/AIChE.
9. Mulligan, K., et al., *Reaction Effluent Analysis Using Adiabatic Calorimetric Experiments*, in *2018 AIChE Spring and 14th Global Congress on Process Safety.* 2018: Orlando, FL. p. 25.
10. Kaisersberger, E. and E. Post, *Practical aspects for the coupling of gas analytical methods with thermal-analysis instruments.* Thermochimica Acta, 1997. **295**(1–2): p. 73-93
11. Xie, W. and W.P. Pan, *Thermal Characterization of Materials Using Evolved Gas Analysis.* Journal of Thermal Analysis and Calorimetry, 2001. **65**(3): p. 669-685
12. NETZSCH. *TGA/STA-GC-MS Coupling: Highest Resolution in Evolved Gas Analysis.* 2017 [cited 2017 September 2]; Available from: https://www.netzsch-thermal-analysis.com/us/products-solutions/hyphenated-techniques-evolved-gas-analysis-ega/tg-gc-ms-coupling/.
13. Crowl, D.A., *Minimize the Risks of Flammable Materials.* Vol. 108. 2012. 28-33.
14. Crowl, D.A., *Understanding Explosions.* 2003: Wiley.

15. Grossel, S.S., *Design and operating practices for safe conveying of particulate solids.* Journal of Loss Prevention in the Process Industries, 2012. **25**(5): p. 848-852

8. Disclaimer

This information was developed solely for The Dow Chemical Company's internal use. While believed to be accurate, Dow disclaims all warranties and any user of this information must assume all liability for any such use. Neither The Dow Chemical Company nor any of its employees or affiliates, makes any warranty, express or implied, or assumes any legal liability or responsibility for the accuracy, completeness, or usefulness of any information, or represents that its use would not infringe privately owned rights. Reference herein to any specific commercial product, process, or service by trade name, trademark, manufacturer, or otherwise, does not necessarily constitute or imply its endorsement, recommendation, or favoring by The Dow Chemical Company. The views and opinions of authors expressed herein do not necessarily state or reflect those of The Dow Chemical Company, and shall not be used for advertising or product endorsement purposes.

Appendix A- GC/MSD EXPERIMENTAL CONDITIONS FOR ARCXGC/MS EXPERIMENTS

Sample Inlet: GC
Injection Source: Manual
Mass Spectrometer: Enabled
6890 GC METHOD

OVEN
Initial temp: 60 C (On)   Maximum temp: 250 C
Initial time: 5.00 min   Equilibration time: 0.50 min
Ramps:

#	Rate	Final temp	Final time
1	20.00	230	50.00
2	0 (Off)		

Post temp: 0 C
Post time: 0.00 min
Run time: 63.50 min

FRONT INLET (UNKNOWN)

BACK INLET (SPLIT/SPLITLESS)
Mode: Split
Initial temp: 150 C (On)
Pressure: 14.7 psi (On)
Split ratio: 10:1
Split flow: 47.0 mL/min
Total flow: 54.6 mL/min
Gas saver: Off
Gas type: Helium

COLUMN 1
Capillary Column
Model Number: CP-PoraBond Q
Description: CP-PoraBond Q, 25mX0.32
Max temperature: 250 C
Nominal length: 25.0 m
Nominal diameter: 320.00 um

Nominal film thickness: 5.00 um
Mode: constant flow
Initial flow: 4.7 mL/min
Nominal init pressure: 14.7 psi
Average velocity: 87 cm/sec
Inlet: Back Inlet
Outlet: MSD
Outlet pressure: vacuum

FRONT DETECTOR (FID)
Temperature: 250 C (On)
Hydrogen flow: 1.5 mL/min (On)
Air flow: 350.0 mL/min (On)
Mode: Constant makeup flow
Makeup flow: 5.0 mL/min (On)
Makeup Gas Type: Helium
Flame: On
Electrometer: On
Lit offset: 1.0

BACK DETECTOR (NO DET)

SIGNAL 1
Data rate: 20 Hz
Type: front detector
Save Data: On

SIGNAL 2
Save Data: Off

THERMAL AUX 1
Use: AED Transfer Line Heater
Initial temp: 293 C (On)

THERMAL AUX 2
Use: MSD Transfer Line Heater
Initial temp: 200 C (On)

AUX PRESSURE 3
Gas Type: Air
Initial pressure: 1.7 psi (On)
Initial time: 0.00 min

#	Rate	Final pres	Final time
1	0 (Off)		

AUX PRESSURE 4
Gas Type: Hydrogen
Initial pressure: 22.0 psi (On)
Initial time: 0.00 min

#	Rate	Final pres	Final time
1	0 (Off)		

AUX PRESSURE 5
Gas Type: Helium
Initial pressure: 30.0 psi (Off)

VALVES
Valve 1 Switching Off

POST RUN
Post Time: 0.00 min

TIME TABLE

Time(min)	Parameter & Setpoint	
0.00	Valve 1:	Off
0.10	Valve 1:	On
0.30	Valve 1:	Off

MS ACQUISITION PARAMETERS

Solvent Delay : 0.00 min
EMV Mode : Relative
Relative Voltage : 0
Resulting EM Voltage : 1529

[Scan Parameters]

Low Mass: 14.0
High Mass: 300.0
Threshold: 0
Sample #: 2 A/D Samples 4
Plot 2 low mass: 20.0
Plot 2 high mass: 300.0

MS Source: 230 C maximum 250 C
MS Quad: 150 C maximum 200 C

APPENDIX B- ARC RUN DETAILS

ARC tests do not necessarily duplicate the process or storage conditions that tested material(s) may be subjected to. The test mode can influence the results obtained. Do not use these data for engineering or design purposes without consulting the original data/reports and reactive chemicals experts.

Experimental Conditions	
Container Material:	316 SS
Container Weight (Mb):	17.549 g
Sample Weight (Ms):	5.012 g
Total System Volume (Vb):	9.52 ml
Sample Volume (Vs):	5.01 ml
Assumed Sample Density:	1.00 g/ml
Head Space Volume (Vg = Vb – Vs):	4.51 ml
Head Space Atmosphere:	Nitrogen
Container Specific Heat (Cb):	0.50 J/g/ °C
Sample Specific Heat (Cs):	2.00 J/g/ °C
Starting Temperature:	300.0 °C
End Temperature:	450.0 °C
Heat Step:	10 °C
Wait Time:	20 min
Isothermal Wait Time:	0 min
Calorimetric Rate Threshold:	0.02 °C/min
Thermal Ratio (PHI = 1 + (Mb x Cb)/(Ms x Cs)) :	1.88
Stirring Rate:	N/A

Figure A1. Experimental Conditions for ARC analysis of Polymeric Resin (Compounded)

APPENDIX C- FTIR EXPERIMENTAL DETAIL

FTIR-ATR spectrum of solid antioxidant additive was acquired on a Nicolet 6700 FTIR spectrometer equipped with a single bounce Czitek MicromATR Vision accessory.

Experimental Parameter	Parameter Value/Description
Sample presentation	Diamond ATR (single bounce)
Detector	DTGS
Spectral Range	4000 – 525 cm^{-1}
Resolution	4 cm^{-1}
Co-added Scans	64
Velocity	0.6329 cm/s
Gain	Auto
Phase Correction	Mertz
Zero-filling	Two levels
Apodization	Happ-Genzel

Figure A2. Experimental Detail for FTIR analysis

How Effective Are Safety Gaps at Mitigating Explosions: Large Scale Testing of Safety Gaps to Prevent DDTs

Scott Davis
Gexcon US
4833 Rugby Avenue Suite 100
Bethesda, MD
sgdavis@gexcon.com

John Pagliaro, Drew Botwinick
Gexcon US
Bethesda, MD USA

Kees van Wingerden
Gexcon AS
Bergen, Norway

Prepared for Presentation at
American Institute of Chemical Engineers
2019 Spring Meeting and 15th Global Congress on Process Safety
New Orleans, LA
March 31 – April 3, 2019

How Effective Are Safety Gaps at Mitigating Explosions: Large Scale Testing of Safety Gaps to Prevent DDTs

Scott Davis
Gexcon US
4833 Rugby Avenue Suite 100
Bethesda, MD
sgdavis@gexcon.com

John Pagliaro, Drew Botwinick
Gexcon US
Bethesda, MD USA

Kees van Wingerden
Gexcon AS
Bergen, Norway

Keywords: deflagration to detonation transition, explosion modeling, DDTs

Abstract

A large vapor cloud explosion (VCE) followed by a fire is one of the most dangerous and high consequence events that can occur at petrochemical facilities. As the size and complexity of facilities increase, designs must consider the potential adverse effects associated with vapor cloud explosions in large congested areas. In fact, if a vapor cloud within a congested region is large enough, the resulting flame or deflagration can continually accelerate within this region and ultimately reach the point where the deflagration can transition to a more devastating detonation or deflagration-to-detonation transition (DDT). As the consequences of DDTs can be orders of magnitude larger than deflagrations, facilities that are at risk need to implement mitigation measures to prevent flames from continuously accelerating in these regions. One method is to provide "safety gaps" or open areas (gaps) within the congested region to interrupt the flame acceleration and provide regions where the flame will actually slow down prior to arriving at the next congested area. The key to this design is to provide a wide enough gap or region between the congested areas so that the flame exiting the first region will decelerate or slow down to adequate levels prior to arriving at the second region.

The problem is that almost all of the safety gap tests have been performed with narrow congested regions where the width is on the order of 3 meters. Unfortunately, most practical congested areas in petrochemical facilities are much wider than 3 meters and recent tests have shown that wider congested areas may be more susceptible to DDTs. Hence, the present paper will present preliminary results of large-scale testing. The net goal of the study is to assess explosion development in both narrow (~3.5 m) and wide (~7 m) congested areas, and the effectiveness of safety gaps in mitigating DDTs. The results will be a valuable reference to facilities owners and designers in helping to provide inherently safer designs at practical scales.

1 Introduction

A large vapor cloud explosion (VCE) followed by a fire is one of the most dangerous and high-consequence events that can occur in petrochemical facilities. The Piper Alpha incident in 1988 was one of the most critical turning points in offshore safety, whereby a relatively minor explosion in the compression module escalated to a massive fire that killed 167 people and destroyed the platform (see Figure 1). This incident demonstrated the need to not only understanding explosions, but also the importance of avoiding minor incidents than can lead to escalating consequences. Despite the Piper Alpha incident being a key driver in explosion research and the development of modern risk assessments for petrochemical installations, incidents continue to occur.

Figure 1: Images of the 1988 explosion and ensuing fire on the Piper Alpha platform.

Designing facilities to withstand a maximum credible event (MCE) is an essential part of the path toward safer designs which can be achieved by avoiding unacceptable escalation of events and damage to safety critical equipment. This is especially true as the size and complexity of facilities increases [1]. Designs must consider the potential adverse effects associated with vapor cloud explosions in large congested areas and understand the potential for more devastating deflagration-to-detonation transitions (DDTs) on these facilities. As the DDT phenomena and acceleration mechanisms have been previously observed experimentally [2][3], it is critical to understand how a facility's geometry or equipment layout can affect explosion consequences and assist in their mitigation and/or prevention,.

While the likelihood of DDTs is lower than deflagrations, they have been linked to certain industrial explosion accidents [4],[5], whereby damage patterns can be readily explained by the deflagration to detonation overpessure generation mechanisms. More specifically, DDTs have been identified in some of the most recent large-scale explosion incidents including: 2005 Buncefield explosion [6], 2009 Jaipur event [7], and 2009 San Juan explosion [8]. The consequences of DDTs can be orders of magnitude larger than deflagrations alone because they have the ability to self-propagate outside the region of high congestion/confinement. Due to the difficulty in predicting such devastating phenomena on the large scale, owners and designers cannot easily evaluate installations for risk of DDTs and provide "inherently safer" layout or mitigation measures to significantly reduce or eliminate such hazards. Recent research has highlighted two main areas that could greatly improve designs from the perspective of reducing the severity of the MCE: (1) more experiments and data at the large scale to determine what factors influence the onset of deflagration-to-detonation transition (DDT), especially evaluating the width of the congestion regions; and (2) the need to properly identify the driving mechanism of these high-consequence events so that effective mitigation measures can be determined and appropriately implemented in the industry [9].

Determining under what conditions a DDT can occur and what mitigation measures will work are crucial in calculating MCE for petrochemical facilities. Gexcon performed numerous large-scale tests with near stoichiometric propane/air mixtures ignited within a high-congestion test rig made up of rigid steel beams in the RPSEA Phase I work [10] and found that DDTs may occur easier than originally anticipated for wide congestion areas (~ 7m). Hence, the goal of this paper is to present preliminary results of continued large-scale testing designed to evaluate under what conditions a DDT can occur and what mitigation measures may be effective; this work is referred herein as RPSEA Phase II. The net goal of the study is to assess explosion development in both narrow (~3.5 m) and wide (~7 m) congested areas, and the effectiveness of safety gaps in mitigating DDTs.

2 DDT Test Data and Results

Testing Setup

Gexcon developed a modular test assembly that allowed for extensive flexibility in test rig configuration, whereby congestion levels and congestion orientations could easily be modified within the rig. In RPSEA Phase I and II, the modules were placed in an array for testing with 2 modules across (~7.3 m), 12-15 modules deep (~44-55 m), and 1 module high (~3.7 m), as shown in Figure 2. The total enclosure was covered with a polyethylene sheet and the rig volume was 1200 m^3 (42,000 ft^3) for 24 modules and approximately 1500 m^3 for 30 modules.

The base configuration (no pipes added) was labeled "low" congestion with an area blockage ratio of 4-5.8%, where only the base structure was included and the structural elements were spaced approximately 1 meter apart.

Figure 2: RPSEA test rig composed of 2 modules across, 12-15 modules long

A single polyethylene sheet covering the modules was used to retain the gas/air mixture prior to ignition. The polyethylene sheet was secured directly to the modules. Tests were performed with a detonation cord cutting apparatus to cut the entire sheet nearly instantaneously to evaluate the initial confinement the sheeting provided. An electric spark was used as the ignition source and was located approximately half a module length within the structure, centered and at mid-module height. Gas was introduced into the rig at four separate locations and mixed thoroughly with numerous box fans. Samples were extracted at 8 locations throughout the rig, with half the sample points in the upper section and half near the ground. The extracted samples were fed to an IR gas analyzer (IR4800D three channel), which was calibrated prior to testing. Tests were performed with experimental grade propane (>99.5% purity).

Preliminary Test Results

<u>DDT sensitivity to stoichiometry</u>

In RPSEA Phase I, DDTs were observed during tests with near stoichiometric mixtures ignited in the "low" congestion level rig configuration (Tests 5-1, 6-1, 7-1, 8-1 and 12-1), and in these tests equivalence ratios ranged from 1.05 to 1.12. Note tests from Phase I will have *-1 after the test number, while those of Phase II will have *-2. The first set of the RPSEA Phase II tests aimed to evaluate how sensitive the potential for DDT is on the fuel/air stoichiometry. Therefore Test 3-2 of RPSEA Phase II was estimated to be slightly fuel rich (equivalence ratio near 1.20-1.25) but the mixture concentrations could not be precisely quantified. This off-stoichiometric mixture did not transition to detonation within the length of the test rig. However, when a near-stoichiometric mixture with an equivalence ratio of 1.10 was tested next (Test 4-2), DDT occurred similarly to what was observed in the RPSEA Phase I tests.

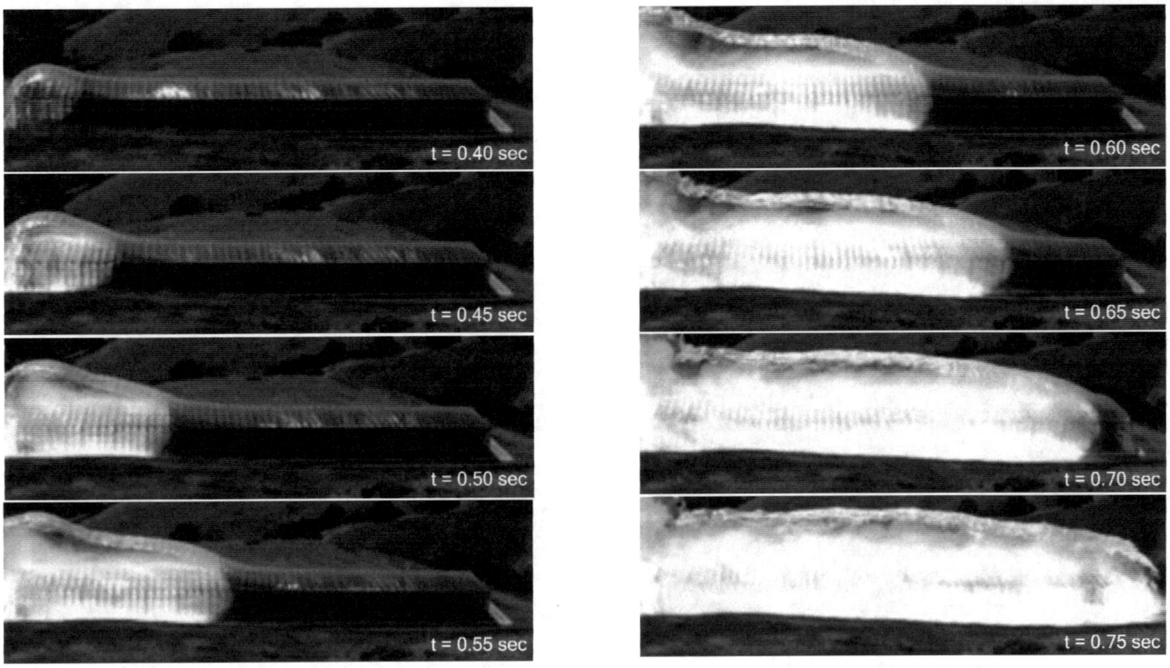

Figure 3: Image sequence of deflagration along the rig for Test 3-2

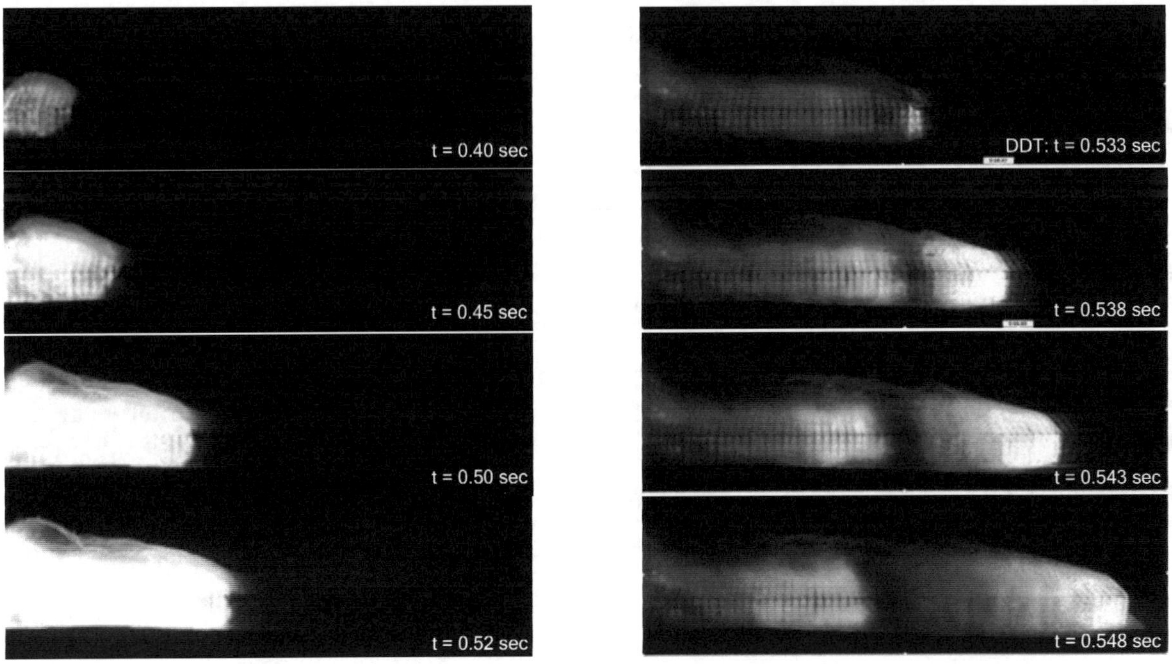

Figure 4: Images of flame propagation before (left) and after transition to detonation (right) in Test 4-2. Note a longer camera exposure setting was used before transition (left) and shorter exposure was used after (right)

<u>DDT sensitivity to tent cutting method</u>

The next set of tests evaluated the effects of tent cutting on the development of flame speed and the influence of the confinement of the plastic. Test 4-2 was performed with the first tent cutting technique whereby a mortar connected to a cable was launched ahead of the flame propagation to cut the tent as the flame propagates along the rig. Another test (Test 6-2) was conducted with the second tent cutting technique whereby a detonation cord was run along the top edge around the entire rig and was triggered just prior to ignition. In this setup, the entire top sheet was cut from the sides just prior to ignition as shown in Figure 5.

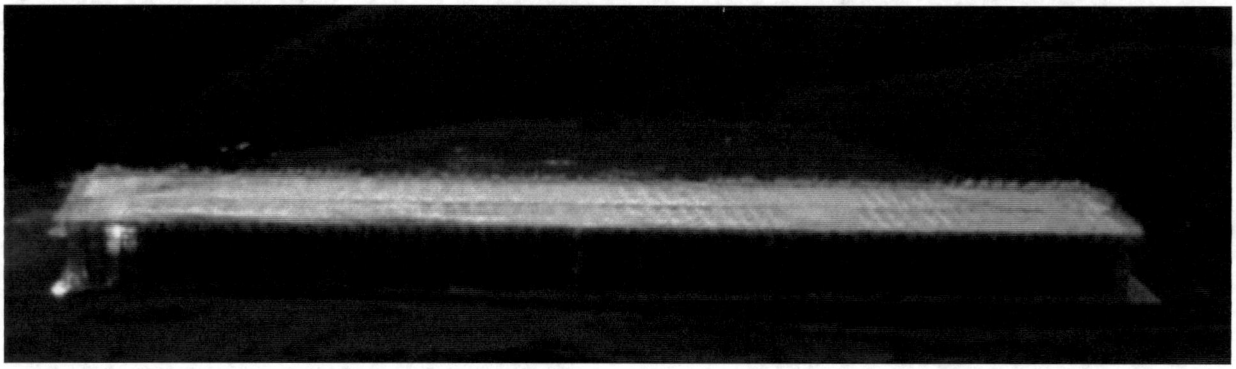

Figure 5: Images of detonation cord cutting the top edge around the modules in Test 5-2.

3 Results and Analysis

Flame front visualization and tracking

Recently developed in-house software provided an advanced method for tracking flame fronts in the high-speed videos from each test. In the software, the user defines a specific row of pixels at any fixed height and the software extracts the row from each frame. Each row is then rotated 90° and stacked sequentially. Pixel widths (before rotation) are determined from the specified start and end points of the rig and its length. The height of each pixel (before rotation) represents a fixed duration based on the video frame rate. Hence once the pixel rows are rotated and combined they form a position versus time plot (see Figure 6 for examples), which we refer to further as an x-t diagram. The software can generate figures in true color, gray scale, or with a color map (see Figure 7). The software can also take the gradient of the light intensity to highlight areas with sharp light intensity changes such as the interface between the flame front and darker background.

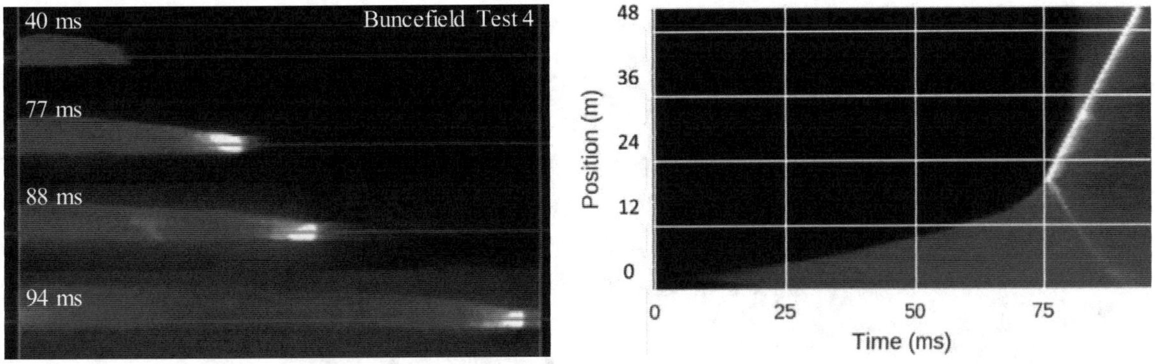

Figure 6: A single horizontal row of pixels (red line in left image) is extracted from each frame, rotated 90° counterclockwise, and stitched together to create x-t diagram (right image). Note that this test data shown in the figure is from another experimental campaign.

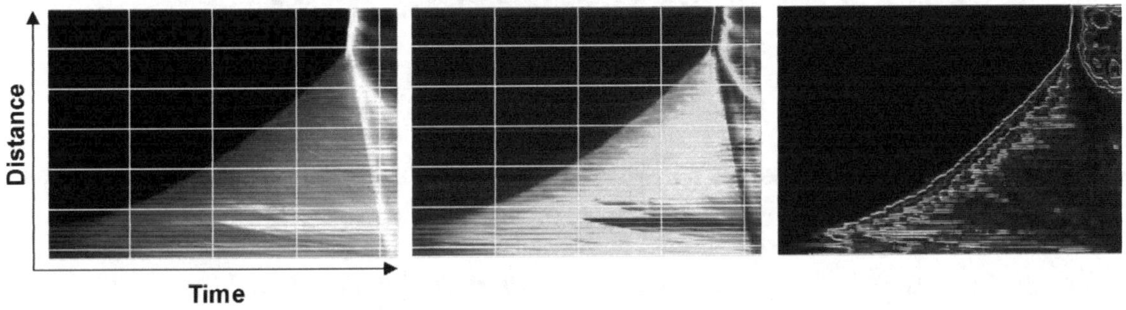

Figure 7: Example color image view (left), colormap view of the intensity (middle), and color map view of the gradient of the intensity (right).

The in-house flame tracking technique proved to be extremely reliable as the user could view the video simultaneously and manually filter out noise such as the light reflecting off of the containment sheet ahead of the flame. Another major source of noise, which could be accounted for using this method, was the influence of the test rig structure. For certain tests, depending on the ambient lighting conditions and camera exposure settings, the flame front would clearly disappear behind the structural elements along the rig creating a step-like flame front position in the x-t diagrams. With this method, the flame front position could be identified when it was in between structural elements and interpolated when behind structural elements. The red circles in Figure 8 shows the points extracted as the flame front which track very well both the deflagration and detonation fronts.

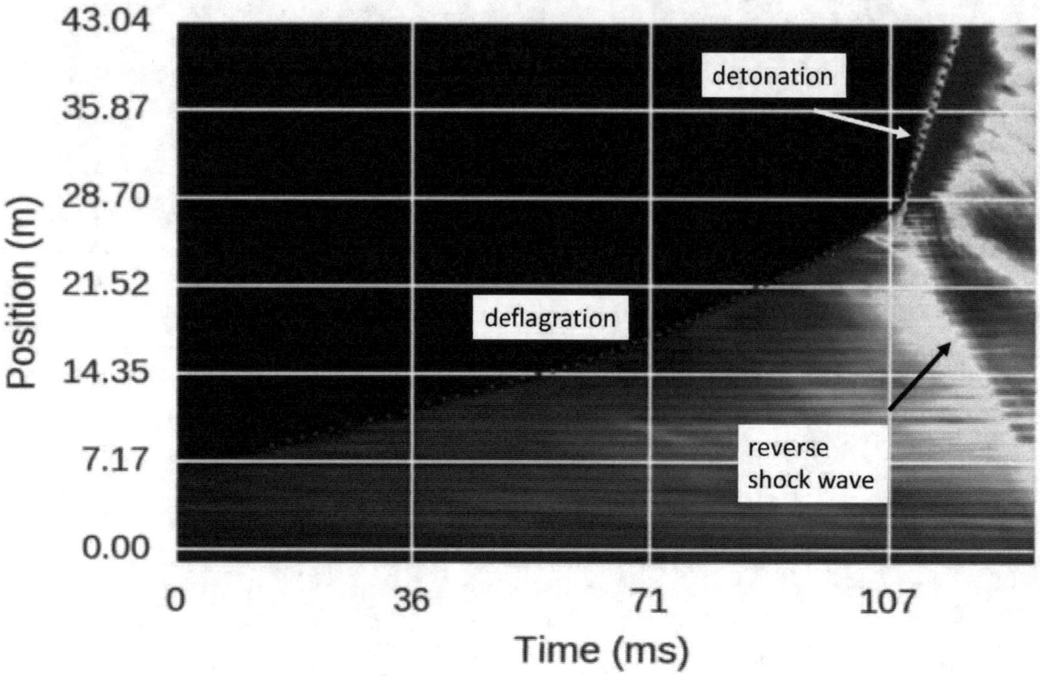

Figure 8: Example tracking methodology - The red dots show the extracted flame front position as a function of time.

Flame front and flame speeds for stoichiometric and off-stoichiometric tests

The flame tracking for the off-stoichiometric tests shows a slight monotonically increasing slope of flame position versus distance, especially near the end of the rig. This is consistent with the videos showing DDT did not occur. In contrast, the near-stoichiometric test resulted in DDT and shows a sharp change in slope at the transition. When plotting the flame speed versus distance, the results show that the near-stoichiometric test transitions to detonation at flame speeds around 300-400 m/s, which is similar to the test results from RPSEA Phase 1. In addition, the flame speeds of the off-stoichiometric tests were close but never transitioned at the end of the test rig. These results show how that DDT is highly sensitive to fuel stoichiometry, and should be considered when considering actual releases where the concentration gradients will exist in congested regions.

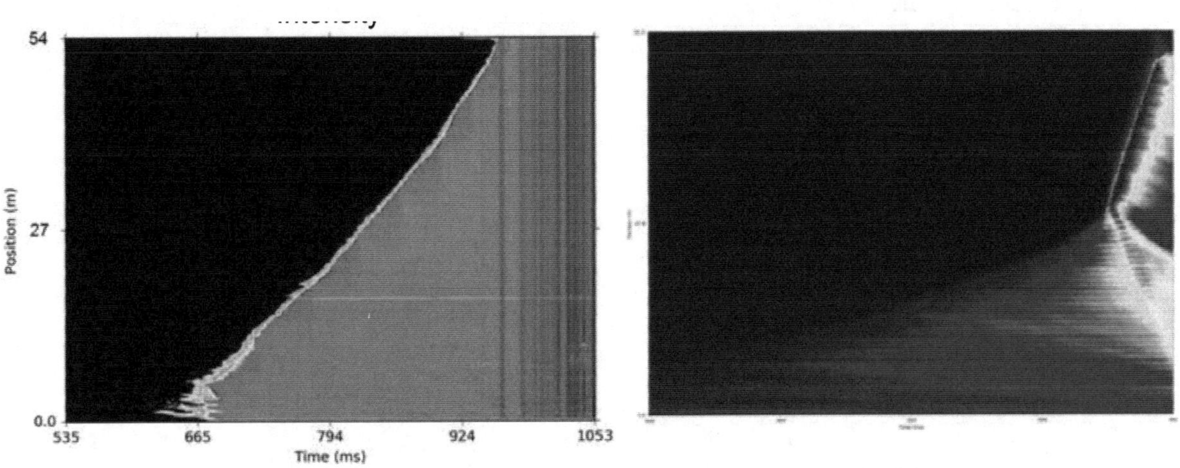

Figure 9: Flame tracking for the off-stoichiometric (left) and the near stoichiometric mixtures (right).

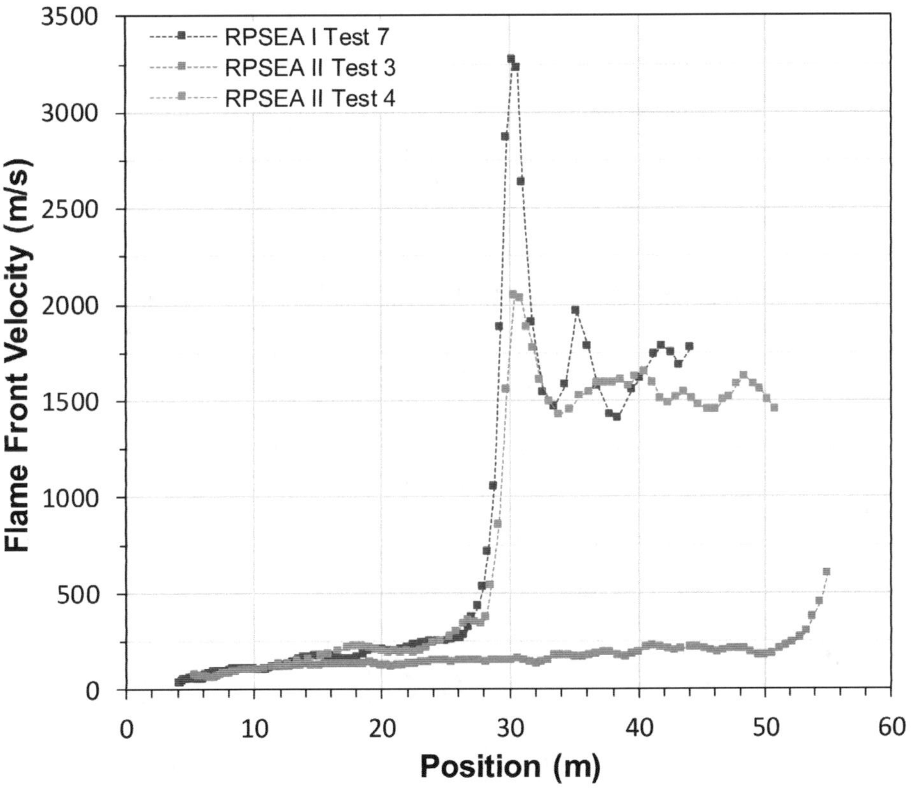

Figure 10: Flame speeds for off-stoichiometric (green 4-2) and near stoichiometric (blue 3-2), compared to the previous near stoichiometric RPSEA Phase I test 7-1 (red).

Effect of tent cutting on flame speeds

A closer examination of flame speeds obtained for tests where the tent was not cut, where the tent was cut using the mortar and wire cutting, and where the tent was cut using the detonation cord show that all results are essentially insensitive to the confinement provided by the tent.

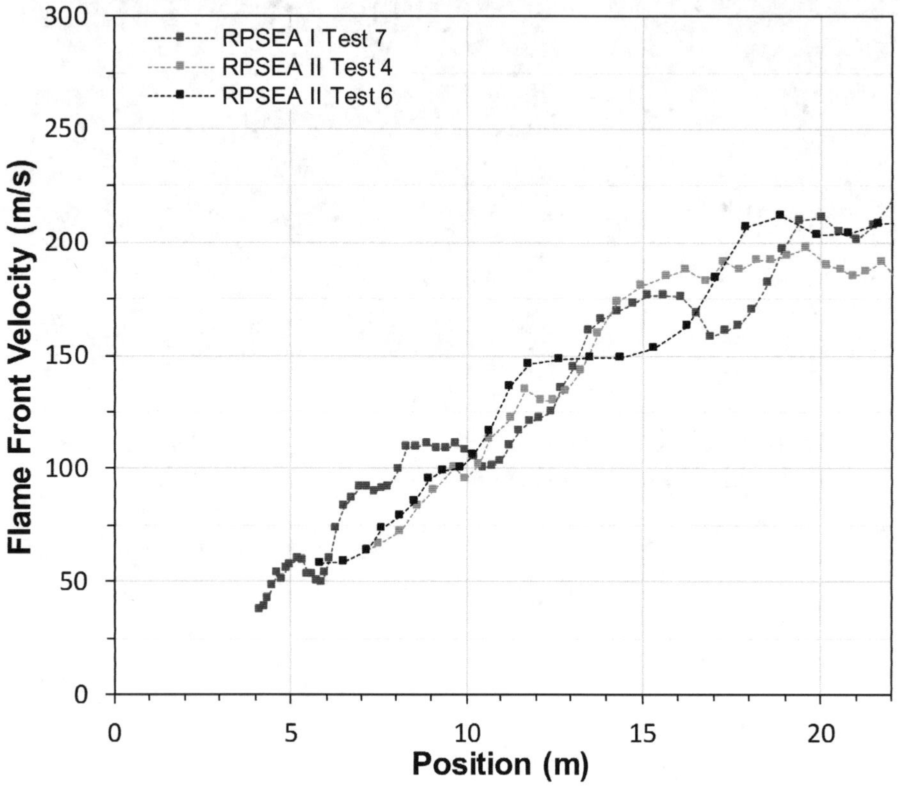

Figure 11: Flame speeds no tent cutting (7-1 red), mortar-wire cutting (4-2 blue), and detonation cord cutting (6-2 black).

4 Conclusion

The present results confirm the results of the previous testing, that the width of the congested region plays a critical role in DDT as the transition occurs at flames speeds much lower than narrower test geometries. The results also show the sensitivity of this parameter to mixture composition, where near-stoichiometric mixtures can more easily transition to DDT than off-stoichiometric mixtures. This result may play a key role in actual release scenarios that do not form "ideal" stoichiometric clouds in congested areas. Finally, the transition to DDT appears to be relatively insensitive to the confinement of the plastic for near-stoichiometric propane/air mixtures as similar flame speeds and transitions were observed for no tent cutting, cutting with a mortar attached to a wire, and cutting with a detonation cord.

Upcoming tests as part of RPSEA Phase II will evaluate the effectiveness of methods for mitigating DDTs when these worst-case and ideal propane/air mixtures are present.

5 Acknlowledgments

We would like to thank the guidance and support provided by our sponsors Chevron, CCS, BP, ExxonMobil, Shell, and Total, without whom this work would not be possible.

6 References

[1] Burgan, B.A., Dispersion and Explosion Characteristics of Large Vapour Clouds, Volumes 1 and 2, *The Steel Construction Institute*, 2014, document number SCI ED024.

[2] Harris R.J. and Wickens M.J. (1989). Understanding vapour cloud explosions - an experimental study, Institute of Gas Engineers communication 1408.

[3] Pekalski A., Puttock J., Chynoweth S. (2015), Deflagration to Detonation Transition in a Vapour Cloud Explosion in Open but Congested Space: Large Scale Test, Journal of Loss Prevention in the Process Industries, vol 36, p 365-370.

[4] Chamberlain G., Oran E., Pekalski A., Detonations in industrial vapour cloud explosions, 12th International Symposium on Hazards, Prevention and Mitigation of Industrial Explosions, Kansas City, USA - August 12-17, 2018. In publication in Loss Prevention Journal

[5] Johnson D.M., Pekalski, A., Tam V.H.Y., Burgan B.A., The Importance of Deflagration to Detonation Transition in Explaining Major Vapour Cloud Explosion Incidents, 12th International Symposium on Hazards, Prevention and Mitigation of Industrial Explosions, Kansas City, USA - August 12-17, 2018.

[6] Buncefield Major Incident Investigation Board, The Buncefield Investigation: Third Progress Report, 2006.

[7] Johnson, D.M. Characteristics of the Vapor Cloud Explosion Incident at the IOC Terminal in Jaipur, 29th October 2009. 8th Global Congress on Process Safety, Houston, TX, April 1-4, 2012.

[8] Investigation Report Summary, Drilling Rig Explosion and Fire at the Macondo Well, US Chemical Safety and Hazard Investigation Board, April 12, 2016.

[9] Davis S., Pagliaro J., Botwinick D., DeBold T., van Wingerden K., Allason D., Johnson M, (October 2018). "Do Not Believe the Hype: Using Case Studies and Experimental Evidence to Show Why the HSE is Wrong about Excluding Deflagration-to-Detonation Transitions", AIChE, Process Safety Progress, pp 1-12.

[10] Davis S., Pagliaro J., Botwinick D., Engel D., van Wingerden K., Merilo E., Ziemba A., Groethe M. Large Scale Testing Confirms: A Deflagration to Detonation Transition Needs to be Considered When Facility Siting, 13th Global Congress on Process Safety, San Antonio, 2017.

Role of Fire Codes and Standards in Hazard Management for Fire, Explosion and Reactivity Hazards

Guy R. Colonna, PE
NFPA
1 Batterymarch Park
Quincy, MA 02169 USA
gcolonna@nfpa.org

Laura E. Moreno
NFPA
1 Batterymarch Park
Quincy, MA 02169 USA
lmoreno@nfpa.org

Prepared for Presentation at
American Institute of Chemical Engineers
2019 Spring Meeting and 15th Global Congress on Process Safety
New Orleans, LA
March 31 – April 3, 2019

Role of Fire Codes and Standards in Hazard Management for Fire, Explosion and Reactivity Hazards

Guy R. Colonna, PE
NFPA
1 Batterymarch Park
Quincy, MA 02169 USA
gcolonna@nfpa.org

Laura E. Moreno
NFPA
1 Batterymarch Park
Quincy, MA 02169 USA
lmoreno@nfpa.org

Keywords: fire codes; fire and explosion incidents; reactive chemicals

Abstract

The hazard nature of materials found in chemical processes, whether raw materials or finished products, begins with the inherent chemical and physical properties of each material. Within those two broad categories lies the physical hazards of flammability and instability/reactivity. It is from that point of departure that the fire codes apply a mostly traditional approach to classification of materials, which then enables the fire code to assess the hazard degree where storage, handling, or use is intended within a specific occupancy. While this is fairly well understood in mixed-use occupancies where materials are used or stored, the awareness and understanding changes when the concepts are applied to a chemical process setting. Using examples in the form of case studies, incident reviews, or incident investigations, the hazard assessment process for fires, explosions, and reactivity will be demonstrated using applicable NFPA codes and standards. This approach to assessing the hazards will be demonstrated specifically using NFPA 400, *Hazardous Materials Code*. Objectives for this presentation include awareness of fire safety codes and standards and ability to apply the requirements to a variety of chemical process hazards within a safety management system.

The NFPA Fire & Life Safety Ecosystem (see Figure 1) offers another way to describe a safety management system. A goal for working within management systems applicable anywhere is that they drive consistent and reproducible performance. Process safety success requires consistent performance and reproducible results within the selected safety metrics that are applied to a given process. In a manner similar to chemical process safety management, the ecosystem illustrates the various stakeholders and shows their connectivity to other stakeholders. It is important to recognize that, depending on the process and where one lies in that process, the impact and influence of various stakeholders varies. Crucial for application of this model is the understanding that if any stakeholders are bypassed in that application, unwanted consequences result.

Figure 1. NFPA Fire & Life Safety Ecosystem

1 Introduction

An outcome of following process safety strategies is an awareness of what is being done well so that management of similar hazards follows. One only has to apply established controls that are already in place and working effectively to other parts of a process in order to expect similar quality results. Those interviewed after an incident commonly are heard to remark, "we always did it that way" and "we never had a problem like this before" [this connects to the *Informed Public* cog of the ecosystem model].

When a facility relies on past practices including a "lack of a serious incident" as justification for not taking proactive risk-based process safety (RBPS) measures, they have no benchmark for what's working and what's not. The incident might suggest something that is not working, but without a formal process safety assessment any conclusions lack substance and thus are of no more value than doing nothing because "we have never had a problem."

National and international standards can be demonstrated as supporting the strategies of risk-based process safety – they provide a consistent way of managing certain well-identified hazards – and standards can be relied upon to frame best practices within industries and show how to consistently implement those practices.

The RBPS (as outlined by the Center for Chemical Process Safety) approach begins with four foundational blocks or pillars which are then supported by a total of 20 elements distributed among the pillars. When a facility adopts and incorporates national and international standards into the safety structure, it demonstrates a commitment to process safety which is the first pillar; those standards are built on the next three pillars:

1. They are developed based upon an understanding of the relevant hazards
2. While not always preventing incidents, they certainly serve to reduce or minimize the consequences, so they contribute to managing the risk
3. While most commonly they are developed and revised based on retrospective assessments (following incidents, new requirements are common), developers frequently seek proactive and forward thinking additions to both learn from experience but also predict and anticipate the next point of vulnerability from a hazard

While national laws and regulations are one form of application, standards that are often voluntary norms of best practice among industries provide both the 'what SHOULD I do as well as what DO I HAVE to do.' Let's look at the ecosystem now to see how and where it connects with these fundamental elements of Risk-Base Process Safety (RBPS) – development and use of current standards, reference standards, code compliance, and investment in safety.

Taking the pillars one at a time, a commitment to process safety results when consistent and repeatable behaviors exist – worker safety, proactive inspections, and preventative maintenance. These behaviors reflect a process safety culture that includes compliance with standards. As already noted, a strong bond exists between standards and process safety as the standards provide a forum for dependable practice (repeatable results) that leads to a strong sense of compliance within all work activities. In addition, the standards offer a template through which decisions, actions and use of compliance results can all be matched for consistency.

The system that results from this commitment and consistent practice only works when the stakeholder is engaged through outreach; both NFPA and CCPS "develop and disseminate information" to prevent incidents.

In two key process safety regulations within the US – OSHA's Process Safety Management (PSM) standard and the EPA's Risk Management Program (RMP) – companies are permitted to use as part of the safety management programs industry safe practices, national standards, other tools of practice to achieve the goals for either PSM or RMP. When such non-regulatory practices are employed, it is under the provisions known as RAGAGEP – recognized and generally accepted good engineering practice. NFPA's fire protection codes and standards comprise an element of RAGAGEP for many companies, just for the reasons outlined.

The 2nd pillar of the CCPS system connects to the NFPA process when our committees receive lessons learned in multiple forms and from multiple sources – state and local as well as federal governmental investigative agencies, research organizations (including the Fire Protection Research Foundation), and industry. The investigations share documentation on causal factors along with a recipe for corrective actions. Taking one government body in the US – the US Chemical Safety and Hazard Investigation Board (or CSB) – investigates incidents and in some instances issues recommendations for amending code requirements. The NFPA standards development process encourages openness and full participation by the affected stakeholders, so this marks another area of standards development that is consistent with and complements the pillars of RBPS.

2 Standards and Process Safety

NFPA's earliest published standards involving chemical substances and industrial processes applied to safeguarding flammable liquids, plastics, and oxidizers and date from the 1930's and 1940's. Through the years as the use of chemical substances increased, so was the case with specific standards that were created to address the unique aspects of respective industries and commodities in production, use, storage, and transportation.

In 2010, four specific standards covering oxidizers, organic peroxides, pesticides, and ammonium nitrate, merged to become NFPA 400 – a comprehensive hazardous materials code created to cover all the physical and chemical characteristics managed for worker safety (such as addressed by the Occupational Safety and Health Administration or OSHA in its general industry standards) and building codes and fire codes.

NFPA 400 establishes basic requirements that become specific depending on both the commodity and occupancy. These requirements are: material classification (flammable liquid, flammable gas, flammable solid, oxidizer, organic peroxide, unstable reactive, explosive); occupancy (such as industrial, storage or miscellaneous use, and mercantile); storage, handling or use quantity; and MAQ. The MAQ stands for Maximum Allowable Quantity and actually is a trigger threshold above which additional protective safeguards become applicable. It is permissible to remain below the MAQ and not incur further protective measures; however, in many applications, the MAQ quantity will be exceeded and the requirement for additional safeguards will be imposed.

Building codes and fire codes point to NFPA 400 for specific protections. NFPA 1, *Fire Code*, as well as the *International Fire Code (IFC)* require hazardous materials occupancies to comply with NFPA 400. While NFPA standards, such as NFPA 1, NFPA 30, *Flammable and Combustible Liquids Code*, and NFPA 400, have been around for quite some time, their approach to these hazards has remained the same:

- Identifying the hazard classification and characteristics
- Identifying the types of containers used for storage
- Limiting the quantities of the more volatile materials
- Using construction to manage some of the storage (separate buildings or structures for certain materials to keep them segregated from incompatible materials) and enforcing separation distances from other structures, people and public areas

Each of these documents, as well as the *International Fire Code*, sets a baseline or framework for hazard management and also creates a manner by which greater amounts of various chemical substances can be increased, provided that requisite increased levels of safety accompany the increased material volumes. Case studies might put these factors into perspective more clearly.

3 Case Study Examples

3.1 Tonawanda, NY: August 18, 1995

A fire at a chemical manufacturing facility in Tonawanda, NY, killed one worker, injured five other employees plus two emergency responders and one bystander. The fire completely destroyed the chemical storage warehouse despite being protected by a dry-pipe automatic sprinkler system.

The plant manufactured three types of persulfates along with peracetic acid and calcium phosphate. The persulfates included ammonium, potassium, and sodium, which are all classified as Class I oxidizers per NFPA 430, *Code for the Storage of Liquid and Solid Oxidizers* (the applicable commodity specific standard in 1995, but which is now covered as Chapter 15 of NFPA 400). The finished persulfates were packaged in bags, drums, and Intermediate Bulk Containers (IBCs). Ammonium persulfate comprised the largest quantity of persulfate and was stored in both drums and IBCs. The investigation, while not determining a specific cause of ignition, suggested that the cause was most probably decomposition and spontaneous heating of the ammonium persulfate following manufacture and packaging in IBCs. The combustible packaging of the IBC and the wooden pallets on which the IBCs were loaded for movement within the warehouse and subsequent storage provided the apparent fuel for the fire. Though protected by a dry-pipe automatic sprinkler system, the growth of the fire and its spread throughout the warehouse overwhelmed the fire protection and the warehouse was completely destroyed. As part of the investigation, a review of the sprinkler design using hydraulic calculations demonstrated that the system was inadequate for protecting the stored persulfate materials.

A series of losses involving oxidizers during the mid-1990's prompted a series of full-scale tests to further study the nature of oxidizer fires and to evaluate protection strategies, particularly sprinkler system design.

3.2 Albany, GA: April 16, 1996

A fire at a fully sprinklered bulk retail store specializing in home improvement products completely destroyed the store without any injuries. Among the products in this facility were oxidizers in the form of swimming pool treatment chemicals (some Class I, Class II, and Class III oxidizers). As part of the fire safety plan at this retail location, the pool chemical manufacturer prepared plans detailing storage requirements for the chemicals in storage racks. One plan highlighted an arrangement of three bays along a single aisle with a recommendation to separate the storage bays by space, walls, partitions, or intervening inert materials. Once again dealing with oxidizers, NFPA 430 forms the primary source for fire and life safety protection requirements applicable to oxidizing materials, whether solid or liquid. NFPA 430 states that no oxidizing material storage should be located adjacent to incompatible materials, particularly flammable or combustible liquids. The rack design for this installation utilized deep racks that permitted storage that would be accessible from adjacent aisles (either side of the rack). With this design, the plan didn't specify what materials should be permitted across the rack on the parallel aisle, and so charcoal and lighter fluid were among the commodities being stored.

Another complication at this site occurred due to the occupancy classification for this store. With the warehouse-type features coupled with the more traditional mercantile components, this store represented a hybrid of mercantile and warehouse occupancy classification and resulted in the chemical products being stored higher in the racks than permitted by the oxidizer code. Without exactly knowing the ignition sequence, the documented time between the fire starting and notification of the fire department and its arrival on scene was minimal. Yet upon arrival of the responders observed that the building was already involved in fire and releasing significant smoke to the outside. The fire protection included two, wet-pipe sprinkler systems and one, dry-pipe system, yet the increased storage arrangement and the lack of in-rack sprinklers as required by the code proved to be significant. The code also required vertical barriers to be placed between the oxidizer materials and any incompatible materials; the vertical barriers were not integrated into this store's installation. Sprinklers were also designed to densities and areas of coverage that fell below current NFPA requirements for oxidizer storage.

Because of this incident and a similar one occurring in May 1995 in Quincy, MA (very near NFPA Headquarters) involving oxidizers (specifically pool chemicals) the manufacturers and the Fire Protection Research Foundation (FPRF) conducted full scale fire tests in 1997 to better inform the lessons from these incidents with specific data on sprinkler design for oxidizer protection. Today, NFPA 430 is found in NFPA 400 as Chapter 15.

3.3 West, TX: April 17, 2013

Around 7:30 pm, April 17, 2013, in the small rural community of West, TX, a fire of unknown origin started and caused ammonium nitrate (AN) to react, become unstable, and eventually explode. The community, served by an all-volunteer fire department not fully familiar with the specific hazards of AN, and without any specific incident pre-planning or emergency response strategy, suffered the loss of 12 emergency responders among 15 who perished in this incident.

This incident raised questions aimed at land use planning, zoning, and safe separation distances from industrial facilities. NFPA's codes and standards addressing safe storage, handling, and use traditionally include measures that ensure 'protection for exposures,' meaning defining safe distances between the hazard (flammable liquids, flammable gases, or reactive materials) and other property. Included in the exposures are other structures (that could include different industrial or storage, handling, or use occupancies, or non-industrial occupancies), adjacent property, and public rights-of-way. The magnitude of the damage from West caused by the ultimate detonation of the AN, a material that is highly reactive when contaminated and unstable when exposed to heat, yet noncombustible, prompted considerable attention from industry, emergency preparedness and response personnel, and regulatory agencies. Similar fertilizer plant operations to those conducted at West were closely reviewed across the nation. West prompted interest in facility siting that was similar to the questions raised in the United Kingdom due to the Buncefield explosion in December 2005. Separation distances in the UK were heretofore predicated on an understanding of the hazard associated with bulk flammable liquid storage (gasoline) and the prior belief that no Vapor Cloud Explosion (VCE) was possible.

When viewed in the context of the ecosystem, a number of the elements or cogs come into focus in the following ways:

- Government responsibility contributes through decisions associated with the placement of developing infrastructure (schools, hospitals, play yards, housing, elderly care centers) relative to an industrial facility
- Application and compliance with primary codes (like building or fire codes) and reference standards (like NFPA 400) – among contributing factors was the AN stored in close contact with incompatible materials (organic materials)
- Emergency preparedness and responder safety – ensuring that the unique hazards attendant to all oxidizers is fundamental for both workers at the plant and for the responders
- An informed public also needs to know about such process hazards wherever conducted in their communities

Because the NFPA technical committee on hazardous chemicals worked closely with the CSB during its investigation, NFPA 400 was revised to reflect critical lessons learned from this tragedy. The consensus standards development process of NFPA benefitted from the timely input from all the stakeholders – community, fertilizer industry, enforcing officials, researchers, government officials – to address the hazards found in such a process facility and eliminate or control them through fire and life safety control measures. The NFPA process encourages openness and full participation by the affected stakeholders, so this marks another area of standards development that is consistent with and complements the pillars of RBPS.

3.4 Tianjin, China: August 2015

A massive series of explosions with additional fires focused attention on Tianjin, China as a warehouse and logistics depot suffered significant damage and approximately 170 fatalities in

August 2015. The specifics from this incident remain unclear, however, an original storage plan developed in 2014 that mixes incompatible materials and lacks additional safeguards that should accompany the large quantities in place highlights a number of failures within the ecosystem model. Reportedly the facility included large quantities of sodium cyanide, ammonium nitrate, toluene diisocyanate, among other flammable and reactive materials. While an initial plan outlined which commodities would be stored and where, over time the plan was not maintained and materials became stored with incompatible materials and in quantities that exceeded safe practices.

Applying the ecosystem shows a system that broke down in regards to following established plans – lack of change management – and a significant gap in terms of emergency preparedness and responder safety. Because the Tianjin Port Public Security and Fire Department was located immediately across from the warehouse complex, response to the incident was swift, but due to the lack of preplanning for specific hazard conditions that had changed over time, the responders were in jeopardy almost immediately. Of the fatalities, reports indicated as many as 100 could have included first responders. What resources are available for those faced with developing process safety guidance and best practices? Besides investigation reports prepared by federal and local agencies, research offers insight into improvements for process safety.

4 Resources – Fire Protection Research Foundation (FPRF) projects

The Fire Protection Research Foundation (FPRF) plans, manages, and communicates research on a broad range of fire safety issues in collaboration with scientists and laboratories around the world. The Foundation is an affiliate of NFPA.

Several projects completed by the Foundation speak directly to the issues raised by the various case studies highlighted in this paper. Completed reports for each of the projects can be accessed at the NFPA web site.

4.1 Separation Distances

NFPA 400 specifies separation distances for hazardous material storage and processes from other equipment and occupied buildings. To satisfy these requirements, standards and industry practices have come from a multitude of sources. The basis of the recommended separation distances in NFPA 400 is challenged by recent accidents, such as the April 2013 ammonium nitrate explosion reviewed in the case studies. Though this is an important specific example, the chemical safety field as a whole needs to study many such hazards and how they could affect one another. The separation distance specifications can be based on historical information, investigation of an accident, near-misses, modeling, or experimentation; all of which can lead to different codes and regulations due to the differences in the information considered in their development. Collecting and distilling information on separation distances will enable a better understanding of both the basis for the current methodologies and the knowledge gaps for storage of hazardous materials.

4.2 Oxidizers

The purpose of this research project was to develop a scientific and reliable method to classify chemical oxidizers based on the degree to which an oxidizer increases the burning rate of combustible materials with which it comes into contact. A test plan was developed to determine the relationship between oxidizer chemical structure, composition, and packaging on the burning rate of oxidizers for classification and fire protection in storage environments. The resulting classification scheme for oxidizers is practical and includes a bench scale screening test method and medium scale burn testing of the oxidizer in its packaging with criteria for classification.

5 Conclusion

The ecosystem provides a framework through which to examine how fire and life safety standards align with objectives for chemical process safety. Using several case studies to illustrate the ecosystem elements or cogs, common areas where connectivity is broken include government responsibility, application of current standards (both primary and reference), emergency preparedness and responder safety, developing and promoting a skilled workforce, and communicating the value of process safety to create an informed public.

The case studies provide harsh but essential lessons that must be applied to current codes and reference standards. Chemical process safety has long focused on the importance of an investment in safety, an element within the ecosystem that bridges the application of the standards to the skilled workforce. Management of change emerges from the ecosystem as illustrated by several of the case studies – those incidents resulted when changes in protection compromised safeguards either planned or in place.

Challenging safety, whether fire and life safety or chemical process safety, is to ensure that lessons learned are shared across all stakeholders in the ecosystem – our governments, the chemical process industry, the public and the workers. NFPA codes and standards are recognized widely as meeting the intent of the RAGAGEP provisions, just like industry best practices. Because each of these procedural sources already have stakeholder buy-in and input from the workers, it ties into the skilled workforce element of the ecosystem as well.

Standards impact risk-based process safety goals in various ways – offer a framework on which to build a program, drive consistent behaviors, react to changes stemming from an incident, fill a gap often left where regulatory provisions do not go, and deliver some buy-in from the affected community. One final takeaway from the ecosystem should be that success in process safety – chemical process and fire and life safety – requires collaboration of many stakeholders.

References

[1] NFPA 400, *Hazardous Materials Code*, NFPA, Quincy, MA 2019

[2] CCPS, *Risked Base Process Safety Overview*, Center for Chemical Process Safety, American Institute of Chemical Engineers, New York, NY, 2014

[3] Isner, Michael, *NFPA Fire Investigation Report, Persulfate Warehouse Fire*, Tonawanda, NY, August 18, 1995

[4] Isner, Michael, *NFPA Fire Investigation Report, Bulk Retail Store Fire*, Albany, GA, April 16, 1996

[5] US Chemical Safety Board, West Fertilizer Company Fire and Explosion, CSB, Report 2013-02-I-TX

[6] Argo, Ted and Evan Sandstrom, *Separation Distances in NFPA Codes and Standards*, Fire Protection Research Foundation, Quincy, MA December 2014

[7] Buc, Elizabeth, *Oxidizer Classification Research Project: Tests and Criteria*, Fire Protection Research Foundation, Quincy, MA, November 2009

Assessment of Battery Energy Storage Systems

Bjorn Nilberg
DNV GL
1400 Ravello Dr
Katy, Texas 77449
bjorn.nilberg@dnvgl.com

Cynthia Spitzenberger
cynthia.spitzenberger@dnvgl.com

Nicholas Warner
Nick.warner@dnvgl.com

Hong Wu
hong.wu@dnvgl.com

Marisa Pierce
Marisa.pierce@dnvgl.com

Tone Rice
tone.rice@dnvgl.com

Trond Elvehøy
trond.elvehoy@dnvgl.com

Prepared for Presentation at the
American Institute of Chemical Engineers
2019 Spring Meeting and 15th Global Congress on Process Safety
New Orleans, LA
March 31 – April 3, 2019

Assessment of Battery Energy Storage Systems

Bjorn Nilberg
DNV GL
1400 Ravello Dr
Katy, Texas 77449
bjorn.nilberg@dnvgl.com

Cynthia Spitzenberger
cynthia.spitzenberger@dnvgl.com

Nicholas Warner
Nick.warner@dnvgl.com

Hong Wu
hong.wu@dnvgl.com

Marisa Pierce
Marisa.pierce@dnvgl.com

Tone Rice
tone.rice@dnvgl.com

Trond Elvehøy
trond.elvehoy@dnvgl.com

Keywords: Battery Energy Storage Systems, BESS, Lithium-ion, risk assessment, hazard assessment

Abstract

Current and future power grids are characterized by a high share of renewable energy sources. This leads to fluctuations in the power injection that need to be balanced. A Battery Energy Storage System (BESS) allows for the balance of supply and demand of electrical energy, utilizing stored energy during "peak demand" times and storing energy during times of "low demand".

From a technological perspective, battery storage is under continuous development and there are hundreds of suppliers providing systems. As such, several barriers must be overcome before Battery Energy Storage Systems (BESS) are fully integrated as a mainstream option in the power sector. These include performance and safety issues, regulatory barriers, and utility acceptance.

As BESS technologies are becoming more common in modern life, the size of installations will increase and they will be installed in occupied buildings. From this arise a concern on the risks associated with these technologies. Hazard identification can provide information for facility siting and proper design of BESS, and establish emergency response procedures for the battery site and local community.

The main core of the energy storage systems are the battery cells. Lithium-ion cells are sealed units, and under normal usage conditions, venting of electrolyte should not occur. However, if exposed to abnormal heating or other abuse conditions, electrolyte and electrolyte decomposition products can vaporize and be vented from cells. Vented electrolyte is flammable, and may ignite on contact with an ignition source, such as an open flame, spark, or a sufficiently heated surface. Batteries themselves can also become hot enough to cause auto-ignition of vented gases or self-ignite.

Some protection strategies for fires can, in some cases, lead to increased explosion risk (e.g., when small fires are suppressed). Selecting a protection strategy that will give the overall lowest risk is therefore an important activity in risk management of BESS. Based on experimental tests DNV GL initially perform phenomenological and risk analyses which in turn are supplemented by structured Computational Fluid Dynamics (CFD) simulations and experiments as needed. Experimental results from a large set of different tests are essential to understand the potential off-gassing, explosion and fire scenarios setting the limits of the energy release into the scenario.

DNV GL has carried out a series of experiments on lithium-ion battery fires at the cell level with different release conditions and ventilation rates to assess the off-gassing associated with a battery exposed to a fire condition or undergoing a runaway reaction. Based on the battery fire testing data, we present a hazard assessment associated with battery failure mechanisms of BESS in this paper. Various scenarios are discussed regarding different battery chemistries, capacities and site designs.

1 Introduction

Energy storage systems are becoming more attractive options for utility operators and providers to improve reliability and efficiency on the grid while reducing emissions. Various countries, including the US, have established strategies to increase the use of energy storage by addressing cost competitiveness, performance, market and pricing regulations, and industry acceptance [1].

Individual states have adopted renewable portfolio standards (RPS) and zero net emissions standards, as well as developed incentive programs, which support the increased penetration of energy storage. Thirty-seven states have an RPS or voluntary clean energy goals, with California as a major leader, and many states are expanding their goals. For instance, New York Governor Cuomo announced in 2017 that New York state has a target of 1500 MW of storage by 2025 [2]. More work is required to fully address the safety concerns related to BESS.

The energy storage market offers various options for consideration including pumped hydro, compressed air energy storage (CAES), flywheels, and batteries [3]. Currently, based on the US Department of Energy (DOE) energy storage database (which, as DOE has disclosed, is self-reported and researched, and may not maintain pace with all installations), there are 24.5 GW of energy storage systems of all types installed and operational in the US. Of that amount,

Electrochemical batteries represent 685 MW to US capacity. Of this subsection, Li-ion chemistries alone count for over 95% of US electrochemical installations (670 MW) [4].

As such, due to their growing market share, Li-ion batteries are the focus of this paper. Li-ion's high-energy density and rapidly decreasing capital costs support their utilization in applications ranging from portable personal electronics to transportation to grid-scale capacity support and beyond. While Li-ion batteries' capabilities are valuable, recent incidents such as the Union Pacific Train Car Explosion in 2017 [5]] and the Boeing 787 Dreamliner fires in 2013 [6]] have caused Li-ion batteries to come under scrutiny. The causes of these incidents are still under investigation. Lithium-ion batteries possess the perfect mixture of features required to create highly flammable and explosive environments and it is these risks that have driven concerns regarding permitting and code development over the last several years, ultimately delaying deployment of these systems in potential multi-billion dollar markets such as New York City[7].

The discussion within this paper is based on significant experience in battery testing, going back years to the Advanced Research Projects Agency-Energy, Advanced Management and Protection of Energy Storage Devices program (ARPA-E AMPED) and more recently through testing for utility companies, battery suppliers and other confidential clients. DNV GL has previously leveraged this data to advise fire fighters on safety measures during ESS emergencies, provide feedback and review to manufacturers regarding ESS safety designs, and develop new testing programs and protocols for stakeholders to gain a better understanding of failure modes and safe operation.

More recently, DNV GL's Maritime Group in Oslo has also begun, as part of an energy safety joint development project (JDP) in conjunction with several entities in Scandinavia, CFD modeling of the explosion risks associated with failure of these systems in confined shipboard battery rooms. This work has been further informed by additional testing work continually underway by both DNV GL and other entities as well as members of the consortium and other published literature. These efforts have given birth to additional internally-funded work within DNV GL to develop comprehensive gas analysis services using CFD at the small scale and Process Hazards Analysis Software Tool (Phast) at the larger scale to understand plume and environmental risks related to the failure of large containers or purpose-built buildings, which house potentially tens or hundreds of tons of these batteries.

The scope of this paper is to discuss the findings of various safety studies conducted by DNV GL for Battery Energy Storage Systems (BESS) connected to onshore grid power systems. The paper focuses mainly on the results of two-dimensional consequence modeling using Phast to determine the safety and environmental impacts from fires, explosions and toxic releases.

2 Batteries

DNV GL has conducted battery studies for Lithium Nickel Manganese Cobalt Oxide (LiNiMnCoO2 or "NMC") and Lithium Iron Phosphate (LiFePO4 or "LFP"). The BESSs studied have been installed in purpose-built structures, ISO-containers, trailers and outdoor cabinets. The illustration in Figure 2-1 shows an example of a Lithium NMC (similar to the batteries that have been studied in this paper) which shows the mass and energy rating of each level.

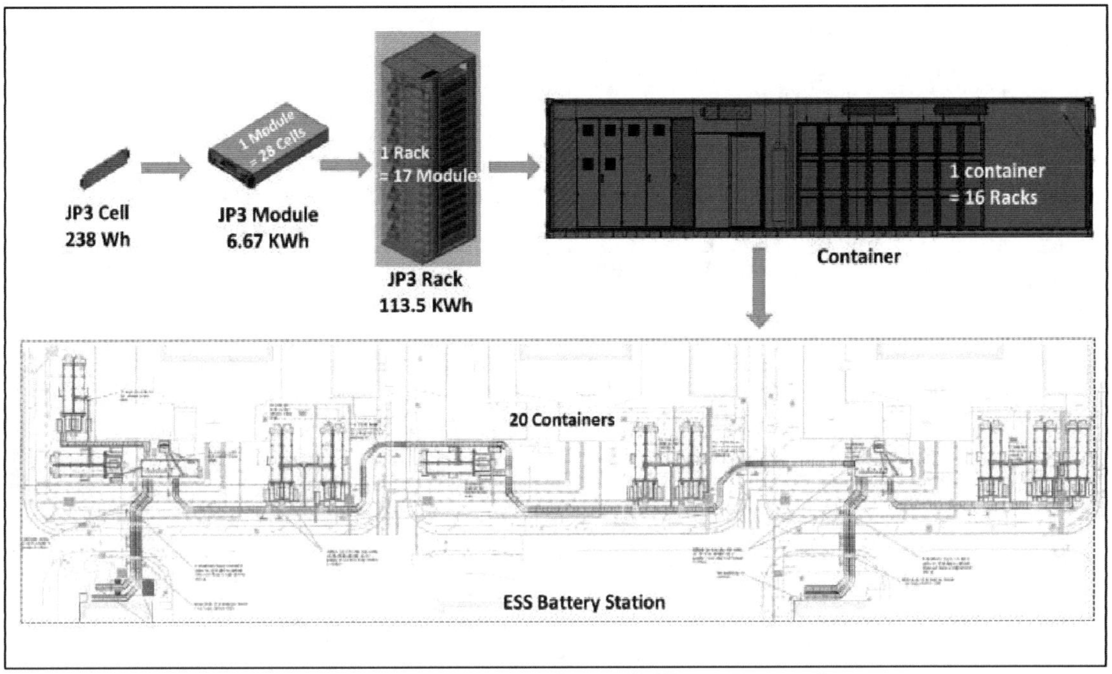

Figure 2-1: Scale up for LGChem Batteries

DNV GL has studied the impacts of thermal runaway and subsequent fires at the various levels of the BESS which are shown and described in Table 2-1.

Table 2-1: Levels and Descriptions

Level	Description
Battery cell	Base unit
Module	Multiple cells in series and parallel
Rack	Multiple modules in series or stacked vertically
Container	Contains multiple racks
Outdoor cabinet cluster	Multiple racks (each in a cabinet) controlled by one controller
Building or warehouse	1) Contains multiple racks controlled by one control system or 2) Several banks of multiple racks each controlled by one controller
Site	1) Multiple outdoor cabinet clusters, 2) Multiple containers or 3) Building or warehouse

3 Hazardous Scenarios

As mentioned before, if the battery cell is exposed to abnormal heating or other abuse, electrolyte and its decomposition products could be vented. A single battery cell fire is assumed to occur due to thermal runaway via various possible causes as described in Table 3-1.

Table 3-1: Possible Failure Mechanisms

Mechanism	Description
Cell Overheat	Battery cell overheats reaching a hazardous temperature which may impact neighboring cells and modules.
Inverter Failure	Inverter failure leading to hazardous current condition within BESS resulting in overcurrent to the cells, impacting multiple modules and potential escalation.
Internal Impact	Impact of batteries within BESS during shipping, construction or maintenance but realization of damage does not occur till normal operation, resulting in overcurrent to the cells, impacting multiple modules and potential escalation.
External Impact	Impact of batteries due to external cause; vehicle impact (for example), leading to overcurrent and potential escalation.
Startup / Commissioning / Construction	Impact of batteries during shipping, construction and commissioning leading to hazardous condition within BESS resulting in overcurrent to the cells, impacting multiple modules and the container / building.
Earthquake	Earthquake which leads to a hazardous condition with resulting damage to the cells, impacting multiple modules and potential escalation.
External Fire	External fire which leads to a hazardous condition at the site resulting in damage to the cells, impacting multiple modules and potential escalation.
Abnormal Grid Situation	Abnormal grid condition due to lightning strike or generator phase overvoltage which leads to a hazardous condition in the system with resulting damage to the cells, impacting multiple modules and potential escalation.

Of note as well from testing is the difference in results from cells that, all things being equal, combust during failure as opposed to simply going into thermal runaway and venting. Thermal runaway, as typically defined, is the exothermic generation of heat within a cell beyond which the cell can dispel. This results in a buildup of gas resulting in venting of the cell and failure. However, thermal runaway does not always lead to combustion, and the difference in thermal runaway temperatures and combustion (and thus gas generation results as well) may be on the order of 400-500°C (~350-400°C without combustion, 750-850°C with combustion). This can make repeatability of tests difficult and, as states of charge (SOC) and energy capacity of the cell tend to determine likelihood of combustion, can lead to a range of results based on numerous factors.

During the failure, the toxic materials are quickly discharged from the battery to the surroundings due to the occurrence of a series of exothermic reactions. A single battery cell fire is very likely to escalate to the neighboring cells, which may cascade to the whole module, and eventually fail the rack.

Burn tests on NMC LGChem JP3 Batteries were conducted to provide chemical discharge data for consequence evaluation applied in this report. Potential toxic, fire and explosion hazards are evaluated. Based on past burn tests, the primary gases of interest, in order of descending approximate quantity, are Carbon Monoxide (CO), hydrogen (H2), methane (CH4), ethylene, Hydrogen Chloride (HCl), ethane, methanol, ethanol, Benzene (C6H6), Toluene

(C7H8), Hydrogen Fluoride (HF), and Hydrogen Cyanide (HCN). Some of the gases evolve from the smoldering of plastic and may present before the actual cell failure as the plastic wrappers of the cells begin to heat. Many of the gases end up in small enough quantities to be discarded, with CO, H2, CH4, and ethylene presenting the bulk of the potential explosion risk.

4 Consequences

Hypothetical gas release leading to potential health impacts, thermal, and overpressure effects from the BESS are evaluated. The objective of this section is to discuss the hypothetical consequences if none of the facility safety barriers were in place or if all barriers failed, which is extremely unlikely. Note that hypothetical hazards related to any improper firefighting strategies are not covered in this analysis.

The consequence assessment of a 60 MWh NMC battery facility is performed with DNV GL's modeling software Phast. Phast is a comprehensive hazard analysis tool applicable to all stages of design and operation across a wide range of process industries. Its theory and performance have been independently reviewed as part of the European Commission (EC) funded project – Scientific Model Evaluation of Dense Gas Dispersion Models (SMEDIS), and it has excelled in both areas.

Two weather conditions are modeled: F stability 2 m/s (F2) and D stability 5 m/s (D5). F2 conditions typically produce more severe dispersion consequences due to stable wind conditions which results in less turbulence and dilution of the vapor cloud. D5 conditions represent neutral atmospheric conditions, either during day or night; typically these conditions produce less severe dispersion consequence zones as the greater wind speed promotes more mixing.

NMC chemistry is represented as LGChem JP3 Batteries. Each module consists of a total of 28 battery cells per module. There are 17 modules per rack and hence 476 cells per rack. Each container contains up to 24 racks, organized in three rows. The configuration of the example battery system is schematically shown in Figure 4-1.

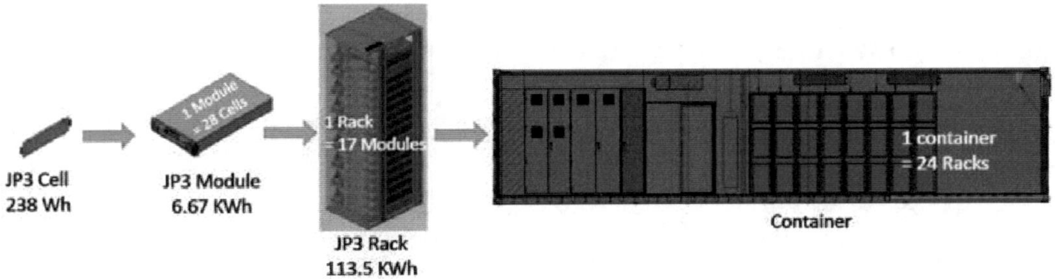

Figure 4-1: Assumed NMC Battery System Configuration

For simplicity, the Energy Storage System (ESS) is considered by racks. For modeling purposes the racks are considered stored in ISO-containers.

ISO-Container Configuration

The ISO-containers are assumed to be 40 ft in length by 8 ft wide and 8.5 ft high. It is assumed that one ISO-container can hold 24 racks. Ventilation from the container is assumed to be on top at the height of 8.5 ft. The containers are assumed to be 10 ft apart.

The escalation between racks inside the containers is assessed. Escalation between ISO-containers outdoors is assumed not to occur ([8]) and is only qualitatively discussed.

The energy capacity of a NMC rack is assumed to be 113.5 kWh. One container has an energy capacity of up to 2.7 MWh. For a 60 MWh purpose-built facility the number of containers required would be approximately 22, consisting of 529 racks.

Modeling

Representative failure cases at different escalation scales are selected to present the possible spectrum of battery fire development:

- Single module failure
- Single rack failure
- 50% of Container (12 racks) failure
- 100% of Container (24 racks) failure

Note that a single cell failure is not modeled as the hazard zone would be very localized.

The fire thermal radiation consequence results are provided at downwind distances from the release source at the following radiation levels, in accordance with CCPS and DNV GL general practices [9], [10].

Radiation Intensity (kW/m^2)	**Effect**
37.5	Sufficient to cause damage to process equipment, 100% lethality
9.5	Pain threshold reached after 8 seconds; second degree burns after 20 seconds
5.0	Sufficient to cause pain to personnel if unable to reach cover within 60 seconds, blistering of the skin (second degree burns) is likely; 0% lethality

Flammable dispersion results are reported at downwind distances from the release source at the lower flammability level (LFL) concentration and half-LFL concentration.

Overpressure consequence distances from the release source are provided at the following levels [11].

Overpressure Level	**Effect**
>7.3 psi (0.5 bar)	Damage to plant and process vessels; some substantial, with total destruction likely above 1bar. Unstrengthened buildings would be demolished.
4 - 5.1 psi (0.3 - 0.35 bar)	Threshold of significant damage to process vessels and pipework. Unstrengthened buildings likely to be significantly damaged/partly demolished.
<2.2 psi (0.15 bar)	Plant damage unlikely and only slight superficial damage to buildings of brick construction. Most windows broken and the glass are likely to cause injury to some people within the buildings. ISO containers fail at 2 psi.

The following toxic components are modeled: carbon monoxide (CO), hydrogen cyanide (HCN), benzene (C6H6), toluene (C7H8), hydrogen chloride (HCl), and hydrogen fluoride (HF). Toxic dispersion results are reported for downwind distances from the release source at the ERPG-3 threshold. Emergency Response Planning Guideline (ERPG) values are adopted as the modeling endpoints. ERPGs are developed by the Emergency Response Planning committee of the American Industrial Hygiene Association. ERPG-3 is the maximum airborne concentration below which it is believed that nearly all individuals could be exposed for up to 1 hour without experiencing or developing life-threatening health effects. Table 4-1 summarizes the values used as modeling endpoints for the different toxic materials in this study.

Table 4-1: Toxic Materials Threshold Values as Modeling Endpoints

Materials	EPRG-3 (ppm)
CO	500
HF	50
HCN	25
Benzene (C_6H_6)	1000
Toluene (C_7H_8)	1000
HCl	150

Results

Table 4-2 presents the toxic hazard distances generated by the failure scenarios. No results are generated for toxic concentrations at ground level given the release scenarios are modeled vertical at height at 8.5 ft. To be conservative, results are reported for distances at the release elevation of 8.5 ft. Greater downwind distances are reached; however, they are at elevations greater than the height of the container.

Table 4-2: Toxic Hazard

Maximum Downwind Distance (ft) at 8.5 ft elevation					
Material	**Endpoint**	**Single Module**	**Single Rack**	**Half Container (12 Racks)**	**Entire Container (24 Racks)**
CO	EPRG-3 (500 ppm)		5	20	30
HF	EPRG-3 (50 ppm)		5	15	20
HCl	EPRG-3 (150 ppm)		5	15	20
HCN	EPRG-3 (25 ppm)				5
C_6H_6	EPRG-3 (1000 ppm)		5	10	15
C_7H_8	EPRG-3 (1000 ppm)				

Table 4-3 presents the fire thermal radiation hazard distances generated by the failure scenarios at height at 3.3 ft.

Table 4-3: Fire Hazard

Event Name	Fire Radiation Levels at 3.3 ft elevation		
	Distance to 5 kW/m² (ft)	Distance to 9.5 kW/m² (ft)	Distance to 37.5 kW/m² (ft)
Single Module			
Single Rack	20	15	10
Half Container (12 Racks)	50	40	25
Entire Container (24 Racks)	50	40	25

Table 4-4 presents the flammable dispersion hazard distances generated by the scenarios. The results are only presented at the 8.5 ft elevation. No results are generated for the 3.3 ft elevation, as the release location is assumed to be 8.5 ft above the Container roof.

Table 4-4: Flammable Dispersion

Event Name	Flammable Dispersion at 8.5 ft elevation	
	Distance to 1/2 LFL (ft)	Distance to LFL (ft)
Single Module		
Single Rack		
Half Container (12 Racks)	3	2
Entire Container (24 Racks)	5	2

Table 4-5 presents the overpressure hazard generated by the ISO-container. The full volume of the container is assumed filled with flammable gas to generate the explosion hazard zones.

Table 4-5: Overpressure Hazard

Event Name	Explosion Overpressure Distance (ft)		
	0.15 bar	0.3 bar	0.5 bar
Entire Container (24 Racks)	86	47	21

Escalation

For the Container configuration, the 50% and 100% Container failure scenarios generate a 37.5 kW/m^2 thermal radiation hazard zone of approximately 25 ft depending on the wind conditions. This distance is from the center of the fire. This radiation level could reach a neighboring container given that the assumed separation for the containers is 10 ft. Note that Phast modeling is free field and this result is conservative given there would be some shielding expected from the originating container walls until they fail. Based on this result, the neighboring container may be impacted by this thermal radiation or greater, but it would take time for the container structure to fail. Effective engineering designs should be considered to further reduce the risk of an event escalating between containers.

5 Safeguards and Best Practices

Safeguards incorporated into BESSs should reduce the likelihood and/or mitigate the consequences of events before a battery fire escalates. Table 5-1 lists some of the most commonly used ESS safeguards and best management practices.

Table 5-1: Common Safeguards and Descriptions

Safeguard Type	Safeguard Description
Inherently safer design	Meeting UL 1973 Battery Abuse Tolerance criteria Facility Heating Ventilation Air Conditioning (HVAC) design for containers or buildings to maintain temperatures HVAC Redundancy
Basic process control system with safety control loops	Thermal management system which regulates the temperature within racks or modules to keep the cells from overheating Battery Management System (BMS) which controls temperature, voltage and current at various levels of the installation; separate from thermal management system BMS which can isolate battery systems and medium voltage equipment external to the ESS Energy Management System can detect loss of communication and disconnect at the inverter and/or disconnect at each bank and rack Remote monitoring of building and batteries from offsite control room
Fire suppression, protection and ventilation system	Fire and gas detection system and associated clean agent and water sprinkler in accordance with NFPA 13, NFPA 15, or NFPA 2001 as applicable Ventilation systems in accordance with NFPA 69 or deflagration protection systems in accordance with NFPA 68
Electrical components	Fuses and circuit breakers on main power lines Ground fault detector and interrupter (GFDI) Inverter voltage trips for inverters installed between the batteries and the main power lines
Shipping, installation and pre-commissioning procedures	Shipping/protection procedures and certification during prior to installation Equipment in safe state/batteries at or below 30% SOC during transport and installation on site Facility HVAC and Emergency Management Systems in place during construction and commissioning before power is connected to battery system Safety procedure and emergency response plan

Safeguard Type	Safeguard Description
Operating procedures	HVAC failure alarm and with procedure HVAC monitoring and maintenance Areas with controlled access and restricted speed (for roadways near the BESS) Emergency response plan for offsite responders with appropriate PPE to protect against chemicals or heat that may be released.
Passive barriers and siting	Spacing between battery containers or outdoor cabinets Buildings or containers have safety alarms, designed to allow time for personnel to exit the BESS Fences or barriers installed around BESS to prevent vehicle impact Distance from the road/railroad Lightning protection

6 Challenges/Opportunities/Further Developments or Testing

While much of the testing done to date has been performed at the cell level, numerous module and pack scale tests have been performed which have provided some insight into the scaling of test data to real world systems; this testing has not provided a complete picture of large scale failure though. Extrapolating the results from the test bench to a full-scale, practical application therefore introduces some additional unknowns that should be further investigated.

Current challenges with battery testing and extrapolation of test results include the following:

- Tests performed outside vs. chamber require more assumptions.
- Ventilation rate within the chamber can impact the burn test results.
- Based on the scenario (overheating, overcharging or burning), the types and compositions of chemicals released from the batteries vary.
- Test conditions are different than real-life operating conditions
- The battery burn tests are not fully reproducible.
- Most burn tests have not occurred beyond the module level. Therefore, the rates at which additional cells or modules burn is not fully known.
- The speed of escalation from cell to module, from modules to racks, and from rack to adjacent racks is not fully understood. System designs further complicate this.
- Chemistries may differ between tests (HSL testing for EDF).
- Toluene readings may zero out or max out. (The zero readings also need to be considered and potentially assigned values.)

7 Conclusion

This analysis, which extrapolated the cell level results based on DNV GL's experience with medium scale fire testing, battery cell failure analysis and propagation, was used as the basis for an environmental consequence assessment to the surrounding area which determined required impact and evacuation zones should the system fail. Safeguards and best industry practices should limit the potential occurrence and impact from BESS failures. Additional testing and scaled experiments will provide greater understanding of the potential hazards and escalation potential from these systems. With potential hazards managed, battery energy storage systems can provide utilities balanced electricity supply and demand.

8 References

[1]	U.S. Department of Energy. *Grid Energy Storage*. Washington, D.C.: Office of Electricity Delivery and Energy Reliability, 2013.
[2]	Spector, Julian. "The Best News Yet for Energy Storage in New York". Greentech Media. https://www.greentechmedia.com/articles/read/new-york-storage-industry-cash-target#gs.Ggh75Us. Accessed 10 January 2018
[3]	U.S. Department of Energy. *Energy Storage Safety Strategic Plan*. Washington, D.C.: Office of Electricity Delivery and Energy Reliability, 2014.
[4]	*Energy Storage Database*. US Department of Energy (DOE). https://www.energystorageexchange.org. Accessed 13 January 2019.

[5]	Smith, M. (2017, April 27). Union Pacific Train Car Carrying Used Lithium Ion Batteries Explodes and Catches Fire Near Downtown Houston, Texas. Retrieved 2017, from Metropolitan Engineering Consulting & Forensics - Expert Engineers: metroforensics.blogspot.com/2017/04/union-pacific-train-car-carrying-used.html
[6]	Kolly, J.M., Panagiotou, J., & Czech, B. A. (2013). The Investigation of a Lithium-Ion Battery Fire Onboard a Boeing 787 by the US National Transportation Safety Board. Boston: National Transportation Safety Board.
[7]	Spector, Julian. "Questions Shaping the Future of Energy Storage in New York". September 1, 2017. Greentech Media. Retrieved from https://www.greentechmedia.com/articles/read/five-questions-shaping-the-future-of-energy-storage-in-new-york#gs.7QVkkSM
[8]	Hutchison, Victoria. "LI-ION Battery Energy Storage Systems: Effect of Separation Distances based on a Radiation Heat Transfer Analysis". Worcester Polytechnic Institute. 12 June 2017. https://www.wpi.edu/sites/default/files/docs/Departments-Programs/Fire-Protection/Final_ESS_Report.pdf
[9]	Center for Chemical Process Safety (CCPS). *Guidelines for Chemical Process Quantitative Risk Analysis.* Second Edition, New York City, NY, USA.: AIChE CCPS, Inc. 2010
[10]	International Association of Oil & Gas Producers, 2009. Evacuation Escape and Rescue, DNV report no. 32338533, rev2, August 2009
[11]	Mannan, S. Loss Prevention in the Process Industries, Rev. 3, 2005

CFD Modeling of Liquid Jet Fire

Jin Zhao, Ph.D.
Baker Engineering and Risk Consultants, Inc. (BakerRisk)
11011 Richmond Ave, Suite 700, Houston, TX, 77042
jzhao@bakerrisk.com

Jef Rowley, Ph.D.
Baker Engineering and Risk Consultants, Inc. (BakerRisk)
11011 Richmond Ave, Suite 700, Houston, TX, 77042
jrowley@bakerrisk.com

Prepared for Presentation at
American Institute of Chemical Engineers
2019 Spring Meeting and 15th Global Congress on Process Safety
New Orleans, LA
March 31 – April 3, 2019

CFD Modeling of Liquid Jet Fire

Jin Zhao, Ph.D.
Baker Engineering and Risk Consultants, Inc. (BakerRisk)
11011 Richmond Ave, Suite 700, Houston, TX, 77042
jzhao@bakerrisk.com

Jef Rowley, Ph.D.
Baker Engineering and Risk Consultants, Inc. (BakerRisk)
11011 Richmond Ave, Suite 700, Houston, TX, 77042
jrowley@bakerrisk.com

Keywords: Jet Fire, CFD, Liquid Fuel, FDS, Thermal Radiation

Abstract

Although jet fires may pose a major hazard in the chemical processing industry, they are typically modeled with semi-empirical models that were developed based on experimental results of natural gas. In most semi-empirical models, it is typically assumed that all liquid is instantly vaporized upon discharge and combusted fully in the jet. In reality, depending on discharge fuel momentum, elevation, temperature and liquid fuel properties, a significant percentage of liquid fuel may fall to the ground and burn as a pool rather than contribute to the jet fire. As a result, semi-empirical models may predict overly conservative results when used to model liquid spray jet fires.

In this study, a Computational Fluid Dynamic (CFD) code was used to model gasoline jet fires. Results show that the CFD code correctly predicts liquid jet flame shape, dimension, rainout condition, and thermal radiation level.

1 Introduction

Since jet fires represent a major hazard for the chemical processing industry [1, 2], it is important to accurately model the consequences of jet fires when performing risk analyses and fire protection design. Semi-empirical models such as the Chamberlain model [3] and Johnson model [4] are widely used to predict free field vertical and horizontal jet fires, respectively. Although both models were developed based on natural gas data, they are frequently used to estimate the hazard associated with other fuels, including liquid jets. Liquid jet fires are typically estimated with such models by assuming that all liquid is vaporized and burned in the jet; in reality, depending on the discharged fuel momentum, elevation, temperature, and liquid properties, a significant percentage of the liquid fuel may fall to the ground and burn as a pool rather than contribute to the jet fire.

In this study, a readily available CFD code was used to model the fuel evaporation and combustion process of gasoline jet fires. Predicted flame shapes, trajectories, and radiation profiles are compared with actual test results.

2 Experimental Method

A series of gasoline jet fire tests were performed at BakerRisk's testing facility. The test cases was performed in two series: one series was performed at night with no measurable wind, and the other series was performed during day time with wind speed about 3m/s blowing toward the same direction as the release direction. Wind speed was measured at 20meters upstream of release point, 1.8m above ground. Pressurized gasoline was released horizontally through an orifice approximately 1.8 m above ground level. Pressure and mass flow rates were measured near the orifice using a Delta Controls 0-3500 kPa pressure transducer and a 25 mm MicroMotion Coriolis flow meter. The release pressure ranged from 150 kPa to 1700 kPa, and the release orifice diameter ranged from 6.3 mm to 12.6 mm. Radiant heat flux was measured using an array of 9 radiometers downstream of the discharge point. All radiometers were set at the release elevation, facing the release axis. Figure 1 shows a side view of the experiment as it was set up and Figure 2 provides a schematic of the top view with locations of the radiation monitors marked. The maximum and minimum flame lengths observed in these series of tests were also marked, as shown in Figure 2, to demonstrate the relative location of thermal radiation monitors to the flame. Note that thermal radiation monitor HS1 was inoperable for the majority of the tests; consequently, no results were reported from that monitor. The physical barriers shown at the north end of the test pad - 20 meters from the orifice - were necessary to prevent a potential grass fire.

Video footage of the tests was processed using the software ImageJ [5]. The last 30 seconds of each test was processed to yield images of the median, maximum, and average flame shape.

Figure 1: Experiment Setup, Side View

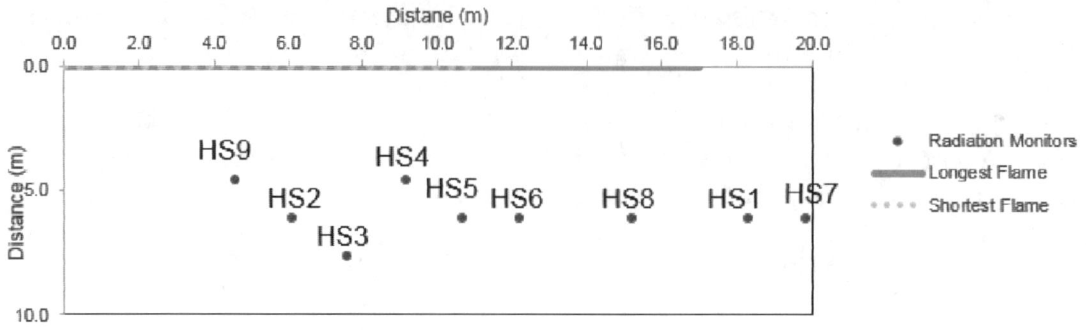

Figure 2: Experiment Setup, Top View

3 Computational Details

Simulation of the jet fires was performed using NIST's Fire Dynamics Simulator (FDS) [6, 7]. A sprinkler object, with a single horizontal release stream, was used to simulate the orifice. The flow rate was specified to match experimental measurements. Release velocity and droplet diameter were calculated with BakerRisk's SafeSite$_{3G}$© consequence analysis software, based on the measured release pressure and orifice diameter. A small flare is first modeled at 0.1m below the release point to ignite the discharged gasoline and was subsequently removed after 1.5 seconds so that it does not affect the jet behavior. The ground is modeled as 0.1m thick concrete.

Depending on the size of the fire, the simulation domains ranged from 24 m × 19 m × 9 m (length × width × height) to 72 m × 19 m × 23 m. Cell size ranged from 12.5 mm to 400 mm. Small cells were used in spaces where liquid droplets may be present and especially near the ground surface where liquid droplets rain out, while larger cells were used in other areas.

All calculations were performed twice: first, using a radiation factor of 0.4, and then with the radiation factor set to 0.7. Results are discussed in Section 4.3.

The flame shape was taken as the combustion surface within which 99% of the fuel is consumed. In all cases, at least 10 seconds of simulation were obtained, although the jet flame was typically predicted to reach steady state within 4-6 seconds of simulation time. When liquid rainout was predicted, the flame shape and dimension of the subsequent pool fire was strongly dependent on the timing of ignition and ground surface condition (temperature, heat capacity, surface roughness, etc.) and required a much longer time period to reach steady state. All jet portions of the flame reached steady state within 6 seconds, even as the pool fire continued to evolve. The simulation duration was insufficient in many cases for the pool fire to reach steady state; however, the primary focus of this study was the jet portion, which reached steady state in all cases.

4 Results and Discussion

4.1 Flame Shape

In all cases, the predicted flame shapes from FDS are adequate approximations of the median observed flame.

Figure 3 (a) and (b) shows the typical rainout flame shape under no-wind conditions. The lower edge of the flame appears to be defined by the predicted liquid trajectory.

Figure 3 (c) shows typical non-rainout flame shape under no wind condition/low pressure condition. Again, the lower edge of the flame corresponds to the predicted liquid trajectory. Vaporized fuel forms the flame above the liquid trajectory. The observed flame length is similar to the predicted liquid trajectory length.

Figure 3 (d) shows typical non-rainout flame shape under significant wind and high pressure conditions. Again, the flame's lower edge appears to be defined by the liquid trajectory. Due to the wind and high initial fuel velocity, the horizontal velocity of vaporized fuel is not dominated by buoyancy; as a result, the flame is longer than the persistence of the liquid spray.

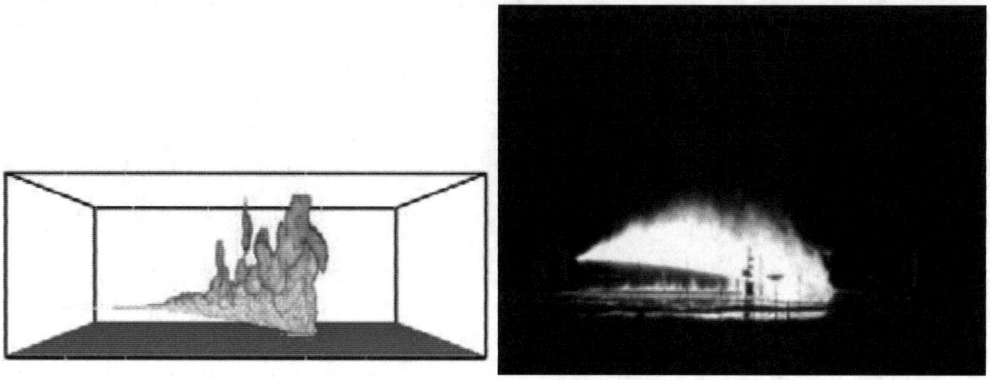

(a) 12.7 mm, ~150 kPa

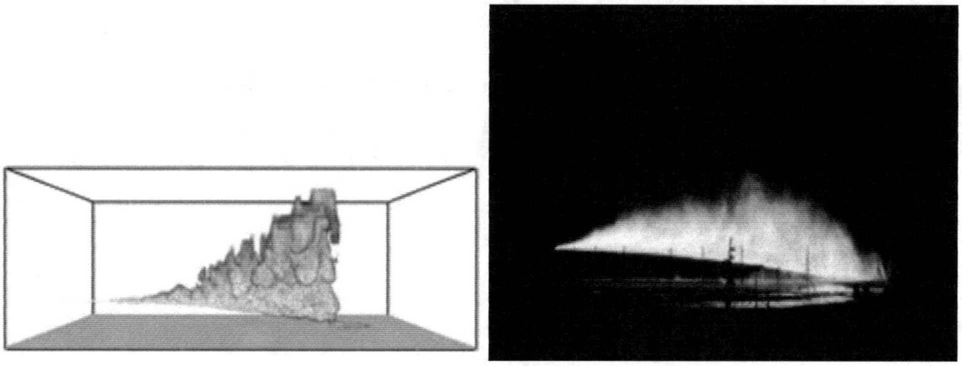

(b) 12.7mm, ~250 kPa

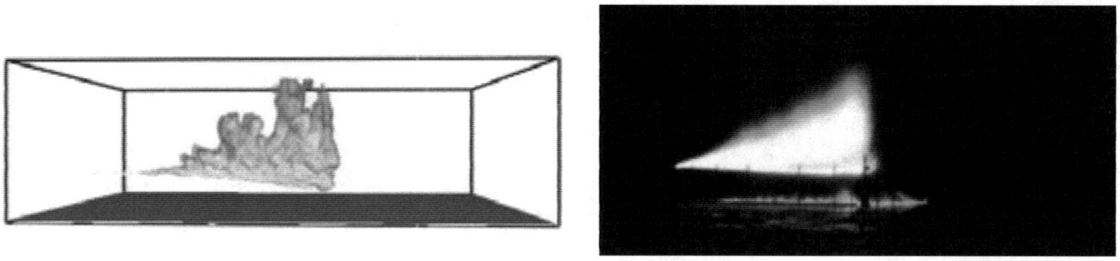

(c) 9.1mm, ~200 kPa

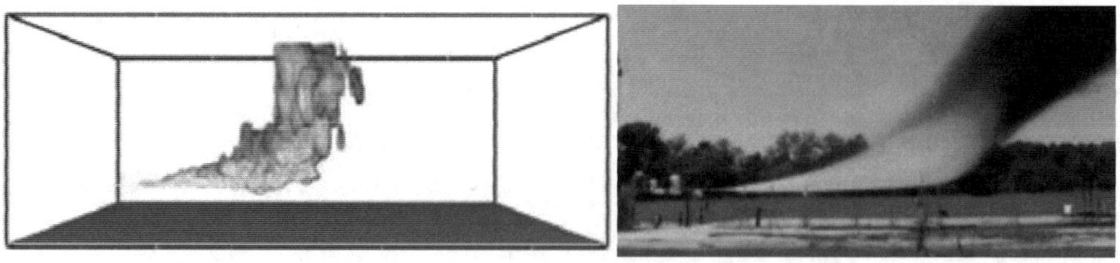

(d) 6mm, ~1,500 kPa

Figure 3: FDS Predicted Flame Shape vs. Median Experimental Observation

4.1 *Rainout Distance*

As the orifice diameter is increased, higher pressure is required to prevent liquid fuel from falling to the ground prior to full combustion. As shown in Figure 4, rainout distances predicted by FDS are very close to observed distances (less than 10% difference). For comparison, rainout distance (s) is also calculated by the initial discharge velocity (v) and elevation (h), using

$$s = v\sqrt{\frac{2h}{g}}$$

which assumes the drag on the liquid droplets is negligible. As shown in Figure 4, neglecting drag results in about 10~40% overestimation of the rainout distance. Therefore, the above equation neglecting drag may be used as a conservative approach to quickly estimate rainout distance as a precursor, or when a more detailed analysis cannot be performed.

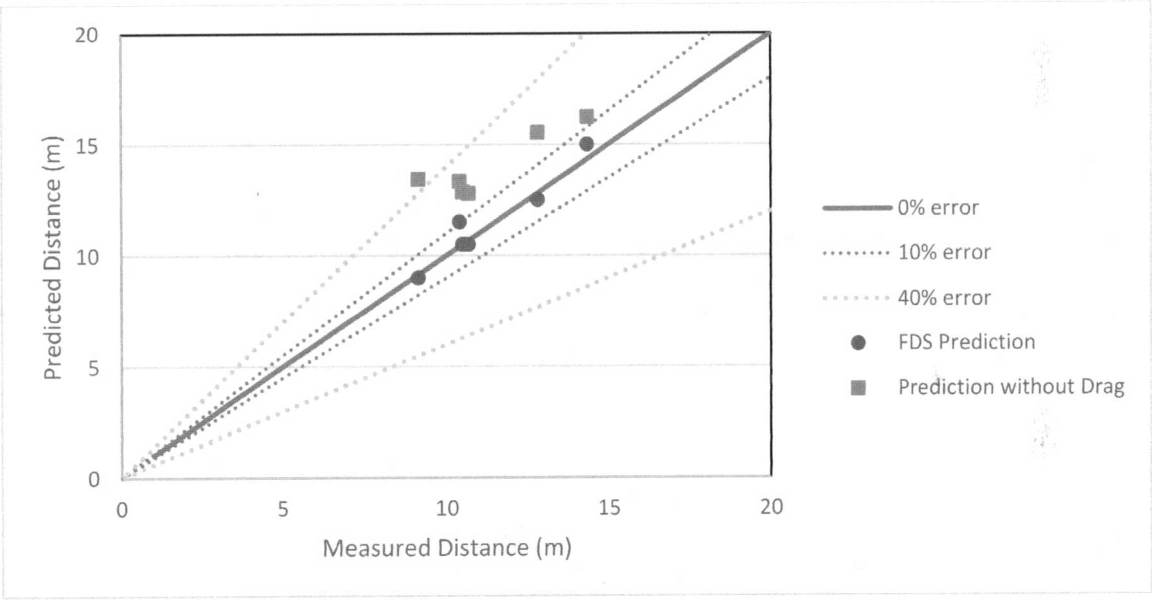

Figure 4: Rainout Distance

4.2 *Flame Length*

The flame length predicted by FDS is compared to the observed length of the median flames in Figure 5. The differences between FDS predictions and observed data are within 20%.

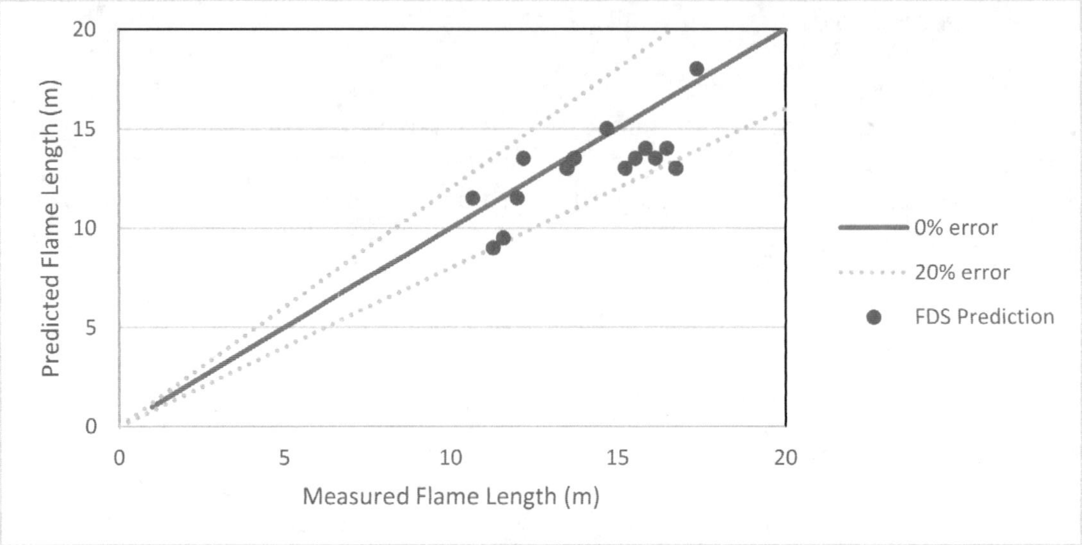

Figure 5: FDS Predicted Flame Length vs. Experiment Results

4.3 Radiation Level

Radiative fraction is a key factor that affects thermal radiation calculation results. In general, documentation found in the FDS release notes recommends a thermal radiation factor of 0.4 for heptane and heavier hydrocarbon [8]. However, gasoline pool fires are reported to have a thermal radiative fraction of 0.6~0.1 depending on the pool size and subsequent soot generation [9]. In this study, all test fire cases exhibited bright flames, with insufficient soot formation to obscure the flame. Based on these observations, a reasonable approximation of the average radiative fraction for a gasoline jet flame may be 0.5~0.6. However, since radiative fraction is a function of flame temperature and chemical composition [10], it seems unlikely that a single radiative fraction is applicable across the entire flame. Cited values in literature are typically derived using a single row of receptors near the flame's center, assuming an idealized flame geometry [10]. Thus, radiant fraction value cited in literature should be applied with caution. In this study, all tested cases were modeled twice: once with radiative fraction assigned to 0.4, and once assigned to 0.7, in order to investigate the impact of a range of values on the predicted radiant flux.

As shown in Figure 6 (a), when radiative fraction is set as 0.4, the calculated radiant flux near the flame tip (monitor HS6, HS7 and HS8, shown in Figure 2) is significantly lower than measured values, with the difference exceeding 60% in some instances. Additionally, calculated fluxes at or near the middle part of flame (HS4 and HS5) are almost always lower than measured value.

As shown in Figure 6 (b), when the radiative fraction is set as 0.7, calculated radiant fluxes near the flame tip (monitor HS6, HS7 and HS8) are still lower than measured values, but most cases are within 20~40% of observed values. Radiant heat calculated in two locations near the flame base (HS2 and HS3) exceed measured values by more than 60%. Radiant heat at the receptor closest to the release point (HS9) is not very sensitive to the radiation factor; it is slightly over-estimated when the radiation factor is 0.7 and slightly under estimated when the radiation factor is 0.4.

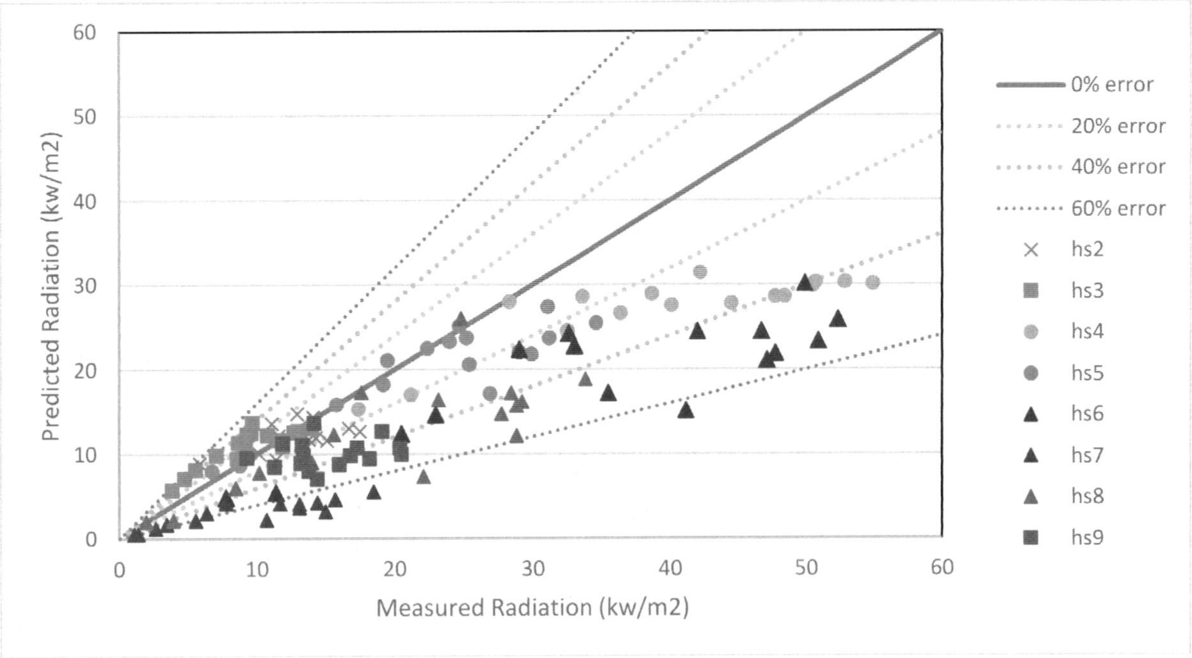

(a) FDS Prediction with Radiative Fraction = 0.4

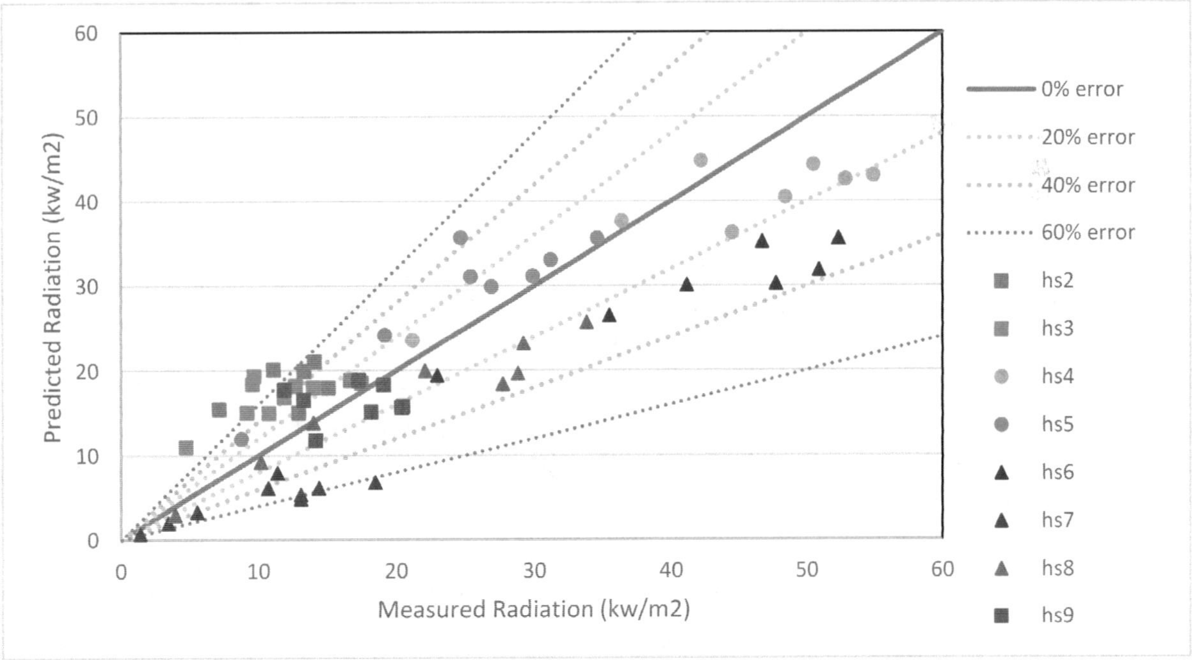

(b) FDS Prediction with Radiative Fraction = 0.7

Figure 6: Predicted vs. Measured Radiant Heat Flux

5 Conclusion and Future Work

Comparison between FDS predictions and experimental gasoline jet fire observations indicates that FDS is able to adequately predict liquid jet fire flame geometry, length, rainout conditions and distance, and radiant heat flux.

Test results and FDS modeling show that:

- The lower bound of the jet flame is based on the liquid trajectory.
- In some instances, liquid fuel will not completely vaporize in the jet; in such cases, liquid forms a burning pool on the ground.
- For low pressure releases in low wind, the jet flame length is determined by the length of liquid evaporation.
- For high pressure releases in high wind, the jet flame may extend beyond the length of liquid evaporation.
- Within the observed range (pressure ranged from 150 kPa to 1700 kPa, release orifice diameter ranged from 6.3 mm to 12.6 mm.), estimating the rainout distance by ignoring drag yields rainout distances slightly longer than observed distances.
- A radiative fraction of 0.7 (or even higher) provides an adequate estimate of the emissive power near the flame tip, while the emissive power near the flame base corresponds to a radiative fraction closer to 0.4.

Future work will investigate parameters that determine the rainout fraction and establish an engineering-level liquid jet fire model capable of incorporating rainout.

6 References

[1] J. Casal, M. Gomez-Mares, M. Muñoz, A. Palacios. "Jet Fires: a 'Minor' Fire Hazard?" Chemical Engineering Transactions, Vol. 26, pp. 13-20, 2012.

[2] N. Gopalaswarmi, Y. Liu, D.M. Labourer, B. Zhang, M.S. Mannan. "Experimental study on propane jet fire hazards: Comparison of main geometrical features with empirical models," J. of Loss Prevention in the Process Industries, Vol. 41, pp. 365-375, May 2016.

[3] G. A. Chamberlain. "Developments in design methods for predicting thermal radiation from flares," Chemical Engineering Research and Design, 65, pp. 299-309, July 1987.

[4] A. D. Johnson, H.M. Brightwell, A.J. Carsley. "A model for predicting the thermal radiation hazard from large scale horizontally released natural gas jet fires," Trans. IChemE. 134, pp. 157-166, August 1994.

[5] Public Domain Code – National Institute of Health (NIH), http://imagej.nih.gov

[6] K. McGrattan (editor). Fire Dynamics Simulator (Version 6), Technical Reference Guide. NIST Special Publication 1018, National Institute of Standards and Technology, Gaithersburg, Maryland, September 2014.

[7] K. McGrattan (editor). Fire Dynamics Simulator (Version 6), User's Guide. NIST Special Publication 1019, National Institute of Standards and Technology, Gaithersburg, Maryland, September 2014.

[8] FDS Release Notes. (2017). Retrieved February 10, 2018, from https://github.com/firemodels/fds/wiki/FDS-Release-Notes

[9] J. G. Quintiere. Principles of Fire Behavior, Second Edition, Taylor & Francis Group, LLC, Boca Raton, FL, 437 pages, 2017.

[10] G. Hankinson, B. J. Lowesmith. "A consideration of methods of determining the radiative characteristics of jet fires," Combustion and Flame 159(3), pp. 1165-1177, March 2012.

Ignition Phenomenon: From Electrostatics to Electrical Discharge

Eric Peterson, Ph.D.
Physical Safety

Morgan Reed, D.Eng.
Physical Safety

Prepared for Presentation at
American Institute of Chemical Engineers
2019 Spring Meeting and 15th Global Congress on Process Safety
New Orleans, LA
March 31 – April 3, 2019

Ignition Phenomenon: From Electrostatics to Electrical Discharge

Eric Peterson, Ph.D.
Physical Safety

Morgan Reed, D.Eng.
Physical Safety

Keywords: Ignition Source, Combustion, Electrostatic

Abstract

Worldwide in many industries, flammable materials are flowing, spraying, vaporizing, and condensing in controlled systems on daily basis. Occasionally, incidents occur where the flammable material is released from its controlled environment and has the ability to mix with an oxidant creating a potentially explosive mixture. If the explosive mixture finds a sufficiently strong ignition source, severe consequences may ensue. One of the most sought after questions in the industry is the quantification of the ignition source.

Historically, reducing or eliminating an ignition source for flammable releases has been paramount. Countless hours have been spent studying the elusive "where did the ignition source come from" and "how can we prevent ignition from occurring". Many avenues exist for creating an ignition source, but none more treacherous than the realm of electrostatics. Electrostatic potentials can develop in conjunction with upset conditions or normal operations, including vessels and piping that have been bonded, grounded, or both.

To our knowledge, this provides an initial look which includes quantum mechanical principals as a mechanism that motivates an electrostatic discharge, specifically relying upon quantum electrodynamics to explain the source of ignition. This paper focuses on the ability of insulating media to develop electrostatic buildup and a subsequent discharge upon the breakdown of a dielectric, including air.

1 Introduction

All around us, in every aspect of life, elements are the building block of life. None more prolific than the carbon atom. From clothes, to plastics, to the fuel we burn in our cars, hydrocarbons are arguably the most widely used substances in this day and age (aside from water). At its most elemental state, a hydrocarbon chain is stitched together as the fabric of life. This hydrocarbon production process requires a great deal of work. The work performed on developing a hydrocarbon, and many other molecular compounds, leads to a stored potential energy of the newly formed system. That potential energy may be transformed into a second kind of mechanical energy, kinetic energy.

The stored energy in the molecular system can be thought of as a minimum, that is, energy in excess of the potential energy is necessary to overcome the forces holding the molecule together. The forces are the bond strength or bond energies of the system and the bond energies are what keep molecular structures intact.

2 Ignition Sourcing

At the heart of this paper is the ignition source. Ignition sources surround us in our daily lives including electrical sparks from electrical outlets, open flames from cigarette lighters, hot surfaces like stoves, and car engine blocks. These are just a few sources we encounter on a regular basis. For example, in the process of ignition, once a developing flame is provided a sufficient energy source, a positive feedback mechanism can be achieved and the system achieves a self-sustaining flame.

It was previously discussed there is an additional energy that needs to be overcome, or required for the energy conversion to take place. That energy is commonly referred to as the activation energy and is attributed to Svante Arrhenius towards the end of the 19th century. The Arrhenius equation allows us to understand the rate at which many of these types of reaction will progress:

$$k = Ae^{-\mathrm{Ea}/RT}$$

where k is the rate of reaction, A is a constant, E_a is the activation energy, R is the universal gas constant, and T is the absolute temperature (in kelvin). As can be inferred from the Arrhenius equation, the rate constant rapidly increases with increasing temperature. This concept has led to an understanding that for a reaction to proceed, an activation energy must be supplied. It is also known that for specified concentration levels for a given temperature, the rate of a reaction is inversely proportional to the energy of activation. That is, there exists a barrier, an energy threshold that must be overcome for a chemical reaction to proceed. This energy/temperature relationship provides an essential information to the elemental collisions and molecular motion that influences and possibly drives the exceedance of the activation energy threshold.

2.1 Electrostatics

Classical electrostatics introduces the electric potential to our discussion on ignition sources, where the phenomena of influencing electron motion or movements from one point to another, therefore leaving behind a positively charged entity in its wake. In doing such, the magnitude of the potential describes a property of the electric field. Our reality is anything but stationary, and as such, electrodynamics allow us to account for the continuous fluctuation of charge in the universe. The secondary related field to help us understand our accumulation and expulsion of electric charges is the magnetic field, each with the ability to be analyzed on separate terms. The discussion of flowing, or dynamics is a bit of a misconception when using the term *statics*. Statics gives one the notion of something unchanged, independent of time itself. When discussing electricity and magnetism, they are two distinct phenomena as long as charges and current are *static*. Life gets more complicated as charges begin to move or transfer in position. This complication is where the interdependence of the electric and magnetic fields is accounted for as the charges move [1][1].

Consider a case where a nonconductive flammable fluid is flowing through a network of piping and pouring into void a space. The flow of fluid essentially sluffing off electrons causing a flux of charge (positive and negative) movement at the surface of a pipe or vessel, but also redistributing the charge at or near the surface of the fluid. All of these charges must be accounted for, which leads to the complexity of the topic. Accounting for charges, their locations, and distribution alludes to the charge density and the components of Maxwell equations for electromagnetics. The culmination reaches a peak where we use the electric field to provide some understanding or glimpse of charged particle movement when under the influence of electric/magnetic fields.

2.1 Electrostatic Discharge

When discussing items in terms of the quantum realm, or intermolecular connections and forces, it is the electron that is often overshadowed by the proton, both literally and metaphorically, as the proton is significantly more massive than the electron. The electron(s) swarms about the positively charged nucleus giving rise to an electron cloud. Therefore, when an object has a partial charge due to intermolecular forces, or a natural polarity, experiences an interaction with another charged force the surface charges will either attract or repel one another causing displacements of those charges. The surface energy effects and displacement of electron clouds is the underlying influence of charge transfer from one atom to another, leading to the "charging" of a molecular structure and any subsequent "discharging".

We speak of an electrostatic discharge when the charge build-up is neutralized. If an electric field is too great, air will break down, so we get a discharge. In Richard Feynman's seminal Lecture Series, he once described the breakdown as:

"Air will break down if the electric field is too great. What happens is that a loose charge (electron, or ion) somewhere in the air is accelerated by the field, and if the field is very great, the charge can pick up enough speed before it hits another atom to be able to knock an electron off that atom. As a result, more and more ions are produced. Their motion constitutes a discharge, or spark. If you want to charge an object to a high potential and not have it discharge itself by sparks in the air, you must be sure that the surface is smooth, so that there is no place where the field is abnormally large."

Electrostatic discharge is the rapid neutralization of charge between two (or more) electrically charged items (molecules, compounds, or objects) caused by intermolecular interaction and/or the breakdown of a dielectric. In our everyday lives, this may commonly be observed in conjunction with the generation of a visible spark. A substance that is normally a good insulator becomes conductive when its breakdown voltage (dielectric strength) is exceeded.

Static electricity is often generated by a separation of electric charges when two materials are brought into contact and then separated. Examples include: walking on a carpet; using a comb to groom one's hair; rubbing a balloon against one's hair; exiting a car form a fabric seat; removing some types of plastic packaging, etc. In these cases, the breaking of contact between two materials creates a difference of electrical potential that can lead to an electrostatic discharge. Figure 1 shows an image from the Barton Solvents facility where the Chemical Safety Board determined an electrical discharge, spark, contributed to the ignition of nonconductive flammable liquids [2].

Figure 1: Barton Solvents Facility – Fire Attributed to Electrical Discharge

The methods most often deployed to prevent static electricity becoming an ignition source are bonding, grounding, or both. When materials are not bonded or grounded, they could amass a sufficient electrical charge capable of producing an electric field strong enough to breakdown and provide a spark. Several conduits capable of producing a static discharge are:

- Liquids flowing through pipes
- Settling of a solid through a liquid

- Expulsion of particles or mists through constricted orifices
- Intense rubbing together and subsequent separation of certain materials

Generally, significant charge accumulations can only persist in regions of low electrical conductivity (large electrical resistivity). Air is considered a very poor conductor of electricity, where very few charges are free to move about the environment. Therefore, the neutralization of charge often occurs when neutral species are ripped apart forming positive and negative species (whether atoms or molecules). This charge separation propagates as an electric current to "discharge" or expel much of the original charge accumulation.

Electrostatic discharge, as a spark, occurs when a strong electric field generates an ionized conductive conduit in air. The spark can lead to severe consequences of fire and explosions, if the surrounding atmosphere contains flammable vapors or dusts. A spark can be generated when the electric field strength exceeds approximately 3 MV/m (normal atmospheric conditions), which is considered the dielectric field strength of air [3]. This process is often termed, the dielectric breakdown, and the discharge causes the sudden and rapid increase in particle motion. The rapid movement of the molecules causes the temperature to rise drastically over short durations (superheats the immediate atmosphere). During this time, the electrons are absorbing finite quanta of energy and transition to elevated energy states. Often described as the electrons excited state. As the electrons relieve themselves of the energy and return to their normal energy states visible light is produced and on grand scales (lightning), audible sound wave. A lightning bolt is a loud and large example of sparks observed in more pedestrian static discharges. The flash occurs because the air in the discharge conduit is superheated to such a high temperature that it emits light by incandescence.

2.3 Energies involved

There are a few types of electrical discharge mechanisms recognized including: the spark, arguably the leading mechanism of ignition source when static electricity is present; brush; propagating brush, which is a high energy discharge and considered dangerous; cone; and corona discharge. The energy released during static discharge can vary over a wide range. The energy (E) can be calculated from the capacitance (C) of the object and the static potential (S) by the formula:

$$E = \frac{1}{2}CS^2$$

There is a relative maximum, about 35–40 kV, for the potential (most often associated with corona discharge) as the charge is dissipated at higher values. In the oil and gas industry, a relatively small amount of energy, as little as 0.2 millijoules, is needed to ignite a flammable mixture of a fuel (methane) and air [4]. Industries working with hydrocarbon gases and solvents, the minimum ignition energy (MIE) required for ignition of a vapor–air mixture is at its minimum when the concentration is approximately at a stoichiometric level, rapidly decreasing as the concentration deviates from this optimum. Though some research [5] shows variations around the stoichiometric value, but it is non-trivial process to determine these values commonly referred to as the lowest minimum ignition energy.

3 Discussion

Electrostatic discharges continue to be an ignition source for flammable material. Many of the industry accidents that have tracked the cause to electrostatic discharge have identified the spark as the ignition source. The industry wide idea of electrostatic discharge remains at the macro-level, where two objects come in contact and then move away from each other causing an electrostatic charge generation. Many times, these ignition sources are considered inherent to the process and determined to be un-mitigatable, since bonded and grounded mitigation options were in place. Thus, leading companies to make choices between less efficient processes and materials at a heavy cost. We believe a deeper understanding of the electrodynamics associated with the electromagnetic fields will provide insights to future design and mitigation options.

Industries should strive for continuing to improve materials of construction, characteristics of flowing material, and means of dissipating charge buildup.

1 References

[1] Feynman et al., “The Feynman Lectures on Physics, Volume II”, 6th Printing February 1977.

[2] Chemical Safety Board, “Barton Solvents – Static Spark Ignites Explosion Inside Flammable Liquid Storage Tank”, No. 2007-06-I-KS.

[3] Glor and Thurnherr, “Ignition Hazards Caused by Electrostatic Charges in Industrial Processes”, The ExplosionProofing Company, February 2015.

[4] Hankinson et al., “Ignition Energy and Ignition Probability of Methane-Hydrogen-Air Mixtures”, June 2009.

[5] Laurence Britton, “Avoiding Static Ignition Hazards in Chemical Operations”, A CCPS Concept Book, 1999.

19 AIChE Spring Meeting
15TH GLOBAL CONGRESS ON PROCESS SAFETY
NEW ORLEANS • 2019

Understanding the Potential for Detonations in Large-Scale Vapor Cloud Explosions

C. Regis L. Bauwens
FM Global, Research Division
1151 Boston-Providence Turnpike, Norwood, MA, USA
carl.bauwens@fmglobal.com

Sergey B. Dorofeev
FM Global, Research Division
1151 Boston-Providence Turnpike, Norwood, MA, USA
sergey.dorofeev@fmglobal.com

Prepared for Presentation at
American Institute of Chemical Engineers
2019 Spring Meeting and 15th Global Congress on Process Safety
New Orleans, LA
March 31 – April 3, 2019

Understanding the Potential for Detonations in Large-Scale Vapor Cloud Explosions

C. Regis L. Bauwens
FM Global, Research Division
1151 Boston-Providence Turnpike, Norwood, MA, USA
carl.bauwens@fmglobal.com

Sergey B. Dorofeev
FM Global, Research Division

Keywords: vapor cloud explosions, detonations, consequence modeling

Abstract

Large-scale unconfined vapor cloud explosions (VCEs) can generate significant damage and destruction over extremely wide areas. The most common models used to assess the consequence of a VCE are flame acceleration methods, which generally limits the contribution of explosion energy to the portion of the cloud where congestion and/or confinement is present. If a deflagration-to-detonation transition (DDT) occurs, however, it has the potential to involve the entire flammable cloud, significantly increasing the consequences of a VCE event.

In this work, prior detonation studies are reviewed to illustrate the inherently unstable nature of detonations, with a focus on the critical cloud dimensions and concentration gradients which can support a propagating detonation wave. These criteria are then applied to representative flammable cloud concentration distributions. By evaluating these limits, it is found that a detonation propagating throughout the entire flammable cloud is extremely unlikely in a realistic unconfined release. Nevertheless, these limits do not rule out the possibility of local detonations, occurring and propagating far from the source of congestion. This work demonstrates how the critical conditions for detonation propagation can be used to estimate the portion of the cloud that participates and how these criteria can be readily incorporated into existing dispersion modeling approaches.

1 Introduction

While rare events, large-scale unconfined vapor cloud explosions (VCEs) have the potential to generate widespread damage over extremely large areas. To estimate the consequences of VCEs, industry has primarily adopted flame acceleration methods, which are models that consider how flames propagate through regions of congestion and confinement. As flames rapidly decelerate in open unconfined areas, these approaches typically neglect the contribution of the flammable cloud where no congestion is present.

Recent studies have suggested, however, that a number of recent large vapor cloud explosions may have in fact transitioned to detonation [1, 2, 3]. This is due to observations of high levels of damage far from congested regions, in otherwise open space [2]. In addition, recent large-scale experiments have demonstrated that a transition to detonation can occur in fuels that are not considered to have high reactivity, with only modest levels of congestion [4]. These studies raise two important questions, should the possibility of DDT be considered in the consequence analysis of VCEs and, if so, how should the consequence of such an event be modeled?

As detonations have the ability to propagate rapidly through open space, it has been widely accepted that, for industrial safety purposes, "If DDT occurs the detonation may spread to all of the cloud which has a concentration within the detonable range" [2]. If this is the case, considering DDT scenarios would significantly increase the expected consequences of a vapor cloud explosion event, compared to traditional flame acceleration models.

On a fundamental level, however, detonation physics has been extensively studied [5, 6]. In particular, there is considerable literature examining the stability of detonation waves. As this paper will show, these studies have identified critical conditions that must be satisfied in order to sustain the propagation of a detonation wave. These limits provide insight into how the consequences of DDT in a VCE can be estimated and suggest that, for a real release of fuel, the non-uniformity of the flammable cloud would significantly reduce the mass of fuel that could support a detonation.

In this work, the basic features of deflagrations and detonations are highlighted, and prior studies that have examined detonation propagation limits are reviewed to illustrate the inherently unstable nature of detonation waves. These criteria are then applied to representative flammable clouds of varying size, shape, and concentration distribution, to demonstrate the implications of these limits on realistic releases and estimate the detonable portion of the cloud.

2 Review of Detonation Propagation Limits

2. 1 *Deflagrations, detonations, and deflagration-to-detonation transition (DDT)*

Deflagrations and detonations represent two distinct propagation mechanisms through which a combustion wave consumes a flammable mixture. When a flammable fuel-air mixture is ignited by a relatively weak source, such as an electric spark, the combustion wave typically propagates as a deflagration, which is characterized by a subsonic propagation velocity relative to the unburned mixture ahead of the flame. In a deflagration, this propagation velocity, commonly referred to as the laminar burning velocity, is controlled by diffusion. Diffusion of heat and radicals from the burned gas into the reactants. While laminar burning velocities of stoichiometric fuel-air mixtures can vary significantly, they are generally on the order of 0.2 – 2 m/s.

If a flammable fuel-air mixture is ignited by a strong ignition source, however, such as a high explosive, a detonation can be initiated in the mixture, which is characterized by a supersonic propagation velocity relative to the unburned mixture ahead of the flame. The mechanism through which a detonation propagates is adiabatic ignition of the unburned mixture as it is compressed and heated by a leading shock wave. As a result, the detonation propagates as a complex of the leading shock wave and the trailing chemical reaction zone. This mechanism results in extremely high fuel consumption rates, with typical detonation velocities on the order of 2,000 m/s for stoichiometric fuel-air mixtures.

In most industrial accidents involving fuel-air explosions, the ignition source is relatively weak and initially results in a slow deflagration. As the flame propagates and grows in size, however, a laminar flame eventually becomes unstable, resulting in an increase of flame surface area and burning rate. Moreover, the deflagration generates a flow field in the unburned gas ahead of the flame, due to the expansion of the combustion products. This flow has the potential to interact with obstructions upstream of the flame, resulting in the generation of turbulence and additional flame surface growth. These factors further increase the effective burning velocity and flow speed ahead of the flame, establishing a positive feedback loop that results in flame acceleration. As a result, regions where obstructions and confinement are present provide the most effective environment for flame acceleration. The role of obstacles in the flame acceleration process has been extensively studied in the literature. In studies of VCE potential, the areas of congestion and confinement are generally considered as the primary source of the blast wave energy, and are considered to be responsible for the majority blast damage [7, 8, 9].

When a flame accelerates through a sufficiently large run-up distance, a shock wave may develop ahead of the flame front. This high speed deflagration can potentially transition to a detonation, through a DDT, due to strong shock reflections off obstacles, or the onset of detonation within the flame brush, and propagate through the unburned mixture. The possibility of a specific mixture undergoing DDT depends on whether certain conditions are met in relation to the flame speed and the size of the flammable mixture [5]. The role of the overall size of the flammable mixture in determining whether DDT can occur has been emphasized in [5, 10] and other studies. It has been shown that even low sensitivity mixtures, such as methane, can undergo DDT given a sufficiently large volume.

The sensitivity of a combustible mixture to detonation is generally characterized by the detonation cell size, a measure of the scale of detonation instabilities, which will be outlined in the following section.

2.2 Detonation instabilities and wave structure.

Since the 1960s, it has been demonstrated by numerous investigators [11, 12, 13] that planar detonations are intrinsically unstable. The instability develops through the formation of secondary waves, more commonly known as transverse waves, propagating back and forth in a direction perpendicular to the detonation propagation. If one were to track the trajectories of the intersection between the transverse waves with the main detonation front, also referred to as triple-points, it would form a diamond-shaped pattern, where the width of the cell, λ, is called the detonation cell width, or more commonly the detonation cell size, as illustrated in Fig. 1.

The detonation cell size is considered a material property of a given mixture and is determined experimentally by analyzing the cellular pattern generated on a soot foil, a foil that has been covered by a thin layer of carbon soot and is placed within the detonation channel. For fuel-air mixtures, the cellular structures that are generated are typically irregular in nature, as shown in Fig. 2, and some interpretation is needed to extract consistent cell sizes. This results in a large source of uncertainty when quantifying the cell size, as these measurements can often vary by a factor of two between different groups.

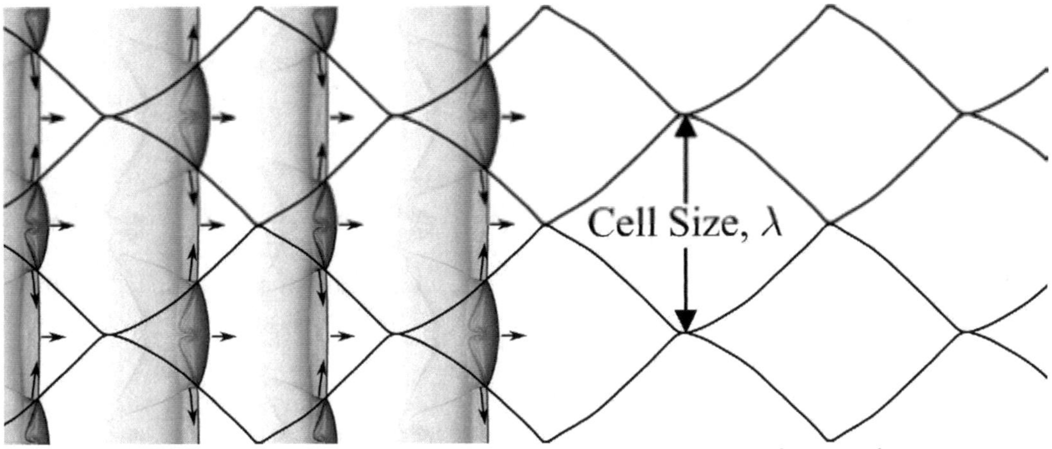

Figure 1. *Schematic of the propagation of a cellular detonation front, showing the trajectories of the triple points with consecutive positions of the main detonation front obtained from numerical simulations* [14].

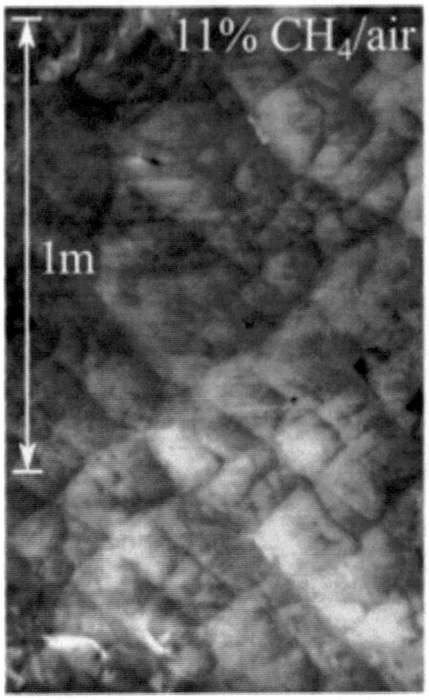

Figure 2. *Example of the cellular structure left by a detonation on a soot foil.* [15]

The instability of propagating detonation waves create a situation where the leading shock and the chemical reaction zone are continuously decoupling. In order for the sustained propagation of a detonation wave, collisions between the transverse waves provide a local ignition source that recouple the shock and chemical reaction. As a result, these transverse waves are a crucial feature of a freely propagating detonation wave. If, for example, the transverse waves were eliminated by an acoustically absorbing material on the channel walls, which provides no reflections, a previously propagating detonation will decay. Therefore, the spacing between the transverse waves, i.e. the detonation cell size, is a critical parameter that determines the ability of a detonation to propagate, where mixtures with a tighter spacing of transverse waves, i.e. a small cell size, produce more stable detonations that are easier to initiate.

The detonation cell size is a function of the chemical reaction rate in a detonation, where small cell sizes correspond to faster rates of chemical reaction. Historically, a mixture's sensitivity to detonation is usually characterized by the energy of the ignition source that is needed to directly initiate a detonation. The smaller the detonation cell size, the more sensitive the mixture, and the less energy needed to initiate detonation. In addition, the sensitivity of the mixture is also important for DDT, where more sensitive mixtures can transition from deflagration to detonation more readily. One of the most sensitive hydrocarbon fuel-air mixtures is acetylene-air (with a cell size $\lambda \approx 5$ mm in a stoichiometric mixture), and one of the least sensitive mixtures is methane-air ($\lambda = 35$ cm) [16]. It is important to note that the cell size increases sharply for mixture compositions away from stoichiometry on both lean and rich side, as shown in Fig. 3.

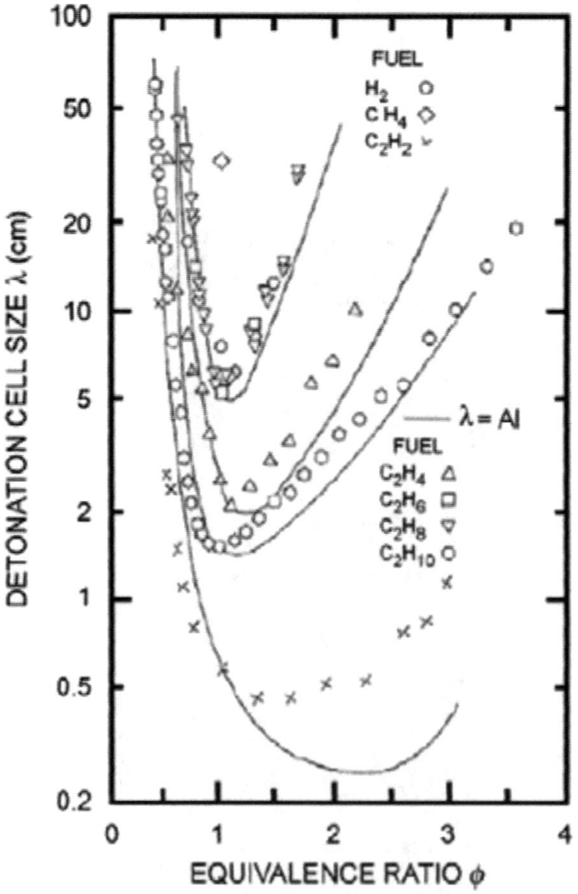

Figure 3. *Detonation cell size of fuel–air mixtures at atmospheric pressure, curves represent cell sizes estimated from the reaction zone width, l, with a coefficient, A, anchored at stoichiometric conditions.* [16]

2.3 Limits of detonation propagation

The importance of the transverse waves to detonation propagation makes the detonation cell size, λ, an important parameter for defining conditions where detonations are possible. For a detonation propagating in a confined channel, its size should, at a minimum, be able to accommodate at least one transverse wave. In a circular tube, the limiting tube diameter corresponds to the situation where one transverse wave rotates along the cylindrical surface of the tube such as that the limiting tube diameter $d_l = \lambda/\pi$.

When a detonation propagates from a confined to unconfined space, e.g., when exiting from a tube to an unconfined combustible mixture, there is a minimum tube diameter that can support the transmission of the detonation to unconfined space, as shown in Fig. 4. This diameter has been traditionally called the critical tube diameter, d_c. This phenomenon was first reported by Zeldovich and coworkers in 1956 [17] in a study of spherical detonation initiation. Further studies [18, 19] concluded that the minimum tube diameter for most hydrocarbon-air mixtures is $d_c = 13\ \lambda$. For square channels, this minimum diameter reduces to $d_c = 11\ \lambda$, and for sufficiently wide rectangular channels, the critical height for the detonation transmission approaches a value of $h_c = 3\ \lambda$.

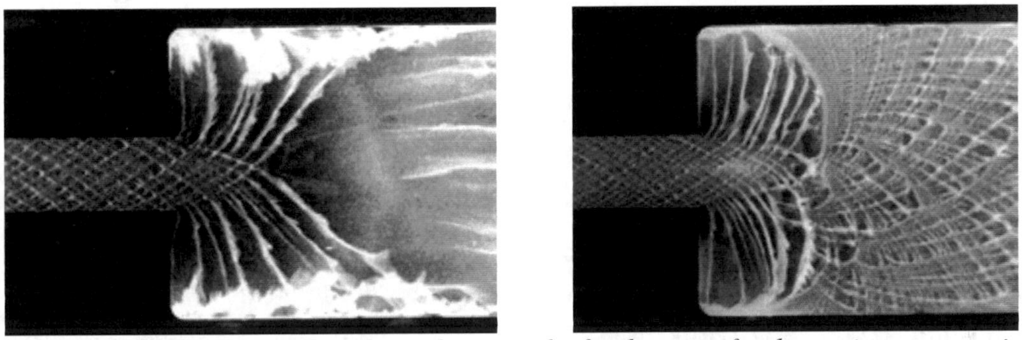

Figure 4. *Open shutter images illustrating the critical tube diameter for detonation propagation.* [20]

The critical diameter problem combines several important phenomena that limit detonation propagation, including the need for transverse waves, the detonation decay due to excess curvature, and the role of expansion waves. For this reason, the value of d_c is often used as an alternative parameter to the cell size λ for characterizing mixture sensitivity.

When considering the potential for detonations in unconfined vapor cloud explosions, it is important to understand the limits for detonation propagation in a free charge of a combustible mixture and in combustible clouds on the ground. A series of detailed experiments have investigated this problem, where novel experimental techniques were used to create conditions suitable for studying detonations in unconfined and semi-confined fuel-air mixtures. Some of these studies have examined the minimum diameter of a free cylindrical charge, d_f, for detonation propagation [20, 21, 22, 23]. In general, it was found that d_f is about two to six times d_c (i.e. $d_f \approx 20\ \lambda$ to $60\ \lambda$) depending on the fuel and the diffusion layer at the boundary of the mixture [21]. Throughout these studies, it was found that any confinement of the charge decreases the critical diameter, while diffusion on the boundary was shown to increase d_f by up to a factor of three [22]. The limits for the critical diameter of a free gaseous explosion charge is illustrated in

Fig. 5. If the free charge is located on the ground as a semi-cylinder, the ground acts as a plane of symmetry, and the critical height of the semi-cylindrical charge is estimated to be $h_f > 10\lambda$.

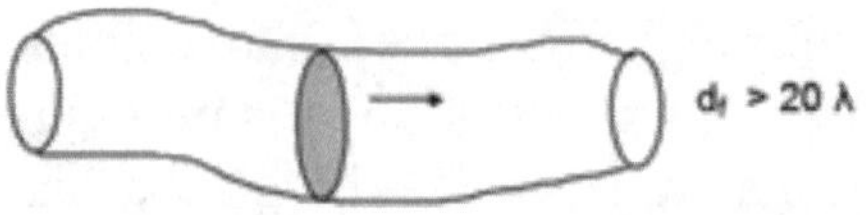

Figure 5. *Illustration of the limit for free gaseous charge diameter needed for detonation propagation in an unconfined cylindrical cloud.*

In addition to free unconfined gas clouds, the critical conditions for detonation initiation and propagation were also studied in heterogeneous fuel-air mixtures of gasoline and kerosene [24]. Semi-cylindrical clouds with volumes up to 1,500 m^3 and heights up to 8 m were created using a special dispersion system for the liquid fuels. The dispersion system was designed to create a uniform concentration distribution within the clouds, although concentration gradients were unavoidable at the cloud boundaries. A high explosive charge was then used to directly initiate detonations within the cloud, and it was found that the minimum height of the gasoline cloud that could support detonation propagation was about 2.5 – 3 m in these tests. The detonation cell size for the gasoline-air mixtures created in these experiments were later measured and found to be roughly 0.05 m [25]. For these tests, the critical diameter of the semi-cylindrical charge can be estimated as $d_f \sim 50\lambda$. The increased critical height compared to previous studies, such as [19], can be attributed to the diffusion layer on the cloud boundary, suggesting that $d_f \sim 10\lambda$ is a conservative estimate for the minimum cylindrical cloud height.

High aspect ratio unconfined explosive gas layers have also been studied [20, 23]. In these studies, it was found that the critical height of the free layer is $h_f \approx d_f/2 \approx 10\lambda$. As the aspect ratio of the layer cross section approaches unity, the value of h_f approaches that of a free cylindrical charge, $h_f \approx d_f \approx 20\lambda$. If the cloud were located on the ground, this corresponds to a critical height of the free layer being within $h_f \approx 5 - 10\lambda$, depending on the aspect ratio of its cross section. This is illustrated in Fig. 6. It should be noted, that these values are higher than those obtained in a recent study conducted in propane-air mixtures that estimated the critical height to be 3λ [26]. In those experiments, however, the initial mixture was contained using a plastic film, and it is probable that the lower value of the critical height obtained was entirely due to the strong effect of confinement that has been demonstrated in the earlier studies [20, 21, 22, 23].

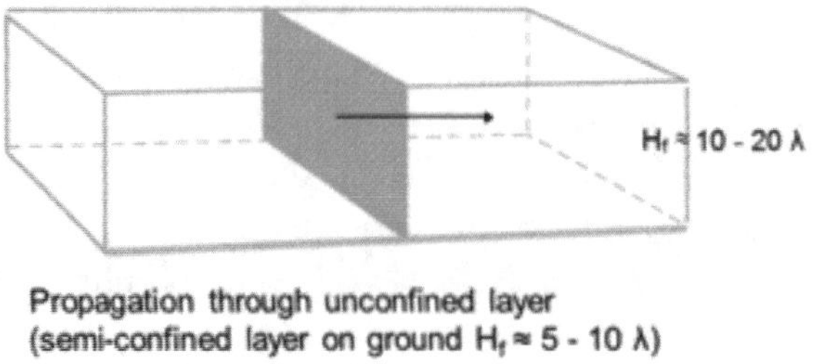

Figure 6. *Illustration of the height limit for detonation propagation in a free gaseous layer.*

2.4 Non-uniform mixtures

The limits for detonation propagation described above were all obtained for well mixed uniform gaseous mixtures. For a cloud generated by the accidental release of a large quantity of flammable material, the fuel distribution will be highly non-uniform, and this fact imposes additional limitations on the potential for detonation propagation. One of these limitations considers detonation propagation through a mixture with decreasing reactivity (or reaction rate). This situation occurs, for example, when a detonation propagates from a nearly stoichiometric mixture to a leaner or richer mixture. It has been shown theoretically, numerically [27], and experimentally [28], that, if the gradient of reactivity is too sharp, the detonation will decay to a deflagration. This critical reactivity gradient can also be expressed in terms of a gradient in detonation cell size, which has been found to be approximately $d\lambda/dx \approx 0.1$. This criterion is equivalent to saying that, if the cell width increases by more than 10% over the course of propagating a single cell length, the detonation wave will fail. An illustration of a decaying detonation wave propagating through a layer with decreasing reactivity, obtained experimentally, is shown in Fig. 7.

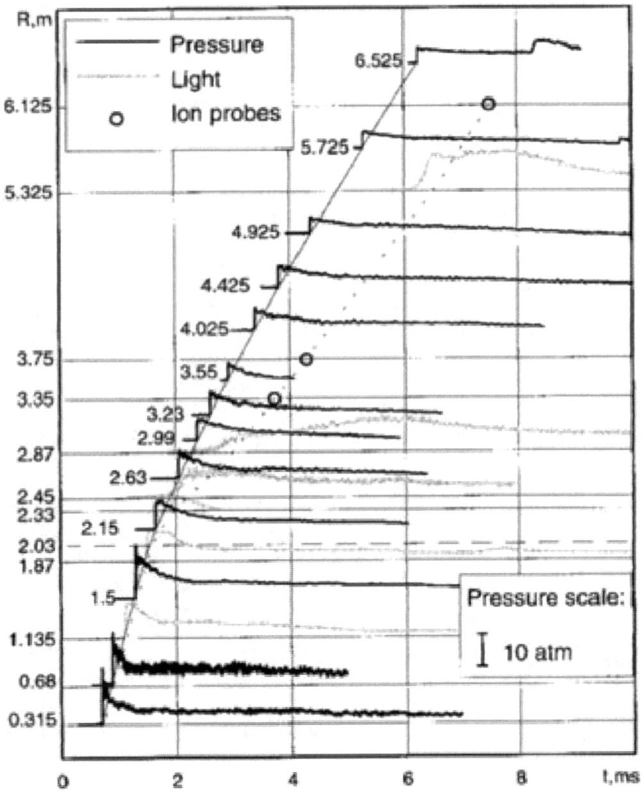

Figure 7. *Diagram of decaying detonation propagating through a concentration gradient* [28] *showing pressure transducers readings at different distances, R, along the detonation tube (bold lines), and the trajectory of the leading shock (thin line), flame trajectory (dotted line), and the center of concentration gradient layer (dashed line).*

In addition to the direct effect of concentration gradients on the ability of a detonation to propagate, non-uniform concentrations also present a number of challenges on how to define the critical layer height limit of an unconfined gaseous mixture. There is the question of how to define the layer thickness to consider when concentration varies at the boundary of the cloud. Furthermore, there is the challenge of specifying the effective reactivity, or cell size, of a non-uniform mixture. While there are no experimental data available that can be directly applied, the following two assumptions can be made to address this problem. The first is to assume that the critical reactivity gradient, $d\lambda/dx \approx 0.1$, represents a boundary in the non-uniform cloud where detonations can propagate in the region where $d\lambda/dx < 0.1$ and detonations would decay in the region where $d\lambda/dx > 0.1$. Thus, these locations form a virtual boundary of the detonable cloud from the standpoint of the reactivity gradient. One can then consider only the thickness of the flammable mixture layer between the critical gradients that occurs on the lean and rich sides of stoichiometry, h_g.

Having defined the effective thickness of the flammable layer within the critical gradients, one then needs to evaluate whether it is sufficiently large for detonation propagation. To do this, the number of detonation cells that fit within the layer height must be estimated. While this is straightforward for uniform mixtures, where λ is constant and the number of cells within the layer simply given by $n_\lambda = h_g/\lambda$, in a non-uniform mixture the cell size varies with height. Instead, the number of detonation cells that fit within the layer height can be estimated as:

$$n_\lambda = \int_0^{h_g} \frac{1}{\lambda(h)} dh \,. \tag{1}$$

Applying the critical layer height criterion described earlier, the critical value for n_λ will vary from 5 to 10, depending on the location of the layer with respect to the ground. For the purpose of the analysis below, a threshold of $n_\lambda = 5$ will be used as a conservative estimate for this value. This definition is equivalent to the critical conditions for uniform mixtures (Fig. 6) and keeps the number of transverse waves across a layer, which are essential to support the detonation, the same between uniform and non-uniform mixtures.

3 Application of Detonation Limits to Representative Clouds

The work described above can be directly applied to estimate the detonable mass of fuel for a given concentration distribution. In this section, a simple cloud dispersion model is used to evaluate the implications of these detonation propagation limits over a range of configurations. Two fuels are examined, propane-air mixtures, representative of a typical straight chained hydrocarbon, and ethylene-air mixtures, representative of a more reactive mixture with increased sensitivity to detonation and a higher likelihood of undergoing DDT. For this study, the effect of concentration distribution, cloud size, and the level of turbulent perturbation is investigated.

3.1 Flammable cloud concentration distribution

The cloud compositions considered for this work assumes a Gaussian distribution, where the fuel concentration decays from a maximum value, c_{max}, in the cloud center, to a maximum cloud height and radius defined by the LFL concentration, c_{LFL}. Using these assumptions, the fuel concentration at any height, h, and radius, r, is given by:

$$c(r,h) = c_{max}\left(e^{-\frac{r^2}{\sigma_r^2}} \cdot e^{-\frac{h^2}{\sigma_h^2}}\right), \tag{2}$$

where the values of σ_r and σ_h are:

$$\sigma_r = \left(-\frac{r_{LFL}^2}{ln\left(\frac{c_{LFL}}{c_{max}}\right)}\right)^{1/2}, \tag{3}$$

$$\sigma_h = \left(-\frac{h_{LFL}^2}{ln\left(\frac{c_{LFL}}{c_{max}}\right)}\right)^{1/2}, \tag{4}$$

such that the concentration is equal to c_{LFL} at the maximum cloud height, h_{LFL}, and radius, r_{LFL}. While the Gaussian distribution has a tail of infinite length, it has a finite integral, and the total mass of fuel within the cloud, m_{tot}, can be calculated:

$$m_{tot} = \iiint \rho_f \, c(r,h)dV = \frac{\rho_f c_{max} \pi^{\frac{3}{2}} \sigma_r^2 \sigma_h}{2} \tag{5}$$

where ρ_f is the fuel density at atmospheric conditions. Using this distribution, representative clouds were generated in 2D, assuming circular symmetry, on a 2.5 cm resolution grid, which was found to be sufficient to converge the results.

To estimate the detonation cell size, as a function of mixture composition, experimental values for λ were obtained from the CALTECH Detonation Database [29]. These values were fitted using a cubic spline interpolation, as shown in Fig. 8, to ensure a smooth variation of the computed cell size gradients.

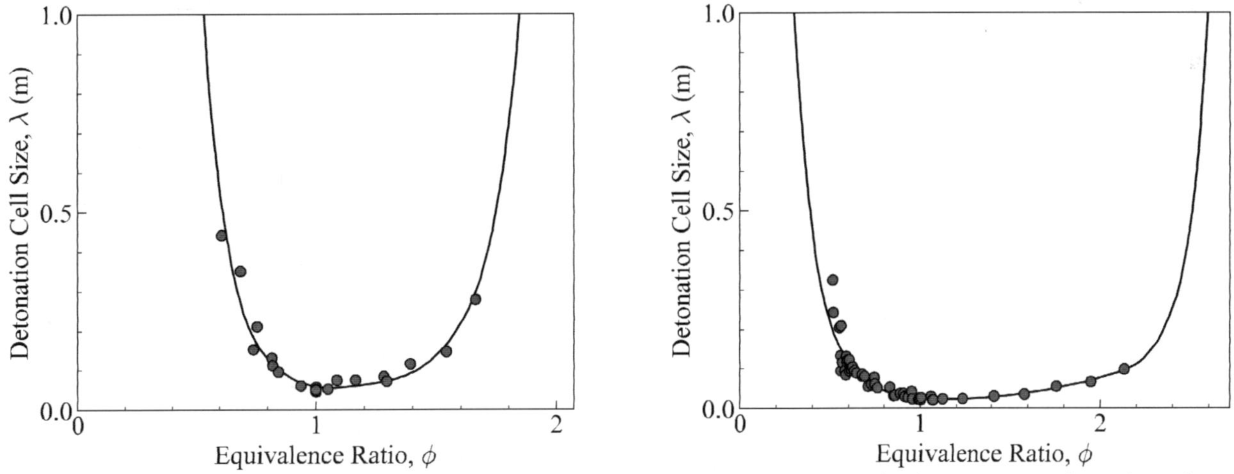

Figure 8. *Detonation cell size fits for propane-air (left) and ethylene-air (right) mixtures, based on data from the CALTECH Detonation Database.* [29]

An example of the results of this model is shown in Fig. 9 for an 80 m diameter, 4 m tall, propane-air cloud, with a center concentration, c_{max}, equal to the upper flammable limit, c_{UFL}, of propane. From top to bottom, this figure shows: the spatial distribution of concentration, with the stoichiometric concentration, c_{st}, in black; detonation cell sizes, in meters; dimensionless cell size gradients; and the portion of the cloud that is within the detonable limits described above.

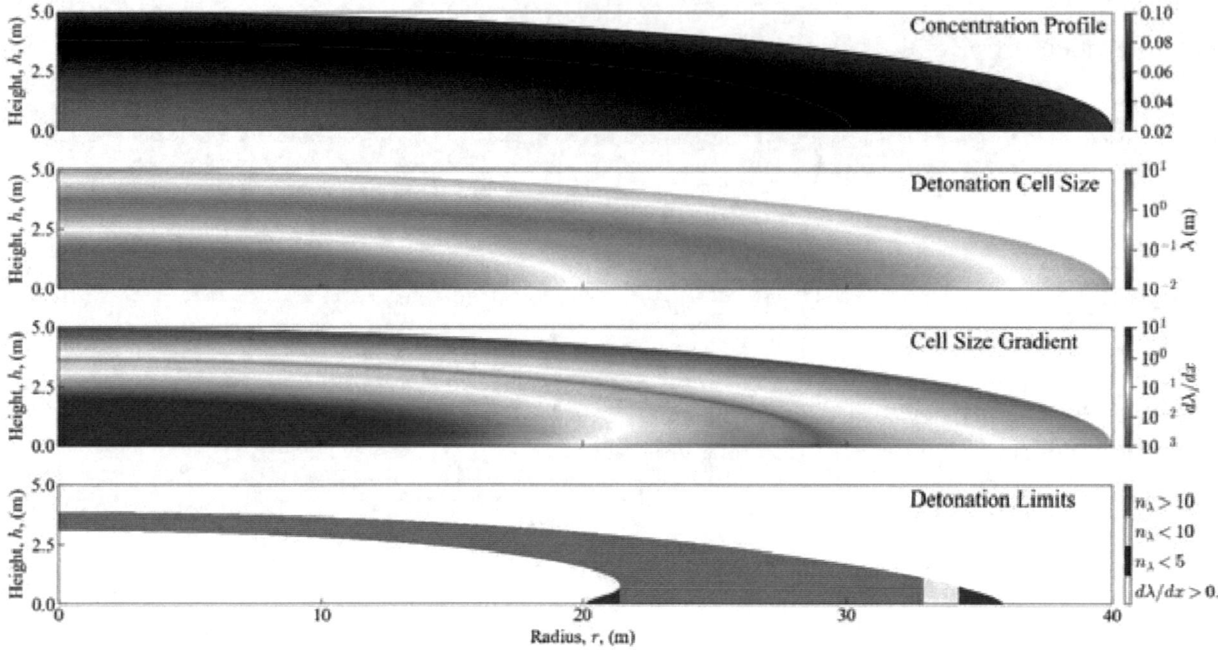

Figure 9. *Sample results for an 80 m diameter and 4 m tall propane-air cloud where* $c_{max} = c_{UFL}$.

For the plot of detonation limits; the white area shows the region where the cell size gradient $d\lambda/dx > 0.1$, which the prior studies suggest is too steep to support the propagation of a detonation; the blue area shows regions where the layer height is less than 5 detonation cells, where a detonation is unlikely to propagate; the yellow area shows the region where the layer height can fit between 5 – 10 cells, and a detonations propagate; and the red area shows the region where more than 10 detonation cells can fit into the layer and a detonation is more likely to propagate. For the summary plots shown in this paper, the region where $n_\lambda > 5$ will be considered to be within the detonable limits, a conservative assumption, that would be appropriate for regions close to the ground surface.

Figure 9 clearly shows that the gradient and cell size limits significantly reduce the detonable region, even for a relatively large cloud. In general, most of the detonable region is located on the rich side of stoichiometry, primarily due to stronger cell size gradients on the lean side. Nevertheless, near the center of the cloud only a thin strip of mixture would be expected to participle in a detonation event for this concentration profile.

3.2 Effect of concentration distribution

In this dispersion model, the center concentration of the cloud has a strong impact on the character of the release and the mass of fuel that can be consumed in a deflagration. This effect is even more pronounced for the detonable limits, as strong concentration gradients lead to steep cell size gradients, reducing the detonable range. As a result, the total mass of fuel in the flammable cloud that could be consumed in a detonation, m_{det}, is a strong function of the fuel concentration at the center of the cloud.

It is also important to account for the fact that, in fuel rich regions, the full mass of fuel is not consumed. In these regions, the mass of fuel that can be consumed in a deflagration, m_{def}, is limited by the remaining oxidizer and is estimated as:

$$dm_{\text{def}} = \rho_{\text{f}}\left(\frac{1 - c_{\text{f}}}{1 - c_{\text{st}}}\right) c_{\text{st}} dV$$

where c_{f} is the local fuel concentration. An equivalent procedure is used to account for incomplete combustion when estimating the detonable mass, m_{det}.

To illustrate the effect of varying c_{max}, the portion of the flammable cloud that can be consumed following a deflagration, $m_{\text{def}}/m_{\text{tot}}$ is shown in the left panel of Fig. 10. It is important to note, that because a Gaussian distribution is used, $m_{\text{def}}/m_{\text{tot}}$ is independent of cloud geometry and only depends on the center concentration of the cloud. The detonable mass on the other hand, is sensitive to the size of the cloud, both because the cell size criterion requires a certain physical layer height, and because the cell size gradient decreases as the cloud size increases. This is shown in Fig. 10, right panel, for three different cloud sizes, where it is clearly seen that smaller clouds reduce the mass fraction of fuel that is within the detonable limit. This figure also illustrates how the increased sensitivity to detonations of ethylene-air mixtures results in a much larger portion of the cloud remaining within the detonable limits for the same size and concentration distribution.

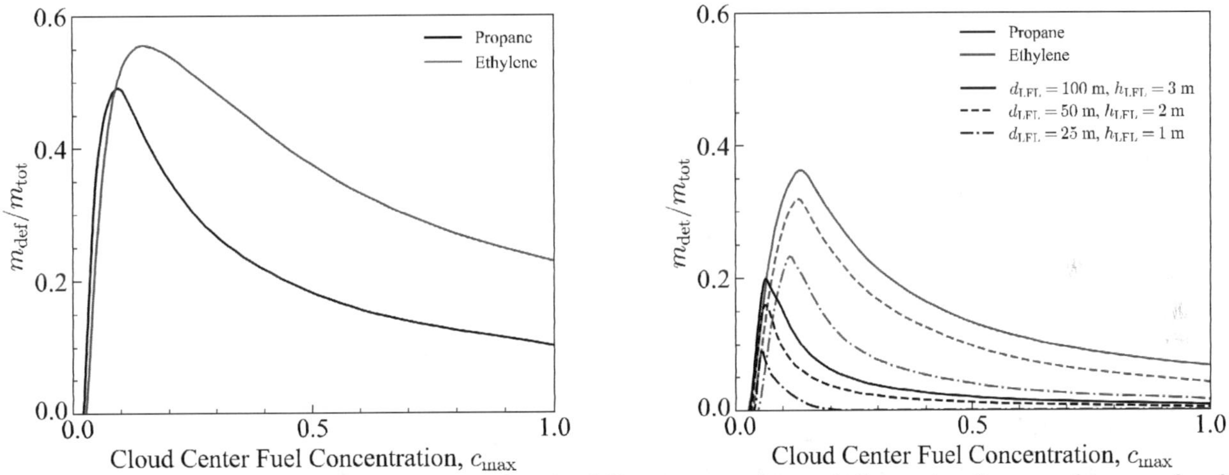

Figure 10. *The portion of the total mass that could be consumed in a deflagration for an arbitrary cloud (left), and the portion of the total mass within the detonable range for three cloud sizes (right), as a function of the cloud center fuel concentration.*

In order to interpret these results, it is important to consider that a real release will not be static, and for a given mass of fuel the cloud will grow in size over time, reducing the concentration at the cloud center. That being said, for ease of comparison in the following sections, it is necessary to select a common concentration distribution. Given that the center cloud concentration that maximizes the detonable fraction varies with cloud size, the cloud center concentration that maximize the total fuel that could be consumed in a deflagration will be used for the following sections. These values correspond to $c_{\text{max}} = 0.096$ for propane-air mixtures and $c_{\text{max}} = 0.146$ for ethylene-air mixtures and maximizes the total energy released for a given mass of fuel. It should be acknowledged, however, that this is close to the worst-case scenario, and is a conservative assumption, as this would assume that ignition and DDT were to occur at the worst possible time in the release.

3.3 Effect of cloud diameter and height

Figure 11 illustrates how the model responds to changes in the height and diameter of the cloud. It can be seen that the amount of fuel within the detonable limits increases monotonically with cloud size, with a stronger initial sensitivity to height than diameter, as the height tends to limit both the layer thickness criteria, as well as the gradient in cell size. For a 100 m diameter cloud, this effect greatly diminishes at a height of approximately 3 m in propane-air mixtures and 1 m in ethylene-air mixtures, where the curve flattens out. These results again demonstrate how a release of ethylene present a much more severe detonation hazard, particularly for shorter cloud heights, having considerably more fuel in the detonable range for the same cloud dimensions.

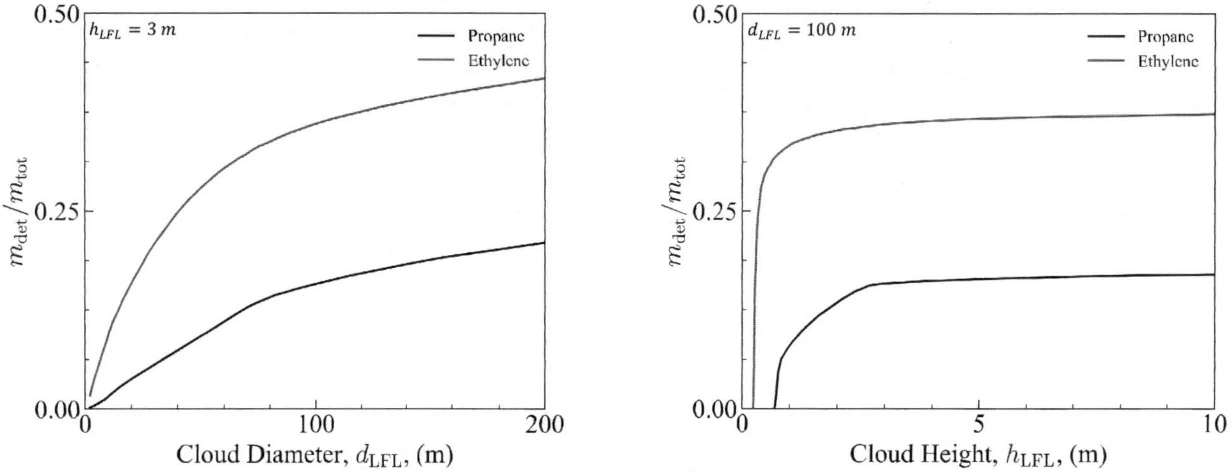

Figure 11. *The portion of the cloud mass that is within the detonable limits as a function of cloud diameter for a 3 m tall cloud (left) and as a function of cloud height for a 100 m diameter cloud (right).*

3.4 Effect of turbulent perturbations

The previous discussion considers perfectly smooth concentration gradients that are not a realistic configuration for a real release. The closest analogue would be a long duration release where the dispersion is primarily controlled by molecular diffusion in perfectly still conditions. In reality, however, the concentration distribution within the cloud will be heavily affected by: the nature of the release itself, particularly when the source is a high pressure jet; ambient conditions; and terrain effects. To illustrate the effect of modest turbulent mixing, spatially correlated random perturbations were superimposed on the concentration field, containing a spectrum of length scales consistent with turbulence, from 0.05 - 0.5 m. The overall strength of these disturbances is quantified by the standard deviation of the turbulent concentration field compared to their unperturbed values.

Figures 12–14 illustrate the effect of these turbulent perturbations for the same scenario previously examined in Fig. 9. From these results, it can be seen that turbulent perturbations generate significantly stronger gradients, which greatly reduce the mass of fuel within the detonable range. Furthermore, as the intensity of these perturbations increase, the detonable regions are no longer contiguous. This strongly suggests that, in many real releases, a detonation involving of all of the fuel within the detonable limits is unlikely to occur, and would be isolated to individual pockets that are adjacent to the region where DDT occurs. These results are consistent with observations of real world events, where small regions of high pressure damage have been observed away from congestion [2], consistent with a small region of fuel being consumed rapidly.

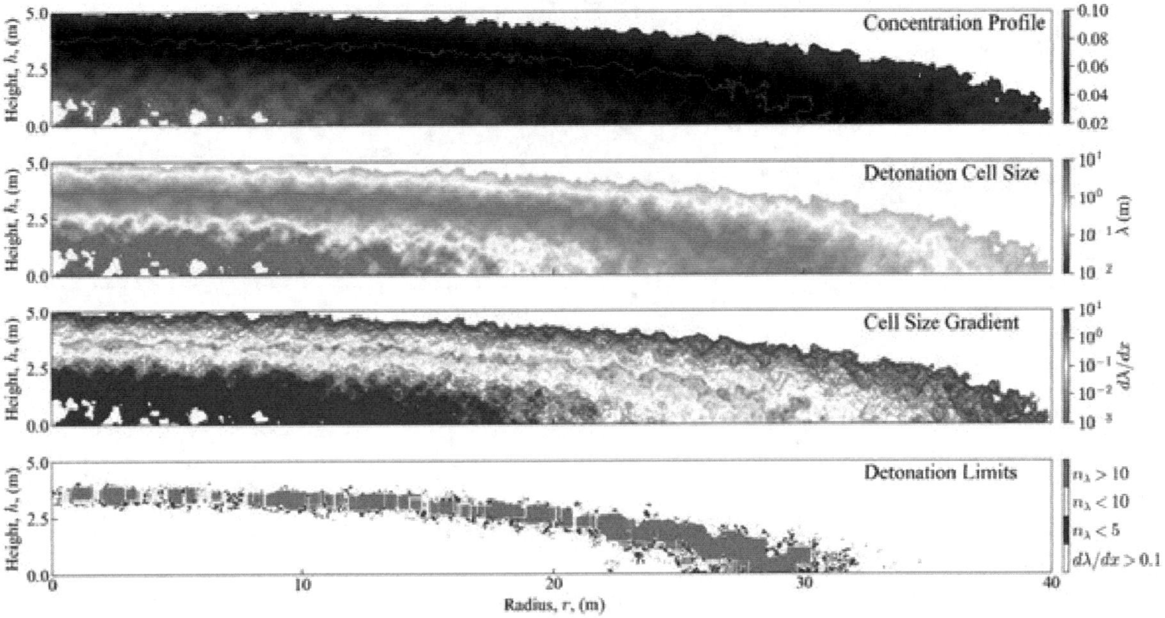

Figure 12. *80 m diameter and 5 m tall propane-air cloud, turbulent perturbation amplitude of 5%.*

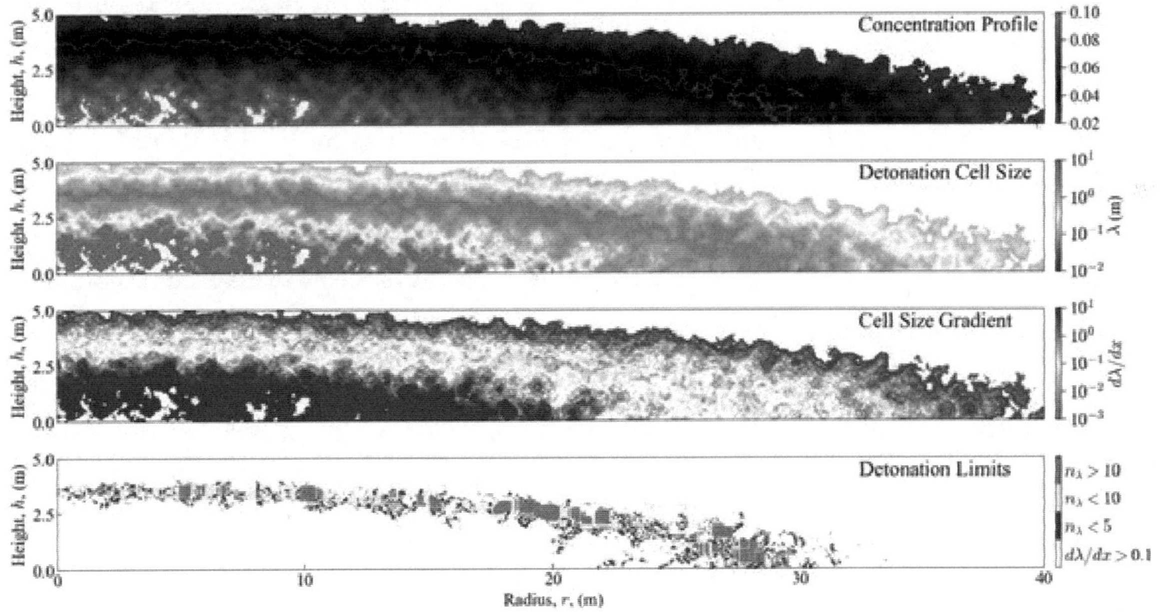

Figure 13. *80 m diameter and 5 m tall propane-air cloud, turbulent perturbation amplitude of 7.5%.*

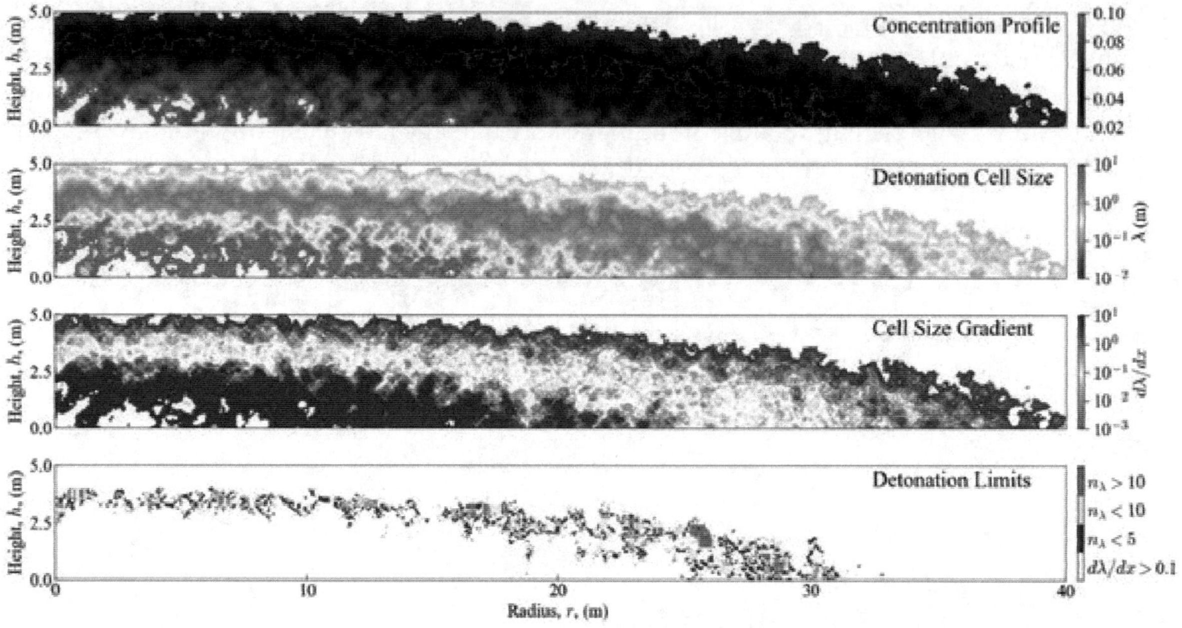

Figure 14. *80 m diameter and 5 m tall propane-air cloud, turbulent perturbation amplitude of 10%.*

The effects of these turbulent perturbations are summarized in Fig. 15. Due to the stochastic nature of these turbulent perturbations, the value for each amplitude was averaged over 100 repeated calculations. It can be clearly seen that the mass of fuel within the detonable range decreases significantly when turbulent perturbations are introduced. These results suggest that the type of release, and ambient wind conditions, plays a critical role in determining the mass of fuel within the detonable range.

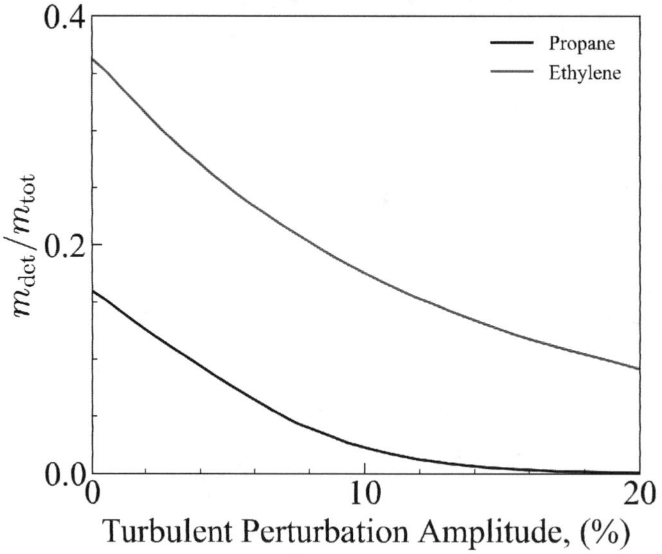

Figure 15. *The portion of the cloud mass that is within the detonable range for a 100 m diameter and 3 m tall cloud, as a function of turbulent perturbation amplitude for propane-air and ethylene-air mixtures.*

3.5 Uncertainty of the results

While uncertainty exists in both the critical cell size gradient as well as the critical layer height, the primary source of uncertainty is in the experimental determination of cell size, as any errors in these values propagate through to both of these criteria. To illustrate the magnitude of this effect, Fig. 16 shows the uncertainty bands for a factor of two variation of detonation cell size, in each direction, for a range of cloud center concentrations of the largest cloud size shown in Fig. 10. These error bands are conservative, as they show the effect of both a factor of two increase and decrease of the detonation cell size, a factor of four difference between the largest and smallest sizes, compared to the more commonly accepted experimental uncertainty of a factor of two. When considering the error band as a fraction of the baseline values, the uncertainty is largest for mixtures far from the center concentration that produces a maximum detonable mass, where the uncertainty is roughly $\pm 30\%$ of the baseline value.

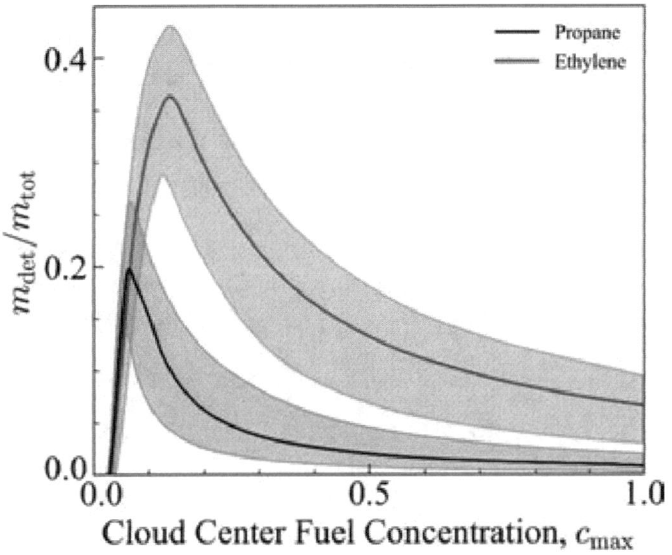

Figure 16. *The portion of the total mass within the detonable range for a 100 m diameter, 3 m tall cloud, as a function of the cloud center fuel concentration, with error bands showing the effect of a factor of two larger, and smaller detonation cell size.*

It is important to emphasize that the results presented in this work are based on the detonation limits described in the literature cited, much of which was obtained at a significantly smaller scale than a typical vapor cloud explosion and some for a different geometry. While those studies clearly demonstrate that these limits exist, it is conceivable that the values for these limits may vary significantly from those studies, in either direction, at the scale of a vapor cloud explosion. As a result, large-scale experimental validation of these limits is essential to more accurately quantify the potential consequences of detonations in a vapor cloud explosion, given a DDT event. Nevertheless, when considering the historical accidents where DDT has recently been proposed to have occurred, and the extreme levels of damage that would have been generated had a significant portion of the vapor cloud detonated, it is likely that these criteria provide a conservative estimate for the detonable portion of the overall cloud.

4 Conclusions

This work reviews and summarizes previous studies that have been performed describing the inherently unstable nature of freely propagating detonation waves. Those studies have clearly demonstrated that the ability of a detonation wave to propagate has additional limits beyond the conventional "detonable concentration range" concept that has been commonly adopted for industrial safety. Those studies emphasize the role of the detonation cell size as a measure for the ease at which a deflagration can transition to a detonation, as well as the mixture's ability to sustain the propagation of a detonation wave. For the detonation propagation limits, the prior studies have identified constraints on both the critical number of cells that must fit within the detonable layer, as well as the critical concentration gradient through which a detonation wave can survive. Using these limits, it is possible to estimate the portion of the cloud that could potentially participate in the event of a detonation.

To illustrate how this can be performed, a simple Gaussian concentration distribution was used to examine how cloud concentration distribution, size, and the level of turbulent perturbations affect these limits. It was found that the specific concentration distribution plays a strong role in the mass of fuel that could potentially participate in a detonation and, as clouds get larger, a larger fraction of the cloud will be within the detonable limits. It was also shown how these limits are extremely sensitive to the concentration distribution within the cloud, and how a uniform concentration field would result in an extremely conservative assumption. In addition, it was demonstrated how the presence of turbulent perturbations can significantly reduce the mass of fuel within the detonable limits, and, more importantly, isolate the detonable regions from one another, limiting the detonation to a smaller pocket of fuel.

This work demonstrates how the detonation propagation limits can be readily incorporated into existing dispersion models, in order to estimate the maximum quantity of fuel that could potentially be involved in the event of a deflagration-to-detonation transition during a large unconfined vapor cloud explosion. Nevertheless, additional experimental work is essential to better quantify these detonation limits at larger scale.

References

[1] D.M. Johnson and V.H.Y. Tam. "Why DDT is the Only Way to Explain Some Vapor Cloud Explosions," Process Safety Progress, Vol. 36(3) pp. 292-300, 2017.

[2] G. Atkinson, J. Hall, and A. McGillivray, "Review of Vapour Cloud Explosion Incidents," Health and Safety Executive, Research Report RR1113, 2017.

[3] G. Chamberlain, E. Oran, and A. Pekalski. "Detonations in Industrial Vapour Cloud Explosions," Proceedings of the 12th International Symposium on Hazards Prevention and Mitigation of Industrial Explosions, Kansas City, MO, USA, August 12-17, 2018.

[4] S.G. Davis, J.L. Pagliaro, D. Botwinick, D. Engel, K. van Wingerden, E. Merilo, A. Ziemba and M. Groethe. "Large Scale Testing Confirms: A Deflagration to Detonation Transition needs to be Considered When Facility Siting," AIChE 13th Global Congress on Process Safety, San Antonio, Texas, March 2017.

[5] G. Ciccarelli and S. Dorofeev. "Flame Acceleration and Transition to Detonation in Ducts," Progress in Energy and Combustion Science, Vol. 34(4), pp. 499-550, 2008.

[6] J.H.S. Lee. "The Detonation Phenomenon," Cambridge University Press, Cambridge, UK, 2008.

[7] The Netherlands Organization for Applied Scientific Research (TNO), TNO yellow book, 1st ed., TNO, Apeldoorn, The Netherlands, 1997.

[8] Q.A. Baker, M.J. Tang, E.A. Scheier, and G.J. Silva, "Vapor Cloud Explosion Analysis," Proc. 28th AIChE Annual Loss Prevention Symposium on Risk Analysis and Process Safety Management, Atlanta, GA, 1994.

[9] W.P.M. Mercx and A.C. van den Berg, "The Explosion Blast Prediction Model in the Revised 'Yellow Book'," Proc. 31st AIChE Annual Loss Prevention Symposium, Houston, TX, March 10-13, 1997.

[10] S.B. Dorofeev, A.A. Efimenko, A.S. Kochurko, and B.B. Chaivanov. "Evaluation of the Hydrogen Explosions Hazard," Nuclear Engineering and Design, Vol. 148(2-3), pp. 305-316, 1994.

[11] J.J. Erpenbeck. "Stability of Steady State Equilibrium Detonations," Physics of Fluids, Vol. 5(5), pp. 604-614, 1962.

[12] D.R. White. "Turbulent Structure of Gaseous Detonations," Physics of Fluids, Vol. 4(4), pp. 465-480, 1961.

[13] Y.N. Denisov and Y.K. Troshin. "On the Mechanism of Detonative Combustion," Proceeding of the Combustion Institute, Vol. 8(1), pp. 600-610, 1961.

[14] Z. Liang, and L. Bauwens. "Detonation Structure with Pressure-Dependent Chain-Branching Kinetics." *Proceedings of the Combustion Institute* 30(2), pp. 1879-1887, 2005.

[15] M. Kuznetsov, G. Ciccarelli, S. Dorofeev, V. Alekseev, Yu Yankin, and T.H. Kim. "DDT in Methane-Air Mixtures," Shock Waves 12(3), pp. 215-220, 2002.

[16] J.H.S. Lee. "Dynamic Parameters of Gaseous Detonations," Annual Review of Fluid Mechanics, Vol. 16(1), pp. 311-336, 1984.

[17] Y.B. Zeldovich, S.M. Kogarko, and N.N. Simonov. "An Experimental Investigation of Spherical Detonation in Gases," Soviet Physics: Technical Physics, Vol. 1, pp. 1689-1713, 1956.

[18] V.V. Mitrofanov and R.I. Soloukhin. "The Diffraction of Multifront Detonation Waves," Soviet Physics: Doklady, Vol. 9, pp. 1055, 1964.

[19] R. Knystautas, J.H.S. Lee, and C. Guirao. "The Critical Tube Diameter for Detonation Failure in Hydrocarbon-Air Mixtures," Combustion and Flame, Vol. 48, pp. 63-83, 1982.

[20] A.A. Vasil'ev. "Critical Diameter of Gas-Mixture Detonation," Combustion, Explosion, and Shock Waves, Vol. 18(3), pp. 349-355, 1982.

[21] S.B. Murray and J.H.S. Lee. "The Influence of Yielding Confinement on Large-Scale Ethylene-Air Detonations" Progress in Aeronautics and Astronautics, Vol. 94, pp. 80-103, 1984.

[22] A.A. Vasil'ev and D.V. Zak. “Detonation of Gas Jets,” Combustion, Explosion, and Shock Waves, Vol. 22(4), pp. 463-468, 1986.

[23] A.A. Borisov, S.V. Khomik, and V.N. Mikhalkin. “Detonation of Unconfined and Semiconfined Charges of Gaseous Mixtures,” Progress in Aeronautics and Astronautics, Vol. 133, pp. 118-132, 1991.

[24] V.I. Alekseev, S.B. Dorofeev, V.P. Sidorov, and B.B. Chaivanov. "Experimental study of large-scale unconfined fuel spray detonations," Progress in Astronautics and Aeronautics, Vol. 154 pp. 95-95, 1993.

[25] V.I. Alekseev, S.B. Dorofeev, and V.P. Sidorov. "Direct Initiation of Detonations in Unconfined Gasoline Sprays," Shock Waves Vol. 6(2), pp. 67-71, 1996.

[26] “Dispersion and Explosion Characteristics of Large Vapour Clouds - Volume 1: Summary Report,” Steel Construction Institute, SCI Document ED023, 2014.

[27] S.I. Kryuchkov, S.B. Dorofeev, and A.A. Efimenko. “Critical Conditions for Detonation Propagation through Mixture with Decreasing Reaction Rate,” Proceeding of the Combustion Institute, Vol. 26(2), pp. 2965-2972, 1996.

[28] M.S. Kuznetsov, V.I. Alekseev, S.B. Dorofeev, I.D. Matsukov, and J.L. Boccio. “Detonation Propagation, Decay, and Reinitiation in Nonuniform Gaseous Mixtures,” Proceeding of the Combustion Institute, Vol. 27(2), pp. 2241-2247, 1998.

[29] M. Kaneshige and J.E. Shepherd. “Detonation Database,” GALCIT Technical Report FM97, available at http://shepherd.caltech.edu/detn_db/html/db.html, accessed on January 21, 2019.

Applying Risk Principles to Combustible Dust Operations at a Multinational Manufacturer

Jeremy Lebowitz, PE
Jensen Hughes, Inc.
1661 Worcester Rd, Framingham, MA 01701
jlebowitz@jensenhughes.com

Kamal Aljazireh, PE
Jensen Hughes, Inc.
600 W Fulton St, Suite 500, Chicago, IL 60661
kaljazireh@jensenhughes.com

Chris Unangst, CSP
Jensen Hughes, Inc.
500 Piedmont Rd NE, Suite 750, Atlanta, GA 30305
cunangst@jensenhughes.com

Marc Hodapp, PE
Jensen Hughes, Inc.
3610 Commerce Dr, Suite 817, Baltimore, MD 21227
mhodapp@jensenhughes.com

Prepared for Presentation at
American Institute of Chemical Engineers
2019 Spring Meeting and 15th Global Congress on Process Safety
New Orleans, LA
March 31 – April 3, 2019

Applying Risk Principles to Combustible Dust Operations at a Multinational Manufacturer

Jeremy Lebowitz, PE, Kamal Aljazireh, PE, Chris Unangst, CSP, Marc Hodapp, PE
Jensen Hughes, Inc.

Keywords: Combustible Dust Hazards, Hazard Identification, Risk Analysis and Regulations, Quantitative Risk Assessment (QRA)

Abstract

Jensen Hughes, Inc. performed a risk-based analysis for three combustible dust handling processes that are prevalent throughout our manufacturing client's enterprise. These systems were determined by our client to present relatively low risk compared to other combustible dust handling processes within the enterprise.

The client has expressed the goal to manage combustible dust hazards in accordance with the industry standard of care with the objectives of protecting life, property, and business continuity. The National Fire Protection Association (NFPA) standards on combustible dust outline several options for meeting these objectives. One approach is to apply prescriptive fire and explosion protection measures where combustible dust fire, flash fire, or explosion hazards are determined to exist. An alternative approach is to apply a risk-based methodology to determine appropriate safeguards for achieving acceptable risk. Naturally, the prescriptive approach is intended to be conservative and broadly applicable. When the prescriptive method is applied to the systems in question, substantial operational and capital resources would be required regardless of whether such measures contribute to an appreciable reduction in risk. A risk-based approach, on the other hand, provides a method for selecting optimized protection measures to achieve industry accepted levels of risk. A risk analysis may also illustrate that no further safeguards are necessary in situations where risk is already low enough to be considered broadly acceptable.

Using a risk assessment methodology, we were able to establish that through cost-effective means, most facilities could comply with industry risk tolerance thresholds, without meeting all prescriptive measures of the NFPA standards.

1. Introduction

Because of growing awareness of combustible dust hazards and broader adoption of regulatory requirements in recent years, new and existing manufacturing facilities are increasing efforts to implement hazard prevention and mitigation measures. As a starting point, facilities are expected to comply with current codes and standards that have been developed and enhanced over time. A common technique used in industry, known as a dust hazard analysis (DHA) [1], aims to identify combustible dust hazards present at a site, identify gaps in compliance with codes and standards, and determine solutions to reduce the risk associated with the identified hazards. Since compliance with prescriptive requirements implies that the facility has provided an acceptable level of safeguards or risk reduction measures, the DHA identifies areas for improvement where compliance gaps exist.

A challenge emerges where these deficiencies are difficult to reconcile, either due to operational conflicts or capital needs. For example, if a multinational manufacturing corporation conducts enterprise-wide dust hazard analyses, a deficiency in explosion protection for similar vessels handling combustible dusts across all manufacturing plants might be identified. Correcting these prescriptive deficiencies may be relatively easy on a small scale; however, "simple" upgrades replicated across several hundred facilities can require a vast capital investment. As a result, companies may not immediately implement safety measures, leaving personnel exposed to known hazards.

For situations such as the one described above, an alternate approach to prescriptive code compliance is to apply a risk-based methodology to determine optimized risk-reduction measures [2]. Our focus is narrow to risk of injury/fatality, consistent with the primary focus of NFPA standards. A risk assessment required for this approach should be a thorough engineering analysis that enables a deep understanding of the process elements contributing to higher risk levels. This method ultimately results in the determination of safeguards and upgrades that provide cost-effective reduction in overall risk and may also illustrate that no further safeguards are necessary in situations where risk is already low enough to be considered broadly acceptable. Such a tool can be valuable for a company to demonstrate its commitment to safety culture.

1.1 Objective

This paper focuses on the framework of the risk assessment approach and presents a methodology to developing an effective quantitative risk assessment (QRA) for similar combustible dust handling equipment or systems across multiple facilities. Through this approach, many of our client's facilities could make substantial gains in reducing exposures consistent with industry risk tolerance thresholds, without necessarily meeting all prescriptive measures of the NFPA combustible dust standards.

Prescriptive codes cannot address all possible equipment and process configurations (most notably in not accounting for variables such as equipment failure rate, process area occupancy and proximity to other processes); therefore, an allowance is provided for alternate approaches that provide comparable or superior performance. A risk-based approach allows for flexibility where the prescriptive method results in disproportionate risk reduction measures relative to generally accepted industrial and societal risk tolerance.

2. Risk Analysis Methodology

This section presents the basis for the first part of the risk assessment: developing a risk estimation method using engineering evaluations. This is also known as a risk analysis, or a study of risk associated with a set of activities or list of potential accident scenarios that can provide qualitative or quantitative results. Once the risk is estimated, it is compared to industry accepted levels in the second step of the risk assessment (summarized in Section 3 of this paper).

2.1 Basis for Risk Analysis and Selection

There are several methods that can be utilized to identify risk ranging from a simple safety review to a complex engineering analysis. Risk is a function of frequency and consequence. Our model analyzes each component separately. The characteristics of the system to be analyzed are used to develop the conditions (frequency) leading to a possible loss event and the measures to prevent or reduce the likelihood of these conditions, as well as the outcomes (consequences) from the top event and the measures to mitigate or reduce the severity of the consequences. The figure below can be used to illustrate this concept, with the left side representing frequency and the right side representing consequence:

Figure 1. Illustration of risk analysis components

A successful risk analysis requires adequate identification and documentation of risk information, high quality results which stakeholders and decisionmakers can easily apply and is completed efficiently [3]. An appropriate methodology and technique must be selected for the scope for a risk analysis to be successful. The process of selecting a technique is complex, with several factors influencing the decision, including:

- Motivation for the study (reason, method, and stakeholders)
- Type of target data needed by stakeholders
- Information and data available to perform the study
- Characteristics of the analysis problem, to ensure that the selected methodology can meet the needs of the desired outcomes
- Perceived risk associated with the subject processes, as processes with higher perceived risk should use risk analysis techniques that minimize the chance of missing an important incident scenario [3]
- Resource availability and management/analyst preference

2.1.1 Required Information for Effective Risk Analysis

The risk analysis should be applicable to processes both in the design phase or existing processes. To perform a quantitative risk assessment, precise data must inform the risk analysis, including:

- Material explosibility and fire properties (e.g. K_{St}, P_{max}, MEC, MIE, etc.)
- Process and plant data: process flow diagrams (PFDs), piping & instrumentation diagrams (P&IDs), design data, site plans, operational procedures, shutdown procedures, testing and maintenance procedures.
- Historical incident data at site or similar facilities, including near-misses
- Equipment reliability data from on-site records and/or industry data
- Human reliability data for routine and non-routine tasks

2.1.2 System Configurations

A schematic should be referenced to help the safety professional identify sequences of events, locations of safeguards, required inputs, and other elements that influence the risk analysis. To apply the risk analysis across multiple locations, the base system configuration must be relatively consistent; distinct risk analyses are necessary if there are significant system differences, (e.g. mechanical conveyance in place of pneumatic). However, several basic process characteristics (inputs) can be addressed by the risk analysis, such as pipe diameters or lengths, material flow rates, materials of construction, system uptime, presence of operators, and presence of building and equipment safeguards such as explosion protection and isolation systems.

2.1.3 Selecting a Risk Analysis Technique

For many complex processes, a quantitative risk assessment documents the necessary substantiated information, clearly defining the risk associated with a process. A common risk analysis methodology that can support a quantitative risk assessment combines two techniques: Fault Tree Analysis (FTA) and Event Tree Analysis (ETA). As presented in this paper, the FTA identifies initiating events that lead to an undesirable consequence (frequency of failure), and the ETA identifies resulting outcomes (severity of failure). Both components lead to the overall risk determination for a selected failure scenario.

The scope of the fault and event tree analyses is limited to top events that involve combustible dust fires, deflagrations, and explosions within those systems.

2.2 Development and Results of the Risk Analysis

The risk analysis methodology assesses risk associated with typical system arrangements, so that the method can be applied globally across a business enterprise, industry, or organization. The method itself can be broadly applicable, while still enabling a safety professional to determine site-specific risk.

To begin the analysis, fault and event trees were developed for the specified system configuration. The fault and event trees account for differences among similar systems as inputs to the method. For example, the same fault and event trees address comparable storage systems with pneumatic conveyance, with or without explosion isolation on the inlet conveyance pipe to a storage vessel.

The risk for each top event was calculated from the product of the fault tree and event tree outputs and combined with the risk of other top events of the system (as necessary).

2.2.1 Fault Tree Analysis – Frequency Component

Fault tree analysis was used to determine the number of occurrences per unit time of a top event, e.g., the number deflagrations within a storage vessel per year. The frequency of the top event is dependent on the frequencies of initiating and intermediate events. The frequency/likelihood of intermediate events was calculated by following the Boolean logic to manipulate the events that contribute to its occurrence. Frequency data for initiating events was compiled from sources such as operational or industry data. An example initiating event, a motor failure, was assumed to fail at a frequency of 0.1 per year (from literature operational data). The effect of preventive safeguards and condition modifiers (e.g., ignition source control, probability of ignition, robust preventative maintenance, etc.) was factored into the overall frequency of the top event.

In the fault tree, the top event is deduced until basic events are found. For the example in Figure 2 below, the components required for a deflagration were accounted for, and from there, each component was examined. Those components include an ignition source, combustible dust in suspension, and at a concentration at or above the Minimum Explosible Concentration (MEC). (Oxygen is assumed to be present where inerting systems are not used.) The basic events used in this methodology are functions of inputs, which is fundamental to applying the same method on a broad scale. For example, normal concentration is a function of dust loading rate (input) and volumetric air flow (input).

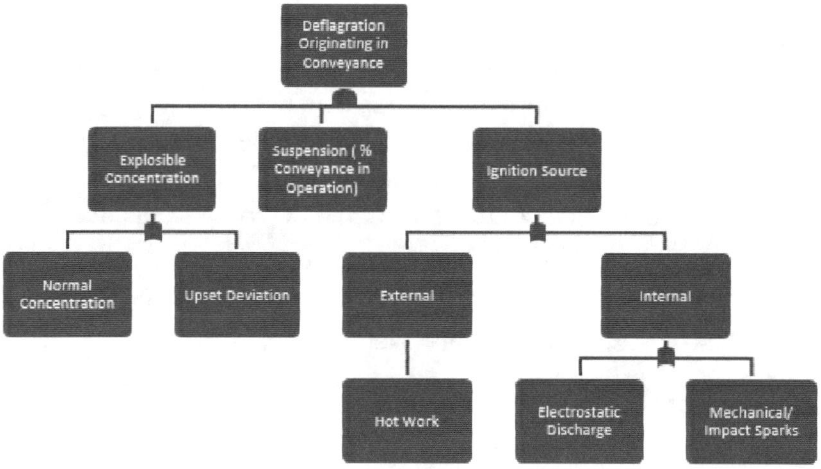

Figure 2 – Example Fault Tree for a Deflagration in Pneumatic Conveyance

2.2.2 Event Tree Analysis – Consequence Component

The consequence component of risk is addressed in the final step of the risk analysis, by identifying all consequences of a top event through an event tree analysis. The event tree analysis was used to determine consequence per event, e.g., the probability of a fatality per deflagration in a vessel. From the top event, branch points were developed for each layer of mitigative safeguards and conditional modifiers. Each branch point has two paths: one representing success of a safeguard (modifier best-case scenario), and the other representing failure of the safeguard (modifier worst-case scenario). Examples of safeguards include explosion protection, fire suppression, PPE, etc. An example of a conditional modifier is the probability of occupancy in the area affected by the top event.

Figure 3 below demonstrates the initial branch points of an event tree for a deflagration in a vessel (the entire event tree is not shown, as there could be a substantial number of branch points). Each branch point can include safeguards, such as explosion suppression and explosion isolation and conditional modifiers, such as location of the system and whether the area is occupied. Each branch point can either be a direct input (e.g. "Is vessel outdoors?" can either be "yes" or "no") or a function (e.g. a safeguard may have a probability of failure on demand). Figure 3 below illustrates how multiple possible conditions are captured; Safeguard 1 is deflagration suppression, Safeguard 2 is deflagration venting, and Safeguard 3 is explosion isolation on the vessel inlet side. While it is doubtful that the same vessel would have both suppression and venting, it is likely that both approaches are used within the same system across a client's portfolio of sites.

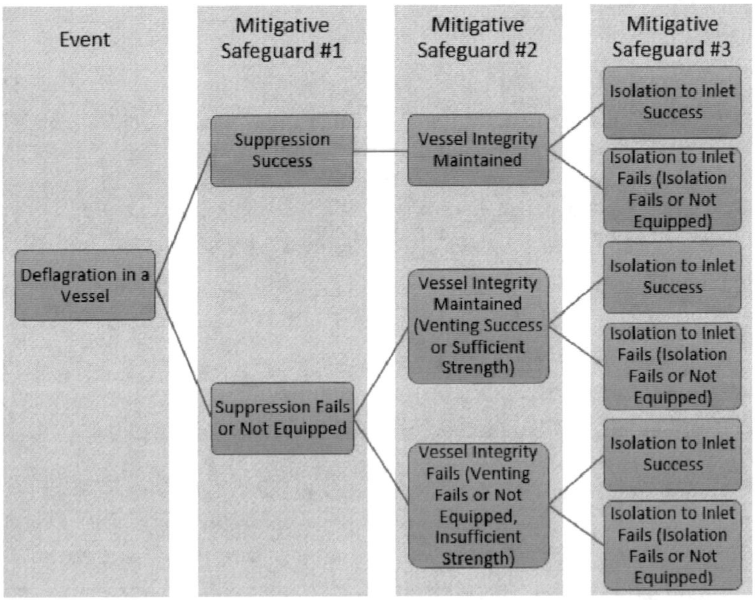

Figure 3 – Event Tree Example for a Deflagration in a Storage Vessel

2.2.3 Determining Risk and Developing User Interface

As discussed above, risk associated with specific system attributes is evaluated based on site-specific input parameters. Inputs provided by the user are fed into functions that provide data to the fault trees and event trees (e.g. frequency of ignition due to ESD or mechanical sparks, frequency of unit in operation, probability that settled material is present in ductwork, probability that an occupant is present in the process area, and any other function that would impact the frequency or consequence of an analyzed scenario).

Once functions for all fault trees and event trees were determined, a list of required inputs was assembled. All necessary input parameters and details that the functions and calculations use in determining probabilities or frequencies were included (see examples below). This list was converted into a site-specific checklist or questionnaire for ease of use by plant personnel.

- Process parameters, e.g. Does process use insulative hoses / couplings? (Yes/No); Days per year equipment is in operation (# days); Volume of each bag filter (ft^3)
- Equipment safeguards, e.g. spark resistant fans, explosion isolation, or bonding/grounding

- Administrative safeguards, e.g. housekeeping procedures, PPE requirements, and operational procedures

The fault trees and event trees were coupled with a user interface and developed in Microsoft Excel to facilitate ease of gathering input data. The user interface, in effect, accepts data and returns outputs with a level of risk that is represented by the product of top event frequency of fault trees and consequences described in event trees. This format ultimately results in a risk model that can be modified by changing user inputs and observing changes to risk levels in real-time.

2.3 Verification and Validation

As this is a complex model, the overall construct must be verified and validated to eliminate errors and ensure it accurately represent the system being evaluated.

The verification step is essential to debug the system, to address errors, and ensure that the method works as intended. To verify the outputs of the calculations, risk levels can be verified with outputs from commercially available risk analysis software.

To correctly validate the methodology, several iterations should be conducted to evaluate sensitivity to input parameters. The results of these iterations should be logical and align with expectations (e.g. providing adequate bonding and grounding in conveyance piping reduces the frequency of occurrence, does not impact consequence, and ultimately reduces the risk of injury/fatality associated with a system).

3. Evaluating Results and Risk Assessment

3.1 Risk Acceptance Criteria

3.1.1 Determine Tolerability of the Risk

Upon completion of the risk analysis model, the next step is to determine whether the existing conditions present a tolerable level of risk. A common technique used to determine risk tolerability is a risk matrix. The risk matrix is a qualitative approach that represents risk as a function of multiple levels of consequence and frequency; the output of the risk matrix indicates whether the evaluated event presents an unacceptable risk. The consequence and frequency levels are qualitative or semiquantitative in nature (e.g. a "Level 2" on the frequency scale may be an event that could occur once every 10 years). Applying a risk matrix can be challenging because it lacks the detail necessary to extract a sound basis for decision-making; slight changes to the process equipment that ultimately reduce risk may not be realized due to insensitive and coarse severity/frequency levels. Qualitative analysis also introduces subjectivity that can cause variation across sites. Therefore, a quantitative method to evaluating risk tolerability should be employed.

To establish risk acceptance criteria, a quantifiable risk level must be defined. Available sources of risk tolerance include statistical data from the target industry or comparable industries, or risk levels associated with common activities such as driving a car or rock climbing. By studying this available data, a "green line" can be determined which will identify a threshold level for acceptable risk.

While it is important to set a threshold for broadly acceptable risk, risk in a quantitative analysis represents a continuum reflective of site specific hazards, conditions, and safeguards. Therefore, structuring risk acceptance as a go-no/go condition can be problematic for decision making, especially when evaluating existing systems where certain risk reduction measures cannot be implemented feasibly. Rather, risk can be viewed as a continuous range that allows for the prioritization of the worst systems/conditions that pose the greatest risk. Systems that present risk marginally above the broadly acceptable risk criteria may be considered tolerable in some instances. The time, effort and capital required to reduce the system risk below the "acceptable line" should be considered in determining whether additional risk reduction measures should be taken to mitigate risk falling in the "tolerable range." In other words, what is the feasibility of required investments to implement risk reduction measures? Situations may arise where significant capital investments are required to reduce the risk level by a marginal amount. This concept of risk management is applied in industry standards, such as NFPA 59A, *Standard for the Production, Storage, and Handling of Liquid Natural Gas* and is often referred to as the "As Low as Reasonably Practicable" (ALARP) principle. Figure 4, which shows acceptability regions of societal risk [4], and Figure 5 illustrates the concept of unacceptable, tolerable, and broadly acceptable risk.

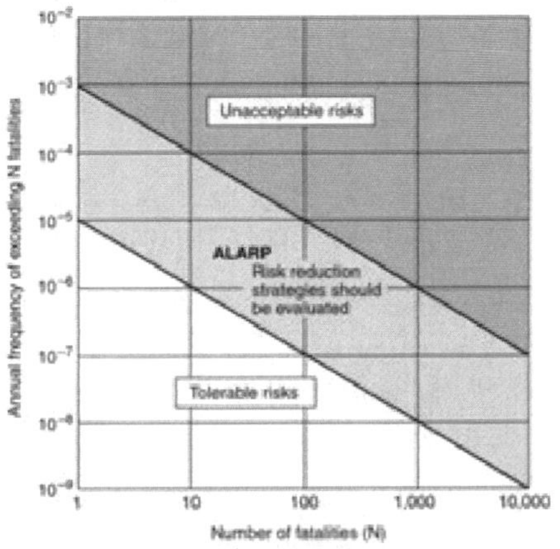

FIGURE 15.10.2 Acceptability Regions of Societal (Injury) Risk in the F-N Domain.

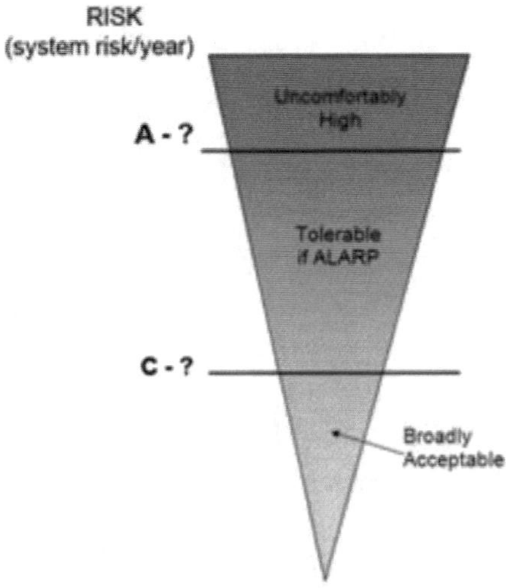

Figures 4 and 5 – Acceptability Regions of Societal Risk (Example from NFPA 59A) on left; Conceptual representation of acceptable, tolerable, and unacceptable risk range on right

In Figure 5, risk level "C" is defined as the threshold for risk acceptance (green line) determined through statistical analysis and collaboration with the involved parties. Risk level "A" sets the upper boundary for the tolerable risk range. Any higher risk is unacceptable.

The risk regions identified in Figure 5 are suggested to be applied as follows:

- Risk within the unacceptable or "red" region cannot be tolerated, and additional mitigation is required.
- Risk within the tolerable or "yellow" region requires further evaluation. Additional risk reduction measures should be applied to the extent practicable to reduce risk to broadly acceptable risk levels. Final acceptance is subject to approval by the authority having jurisdiction (AHJ).
- Risk below the risk acceptance threshold (i.e. below green line in the broadly acceptable "green" region) is acceptable and no further mitigation is necessary. Final acceptance is subject to approval by the AHJ.

3.1.2 Risk Acceptance Criteria

Risk acceptance in the scope of this analysis was focused on life safety. Accordingly, risk of injury or fatality resulting from the consequences associated with a combustible dust fire, flash fire, or explosion were evaluated. Risks associated with property damage, environmental impact, business continuity or reputational harm can also be assessed, but are not the primary goal of this assessment.

The life safety risk acceptance criteria that was used in this assessment is broadly acceptable based on the current industry standard of care. The first step in determining this level of risk was to conduct a statistical analysis to determine current risk levels in industry. Publicly available data from the United States Department of Labor – Bureau of Labor Statistics (BLS) and the Chemical Safety Board (CSB) was referenced. The BLS data search functions allow a user to extract compiled data related to labor in the United States, including injury and fatality numbers and rates for specific industries, while the CSB data provides additional insight into the nature of the incidents resulting in injuries or fatalities (e.g. fatalities related to combustible dust incidents). Based on statistical analysis, the individual injury and fatality rates can be found for a specific industry.

It is important to note that the major discrepancy between the two data sets (particularly in the injury rates) is due to the nature of the incidents that occurred. The BLS Data shows injuries that can include occupational hazards such as slips/trips/falls, pinch points, and electrical hazards, in addition to those from fires/explosions. Therefore, the injury and fatality rates are comparably high in the BLS data and are not appropriate for this risk assessment method. The higher calculated rates would result in a less conservative risk acceptance threshold. As a result, CSB data were used as the benchmark for this methodology.

Fatality statistics available for common nonindustrial activities were also reviewed. Most of the data was taken from the National Safety Council and based on U.S. statistics. The data showed that the individual risk (borne both voluntarily and involuntarily) associated with accidents occurring in daily routine activities is approximately 1×10^{-5} fatalities /year. Since these common activities are undertaken by most members of society in general, for the purposes of this study the risk associated with such activities is considered as a broadly acceptable level of risk.

3.1.3 Comparing the Risk Acceptance Criteria to Common Activities

The guidance provided by CCPS [5] and UK HSE [6] indicates that, for developing risk criteria, a "broadly acceptable" level of risk is typically considered by regulators as "comparable to those that people regard as insignificant or trivial in their daily lives." Thus, it should be considered acceptable when employees working in industrial facilities face on-the-job risks comparable to other activities undertaken in daily life or accidents caused by common events.

Selected thresholds determined for the analysis were consistent with the magnitude of acceptable risk specified in NFPA 59A (see Figure 4) for societal risk, which is more stringent than that allowable for losses limited to a facility and its employees, as is assumed in this analysis. When compared to the risk individuals are routinely willing to accept during common activities, it also must be evident that the selected risk acceptance criteria provide a conservative risk framework. The individual risks of death from traveling in a car is $1.7E^{-4}$, from rock climbing is 4E-5, and from being struck by lightning is 1E-7 [7]. The data show that the risk acceptance criteria framework applied in this analysis can result in at least an order of magnitude less risk than people are commonly willing to accept when participating in common activities.

3.2 Baseline Risk and Statistical Analysis

As noted throughout the paper, the methodology developed applies to like processes throughout a business enterprise, organization, or industry. Once the risk analysis methods were created, site-specific system configurations were validated, and site-specific input information was gathered. Each site's input information was entered to determine a baseline level of risk. These risk levels were plotted to determine the level of risk at each facility. Figure 7 in Section 3.4 is an example of how this summary may look for an organization that has dust handling equipment at 15 locations. The figures are scatter plots of the risk of injury or risk of death for each in relation to the risk regions as determined for the applicable processes (broadly acceptable, tolerable, and unacceptable). The broadly acceptable risk region is to the left of the green line, the unacceptable region is to the right of the red line, and the tolerable region is between the green and red lines.

3.2.1 Analysis of Results

For the example given, the baseline results illustrate that most of the studied dust handling systems fell within the tolerable or acceptable regions of risk for fatality. To validate the results of the analysis, a statistical analysis was completed using a confidence interval acceptable to stakeholders.

3.3 Selection of Optimized Risk Reduction Measures

The outcome of the analysis can be analyzed to determine the major contributions to risk for a specific site/facility. To determine major contributions to the frequency of the top event, intermediate and basic event values can be reviewed. Similarly, one could use a software program to determine minimum cut sets of the FTA. By combining the safeguard and conditional modifier probabilities using AND logic, the probability of each consequence per top event can be determined. Injury or fatality consequences can be evaluated to identify the event sequence that is most probable to result in the undesirable consequence. By identifying the system characteristics that most dominantly contribute to risk, the site/facility can select the mitigation techniques that realize the optimal reduction in risk with respect to resources expended.

Another means to optimize mitigation techniques is to employ a "guess and check" method. Inputs of a system can be changed to see their effect on the overall risk of the system, while keeping all other inputs constant. Another analysis strategy can be to select safeguards and conditional modifiers that could be readily implemented at a facility (e.g. good housekeeping, ensure bonding and grounding, following a hot work program, etc.), to see the cumulative effect on risk. From there, it may be realized that no further measures are needed, or additional measures can be selected to lower the risk of the process to tolerable or acceptable levels. By selecting all hazard mitigation techniques, the effect of diminishing returns will be realized; the incremental effect of each subsequent risk reduction measure can be plotted, as shown in Figure 6 below.

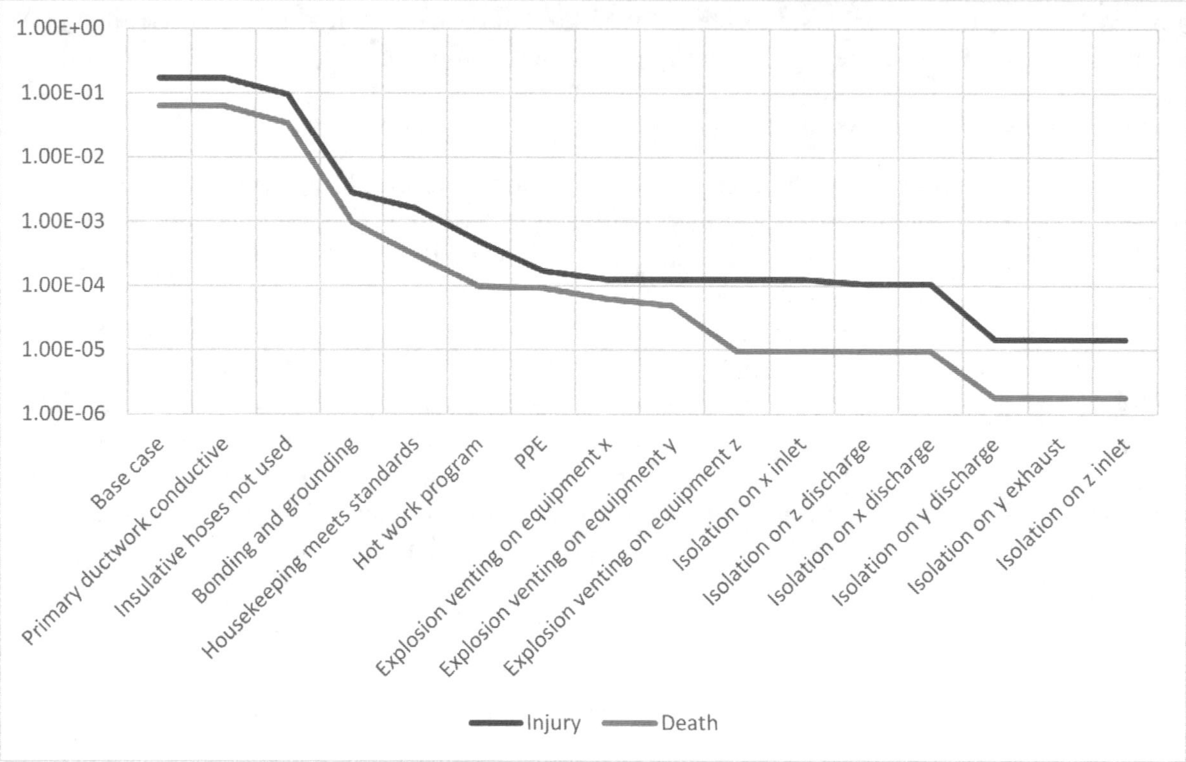

Figure 6 – Example plot of mitigation measures vs risk (cumulative)

The techniques of optimizing risk-reduction measures can also be accomplished on a macro scale, for all comparable system configurations. Site-specific system risk analyses can be completed side-by-side. In this manner, a unified approach for the implementation of hazard mitigation techniques can be developed enterprise- or industry-wide.

In addition to the mathematical and guess-and-check techniques described above, a common-sense approach can be used when evaluating risk of a group of facilities. Some considerations may include:

- Evaluate common system characteristics or administrative and engineering controls for facilities presenting highest risk. These elements could be contributing to the higher risk levels (e.g. it may be revealed that facilities with highest risk do not have adequate housekeeping procedures or similar equipment was installed without explosion isolation).
- Review the calculations and the construct of the fault trees and event trees to determine elements that are driving higher risk. There may be a situation where a high likelihood of an event is driving risk of injury or fatality, so the focus should be on preventive barriers that would reduce the frequency component of risk. A deeper inspection may reveal that ignition sources are the main component driving higher frequency and should be addressed by measures such as bonding and grounding.

3.4 Final Risk Levels

Using the assessment techniques above, an organization can make optimal risk reduction decisions. The organization may realize an acceptable level of risk without employing all prescriptive measures.

Returning to the example summaries previously referenced, an acceptable risk level was achievable by the evaluated facilities by implementing the following risk reduction measures: following housekeeping procedures as outlined by NFPA 654, creating a hot work program, bonding and grounding the system, and protecting the air-material separator with automatic sprinklers. See Figure 7 below of an example of a risk summary after implementation of selected risk reduction measures. Like the baseline risk analysis, the final risk should be statistically analyzed to ensure confidence in results.

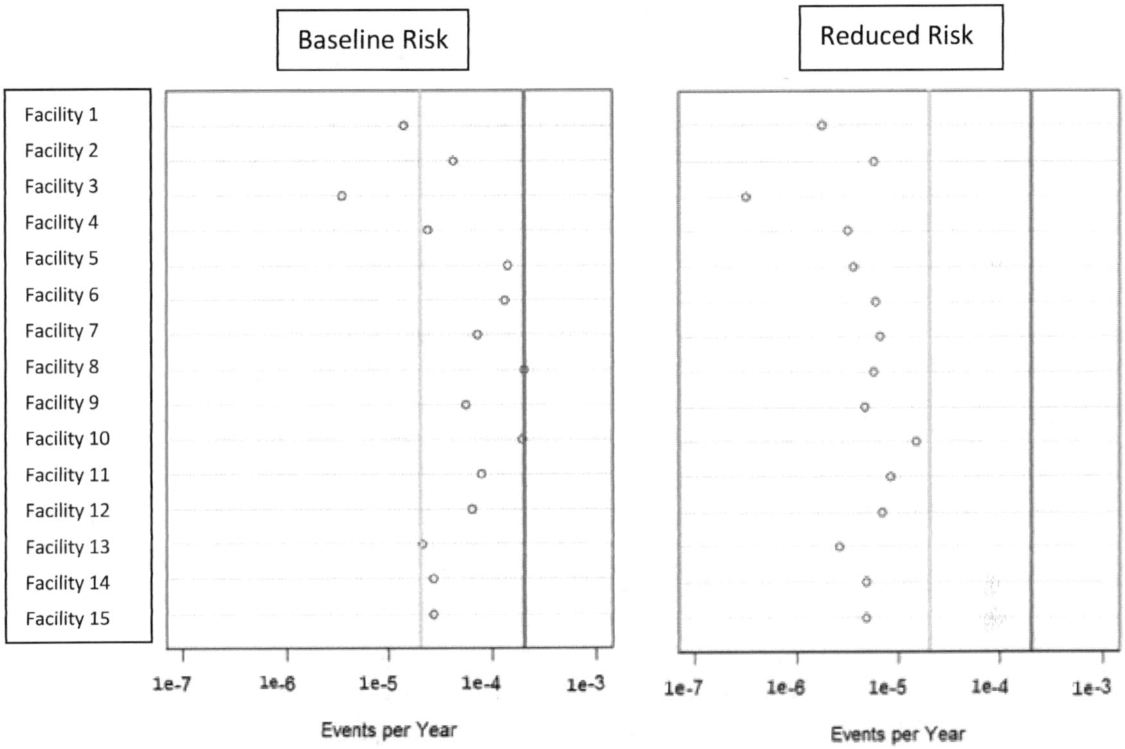

Figure 7 – Example system: baseline and reduced risk of fatality based on selection of optimized risk reduction measures

4. Conclusion

4.1 Discussion of Results

With a well-developed risk assessment approach, manufacturers with large portfolios could strategically reduce combustible dust related risks to acceptable levels without meeting all prescriptive code requirements. The approach evaluates the risks of injuries and fatalities and can be adapted to evaluate other parameters of interest, such as environmental or reputational impact. This approach is also in alignment with NFPA standards on combustible dust and therefore allowable through the regulatory framework.

An additional benefit of the risk analysis is as a platform to optimize risk-reduction recommendations. By comparing risk reductions of each improvement project, the user can prioritize upgrades based on options that provide the largest reduction in overall risk compared to cost of implementation.

4.2 Key Considerations

Several aspects of this methodology warrant scrutiny:

- The system configurations must be carefully decided and robustly modeled to capture all possible variants that may impact frequency of failure or magnitude of consequence.
- Establishing a singular set of risk acceptance criteria is critical to the entire assessment; all stakeholders should be involved in the final approval of these criteria before comparing to site specific data.
- For this approach to be successful, the AHJs must agree to all aspects of the risk assessment (including the risk acceptance criteria). Therefore, there must be close alignment of the risk acceptance criteria with enforceable codes and standards in all jurisdictions of the owner's facilities.
- If a model is developed, it must be verified and validated. The verification step is an essential step in debugging the system to address errors and ensure that the model works as intended. Validation demonstrates that the model is an accurate and reasonable representation of the actual system.

4.3 Further Studies

Although this framework exists loosely in other industries, it is not well established in the combustible dust industry due to a higher level of complexity in the dynamics of combustible dust systems. This complexity is clear when developing the equations and calculations that are used in the fault trees and event trees; approximations were made for concepts such as concentration gradients of combustible dust present in conveying lines, probabilities of ignition, and effects of pressure piling through several interconnected vessels without explosion isolation. Further research and experimental testing in the understanding of combustible dust systems and explosion characteristics could enhance the applicability and robustness of risk assessment studies.

References

[1] National Fire Protection Association, *NFPA 652, Standard on Fundamentals of Combustible Dusts – 2016 Edition*, National Fire Protection Association, Quincy, MA, 2015

[2] National Fire Protection Association, *NFPA 654, Standard for the Prevention of Fire and Dust Explosions from the Manufacturing, Processing, and Handling of Combustible Particulate Solids – 2017 Edition,* National Fire Protection Association, Quincy, MA, 2017

[3] CCPS, *Guidelines for Hazard Evaluation Procedures,* American Institute of Chemical Engineers, Inc., Center for Chemical Process Safety, John Wiley & Sons, Inc., Third Edition, 2008

[4] National Fire Protection Association, *NFPA 59A, Standard for the Production, Storage, and Handling of Liquefied Natural Gas (LNG) – 2016 Edition*, National Fire Protection Association, Quincy, MA 2016

[5] CCPS, *Guidelines for Chemical Process Quantitative Risk Analysis,* American Institute of Chemical Engineers, Inc., Center for Chemical Process Safety, John Wiley & Sons, Inc., Second Edition, 2000

[6] UK HSE, *Reducing Risks, Protecting People. HSE's Decision-Making Process*, London, 2001

[7] Lees, Frank (editor Sam Mannan), *Lees' Loss Prevention in the Process Industries – 3rd Edition,* Elsevier, Inc., Burlington, MA, 2005

548428 3D Printing- Unique Hazards When Metal Dust Is Created

Over the past several years, manufacturers in the additive industries have employed 3D printing to increase efficiency on plant floors. Materials used in the additive industries cover a wide variety of plastics and metals. The metals used include alloys such as stainless steels, brass, and copper. The nature of the 3D printing process tends to create fine dusts and typically, dust from metal alloys is not a high-risk concern. However, metals such as aluminum and titanium can pose significant fire risks. The particle size of a dust is a major contributing factor associated with the hazards they may pose. Fine aluminum and titanium dust can be extremely sensitive to ignition, and can produce tremendously high temperatures and overpressures. Typically, an additive can range from 10 to 70 microns, at which size aluminum and titanium dusts are highly dangerous. In these particle ranges, aluminum has an explosion severity (Kst) greater than 400 bar m/sec. Fine metal dusts have also shown to have low ignition sensitivity, such that a static discharge (less than 25 mJ) can ignite them. Such characteristics of a dust indicate potential for a catastrophic event under the right conditions. 3D printing can also create aluminum and titanium dust in particle sizes in the nanometer range. Titanium dust in the nanoparticle sizes has been shown to display pyrophoric tendencies. 3D printing can create hot dusts but, even without a credible ignition source, metal dusts such as titanium can catch on fire. It is important that industries employing 3D printing with metal additives are aware of the hazards associated with them. Metal dusts such as aluminum and titanium can be extremely dangerous and pose a high risk for a fire or explosion. In addition to having adequate knowledge of the risks, actions should be taken to alleviate these risks. This work seeks to present a detailed analysis of the hazards associated with 3D printing, as well as ways to prevent and mitigate these hazards. The work further provides three fire and explosion case studies resulting from 3D printing metal dust.

19 AIChE Spring Meeting
15TH GLOBAL CONGRESS ON PROCESS SAFETY
NEW ORLEANS • 2019

COMBUSTIBLE DUST - HAZARDS / MITIGATIONS / PITFALLS

David Rowe
Global Risk Consultants
Conroe, Texas
David.rowe@tuvsud.com

Prepared for Presentation at
American Institute of Chemical Engineers
2019 Spring Meeting and 15th Global Congress on Process Safety
New Orleans, LA
March 31 – April 3, 2019

COMBUSTIBLE DUST - HAZARDS / MITIGATIONS / PITFALLS

David Rowe
Global Risk Consultants
Conroe, Texas
David.rowe@tuvsud.com

Keywords: Combustible dust, dust hazards, DHA,

Abstract

Combustible dust presents a real problem and while the hazard is well known, many facilities either do not believe it can happen to them or do not believe that they have a problem. A number of recent combustible dust incidents have led to a heightened awareness of the hazards, modification of standards and regulations, as well as enforcement actions. This paper will touch on sampling and testing to understand the dust problem a facility may have as well as mitigations to prevent disasters. It will also address, pitfalls that many facilities face when dealing with combustible dust.

1. Introduction

Combustible dust is and has been a serious issue in industry for decades. While not new combustible dust has been more in the spotlight in recent years with accidents being taken more seriously, incidents being published and legislation coming into effect. Companies and their employees need to understand the hazards of combustible dust as well as the threat to their industry, their facility and their lives. There are several recent examples of incidents related to combustible dust that brought it to the forefront.

In addition to understanding that there is a hazard, and what this hazard may present, companies need to understand how this specifically relates to their facility. This is accomplished through sampling and testing and educating employees. Once the hazards of dust are understood, a facility must know how that applies to them in order to correctly address the hazard and develop prevention and mitigation strategies. Prevention strategies may include housekeeping, electrical classifications as well as dust collection systems. Mitigation strategies may include any or all: isolation, venting, and / or suppression.

2. Combustible Dust What is It?

Combustible dust is defined as a solid material composed of distinct particles or pieces, regardless of size, shape, or chemical composition, which presents a fire or deflagration hazard when suspended in air or some other oxidizing medium over a range of concentrations. Typically, it is a material that is 420 microns or smaller in diameter (material passing a U.S. No. 40 diameter Standard Sieve). Above that size and the chance for ignition is greatly reduced.

Given the proper circumstances combustible dust can have extreme consequences. For example, on February 7, 2008 there was an explosion of sugar dust that leveled the Imperial Sugar facility, killed 13 and sent many more to the hospital many with serious injuries. What exploded? A minor explosion shook the facility allowing sugar dust that had accumulated to became airborne and explode. Stories from workers indicated that there were large accumulations of dust in the facility that contributed to the severity of the secondary explosions. The CSB determined that an effort to prevent contamination provided the concentration necessary for an explosion.

3. Regulatory Influences

Over the years, the CSB has issued several recommendations for OSHA to develop a comprehensive general industry standard for combustible dust but as of this time there is no OSHA standard on combustible dust. That does not mean that there is no enforcement potential. OSHA can and has used the General Duty Clause and can cite issues using several other standards. Under OSHA's Combustible Dust National Emphasis Program (Oct 2007), the agency can issue dust-related fines under 18 different CFR standards as well as the General Duty Clause.

Fire marshals use NFPA codes, including NFPA 652: Standard on the Fundamentals of Combustible Dust, when inspecting workplaces for fire hazards. These codes are also cited in OSHA citations and documentation.

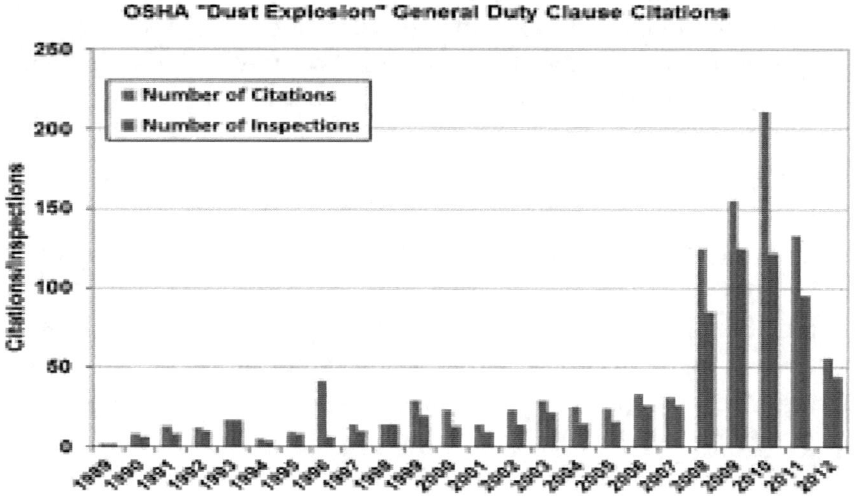

4. Industries in Which Combustible Dust is An Issue

Combustible dust can be found in numerous industries, some we expect and some maybe not. Basically, any industry that uses, produces or transports a fine powder / dust may have a hazard and should at least investigate the potential. A few of industries include: Agricultural, Chemical, Pharmaceuticals, Wood, Plastic, Food, Coal, etc.

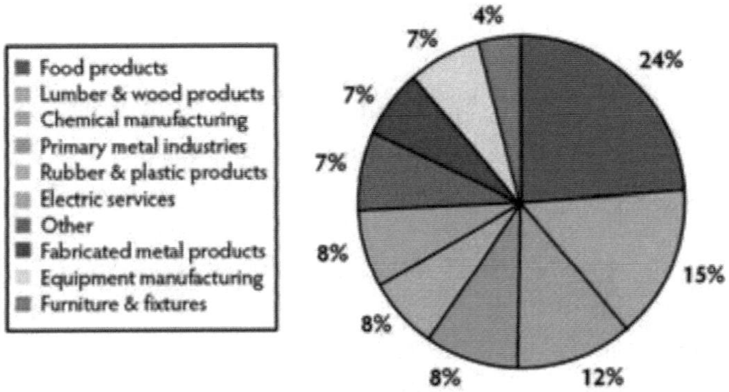

5. How to Determine if Combustible Dust Exists?

Any of the industries above should be aware of potential dust hazards. If equipment or structure has dust settling on it, then care should be taken. Taking samples to determine if there is indeed a combustible hazard is only prudent. If there is a dust hazard, then steps can be taken to ensure a safe environment and if it is not combustible, there is conclusive evidence that the facility is safe (at least from that hazard)

a. Sampling

Knowing that there is potential dust hazard is a start, but a facility must also determine the level of risk. This involves sampling and testing. In order to sample for combustible dust, it is recommended to get a representative sample but keep it to a given area – milling room for example, not the entire ground floor. Samples taken from specific locations can minimize the effort needed to reduce the hazard.

Typical areas of concern include but are not limited to 1) the dirty side of dust collectors 2) elevated areas (don't forget about these as they are excellent sources of fuel for secondary explosions and dust will tend to rise) such as building framework, piping, etc. 3) inside of bins and duct work transporting potentially dusty material. 4) Anywhere that material is transported including belts, conveyors, elevators, bins, hoppers, screens, etc. all could potentially generate dust capable of combustion. Basically, anywhere that dust is accumulating, especially if it is obscuring the color of the material its on, one should consider a hazard.

b. Testing

Dust testing can be expensive and as such all the tests on all the samples is typically not warranted. One of the first tests should be a sieve test and / or Combustibility Screening test (Go / No Go). The sieve test determines the size distribution of the sample. Typically, anything over 420 microns (US No 40sieve) is not considered combustible. The Go/No Go test determines if there is an explosive potential to a given dust sample. If this test comes back negative, then there is no combustible hazard, but if it comes back positive then the sample should be tested to quantify the sample.

At this point the MIE, MIT, MEC and LIT will allow an understanding of what it would take to have an explosion. The P_{max} and K_{st} are tests that explain the severity of a potential explosion.

MIE - Minimum Ignition Energy (MIE) - mJ

The MIE is the minimum amount of energy required to ignite the dust. This test will determine if the dust can be ignited by low-energy sources such as static or if higher energy sources would be required.

Product	MIE (millijoules)
Corn	45-100
Sugar	<10
Wheat	50 – 100
Wood bark	40-60
Zinc	960

MIT – Minimum Ignition Temperature of Dust Cloud - °C

The lowest temperature of a hot surface at which ignition occurs when in contact with airborne dust (cloud). MIT values are generally between 150 and 700°C depending on the material.

MEC - Minimum Explosible Concentration - g/m^3 or vol %

The lowest concentration of a combustible dust in air, expressed in grams per cubic meter that will propagate a flame. This can be equated with LEL in vapors or liquids.

LIT – Layer Ignition Temperature - °C

The lowest temperature at which layered dust will ignite when in contact with a heated surface.

-

K_{st} – Deflagration Index – bar-m/s

This is the normalized speed of pressure rise in created by a dust cloud explosion. Combustible dusts are ranked into one of four classes; St0, St1, St2, St3. The level of explosion violence increases with the class number. Some of the most famous dust explosions case studies come from St 1 class dusts: such as the Imperial Sugar explosion in 2008 in Georgia.

The St class is based on the K_{St} value as follows:
St 0 - K_{st} value = 0

St 1 - K_{st} value less than/or equal to 200 bar-m/sec and greater than 0
St 2 – K_{st} value between 201 and 300 bar-m/sec
St 3 – K_{st} value greater than 300 bar-m/sec

The higher a K_{st} value is the more "energetic" the dust is but that does not change the danger associated with dusts having a lower K_{st}.

Dust Type	K_{st}	P_{max}
Corn Starch	202	10.3
Wood Dust	10	9
PVC	46	7.6
Sugar	138	8.5

P_{max} - Maximum overpressure - bar

This value measures the maximum explosion overpressure generated by a dust cloud explosion.

6. What are the options once Combustible Dust is confirmed?

According to CSB investigations into recent dust explosions several common issues were revealed. 1) Poor level of understanding of the risks of combustible dust 2) Inadequate efforts to manage combustible dust hazards and 3) Inadequate inspection methods to proactively address potential "unsafe" levels of dust.

One of the most important things in addressing combustible dust is awareness and understanding of the associated hazards. There are several issues that affect the level of awareness within a company. During training, the hazards may be glossed over leaving employees truly unaware as to the extent of the potential hazard. Facility history can also make a true understanding of potential hazards difficult. Comments like, "we have worked this way for years and never had an issue", or "I have never seen a fire or explosion" make people complacent to hazards or even not believe that the hazards exist. Training as well as confirmation of understanding should be documented to ensure that employees and management know and understand the hazards that are present

Along with training, other documented programs should include:

- Documented Housekeeping program with detailed inspection and cleaning schedules and procedures.
- Documented Electrical classifications. The philosophy that the work can be performed quicker than having to follow the rules is a dangerous one.
- Maintenance procedures (routine as well as infrequent) need to be documented so that the hazards are understood and mitigated.

- Management of Change (MOC) procedures should be documented and followed to ensure that things like electrical classification on modifications are followed and that employees are trained

7. Combustible Dust Hazard Prevention

a. Dust Hazard Analysis (DHA)

DHA's were initially to be completed as of 2018 but, as of now, that requirement has been postponed to 2021. A DHA is a Dust Hazard Analysis, similar to Process Safety Management's (PSM's) Process Hazard Analysis (PHA) but with the focus on combustible dust. DHAs may be carried out in a manner very similar to the PHA. HAZOP is still a viable approach but can be more detailed than necessary. Checklist / What-if styles are more commonly used but it is always up to the facility.

The DHA will help a facility to understand where the hazards exist, the prevention / mitigation needed and increase the understanding of employees and management within the organization.

b. Housekeeping

As with the goal of Process Safety Management, if the hazard can be maintained in a manageable location then the effects on a facility can be minimized. If the dust can be kept from accumulating, then there is no hazard. 1/32 inch is the calculated minimum dust layer that is sufficient to form a combustible dust cloud.

Moreover, dust accumulation on, in, or in the vicinity of the electrical equipment could be sufficient to interfere with the safe dissipation of heat from the electrical equipment or the dust could become ignitable by abnormal operation or failure of electrical equipment.

Many times, air wands or lances are used for cleanup. They are able to blow debris, similar to sweeping. This practice is highly discouraged if you have or suspect you have a combustible dust issue. The pressurized air can work to create the MEC (minimum explosible concentration), thereby actually increasing the risks during housekeeping activities.

c. Dust Collection systems:

Dust collection systems aid in housekeeping and are a vital part of any hazard prevention program. If properly designed, they minimize the amount of free dust in an area by controlling the amount of dust at its source and not allowing it to disperse throughout a facility. It is imperative that electrical ratings be followed and any mechanical equipment in the potentially dust producing areas are rated properly. Part of the design must be to ensure that the draw on the collection ends is sufficient to keep dust from becoming airborne.

Types of dust collection systems include cyclones, bag houses and water wash or wet systems. While different in operation they all collect potentially explosive dust and confine it. While the collection system can be considered part of the prevention program, explosion mitigation required on the collectors. Most dust collectors are located outside of a facility but if proper protections are taken, they can be located inside a facility. To locate the collector inside it must have one or more of the following 1) Listed flame arresting device through which an explosion would be vented, 2) dilution with non-combustible dust, 3) Explosion suppression, 4) Explosion pressure containment, 5) Explosion venting or 6) Oxidant concentration reduction. The purpose of a dust collection system is to collect the dust, keep it from getting airborne and provide a means of disposal. As such, there are often explosive concentrations of dust within them and they have been the source of dust explosions. Thus, mitigations are necessary on these devices.

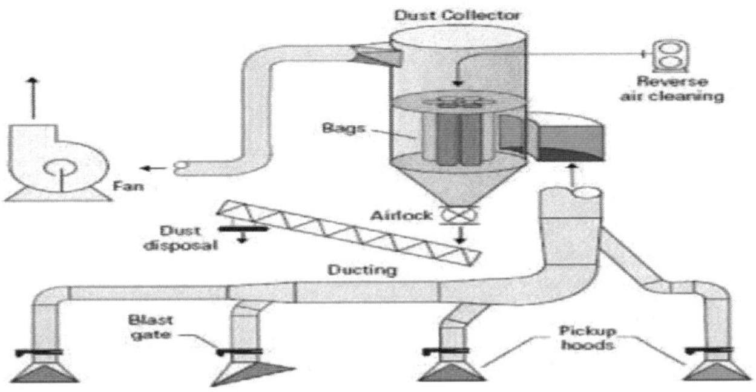

d. Electrical Classification

NEC Classifications, Article 500 states that combustible dust is a Class II material. From there it is further subdivided:

- Division 1 Combustible dust in air under normal operating conditions in quantities sufficient to produce an explosive or ignitable mixture.
- Division 2 Combustible dust due to abnormal conditions in quantities sufficient to produce an explosive or ignitable mixture.

It can be further delineated by the following groups based on the nature of the dust.

Group E (Class II Div I Group E for example) - Atmospheres containing combustible metal dusts, including aluminum, magnesium, and their commercial alloys, or other combustible dusts whose particle size, abrasiveness, or conductivity present similar hazards in the use of electrical equipment.

Group F - Atmospheres containing combustible carbonaceous dusts that have more than 8 percent total entrapped volatiles or that have been sensitized by other materials so that they present an explosion hazard.

Group G - Atmospheres containing other combustible dusts, including flour, grain, wood flour, plastic, and chemicals.

NFPA has developed recommendations for classifying an area based on the availability of a dust layer. This is a quick visual determination.

Table A.5.2.2(a) Division Determination Guidelines Based on Dust Layer Thickness

• Greater than 1/8-inch (3.0mm)	Division 1
• Less than an 1/8 but surface color is not discernable	Division 2
• Surface color is discernable under the dust layer	Unclassified

There are everyday items that can defeat even the best electrically classified areas. These may due to non-routine operations, or a poor understanding of the full extent of the dust hazard, thereby using non-rated tools in a classified area. These tools may include:

- Unrated Forklifts
- Unclassified Vacuum units
- Spark producing Power tools for maintenance or production

8. Combustible Dust Hazard Mitigation

a. Venting

Some form of vent is required. Should an explosion occur, this option allows for the venting of the explosion. It may be ductwork that is attached to the equipment and vented to a safe location. Venting can be accomplished through duct work or blow out panels from a dust collector, or hammer mill, etc. Complications with this mitigation are that it allows the explosion to happen. Theoretically venting will relieve the over pressure and explosive material to a safe location however, damage will likely be done to the equipment.

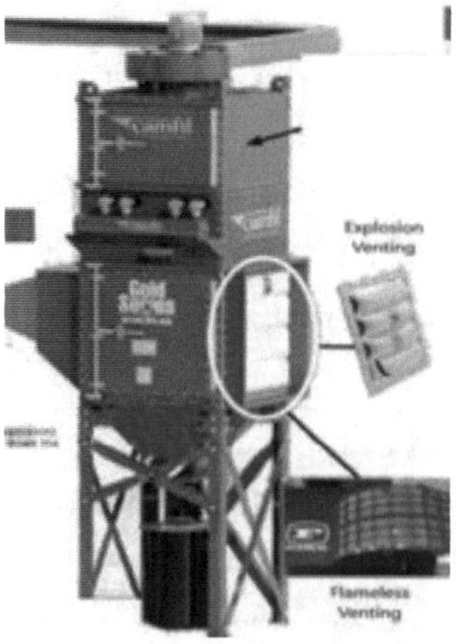

This picture shows the explosion vent as well as a flameless vent. The upper explosion vent typically requires the device to be outside while the flameless vents allow the device to be located inside as the flameless vent extinguishes the flame as well as eliminates the pressure wave.

Flameless venting is a good option whether the collector is to be located indoors or outdoors. The vents are designed as flame arrestors, they knock down the pressure wave and collect the dust knocked out of the explosion.

These pictures show an explosion without a flameless vent and the same explosion with a flameless vent. This type of venting must be certified.

b. Suppression

This system allows an explosion to start (ignition to occur) but as soon as the sensor detects an ignition it signals a device to introduce a suppressant into the system and stop the ignition before the explosion can happen. While suppression systems are expensive options, they are also designed to eliminate the explosion thus removing the potential damage to the equipment. These systems require routine maintenance to ensure that the devices work when needed. They must also be inspected and certified by a qualified 3rd Party to verify that the suppression system is still valid.

IR or UV/IR technology is utilized to detect sparks or hot embers. Smoke detectors can be used to detect smoldering material. Temperature sensors are used as well to monitor for temperature rises that may be associated with fires. Any or all of these signals can be used to initiate a suppression device that will extinguish the heat source or spark.

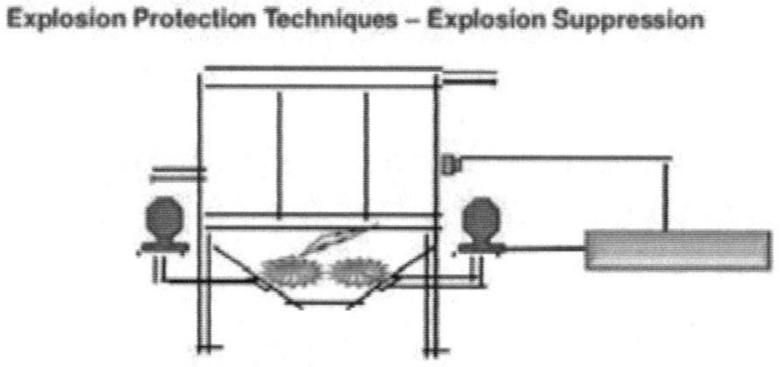

c. Isolation

Isolation system are designed to isolate the explosion to prevent it from spreading. Typically, this is a mechanical device but there are chemical isolations as well. The intent here is to use a fast-acting device to stop the flame front from propagating to further fuel sources. This does mean that a fire has occurred and if an unexpected explosion happens, they could be quickly rendered ineffective.

Explosion isolation devices are defined as active or passive. Active devices need a detection device to sense pressure increase and signal the isolation device to actuate. Passive isolation devices or flap valves are fully operational with airflow and close shut in reaction to a deflagration pressure increase.

Chemical Isolation includes a bottle charged with sodium bicarbonate which discharges an explosion suppressant into the duct to suppress a fireball, preventing it from reaching inter-connected equipment or plant areas. As an active device, it is activated by a signal from a rate of rise pressure detector thru a control and monitor panel.

Mechanical Isolation includes an isolation valves that is activated when a pressure detector senses the pressure wave from the deflagration and triggers the release of a high-speed knife gate to close, forming a mechanical barrier in the pipeline. The barrier prevents propagation of the fireball back into the interconnecting duct and processes; isolating and protecting the return air duct from a secondary explosion. High speed valves are an active device, and a control panel and pressure detector are required as part of the package.

9. Common Pitfalls when dealing with Combustible dust

a. Status quo

The concept of, we have run this way since I can remember without a problem, encourages apathy and complacency. It's always nice to have employees that want to learn and improve but not always feasible. Training and thorough risk assessments should help to show why status quo is a bad idea.

b. Inadequate Housekeeping

There must be a documented program, not only because the regulation requires it but because without a documented program, then housekeeping will not get completed or at least adequately completed. Cleaning and inspection frequencies will be modified to fit the mood of the staff on hand at that moment. Also, an adequate documented program will reinforce housekeeping at the harder to reach locations or those location people may not think of.

c. Not conducting adequate training on the hazards

The news and social media can desensitize people to the hazards of what can really happen and without a good understanding of the hazards and how they apply may believe that it cannot happen to them. It was understood at a management level at Imperial Sugar that sugar dust posed a problem, but the workers did not know that sugar even presented that kind of hazard. If workers understand the issue and what threat it poses to them and have some level of buy in to ensuring that the hazard is minimized, then they will be more likely to do the things necessary to prevent a dust issue.

d. Not performing a Risk Assessment

Risk assessments should be done to some extent on all process whether for dust, or general hazards or PPE. If there is dust present, a dust hazard assessment is the prudent thing to do. Complete a formal DHA (Dust Hazard Analysis) to ensure that you not only understand the hazard but have addressed any potential hazards. Once complete, share the results with the employees so that they understand the hazards as well as what the facility is doing to protect the safety of its employees. NFPA is recommending that all facilities with dust hazards complete a DHA by 2021.

e. You get what you pay for

That term is well used and generally applicable. There are companies that buy spare parts off eBay to save money. While the equipment may meet the concept of the protection, cheap equipment may not have the proper mitigations or adequate safeties for a specific dust. Also, buying cheap may result in installing inferior equipment that may fail when needed.

f. Buying the top of the line but not having the time or expertise to understand how to work it.

The opposite of going super cheap is buying the top of the line when it is not needed. A company purchases something that is either more than is realistically needed and then not maintaining or even understanding it. Companies will also buy the top of the line but not investing the time and resources to understand how it should work or how it should be maintained.

g. Conclusion

Dust hazards, while known to be a problem, are often still not given the respect they deserve when it comes to risk assessments and facility operations. Combustible dust is a serious hazard and needs to be considered, understood and the hazards prevented or mitigated. Housekeeping, electrical classification and education are critical to preventing dust hazard issues. Dust collectors, as well as suppression, venting and isolation systems ensure adequate mitigation of the hazards.

10. References

[1] Christopher A. Jarc, P.E., PMP, LEED A.P. "Combustible Dust Basics" Hixson Engineering

[2] Chemical Safety Board Investigations http://www.csb.gov/investigations

[3] OSHA "Combustible Dust National Emphasis Program (Reissued) CPL 03-00-008 (2008)

[4] Ann Marie Fauske "Weighing in On Balancing Knowledge and Action to Combat Dust Hazards"

[5] Donaldson Company Inc “DUST COLLECTORS & COMBUSTIBLE DUST STRATEGIES” 2011-2018

[6] Glacier Technology “Engineered Solutions for Industrial Fans & Air Filtration”

[7] Dust Explosion Hazard Assessment, Including OSHA Combustible Dust NEP Vahid Ebadat, PH.D

[8] Jef Snoeys “Dust Explosion Protection Using Flameless Venting” Fike Corporation 2015

[9] Cooldo Electromechanical Equipment Co LTD “Industrial Dust Collection”

Driving Transformation in Process Industries through Materials and Chemicals Informatics

Daniel Christe
Elsevier
d.christe@elsevier.com

Matteo Caligaris
Elsevier
230 Park Ave Suite 800, New York, New York, 10169
m.caligaris@elsevier.com

Prepared for Presentation at
American Institute of Chemical Engineers
2019 Spring Meeting and 15th Global Congress on Process Safety
New Orleans, LA
March 31 – April 3, 2019

Driving Transformation in Process Industries through Materials and Chemicals Informatics

Daniel Christe
Elsevier
d.christe@elsevier.com

Matteo Caligaris
Elsevier
230 Park Ave Suite 800, New York, New York, 10169
m.caligaris@elsevier.com

Keywords: Materials Informatics, Natural Language Processing, Data Science, Knovel

Abstract

Connecting data, people, and processes is the key to achieving true digital transformation. Process safety intelligence is derived from a vast corpus of structured and unstructured data sources - historical reports, equipment data sheets, materials and thermophysical property data, safety data sheets, best practice guides, handbooks, textbooks, and real-time feedback from instrumented equipment. Navigating this emergent complexity demands dedicated platforms to enable insight discovery across all design stages in a continuous digital thread from inception, to design and manufacturing. Edge computing coupled with machine learning approaches will further enable predictive and prescriptive analytics for process design, safety, and intensification.

The talk will share a perspective on materials, chemicals data management and digitization, state-of-the-art tool and approaches on the market today to manage, and process safety information competencies that engineers entering the profession need to have.

1 Introduction

The chemicals industry faces considerable pressures to innovate products and processes to drive growth and maintain competitive edge in an increasingly volatile economic landscape[1]. As product cycles compress, the challenge for chemicals companies today lies in optimizing innovation processes and infrastructure to shrink time-to-market, while managing risk to acceptable levels and maintaining or exceeding environmental health, safety, & sustainability goals. A growing fraction of chemical companies' revenues come from product lines that did not exist even a few years ago. For instance, in 2010 40% of Dow's revenues came from product lines invented in the last five years [1].

The chemicals industry is forecast to spend $30 billion on R&D in 2020 [2], with half of this cost is directed toward experimentation. Nearly a third of those experiments will be duplicate experiments, partially a consequence of fragmented data and information systems. While some duplicate experimentation is inevitable, much of this cost can be avoided through improved access and visibility to high resolution, clean datasets across the enterprise.

Sound data management is not a goal in itself – it is a conduit to accelerated discovery, new product introduction, and increased revenues. Chemicals and materials data is inherently fragmented in nature, posing challenges for data management – which forces many chemicals professionals need to "jump" between various applications to find the appropriate data, insight, and knowledge for path-critical decision-making. Ideally, a single platform would be able to surface all known materials & chemicals data on a particular material or process to avoid accidental duplication and streamline the path to discovery.

Addressing this challenge, Elsevier aims to harmonize all scientific, commercial, and company data on a single platform designed to streamline decision-making, reduce risk, and improve collaboration across the product lifecycle (**Figure 1**), paired with contextual search and machine learning capabilities.

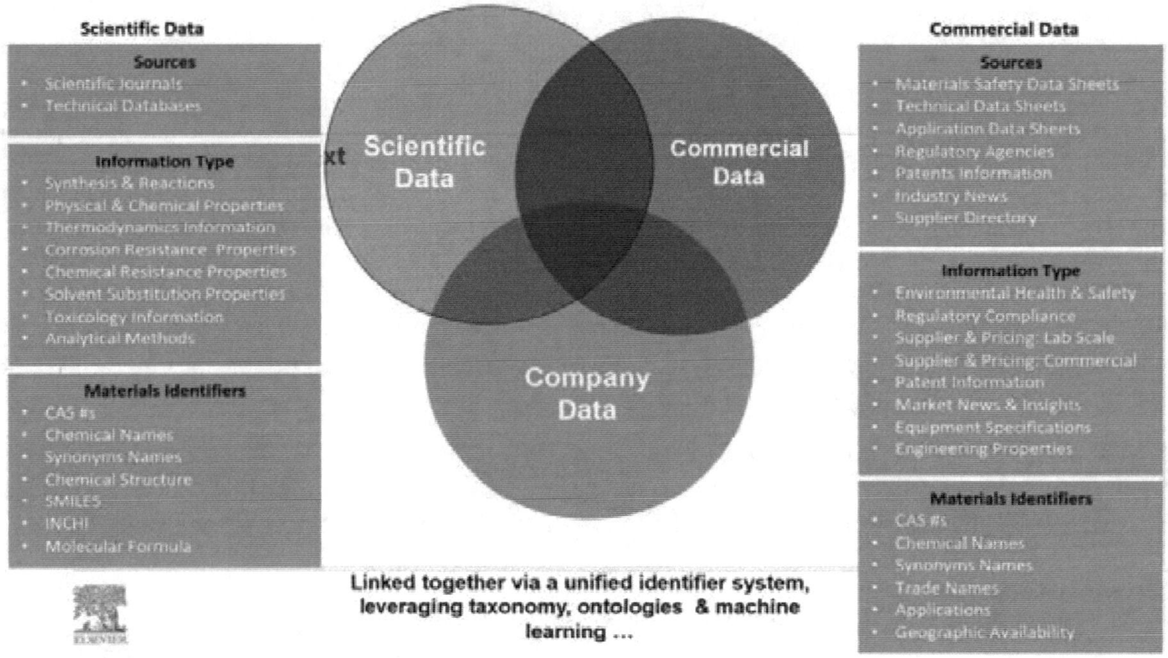

Figure 1: Scientific, commercial, and company internal data harmonized in an interconnected innovation platform to enable collaboration

FAIR (Findable, Accessible, Interoperable, Retrievable) principles are a guiding philosophy to structure and management of research datasets. The focus of FAIR is on "enhancing the ability of machines to automatically find and use the data, in addition to supporting reuse by individuals" [3].

Data should be visible across the entire enterprise, and most if not all employees across functions should be able to view the available data. Those who take advantage of a data-first process innovation will enjoy the business benefits of accelerated new product introductions, cost reductions, increased revenue and higher margins. Elsevier's Knovel team is always interested in development partners and the authors may be contacted to discuss particular needs in your organization.

2 References

[1] J. Boringer and T. Jan Simons, "Commoditization in chemicals: Time for a marketing and sales response," 2016. [Online]. Available: https://www.mckinsey.com/industries/chemicals/our-insights/commoditization-in-chemicals-time-for-a-marketing-and-sales-response.

[2] A. Booth, J. Hart, and S. Sim, "Building a great data platform," 2018. [Online]. Available: https://www.mckinsey.com/industries/electric-power-and-natural-gas/our-insights/building-a-great-data-platform.

[3] M. D. Wilkinson *et al.*, "The FAIR Guiding Principles for scientific data management and stewardship," Sci. Data, vol. 3, p. 160018, Mar. 2016

CAROL Robotic Catalyst Unloading – Results from Field Trials

Christopher Jansen
Program Manager, WorleyParsons
12 Creek St, Brisbane, QLD 4000
christopher.jansen@worleyparsons.com

Andrew W. Sloley
Principal Consultant, Advisian (WorleyParsons Group)
Energy Center II, 575 N Dairy Ashford St, Houston, TX 77079
Andrew.Sloley@advisian.com

Scott Schroeder
Senior Technical Consultant, Advisian (WorleyParsons Group)
Energy Center II, 575 N Dairy Ashford St, Houston, TX 77079
Scott.Schroeder@advisian.com

Prepared for Presentation at
American Institute of Chemical Engineers
2019 Spring Meeting and 15th Global Congress on Process Safety
New Orleans, LA
March 31 – April 3, 2019

CAROL Robotic Catalyst Unloading – Results from Field Trials

Christopher Jansen
Program Manager, WorleyParsons
12 Creek St, Brisbane, QLD 4000
christopher.jansen@worleyparsons.com

Andrew W. Sloley
Principal Consultant, Advisian (WorleyParsons Group)

Scott Schroeder
Senior Technical Consultant, Advisian (WorleyParsons Group)

Keywords: Catalysts, Safe Work Practices, Product Design

Abstract

In the 1970's, the first commercial Remote Operated Vehicles (ROVs) for the offshore oil & gas industry were developed. Prior to this, underwater inspections and maintenance operations were performed by personnel wearing heavy underwater suits equipped with breathing apparatus. A catalyst handling technician faces a similar scenario when they enter vessels filled with nitrogen, to vacuum the spent catalyst material. The number of fixed-bed catalytic vessels in the global refining and petrochemical industry is estimated to exceed 50,000 (average of 50-60 vessels per facility)[1] and yet the process for catalyst removal has remained essentially unchanged for the past 75 years.

In 2018, the industry's first commercial robot for catalyst unloading was used to remove molecular sieve adsorbent from a dehydrator at an LNG plant in Australia. The development of CAROL™ (Catalyst Removal Amphirol) responds to increasingly stringent requirements to reduce human risk in inert confined-space entry. It is the culmination of a three-year development effort from conception and design to prototyping and testing and represents a potential game-changer in the industry.

This paper showcases the challenges associated with current catalyst removal techniques, and the inherent safety advantages that CAROL has shown in its pilot operations to date. One case study demonstrates CAROL eliminating the need for confined space entry under inert conditions during adsorbent unloading from a dehydrator. The second case study examines the use of CAROL on a wet catalyst bed under water flood conditions. The final case study illustrates the use of CAROL in removing material built-up at its angle of repose, at temperatures not suitable for human entry.

1 Introduction

Based on current technology a rough estimate of the number of worker days of risk exposure due to confined-space entry during catalyst unloading (Figure 1) exceeds 10,000 days per year globally (assuming catalyst removal every 5 years with an average confined space entry duration of one day per vessel)[1]. Catalyst removal is most commonly achieved by a worker standing on the catalyst and manipulating the end of a high-volume vacuum hose (Figure 2).

The challenges faced by a catalyst handling technician are not too different to that of a deep-sea diver. The working conditions require supplied breathing air to sustain life and one mistake can lead to catastrophic consequences. While the off-shore industry has seen the development of the Remote Operated Vehicle (ROV) to greatly reduce the requirement for underwater diving at oil platforms, the safety improvements in the catalyst handling industry have primarily focused on improved personal protective equipment (PPE) and procedures.

Many vessels require inert conditions throughout the catalyst unloading process. This is typically because pyrophoric scale on the catalyst can auto-ignite in the presence of oxygen. Anti-panic helmets have been developed that consist of a clam-shell design bolted to the occupant's head. The helmet cannot be removed by its occupant, a design that has become standard best practice, after fatalities occurred due to technicians removing their helmets in the vessel. It is likely they had become claustrophobic and were not thinking clearly, potentially influenced by heat exhaustion. In operation, breathing air is supplied to keep the helmet at a slight over-pressure.

Figure 1: Entering a vessel[2]

Figure 2: vacuuming catalyst[3]

Fatalities have occurred in the catalyst handling industry due to asphyxiation, exposure to heat, fire, falling from heights, pressure build-up, and engulfment under catalyst[4]. A study conducted by the U.S. Chemical Safety Board identified 85 nitrogen-exposure incidents in the U.S. between 1992 and 2002, resulting in 80 deaths and 50 injuries[5]. Whilst improvements have been implemented, fatalities continue to occur.

In 2014, a catalyst handling worker was killed at a refinery in Germany. The worker was vacuuming catalyst from the bottom of the vessel when he reported hard sections of catalyst via the intercommunication system. Ten minutes later the worker called for help and the safety crew initiated rescue activities. The worker reported that a wall of catalyst had collapsed on him and that he could feel heat coming from within the catalyst. Outbreathing from the helmet was enough oxygen to cause pyrophoric scale on the catalyst to ignite. Communication with the entrant gradually slowed and then stopped. The burned body was only recovered after 350 cubic feet of catalyst had been removed.

Whilst fatalities present the ultimate risk, the health impact on catalyst handling technicians when performing manual vacuum operations is also significant. The ergonomic position taken by the worker is to crouch over in their hot protective clothing, meanwhile maneuvering the heavy vacuum hose around the vessel. They cannot drink water while in the vessel due to the air supply system. To get out of the vessel for a break they may need to climb 40 feet of ladder to get to the manway which could be as narrow as 18 inches.

2 Development of CAROL

One of the earlier attempts to create a catalyst removal robot was by Greg Kraus and Ray Arnold of Catalyst Services, Inc., Texas. With the patent published in October 2008, the design involved a vacuum line connected to an articulating nozzle, supported against the sides of the vessel using stabilizing arms[5]. One of the key challenges the inventors likely faced was achieving a catalyst removal efficiency comparable to that of a person. Catalyst change-outs are often critical path activities during a plant turnaround meaning delays are very costly for the operator. It is likely that this challenge is what prevented the robotic arm concept from being commercialized.

The CAROL development team went for a different approach. The aim was to develop a machine that sits on the catalyst, moving around on the material as it was vacuumed. The characteristics of the catalyst, sometimes referred to as being like quick sand, was the critical parameter to overcome. The amphirol (screw-propelled vehicle) was selected (Figure 3).

CAROL 1.0

CAROL 2.0

CAROL 2.1

CAROL 3.0

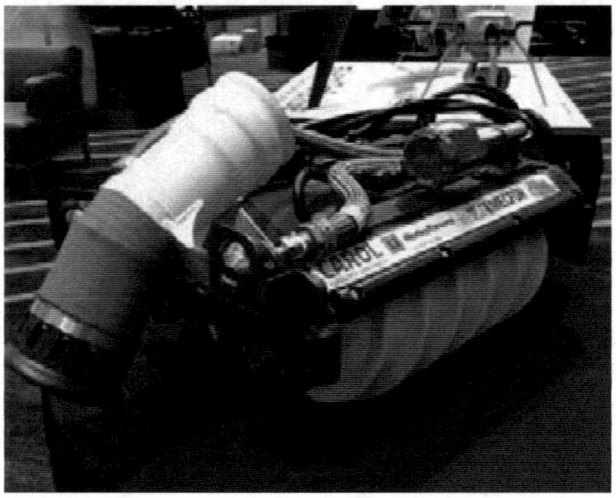

Figure 3: CAROL prototype development

The development of the amphirol catalyst unloading robot was adapted from a technology the manufacturer, Mecfor, had previously implemented for the red mud lakes of alumina refineries. The technology (using worm gear propulsion) allows the device to move on a semi-liquid substance. The combination of a lightweight aluminum frame and large diameter screws provides a buoyancy effect on the catalyst (Figure 4). CAROL has been tested on the full range of media density (40 – 80 lb/ft^3) and particle size (0 – 1 inch) to be expected in catalyst unloading.

The amphirol is controlled remotely from the top platform or at grade (Figure 5). Movement around the catalyst bed is achieved via remote-operated direct drive hydraulic motors. Meanwhile the vacuum head, connected to the robot via a hydraulic cylinder, can be raised and lowered remotely to achieve an optimum catalyst-to-air/nitrogen efficiency. The CAROL in-vessel electrical equipment is rated for use in Zone 1 and Class 1 Division 2 hazardous areas (ATEX, IEC Ex and AEx certification on the entire system). The equipment is operational up to a temperature of 167 degrees Fahrenheit. For manual unloading, the vessel will be cooled to less than 100 degrees Fahrenheit for human entry and atmospheric pressure is maintained. CAROL provides productivity advantages since it can be deployed while the vessel is still hot. It can also operate continuously throughout the shift meaning the support crew can have a staggered lunch break and 12-20% of downtime can be eliminated.

Figure 4: CAROL Testing

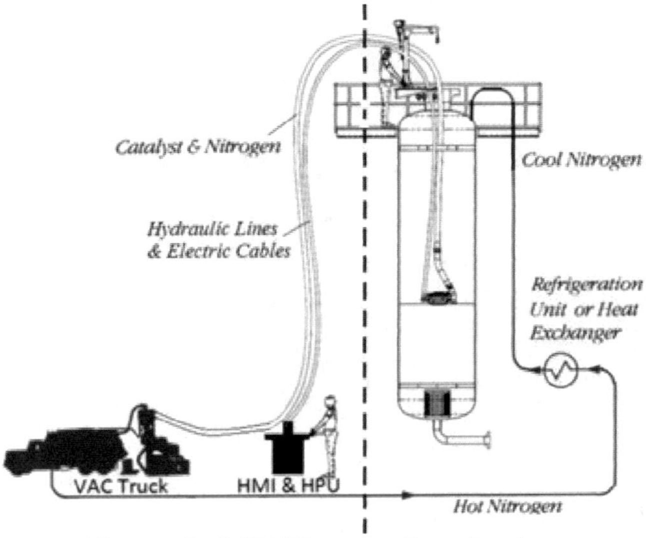

Figure 5: CAROL operational set-up

Risk mitigation is a critical aspect of any project and WorleyParsons has approached the CAROL development using the same rigor as would be applied on a major greenfield project. Examples of outputs from the risk assessment process include: hydraulic oil low level trip to minimize potential leakage from a broken hose; continuous monitoring of vessel lower explosive limit, oxygen content and temperature to minimize the risk of fire and/or explosion; and operating procedure with step-by-step guidance and sign-off to ensure effective operation of the equipment.

3 Case Studies

The case studies presented highlight the use of CAROL under a range of challenging circumstances typically encountered during catalyst unloading.

3.1 Dehydrator unloaded under inert conditions

CAROL was used to remotely remove adsorbent from a dehydration vessel, avoiding the need for worker confined space entry under inert conditions.

Figure 6: Dehydrator Vessel

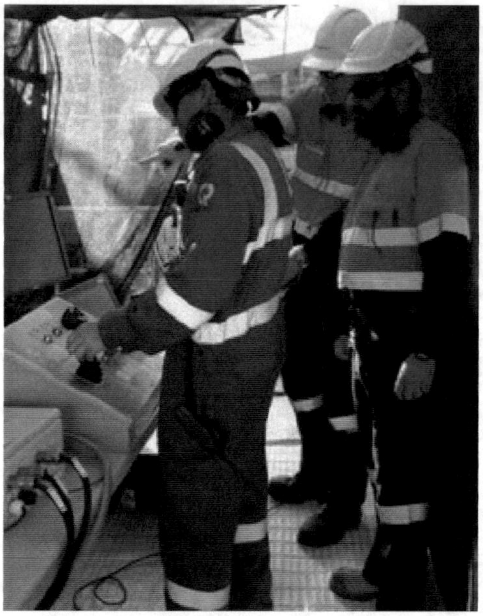

Figure 7: Control Station

Customer Challenge

The dehydration vessels, which remove water from gas before it is liquified, consist of 2000 cubic feet of molecular sieve adsorbent which is low density catalyst. When unloaded under a nitrogen blanket the catalyst is dry and free flowing. When unloaded manually, the technicians boot will sink into the material up to their knee. The plant is located in a tropical climate with temperatures in the vessel reaching up to 120 degrees Fahrenheit. The catalyst sits between a lower and upper steel mesh screen with ceramic balls at the top. The vessels are of a large diameter with no gravity dump capability. The client was seeking to eliminate confined space entry under inert conditions.

Our Solution

CAROL was used to unload the dehydration vessels under inert conditions with no requirement for human entry in the nitrogen environment. This was the first time that CAROL had been used in a live operating plant. The robot was first deployed to remove 1-inch ceramic balls from above the top mesh screen. The screen was speared and removed from the top manway. CAROL was then used to remove the molecular sieve adsorbent, navigating to all parts of the vessel, including in and around the moisture probes located towards the bottom of the bed. The vessel was then opened to air so that the bottom internals could be removed.

Value Delivered

CAROL remotely removed more than 95 per cent of the molecular sieve adsorbent from the vessel including the ceramic support material at the top of the bed. Inert conditions were maintained throughout and the requirement for confined space worker entry in the oxygen deficient atmosphere was eliminated.

3.2 Dehydrator unloaded under water flood conditions

CAROL was successfully used to remove wet material from another dehydration vessel following a water flood operation (part of the same plant as the previous case study).

Figure 8: Camera on robot

Figure 9: CAROL post job

Customer Challenge

Some operators have sought to move entirely away from confined space entry under inert conditions. This can be achieved by flooding the vessel with water several times to flush away the residual hydrocarbons. In the case of pyrophoric material, then the catalyst is kept relatively wet by lowering the water level in the bed slowly as the catalyst is removed. Although the catalyst can then be vacuumed in an open-air environment, the water introduces additional hazards for the occupants, and full suits with breathing air are still required.

Our Solution

CAROL was used to vacuum the wet adsorbent. The screws were effective in allowing the robot to maneuver on the catalyst bed, maintaining an even level of catalyst during removal, important because the catalyst was not free flowing.

Value delivered

The use of CAROL on the wet adsorbent illustrated that the technology can be used in tandem with the water flooding method. The time at risk for confined space entry was significantly reduced.

3.3 Gas shift reactor unloaded post gravity dump

CAROL was used to remotely remove built-up catalyst at its angle of repose with vessel temperatures exceeding 120 degrees Fahrenheit. This was the first time that CAROL was used in a live operating plant in the USA.

Figure 10: Control system trailer

Figure 11: Camera top of vessel

Customer Challenge

Catalyst from the gas shift reactor vessel was removed in the first instance using a dump nozzle located at the bottom of the catalyst bed. This allowed for catalyst to be gravity dumped from the vessel, however not all catalyst was removed. The residual catalyst can be up to 40% of the total vessel volume. Workers are typically required to enter the vessel to vacuum the remaining material and must be wary of the built-up catalyst around the vessel walls and the associated risk of being buried. The pyrophoric material also tends to heat up if the nitrogen purge is insufficient to maintain inert conditions. At this plant the supply of nitrogen was low and vessel temperatures were difficult to control.

Our solution

CAROL was used to unload the residual catalyst material (post dumping) in the gas shift reactor vessel. CAROL was operated in temperatures not suitable for human entry. CAROL successfully removed the catalyst until a hot spot on the catalyst bed caused the plastic prototype screws to melt. Hot catalyst beds are not uncommon, and workers will on occasion experience their boots getting hot, at which point they must get to the ladder before the soles of their shoes melt.

Value delivered

This use of CAROL demonstrated that the robotic catalyst removal technology can be used effectively in combination with gravity dumping, to minimize the time spent by workers inside confined spaces during catalyst unloading. The vessel temperatures following the catalyst dumping operation were still too high for human entry, however the use of CAROL allowed for the job to continue. Although a set of screws were destroyed due to a hot spot in the catalyst bed, this highlighted the reason robotic catalyst removal is being introduced as an alternative to manual vacuum unloading. A new set of 3D printed aluminum screws are currently in production which will be able to withstand catalyst temperatures of greater than 200 degrees Fahrenheit.

4 Future Work

In some cases, catalyst can be fused, normally due to a process upset prior to the shutdown or extended runs between change-out. During in-house tests, CAROL has broken through a layer of crust created using a carbon and mortar mixture. The screws on the current unit are effective in catalyst conditions with mild agglomeration or 'shelving' (build-up of sticky material against the walls after dumping). However additional implements are being developed to extend the capabilities to heavily agglomerated or coked beds, typically encountered in heavy oil applications.

The CAROL robot is currently remotely operated using a joystick and video camera footage from inside the vessel. A joint industry project with the Queensland University of Technology is underway to implement automation of the CAROL movement around the bed. The goal is for the CAROL amphirol to automatically move around the bed to maintain an even catalyst service and to automatically adjust the vacuum head position to optimize catalyst removal efficiency.

5 Conclusion

WorleyParsons and its partners have developed the industry's first commercial fixed bed catalyst unloading robot. CAROL (Catalyst Removal Amphirol) aims to reduce the associated risk with placing people inside vessels, hopefully kick-starting a revolution for the onshore oil and gas industry, like that which has occurred with ROVs in the offshore industry, and the associated phasing out of the current manual-labor-intensive process.

CAROL has a straightforward design with few moving parts. By using screws for propulsion, CAROL moves freely on the surface of loose material. It is light weight and easy to maneuver inside the vessel using a joystick and live video feedback. CAROL facilitates effective vacuum removal of catalyst with no requirement for human entry during the bulk catalyst unloading phase of vessel change-out.

WorleyParsons have challenged the status quo that catalyst unloading must rely entirely on personnel–CAROL has the potential to radically alter what has been done the same way for almost 75 years. This represents one example of the way that WorleyParsons is looking to develop and support modern technologies, in line with its strategy to be a key player in the New World.

6 References

[1] Oil & Gas Journal online research center, "2017 Worldwide Refining Survey", Data as of Dec 5, 2016.

[2] Mourik Catalyst Handling Services, Available at http://www.mourik.co.uk/catalyst-handling-services.php Accessed on January 14, 2019.

[3] Catalyst Handling Resources, Available at http://www.catalysthandling.com/ Accessed on January 14, 2019.

[4] Malewitz et al (2015). “A Deadly Industry,” Available at: http://www.ehstoday.com/safety/deadly-industry Accessed Jan 10, 2018

[5] U.S. Chemical Safety and Hazard Investigation Board, “Hazards of Nitrogen Asphyxiation”, Safety Bulletin No. 2003-10-B, Jun, 2003

Using Unmanned Aerial Vehicles and Robotics in Hazardous Locations *Safely*

Kathleen A. Kas, PE
The Dow Chemical Company
4460 State Highway 225, Deer Park, TX 77536
KKas@dow.com

Gary K. Johnson, PE
The Dow Chemical Company
21255 LA Hwy 1 P.O. Box 150, Plaquemine, LA 70765
JohnsonGaryK@dow.com

Prepared for Presentation at
American Institute of Chemical Engineers
2019 Spring Meeting and 15th Global Congress on Process Safety
New Orleans, LA
March 31 – April 3, 2019

Using Unmanned Aerial Vehicles and Robotics in Hazardous Locations *Safely*

Kathleen A. Kas, PE
The Dow Chemical Company
4460 State Highway 225, Deer Park, TX 77536
KKas@dow.com

Gary K. Johnson, PE
The Dow Chemical Company
21255 LA Hwy 1 P.O. Box 150, Plaquemine, LA 70765
JohnsonGaryK@dow.com

Keywords: Unmanned Aerial System, UAS, Unmanned Aerial Vehicle, UAV, Drone, Robotics, Robot, Crawler, Submarine

Abstract

Confined space entries and elevated work are two of the most hazardous practices in the chemical industry. According to 2017 data collected by the U.S. Department of Labor, over 160 fatalities occurred during confined space entries and over 800 workplace fatalities were attributed to falls [1]. As part of its breakthrough Manufacturing 4.0 initiative, The Dow Chemical Company has been employing devices such as Unmanned Aerial Vehicles (UAVs or "drones") and robots to eliminate the need for employees to enter confined spaces or conduct elevated work, while allowing effective, remote inspections. UAVs and robots have been equipped with cameras to allow remote inspection of areas that before required placing a person in confined spaces and other hazardous locations. Since piloting the use of UAVs and robots in 2016 at Dow, over 400 confined space entries have been avoided. UAVs are also being used to inspect distillation columns, flares, transmission towers, and pipelines – avoiding the need for over 2,200 elevated work inspections by humans. Robotic miniature submarines are also being used to allow remote sensing for underwater inspections. The role of UAVs and robots in safety is continuing to expand beyond gathering visual data; robots are now equipped with sophisticated tools for gathering samples and non-destructive data, enhancing the ability to monitor environmental performance and ensure mechanical integrity. They are also being used to perform functions such as dismantling and removing internal equipment, cutting/ sawing, hydroblasting, vacuuming sludge, and performing vessel wall/pipe thickness, ultraviolet, gauss (magnetic), and infrared scans.

The benefits of UAVs and robots in reducing human risk, enhancing data collection, and improving productivity are quite clear; however, the hazards of these technologies need to be carefully evaluated to ensure that the devices can be used safely in specific applications. UAVs can present collision hazards, and a poorly chosen flight path could result in human injury, significant damage to process equipment, or the loss of the UAV. During internal equipment inspections, UAVs and robots can be made of incompatible materials of construction that could contaminate or damage the equipment being inspected or the UAV or robot itself. Also, during internal inspections, UAVs and robots can become disabled or lose power inside of the equipment, requiring emergency extrication. Another key consideration is hazardous area classification. UAVs and robots are energized devices that could potentially present ignition sources to flammable materials and combustible dusts, so consideration of the hazardous area classification of the work area (internal and external to the equipment) – as well as the certification of the robotic equipment for the specific application - is vital when employing these technologies.

Manufacturing 4.0 provides a compelling vision of safer production operations. The use of UAVs and robots can significantly reduce human exposure to process and personal hazards. However, to capitalize on this potential, it is critical that organizations employing these technologies perform thorough hazard analyses, supply adequate guidance, and provide tools that can be leveraged at the site level to facilitate the development of safe work plans for specific applications. This paper discusses potential hazards that should be considered when evaluating the suitability of a UAV or robot for a particular application and provides guidance on leveraging hazard recognition tools and training to the site level.

1 Introduction

Unmanned Aerial Vehicles (UAVs) and robots have been in existence for decades, primarily for military applications. Over the past 10 years, advances in technology and decreasing costs have enabled commercial and even private use. As a result, UAVs and robots have been used for a variety of purposes within industry. These include:

- Photography and video
- Mapping
- Security/surveillance
- Emergency response
- Search and rescue
- Construction monitoring
- Accident investigations
- Facility/structural inspections
- Sampling
- Non-destructive testing
- High-pressure cleaning
- Cutting, drilling, and other mechanical activities

This paper shares aspects of Dow's experience in this area. While considered correct, one must confirm applicable legal constraints and the appropriateness of a specific application before using any UAV or robot.

Figure 1 provides some examples of UAVs and robots in use within Dow.

Hydraulically powered crawler with grab arm used to place jack stands under a failed internal floating roof. This same robot was later outfitted with a cutting saw to remove the floating roof in pieces.

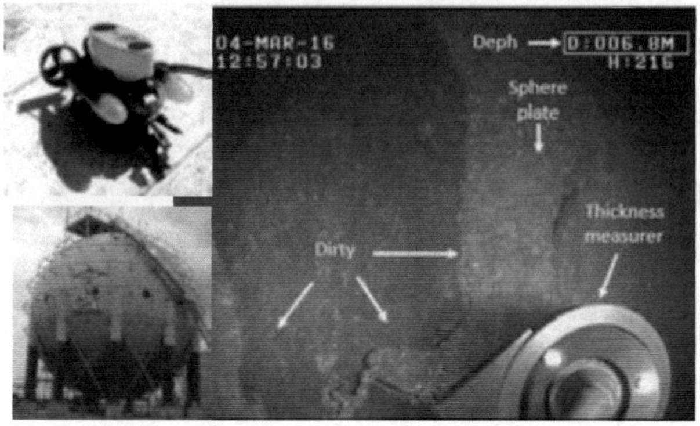

Submarine with camera and ultrasonic thickness detector used to inspect the internals of a spherical gas storage tank (evacuated and filled with water).

UAV (drone) performing a visual inspection of a pipe rack system with high resolution camera.

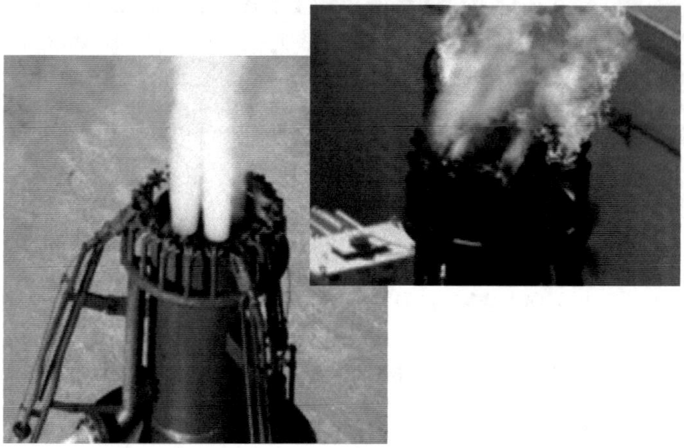

UAV (drone) photos taken while performing flare tip inspections.

Figure 1. Photos of UAVs and Robots in Use within Dow

Benefits of the use of remote-operated technologies in process operations include:

- Reduced health and safety risks to humans
- Access for hard-to-reach equipment
- Real-time image viewing
- Enhanced maintenance inspection capabilities
- Reduced time and cost for maintenance

Dow recognized these benefits and began a corporate-wide Unmanned Aerial Vehicle (UAV) program in 2016. Over the past three years, use of UAVs has increased significantly. Dow later expanded the program to include internal inspections of process equipment. In 2018, Dow completed over 1,500 UAV flights in the United States; approximately 10% of those flights consisted of internal process equipment inspections.

Since implementing the use of UAVs, Dow has achieved tens of millions of dollars in savings as a result of decreased costs for inspection/maintenance labor and material/contracted services (such as scaffolding), as well as productivity savings due to reduced downtime. Most importantly, Dow has avoided over 400 confined space entries, eliminating the need for workers to manage the hazards associated with these activities.

In addition to UAVs, Dow is also using robots for a variety of applications, including inspection, cleaning, and other mechanical activities. Examples of robotics include submersibles and crawlers. In addition to cameras, robotics can also be equipped with a variety of attachments, such as specialized sensors, high pressure cleaning nozzles and cutting blades. Figure 2 provides examples of robotic devices.

Track-driven submersible robot, small and lightweight, used for pipe and sewer inspections

Rover with camera and grab arm

Hydraulically powered robot with sludge vacuum

Submarine operated by an umbilical cord

Battery-powered multifunction track robot

Pipe inspection robot - submersible, magnetic track

Magnetic wheel robot for tanks, pipes and vessels

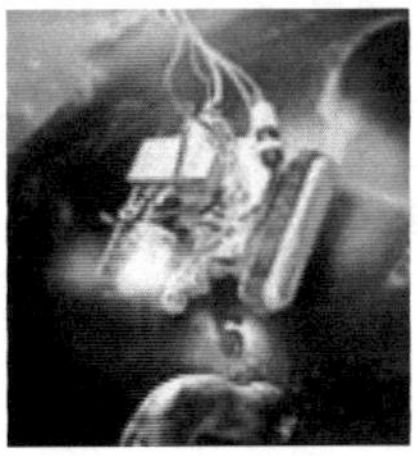

Tank inspection robot

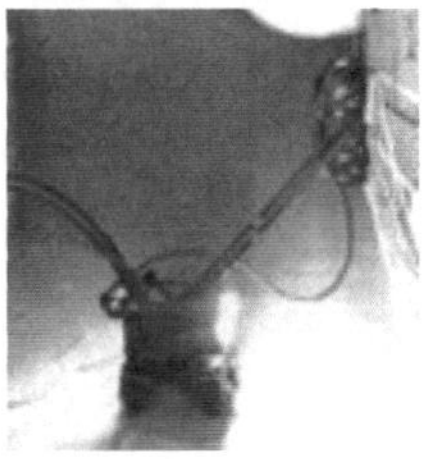

Tank robot with attachable side wall ultrasonic thickness sensor

Figure 2. Examples of Robotic Devices

The use of UAVs and robots has been shown to save money and improve worker safety. However, there are risks associated with these technologies that need to be managed in order to obtain the desired benefits safely. UAVs and robots are often electrically-powered devices and represent potential ignition sources. Battery-powered devices contain a power source which cannot be deenergized. Intrinsically safe devices operated by an umbilical cord have limitations on cable length; thus, controller placement relative to hazardous area extents is impacted. Pneumatically-operated or purged devices using air may not be compatible with certain process materials. UAVs in combustible dust areas can create a dust cloud from their propeller ("prop") wash and can also create static electricity from the particulate movement of non-conductive dusts. Robots can collect and track dust layers like footprints, expanding dust areas. Even hydraulically-driven robots may also need to be evaluated for autoignition potential where elevated oil temperatures may develop. UAVs have collided with towers, power lines, birds, and even planes. Other potential consequences include power outages, process upsets, and process safety events.

Like any activity in a chemical facility, it is critical to ensure that the appropriate technology is selected for the application, job pre-planning and safety analyses are completed, and plans are in place for safe task execution – including plans for how to respond in the event of equipment failure or other unexpected conditions.

2 Definitions and Key Regulations

2.1 Definitions

Before embarking upon a detailed safety discussion, it is important to first have a common understanding of the terminology associated with these technologies. For the purposes of this paper, the following definitions will be used:

Unmanned Aerial Vehicle (UAV) – The flight operable device used to conduct photography, inspection, or testing (often referred to as a "drone").

Robot – A device, such as a crawler or submarine, used to remotely perform a task.

Nationally Recognized Testing Laboratory (NRTL) - A governmentally-approved testing facility authorized to confirm the suitability of design, most common to the Occupational Safety and Health Administration (US) and the Canadian Centre for Occupational Health and Safety.

Notified Body (NOBO) - A governmentally-approved testing facility authorized to confirm the suitability of design, most common to ATEX regulations in the European Union.

Pilot – The individual who controls the UAV or robotic device. Depending on the jurisdiction, pilots may require certain certifications or training.

Visual Observer – A second set of eyes, helping to monitor the UAV and communicate with the pilot to provide improved situational awareness.

Note: The term "UAV" will be used in this paper to represent devices that fly and the term "robot" will be used to represent other devices.

2.2 Key Regulations and Standards

In addition to using common terminology, it is also important to understand the key regulations and standards that apply in the jurisdiction in which the UAVs will be operated. In addition, legal requirements applicable to UAV and robot use are rapidly changing and should be monitored. Some current key rules and standards include:

2.2.1 United States Federal Aviation Administration (FAA) Rule, 14 CFR Part 107 [2]

This regulation provides guidance for the use of UAVs in the United States. The regulation classifies UAVs by size. Specific regulations apply, depending on the size of the UAV, including maximum altitude permitted. This regulation also requires that small UAVs:

- be registered,
- weigh less than 55 pounds,
- be operated with visual line-of-sight, and
- be operated during daylight or civil twilight below 400 ft in elevation.

FAA approval is required to operate a UAV within a controlled air space in the United States. Use of a visual observer may also be advised in certain cases. Other regional aviation authorities may have similar requirements.

2.2.2 National Fire Protection Administration (NFPA) 2400, *Standard for Small Unmanned Aircraft Systems (sUAS) Used for Public Safety Operations* [3]

NFPA 2400 provides guidelines on the implementation of programs for the use of small UAVs, including professional qualifications for personnel, UAV maintenance, and the safe operation of UAVs. Although intended for public safety applications (such as fire and police use), general guidance on the use of UAVs is included.

2.2.3 Other References

Other regional rules, standards, and useful guidance include:

- *Verein Deutscher Ingenieure (VDI) 2879* [4] – This German standard covers the inspection of installations and buildings using UAVs.
- *Underwriters Laboratories (UL) Subject 3030 – Standard for Unmanned Aircraft Systems* [5] – UL 3030 addresses the electrical system design of UAVs and the risks of electrical shock, fire, and explosion hazards associated with inherent features of UAVs, batteries, and charging systems.
- *ABS 242 – Guidance Notes on Using Unmanned Aerial Vehicles (UAVs)* [6] – This document provides the authors' guidance and best practices regarding the use of UAVs, with an emphasis on marine and offshore applications.
- *International Organization for Standardization (ISO) 19901-9 Petroleum and natural gas industries– Specific requirements for offshore structures Part 9* (draft) [7]. This standard (currently in draft) provides guidance addressing the use of UAVs around flammable materials and combustible dusts.

3 Use of UAVs and Robots External to Process Equipment

3.1 Planning

Planning the safe use of a UAV or robot requires evaluation of a number of important safety and logistical considerations, such as:

1. Objectives – The initial step in selecting the appropriate technology is to consider the objectives of the activity. UAVs and robots have been used for a number of activities, including photography and video, facility/structural/equipment inspections, sampling, non-destructive testing, high-pressure cleaning, cutting, and drilling. It is first necessary to describe what needs to be achieved as well as the environment in which the operation will take place so that the correct technology can be selected. In some cases, more than one device may be needed to achieve the objectives.

2. Pilot qualifications – In the US, the use of an FAA-certified UAV pilot is required for commercial operations. Outside of the US, pilot qualifications may be more or less stringent. Some organizations may choose to specify a minimum level of certification for their corporate UAV program above that required by local regulations.

3. UAV registration – In the United States, the FAA requires that all commercially operated UAVs be registered [2]. Other regional aviation authorities vary in their expectations for registration. If using a contractor, it may be prudent to verify that the UAV to be used is properly registered where required.

4. Data transmission – Signals are transferred between the UAV and controller wirelessly, the same way that data is captured from the UAV. Frequently, in process operations, other devices also transmit data wirelessly, creating the potential for interference that could affect process operations as well as the operation of the UAV. It is important to verify that the UAV communication/control frequency is not the same wavelength that is used by the site's internal wireless communications and controls, so that interference would not result. In addition, some regional regulatory authorities (e.g. Health and Safety Executive in the United Kingdom [8]) have limitations on the emission energy of wireless devices in proximity to hazardous (classified) locations.

3.2 Hazardous Area Classification (HAC)

One of the most important safety aspects to consider when evaluating UAV and robotic technology is the hazardous (classified) locations where flammable materials or combustible dusts exist. UAVs and robots represent potential ignition sources; motors may have brushes that generate sparks, impact or crashes could dislodge electrical connections to the battery, or the battery could fail catastrophically. To avoid a potential ignition of flammable materials or combustible dusts in the area, proper evaluation and planning for operations in hazardous (classified) areas are some of the most critical – and most challenging – aspects of safely using these technologies. Options include managing the potential for ignitable materials near the device and reducing the potential for spark generation. This can be done in several ways:

1. Selecting a system that is rated for the HAC of the travel route and inspection area
2. Removing flammable materials and combustible dusts from the process area to temporarily change the HAC rating
3. Utilizing a Hot Work Permit and safe operating discipline

Each approach has advantages and disadvantages. Some combination of approaches may be needed to maintain a tolerable level of risk during a remote operation. These three potential approaches are discussed in more detail below.

1. Select a system that is rated for the HAC of the travel route and the area to be inspected

Regulatory authorities require the utilization of fixed equipment that is rated for the hazardous classification of the area in which the equipment will be used. It would therefore be logical to select a remote-operated device that has been certified for the hazardous (classified) location. However, this can sometimes prove to be a challenge. The majority of UAVs and robots available today include no certifications for use in hazardous (classified) locations. Those that are available with HAC certifications are rarely certified for use in all geographic locations and for all process materials (e.g. approved for all group ratings and autoignition temperatures).

NOTE: Various suppliers of UAVs and robots have represented their equipment as suitable for hazardous (classified) locations. However, some of these suppliers have not recognized that the assembly of the UAV or robot from certified components does not make a system certified. Systems that are claimed to be certified but do not have NRTL or NOBO validation may not be suitable for hazardous (classified) locations and may not be aligned with standards and requirements of the appropriate jurisdictions.

> ***Real Example:***
> A vendor indicated that a robot was certified for Class I, Division 2 service because the camera component mounted on the robot was certified. However, the assembled robotic unit was never certified as a system - and was actually not suitable for use in a hazardous (classified) location since the wiring and power management methodologies were not compatible with the HAC service.

It should also be noted that certification requirements are not globally uniform. In the US, OSHA expects that equipment designed for service in a hazardous (classified) location have an independent NRTL certification [9]. For ATEX-aligned countries, ATEX requires a NOBO certification for equipment to be used in a Zone 0/1 or Zone 20/21 location. However, equipment certified for use for Zone 2/22 locations are allowed by ATEX to be certified by the manufacturer by a Declaration of Conformity – a certification that the device meets a minimum set of technical and regulatory requirements [10].

2. Remove flammable materials/combustible dusts from the process to temporarily change the HAC classification

Using this approach, flammable materials and combustible dusts are removed from the area to temporarily change the HAC rating. There are certainly some disadvantages to process material removal – including downtime, cost, and waste generation. However, removing the fuel source remains an effective way of using remote equipment or devices which have not been certified for hazardous (classified) service.

3. Utilize a Hot Work Permit with safe operating discipline

In many cases, it is not possible to provide UAVs or robots that are certified for the hazardous (classified) location. Except during major turnarounds or maintenance activities, it also may not be possible to de-inventory the equipment being inspected or adjacent equipment or tanks. In this case, use of Hot Work Permitting in conjunction with safe operating discipline may be an appropriate way to manage the risk.

Standard Hot Work policies typically require a portable gas detection test in the area where electrical devices are to be used. This practice would generally work equally well for most robotic activities. However, for a UAV that is operating 60 meters (200 feet) in the air, a gas detection test is generally not possible. Therefore, the safe work evaluation also needs to consider material properties, operating conditions, maintenance activities, and ongoing work. In a facility where the flammable vapors/gases are heavier than air, approaching the hazardous location at elevated heights can help mitigate the risk. Where vapors and gases are lighter than air, areas directly above potential release points need to be avoided or observed at a distance using a high-quality camera.

3.3 Failure Modes

It is also important to consider how the UAV or robot might fail and whether it is necessary to ensure that the device has certain safeguards incorporated into the design to mitigate consequences from a failure. Potential failure modes include:

1. Collision with structures – The UAV flight path should be pre-planned and mapped in three dimensions. The UAV flight plan can be controlled with operational limitations that prevent it from running into objects and obstructions.

   Collisions are also a concern for robots as well. Some robots are heavy and powerful, and they do have the potential to inflict damage on plant equipment in the event of impact. A maximum robot speed should be specified to maintain control, and work paths should be defined to avoid contact with process equipment. Robot travel paths should be as direct as possible and ensure that cables, tubing, and hoses do not get snagged on structures or equipment.

2. Loss of power – Some robots operate with an electrical cord, providing a continuous source of power. Heavy duty service cords with outer protective jackets that meet the hazardous area classification needs are recommended. Cabling should

be routed to prevent the cable from becoming a tripping hazard and to avoid potential damage to the cable from pinching or sharp edges.

Some robots - and most UAVs - operate on battery power with a limited battery life. It is necessary to determine the battery life of the unit and define limits for UAV flight time/robot service time to ensure that recovery can be completed.

3. Controller failure - Loss of control of the UAV is a significant concern. Some UAVs are designed with self-diagnostics and recognize failures such as loss of signal, excessive yaw, excessive vibration, and low battery voltage. Upon controller failure, some UAVs can return to a pre-programmed safe altitude and return to a home position. For those UAVs that do not have these features, select flight paths which limit the risk of equipment damage or personnel injury. Consider restricting personnel access to the flight path.

   Although some robots have power cords, they are not often rated to be used to support the weight of the robot. Pull chains/cords may be beneficial for device retrieval purposes if the work path allows for their use without interfering with safe device operation. Otherwise, alternative plans for extrication are needed, which may also include use of a crane or lifting device.

4. Uncontrolled descent – An uncontrolled descent of a UAV can happen as a result of a number of failures including loss of a propeller motor, drained battery, collision, or electrical failure. A good practice is to restrict the area beneath the UAVs flight path, to prevent injury in the event of an uncontrolled descent and to avoid process areas that might be adversely impacted by an uncontrolled UAV impact.

3.4 Route Selection

The selection of the travel route for the UAV or robot requires careful evaluation. A key aspect of route selection involves an assessment of any hazardous (classified) areas through which the device will travel or in which the device will conduct work activities. In addition to hazardous area classification, other factors should be considered when determining the travel route. This section discusses hazardous area classification and other factors to evaluate when selecting the most appropriate route for the remote-operated device.

3.4.1 Route Selection: HAC Considerations

When planning the travel route of a UAV or robot, it is important to consider the hazardous classification(s) of the locations(s) above/through which the device must travel. Remote-operated devices could be potential ignition sources, and both the certification of the device and the hazardous classifications of the locations need to be evaluated to prevent the potential for an ignition source in areas which may have an ignitable atmosphere. This section discusses area classification considerations when selecting a travel route.

It is certainly preferable to choose a route for a UAV or robot which does not enter a hazardous (classified) location. However, this may not be feasible. Though it is not appropriate to fly an uncertified UAV directly though Class I, Division 1/Zone 1 location, it might be suitable to fly near/over such areas in transit where they typically represent a small footprint and the reasonable risk for falling into such areas is slight. These types of areas include the following:

- Sumps
- Trenches
- Tank breather vents
- Pressure relief devices
- Process vents

However, where hazardous (classified) locations rated Class I, Division 1/Zone 1 do represent a larger footprint, such as above an open external floating roof tank, it would be preferable for a UAV to avoid passing over, or hovering over, these areas.

It is recommended that there be a review of operations and, if needed, actions taken to minimize the likelihood of releases from relief valves and process vents. Should process conditions change, flight activities should be ceased immediately.

Combustible dusts can become hazardous when suspended, and UAVs present a significant risk of creating a suspended cloud even when only small quantities of dust are present. UAVs are likely not suitable in the presence of combustible dusts, even with surface levels of dust that would normally support a classification of Zone 22 or Class II, Division 2. Likewise, robotics may not be suitable in such conditions if the overall travel of the robot – and the cumulative dust layer thickness collected on the robot – could exceed levels that would allow the area to be considered unclassified.

3.4.2 Route Selection: Other Factors

In addition to consideration of the hazardous area classification(s) throughout the route of travel and in the work area, other factors should be considered when defining a route.

1. Birds – Birds in flight can represent a significant collision risk to UAVs. The hazards of nesting or resting birds on transmission lines and pipe racks should be considered, especially if there is a potential for a flock to take flight simultaneously, perhaps if frightened by the UAV.

2. Flight instability - Avoid flying above cooling towers, fin-fans, flares, and other equipment which could create air turbulence and disrupt the UAV's flight stability.

Real Example:
A UAV flight intended to inspect a flare passed by an elevated pipe rack where a large number of doves were roosting. As the UAV approached, over a dozen doves disturbed by the UAV took flight directly in the UAVs path. No collision occurred, but the doves flew within feet of the UAV.

3. Elevated structures – Planning a UAV flight path at any point in time is a three-dimensional exercise. Elevated objects need to be identified and avoided, especially if impact could lead to damage to important plant infrastructure, loss of power, process upsets, and process safety events. Figure 3 includes a number of examples of elevated structures typically encountered in an operating facility.

230kv Transmission line

Switchyards

Overhead Pole Lines and Cable Trays

Vent Stacks

Elevated Flares

Distillation Columns

Figure 3. Examples of Potential Elevated Obstacles for a UAV Flight Path

A best practice is to mark the flight path – including the "no fly zones" – on a copy of the Hazardous Area Classification drawing, including an indication of intended flight elevation. Include the time or conditions during which the flight path has been planned. Ensure that the process owner reviews the plan and supports the indicated flight path.

3.5 Planning and Execution

Once a route is selected, additional factors should be evaluated to ensure safe operation of the remote-operated equipment, including contingency planning for unexpected events or changing condition. These factors include:

1. Flight planning – Activities involving remote-operated devices should be pre-planned prior to the start of the operation. A key aspect of planning includes a hazard assessment and identification of any necessary safeguards to mitigate the risks. Plant

personnel may not perform these operations often, and they may not have expertise in identifying the potential hazards associated with these technologies. Use of a standard UAV checklist developed by Subject Matter Experts (SMEs) is recommended to ensure that plant personnel consider all relevant factors when planning the operation. The need to obtain a Safe Work Permit (and possibly a Hot Work Permit) should be identified during the planning phase.

2. Approval – If the UAV is to fly over private land, the flight should be pre-approved to respect the privacy of the owners and avoid photographing intellectual property belonging to others.

3. Pilot location - The pilot should be located where there is a direct line-of-sight of the flight path. It is also important for the pilot to have a safe egress path in the event of a plant emergency. It is preferable to locate the pilot in an unclassified location since UAV controllers are not always rated for hazardous (classified) locations. When the pilot needs to be located in a hazardous (classified) area in order to have direct line-of-sight of the UAV, certified controller equipment or appropriate safety permitting procedures should be followed.

4. Equipment inspection – The UAV or robot should be inspected prior to use for potential defects such as loose wiring, damaged batteries and components, or contamination from previous applications.

5. Location of batteries and the battery charging unit – UAVs and remote-operated robots have finite battery life, and a battery charging station is typically required. Spare batteries should be available near the flight/operational base. While in use, the battery station should be located in a continuously attended unclassified location. During the charging of lithium-based batteries, a bucket of sand and a dry fire extinguisher should be available in the event of a battery fire.

6. Obstructions – The take-off/landing location of UAVs should be free from obstructions and allow for the pilot to have line-of-sight. Where the pilot cannot maintain line-of-sight, use of a visual observer should be considered. The route chosen for the UAV or robot should be walked or otherwise reviewed just prior to flight to ensure that the path is clear and unexpected obstacles have not been introduced.

7. Weather – UAVs and other robotics should be operated in weather conditions which provide sufficient visibility and a stable path. Define the weather parameters (winds, rain/snow, temperature, etc.) for which the device is rated and specify under what conditions a flight is to be discontinued.

8. Communication – A communications protocol should be developed between the pilot/operator, the visual observer (if used), and plant personnel. The pilot/operator, who may not be familiar with the facility, should be trained on how to respond in the event of an alarm or be stationed with plant personnel who can give direction in case of alarm. In addition, the pilot/operator and visual observer (if used) should be in

communication with the plant so that they are aware of changing conditions or operations that might result in the need to abort the operation. (If the area is noisy, hand signals might also be necessary and should be coordinated in advance.) It is also important to notify the plant personnel at the beginning and end of an operation; the unexpected appearance of a UAV in the unit could create a distraction that could result in a personal injury.

9. Cessation of Work – The flight or travel plan may require that certain activities be suspended during the remote operation. This may include cessation of material transfers, sampling, sump/trench operations, or maintenance involving line and equipment opening associated with flammable materials or combustible dusts.

10. Discontinuation of UAV or robot activity - Upon any change in operations (such as an unplanned event), operating discipline should ensure that all remote-operated device activities immediately cease.

Selecting a travel route, completing a job plan, and safely executing the activity according to the plan requires the input of multiple individuals with various areas of expertise. Input may be needed from the device pilot/operator, plant owner, Environmental, Health and Safety (EH&S) expert, and someone knowledgeable in hazardous area classification principles as well as the specific hazardous area classification(s) of areas within the facility. Additional notifications/approvals may also be required by certain roles within the facility or organization as well authorities, neighbors, and other stakeholders. It is critical to obtain the necessary input to effectively plan for the safe execution of a UAV/robotic operation.

4 Use of UAVs and Robotics Inside Process Equipment – Special Considerations

Use of remote-operated equipment inside process equipment presents unique challenges. This equipment may contain, or have contained, combustible dusts or flammable, pyrophoric, water-reactive, oxygen-reactive, or toxic materials. The potential for damage or contamination of process equipment, and the need to decontaminate the UAV/robot after use, require careful evaluation and planning. The remote devices may not be within the line-of-sight of the pilot, and internal obstacles may present a challenging flight/travel path. A retrieval plan should be defined prior to the activity in case the event the UAV/robot becomes disabled inside the equipment. These special considerations are discussed below.

4.1 Hazardous Area Classification

To prevent an ignition of flammable materials or combustible dusts, at least one of the three sides of the "fire triangle" must be removed:

1. Fuel Source
2. Oxidant
3. Ignition Source

This section covers special considerations for evaluating the potential use of UAVs/robots in hazardous (classified) locations inside process equipment.

1. Fuel Source

One means of preventing a fire or explosion is to remove combustible dusts and flammable materials from the equipment prior to introducing the UAV or robot – in essence, reducing the internal hazardous area classification. This can be challenging, especially when the drain is not located at the equipment low spot, leaving residual pools of material.

However, draining the flammable material alone will likely not change the internal hazardous area classification and make the environment suitable for introducing UAVs or robots. Generally, the equipment should then be cleaned, flushed, and ventilated until the concentration of vapors is less than 10% of the Lower Flammable Limit (LFL). Also, there may be circumstances where additional steps or precautions could be necessary. For example, for toxic or carcinogenic materials, ventilation thresholds more restrictive than that needed for 10% LFL may be required.

Removal of process materials and residues can be especially challenging in systems containing packing and other surfaces that cannot be easily cleaned or inspected. Special precautions should be taken when ventilating systems with the potential to create pyrophoric residues when wetted material is dried.

"Prop wash" can create several hazards when a UAV is introduced into process equipment that has not been thoroughly cleaned:

1. A hazard can continue to exist in systems where a sludge or solid layer remains after draining, cleaning, and ventilation. In some cases, the internal concentration of flammable vapors might be verified as suitable using a portable gas detection unit. However, robotic activities such as hydroblasting, vacuuming, or even the air movement created by a UAV can create mixing/movement of residual material, releasing trapped ignitable material or combustible dust.
2. Materials with low Minimum Ignition Energy (MIE) levels require special scrutiny when UAVs are used inside process equipment. The particulate movement or atomization of residual material from "prop wash" can increase the internal concentration of flammable materials from heel material, sludge, or residues trapped in internal structures such as internal packing while simultaneously generating static electricity.
3. UAV air movement can create a combustible dust cloud. Since commercial UAVs are not typically rated for Class II Division 1/Zone 20 service, it is important to clean the area sufficiently to prevent dust cloud formation.

Continuous ventilation is one approach that is used to remove flammable vapors and temporarily alter the hazardous area classification inside the equipment. However, ventilation with high air flow rates may create air instability and make control of a UAV difficult. The equipment owner can provide input on locations where air flow or turbulence may be increased within specific locations of the equipment, particularly in narrow areas. Consult with the UAV pilot regarding specific air flow restrictions for the UAV that are necessary to maintain controlled flight.

2. Oxidant

Another means of preventing ignition of flammable materials and combustible dusts is to remove the oxidant, generally air, from the system. By reducing the oxygen concentration to below the Minimum Oxygen Concentration (MOC), the concentration inside the equipment no longer supports combustion.

Since the process equipment must be opened to introduce a remote-operated device, individuals working near the equipment opening may need to wear supplied air respirators to avoid the risk of asphyxiation. Also, if the device needs to pass through a short transitional region with the potential for ignition during insertion/removal of the device, additional safety measures may be needed. For example, testing the interface region for % LFL might be appropriate, and the power supply to the remote-operated device might need to be de-energized until the device is located in a sufficiently inerted location.

Though UAVs and robots might be used inside evacuated or inerted process equipment, the hazardous area classification external to the vessel should also be considered. Certification or safe work permitting may still be needed for activities external to the vessel.

It is possible that the use of robotics might be considered inside tanks containing combustible dusts or flammable materials which have not been emptied of their contents. This practice could result in significantly less downtime and waste. In these cases, robotics certification to Division 1/Zone 0/Zone 20 is necessary in the absence of other mitigation measures. In addition, use of a combination of safeguards might be necessary, such as LFL and oxygen monitoring at the tank opening, use of proper Personal Protective Equipment (PPE) by employees at the equipment opening, and de-energizing the robot during introduction and removal. A hazard review and risk mitigation plan should be developed for these activities, involving representatives from the plant as well as Process Technology, Process Safety, EH&S, and Industrial Hygiene Experts.

Note: Most standards generally consider the area inside process equipment below the liquid level to be unclassified. The American Petroleum Institute (API) Standard 2000 [11] provides guidance on conditions that could be used to allow nitrogen inertion as a means to consider the internal headspace above liquid level inside of a tank as unclassified while flammable liquid remains. This approach could be a means to satisfy conditions where robots cannot be certified as suitable for Division 1/Zone 0 service.

3. Ignition Source

Often, UAVs and robots contain electrical components that could represent ignition sources. Some robots use hydraulics or electrical components which can generate temperatures in excess of material autoignition temperatures. As such, it is important to use a device that is certified for the hazardous (classified) location in which it will be used and is validated to meet material temperature limitations, or is operated under administrative controls.

Commonly, for equipment storing or processing flammable material, the internal classification of the equipment is Division 1 or Zone 0. This classification does not support the insertion of non-certified equipment without measures being taken to temporarily declassify the equipment. A few commercially available UAVs and robots are certified for service in hazardous (classified) locations rated Division 1 or Zone 0. Often, the certification documents also have restrictions of use which must be followed, such as when to energize the device and operating temperature requirements. However, these certifications are limited to addressing hazardous area classification needs only and could introduce additional hazards that need to be managed.

Real Example:
A robot, remotely operated using an umbilical, was under consideration for use in several tanks containing flammable materials. The robot did have an ATEX/IECex certification. However, it was not initially recognized that the certification permitted energization of the robot only after it was lowered through the Zone 0 tank headspace and positioned below the surface of the material. A further review of the certification indicated that the robotic assembly was only certified for Group IIA materials. Since one of the tanks contained a Group IIB material; the robot would not be suitable for inspecting this particular tank.

4.2 Other Considerations

While management of flammable and combustible dust hazards is a critical consideration for the safe use of UAVs and robots, there are other aspects of their internal use that merit consideration.

1. Compatibility – The device should be evaluated for compatibility with the materials normally contained within the process equipment. Acids, bases, and solvents could damage the remote-operated equipment. Likewise, it is possible that contamination from the remote-operated device could contaminate process equipment. For example, some process materials might be incompatible with metals such as copper and aluminum, often found in robotic devices. Greases, oils, lubricants, and contamination from previous equipment inspections could also create product quality or product instability issues.

> ***Real Example:***
> A plant was evaluating the use of a robot to inspect a tank without removing the process chemical. The metallic components were evaluated and found to be compatible with the process chemical. However, it was subsequently determined that the soft, elastomeric components of the robot would have dissolved while immersed in the process material.

The remote-operated device should also be evaluated to ensure that it will not damage the process equipment in which it will operate. For example, UAVs and robots could potentially damage softer materials, such as non-metallic vessels and lined equipment. The UAV/robotic device weight, available motive force, and energy should be considered when evaluating the remote-operated devices for suitability.

2. Introduction/removal of the device from the equipment – The use of UAVs typically require the operator to insert the UAV and recover the UAV by hand. As a result, the operator might be exposed to the atmosphere inside the equipment as well as to any residue that has been picked up by the UAV via contact with the equipment surfaces. The operator should wear appropriate PPE to protect from exposure to chemicals and to an oxygen-deficient atmosphere, where required.

   Often, the remote-operated device is introduced into the equipment at grade level; sometimes, ramps are built so that robots can crawl directly into the equipment. For heavier robots, this can reduce the need for lift equipment such as cranes. In other cases, remote-operated devices need to be introduced to process equipment in less convenient locations, such as from small platforms at elevation. Consideration should be given to lifting, fall protection, and other hazards associated with the means of introducing and removing the device. For example, for device entries using manways on the top of the vessel, some robots require hoist assemblies mounted on the top of the tank to lower and retrieve them. Some robots weigh in excess of 100 kg (220 lbs), and validation of tank roof structural integrity may be required for tank-mounted hoists. Also, when lifting a device to the top of a tank or vessel platform, cranes and manlifts may be required. A lift plan is recommended.

3. Contamination/decontamination – Robots or UAVs in flight inside the vessel will likely contact the vessel walls and internal components, contaminating the device. In most cases, the owner would like to decontaminate and re-use the device, requiring a decontamination plan. Additionally, robotic devices and UAVs are often transported by commercial transportation, so failure to decontaminate them properly might impact not only the owners, but also the transportation firm and potentially the general public. Decontamination may involve providing secondary containment, the use of specific cleaning chemicals, disposal of spent cleaning materials, and defining a procedure and appropriate PPE for the decontamination activity. In some cases, it may be safer to plan to dispose of the device after use rather than to try to decontaminate it. One should also consider decontamination of any ramps or lifting devices.

Note that some remaining/residual process materials inside (or on the flange/opening area of process equipment) may be classified as Naturally Occurring Radioactive Material (NORM). In these cases, consult with Industrial Hygiene specialists to determine if radioactive decontamination procedures should be developed.

4. Visibility – Unlike external operation, internal operation is typically done without the remote-operated device being within line-of-sight of the pilot/operator. Maneuvering of the device is done by means of cameras mounted on the device with visual operation performed from the controller's monitor.

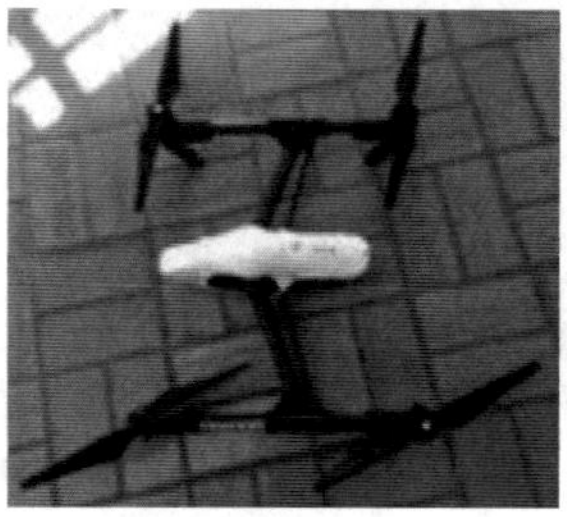

External UAV

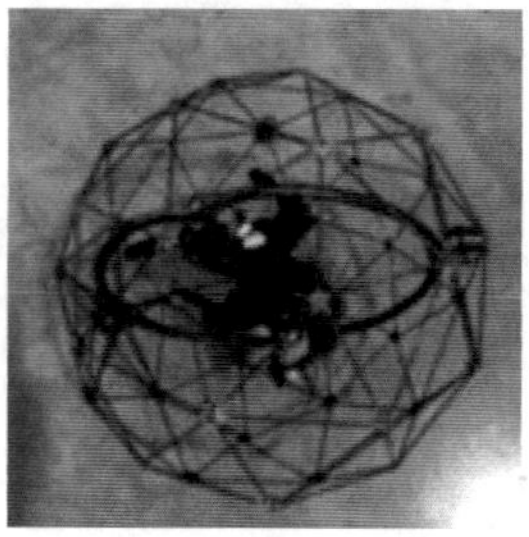

Internal UAV

Figure 4. UAVs Used External to and Inside Process Equipment

As shown in Figure 4, some UAVs are designed with cages, to protect rotors and electrical equipment from contacting process equipment. Technology is also being developed with "sense and avoid" capability. In 2018, the North Atlantic Treaty Organization (NATO) approved its first standard on this topic: *Guidance on Sense and Avoid for Unmanned Aircraft Systems*, STANREC 4811/AEP-101[12]. It is expected that this technology will reduce the frequency of contact between the device and the internal surfaces of process equipment.

5. Emergency retrieval plans – If a UAV or robot fails during an external inspection, the device can typically be retrieved from the ground where it failed (or fell). However, if a device fails inside equipment, the implications can be more significant. In most cases, it is necessary to remove the device to safely restart the equipment. This requires the development of a retrieval plan. In some cases, the device may be tethered and can be pulled out of the equipment in the event of failure. In other cases, a second device might be introduced to retrieve the failed device.

Real Example:
A UAV was in use inside a process vessel. However, it was difficult for the pilot to maneuver around obstacles, and the operation required more flight time than expected. When the low power alarm was received, the pilot immediately began the extrication flight. Unfortunately, because of the complicated return flight path, the UAV ran out of power before it could reach the manway, and it fell into the bottom of the vessel. Retrieval was later successful by "hooking" the failed UAV with a "rescue" UAV.

6. Scheduling contingency plan – When inspections or maintenance are planned using traditional confined space approaches, the activity is generally well-understood and the required time to complete the work is incorporated into a maintenance or turnaround schedule. The use of UAVs and robots can prevent the need to place humans inside confined spaces and save a significant amount of time – if the activity is successful. However, what if the remote-operated device fails or the activity is not successful? Can the schedule be delayed, or could the activity be deferred? Consideration should be given to what should be done in the event that the use of remote-operated technology is unsuccessful.

Conducting remote-controlled operations inside process equipment can provide significant benefits. However, the use of UAVs and robots inside equipment also presents special challenges that need to be managed so that the work can be performed safely and effectively. Section 4 highlighted some of these challenges and the importance of engaging experts when needed to develop safe work plans.

5 UAVs and Robotics – Organizational Implementation

Dow has achieved a significant benefit by implementing a UAV/robotics program. Other organizations may also have an interest in developing their own program to capture benefits in safety, cost, and productivity. The implementation of a UAV/robotics program can be done in a number of ways. Some organizations purchase and operate their own equipment. This approach helps ensure that equipment is suitable for the specific application, maintained, and properly registered, and that pilots are properly certified and understand corporate requirements. Although this approach can have a higher initial start-up cost, it can be quite cost-effective for regions where UAVs/robotics will be used frequently. Even one large turnaround requiring multiple inspections could justify the purchase of a UAV instead of hiring external contractors. For locations where utilization would be infrequent or where the application is specialized, the use of an experienced contractor could be more cost-effective. This approach avoids the capital outlay to purchase equipment and the cost of training internal staff to operate it. However, the contractors, their equipment capabilities, and safety rules should be evaluated and approved by the organization prior to use. Some organizations use a hybrid approach; they operate their own equipment in regions with frequent use and hire contractors in areas where the need is less frequent or where applications require specialized equipment.

Regardless of whether the equipment and pilots/operators are part of the organization or are contracted resources, corporate guidance is essential to ensure that the devices are employed safely. To ensure that best practices are captured and leveraged consistently throughout the company, training and operational discipline should be developed at the corporate level. Elements of a UAV/robot corporate program may include:

Corporate Policy or Standard

A corporate policy or standard can create a framework for the use of these technologies within the organization. Key elements may include:

- Role definition – The policy may define who needs to provide input and/or approval for UAV/robotics activities.
- Training requirements – The standard may define the minimum qualifications or training necessary to fill the role responsibilities identified by the standard.
- Legal requirements – As UAVs and robotics are rapidly developing technologies, legal requirements continue to evolve. Check with appropriate legal experts often to ensure that corporate standards remain up-to-date with regulatory requirements.
- Limitations/boundaries – The organization may have evaluated and approved the technology for certain uses; however, individual plants could potentially expand the use of the technology beyond the activities for which the technology has been assessed. A corporate standard is one means to define applications for which the technology has been approved and require a review by corporate Subject Matter Experts for proposed applications beyond those previously approved.
- Expert contacts – Providing the names of expert contacts for questions makes it more likely that facilities will consult with experts when appropriate.

Checklists and Other Tools

Checklists are useful tools to ensure that all necessary steps are consistently taken for a successful operation. Checklists may include elements such as:

- Verification that the device and pilot/operator are appropriately certified
- Documentation of the flight path, including the identification of obstacles and precautions to be taken when flying above/through a hazardous (classified) area
- Flight restrictions (e.g. weather, elevation and areas to avoid)
- Hazard identification/mitigation
- Safe work permit/hot work permit requirements
- Identification of appropriate personal protective equipment
- Emergency retrieval plan
- Communication plan
- Documentation of approvals to conduct the activity

Approved Device/Contractor List

To facilitate selection of appropriate devices and vendors at the plant level, an organization may choose to develop a list of approved devices and contractors. The list may include the types of devices approved by the corporation and the applications for which they are approved (e.g. internal inspections, external use only). If possible, devices and contractors should be identified for all key geographic regions in which the organization operates.

Training

Planning a remote-operated device activity can require specific knowledge, especially when the activity involves a hazardous (classified) location. Organizations can develop training targeted at the specific roles who will be executing responsibilities associated with these activities. Training may need to be provided in multiple languages to ensure that the training is effective in all geographies.

Communication

Communicating the "success stories" may raise the awareness of the benefits of these technologies and encourage plants to begin taking advantage of them. Likewise, communicating near misses and "lessons learned" will help to build hazard awareness and improve the safety and effectiveness of the UAV/robotic program. The use of corporate websites, email, social media, and recognition programs may enhance the awareness and ultimate success of these programs.

6 Conclusions

Since the inception of the program in 2016, the use of UAVs and robots in Dow has saved tens of millions of dollars and eliminated over 400 confined space entries. UAVs are also being used to inspect pipelines, flares, and towers – eliminating the need for elevated work, and submersibles are used to conduct underwater inspections. The benefits of UAVs and robots in reducing human risk, enhancing data collection, and improving productivity have been clearly demonstrated; however, the hazards of these technologies need to be carefully evaluated and managed to ensure that the devices can be used safely.

The planning and execution of a UAV/robotic activity may require various areas of expertise, and this expertise may not be available at all plant locations. In addition, without guidance, plants may use these technologies in applications that have not been approved by the organization. An established corporate governance program is essential to ensure that the full advantages of UAV/robotics can be safely realized.

7 References

[1] Bureau of Labor Statistics (BLS), "National Census of Fatal Occupational Injuries in 2017," US Department of Labor (DOL), December 18, 2018, available at weblink: https://www.bls.gov/news.release/pdf/cfoi.pdf (accessed on January 6, 2019).

[2] United States Federal Aviation Administration (FAA) Rule 14 CFR Part 107, *Operation and Certification of Small Unmanned Aircraft Systems*, Department of Transportation, Docket FAA-2015-0150, Amdt. 107-1, 81 FR 42209, June 28, 2016, available at weblink: https://www.ecfr.gov/cgi-bin/text-idx?SID=deaea7be76e4f4d42e84571a29d9e635&mc=true&node=pt14.2.107&rgn=div5#sp14.2.107.a (accessed on January 6, 2019).

[3] National Fire Protection Administration (NFPA) 2400, *Standard for Small Unmanned Aircraft Systems (sUAS) Used for Public Safety Operations,* 2019 Edition.

[4] Verein Deutscher Ingenieure (VDI), VDI-2879, *Inspection of installations and buildings with UAVs (unmanned aerial vehicles),* September 2018.

[5] Underwriters Laboratories (UL) Subject 3030, *Standard for Unmanned Aircraft Systems,* Edition 1, September 18, 2018.

[6] ABS 242, *Guidance Notes on Using Unmanned Aerial Vehicles (UAVs),* 2018.

[7] International Organization for Standardization (ISO), ISO-19901-9, *Petroleum and natural gas industries– Specific requirements for offshore structures Part 9* (draft), currently unpublished.

[8] British Standards Institution (BSI), *Assessment of inadvertent ignition of flammable atmospheres by radio-frequency radiation,* PD-CLC/TR-50427, April 28, 2006.

[9] Occupational Safety and Health Administration (OSHA), *Electrical*, 29 CFR 1910.303, 2008 requires that conductors and equipment be approved per OSHA *Electrical,* 29 CFR 1910.399, 2015. OSHA 29 CFR 1910.399 requires that such installations or equipment be certified by a nationally recognized testing laboratory per *Definition and requirements for a nationally recognized testing laboratory*, 29 CFR 1910.7, 2011. The Nationally Recognized Testing Laboratory Program is available at weblink: https://www.osha.gov/dts/otpca/nrtl/nrtllist.html (accessed January 6, 2019).

[10] ATEX, Directive 2014/34/EU of the European Parliament and of the Council, *Harmonisation of the laws of the Member States relating to equipment and protective systems intended for use in potentially explosive atmospheres*, February 26, 2014.

[11] American Petroleum Institute (API), Standard 2000, *Venting Atmospheric and Low-Pressure Storage Tanks*, 2014.

[12] North Atlantic Treaty Organization (NATO), STANREC 4811/AEP-101, *Guidance on Sense and Avoid for Unmanned Aircraft Systems*, February 23, 2018.

Application of Visual Flame Detection and Hazard Mapping to Large Atmospheric Storage Tanks

William Pittman
Micropack (Detection) Americas
800 Town and Country Blvd, Suite 300, Houston TX 77041
WilliamPittman@micropackamericas.com

Jim Scoggins
Micropack (Detection) Americas
800 Town and Country Blvd, Suite 300, Houston TX 77041
JimScoggins@micropackamericas.com

Prepared for Presentation at
American Institute of Chemical Engineers
2019 Spring Meeting and 15th Global Congress on Process Safety
New Orleans, LA
March 31 – April 3, 2019

Application of Visual Flame Detection and Hazard Mapping to Large Atmospheric Storage Tanks

William Pittman
Micropack (Detection) Americas
800 Town and Country Blvd, Suite 300, Houston TX 77041
WilliamPittman@micropackamericas.com

Jim Scoggins
Micropack (Detection) Americas

Keywords: Storage Tank, Tank Fire, Fire, Flame Detection, Rim Seal, Large Atmospheric Storage Tanks, Optical Flame Detection,

Abstract

Fires involving large atmospheric storage tanks can be dangerous and expensive. The size of the tanks and the fires make them difficult to extinguish and sometimes difficult even to control once the fire has spread to engulf the entirety of the tank, the bund or just the rim-seal. High-profile events in the United States and Europe have increased the visibility of these incidents, demonstrated the potential severity of the consequences and set off a wave of proposed revisions to codes and standards.

While design standards should be reviewed and improved, it is also important to remember detection and suppression systems intended to mitigate the consequences of these events when and if fires occur. Traditionally, operators have used linear heat detection systems to monitor the rim seal at the top of the tank. The authors believe that the industry should begin to more widely employ optical flame detection systems to monitor large atmospheric storage tanks, using detector layouts designed and validated using 2D or 3D hazard mapping software.

1 Introduction

Fires involving large atmospheric storage tanks - with diameters in excess of 200 ft - have always been a major cause for concern in petrochemical facilities, especially where multiple large tanks are grouped together in such a way that allows for a fire to spread from one large storage vessel to its neighbors. The potential size and severity of these incidents began to receive additional recognition in the wake of the 2005 Buncefield disaster in the UK and the October 2009 Caribbean Petroleum Tank Terminal Explosion that occurred in Puerto Rico.[10] Both incidents resulted in powerful blast waves that cause significant damage to surrounding areas and burned in excess of a day. The incidents have sparked increased interest in the study of tank farm incidents and a major push to improve both the design of the tank farms themselves and of the fire detection / protection systems installed within them. Regulators and operators in the US, UK and beyond have been looking into the causes of these incidents and what engineering options are available to resolve them.

Chang and Lin conducted a review of 242 incidents involving storage tanks that occurred between 1960 and 2003.[7] The study proved, perhaps, timelier than the authors had intended or imagined. The paper was accepted in May 2005, only about 6 months before the Buncefield incident in December 2005. The journal published the study in January 2006.

Most of what they found is not at all surprising given the sources they pulled information from - including Lees' Loss Prevention, CSB reports, and US and UK databases. The vast majority of the incidents reviewed were in the US and Europe with most occurring at refineries, terminals or storage facilities. The vast majority of the fires involved crude oil and oil products and over half of the incidents in which the type of tank could be identified involved floating roof tanks. Roughly 33% of the incidents were attributed to lightning strikes with another 20% being attributed to static electricity, leaks and line ruptures and equipment failure. It is likely not coincidental that the 53% of incidents attributed to these four causes lines up closely with the 60% of incidents that involved fires. This would suggest that lightning was the cause of roughly 50% of the storage tank fires, echoing statements made by Dr Vengasandra of Emerson[2] and others.[9]

Chang and Lin concluded that "most of those tank accidents would have been avoided if good engineering in design, construction, maintenance and operation has been practiced and the safety management program has been implemented and executed." There's more than a little bit of stating the obvious in this conclusion. It's a widely held tenet of process safety and fire protection professionals that "all incidents are preventable," however, this is always followed by, "but that is not to say that all incidents will be prevented." Why? Because people make mistakes and you can't always predict when something will fail and preempt the failure.

Because of this the industry cannot simply rely on good design and maintenance to prevent incidents. Operators must also make provision for monitoring, rapid detection and mitigation. The take away from this study and others like it is that fires involving storage tanks are a relatively common occurrence and there is a well-documented need for fire detection and protection systems in tank farms and at the rim-seals of large storage tanks.

The severity of storage tank incidents can vary widely, with the consequences often being closely correlated with the size and number of tanks involved in the incident. A lesser publicized incident at a United Airlines fuel storage depot involved four tanks and 2 million gallons of fuel. The fire burned for two days and led to localized jet fuel shortages and flight delays.[12] Incidents like Buncefield and the Caribbean Petroleum incidents represent the upper extreme in terms of severity in large part because of the scale of the tanks involved. Providing protection for these extremely large tanks and the tank farms that sometimes include several of such tanks is the focus of this study.

Fire detection and protection on large storage tanks normally focuses on the rim seal at the top of the tank. Flame detection is sometimes provided for large tank isolation valves and transfer pumps associated with the tanks as these are, after the rim seal, the most likely location for leaks and fires. Flame detection is rarely provided for the surrounding bunded area as covering such a large area can be expensive and it is assumed that any fire will start near or include the transfer pumps and associated valves. Coverage of the rim seal is most often provided using linear heat detection systems. Detection using linear cable systems can be accomplished in several different ways. Such systems may use pneumatic tubes, electrical cable,[13] or fiber optic cable.[14] These systems are typically very prone to damage and / or false alarms. These problems have led to growing interest in alternative approaches in general and optical-based flame detection in particular.

Optical-based flame detection has matured significantly in recent decades and now allows operators to detect small test fires at ranges of well-over 100 ft and larger fires at much greater distances. Adoption of the technology for application on large storage tanks has been delayed by concerns that the technology would not provide effective coverage for the roof of very large tanks and fear of false alarms. In the past these concerns were very legitimate. However, as the technology has advanced solutions have been developed including improved detection systems and mapping tools that allow detector coverage to be assessed objectively. Used in concert, the new detectors and mapping systems enable operators to design and install reliable flame detection systems that they can have confidence in.

2 Why Use Optical Flame Detection?

An alternative title for this article and this section of it would be "can optical flame detection on a large storage tank even work?" Conventional wisdom indicates that optical based flame detection cannot work on very large atmospheric storage tanks (LASTs). While there is some truth to this, the fact is that optical flame detectors have been used to protect the roofs of LASTs by clients of Micropack and others[2] - sometimes on tanks with diameters of up to 100 m (330 ft).[3] LASTs usually have a diameter of 250-300 ft (76-91 m).

Traditional multi-spectrum IR (IR3) flame detectors - which are the de-facto industry standard and the type most often seen installed in oil and gas facilities - usually only have a range of 30 m (98 ft) under ideal conditions when using a 1 ft^2 n-heptane pool fire - the standard test fire used in FM3260 testing. However, this 1 ft^2 test pan fire is relatively small - much smaller than the fires that would be expected shortly after ignition at the top of a LAST.

The inverse square law can be applied to the detection range of the flame detectors within reasonable limits - extrapolation beyond 1000 kW may be unwise. Based on this, optical flame detectors can be expected to detect larger fires at greater distances - potentially much greater. A detector that can detect a 40 kW fire at 30 m can be reasonably expected to detect a 160 KW fire at 60 m and a 640 kW fire at 120 m.

Desensitizing environmental factors, interference sources and dirty optics can significantly shorten the effective range of an optical flame detector. The extent to which the detector is expected to be desensitized can be estimated using test values from the detector's FM3260 report and a well-established formula with broad acceptance in the industry.[4,5] Some IR3 units can only reliability detect the standard test fire at 59-62% of their ideal range when subjected to real-world conditions present in most oil and gas and chemical manufacturing facilities. This gives most IR3 detectors an effective range of only about 18 m (58 ft) for the 1 ft^2 test fire when installed in the field. However, while this desensitization can sometimes be quite severe, it is not necessarily so severe as to prevent the detector from meeting the needs of this application. A detector that effectively detects a 40 kW fire at 18 m should be able to detect a 640 kW fire with a similar response time at a range of 72 m - sufficient to cover many large tanks.

It is also worth noting that most IR3 detectors have between 2 and 4 different sensitivity settings and the range values used in the forgoing discussion are based on the default sensitivity setting used “out-of-the-box.” This is usually called the “medium” sensitivity setting. Most IR3 units also have a High and / or “Very High” sensitivity setting.

These higher sensitivity settings sometimes allow the detectors to achieve detection of a 1 ft^2 n-heptane pool fire at 60-80 m. However, optical flame detectors are typically expected to have a mean time to detection (MTD) of <10 seconds. At higher sensitivity settings, some IR3 detectors have MTDs in the range of 17-22 seconds. This is not considered acceptable by many operators. After desensitization, these higher sensitivity settings only extend the effective range for detecting a 40 kW fire out to 30-40 m. The operator will have to verify that the installed detectors are using the correct sensitivity setting and not the default setting if they wish to claim this range. Still, if an operator is willing to accept a 15-30 second average response, this introduces the prospect of a IR3 detector being able to detect a 640 kW fire at a range of 100-150 m. Relatively few storage tanks of such enormity exist, if any.

These range values for specified fire sizes apply to typical IR3 detectors, which tend to be the most common in oil and gas facilities. Some other, newer, technologies can provide significantly better range and performance.

Some might reasonably point out that the optical flame detectors are significantly more expensive than the linear heat detection systems. A single optical flame detector can cost $2,000-4,000 and the installed cost of each unit can be $20,000. There are two responses that can be offered here.

In the immortal words of Trevor Kletz, “If you think safety is expensive, try having an accident.” Large storage tanks can be expensive to build but the true value to be protected is the product inside - which can be worth tens of millions of dollars - to say nothing of the lives of emergency responders or the lives and property of those living and working around the facility. While the optical flame detectors are easily more expensive to purchase and install than linear heat detection systems, the optical flame detection systems can be significantly easier and cheaper to operate and maintain.

Linear heat detection systems are easily damaged where optical detectors are highly robust. While all optical flame detectors are not created equal - some are far more reliable than others - many are designed to function in harsh environments, subject to high humidity, corrosive salt water and temperature extremes for years with minimal care or maintenance. The labor costs associated with inspection and maintenance of a linear heat detection system can add up quickly and will do so throughout the life of the system. A linear-heat detection system will require repair after any fire or false alarm incident as the system must sustain damage to the cable for detection to occur. Optical flame detectors can identify flames at a safe distance without sustaining damage. Some are designed to survive even direct flame impingement - which is useful when the detector is going to be mounted at the top of a tank, near the rim seal, where the fire is expected to start.

The optical flame detectors may also detect the fire faster and earlier. In this application the optical flame detectors will alarm in response to a fire with a radiant heat output (RHO) of 100-640 kW in 10 seconds or less. Even a 640 kW fire may not be large enough to break a linear heat detection line and cause an alarm. Delayed detection means higher consequences.

3 Potential Benefits of Visual Flame Detection

Intelligent Video Flame Detectors (iVFDs) use 640x480 resolution cameras that are sensitive to near-IR radiation to detect fires. The video is subjected to pixel analysis algorithms which identify fires based on the radiation intensity, flicker frequency and shape of the radiation source. This gives the detector a spatial awareness and allows it to distinguish between black body radiators and genuine flames. This gives the iVFDs significantly better immunity to blinding or desensitization when compared to traditional IR3 units. The iVFDs also have far superior immunity to false alarm stimuli.

iVFDs are not as easily blinded by water, rain, fog, snow, or dirty optics because they detect using near-IR radiation instead of the 4.3-4.5 micron band that IR3 units depend on. As a result, where IR3 units are only expected to be effective at 59-62% of their ideal range under real-world conditions, iVFDs are still effective at 90% of their ideal range when deployed in the field. These desensitization factors are determined using methods with general acceptance in the industry.

Visual detectors have also been shown through independent testing to be able to detect 1 ft^2 n-heptane pool fires at ranges of 44 to 60 m with an average time to detection of under 10 seconds - this range is far superior to the 30 m achieved by IR3 detectors with the same limitations on response time.

Taken together, these advantages leave iVFDs able to reliably detect a 40 kW pan fire in the field at ranges that are more than double that achieved by most IR3 units - 40 m or more. As the case study that follows will demonstrate, this makes it possible for iVFDs to provide better coverage for larger storage tanks, often with fewer detectors when compared to IR3 or UV/IR units.

Because the iVFD is essentially a camera with additional signal processing equipment it also sometimes comes with additional advantages. Some iVFDs - though, admittedly, not all - can transmit a live video feed from the camera to the control room if the detector is installed and wired appropriately. This allows an operator to pull up live video showing what the detector is seeing in the event that an alarm sounds. The operator could then confirm the fire within seconds and initiate mitigative actions as necessary. This could be particularly useful given that some operators will not wish to install automated fire suppression systems on storage tanks. Spurious activations can be expensive and even the best detectors can be fooled into a false alarm under the right conditions. The authors have seen instances where water from rainfall pooled on top of a tank roof and functioned as a mirror, reflecting the image of a flare back to the detector, triggering a false alarm. Even iVFDs cannot avoid false alarms in such situations.

This live video feed may not seem like much of an advantage to a facility owner with security cameras already installed to monitor the tank farm and the surrounding area. However, the security camera feeds may only be accessible to security personnel in the guard shack and not operations personnel. Additionally, the security camera system is unlikely to be designed to monitor the tops of the tanks and the rim-seal. Those cameras were likely installed to guard against intruders, not to allow operators to confirm rim seal fires. Finally, and perhaps most importantly, these detectors are certified for service in explosive atmospheres and designed for use in harsh, extreme and dirty conditions. Are the security cameras?

4 Why Use Hazard Mapping

Using 3D hazard mapping tools will allow designers to quickly determine if adequate coverage is provided for the entire rim seal of the tank, identify any worrisome coverage gaps, and iteratively adjust the layout to resolve coverage gaps. While heuristic approaches can be used to design some optical flame detector layouts, and while it is reasonable to use heuristic methods to propose an initial design, without a Geographic coverage assessment (GCA) the end user of the system cannot be assured of adequate coverage or optimal performance.

5 Case Study: 60 Meter Diameter Floating Roof Tank

The following analysis shows the results of GCAs using a 3D model of a 60-m diameter tank with a bunded area surrounding it. Relative to some of the larger tanks present at oil and gas facilities this tank is somewhat small. However, it is large enough to provide a proof of concept and to allow for the relative advantages of certain detectors and layout design approaches to become apparent. More importantly, this allows for a demonstration of hazard mapping and how it can be used to assess, verify, and improve detector layouts.

Please note, the 3D model used for this analysis does not use a foam system or have a foam dam on the roof of the tank. For tanks with foam dams installed coverage may be somewhat impaired and some small delays in detection may occur, but the impact on system performance is expected to be minor.

The first impulse of many operators when trying to monitor the top of a large tank is to position the detectors to face straight out across the diameter of the tank. Figure 1 shows what this looks like with three IR3 detectors mounted on the top of a 60-m (200 ft) diameter tank, arranged at 120° angles to each other, relative to the center of the tank. The field of view (FOV) cones shown for the detectors in Figure 1 use the expected effective range for detecting a 40 kW RHO fire - roughly equivalent to the 1 ft^2 test fire used for the FM3260 testing. The FOV cones shown have been adjusted to account for expected desensitization as discussed in the previous section. The IR3 units are modeled using a model with a 30 m range and 90-degree horizontal FOV - highly representative of the current technology. The results obtained here will closely match what will be achieved by most major IR3 models on the market including the Det-tronics X3301, the Honeywell FS24X, the Spectrex 40/40I, and the Micropack FDS303.

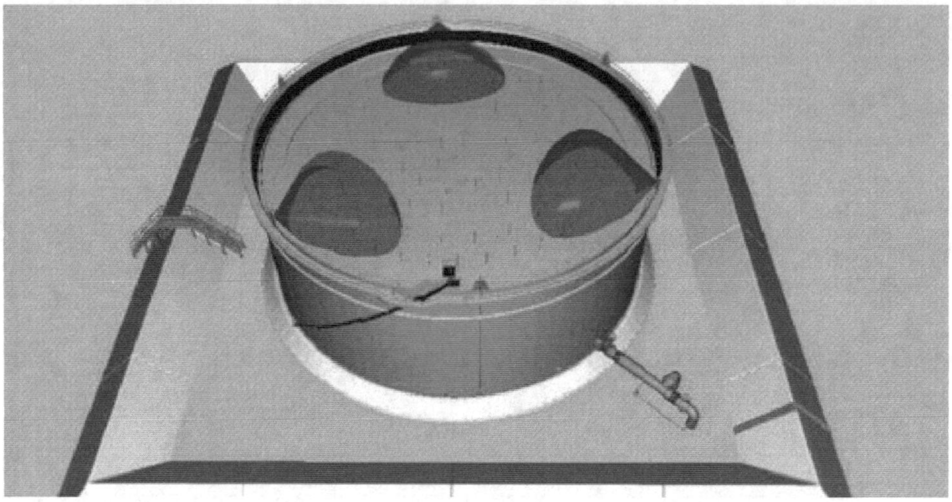

Figure 1: Three IR3 detectors positioned on top of a LAST, pointed towards the middle of the tank.

It makes sense that this is the first thought of many layout designers. Assuming the range of the detector is adequate for the task, the layout achieves complete coverage of a circular area with only three detectors - two detector layouts with the detectors on opposite sides of the roof do not achieve full coverage.[3] This is an approach that Micropack recommends for helipads at normally unoccupied facilities, but not necessarily for LASTs. The helipads are only 15-24 m across, a fraction of the size of a LAST. This approach can work for a LAST,[3] but its appropriateness will depend on the specifics of the application, including the size of the tank, the target fire size specified by the system owner, and the detector specified for the project.

Based on the foregoing information regarding the effective range of most IR3 detectors it should not be surprising that the desensitized IR3 detectors come nowhere close to reaching all the way across the tank's diameter to see the rim on the other side when looking for a 40 kW fire. Looking at Figure 1 makes it all the more obvious that this approach is not going to work for that fire size with these detectors. However, it's useful at this point to assess the coverage achieved with a more reasonable target fire size to establish a baseline.

The authors used Micropack's mapping software - HazMap3D - to assess the coverage provided by these three detectors. The assessment was run using a target fire size of 500 kW RHO. This target fire size was chosen to be consistent with the recommended practice for open, uncongested, onshore facilities detailed in ISA TR 84.00.07[1] and generally accepted industry practice. As can be seen in the literature,[3] operators may choose larger or smaller target fire sizes depending on their internal risk tolerance or their expectations with regard to a specific tank. In either case, a 500 KW target fire size is more than reasonable for this application.

A 500 kW RHO fire is significantly larger than the 1-ft^2 test fire and, because of this - invoking the inverse square law - the detectors are expected to respond to a fire of this size from more than three times the distance at which they would respond to the 1ft^2 test fire. Figure 2 shows the assessment results as they appear in HazMap3D. This assessment assumes that only 1 out of the N detectors (1ooN) is required to see a fire activate an alarm and that no automated control action occurs (alarm only coverage).

Figure 2: Assessment results for the three center-facing IR3 detector layout using a 500 kW RHO target fire.

The coverage shown in Figure 2 may be quite surprising to those not familiar with the technology and industry practice in this area, especially after seeing Figure 1. With the large target fire size specified the IR3 detector comes close to reaching across the tank, but falls short in the end, failing to effectively cover large portions of the rim seal. This highlights the power and importance of leveraging the inverse square law in assessing coverage and detector performance and optimizing detector layouts. This is not to say that an operator should just choose a larger target fire size if they find a tank is too large to be covered by a detector with the current specified fire size - performance targets should be based on risk acceptance criteria, not any limitations of the equipment. This is merely meant to say that, once specified, target fire size is an important consideration in layout design.

The results in Figure 2 show that this approach can achieve complete 1ooN coverage of the tank roof and the rim seal for a target fire of 500 kW RHO for any tank smaller than 56 m (184 ft) in diameter - slightly smaller than the tank in the model. The coverage achieved when applying this method with detectors of this type to larger tanks with diameters in the range of 76-91 m will be significantly worse and will clearly not be acceptable.

Achieving acceptable coverage for such tanks requires a different, better, detector, a different approach to layout design, or both.

2.1 Using iVFDs to Provide Improved Coverage

Figure 3 shows the FOV and expected detection range for a 40 kW RHO fire when using the Micropack FDS301 iVFD, using the same detector layout as was used in the previous assessment.

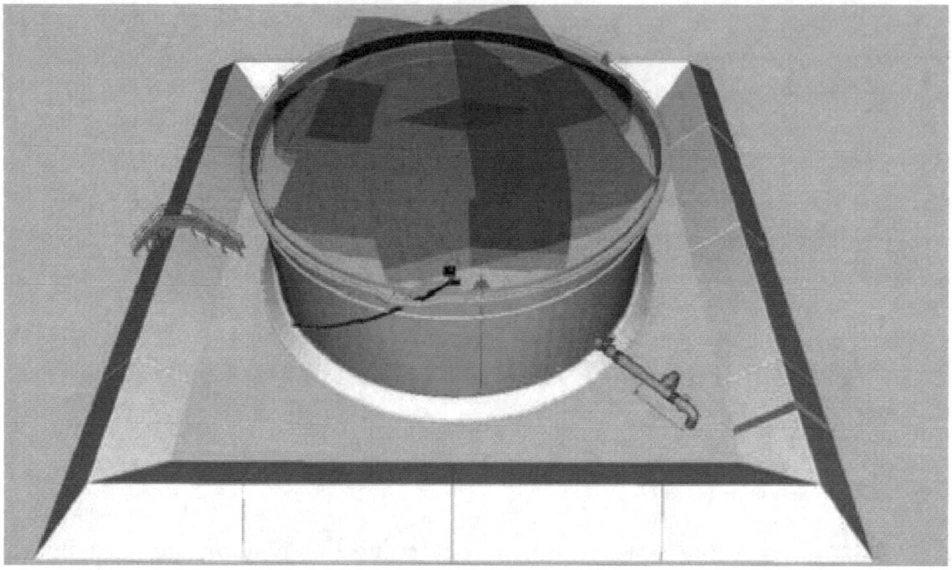

Figure 3: Three iVFDs positioned on top of a LAST, pointed towards the middle of the tank.

Figure 4 shows the results of a coverage assessment, again, using a target fire size of 500 kW RHO, consistent with ISA TR 84.00.07. Perhaps unsurprisingly, the FDS301 iVFD achieves complete 1ooN alarm-only coverage of the rim seal for this 60-m diameter tank where the IR3 unit failed to do so.

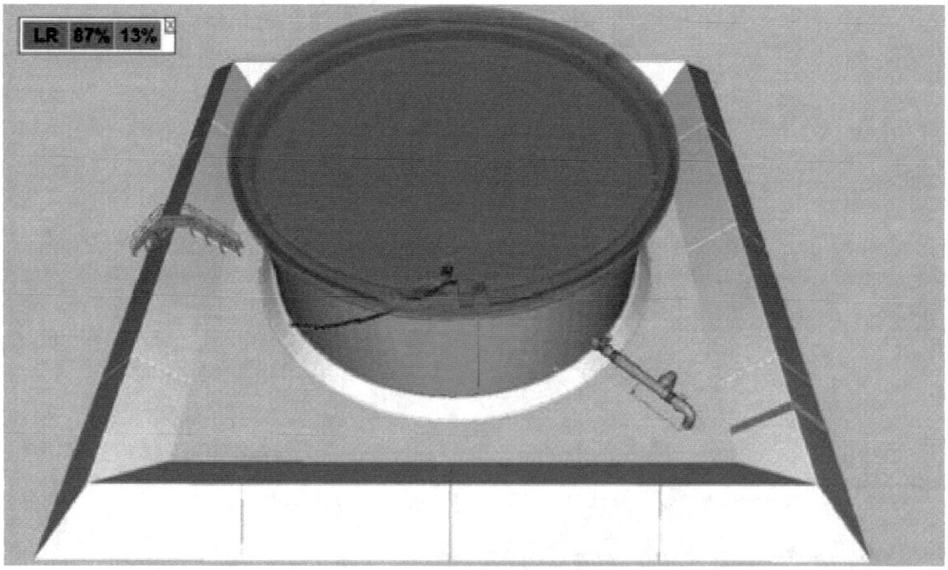

Figure 4: Assessment results for the three center-facing iVFD layout using a 500 kW RHO target fire.

The FDS301 iVFD in this 3-detector configuration can provide complete coverage of a tank roof for a target fire size of 500 kW RHO for tanks that are 100 m in diameter or larger. For this 60 m diameter tank complete coverage is provided for fires of 100 kW RHO or larger.

2.2 Using Angled Detector Layouts and Additional Detectors

Some operators may desire complete coverage for fires smaller than 100 kW, or to achieve such coverage on tanks with diameters larger than 60 m - like the 76-91 m diameter tanks. Other operators may choose to specify an IR3 unit and not use an iVFD - possibly to satisfy internal company standards or approval requirements. Achieving complete coverage of the rim-seal under such conditions is still possible but doing so requires a change to the layout design approach.

When the detectors are not capable of "seeing" across the full diameter of the tank and still detecting fires of the expected size, the detectors should be rotated to look out along the edge of the tank rim. Detector spacing can be adjusted and additional detectors can be placed as needed in order to achieve complete coverage of the rim-seal. An example of this, using a layout with three IR3 detectors, can be seen in Figure 5: This is the same approach to detector placement recommended by Emerson for providing coverage for tank roofs with diameters of over 75 m using optical flame detectors.[2] It is worth noting, however, that the Emerson article uses UV/IR detectors to provide coverage in their example - Micropack does not recommend the use of UV/IR units in outdoor oil and gas facilities. Instead, this analysis will continue to look at IR3 detectors.

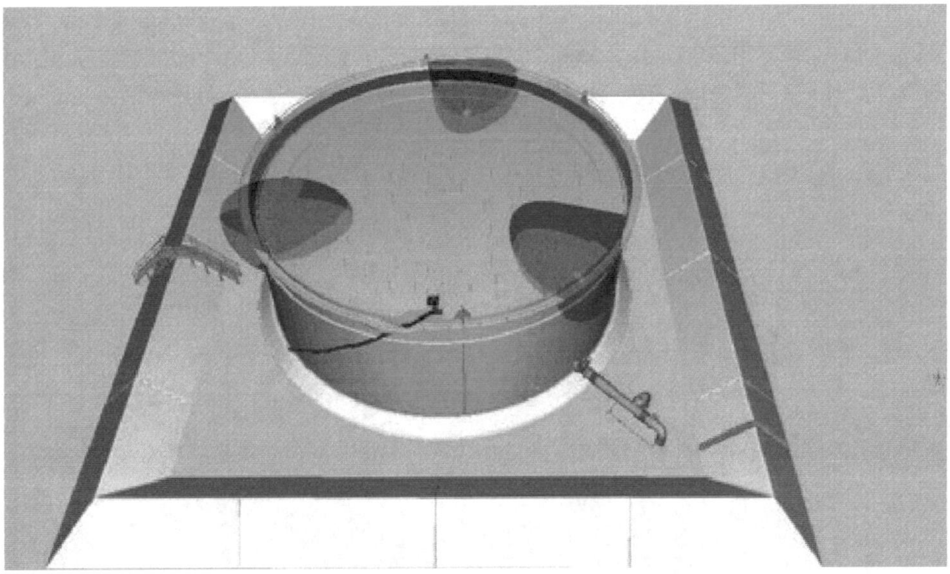

Figure 5: Layout using three IR3 detectors, facing along the rim of the LAST.

Figure 6 shows the assessment results obtained with the layout in figure 5 with a 500 kW RHO target fire size. Where three IR3 detectors previously failed to reach across this 60-m tank to detect a 500 kW RHO fire, this three-detector layout can achieve complete coverage of the rim seal. This approach causes gaps in coverage to open up in the middle of the tank roof, but fires are not really expected in this area, so this gap in coverage is not concerning.

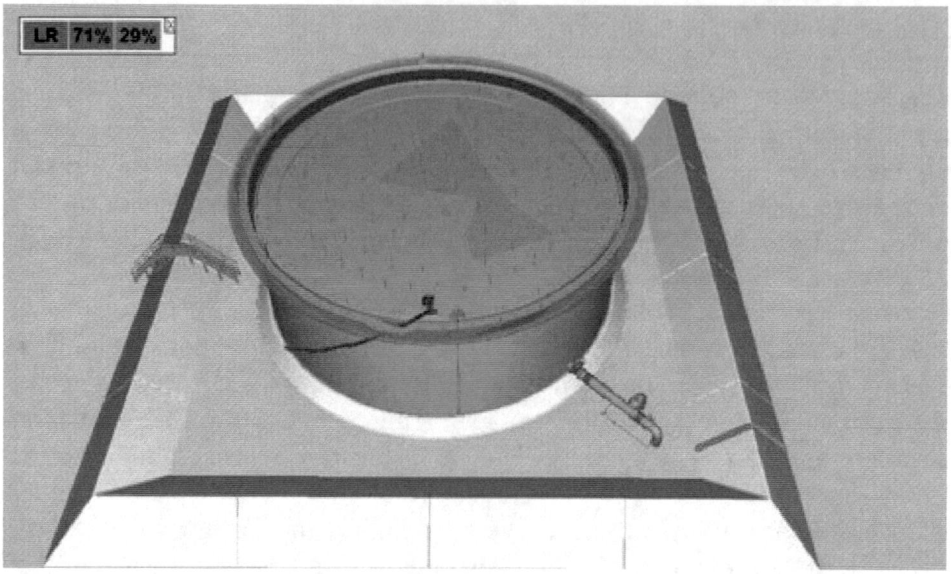

Figure 6: Results obtained three IR3 detectors facing along the rim seal of a 60 m LAST with a 500 kW RHO target fire size.

Figure 7 shows a layout using six IR3 detectors and Figure 8 shows the results obtained with a target fire size of 250 kW RHO instead of 500 kW. The results show that additional detectors can be deployed to allow for coverage of larger tanks or smaller target fire sizes. This is, again, consistent with the layout recommended by Emerson for protecting LAST rim-seals using UV/IR detectors.[2]

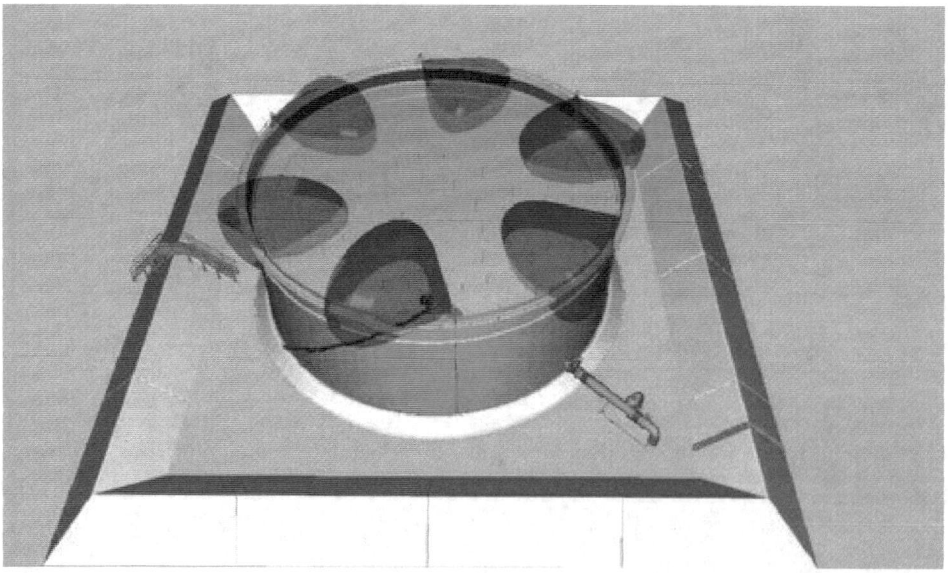

Figure 7: Layout using six IR3 detectors, facing along the rim of the LAST.

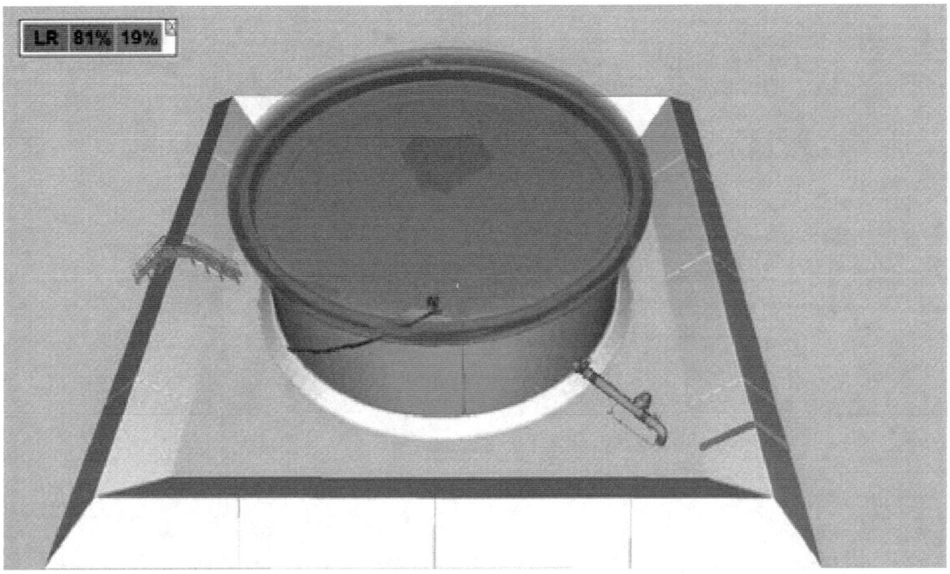

Figure 8: Results obtained six IR3 detectors facing along the rim seal of a 60 m LAST with a 250 kW RHO target fire size.

While both IR3 or iVFD detectors are capable of achieving complete coverage of the rim-seal for any size of tank using this approach to layout design, on larger tanks, the number of IR3 detectors required to achieve complete coverage will be significantly larger than the number of iVFDs required, owing to the superior range that the iVFDs achieve in the field.

To illustrate this point, Figure 9 shows a layout using three iVFDs arrayed in the same way as the IR3 detectors shown in Figure 6. Figure 10 and 11 show the coverage achieved for a 100 kW RHO target fire size using the three iVFD layout shown in Figure 10 and the six IR3 detector layout shown in Figure 8 respectively.

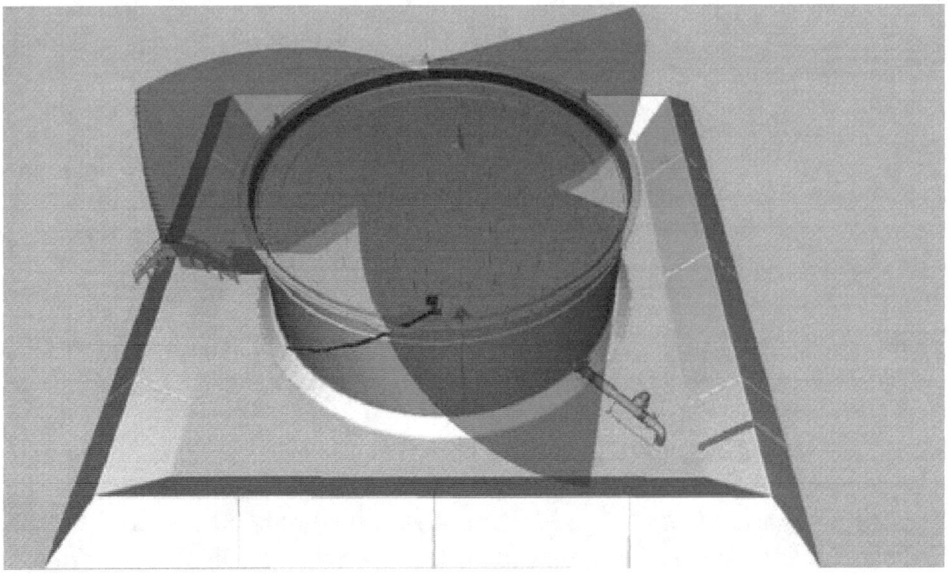

Figure 9: Layout using three iVFDs, facing along the rim of the LAST.

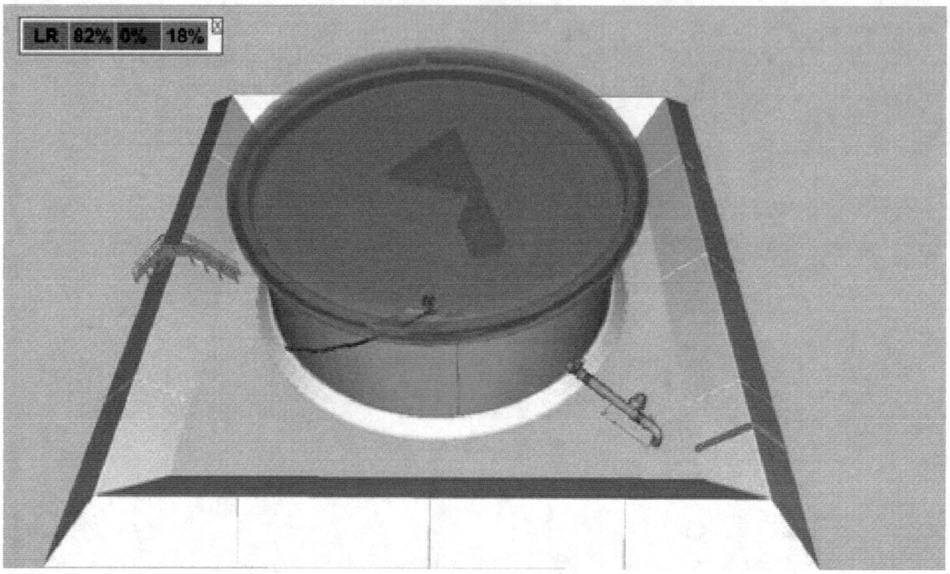

Figure 10: Results obtained three iVFDs facing along the rim seal of a 60 m LAST with a 100 kW RHO target fire size.

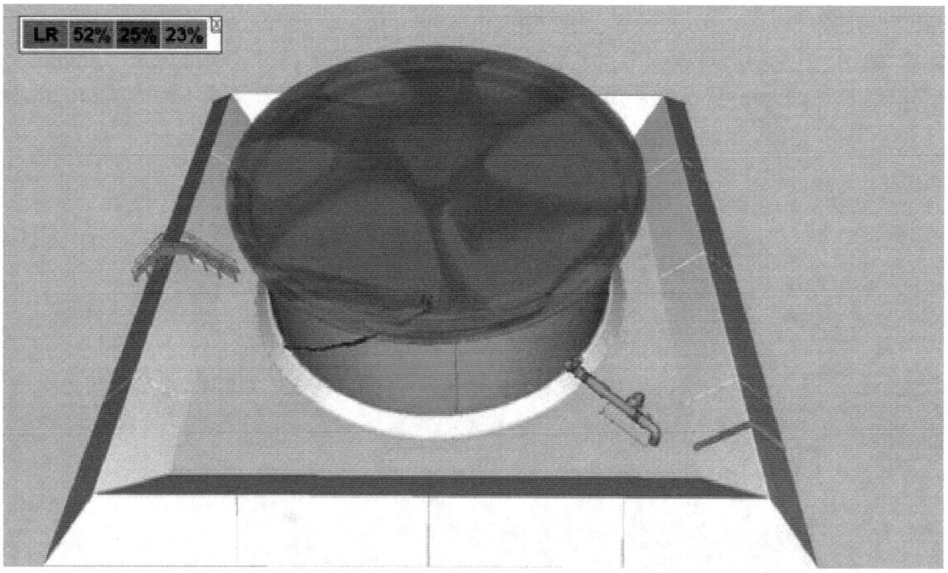

Figure 11: Results obtained six IR3 detectors facing along the rim seal of a 60 m LAST with a 100 kW RHO target fire size.

The results clearly show that, using the same target fire size (100 kW RHO), the iVFDs can provide superior coverage using 50% fewer detectors.

2.3 The Impact of Voted Coverage Requirements

The foregoing analysis has used 1ooN voting, such that, if a single detector could respond to a fire of the specified size in that area, that area was regarded as having acceptable coverage. This works for an application in which the operator is going to use the detectors in an alarm-only role and an employee is required to confirm the fire and manually activate fire suppression systems, if any. If an operator chooses to have automated foam fire suppression systems and require 2ooN voting to confirm a fire before activation, the coverage problem becomes somewhat more nuanced.

Recall the result shown in FIgure 8, using six angled IR3 detectors to provide coverage for a 250 kW target fire. If this assessment is repeated but done so with a requirement for 2ooN voting for control action the result looks considerably different. The result is shown in Figure 12. The 2ooN voted coverage of 35%, which doesn't include much of the rim seal, isn't much to brag about.

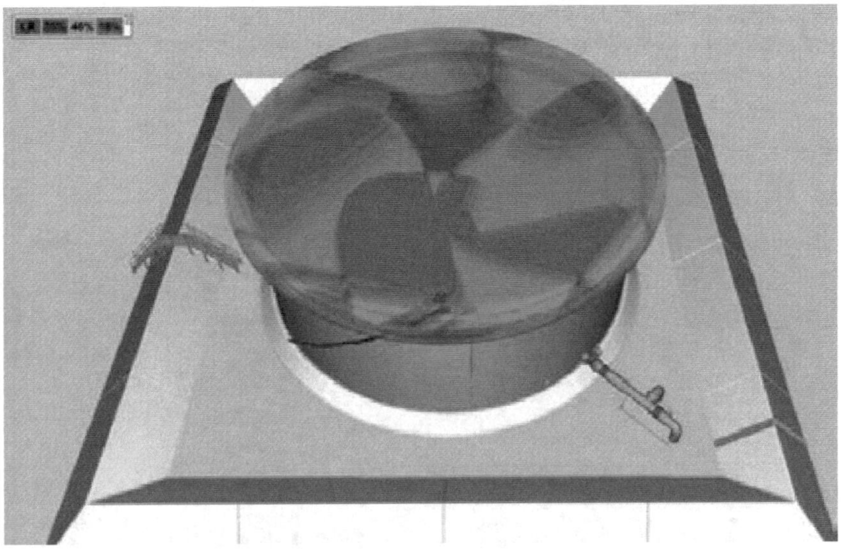

Figure 12: Results obtained six IR3 detectors facing along the rim seal of a 60 m LAST with a 250 kW RHO target fire size and requiring 2ooN voted coverage.

If this experiment is repeated with the iVFD, with the same detector count and placement, the superior range of the iVFD again allows it to provide much better performance - even for much smaller fires (100 kW) - as can be seen in Figure 13. Using 6 IR3 units arrayed as in Figure 7, 2ooN coverage for a 100 kW RHO fire is 0% for a tank of this size.

Figure 13: Results obtained six iVFD detectors facing along the rim seal of a 60 m LAST with a 100 kW RHO target fire size and requiring 2ooN voted coverage.

These results - 81% 2ooN voted coverage, with effectively complete 2ooN coverage of the rim - are with a 60 m diameter tank. With larger tank diameters, even the range of the iVFD will be challenged eventually and cause 2ooN coverage to start dropping. In both cases the operator would be forced to add more detectors positioned closer together along the rim of the tank, or accept a larger target fire size and diminished performance. This will just happen much faster with the IR3 units and require significantly more detectors than what would be needed with the iVFD.

However, in applications where 2ooN voting and automated response are desired, operators might at least consider implementing the layout shown in Figure 3, using 3 iVFDs facing directly out across the tank roof. The results shown in Figure 14 use the layout from Figure 3 but, unlike the results shown in Figure 4, 2ooN voted coverage is now required.

Figure 14: Assessment results for the three center-facing iVFD layout in Figure 3 using a 105 kW RHO target fire, requiring 2ooN voted coverage.

This result is only valid where the iVFD is able to respond to the specified target fire size at ranges equal or greater to the diameter of the LAST. For a 60 m tank, this approach works for a 105 kW fire. For a 100 m tank, the target fire size would need to be about 300 kW RHO or larger. The 67% 2ooN coverage achieved here is significantly lower than what's achieved with the six angled iVFDs and 2ooN coverage is lost for a major portion of the rim, but the layout achieves this coverage with only three detectors instead of six. Using the three iVFD layout shown in Figure 9 provides 2ooN coverage of only 16% - as shown in Figure 15.

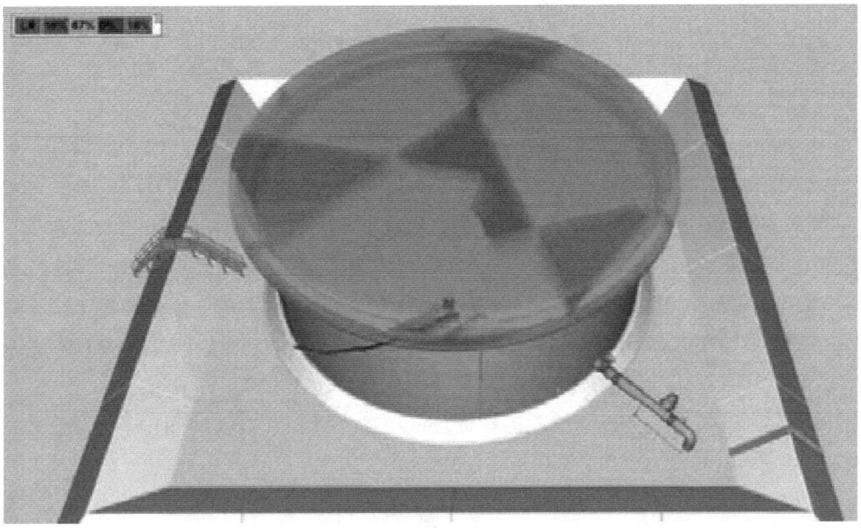

Figure 15: Results obtained three iVFDs facing along the rim seal of a 60 m LAST, as in Figure 9, with a 100 kW RHO target fire size, requiring 2ooN voted coverage.

If four detectors are used, facing inward, at 90-degree angles relative to the center of the tank, the detectors will provide complete 2ooN coverage of the rim seal provided the detectors have a horizontal FOV of at least 90 degrees and sufficient sensitivity to detect a fire of the specified size on the other side of the tank roof.

6 Conclusions

Optical flame detection technology in general - and intelligent visual flame detection (iVFD) in particular - has the potential to significantly improve the speed and reliability of flame detection systems for large atmospheric storage tanks. The technology offers obvious advantages over linear heat detection systems, especially in terms of reliability. The authors have shown that current technology is capable of providing effective coverage for even the largest of LASTs currently in service.

The foregoing analysis has shown that, contrary to popular misconception, different optical based flame detectors can be used to achieve complete coverage of the rim-seal of a LAST if arrayed properly. While most detectors are incapable of detecting a standard test fire at a range of 100 m and are advertised at being effective at ranges far smaller than the diameter of most LASTS, the detectors can still offer coverage for the types and sizes of fires most operators are looking to detect and respond to with LASTs and do so at the necessary ranges. The appropriate detector layout and the number of detectors needed will depend on the size of the tank, the specified target fire size for detection, and - most importantly - the type of detector used. The authors expect that the vast majority of LAST applications can be satisfied using 3-5 iVFDs - most likely three. Providing adequate coverage of a LAST rim-seal using IR3 units may require 6-8 detectors, once the loss of detector sensitivity is accounted for in the assessment.

7 References:

1. ISA TR 84.00.07. Guidance on the Evaluation of Fire and Gas System Effectiveness.

2. Vengasandra, Srikanth. Flame Detection for Crude Tanks. Emerson. Mar 23, 2015. Available at: < https://www.analyticexpert.com/2015/03/flame-detection-for-crude-oil-tanks>.
3. Duncan, Graham; McNay, James. "Restoring Fire Detection Confidence: The use of visual flame detection equipment can provide operators with a proven and comprehensive early warning system. Available at: <http://micropack.co.uk/images/downloads/ Micropack-Visual-Flame-Detection-Tank-Storage.pdf>.
4. McNay, James. "Desensitisation of Optical Flame Detection in Harsh Offshore / Onshore Environments." Proceedings of the 70th Annual Instrumentation Symposium. Jan 2015. Texas A&M University. College Station, Texas.
5. McNay, James. "Desensitisation of optical based flame detection in harsh offshore environments." International Fire Professional. July 2014. No. 9. 12-14.
6. Duncan, Graham. "Helideck Flame Detection." Micropack Engineering Ltd. Available at: <http://micropackfireandgas.com/images/downloads/CAA%20CAP%20437%20 Helideck%20Flame%20Detection.pdf>. Guidance on the Evaluation of Fire and Gas System Effectiveness.
7. Chang, James and Lin Cheng-Chung. "A Study of Storage Tank Accidents." Journal of Loss Prevention in the Process Industries. 19 (2006) 51-59.
8. "Fire Detection at Floating Roof Tanks by fiber optic linear heat detection. linearheatdetection.wordpress.com. Dec 21, 2009. Available at: <https://linearheat detection.wordpress.com/2009/12/21/fire-detection-at-floating-roof-tanks-by-fiber-optic-linear-heat-detection>.
9. "Spark, Arc and Boom!" www.lightningprotection.com. Jun 10, 2015. Available at:<https://www.lightningprotection.com/spark-arc-boom>.
10. "Final Investigation Report: Caribbean Petroleum Tank Terminal Explosion and Multiple Tank Fires." Report No 2010.02.I.PR. US Chemical Safety and Hazard Investigation Board. October 2015.
11. "The Buncefield Incident: 11 December 2005. The Final Report of the Major Incident Investigation Board." Volume 1. 2008.
12. Associated Press. "Denver Fuel Tank Fire is Extinguished at Last." 28 Nov 1990. New York Times.
13. "What is Standard Linear Heat Detection? Safe Fire Detection. Available at: <http://safefiredetection.com/products/linear-heat-detection/>.
14. "Fibre Optic Linear Heat Detection Cable." Patrol Fire Detection Solutions. Available at: <http://safefiredetection.com/products/linear-heat-detection/>.

Improved Modeling of Continuous Gas Releases of Finite Duration

Jessica M. Morris, Ph.D.
Tom O. Spicer III, Ph.D., P.E.
Ralph E. Martin Department of Chemical Engineering
University of Arkansas, Fayetteville, AR
Morri968@gmail.com

Prepared for Presentation at
American Institute of Chemical Engineers
2019 Spring Meeting and 15th Global Congress on Process Safety
New Orleans, LA
March 31 – April 3, 2019

Improved Modeling of Continuous Gas Releases of Finite Duration

Jessica M. Morris, Ph.D.
Tom O. Spicer III, Ph.D., P.E.
Ralph E. Martin Department of Chemical Engineering
University of Arkansas, Fayetteville, AR
Morri968@gmail.com

Keywords: Consequence Modeling, Hazard Identification and Risk Analysis, Research and Development

Abstract

The consequence assessment of toxic or flammable gases released into the atmosphere is typically based on idealization of the gas source as a puff or plume, but continuous gas releases of finite duration are typically considered more realistic. Frequently, operational dispersion models are based on Gaussian concentration profiles which inherently assume symmetric concentration profiles in the along-wind direction. While this may be appropriate for very long distances, near field behavior of finite duration releases are shown to not follow this behavior in a recent experimental program. This paper discusses the wind tunnel experimental program which studied ensembles of continuous gas releases of varying duration conducted at two different wind speeds at multiple downwind distances. The data were analyzed to propose improvements in the near field modeling of finite duration releases.

1 Introduction

Many chemicals are used or stored at industrial facilities that are toxic, flammable, or both. If releases, devastating effects can occur, as seem with previous accidental releases (e.g., Bhopal[1], Flixborough[2], Seveso[3]). While accidents cannot always be prevented, the modeling of where chemicals travel through the atmosphere can reduce the potential impact of the airborne hazardous materials.

Modeling contaminants released into the atmosphere often classifies the release as either instantaneous or continuous, while many real world releases are of finite-duration. Both instantaneous and finite-duration releases are affected by the stretching of the cloud in the along-wind direction, known as along-wind dispersion. Along-wind dispersion, σ_x, is determined from the time distribution coefficient σ_t, and effective wind speed u_e from the established relationship $\sigma_x = \sigma_t u_e$.

The determination of σ_t can be appropriately measured with finite-duration releases. With finite-duration releases, the total impact on the leading and trailing edge can be observed compared to the steady-state period. Hypothetical concentration time-histories are shown below to determine the impact of the time distribution coefficient from the release duration, T_d, and the total travel time of the cloud, ΔT.

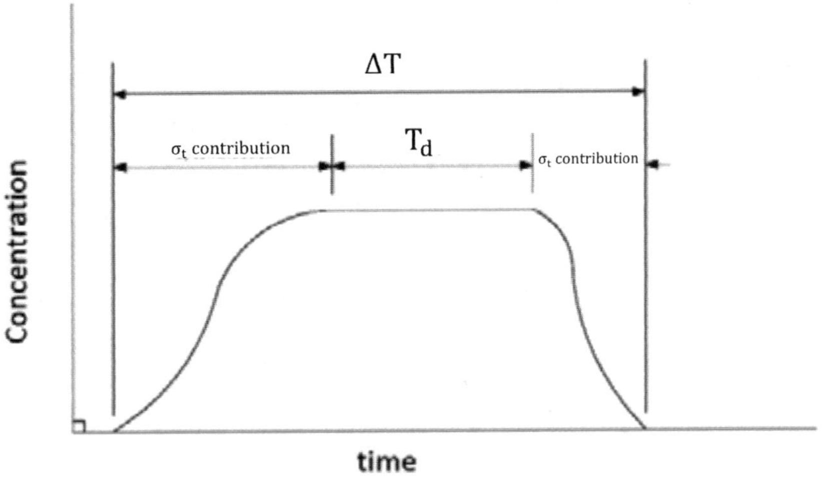

Figure 1. Hypothetical concentration time history for uniform concentration as analyzed by Hanna and Chang[4]

The overall time distribution coefficient has previously been determined from $\sigma_t = (\Delta T - T_{dur})/4.3$. This approach was taken to determine σ_t and σ_x from a single data set and was largely unquestioned due to the lack of data. From this method, the effects of σ_t and σ_x only act on the leading and trailing edges of the cloud and not on the period of steady-state. A proposed concentration time history is shown below.

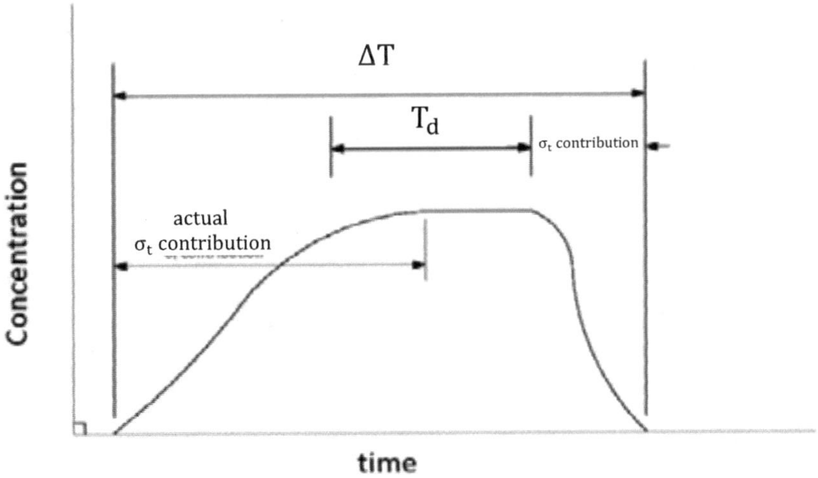

Figure 2. Expected concentration time history for uniform concentration

The expected concentration-time history shows a transfer of mass from the leading and trailing edge resulting in a decreased steady-state period that is less than the release duration resulting in an increased time-distribution coefficient. Ensemble averages of original finite-duration releases allow for an improved method in determining and predicting the time distribution coefficient and subsequently the along-wind dispersion coefficient with continuous gas releases of finite-duration.

2 Equipment and Facility

The Chemical Hazards Research Center (CHRC) houses an ultra-low speed wind tunnel with a working section of 24 m in length, 6.1 m in width, 2.1 m in height, and wind speeds from 0.34 m/s to 1.1 m/s. The facility employed a 3D laser Doppler Velocimeter (LDV) used to ensure uniform flow distribution throughout. A constant stress layer of 40 cm is achieved in the tunnel with a modified Irwin spire and surface roughness arrangement that ensures that all tests performed are within the constant stress layer. A fast Flame Ionization Detector (fastFID) system allows for concentration-time histories to be accurately measured with essentially no time delay. An area source developed specifically for this program provides repeatable release durations at constant flow rates. All continuous releases of finite duration were performed in the ultra-low speed wind tunnel at the CHRC.

2.1 Ultra-Low Speed Wind Tunnel

The ultra-low speed wind tunnel has an established boundary layer of 60 cm and a constant stress layer of 40 cm with two layers of arranged spires. The "push through" tunnel produces wind speeds less than 2 m/s for scalability to atmospheric conditions[5]. A control room adjacent to the wind tunnel provides non-invasive observation and control of equipment during testing without disruption of flow in the wind tunnel located centrally in the room. A floor plan is shown below in Figure 3.

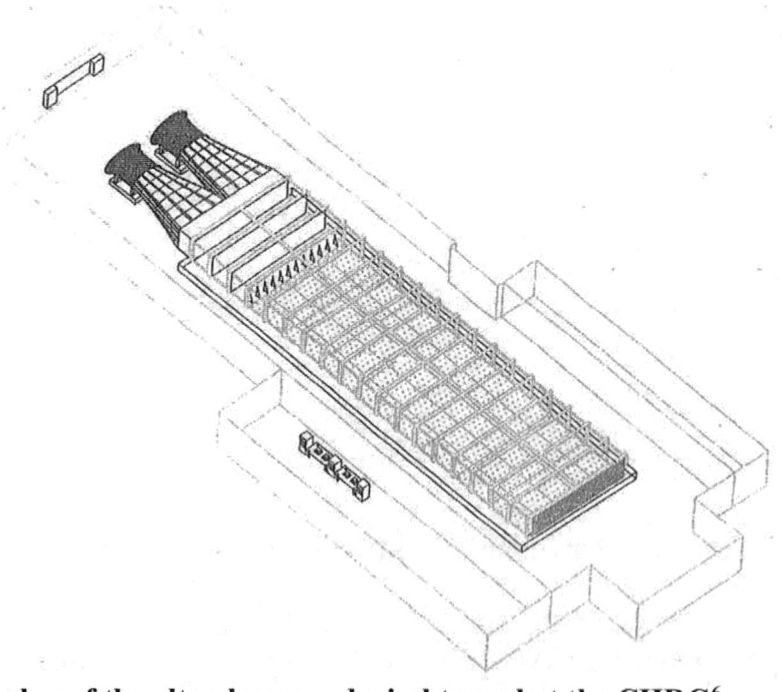

Figure 3. Floor plan of the ultra-low speed wind tunnel at the CHRC[6]

The wind tunnel is powered with two adjustable pitch vane-axial fans (Buffalo Forge Company Model 72D5 ADJUSTAX) built by the Buffalo Forge. The 1.8 m diameter, 75 horsepower fans are controlled with a Fenner M-Trim speed controller optimizing the master-and-slave method (Eaton Company DYNAMATIC Model VT-323). Transition ducts carry wind from the circular output of the fans into a rectangular input at the front of the wind tunnel. A pressure drop and flow straightening is achieved with Honeycomb (1.2 cm cell size with 10.1 cm thick plastic-impregnated Kraft paper) and seamless fiberglass screens (at distances of 1.0 m, 1.1 m, 2.0 m, and 2.1 m downwind) before entering the working section.

Two rows of Irwin spires, comprised of 13 or 14 spires per row, placed at a distance of 30 cm downwind of the last screen, were used to generate a turbulent boundary layer. These spires are comprised of 0.36 cm thick acrylic and are 92.7 cm in height, 13.2 cm at the base, and 23.2 cm in length. In addition to the Irwin spires, aluminum angles of 3.81 cm in L x W x H were placed at a spacing of 30.5 cm between each roughness element to provide uniform surface roughness. A fully developed boundary layer of 60 cm in depth is developed after the boundary layer generation region of 6 spire heights downwind. The remaining downwind portion of the wind tunnel served as the working section.

2.2 Area Source

Ground level releases were chosen for these tests for direct application in modeling. For the experiments within this project, an area source was developed for continuous rate finite-duration releases to improve understanding of the effects of along-wind dispersion. The top of the box was comprised of two sheets of cold rolled steel with 23% perforated holes and one sheet of polypropylene with 23% perforated holes between the two metal sheets. This design allows for the polypropylene middle sheet to be moved pneumatically between the two metal sheets to open and close the holes when aligned and misaligned, respectively, to allow for finite-duration releases of the source gas through the top of the area source. When the top of the box is closed, slots in two the sides of the box are opened for the source gas to be vented under the floor of the wind tunnel allowing for a continuous release of the source at all times either above or below the floor. This configuration allows the mass flow meters providing the source gas to operate at a constant rate which eliminated encountered problems with mass flow meters stopped and started. The area source was tested for uniform distribution of the source gas which was achieved with both visual source gas releases and a measured uniform concentration across the full release area. The box was secured into the floor of the wind tunnel at a distance of 20 ft downwind, and 10 ft crosswind (or centerline), with the top of the area source, mounted flush with the floor. After instillation, the area source and data acquisition was fully automated for accurate release of the source gas for a specified duration and data acquisition beginning the moment the box was opened.

2.3 Concentration Measurements

A Cambustion HFR-400 Atmospheric fast Flame Ionization Detector (FID) was used for concentration measurements. The FID system utilizes a hydrogen flame to burn a hydrocarbon traced sample through a capillary tube inside the instrument. Ions are detected with a metal high DC voltage ion collector. The current produced by the high voltage ion collector is proportional to the rate of ionization which is dependent on the concentration of hydrocarbons being sampled. The FID is a non-specific aspirated probe that utilizes propane, in these tests, as the source tracer gas for concentration measurements. Known concentrations of propane were used for calibration of the FID. The standard deviation of each signal varied from 1.2 mV at 0.15% C_3H_6 to 7 mV at 1% C_3H_6.

3 Experimental Program

The experimental program in the ultra-low speed wind tunnel at the Chemical Hazards Research Center was established to determine an appropriate along-wind dispersion coefficient as a function of the multiple parameters tested. To account for all of the parameters and their relationships, a large test matrix was designed. Two wind speeds of 0.35 m/s and 0.71 m/s were chosen to test. Downwind distances of 2 ft, 4ft, and 8 ft were chosen for the 0.35 m/s wind speed and distances of 4 ft, 8 ft, and 12 ft were investigated for 0.71 m/s. Two release rates were determined for the wind speeds to mitigate entrainment effects of the area source with velocities matching 1/10th of the maximum wind speed. Release durations of 30-, 60-, and 120-seconds were investigated. A constant flow rate of 1% propane in air was used as the source concentration for all release durations. A summary of test parameters are shown below in Table 1.

Table 1. Test conditions performed for finite-duration releases

test	U (m/s)	U_{1cm} (m/s)	u_* (m/s)	distance (m)	release duration (s)
6	0.71	0.224	0.036	1.37	120
5	0.71	0.224	0.036	1.37	60
8	0.71	0.224	0.036	1.37	30
9	0.71	0.224	0.036	2.59	120
10	0.71	0.224	0.036	2.59	60
12	0.71	0.224	0.036	2.59	30
30	0.71	0.224	0.036	3.81	120
31	0.71	0.224	0.036	3.81	60
33	0.71	0.224	0.036	3.81	30
17	0.35	0.102	0.015	0.76	120
18	0.35	0.102	0.015	0.76	60
19	0.35	0.102	0.015	0.76	30
21	0.35	0.102	0.015	1.37	120
22	0.35	0.102	0.015	1.37	60
23	0.35	0.102	0.015	1.37	30
25	0.35	0.102	0.015	2.59	120
27	0.35	0.102	0.015	2.59	60
28	0.35	0.102	0.015	2.59	30

Ensemble averages for each set of test conditions was performed to determine accurate cloud characteristics. If tests had fluctuations in concentration after the 65 ensemble averages, additional tests were performed. With ensemble averages, the experimental matrix resulted in over 2,000 tests. From the varying test conditions, appropriate relationships between the different parameters can be found in order to determine an improved along-wind dispersion coefficient.

4 Analysis of Data

The analysis performed in this experimental program takes into consideration past methods performed for data analysis of experimental releases. Some methods applied in the analysis are new, but of higher caliber due to the quantity of releases with ensemble averages performed. The background concentration, time of arrival, time of departure, steady-state, time averaging, effective wind speed, and determination of the distribution constants are all discussed in this section with respect to prior experimental programs.

4.1 Background Concentration and Time Averaging

Background concentration increased as a function of time due to operation in a closed room. Background concentration was averaged from the beginning of the release and accounted for across the overall concentration-time history. A similar method was performed by Hall[7] in 1996 with the first 20 samples (0.2 seconds) to provide a "sensor zero" for each test. With ensemble averages incorporating cloud meander and concentration fluctuations observed across multiple releases, time averaging was not applied to the tests performed in this program.

4.2 Time of Arrival (TOA) and Time of Departure (TOD)

Time of arrival (TOA) and time of departure (TOD) was determined when the cloud has arrived at or departed from the concentration sensor. The quantification of the concentration exceeding background concentration or noise is quantifiable by different methods in previous analysis. For a consistent analsis across varying maximum concentrations for the different downwind distances, the time in which the ensemble average exceeded or fell below 1% of the maximum concentration for the specified distance resulted in the TOA and TOD, respectivly.

4.3 Time of Steady-State or Peak Concentration Arrival and Departure

To determine the concentration at steady-state, ensembles of 10-minute release durations were performed. The average concentration for the 10-minute releases was preliminarily used for steady-state concentrations. If a cloud reached this concentration, a moving average across the period in which steady-state was visually achieved, covering 50% of the release duration, was moved across 25% of the cloud. If the variation was < 0.1%, steady-state was quantified. This method resulted in steady-state being achieved in all 120-second releases. For shorter release durations, if the concentration reached the concentration that achieved at 120-seconds, it was determined to also reach steady state.

Due to fluctuations in the concentration over time, there was no set period in which the steady-state or peak concentration was exceeded and held constant. Moving averages were applied to the ensemble averages for a smoother transition. When the derivative of the moving average was essentially equivalent to zero (<0.001), the time of steady-state or peak arrival (TOPa) was achieved. For the trailing edge, the sharp drop in concentration was easily determined visually and numerically with binning, or group of the data, by 5 millisecond intervals.

4.4 Effective Wind Speed

Effective wind speeds are commonly derived from the distance traveled divided by the travel time. The distance divided by the time of arrival was found to be most consistent values with the measured wind speeds at 1 cm elevation shown in Table 2.

Table 2. Results of effective wind speeds using time of arrival (TOA)

test	U (m/s)	U_{1cm} (m/s)	Dist. (m)	Dur. (s)	TOA (s)	Ue (m/s)
6	0.71	0.224	1.37	120	8.86	0.15
5	0.71	0.224	1.37	60	8.24	0.17
8	0.71	0.224	1.37	30	8.53	0.16
9	0.71	0.224	2.59	120	11.28	0.23
10	0.71	0.224	2.59	60	13.53	0.19
12	0.71	0.224	2.59	30	12.38	0.21
30	0.71	0.224	3.81	120	15.17	0.25
31	0.71	0.224	3.81	60	15.07	0.25
33	0.71	0.224	3.81	30	14.73	0.26
17	0.35	0.102	0.76	120	10.67	0.07
18	0.35	0.102	0.76	60	10.76	0.07
19	0.35	0.102	0.76	30	8.38	0.09
21	0.35	0.102	1.37	120	12.71	0.11
22	0.35	0.102	1.37	60	11.59	0.12
23	0.35	0.102	1.37	30	11.00	0.12
25	0.35	0.102	2.59	120	16.55	0.16
27	0.35	0.102	2.59	60	16.98	0.15
28	0.35	0.102	2.59	30	16.72	0.15

While the effective wind speeds are similar to the measured wind speeds at the same vertical distance, they are not equivalent at all distances. As the cloud travels downwind, the effective wind speed becomes larger. These effects can be attributed to the effect of vertical dispersion on the cloud. Initially, the cloud has no momentum in the x-direction when releasing from the area source. As the cloud begins to accelerate due to the tunnel flow, the effective wind speed will be less than the wind speed at the measured height (1 cm). As the cloud travels downwind, higher speed tunnel flow will be entrained at the leading edge of the cloud which will ultimately cause the effective wind speed to be larger than the wind speed measured at 1 cm. Similar effects have been observed in other experimental programs from ground level releases[8]. The effective wind speed reaches or exceeds the measured wind speed at the same height for the 2nd and 3rd downwind distances for each wind speed observed.

4.5 Determination of σ_t

The determination of the time distribution and along-wind disperion coefficent from the previous method using the total cloud duration and release duration are discussed previously in the introduction. This method assumes a steady-state period of the same duration as the release. A concentration-time history for the ensemble of 30-second releases at 4 ft downwind of the area source in 0.71 m/s wind speeed is shown below.

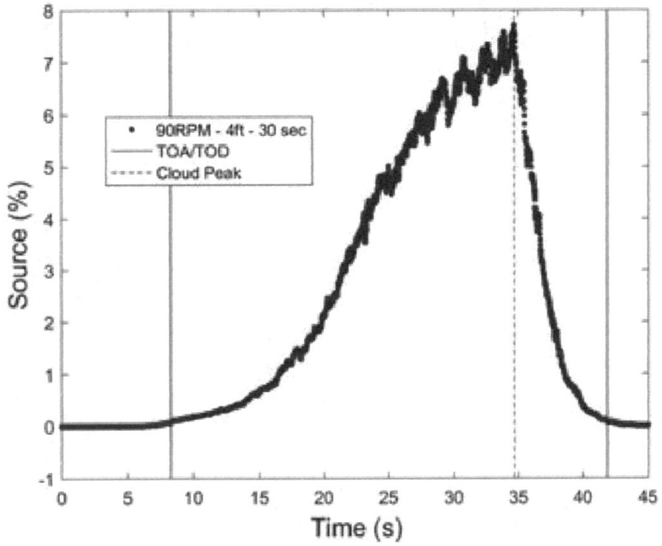

Figure 4. Concentration-time history for 30-second release in 0.71 m/s at 4 ft downwind

In the case shown in Figure 4, there is no period of steady-state. The difference between the TOA and TOD is not much larger than the total release duration, T_{dur}. With the previous method, the estimate of σ_t is around 1.2 s. By contrast, consider Figure 5 of the ensemble concentration-time history for a 120-second release at the same location and wind speed as Figure 4.

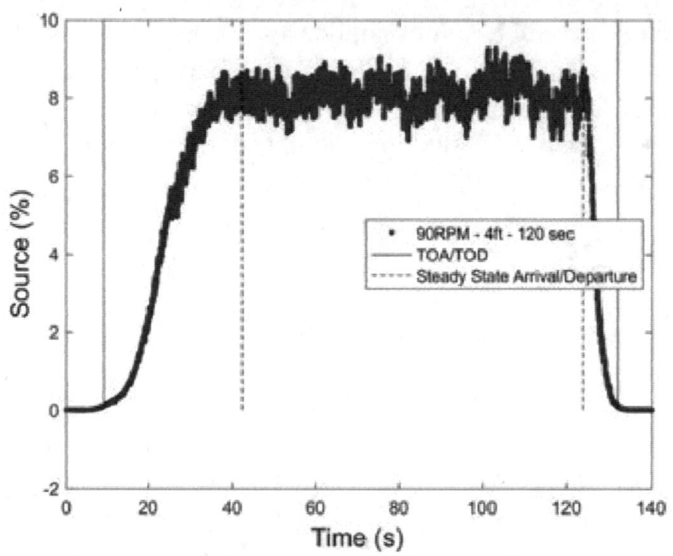

Figure 5. Concentration-time history for 120-second release in 0.71 m/s at 4 ft downwind

This release shows characteristics more similar to Figure 1, but the duration of steady-state is much less than 120-seconds. The ensemble concentration-time histories of 120-second releases are representative of true finite-duration releases and total impact of the leading and trailing edges. The leading edge of the cloud (earliest concentration measurements) shows an increase in concentration that is much slower than the trailing edge, which rapidly decreases. For longer releases, there is a period of constant concentration that is the same as that of a steady-state release, but the duration of this period is significantly less than the total release duration. For shorter release durations, the observed maximum concentration is less than the steady-state maximum concentration and a period of steady-state is never achieved.

In field tests, ensemble averages are impractical due to cost and repeatability with varying test conditions such as wind speed, direction, and stability class. Concentration-time histories captured in a field test would be similar to a single wind tunnel test. An example of a single test in comparison to the ensemble of 100 tests is shown in Figure 6.

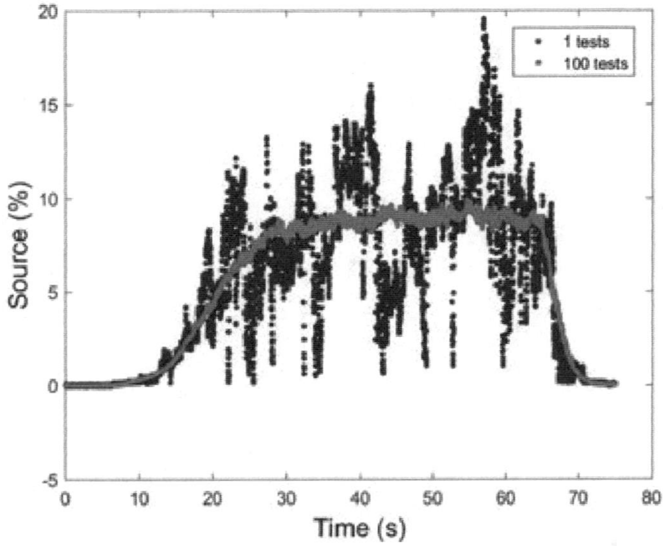

Figure 6. Concentration-time history for 120-second release in 0.71 m/s at 4 ft downwind for both 1 individual test (blue) and 100 ensemble averages (red)

From individual tests, the time of arrival and time of departure can be measured, but determination of any constant concentration or steady-state period would be difficult, if not impossible. The ensemble averages provide a much clearer picture of the cloud behavior which is achieved in this experimental program.

At long distances from the source, the distinction between leading and trailing edges are expected to disappear so determination of σ_t taking into account the both the leading and trailing edges would be advantageous. Consequently, σ_t was evaluated as:

$$\sigma_t = \frac{\textbf{leading edge} + \textbf{trailing edge}}{\mathbf{4.3}} = \frac{(\boldsymbol{TOPa} - \boldsymbol{TOA}) + (\boldsymbol{TOD} - \boldsymbol{TOPd})}{\mathbf{4.3}} \qquad \textbf{(1)}$$

This approach can be applied to all measurements here. For shorter release durations, the TOPa and TOPd are equivalent with one peak.

5 Results and Discussion

Results of the continuous rate finite-duration release experiments are presented. An overall recommendation for the time distribution constant and along-wind dispersion constant is achieved from the continuous rate finite-duration releases.

5.1 Concentration-Time Histories

For the 120- and 60-second releases, the steady-state maximum concentration was achieved. For the 60-second releases, the maximum measured concentration period is significantly shorter for the same test conditions compared to the 120-second releases and is effectively zero for the low wind speed tests. For the 30-second releases, the maximum concentration (peak) did not always reach the maximum steady-state concentration for the same downwind distance and wind speed. All trends can be observed in the ensemble concentration-time histories plotted as a function of the downwind distance and wind speed shown in Figure 7.

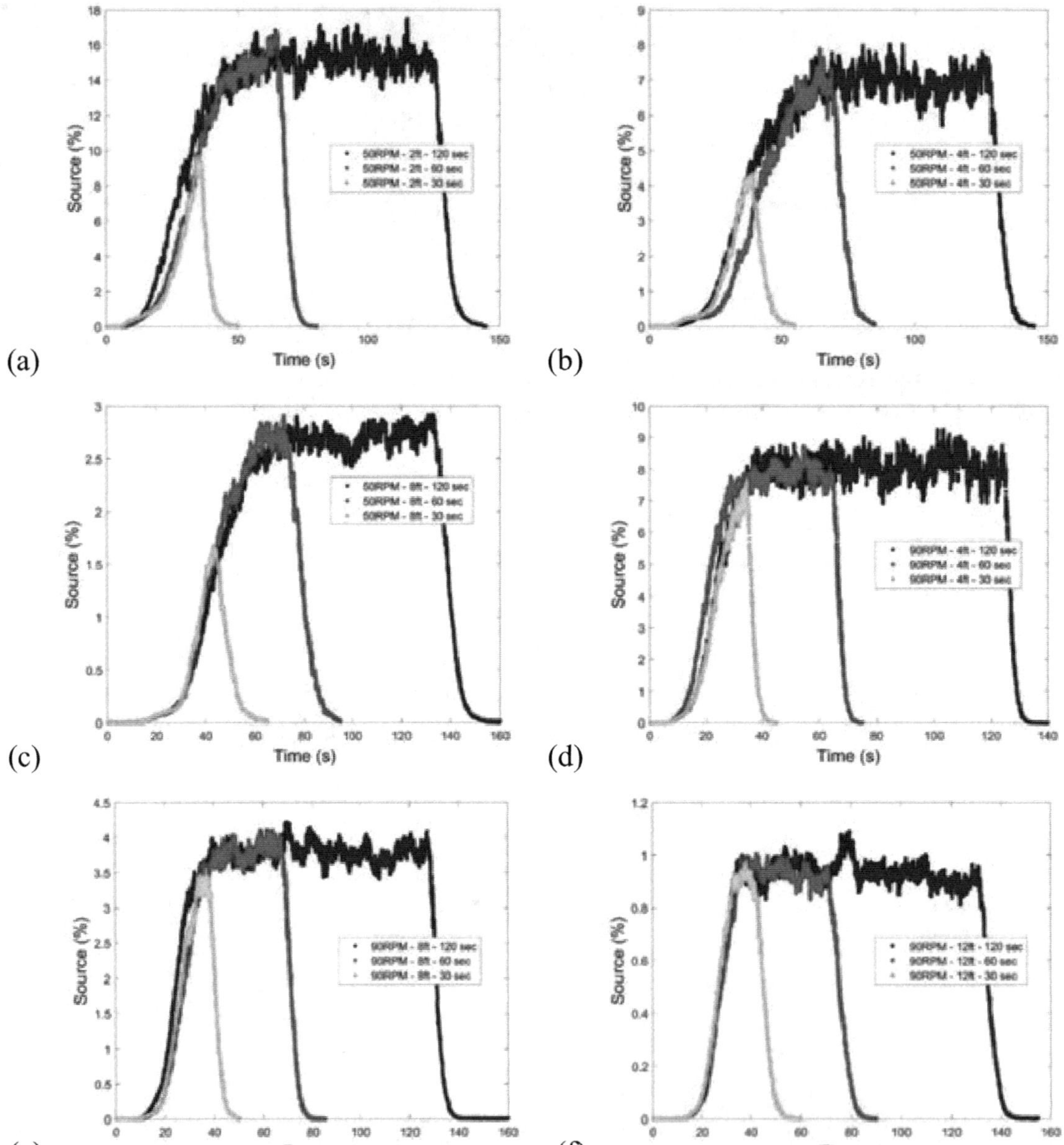

Figure 7. All finite-release durations at downwind distances of 2 ft (a), 4 ft (b), and 8 ft (c) for 0.35 m/s and 4 ft (d), 8 ft (e), and 12 ft (f) for 0.71 m/s

As expected, the effect of along-wind dispersion increases with downwind travel (in both distance and time). For both the 60- and 120-second releases, the impact of along-wind dispersion at the leading edge and trailing edge are very comparable. For the 30-second releases, the impact of along-wind dispersion on the leading edge does not see the full effects as compared to the longer release durations. The 30-second releases do not always reach the steady-state maximum concentration due to the transfer of mass from the steady-state period to the leading and trailing edges, thus decreasing the steady-state duration and maximum concentration achieved. Additional shorter release durations were performed at 4 ft downwind at the 0.71 m/s wind speed. The concentration-time histories of these tests are shown below.

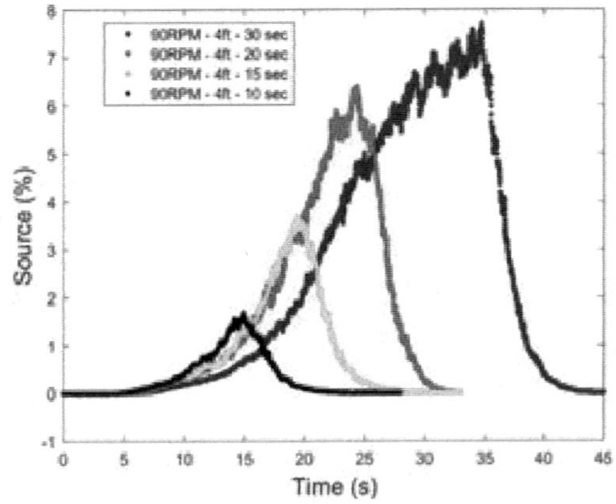

Figure 8. Shortened release durations at 0.71 m/s and 4ft downwind

The 30-second releases paired with the shorter release durations show that as the cloud travels downwind, the leading and trailing edges come together more quickly reducing the maximum (peak) observed concentration. The decreased release duration is shown to have a larger impact on the leading edge as compared to the trailing edge which is essentially consistent across all release durations.

5.2 *Measured* σ_t

The time distribution coefficient and along-wind dispersion coefficient are determined from the ensemble averages using equation (1) and the $\sigma_x = \sigma_t u_e$ relationship. Non-dimensionalized values for σ_t and σ_x are shown below for varying times.

Table 3. Summary of $^{\sigma_t}/_t$ for different times and $^{\sigma_x}/_x$ for the only distance

test	U (m/s)	Distance (m)	Duration (s)	σ_t/TOA	σ_t/TOD	σ_t/TOPa	σ_t/TOPd	σ_x/x
6	0.71	1.37	120	1.09	0.07	0.23	0.08	1.09
5	0.71	1.37	60	1.21	0.14	0.22	0.15	1.21
8	0.71	1.37	30	0.91	0.19	0.22	0.22	0.91
9	0.71	2.59	120	0.91	0.07	0.23	0.08	0.91
10	0.71	2.59	60	0.70	0.12	0.21	0.14	0.70
12	0.71	2.59	30	0.66	0.17	0.22	0.22	0.66
30	0.71	3.81	120	0.76	0.08	0.22	0.09	0.76
31	0.71	3.81	60	0.78	0.14	0.22	0.16	0.78
33	0.71	3.81	30	0.61	0.17	0.23	0.23	0.61
13	0.71	5.03	120	0.99	0.12	0.24	0.13	0.99
14	0.71	5.03	60	0.96	0.19	0.23	0.22	0.96
15	0.71	5.03	30	0.53	0.16	0.22	0.22	0.53
17	0.35	0.76	120	1.38	0.10	0.25	0.12	1.38
18	0.35	0.76	60	1.23	0.18	0.23	0.20	1.23
19	0.35	0.76	30	1.00	0.19	0.24	0.24	1.00
21	0.35	1.37	120	1.23	0.11	0.24	0.12	1.23
22	0.35	1.37	60	1.34	0.18	0.24	0.23	1.34
23	0.35	1.37	30	0.85	0.18	0.24	0.24	0.85
25	0.35	2.59	120	1.08	0.12	0.24	0.13	1.08
27	0.35	2.59	60	0.90	0.16	0.24	0.21	0.90
28	0.35	2.59	30	0.90	0.25	0.24	0.35	0.90

As expected,

$$\frac{\sigma_x}{x} = \frac{\sigma_t}{TOA} \quad (2)$$

because u_e was taken to be $^{x}/_{TOA}$. From the table, it is interesting to note that $^{\sigma_t}/_{TOPa}$ is essentially constant over all wind speeds, distances, and release durations.

5.2 Comparison of σ_t and σ_t with previous data

Past field and wind tunnel data were analyzed[9][10]. The results from this experimental program are added to the previous data in Figure 9 using the TOPa as the time scale.

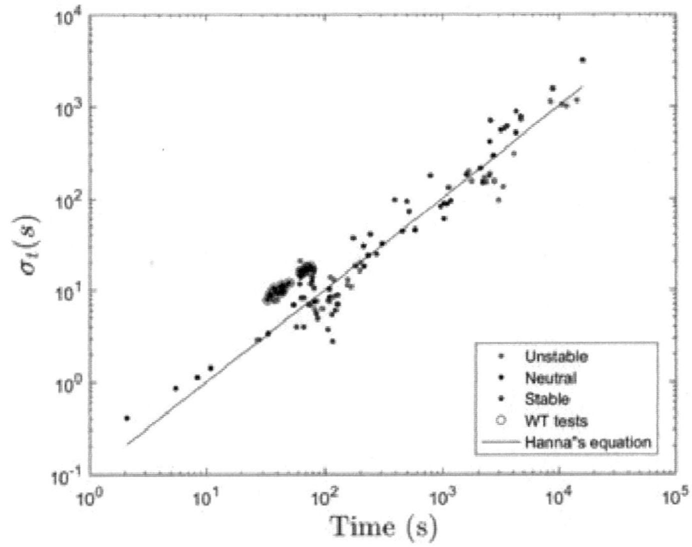

Figure 9. σ_t vs. time for all previous experiments, Hanna's equation, and the new finite-duration releases plotted as a function of stable, unstable, or neutral stability

As expected, the values of σ_t estimated in this program are larger since previous values of σ_t had been estimated using $\sigma_t = (\Delta T - T_{dur})/4.3$. Clearly, the use of ensemble average measurements provides a different perspective on this problem that cannot be obtained from field scale data.

Following the previous discussion, Figure 10 includes data from this experimental program along with previous data. As with Figure 9, the time scale used to represent the data in this work is TOPa.

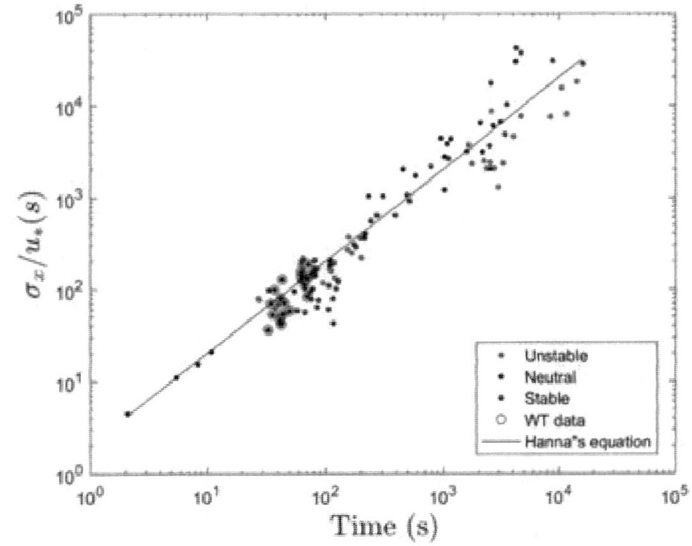

Figure 10. $\frac{\sigma_x}{u_*}$ vs. time for all previous experiments, Hanna's equation, and WT tests as a function of stable, unstable, or neutral stability.

When compared in this fashion, the parameters estimated here are in agreement with past data. It should be noted that none of the CHRC finite-duration data sets are scaled. While the σ_t values are larger in comparison with previous data, the $\sigma_x$values are comparable.

5.3 Modeling Applications

While the experimental program was performed in a wind tunnel, the results and relationships are directly applicable to field tests. From the additional tests performed, the shorter the finite-duration releases result in a decreased concentration, TOPa, and time distribution coefficient. The same effect would be expected in field scale tests. This observation is seen with advanced similarity models. The two main relationships that can be used when scaling are the $\sigma_t = 0.23 * TOPa$ and $TOPa = C_1 * T_{dur}{}^{p} * (\frac{u_e}{u})^q$. From the time distribution coefficient, the along-wind dispersion coefficient can be determined from $\sigma_x = \sigma_t u_e$ relationship.

6 Summary and Future Work

Continuous rate finite-duration releases were performed in an ultra-low speed wind tunnel at the Chemical Hazards Research Center. The primary test program focused on area source releases with durations of 30-, 60-, and 120-seconds. Tests were conducted at freestream wind speeds of 0.35 m/s and 0.71 m/s at downwind distances of (nominally) 2 ft, 4 ft, 8ft, and 12 ft. Ensemble averages of at least 65 tests were performed for each set of test conditions. Previous assumptions for determining the time distribution coefficient from field data may be open to question. Across all wind speeds, distances, and release durations, a relationship of $\sigma_t = 0.23 * TOPa$ was observed. From a modeling perspective, the time distribution coefficient can be reasonably predicted for finite-duration releases that both reach steady-state and for release durations that do not. For the along-wind dispersion coefficient, the experimental results did not provide a simple relationship but can be determined from the total time distribution coefficient and effective wind speed. Overall, throughout the course of the study: 1) true finite-duration releases from a purpose-built area source were accomplished, 2) quality data to be used in validation of models under multiple test conditions were established, 3) a method to predict overall σ_t and impact of release durations was determined and 4) an accurate prediction for σ_x from σ_t was accomplished.

Additional tests are underway at the Chemical Hazards Research Center for the exploration of different release rates under the same wind speed, downwind distances, and release durations. Additional downwind distances are desired for additional data under the same experimental program for additional trends.

7 Acknowledgments

We gratefully thank all of the undergrads working at the Chemical Hazards Research Center who were apart of the data collection. We also thank Henk Witlox and Maria Fernandez from DNV GL who provided insight and expertise that assisted with the research and funding of the project.

8 References

[1] Boybeyi, Zafer, Sethu Raman, and Paolo Zannetti. "Numerical investigation of possible role of local meteorology in Bhopal gas accident." *Atmospheric Environment* 29.4 (1995): 479-496.

[2] Høiset, S., et al. "Flixborough revisited—an explosion simulation approach." *Journal of Hazardous Materials* 77.1 (2000): 1-9.

[3] De Marchi, Bruna, Silvio Funtowicz, and Jerome Ravetz. "Seveso: A paradoxical classic disaster." *The long road to recovery: Community responses to industrial disaster* (1996): 86-120.

[4] Hanna, S.R., and Chang, J.C., "Analysis of dispersion of finite-duration releases of dense gases and recommended revisions to HEGADAS, Earth Tech Report 1411, 196 Baker Ave, Concord, MA01742, 196 Baker Ave, Concord, MA01742, prepared for Exxon Research and Engineering, 180 Park Ave, Florham Park, NJ07932 (1995)

[5] Havens, J., and T. Spicer. "Vapor dispersion and thermal hazard modelling." *Final topical report to Gas Technology Institute under sub-contract K* 100029184 (2006).

[6] Walker, H., and J. Havens. "Comparison of hot wire and laser Doppler anemometry measurements of the turbulence structure of an ultra low speed wind tunnel boundary layer." *Data report, Chemical Hazards Research Center,* University of Arkansas, Fayetteville, AR 72701 (1998).

[7] Hall, D. J., et al. "The Effect of Release Time on the Dispersion of a Fixed Inventory of Heavy Gas – A Wind Tunnel Model Study." Physmod 2007.

[8] Hanna, Steven R., and Joseph C. Chang. "Use of the Kit Fox field data to analyze dense gas dispersion modeling dispersion modeling issues." *Atmospheric Environment.* 35.13 (2001): 2231-2242.

[9] Morris, J.M. and Spicer, T.O. (2018). Improved along-wind dispersion coefficient model from field and wind tunnel data. Annual Meeting for the Global Conference of Process Safety (GCPS), Orlando, FL.

[10] Morris, J.M. and Spicer, T.O. (2017). Generalization of modeling along-wind dispersion. Annual meeting for the Global Conference of Process Safety (GCPS), San Antonio, TX.

BRINGING IT ALL TOGETHER: THE EMERGENCE OF OPERATIONAL RISK MANAGEMENT DIGITAL TWINS - TWO CASE STUDIES

Scott Lehmann
Sphera – VP, Operational Excellence
SLehmann@sphera.com

Simon Jones
Sphera – Director, Solutions Consulting
Simon.Jones@sphera.com

Prepared for Presentation at
American Institute of Chemical Engineers
2019 Spring Meeting and 15th Global Congress on Process Safety
New Orleans, LA
March 31 – April 3, 2019

BRINGING IT ALL TOGETHER: THE EMERGENCE OF OPERATIONAL RISK MANAGEMENT DIGITAL TWINS - TWO CASE STUDIES

Scott Lehmann
Sphera – VP, Operational Excellence
SLehmann@sphera.com

Simon Jones
Sphera – Director, Solutions Consulting
Simon.Jones@sphera.com

Keywords: Operational Risk Management. Digital Twins. ORM. Process Safety Management. Process Safety Barrier health

Abstract

Modern hazardous industries face the challenge of routinely connecting risk to the operational reality. While process safety methods and techniques developed over the years have improved the design of our facilities, many industry operators do not provide systems and solutions for maintenance and operations - where decisions continuously take place that impact well-designed process safety barriers and increase exposure to Major Accident Hazard (MAH) risk.

According to Petrotechnics' 2018 industry survey, 86% of industry leaders report a gap between how process safety is intended and the reality of operations. Only 38% believe industry operators proactively manage process safety risk. Sixty-nine percent say challenges to delivering effective process safety are due to safety culture and overall risk awareness.

The survey indicates gaps in organizations' abilities to develop and make accessible a single, shared view of the operational reality for everyone across an organization. Although companies have a wealth of data, it's often trapped in organizational silos making it difficult to utilize in supporting real-time decision-making.

The dynamic nature of risk at the frontline demands new ways to make MAH risk exposure visible and prominent, so every operational decision is informed by real-time events and managed with care.

There is a perceptible shift in the hazardous industries with connected systems changing how risk and process safety are measured and managed. Industry leaders report the adoption of advanced technologies – such as IIoT platforms, digital twins, predictive analytics, and mobile workforce applications - for managing process safety and operational risk will increase significantly in the future.

An emerging category of Operational Risk Management (ORM) Digital Twins offer the potential to make companies and their work teams safer through their ability to deliver an enterprise-wide understanding of how planned work activities impact process safety barrier and equipment health. This early warning system makes all risk exposure visible, prominent and available to all in real-time, reducing vulnerability to high-potential near-misses and major hazard events.

This paper will set out the principles of this new approach and include case studies sharing how two major operators are using ORM Digital Twin capabilities to connect a real-time view of process safety barrier health to the frontline to support more effective hazard management.

1 Introduction

Process Safety approaches developed and implemented over the past 20 – 30 years have, few would argue, enabled us to improve the design basis for our facilities. The use of a risk-based approach is commonplace and indeed a requirement of many regulatory bodies around the world.

The process safety practices and guidelines for designing, managing and operating the facility are well-known and documented by regulatory and engineering bodies. The rules of how to run facilities, maximize production and manage risks are typically encoded in organizations' Operational Management systems.

But, how do we know good practice is being applied? How do we ensure it is effective, and how do we connect this practice to the frontline? Traditionally, KPIs and audits are our principal tools, yet despite our investments in performance monitoring and continuous improvement, the industry continues to experience catastrophic accidents.

The 25th edition of Marsh and McLennan's report[1] outlines the most significant losses in the Hydrocarbon industry between 1978 and 2017, covering the costs of property damage, debris removal, and clean-up costs – all normalized to 2017 values.

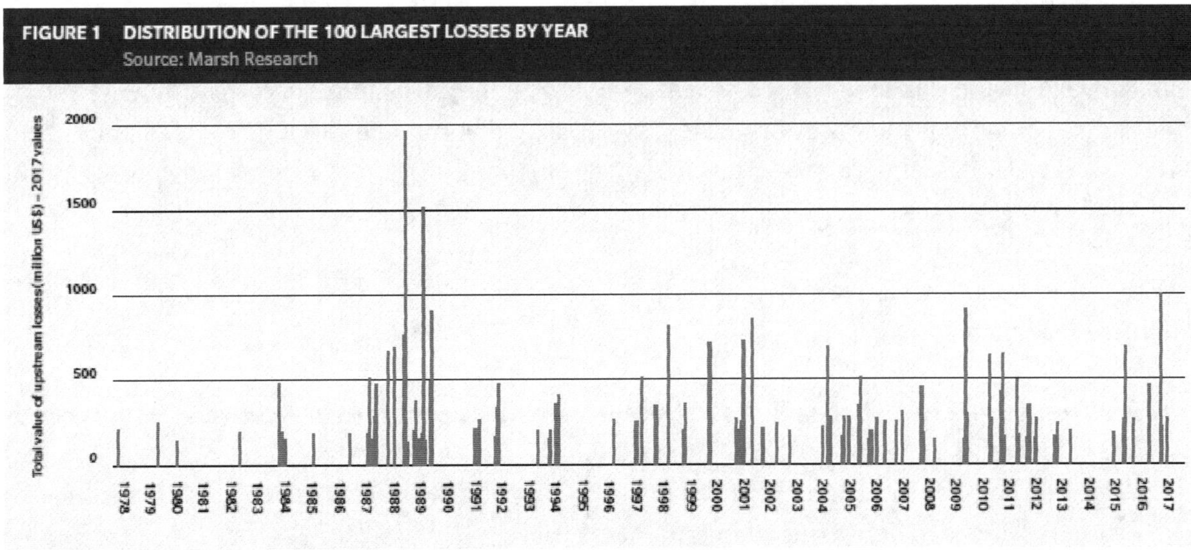

Figure 1: Distribution of the 100 largest losses by year (from Marsh Report[1])

The report authors highlight that in the last two years there has been a spike in the number of high-value downstream oil and gas losses which have had a significant impact on the global energy industry. For the same period, the authors draw attention to the findings from engineering surveys undertaken by the insurance sector. The findings highlight issues with work management systems, such and permit to work (PTW), shift handover, management of change (MoC), and with fundamental inspection processes and practices. The report also references other insurance industry loss analyses,[2] which indicate "mechanical integrity failure" was responsible for 57% of human-influenced losses concluding:

> *"Given these concerns, and combined with high operating rates, reduced staffing levels, and other cost saving programs, operators must maintain high levels of monitoring and vigilance to ensure that asset integrity is being maintained and accidents are eliminated."*

2018 Survey of process safety and operational risk engineers

In 2018, Petrotechnics conducted its third major survey of process safety and operational risk engineers[3]. Over 100 process safety, asset integrity, and operational risk management senior leaders from around the world participated. Fifty-three percent of respondents have more than 15 years' experience, with 61% holding a Corporate or Regional management position. In terms of the sectors represented, 55% came from oil and gas, 20% from chemicals, and the remaining 25% came from a variety of other hazardous industries.

When asked in 2018 about the operational reality of risk from their viewpoint, the following figures emerged:

- 86% believe there are gaps between how process safety is intended and what happens on the plant/asset - up from the 2017 survey figure of 70%
- 56% found an increase in risk on their plant when undertaking periodic process safety reviews (for example, 3-5 year periodic reviews of PHAs, HAZOPs, HAZIDs, LOPA studies, etc.)

This speaks to the challenge faced when a good design enters into service at the frontline; the operational environment causes processes, systems, and equipment to degrade over time, thereby increasing risk. It seems this change in risk is only highlighted when periodic process safety reviews are undertaken.

When asked about the delivery of scheduled maintenance on safety critical elements on facilities, industry leaders reported:

- An average of 73% of scheduled safety-critical maintenance is achieved, and 22% do not think it's practical to achieve 100% scheduled safety-critical maintenance
- When asked why, conflicting priorities (75%) and limited resources (72%) are said to be the main challenges to delivering planned intent

Following up on two key pieces of feedback from Petrotechnics' 2017 survey[4]:

- "It's important that we understand hazards on a real-time basis and that the continual state of barriers is maintained as designed to reduce incidents."
- "Everyone would be more thoughtful on ensuring barriers perform to standards if they truly understood what the barrier was trying to prevent."

The 2018 survey found that only 38% believe industry operators proactively manage process safety. Furthermore, respondents think companies do not have effective systems in place for:

- monitoring and managing impaired process safety barriers (60%)
- monitoring and managing deviations from management system requirements or expectations (64%)

The dynamic nature of frontline operations

The real world of operations is neither simple nor static. In the real world, when our well-designed and well-specified processes and equipment enter service, they are almost immediately subject to change. In reality, day-to-day operations are characterized by a changing dynamic where there are multiple sources of potential risk which typically change daily, hourly and sometimes by the second, such as:

- Process safety barrier impairments - deviations including damage and failure to physical barriers such as corrosion, mechanical failure, temporary repairs, procedure deviations and deferred inspections, testing or preventative maintenance
- Operational activities such as equipment start up or shut down
- Operator actions such as hydrocarbon sampling, filter change out, pig launching and receiving, etc. Excursions outside safe operating limits. Temporary defeats to safety systems. Process isolations which do not meet company standards. Periods of instability on the processing plant
- Working in and around live plant introduces immediate hazards, such as ignition sources from hot work, breaking flanges, heavy lifting, excavation work, and more. These activities are typically risk assessed and controls put in place (such as gas testing while conducting hot work) which mitigates or minimizes the risk, but cannot entirely eliminate it
- Simultaneous operations (SIMOPS) compound operational risk when work teams operate in the same or nearby areas. Some activities may be managed separately from plant operations, such as construction, project work, subsea, drilling or well service. So it is essential that rules be defined for work coordination; however, residual risks will always remain, and more importantly, they will place work crews - possibly unfamiliar with the plant - into hazardous territory
- Environmental conditions such as temperature extremes, wind, visibility, precipitation, electric storms add risk potential and can compromise work programs

- Human factors such as competence, lack of experience, discipline, fatigue, and distraction can compromise the quality of decision-making at the frontline

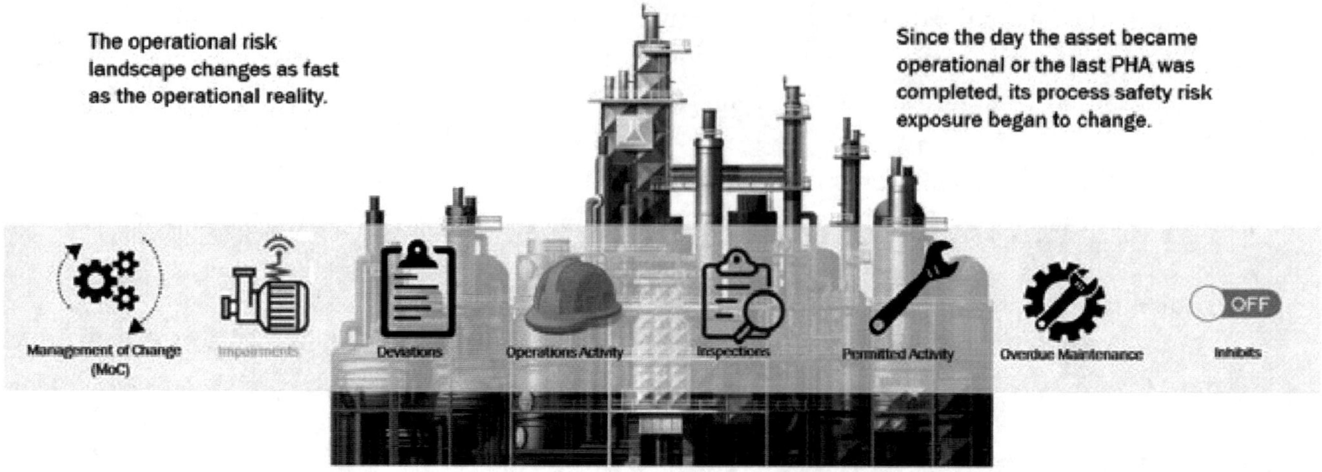

And unavoidably, risks are often managed in different parts of the organisation. Information is siloed. Dots are not connected.

Faced with an array of potential risk sources, decision-makers are faced with many questions, but our information and data systems and approaches to managing risk are not set up to offer real-time insights to support critical operating decisions. Traditionally operational risk management solutions have ended up adding layers of complexity, requiring extensive, manual updates – often disconnected from the operational reality.

A new model for operational risk management

As we have detailed, operational risks arise from a complicated set of interrelated parameters and are viewed and managed in differing ways depending on the role and level in the organization. The challenge lies in simplifying this complexity and enabling all levels of the organization to collectively focus on the elements of risk that are most important.

A new approach to managing the cumulative impact of all these dynamic operational risks is to model their impact on process safety barrier groupings and associate them to the major accident hazards under management. This simple, elegant approach enables operators to predict and better manage the outcome – whether that means to postpone a particular planned activity or accelerate maintenance to address the deviations or non-conformances on the facility.

In many ways, the dynamic nature of the frontline is ripe for technology to better support decision-makers. In fact, according to a Verdantix study, Operational Excellence and Industry 4.0 strategies are among the top factors triggering operational risk management implementation today. Over 73% consider digital technology valuable, if not essential, for effective operational risk management. An emerging category of Operational Risk Management enterprise software seeks to support safe and effective operational decisions by providing critical innovations[7], including:

- An enterprise ORM approach to barrier management, permit to work, management of change, incident management, risk assessments, and process safety management - all accessible via desktop web applications and mobile devices for use in the field
- Wearable technology as a source of data for use in the field
- Industrial Internet of Things (IIoT) as a source of data from critical devices in the field
- Advanced modeling capabilities to create dynamic "digital twin" asset models
- Advanced analytics from big data to provide actionable insights on operational risk status and trends

At the heart of this model (figure 2) is the need to carry out an operational risk assessment for any performance deviation or non-conformance identified on the facility. Examples include:

- Performance standard failure
- Verification inspection finding
- Overdue safety-critical maintenance
- Override of a safety-critical system or device
- Management of hydrocarbon leaks
- Temporary equipment
- The absence or inadequacy of a control system(s)

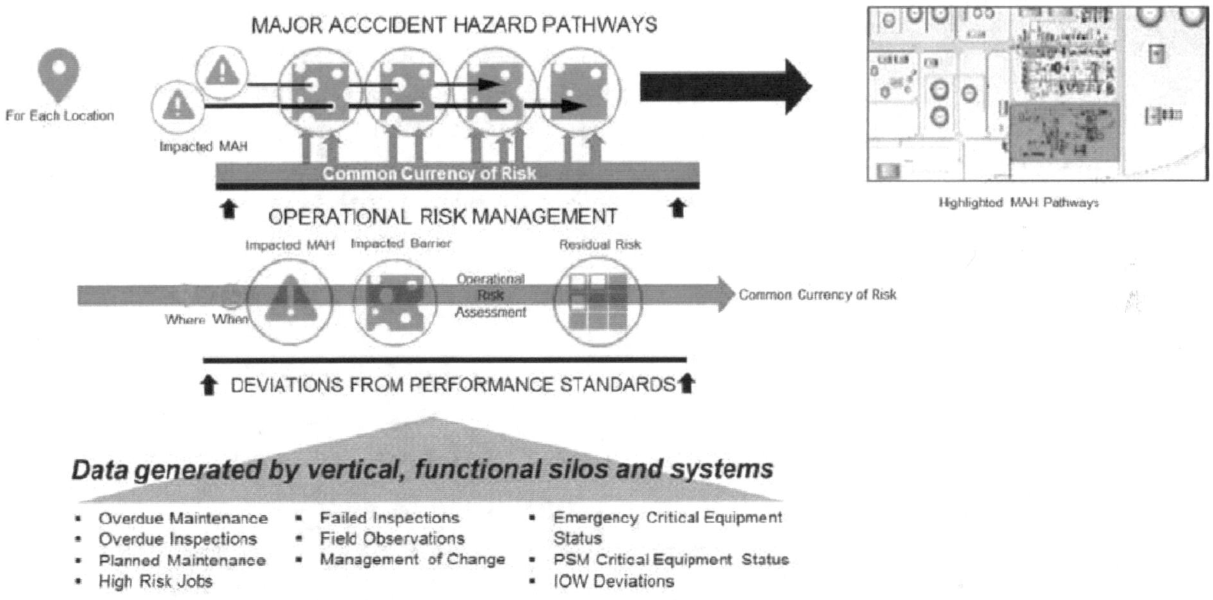

Figure 2: Deviations and non-conformances as a source of risk

An engineering technical authority typically leads the risk assessment process. This process identifies the major accident hazards under management, the fundamental barriers impacted, defines interim mitigation measures and authorizations required and also the resulting residual risk associated with the individual deviation or non-conformance.

Deviations and non-conformances may be tracked and managed through many different business processes and systems, such as asset integrity inspection systems, maintenance management systems, operator rounds or management of change processes, inspection data and environmental control systems. Therein lies the challenge and the opportunity from an Industry 4.0 perspective: if we can map these operational conditions, from whichever business system they arise from, and provide a better illustration of their impact on operational risk and the major accident hazards under management, we help support better decision-making from a risk management perspective.

Can an operational risk management digital twin help close the gap?

There is increasing focus and attention on the potential for new digitalization technologies to deliver increased value and sustainability in the energy and petrochemicals sector. And in particular, Operational Risk Management (ORM) digital twins. Over 90% of industry leaders recognize the power of digitalization to accelerate and provide sustainable operational excellence[5]. A reduction in operating costs, broader operational efficiencies, and fundamental transformation of the business are expected. The promises of data connectivity and analytics suggest continuous uptime, rapid response to risk exposure, incremental revenue gains, opportunities to better utilize assets, coordinate with operating and business needs and improve the efficiency of field service groups.

> *"The effective use of digital technologies in the oil and gas sector could reduce capital expenditures by up to 20 percent; it could cut operating costs in upstream by 3 to 5 percent and by about half that in downstream." – McKinsey*[6]

So, what is an ORM Digital Twin, and how could it help improve the way we manage operational risk on our assets? To be a digital twin, one must have a digital representation or model of a real and specific object such as a plant, factory or physical asset. Other software systems and operators on the plant must be able to interact with it on a direct basis bypassing the real object. From an ORM perspective, the digital twin must be able to bring together human, system, and sensor-derived inputs in a joined-up way and provide a continuously updated picture of operational risk on the specific asset.

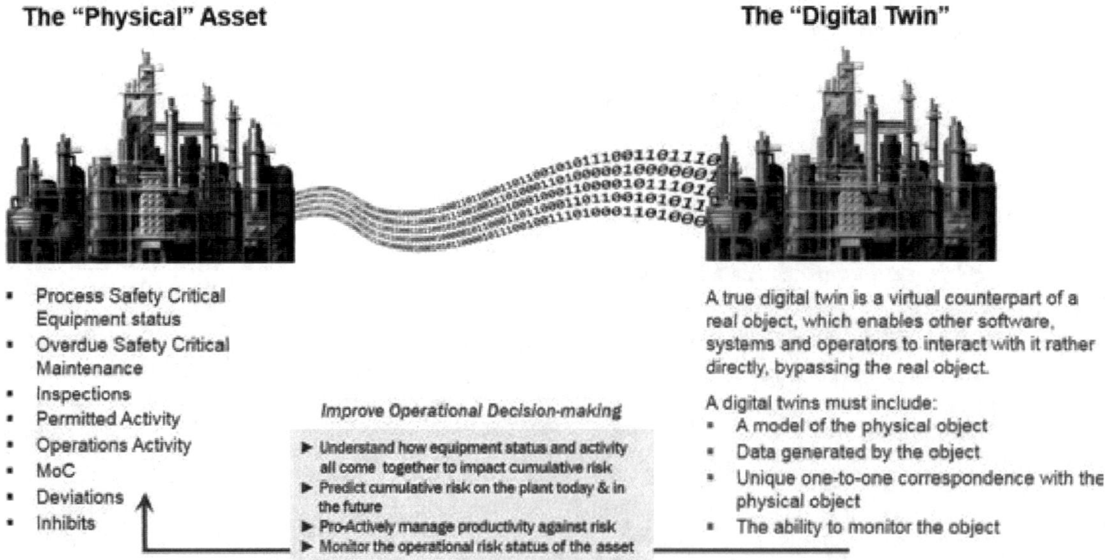

An ORM Digital Twin Connects Operational Risks to a Fundamental Barrier Model

As hazardous industry operators move toward a fundamental barrier model, based on multiple layers of protection/mitigation an ORM Digital Twin can bring all of the risks together to understand their cumulative risk impact in a practical and tangible manner. The idea is simple; if you have impairments in several barrier groups), risk increases because a major accident is more likely to occur. Note that these barrier groups are not singular barriers; each is a collection of equipment, instrumentation or people-driven processes which collectively fulfill the function of the barrier (figure 3).

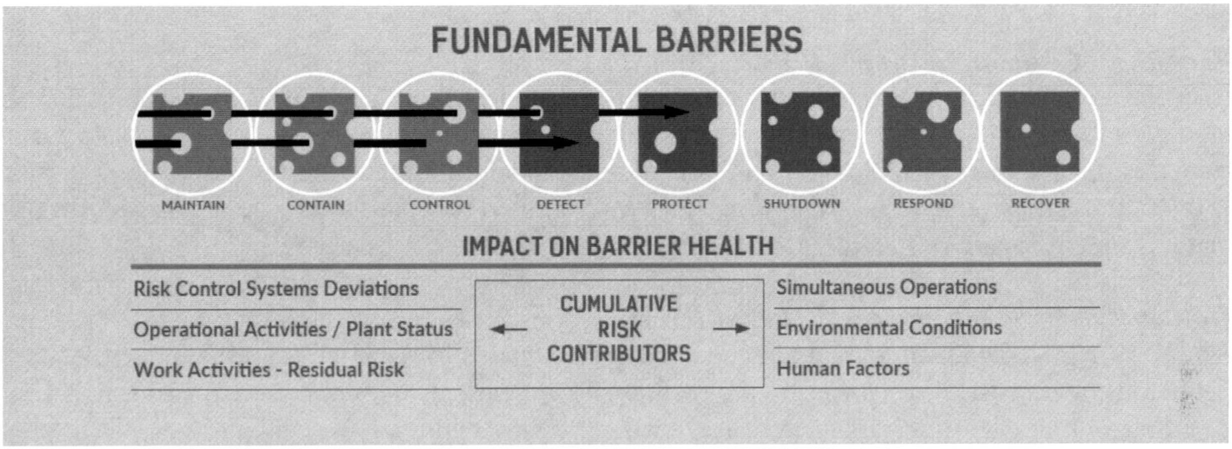

Figure 3: A fundamental barrier model

The fundamental barrier model has its origins in James Reason's Swiss Cheese metaphor; the holes in the barriers represent impairments, and when the holes line up, an event can occur and escalate into a major accident. The degree to which risk pathways develop represents the potential level of risk.

The grouping of the barrier systems is important because it allows work teams and process safety engineers to see how the impairment of barriers can line up sequentially with others, with the potential to compound risk. For example, if there is an impairment on the 'containment' barrier happening at the same time and within the same location as impairment in the 'ignition control' barrier, the risk of having an uncontrolled release of hydrocarbons to the atmosphere is higher, as is the risk of ignition. Combined, they can result in a fire or explosion as the result of the subsequent failure of more than one barrier group. If the 'detect' barrier (e.g., gas and fire detectors) is also compromised, and the ability to protect is also compromised (e.g., say the water deluge system is inoperable), these additional breaches can result in the potential occurrence of a major event. The degree of escalation and the scale of the consequences will depend on the mitigation barriers or the ability to respond to the incident.

An ORM Digital Twin Connects Activity Risks to Fundamental Barriers

Planned activity on the facility can also introduce risk as it often includes the introduction of hazards into a live process plant whether its work involving spark potential, breaking containment, startups, shutdowns, isolations, de-isolations. All of these activities have the potential to increase risk and impact process safety barriers. These activities are typically planned and scheduled in a maintenance management system, and their execution is managed via a work permit processes, supported by task risk assessment or job safety analysis (JSA). In addition, operational activities are managed through a combination of operational procedures and operator rounds practices.

The potential barrier impact of this activity can be modeled – for instance, if a planned positive isolation is needed to prepare for confined space entry, it is reasonable to assume there is a potential impact on the process containment barrier for the period in which first line break is undertaken. Similarly, open flame hot work in a unit represents a degradation of the ignition control barrier for the period the permit is issued.

Enabling a Common Currency of Risk

If we have carried out operational risk assessments for all deviations and non-conformances on the facility, and we also know the potential barrier impairments introduced by planned work, we can have a complete view of all activity and risk and their potential impact on the asset's operational reality. And we can map this to a specific location, a given time/shift, and see the impact on the major accident hazard risks under management.

How does data get into the ORM Digital Twin?

An ORM Digital Twin allows us to connect disparate sources of data that represent all activity, deviations, and non-conformances on the facility and generate a "common currency of risk." The cumulative impact of these risks can be modeled to help everyone understand and assess risk by the same criteria, to make better operational decisions and proactively intervene to prevent major hazard events. Support for diverse integration needs, ranging from direct point-to-point integration with 3rd party systems to robust involvement with integration middleware via RESTful Web Service API, Plugins, Messaging (MQ) and custom connectors is key. The diagram below illustrates how insights can be created with the help of an ORM Digital Twin by extracting, translating and aggregating data from sensors, systems, and human activity.

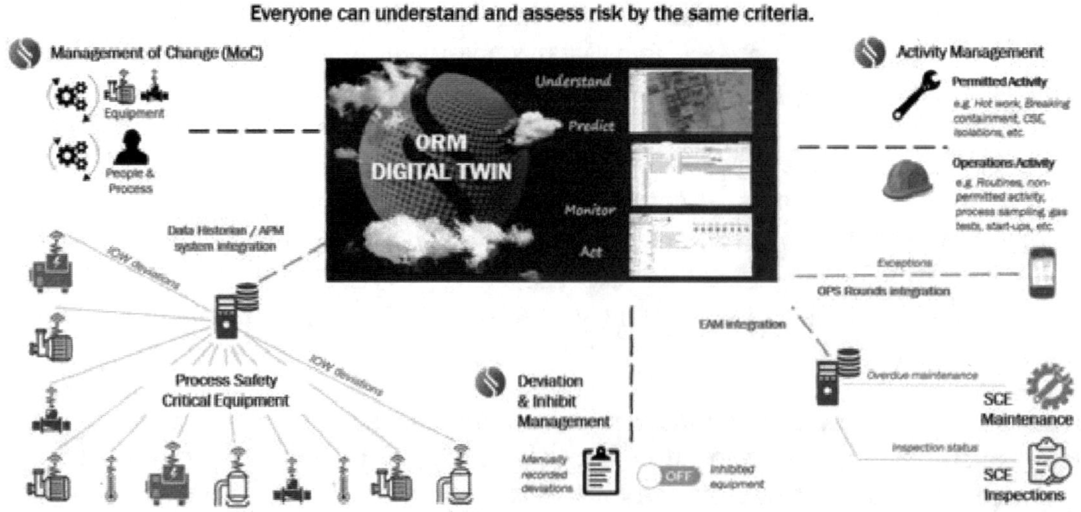

- Make the IIoT operational by connecting the status of process safety-critical equipment sensor (e.g., temperature, pressure, vibration sensors) data to its impact on cumulative operational risk with real-time or near-time integration to Historians and APM systems using time-weighted averages
- Automatically connect overdue and planned maintenance and inspection data with integration to EAM/CMMS and Inspection systems
- Automatically capture process and people related "risk" data with real-time integration of EPTW-CoW, EHS, and other systems
- Automatically capture the operational risk impact of MoC, inspection data, inhibits, emergency-critical and environmental control system statuses
- Manually raise, risk assess and manage performance deviations for impairments, deferrals, overrides, MoCs, or even how well processes, procedures, and drills are followed

An ORM Digital Twin provides a single, shared view of the operational reality

Capturing all the process safety risk data isn't enough. It needs to be connected to the operational reality - in a practical and routine way – for everyone. An ORM Digital twin can do just that by delivering real-time insights to support daily operational decision-making on the ground. With a single, shared view of the operational reality everyone can make better, more informed decisions because they can see:

- What is happening
- Where is it happening
- When is it happening
- What's driving the risk

If we think about the daily activities of the frontline worker, there are many situations where providing a common view of the operational reality of the facility - that is, an understanding of where equipment conditions or planned activities may impact operational risk – will help support effective decision-making.

By bringing all human and sensor-derived inputs together in a joined-up way to see their cumulative risk impact on the operational reality of the plant, operators can see the health status of process safety-critical equipment and the impact of performance deviations and abnormal conditions. With a common view of risk, everyone can understand and assess risk by the same criteria. This provides a holistic and common means of balancing risk against production at all levels of the operation.

Major accident hazard risk exposure can be made visible, prominent and available in real-time. It can be viewed in time, location, and in a dynamic process safety barrier model with drill-down capabilities to quickly understand what's driving the risk. Everyone can understand what activities or conditions drive risk and which process safety barriers are impacted or impaired by specific work activities, actions, and operational requirements. This allows you to proactively manage the health status of safety-critical equipment and the impact of performance deviations and abnormal conditions.

Knowing what is happening and where it is located allows operations staff to understand better how the state of the plant potentially can impact planned activity and how planned activity could potentially impact the state of the plant. Seeing the risk on the asset from a geographic perspective

Diagram: See the risk on the asset

With these visual cues, we can better institutionalize operational risk management standards and expertise in a manner that is tangible and practical for everyone across the organization. We can understand what is happening on the physical plant, what is planned to occur, the state of the plant and through risk shading the risk exposure on a shift by shift basis.

We can then use this insight to understand how process safety risk might potentially impact the operational schedule. No longer do we need to rely on the experience and expertise of operations supervisors to juggle all of the moving pieces in their heads. An ORM Digital Twin can deliver a real-time, 360° view of the operational reality to optimize the schedule against risk.

Diagram: Understand the impact on the schedule

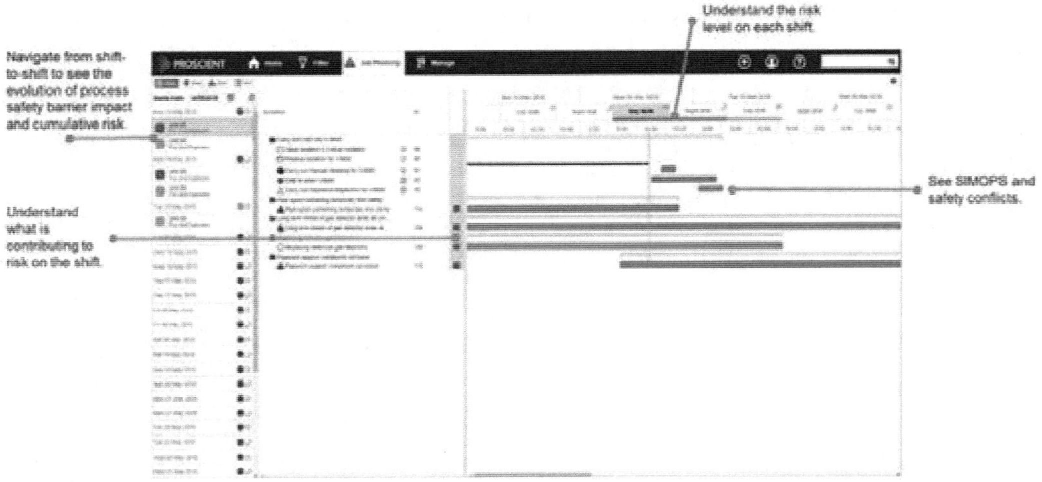

In the ORM Digital Twin, we can see shift by shift basis the cumulative risk and which activities and conditions contributing to the risk. Importantly we can also see SIMOPS and work sequence conflicts. This allows us to make informed decisions to turn SIMOPS from a half-empty proposition of what we can't do at the same time to half full exercise of optimizing productivity against risk. With real-time monitoring and management as the dynamic of each shift unfolds supervisors have insight and the tools to help them move from reactive to proactive decision-making.

A key enabler of proactive decision-making is easy access to real-time information to drill down to the details that matter most for the decision to be made. An ORM Digital Twin allows you to see and understand the health of dynamic process safety barriers and what's driving the risk. It connects the dots and unlocks meaningful operational risk relationships, previously unseen. With this information, everyone can practically understand how the state of the plant impacts live and planned activities and how that activity could impact the state of the plant.

Diagram: Understand what's driving the risk

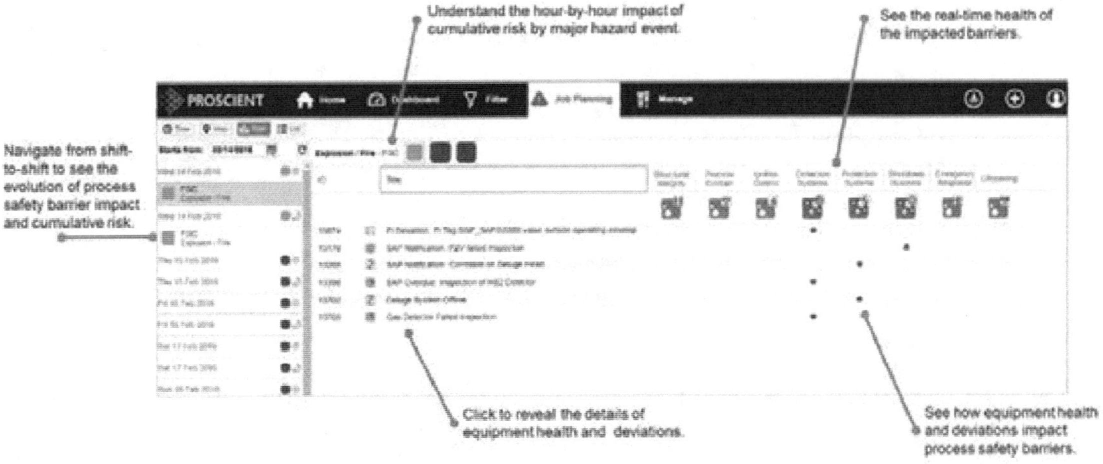

What's critical is an understanding of how specific impairments to equipment and processes combine with activities to impact a specific major accident hazard such as fire and explosion. By seeing how the holes in the barriers line up, we have more insight into how we can make better decisions of how we mitigate and manage risk (e.g., defer planned activity, add further controls, prioritization changes, etc.).

An ORM Digital Twin provides the ability to drill down further into the specific elements causing the risk. It gives us a real-time view of equipment risk and its cause. We can map process safety critical equipment tags held in our Historian and APM system to flag to the ORM Digital Twin when for example the timeweighted average reading for a temperature sensor is outside of its normal integrity operating window. By relating the real-time health of process safety equipment to its cumulative risk impact against a barrier model, the physical world is connected to a decision support model that can inform proactive decision-making.

Diagram: A Real-time View of Risk (Equipment)

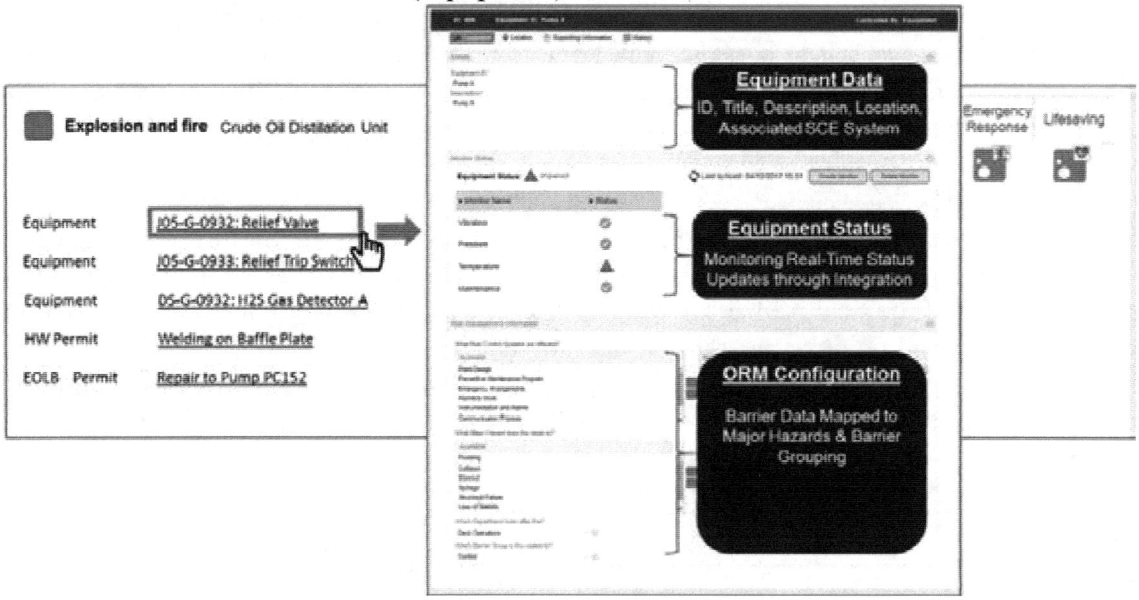

Importantly, an ORM Digital Twin also allows you to drill down into an activity that is either live or planned to understand the details. See,

- The details of the risk assessment
- What hazards have been selected
- What controls have been put in place to manage the risk associated with this activity as well as other relevant details

Diagram: A Real-time View of Risk (Activity)

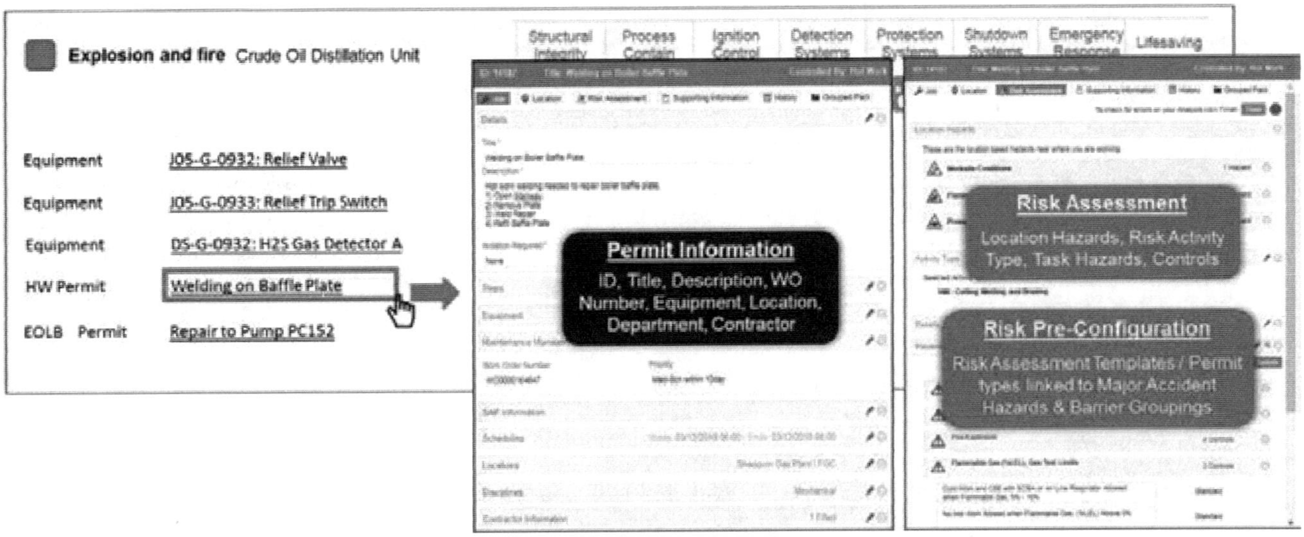

With this real-time access to activity details at the fingertips of an operator, more informed and joined up operational decisions can be made. Importantly, the visibility of how an activity they are about to do or approve impacts the state of the plant. With this understanding, they can be better understood whether they should proceed with the work, add additional controls to reduce risk or even defer it to a later date when the overall risk exposure may be lower.

Managing Process Safety with an ORM Digital Twin

Digital Transformation / Industry 4.0 done right is an ongoing process that will change the way hazardous industries operate. It's fundamentally about new end-to-end business processes empowered by technology to produce positive business outcomes. As we have demonstrated an ORM Digital Twin with its digital representation of the operational reality can unlock a radically different, far more effective way to visualize and manage activity and risk. By connecting people and processes and closing-the-loop between operations, maintenance, engineering, HSE, and other functions – digital solutions can deliver meaningful, actionable insights with powerful visualizations of risk and activity.

An ORM Digital Twin can help connect previously disparate business processes in ways that just haven't been possible until now. It can relate the collective performance of your process safety systems to the real, cumulative risk impact on operations at any given point in time.

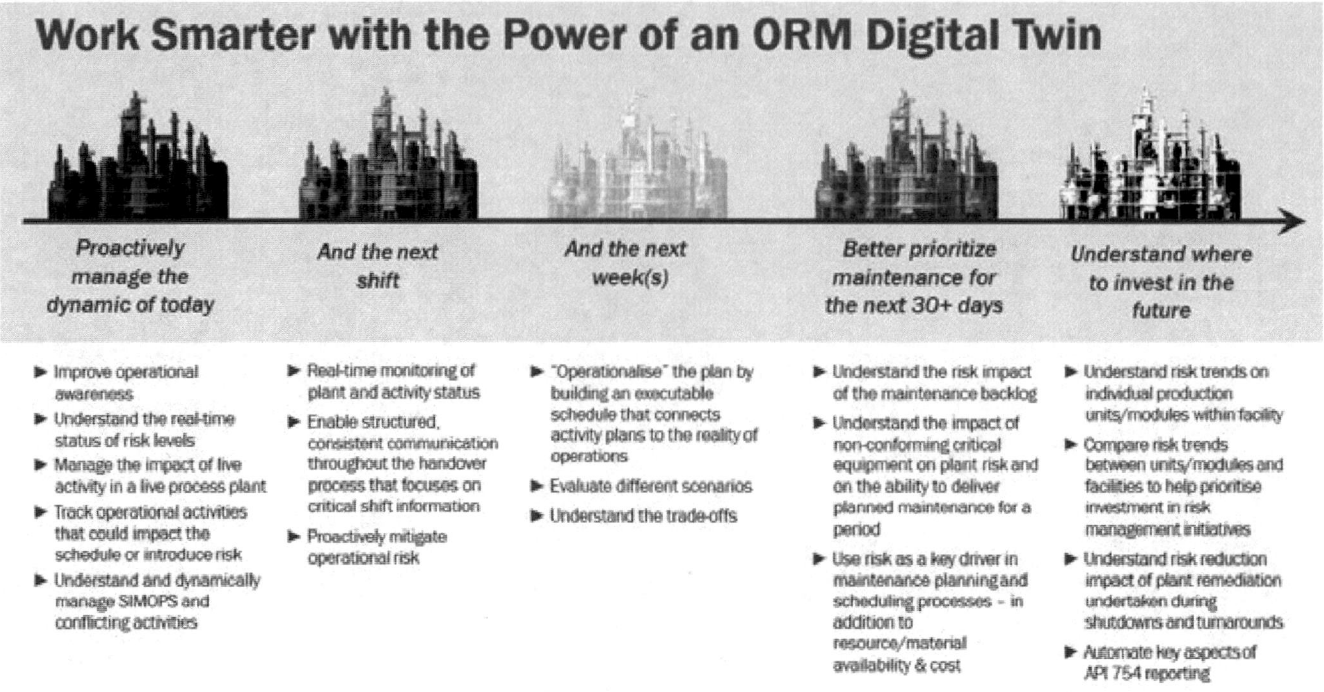

ORM Digital Twins in practice

ORM Digital Twin capabilities are not an aspirational vision of what the future could be, they are already used in production today. Here we share how two major international oil and gas industry operators have implemented ORM Digital Twin capabilities.

CASE STUDY 1: Improving the quality of technical risk assessments and modeling their cumulative risk impact

A major international oil company operates multiple platforms offshore in the UK Continental Shelf (UKCS). This operator has a mature management system and a well-defined approach to process safety, asset integrity management and work control. By implementing their Operational Risk Management platform, a technical manager sought to further improve the risk assessments that are undertaken when critical equipment is not meeting its performance standard.

The existing practice was to immediately carry out an operational risk assessment once such a deviation had been identified from formal inspection, maintenance, or operator activities. This initial risk assessment was approved by the local offshore installation manager (OIM) and would be discussed with the engineering support team onshore.

A general criticism of the UKCS regulator (not specific to the operator) was that such risk assessments in practice rarely identified the true hazard related to the failure of the protective function of the critical equipment.

This operator had a well-defined approach to managing safety-critical elements (SCE) and associated components and equipment. A performance standard was defined for the identified SCEs on each installation, which is related to the risk reduction credit taken in the regulatory Safety Case.

To improve the quality of the initial risk assessment, the operator uses Petrotechnics' ORM for Operations to present the assessor with templated risk assessments based on the type/category of SCE impaired. The templates helped to:

- Define the true hazard that the non-conforming equipment as a class gives rise to
- Provide typical mitigating measures for the assessor to consider minimizing risk based on the equipment class/function
- Present relevant SCE performance standard content as checklists – which encouraged the assessor to,
    - Identify the level of safety or integrity criticality the deviation represents
    - Consider other protective functions that might compound the problem – that is, other deviations that also impact the area and major hazard under management

ORM for Operations also helped the assessor define if the impairment would impact a local area of the facility, or the whole platform – for example single gas detector may have a localized risk impact, whereas firewater pumps unable to deliver the required capacity could impact the entire installation.

ORM for Operations was also used to drive a revised approval process:

- Formally involve the defined functional, technical authority in the approval of the technical risk assessment, based on the equipment class/function; and,
- Based on the level of safety or integrity criticality identified in the assessment, indicate required approvals from asset and business managers, in addition to the local OIM

ORM for Operations is also used to manage all permitted activity on the operator's facilities. This provides a combined view of all equipment risks and all activity risks on the barrier model, highlighting MAH pathways.

CASE STUDY 2: Delivering a real-time view of critical equipment status and its impact on risk

A major national oil company is building and will operate a world-scale refinery in the Middle East. Currently in its greenfield phase, the operator has commissioned a significant Industry 4.0 initiative to develop and deliver a technology-driven approach to integrate a suite of business systems to better support asset and operations processes and management. The operator has a sophisticated safety management system and clear corporate standards and practices. The operator wishes to make a real-time view of operational and process safety risk a central element of decision-making from plant start-up onwards.

The operator is implementing *ORM for Operations* to manage all permitted activity and all deviations on the facility. *ORM for Operations* will integrate with three other business systems to deliver a real-time view of the risk status associated with critical equipment:

- Data historian – for near real-time status of critical equipment (this data historian is itself tracking the operational DCS and critical alarms systems)
- Maintenance Management System (MMS) - for inspection and maintenance records and associated plans and schedules
- Operator rounds system

Since the project is still in the design phase, the project team was able to access design package materials from the EPC contractors responsible for each refinery unit to identify the critical equipment. This includes bringing together a variety of information in useful formats:

- From design phase Hazop and asset integrity studies
- Health parameters associated with specific items of identified safety-critical equipment
- Critical equipment types/categories mapped to the fundamental process safety barrier model
- Records representing all critical equipment was set up in the *ORM for Operations* software
- Through integration, the *ORM for Operations* "listens" for the health of critical equipment, based on the above parameters, from three sources:
    - Near real-time status from the data historian
    - Inspection records for critical equipment from operator rounds and inspection management system
    - Deferred planned maintenance for critical equipment from MMS
- Non-conformances are mapped to the fundamental process safety barrier model

The future is now

Technology has caught up with ambition and aspiration. Organization's are already using ORM Digital twin capabilities live in their operations to reduce risk, improve productivity, and optimize costs. The benefits of digital transformation for process safety management are tangible and make the IIoT possible in an increasing number of organizations. With the single, shared view of the operational reality, an ORM Digital Twin enables organizations to close the gaps between how process safety is intended and the reality of operations.

References

1. The 100 Largest Loss events 1978 - 2017 - Large property damage losses in the Hydrocarbon Industry, 25th edition, Marsh & McLennan, 2018

2. Jarvis, R. and Goddard, A. An Analysis of Common Causes of Major Losses in the Onshore Oil, Gas and Petrochemical Industries, IChemE, Symposium Series No. 162, Hazards 27. Paper 34

3. 2018 Survey on Process Safety and Operational Risk Management, Petrotechnics 2018 – https://www.petrotechnics.com/insights/psm-orm-part-2/

4. 2017 Survey on Process Safety Management and Operational Risk Management. Petrotechnics, 2017 - https://www.petrotechnics.com/insights/petrotechnics-2017-survey-process-safety-and-operational-risk-management-report/

5. Operational Excellence Index 2018/2019

6. Choudhry, H. et al. The next frontier for digital technologies in oil and gas, August 2016 https://www.mckinsey.com/industries/oil-and-gas/our-insights/the-next-frontier-for-digital-technologies-in-oil-and-gas

7. Verdantix, 2018a, Smart Innovators: Operational Risk Management Software, August 2018, London, UK

19 AIChE Spring Meeting
15TH GLOBAL CONGRESS ON PROCESS SAFETY
NEW ORLEANS • 2019

When Emergency Response Goes Fatally Wrong: Lessons Learned from Investigating Two Major Incidents

Scott Davis
Gexcon US
4833 Rugby Avenue Suite 100
Bethesda, MD
sgdavis@gexcon.com

Prepared for Presentation at
American Institute of Chemical Engineers
2019 Spring Meeting and 15th Global Congress on Process Safety
New Orleans, LA
March 31 – April 3, 2019

When Emergency Response Goes Fatally Wrong: Lessons Learned from Investigating Two Major Incidents

Scott Davis
Gexcon US
4833 Rugby Avenue Suite 100
Bethesda, MD
sgdavis@gexcon.com

Keywords: emergency response planning, lessons learned, case studies, major explosion incidents

Abstract

Emergency response plans are an essential, yet oftentimes overlooked, layer of protection for facilities where all other safeguards failed to prevent an incident. While catastrophic accidents such as large releases of chemicals, fires or explosions are devastating for the process industry, experience investigating numerous incidents has shown that a lack of an effective emergency response plan can lead to an unnecessary and tragic escalation of the initial incident. More specifically, investigation of two recent incidents: (1) 2013 West Texas Explosion and (2) 2015 explosion on the FPSO Cidade São Mateus; will demonstrate how fatal consequences resulted from the lack of emergency planning in the first case study and from deficiencies in the emergency response plan or implementation of the plan in the second case study. More specifically, the FPSO explosion resulted in 9 fatalities, eight of which were responding to a leak of a light-end hydrocarbon, and the West Texas explosion that resulted 15 fatalities, most of whom were responding to the fire that preceding the devastating explosion. Lessons will show that the fatal consequences could have been avoided with the proper emergency planning. In addition, while emergency response plans consider "maximum credible" scenarios, past events have shown that low probability high consequences should also be at least considered.

1 Introduction

Emergency response plans are an essential, yet oftentimes overlooked, layer of protection for facilities where all other layers of protection failed to prevent an incident. An emergency plan specifies procedures for handling sudden or unexpected situations, where the ultimate objectives are to be prepared to: (1) prevent fatalities and injuries; (2) reduce damage to buildings, stock, and equipment; (3) protect the environment and the community; and (4) accelerate the resumption of normal operations [1]. Developing the emergency response plan consists of five major steps (see Figure 1). The first step is to assess the vulnerability of the site and determine what hazards pose a threat to the installation. Such an exercise will help determine not only what "credible" events can occur at a given facility, but also what catastrophic or major events can occur. It should be noted that major emergencies are generally rare events.

Next, it is necessary to develop your emergency response plan and determine what means are currently available to execute the plan and what is still needed. At the planning stage, it is important that several groups be asked to participate, such as: subject matter experts and process engineers; the health and safety group; and even appropriate municipal officials should also be consulted since control may be exercised by the local government in major emergencies and additional resources may be required. Once the emergency response plan is developed, the necessary facilities and equipment need to not only be provided but tested and maintained. Communication, training and periodic drills will ensure adequate performance if the plan must be carried out. Lastly, there should be a periodic review of the emergency response plan.

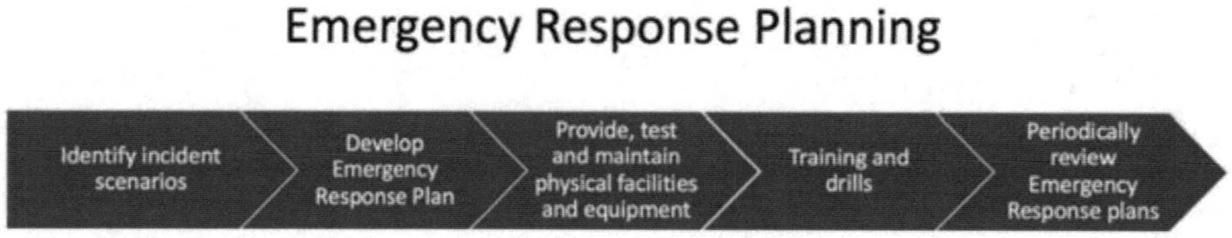

Figure 1: Steps of Emergency Response Planning

While the major benefit is to provide guidance during an emergency, developing the plan has other advantages. For example, previously unrecognized hazardous conditions, which would aggravate an emergency situation, could be discovered during the emergency planning and means put in place to mitigate or eliminate these hazards. The planning process may also identify deficiencies, such as the lack of resources (equipment, trained personnel, supplies), which can be easily corrected before an emergency occurs.

While catastrophic accidents such as large releases of chemicals, fires or explosions are devastating for the process industry, experience investigating numerous incidents has shown that a lack of an effective emergency response plan can lead to an unnecessary and tragic escalation of the incident. Chaos can quickly occur during an emergency if there are lack of resources and trained personnel coupled with an urgent need for rapid decisions and a very short period of time. The stress of the situation can lead to poor judgment resulting in severe losses, and normal channels of authority and communication cannot be relied upon to function routinely. Therefore, it is imperative to eliminate the "it can't happen here" mentality, as emergencies can occur and preplanning is necessary.

The importance of emergency planning will be discussed within the context of two recent incidents: (1) 2013 West Texas Explosion and (2) 2015 explosion on the FPSO Cidade São Mateus. These two case studies will demonstrate how the lack of emergency planning resulted in devastating consequences in the first case study and from deficiencies in the emergency response plan or implementation of the plan in the second case study.

2 Background of Incidents

2013 West Texas Explosion

On April 17th, 2013, an explosion occurred at the West Fertilizer Company storage and distribution facility in West, Texas. The explosion at West Fertilizer resulted from an intense fire in the seed and fertilizer area of the facility that led to the detonation of approximately 30 tons of ammonium nitrate stored inside a wooden receiving bin. The explosion occurred while emergency services personnel were responding to a fire at the facility. Fifteen people were killed, more than 250 were injured, and numerous buildings were damaged or destroyed.

The seed and fertilizer building (Figure 2) was built in 1958 and stored fertilizer and seed, which were distributed to farmers. Figure 3 shows the layout of the seed and fertilizer building, which included an ammonium nitrate (AN) storage bin neighboring the seed room, as well as an empty bin used for ammonium nitrate storage, and bins containing ammonium nitrate, K-Mag (fertilizer mix of potassium, magnesium and sulfur), Potash, ammonium sulfate, DAP (diammonium phosphate), and a bin for potassium chloride along the west wall. This building was capable of receiving rail shipments directly from the rail car by means of a rail spur junction and an underground conveyor belt and bucket elevator; the bucket elevator dumped directly into the northern AN storage bin, or onto a conveyor belt feeding the bins along the west wall.

Figure 2: West Fertilizer Facility – viewed from the west (upper) and east (lower)

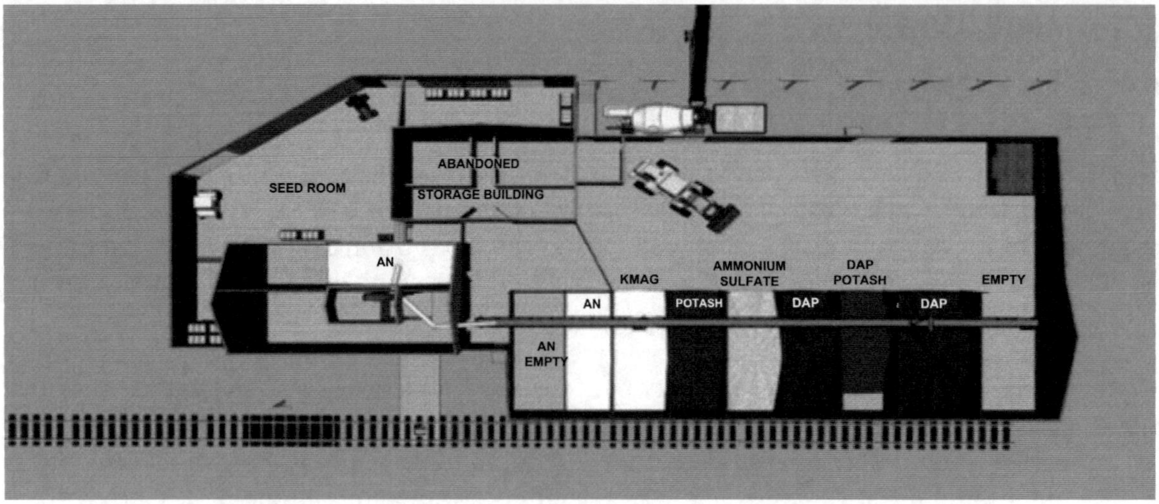

Figure 3: Layout of the Seed and Fertilizer Building.

On April 17th, 2013, at approximately 7:29 pm, a fire was reported, and fire fighters responded to the fertilizer plant [1]. All operations were already stopped for the day and employees had left the plant at about 5:30-6:00 pm. GexCon obtained several videos that recorded the progression of the initial fire [3]. Figure 4 shows the initial fire as viewed from the east and west; note that the highest point on the building represents the location of the bucket elevator by the northern AN storage bin. The image shows that the fire is isolated to the northern section of the building, which includes the seed room and portions of the northern AN storage bin containing about 30 tons of ammonium nitrate.

At approximately 7:50 pm, a massive explosion occurred at the fertilizer plant. Figure 5 shows images from a video taken from the high school parking lot to the southeast and Figure 6 the ensuing blast damage and crater from the detonation. Based on the videos from the day of the incident, there were actually two explosions: one smaller blast followed by a larger blast roughly 300 msec later.

Figure 4: Close up of the Seed and Fertilizer Building during fire as viewed from east (left image) and viewed from the west (right image)

Figure 5 – View of the explosion (viewed from southeast)

Fifteen people were killed as a result of the explosion, which included voluntary firefighters, and over 250 people were injured [3]. Damage was extensive throughout the neighboring community, which included homes, schools, an apartment building and a nursing home.

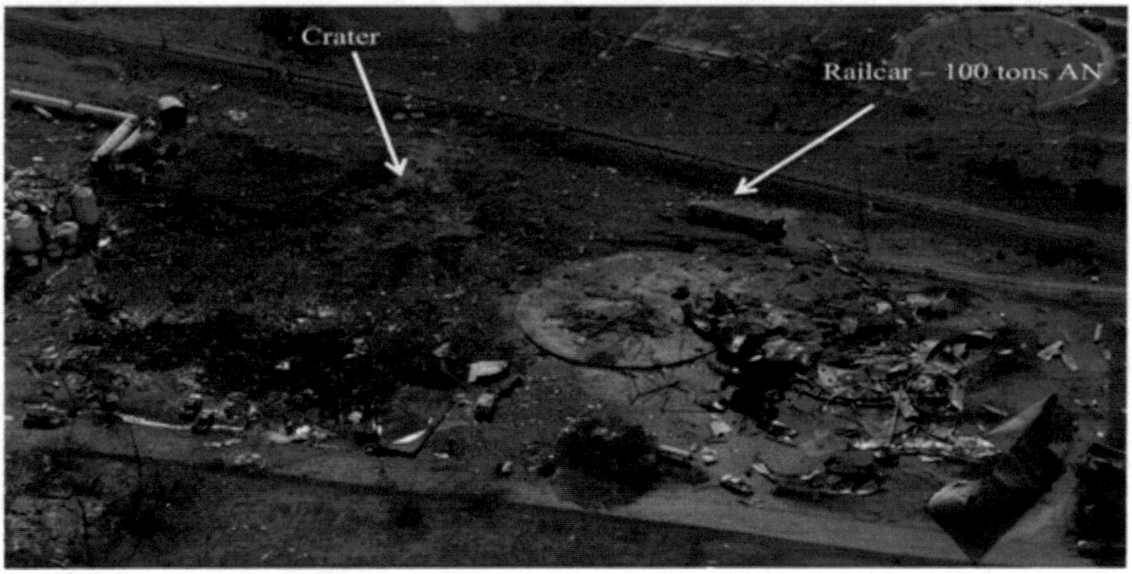

Figure 6: Blast damage and resulting crater

2015 explosion on the FPSO Cidade São Mateus

Due to the confidentiality of matters in this case, all references are made to two publicly available reports regarding the incident [5], [6]. The Cidade São Mateus was a Floating, Production, Storage and Offloading of Oil and Gas Export (FPSO), with a capacity to produce 4,000m³/day of oil and 10,000,000m³/day of gas as shown in Figure 7.

Figure 7: 2015 FPSO Cidade São Mateus.

On February 11, 2015 there was a leak of condensate (light-end hydrocarbon liquid) discovered in the pump room at a spade blind flange connection. Figure 8 shows that the pump room is situated just forward of the engine room and aft of the storage tanks. The layout and levels of the pump room is given in the left image of Figure 9, while the leaking spade blind flange is in the right image.

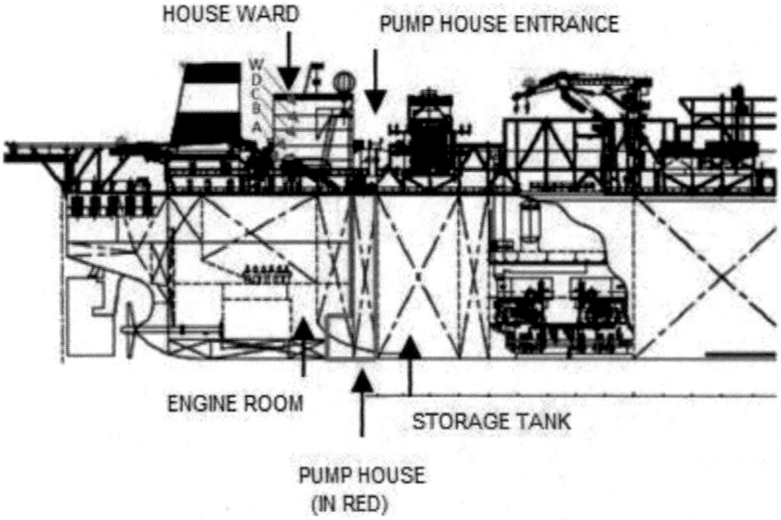

Figure 8: Cutaway side view of the pump room on the FPSO Cidade São Mateus.

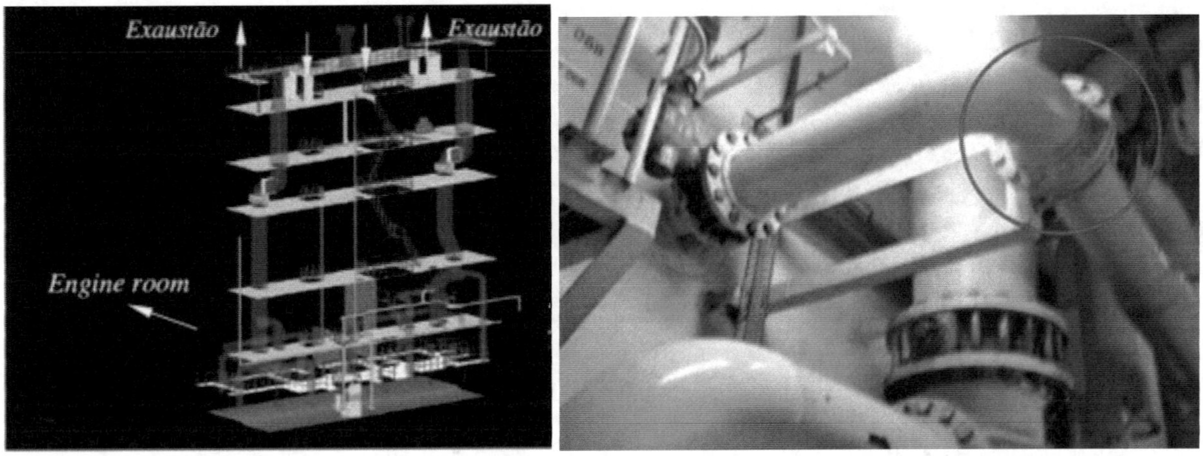

Figure 9: Model of the pump room on the FPSO Cidade São Mateus (left) and spade blind flange (right).

Just prior to the leak, the operations crew was attempting to drain condensate center cargo tank 6 (6C) into an alternate tank and they used an alternate pump alignment (stripping pump). However, the valve on discharge side of stripping pump was closed and they were pumping against a spade blind that was inadequate for such pressures, which ultimately failed the blind and seals. The gas detectors in the pump room began to alarm, and ultimately all three were alarming.

The ventilation was interlocked to shutdown per "confirmed gas" in the pump room. Emergency response teams entered the pump room to attempt clean up and repair the flange. There were three separate entries, and on during the third entry the explosion occurred. Figure 10 through Figure 12 show the level of blast damage to the pump room. Nine people were killed, of which eight were first responders.

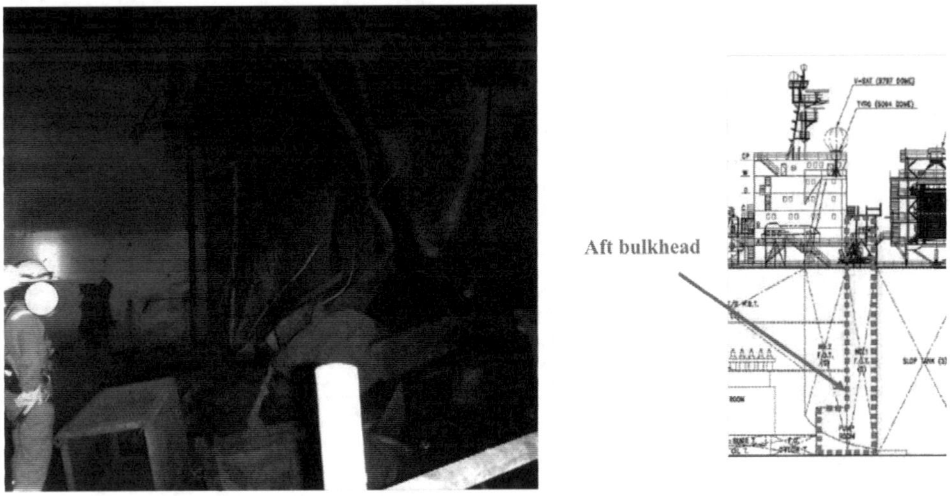

Figure 10: Damage to level above the pumps – complete bulkhead failure.

Figure 11: Damage to the engine room.

Figure 12: Damage to upper level of pump room.

3 Emergency Response

Response during the 2013 West Texas Explosion

As the fire occurred 20 minutes prior to the explosion, fire personnel were deployed to scene. Of the 15 people killed, 12 were first responders (8 volunteer firefighters and emergency services), and the other individuals were good Samaritans who supported the emergency response at the fertilizer plant. One of the volunteer fire fighters, who died in the blast, was employee of West Fertilizer. In fact, he was the plant foreman and knew what was stored in the facility. The emergency response is outlined in Figure 13.

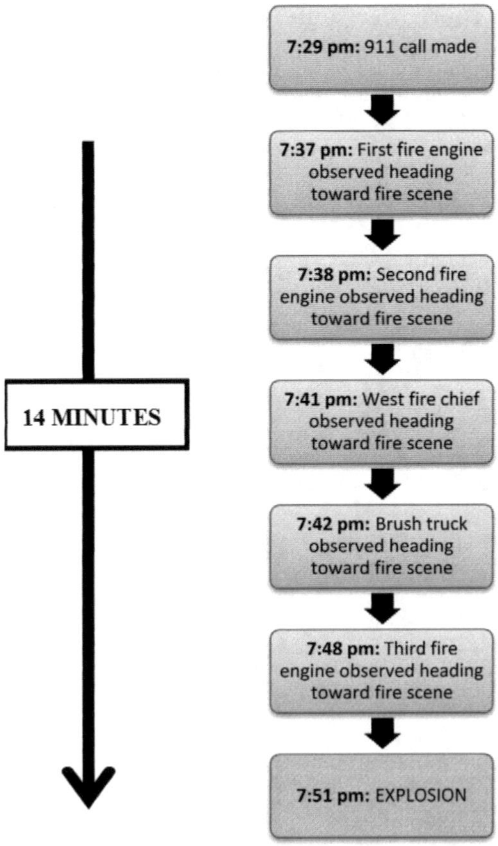

Figure 13: Emergency Response Plan at West Texas

Plant personnel and management did not understand the unique explosive nature of ammonium nitrate. More specifically, while the conditions necessary for detonation are not precisely known, the data suggest that detonations are promoted by severe shock or by heating with confinement, such as the massive fire at the plant. In fact, pools of molten ammonium nitrate may form in the event of fire, and if the molten mass becomes confined (e.g., in the elevator pit) it could explode. These hazards were not known nor understood by plant personnel and emergency responders. Therefore, responders were unaware of the potential of ammonium nitrate to detonate, had no knowledge of how long the fire had been burning, no robust incident pre-planning process in place, and no previous ammonium nitrate-related fire emergency training or drills. Hence, the strategy and tactics utilized by the West Volunteer Fire Department were not appropriate for the rapidly developing situation and they were exposed to unnecessary and extreme risk that proved fatal.

Response during 2015 explosion on the FPSO Cidade São Mateus

The first detector alarmed at 11:31, and by 11:38 all three detectors reached the over limit levels. The 1st team entered the pump room between 11:49-11:54 to assess the gas alarms, and they reported a leak at the flange and a 2-meter diameter pool under the flange. A 2nd team entered the pump room around 12:07 to assess the damage and whether repairs could be made. Despite the confirmed gas on all three detectors, the 2nd team still entered, and a personal gas detector alarmed at 100% LEL. The 2nd team exited around 12:15 and the ship was partially demobilized around 12:20 for lunch. The 3rd team enters around 12:27 to clean and repair, with ladders, wrenches, fire hose and absorbent blankets. The explosion occurred around 12:38 and killed four members in the pump room and four members at the top of the pump room on watch. Due to the failure of the wall into the engine room, the utilities operator was fatally injured in the engine control room.

There was a detailed Emergency Response Plan on the FPSO, along with the muster points for each person involved in the emergency response as shown Figure 14.

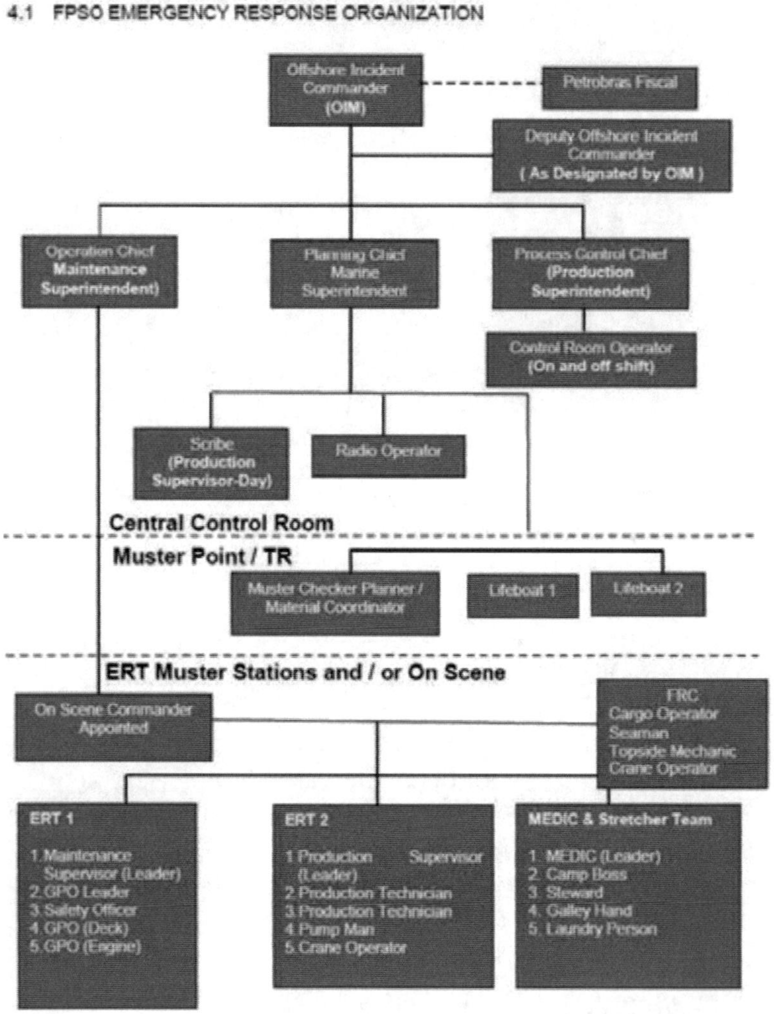

Figure 14: Organization Chart and muster points of the emergency response team

The Offshore Incident Commander is the Offshore Installation Manager (OIM) and would have the roles of implementing a command structure that would satisfy the organizational needs to protect life, environment and the installation, as well as deploying quick and efficiency actions to control incidents, as well as managing response actions to ensure that the escalating potential, risks of injury to personnel and damages to the installation are minimized. In the event of a leak in potentially explosive atmospheres, the emergency response must consider the possibility of the event escalating to a fire or explosion scenario. Thus, systems and procedures should be provided for the crew to control the situation and mitigate the risks of escalation from explosive atmospheres.

Investigation showed that the vessel was not designed for condensate, but for crude and gas. Yet despite the fact that there was confirmed gas detection within the pump room and on personal gas detectors, the crews entered three times. On the third entry, the explosion occurred. It was also reported that an explosive atmosphere in the pump or engine room was not included as an accident scenario in the emergency response plan, and there were no clear instructions of this particular scenario. Despite the knowledge that there was confirmed gas detection on the fixed sensors installed at the bottom of the pump room and also on personal detectors, Emergency Response Teams were sent into the pump room to evaluate the scenario in an explosive atmosphere.

The actions of Command essentially played down the risk of the condensate leak and allowed personnel to enter the pump room on multiple occasions to clean the spill without first performing a risk analysis, and without additional monitoring the room's atmospheric conditions. This noncompliance with the applicable emergency procedure was not only observed during entry operations, but also through the partial or total demobilization of the people irrespective of their muster points with an ongoing confirmed gas detection in the pump room. This demobilization resulted in two members being injured as they were found in a damaged elevator shaft after the blast.

4 Analysis

While deficiencies in the emergency response plans resulted in unnecessary and tragic consequences, the reasons each were lacking are very different and distinct.

For the West Texas Explosion, the Texas State Fire Marshal incident investigation [2] concluded that neither West Fertilizer nor West Fire Department had pre-fire and pre-incident plan at the facility. -Therefore, the recommendations provided were to ensure that local fire departments implement an Emergency Response Plan, with a particular focus on high-hazard occupancies. It was identified that fire departments visit and pre-incident plan responses to commercial structures in their jurisdiction. Further, they concluded that it is imperative that high-risk occupancies containing hazardous materials are identified and appropriate emergency response strategies and tactics are developed through pre-plan documents and SOP directives. Despite the extremely low likelihood of ammonia nitrate detonations, the recommendations state that high-hazard and high-risk occupancies containing heavy fire loads and hazardous materials should be identified by Fire Departments and all members made aware of their potential dangers. This will better prepare them for not only credible hazards, but also those high consequence hazards which may be found in the event of a fire and also provide them with insight as to when it is prudent to attack a fire or evacuate the area. More specific recommendations regarding the emergency response planning included:

- Develop a detailed Emergency Response Plan and Evacuation Plan for the area within 1 mile of existing ammonia nitrate buildings
- Develop standard operating procedures/guidelines to train and educate members on firefighting and emergency scene strategies and tactics for high-hazard occupancies
- Develop pre-incident plans by visiting the commercial and high-risk properties within the response jurisdiction.
- Utilize risk management systems on all emergency responses to determine when to employ alternative tactics.

- Incorporate incident management systems in every emergency incident, training exercise and basic operational process of the fire department to provide appropriate command and control for conducting operations.
- Implement a strategic incident action plan (IAP) for all emergency operations through a coordinated application of appropriate emergency scene tactics for the presented dangers
- Ensure that an incident safety officer (ISO), a person who is independent from the incident commander, is appointed and effectively utilized at every incident that would require an ISO.
- Establish a hazardous materials program and actively train members on the tactics to use when an incident involving hazardous materials occurs

In contrast, for the 2015 FPSO event, there was a detailed Emergency Response Plan for the vessel. Despite this plan, investigations concluded that there were significant deficiencies in the plan that needed to be remedied, and largely included reference to personnel exposure, failure to identify the accident scenarios for the emergency plan, training and mobilization. The execution of the plan failed to properly identify the potential high consequence and catastrophic result of the actions of the emergency responders. More specifically:

- Exposure to personnel
  - In the event of a leak in potentially explosive atmospheres indoors, the plan must consider the possibility of event escalation to a fire and/or explosion. Thus, systems and procedures should be provided once there is the formation of an explosive atmosphere. To ensure the success of mitigation actions, emergency procedures and resources for emergency response should be pre-planned, drawing on the use of all available resources, including the time of scenario assessment.
  - To determine that all functions involved in emergency response, including land-based support, have periodic theoretical training in emergency response procedures.
  - Set in the emergency response plan the available material and human resources to the response of each accidental
- Failure to identify accident scenarios or not included from the risk scenarios
  - Establish criteria and procedures for the migration of risk scenarios identified in the risk analysis and safety studies for emergency response plans. In addition to the selected scenarios, the procedures should contain scenarios that represent challenges to response actions, such as fatalities, and multiple injuries, among others.
  - Establish periodically simulated emergency covering all identified accident scenarios, including scenarios with multiple injuries and fatalities. Such simulations should include the activation of external resources to inform platform that response structure is effectively tested and that corrective measures arising from simulated ensure continuous improvement of the response actions of all entities involved.
- Demobilization of meeting points
  - Periodically coordinate all staff regarding the hazards involved in the assumptions made in the response procedure to the emergency unit.

    - Prohibit the release of people to normal activities before the full control of the emergency.
- Ignition of the explosive environment
    - As the ignition was introduced by the action of the emergency responders within explosive atmospheres, review the procedures related to emergency response and safe work practices considering the possible ignition sources identified in existing rules

5 Conclusion

The lessons from two specific incidents show that critical deficiencies in the emergency response plan or implementation of the plan can result in fatal consequences that could have been avoided. The 2013 West Texas explosion shows how important it is to not only recognize high risk events, but also have the appropriate plans in place to avoid unnecessary escalation of an already tragic incident, where 12 of the 15 people fatally injured were first responders. By preparing first responders for not only credible hazards, but also those high consequence hazards, can provide them with crucial insight as to when it is prudent to attack a fire or evacuate the area. In addition, not properly following or playing down the risk of a high consequence event was observed in the 2015 FPSO event, where eight of the nine crew members killed by the blast were first responders. The failure to recognize the potentially high consequence and catastrophic event, led to an emergency response that unnecessarily exposed responders and ultimately led to an exacerbated event. This study confirms that emergency response plans are an essential layer of protection for facilities where all other safeguards failed to prevent an incident, which could have prevented unnecessary and tragic escalation of the incident.

6 References

[1] Preparedness for Emergency Response - FEMA Training - FEMA.gov

[2] Texas State Fire Marshal, "Firefighter Fatality Investigation FFF FY 13-06," West, TX, 2013

[3] Davis, SG, DeBold, TF, & Marsegan, C (2017). Investigation Findings and Lessons Learned in the West Fertilizer Explosion. Journal of Fire Science. Volume 35, Issue 5, September 5, 2017: pp. 379-395.

[4] "A Public Health Report on Injuries Related to the West (Texas) Fertilizer Plant Explosion," June 24, 2014.

[5] Brazilian Navy Directorate of Ports and Coasts, Maritime Accident Safety Investigation Report, FPSO Cidade São Mateus - Explosion followed by flooding and casualties, February 11th, 2015.

[6] Investigation report of the explosion incident occurred on 11/02/2015 in the FPSO Cidade São Mateus, Brazilian National Agency of Petroleum, Natural Gas and Biofuels, Superintendence of Operational Safety and the Environment, August 2015.

19 AIChE Spring Meeting
15TH GLOBAL CONGRESS ON PROCESS SAFETY
NEW ORLEANS • 2019

Major Explosion at a Chemical Company in China: the Causes and Lessons Learned

Wei Wang
China University of Petroleum (East China)
No. 66, West Changjiang Road, Huangdao District, Qingdao, China
15621097585@163.com

Yi Liu
China University of Petroleum (East China)
No. 66, West Changjiang Road, Huangdao District, Qingdao, China
liuyi@upc.edu.cn

Dongfeng Zhao
China University of Petroleum (East China)
No. 66, West Changjiang Road, Huangdao District, Qingdao, China
zhaodf@vip.sina.com

Yanchun Wang
China University of Petroleum (East China)
No. 66, West Changjiang Road, Huangdao District, Qingdao, China
Wychun25@163.com

Jian Chen
China University of Petroleum (East China)
No. 66, West Changjiang Road, Huangdao District, Qingdao, China
836771528@qq.com

Hanhai Dong
China University of Petroleum (East China)
No. 66, West Changjiang Road, Huangdao District, Qingdao, China
donghanhaiupc@126.com

Prepared for Presentation at
American Institute of Chemical Engineers
2019 Spring Meeting and 15th Global Congress on Process Safety
New Orleans, LA
March 31 – April 3, 2019

Major Explosion at a Chemical Company in China: the Causes and Lessons Learned

Wei Wang
China University of Petroleum (East China)
No. 66, West Changjiang Road, Huangdao District, Qingdao, China
15621097585@163.com

Yi Liu
China University of Petroleum (East China)

Dongfeng Zhao
China University of Petroleum (East China)

Jian Chen
China University of Petroleum (East China)

Yanchun Wang
China University of Petroleum (East China)

Hanhai Dong
China University of Petroleum (East China)

Keywords: Fishbone Diagram; Analytic Hierarchy Process; Process Safety Management; Incident Investigation

Abstract

On July 12, 2018, a devastating explosion incident occurred in Yibin City Hengda Technology Co., Ltd. in Sichuan Province, resulting in 19 deaths and 12 injuries, and impacting an area of about 1,800 square meters. This was one of the most disastrous explosions that occurred in a chemical plant in China in the past five years. The incident reflects the general deficiencies of process safety management in small and medium-scale companies. Therefore, it is significant to analyze the causes of the incident and put forward countermeasures to avoid such kind of incidents.

According to the characteristics of complexity, diversity, and uncertainty about the influencing factors of the explosion, this paper presented a systematic analysis from qualitative and quantitative perspectives. The fishbone diagram model was used to identify influence factors from six aspects of people, design, management, environment, safety prevention, and materials. Furthermore, an assessment index including 26 factors was determined. The analytic hierarchy process method was used to analyze the importance degree of each factor quantitatively, and the key factors leading to the incident were obtained. The results showed that the sum of weights of three main aspects, including people, design, and safety prevention is nearly80%, so they were the dominant causes leading to this incident. Lack of chemical industry experience, non-standard process design, and incomplete automatic control system were the focus of prevention among 26 factors.

Process safety management is the application of management systems to the identification, understanding, and control of process hazards to prevent incidents from occurring. This analysis considered process hazard analysis, pre-startup safety review, operation procedures, training, and emergency plan and response, which were the main elements in process safety management framework. Finally, we proposed safety precautions for the identification, storage, production, and use of hazardous chemicals, as well as recommendations for process equipment management.

1 Introduction to the event

At 18:42:33 on July 12, 2018, an explosion occurred in Yibin Hengda Technology Co., Ltd. in the industrial park of Jiang'an County, Sichuan Province, causing 19 fatalities and injuring 12 people. The explosion caused the second and third workshops to catch fire, and the combustion area is $1800m^2$ [1]. The first workshop, warehouse, drying room, analysis room, office building, and other buildings had been damaged to some extent. The fire department dispatched 42 fire engines and more than 200 firefights. It took more than two hours to extinguish the fire on the spot as shown in Figure 1. It was estimated that the total energy of the explosion in this incident was equivalent to 235 kg of TNT. The direct economic loss was 41.42 million yuan.

At the time of the incident, the company was producing imazamox and triazole. Triazole is a pesticide, and imazamox is an herbicide. The process of the incident was the dehydration of butyramide in the process of synthesis of imazamox. At about 11:30 on July 12, the company collected 2 tons of chemicals labeled as raw materials (actually the explosive chemical sodium chlorate, not the raw material required for the production of butyramide). At around 14:00, the deputy director of the company's second workshop received raw materials from the warehouse. At around 15:30, the workers in the second workshop raised the raw materials for production through the lift to the third floor of the second workshop. At around 16:00, the 2R301 kettle for dehydration of butyramide was completed, and the feed was in an empty state. Before 17:20, the 2R301 kettle was completed. At about 6 pm, the dehydration completed. At 18:42:33, it was due to the on-site handover time. A chemical explosion occurred in the 2R301 kettle on the third floor of the second workshop. The direct cause of the incident was that the operator replaced the unlabeled sodium chlorate with butyramide and put it into the 2R301 kettle for dehydration, which caused the explosion and fire. Sodium chlorate is a kind of explosive chemical, which is easy to form an explosive mixture when mixed with butyramide. The explosion led to the disintegration of 2R301 kettle. The high-temperature toluene vapor rushed out rapidly formed an explosive mixture with the external air and produced a secondary explosion, which also caused the explosion and combustion of the materials such as sodium chlorate, toluene, and methanol stored at the workshop site.

Figure 1. Extinguishing at the scene.

According to the investigation, in this incident, the second workshop was the core area of the explosion, and the overall structure was severely damaged as shown in Figure 2. The platform on the third floor collapsed. The reactor with a height of nearly two meters and a diameter of about one meter fell from the upper level.

Figure 2. Severely damage to the second workshop.

2 Incident evaluation index system based on the fishbone diagram

Fishbone Diagram is also named as cause and branches figure analysis. Kaoru Ishikawa, a Japanese management master, invented it for finding out the root cause of incident [2]. It indicates the relationship between a problem and its underlying causes. It is a qualitative analysis tool to help identify incident causes [3]. With step by step in-depth study to analyze its influencing factors,

Fishbone Diagrams can systematize complicated incident causes. When starting the specific report, we first analyze the factors which affect the safety most and then followed by departure from the overall reason to find the basic reason and to detect and determine the main reason [4,5,6].

After the incident, the government set up a team of experts to identify risk factors from six aspects: human, design, management, environment, safety prevention, and material and establish an incident evaluation index system with 26 elements. The fishbone diagram that induced the explosion is shown in Figure 3.

Figure 3. Fishbone diagram of Yibin explosion incident [7].

2.1 People

Neither the safety managers nor the operators demonstrated proficiency in the requirements of process safety competency. There were two main aspects below:

- Corporate management - The general manager of the enterprise had responsibility for the mechanical processing and production contract. However, he did not have appropriate experience in the chemical industry. The technical director did not have experience with management of process safety and production systems. The corporate legal person responsible for finance and logistics also had no experience in chemical engineering.

- Operators - Among the 19 people who died in the incident, 16 were operators, and most of these operators were local farmers. They lacked the basic knowledge of chemical process safety production, and they did not understand the process safety risks in the production process.

2.2 Design

The second workshop where the incident occurred was not constructed and installed strictly by the design drawings, and it was originally planned to have nine sets of 5000L tanks and one set of 3000L tanks and produce 5-nitroisophthalic acid. As built, the workshop had three sets of 5000L tanks, seven sets of 3000L tanks, and eight sets of 2000L tanks were set up. The product was also different from the original, but imazamox and triazole. A similar situation existed in the third workshop. The production technology of imazamox and triazole used by the company had no formal technical source, and the enterprise also did not entrust professional institutions for process calculation and construction drawing design. The technical director of the enterprise only planned the required production equipment and operation process according to his experience, but he knew nothing about the process safety risks of related technologies. It was during the production of imazamox that the distillation kettle used for the dehydration of butyramide by toluene first exploded. According to the *Notice on Further Strengthening the Safety Design Management of Hazardous Chemicals Construction Project*, Hengda company needed to start construction in April 2018, but the company had already started construction in July 2017. Also, the fire separation design of each building did not meet the requirements of the *Petrochemical Enterprise Design Fire Code (GB50160).*

2.3 Management

The lack of process safety culture is the main cause of management failure. Safety culture is developed through the interactions of factors such as safety policies and safety goals, the discipline of the organization, the behavior and attitude of the company staff, and the competence of workers. Improper operating discipline and procedures can lead to a negative process safety culture and an increased likelihood of process safety incidents. The enterprise had not formulated the post safety responsibility system, safety management rules, and regulations. Furthermore, the enterprise manager and the person in charge of each department only had the oral authorization, and there was no formal appointment document to clarify the post responsibility. The safety management organization was not complete. The company's safety and environmental protection department had only one safety officer who did not have safety management capabilities. There were also no full-time or part-time safety officers in each workshop. There was a lack of management of hazardous chemicals in the process safety management. The technology provider concealed some chemical name marks for confidentiality, and there was no safety specification. Required knowledge and competence connected with safety for workers were the pivotal content of safety training. New employees did not have security education directly work or internship. The government had not strictly regulated the safety departments of local key hazardous chemicals companies.

2.4 *Environment*

The factory built the workshop while producing, and used the polypropylene pipe which was easy to produce static electricity to transport flammable and explosive liquid materials. The building materials, hardware, and chemicals in the warehouse were mixed and piled with hidden dangers. The incident occurred during shift time, resulting in a large number of casualties. In the on-site operation workshop, there was neither operation instructions nor warning signs for hazardous chemicals. At the time of the incident, the second workshop produced two products at the same time. This practice of producing multiple produced at the same time in the same workshop has excellent security risks.

2.5 *Safety prevention*

Compared with the 20 elements of CCPS, China drew on the process safety management experience of the Occupational Safety and Health Administration (OSHA) and further optimized the formation of 12 elements for the production of Chinese chemical companies. After that, a series of process safety technologies were developed and promoted for the key elements of risk management and equipment integrity. In 2013, the SAWS issued *Guidance for enhancing chemical process safety management*. The Ministry of Emergency Management, which was integrated by the State Administration of Work Safety(SAWS) in 2018, attaches great importance to the supervision of hazardous chemicals, hazardous chemical processes, and major hazardous sources, referred to as "Two Priorities and One Major." If an enterprise violates The relevant provisions, it shall be punished according to the upper limit of economic punishment according to law.

In the production process of imazamox and triazole, the nitrification, oxidation, diazotization, hydrazine hydrate, hydrogen peroxide, methanol, and other dangerous chemicals and processes involved were all "Two Priorities and One Major" projects, which needed strict supervision. However, the company's automatic control system, flammable and toxic gas alarm system and fire water system and other safety facilities were not installed. Once there was a danger, the enterprise itself could not deal with an emergency. The lack of an automated control system led to more than ten operators performing manual operations around the reactor, which was an important cause of major casualties.

2.6 *Materials*

According to the investigation, the safety evaluation report made by the enterprise before the incident did not fully consider the process safety information, such as the lack of description of dangerous chemicals and the risk identification of the process, so the enterprise had great potential risks. Workers didn't know the source of raw materials, and many chemical production raw materials were not marked. Workers knew nothing about the risks of these dangerous chemicals, and company officials failed to communicate information about the hazards of the process to employees. The workers compiled codes to replace the names of products and raw materials, and these numbers were not abbreviated as specifications. The raw materials used by Hengda Company, such as hydrazine hydrate, nitric acid, toluene, hydrochloric acid were dangerous chemicals or precursor chemicals, but the company still purchased them illegally without relevant licenses. In the production process, regulators did not supervise the management of hazardous chemicals.

3 Determination of the weights of the incident incentive index based on the analytic hierarchy process

The Analytical Hierarchy Process (AHP) is a decision-making method proposed and used by American operations researcher T.L.Saaty in the mid-1970s [8]. AHP is a process that breaks down a problem in a hierarchical matrix with increasing level of detail to determine the most significant causes of an incident. According to the obtained fishbone diagram of the Yibin explosion incident, an incremental layer structure for analytic hierarchy analysis was constructed, as shown in Fig. 4, where the A, B, and C layers illustrated the primary effect and hierarchy of causes.

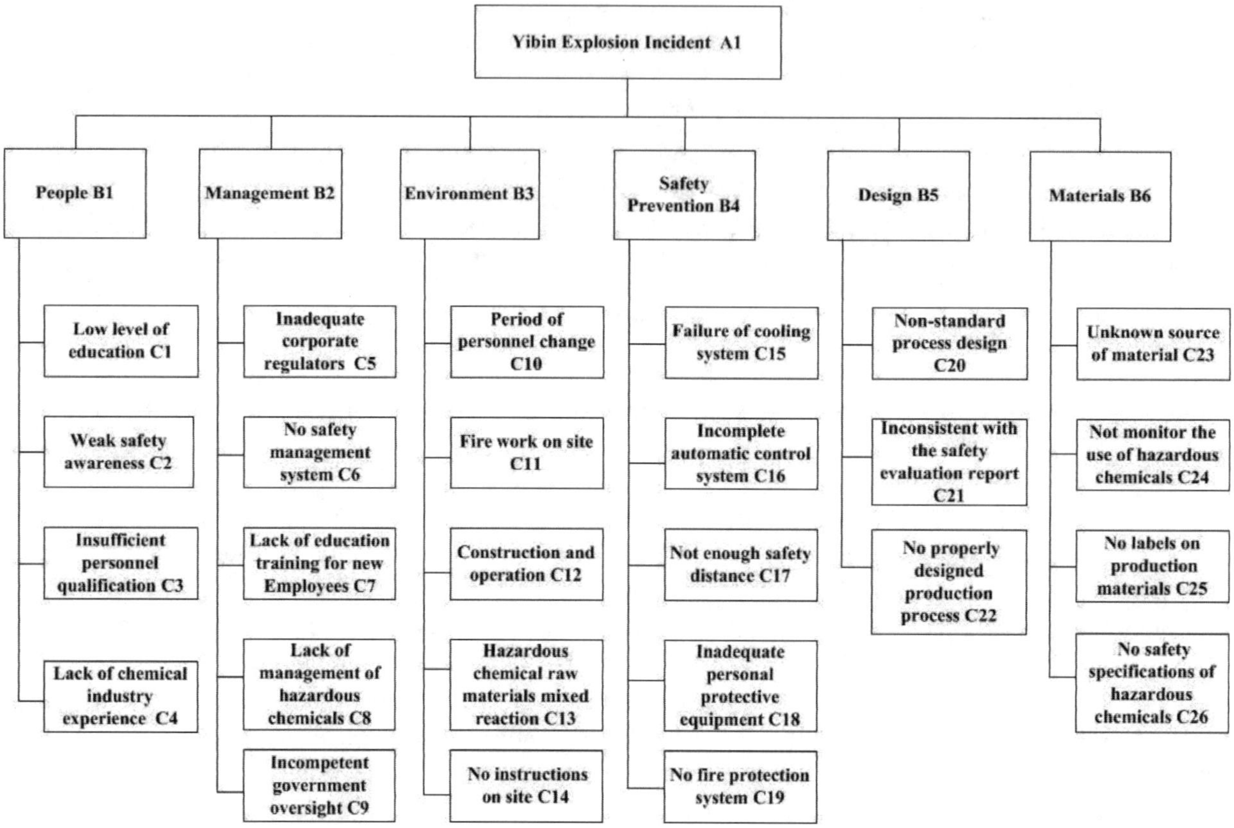

Figure 4. Hierarchical structures of safety evaluation system for Yibin explosion incident.

3.1 The single-layer index weights of the influencing factors

According to the actual data collected and expert opinions, we used 1-9 scale to construct the judgment matrix (A) of the first-level evaluation index, as shown in table 1, and obtained the influence of various factors in layer B on layer A. λ_{max} is the maximum eigenvalue, CI is the consistency index, RI is the average random consistency index, and CR is the random consistency ratio; When CR<0.1, it passed the consistency test [9]. According to table 1, the weights of design is the largest, followed by people and safety prevention, with the sum of the weights of the three as high as 0.8. It can be seen that the leading cause of this incident is due to the low level of process design, operators' insufficient qualification of chemical safety production and non-compliant protection measures. Therefore, enterprises should pay attention to intrinsic safety design, improve the safety awareness of staff by popularizing safety knowledge and operational guidelines, and finally achieve the goal of safe production.

Table 1. Judgment matrix of primary evaluation indicators and weights results

A	B1	B2	B3	B4	B5	B6	Weights	Consistency	Ranking
B1	1	2	3	1/2	1/3	7	0.1645	λ_{max}=6.4698	3
B2	1/2	1	2	1/3	1/4	6	0.1060	CI=0.094	4
B3	1/3	1/2	1	1/4	1/5	5	0.0700	RI=1.26	5
B4	2	3	4	1	1/2	8	0.2545	$CR=0.0746<0.10$	2
B5	3	4	5	2	1	9	0.3810	Satisfy consistency	1
B6	1/7	1/6	1/5	1/8	1/9	1	0.0239		6

3.2 The comprehensive weights of the influencing factors

We built each judgment matrix of layer C to layer B and calculated the single layer weights of layer C, and finally, we obtained the comprehensive weights of incidents caused by bottom influencing factors. We considered the product of these two layers as the comprehensive weights of the sub-factor. For example, the comprehensive weights of C1 is equal to the single-layer weights of C1 to B1 multiplied by the single-layer weights of B1 to A. The comprehensive weights result and ranking of each sub-factor were shown in table 2.

Table 2. Single and synthetic weights of C layer indexes

		Single weights	Consistency	Synthetic	Ranking
B1	C1	0.0843	λ_{max}=4.1375 CI=0.0458 RI=0.9 $CR=0.0515<0.1$ Satisfy consistency	0.0139	16
	C2	0.2120		0.0349	8
	C3	0.1544		0.0254	12
	C4	0.5492		0.0904	3
B2	C5	0.1556	λ_{max}=5.2550 CI=0.0638 RI=1.12 $CR=0.0569<0.1$	0.0165	15
	C6	0.0566		0.0060	21

	C7	0.4735	Satisfy consistency	0.0502	5
	C8	0.2578		0.0273	11
	C9	0.0566		0.0060	22
B3	C10	0.6253	λmax=5.3598 CI=0.08995 RI=1.12 CR=0.0803<0.10 Satisfy consistency	0.0438	7
	C11	0.0901		0.0063	19
	C12	0.0435		0.0030	23
	C13	0.1976		0.0138	17
	C14	0.0435		0.0030	24
B4	C15	0.0781	λ_{max}=5.1717 CI=0.042925 RI=1.12 CR=0.0383<0.1 Satisfy consistency	0.0199	13
	C16	0.5895		0.1501	2
	C17	0.1271		0.0324	9
	C18	0.0781		0.0199	14
	C19	0.1271		0.0324	10
B5	C20	0.6833	λ_{max}=3.0246 CI=0.0123 RI=0.58 CR=0.0236<0.1 Satisfy consistency	0.2604	1
	C21	0.1168		0.0445	6
	C22	0.1998		0.0761	4
B6	C23	0.1242	λ_{max}=4.1891 CI=0.0630 RI=0.9 CR=0.0708<0.1 Satisfy consistency	0.0030	25
	C24	0.2626		0.0063	20
	C25	0.5465		0.0131	18
	C26	0.0667		0.0016	26

As can be seen from table 2, the full weights of C16 (Incomplete automatic control system) and C20 (Non-standard process design) was greater than 0.1. The full weights of C4 (Lack of chemical industry experience) and C22 (No properly designed production process) were very close to 0.1. The combined weights of these four factors reached 0.577, which was more than half of the total. Therefore, these four influencing factors were the primary causes of the explosion incident in Yibin, Sichuan, as strict control and prevention were given priority.

Among them, both C20 and C22 belong to the design aspect, and enterprises should entrust qualified formal companies to design the process and production plan. In the method of safety design, the operation and design should be analyzed in real time to check the status of dangerous materials, effectively reduce the operating temperature and pressure, and control the influence of hazards factors in the process on chemical equipment. The equipment used in production shall meet the strength and anti-corrosion performance required by the process, and shall be designed and manufactured by qualified factories.C4 is a human problem. Therefore, it is necessary to conduct chemical safety education and operation training for operators before entering the plant. Many of Hengda's workers were the relatives and friends of management personnel. They were generally poorly educated and skilled in operation, and had not received any training in chemical safety. The workers did not go through systematic learning, but the workshop director directly taught them operating skills. C16 is a problem of safety prevention. The automatic control system can not only improve the operation efficiency of production equipment but more importantly, it can protect the safety of chemical personnel and reduce the loss caused by incidents. Enterprises should follow the safety production regulations, install automatic control systems, toxic and harmful gas alarm systems, and fire protection systems to effectively prevent incidents.

The weights of C7 (Lack of education training for new employees), C10 (Period of personnel change) and C21 (Inconsistent with the safety evaluation report) were all around 0.05. The weights of C2 (Weak safety awareness), C19 (No fire protection system) and C17 (Not enough safety distance) also exceeded 0.03. These five factors were important causes of incidents and should be effectively controlled.

C7 is a new employee training management problem. Enterprises should pay attention to the educational training of employees before entering the plant and implement the three-level safety training education of the plant, workshop, and team. C10 involves change management. At the time of the incident, it was during the shift time of employees, the change of personnel and the change of materials and equipment, which had great potential risks. Enterprises should establish a change management system, conduct comprehensive training for the staff who play a vital role in the change management process and clarify the position and responsibility of each person in the change system.

4 Lessons learned from the causes of the incident

The Yibin explosion was a pharmaceutical chemical incident. Through the investigation and analysis of historical disasters in the chemical industry in China, it was found that the proportion of pharmaceutical chemical incidents in the chemical industry was huge [10]. It was characterized by a random change of chemical process, small scale, low level of process control, involving dangerous materials and complex chemical reaction. Especially in small and medium-sized chemical enterprises, the lack of process safety management is an essential cause of such incidents. From the perspective of process safety management, this paper summarized and put forward suggestions for preventing such incidents.

- Process Hazard Analysis - In the design stage, the risk identification and assessment tool system should be applied to identify the risks of materials in the process of technology. In the review stage, HAZOP, SIL are used to review the process, layout and material flow. Companies should use scientific models to calculate the explosion-proof grade and safety distance of buildings. By adding the explosion-proof wall and the explosion-proof film, the building close to the installation area can reach the required safety factor. After the device is put into operation, it needs to be compared with the hazard analysis in the design stage, and the hazard that is inconsistent with the design should be analyzed again. Hengda's actual operating equipment, raw materials, and products were inconsistent with the original plan, and the company had not discovered and controlled the real hazards [11].

- Pre-startup Safety Review - The company shall organize a safety inspection team, whose task is to confirm whether the improvement measures and safety guarantee measures in the process hazard analysis report have been implemented as required; Whether adequate training of personnel in operation, maintenance, and management of process equipment has been completed; Whether necessary safety information of process equipment is completed and updated; Whether necessary equipment testing and checking have been completed and confirmed; Whether the process equipment is constructed and installed according to the design requirements [12].

- Operation Procedures - The incident occurred during the change management phase, and there was no review or other risk assessment during the shift. Staff should review the operating procedures as often as possible to ensure that the operating procedures reflect current operating conditions, including changes in chemicals, process equipment, and facilities. The enterprise shall confirm the adaptability and effectiveness of the operating procedures and ensure that the operators understand and master the operating systems [13].

- Training - A complete process safety management is inseparable from the participation of operators, and the human factor is the most unstable factor in process safety management. Therefore, Enterprises should establish strict production operation disciplines and require operators to master process safety information, equipment integrity management points, and emergency response policies and finally improve workers' ability to control equipment safety operations [14].

- Emergency Plan and Response - Enterprises need to establish a set of overall emergency plans, which usually stipulate in written form how the factory should deal with abnormal or urgent situations. Also, enterprises should establish emergency response teams, traditionally composed of enterprise personnel, but also include external staff; The responsibilities of each team member should be clear, ensuring that members have no questions about accountability and delegation. In case of an emergency, the person in charge may determine the safe area according to the emergency response manual and direct the person to evacuate to a safe place [15].

5 Conclusion

The circumstances of the incident described in this article are typical in China, and the incident reflects the common safety problems in small and medium-sized chemical enterprises. Managers should understand these lessons we learned from this incident. By using fish-bone diagram method, the causes of the explosion incident were obtained through qualitative analysis, including six aspects, such as people, design, management, environment, safety prevention and materials, and 26 factors influencing the incident were further analyzed. Then the analytic hierarchy process was used for quantitative analysis, and the results showed that lack of chemical industry experience, non-standard process design, and incomplete automatic control system were the main factors affecting the incident, so the enterprise should strictly control and prevent these factors. Finally, we summarized the causes of such incidents and put forward the measures from five aspects of process safety management, including process hazard analysis, pre-production safety inspection, operation procedures, training, and emergency plan and response plan. An incident brings not an only economic loss but also bad social influence to the factory. It is our hope that this paper will help prevent future accidents.

6 References

[1] Sichuan Provincial Emergency Management Office, Yibin Hengda Technology Co., LTD. "7.12" Major Explosion Fire Accident Investigation Report, Available at http://www.chinasafety.gov.cn/gk/jsts/201902/t20190214_224397.shtml, Accessed on February 14, 2019.

[2] Tongyuan Luo, Chao Wu and Lixiang Duan, Fishbone diagram and risk matrix analysis method and its application in safety assessment of natural gas spherical tank, Ind. Eng. J. 25(02), 2007, 138-141.

[3] Xing, W., QC Team Based Materials, China Society Press, Beijing, China, 2004.

[4] T Deng, X., Chen, Q. and Jiang, D., Application of fishbone chart analysis method in prevention for steam turbine overspeed incident, Guangdong Electr. Power, 20(02), 2013, 73-77.

[5] Changlong Zhu, Jiayong Zhu, Li Wang and M. Sam Mannan. Lessons learned from analyzing a VCE incident at a chemical plant, Journal of Loss Prevention in the Process Industries,50,2017, 397-402.

[6] Dennis P. Nolan, Chapter 5: Specialized Reviews—CHAZOP, EHAZOP, Bow-Tie Analysis, Layers of Protection Analysis, Safety Integrity Level, FishboneDiagram, and Cyber Security Vulnerability Analysis, Safety and Security Review for the Process Industries (Fourth Edition), Gulf Publishing Company,2015,Pages 17-27.

[7] Xinhua News Agency, Emergency management department: 7·12 incident enterprises have no law and discipline, and government supervision is lax, Available at http://www.gov.cn/xinwen/2018-07/17/content_5307196.htm?_zbs_baidu_bk, Accessed on July 17, 2018.

[8] Seong KonLee, GentoMogi and Jong WookKim. Decision support for prioritizing energy technologies against high oil prices: A fuzzy analytic hierarchy process approach, Journal of Loss Prevention in the Process Industries,22(6), 2009, 915-920.

[9] Saaty T.L., A scaling method for priorities in hierarchical structures, Journal of Mathematical Psychology, 15 (3),1977, 234-281.

[10] Heda Zhang and Xiaoping Zheng, Characteristics of hazardous chemical incidents in China: A statistical investigation, Journal of Loss Prevention in the Process Industries, 25(4), 2012, 686-693.

[11] Hao WU, Igor PEÑARRUBIA, Lin CUI and Jinsong ZHAO, Process safety management considerations for biofuel production, Higher Education Press, 4(3), 2017, 357-367.

[12] Yuan Gao, Yunxiao Fan, Jing Wang, Xi Li and Jingjing Pei, The mediating role of safety management practices in process safety culture in the Chinese oil industry, Journal of Loss Prevention in the Process Industries, 57, 2019, 223-230.

[13] Paul R.Amyotte and Cathleen S.Lupien, Chapter Three - Elements of Process Safety Management, Methods in Chemical Process Safety, 1, 2017,87-148.

[14] T. Kontogiannis, M.C. Leva and N. Balfe, Total Safety Management: Principles, processes and methods, Safety Science, 100(B), 2017, 128-142.

[15] Hao WU, Igor PEÑARRUBIA, Lin CUI and Jinsong ZHAO, Process safety management considerations for biofuel production, Front. Eng. Manag, 4(3), 2017, 357- 367.

19 AIChE Spring Meeting
15TH GLOBAL CONGRESS ON PROCESS SAFETY
NEW ORLEANS • 2019

Taming the Wild River Rapids – How Process Safety Can Apply Outside the Process Industries

Trish Kerin
IChemE Safety Centre
Level 7, 455 Bourke St, Melbourne, Victoria, Australia
tkerin@icheme.org

Prepared for Presentation at
American Institute of Chemical Engineers
2019 Spring Meeting and 15th Global Congress on Process Safety
New Orleans, LA
March 31 – April 3, 2019

—

Taming the Wild River Rapids – How Process Safety Can Apply Outside the Process Industries

Trish Kerin
IChemE Safety Centre
Level 7, 455 Bourke St, Melbourne, Victoria, Australia
tkerin@icheme.org

Keywords: lessons learned, process safety, design, maintenance

Abstract

On the 25th October 2016 the Thunder River Rapids ride at Dreamworld on Australia's Gold Coast suffered a pump failure, resulting in the tragic deaths of four riders. This incident was investigated by the police and later a coronial inquest to determine how a tragedy like this could occur. As a process safety professional, it was interesting to follow the coronial inquest findings and then think about how process safety management may have helped to prevent this incident. Assuming that a process safety event can be defined as a high consequence or potential consequence low likelihood loss of control event [1], this incident had all the hallmarks. Therefore, it is possible to apply the IChemE Safety Centre (ISC) framework [2] for process safety to identify where the system failed and understand how it could be averted in future. This paper will explore various elements of the incident and how process safety management techniques may have helped, including that the ride had been modified to be different from all other similar rides, that the staff did not know how to stop the ride in an emergency and that all means to stop the ride required manual detection and intervention, with no alarms or shut down systems installed.

1 The Incident

The Thunder River Rapids ride at Dreamworld in Queensland, Australia, was a water ride, simulating white water rafting. Up to six riders sit in a circular raft, which flows along a turbulent river, including travelling up a conveyor to ride down a waterfall. The water flow is driven by pumps and made turbulent by the installation of underwater obstacles. It was in full operation on 25th October 2016, when the pump managing the water flow tripped out. This resulted in the water level on the ride dropping. This drop in water level caused a raft to become stuck on the rails at the conveyor leading to the disembarkation platform. This raft remained stationary for some 57 seconds while the ride continued to run before it was impacted by the next raft in the sequence. This following raft contained 6 passengers, four adults and two children. When this raft struck the stationary one, it flipped over, killing the four adult passengers. The two children managed to escape the raft [3]. This incident is

—

currently under a Coronial Inquest. As such this paper is written from several media outlets that have been reporting the proceedings and based on the evidence reported, not the findings of the Coroner. The Coronial Report is due to be published in mid 2019.

2 History of the Ride

The Thunder River Rapids ride was completed in 1986 and was likely an in-house design and construction [4]. Sometime after construction, the ride was modified to remove some of the timber slats that form the base of the conveyor [4]. Two in every three slats were removed, though is unclear why this was done. This did create significant gaps in the base of the conveyor belt. The ride was also refurbished some months prior to the incident. [5].

Over the history of the ride there had been a number of incidents that had occurred with the rafts and the pumps. This in itself is not unusual, in that pumps can trip out from time to time, however there had been an incident in 2001 where the pump had tripped, and a raft became stuck [6]. This was then struck by a second raft, which flipped over. In this incident there were no passengers in the raft, and therefore no injuries. The investigation found that the ride operator had failed to adequately monitor the ride and he was dismissed. Details of this incident were not shared broadly and the operators and maintainers of the ride in 2017 had no knowledge of it. At the time an email was written by a park engineer discussing the incident and stated "I shudder when I think if there had been guests on the ride" [3]. This demonstrates that someone at some time recognized the significance of this incident, but it was then lost in history.

3 Technical and Organizational Causes of the Incident

There is one key immediate cause of this incident, however there are many different and wide-ranging causal factors. The immediate cause has been determined to be the tripping of the water pump, which in turn allowed the water level to drop, causing the first raft to become stuck. This provided the obstacle that the second raft, with six passengers, struck, causing it to then flip over. But let us examine the broader aspects of this incident, from the pump reliability to the safety systems and organizational capability. The ISC framework [2] provides a useful template to examine the causes. This framework in represented by Figure 1. Fundamentally leadership underpins all aspects of process safety. There are six functional elements that must be robust in an organization to manage major hazard safety. These are; knowledge & competence; engineering & design; systems & procedures; assurance; human factors and culture. In this incident there were clear failures in each of these elements.

—

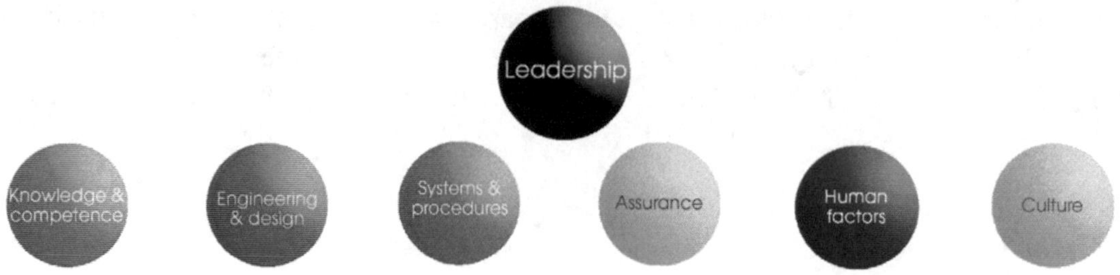

Figure 1. ISC Framework for Process Safety

3.1 Knowledge & Competence

The training of the ride operators took only 90 minutes and was unstructured with no evidence of competence consistently recorded [7]. There was a procedure manual however it was thought to be confusing by the ride operators [8]. The operator running the ride on 25th October 2016 had been trained for 90 minutes that morning, as it was her first day running the Thunder River Rapids. The junior operator stated that the trainer "showed me the front page of the manual and just flipped through it. I had to go through it myself. She said just sign off on it when you are happy." [9] The trainer stated that she sometimes had to "invent" training techniques because of the complexity of the control panel [8]. There was no clear guidance on how and when to use the emergency stop for the ride. The new operator testified that she was told "don't worry about it, you don't need to use it" with regard to the emergency stop button [10]. This is disputed by the person who delivered the training. The new operator believed it was the role of the senior operator to shut the ride down. It should be noted that the emergency stop would shut down the moving equipment with in two seconds, where the normal shutdown took around eight seconds. The ride operator that day stated that she "would have waited for the senior operator to tell her to hit the e stop" [10]. The senior operator that day was "surprised" to learn there was an emergency stop that would have shut the ride down within two seconds [7].

3.2 Engineering & Design

Intamin Amusement Rides designed and built the first river rapids rides in the world, and following this tragedy released a statement stating that the ride was "neither built by Intamin nor supplied by Intamin". The ride at Dreamworld was a custom build, and this was said to be common for rides built in the 1980's in the Gold Coast[4].

3.2.1 The Timber Slats

—

A few years after the ride was opened, a modification was made to the timber slats in the conveyor system [4]. The slats appear to serve at least two safety purposes. The first was to provide a solid platform for a passenger if they fell from the raft, essentially isolating them for the moving machinery and providing a platform to stand on [11]. The second was so that the rafts did not get caught in gaps and become stuck or indeed flip over. The Thunder River Rapids ride had two of every three slats removed. This provided large gaps between the remaining slats. There does not appear to be a theme park anywhere else that has made this change. There also appears to be no evidence as to why the change was made, nor was there a considered Management of Change (MoC) review undertaken. It is not clear if the organization had a structured and systematic MoC process. The safety regulator did not approve the modification and believed it added in additional hazards [4, 12]. In this incident, the raft became lodged in the gap, and as the conveyor continued to move forward, the raft was driven up vertically. When the passengers then fell from the raft, there was no solid base and they landed in the mechanism of the conveyor. There had also been conveyor framework modifications made to the ride in early 2016 in an attempt to stop rafts flipping over [13]. It is unclear what this modification was or indeed why it was made.

Figure 2. Aerial view of Thunder River Rapids showing the large gaps in the timber slat conveyor after the incident [10].

Figure 3. Aerial view of Infinity Falls, Orlando, showing close spacing of the boards on the conveyor [14].

3.2.2 The Pump

The water pumps had failed three times in the days leading up to the incident, and then twice on 25th October 2016, before the final failure that lead to the tragedy [15]. The pump's purpose was to keep the water flowing through the ride to keep the rafts moving. If the pumps stopped the water not only stopped flowing, it also drained back, and the water level dropped such that rafts could become stuck on the underwater obstacles that produced the rapids [16]. The reason for the pumps tripping out was not determined, the protocol seemed to be to just reset them to get the ride operating again [3]. There also is some confusion as to whether an alarm was sounded when the pumps tripped out, with an old procedure having the comment about an alarm crossed out [17]. At the time of the incident

—

the pumps were approximately 10 years old, and the manufacturer stated that they had a life expectancy of about a decade. There did not appear to be any plans to replace the pumps, however there was a maintenance request into the pump supplier in the week prior, but this visit did not eventuate [18].

3.2.3 Low Water

Low water in the ride was a critical hazard, as the rafts could become stuck or indeed flip. There had been previous incidents where low water level hazard led to rafts flipping, notably in 2001. There was no low water level switch that would shut down the ride if the water level became too low [3]. The only indication the operators had was to notice the water stain level on the wall of the ride canal to determine if the water was too low [19]. When the water level dropped on 25th October, one empty raft became stuck on the rails near the base of the conveyor. As the second raft containing four adults and two children engaged with the conveyor it was then driven into the stuck raft. A safety review conducted after the event stated that the primary cause was a lack of a safety rated water level detection system. It was believed that such a device could have been fitted for as little as AUD 3,000. The ride had been refurbished some seven months prior to the incident, but a device was not fitted. The nearby log ride at the park had such a safety device fitted [18].

3.2.4 Emergency Stop

Around two years prior to the incident an external safety report was written suggesting the emergency stop system should be simplified, however this was rejected by the company as the emergency system in place was seen to be an "acceptable risk". The stopping procedure for the ride contained several individual steps. There was a single button which would stop all components of the ride, however this was located at the unloading area and not the control panel where the operators worked. [20]

There was a great amount of confusion regarding the emergency stop button and what it actually did. The senior operator stated that they were "surprised" to learn there was an emergency stop that would stop all components and the junior operator on her first day was unsure what the button did [7]. She stated she was told to not worry about the button because she did not need to use it and testified, she "would have waited for my senior operation to tell me to press it". She also did not know the button would stop the conveyor from moving. [10] The maintenance workers who performed the daily checks on the ride also stated they did not know what the emergency stop button did as they never tested it [21].

3.2.5 Maintenance

Seven months prior to the incident a directive to cut all repairs and maintenance spending was ordered by the owners. At the time of the incident the facility was approximately AUD125,000 over maintenance budget for the year to date. Capital expenditure could continue to be spent during this time [16]. One maintainer stated that he performed comprehensive daily, monthly and annual checks on the ride, which include consumable

—

replacements such as nuts and bolts [22]. When the regulator inspected the ride after the incident, he noted it was in "generally poor condition" with excessive corrosion and rust on the electrical cables [23].

The maintenance planner noted that if spending could be deferred it would be. He also noted that the ride was shut annually for review, but on the nine occasions he witnessed a shutdown, he never saw an engineer inspect the ride nor were any risk assessment carried out [24].

3.2.6 Seatbelts

The Thunder River Rapids rafts had seat belts fitted for the safety of the riders. It should be noted that in May 2017, some seven months after the Dreamworld incident, that an 11 year old girl died on a similar ride, Splash Canyon, at Drayton Manor in the United Kingdom. The Splash Canyon ride rafts had no seatbelts installed at all, and as the child stood to swap seats with a fellow rider she fell overboard and struck her head. She died days later in hospital. This was not the first time a rider had fallen from the Splash Canyon ride [25]. This incident highlights the importance of proper restraints on these types of rides.

The seatbelts provided on the Thunder River Rapids rafts consisted of Velcro straps. It was vital that these were engaged correctly to provide the necessary restraint. Unfortunately, the Velcro was worn, and this made correct engagement more difficult. It is thought that at least one of the four riders may have survived the incident if the rafts had a better quality seatbelt, such as a mechanical buckle type that are on aircraft. This is because as the raft was shaken by the mechanism the belt failed to hold and the rider dropped to the conveyor system below [23, 26].

3.3 Systems & Procedures

Dreamworld has a series of operating procedures for the Thunder River Rapids ride however they were thought to be confusing [8]. During the investigation into the 2001 incident where empty rafts flipped, it was determined that the operator failed to follow the startup and emergency procedures. The investigation also reviewed the operating procedures and determined that the chance of a similar incident occurring with passengers on the rafts "is nil" [6]. A senior engineer believed that the cause of the 2016 incident was the operator not following procedures and that if procedures were followed the "system is a safe system" [27, 28].

There was also a lack of adequate emergency response procedures. Operators had discussed wanting better procedures and undertaking emergency drills, however, even two years after the incident, these have not been conducted at all [29].

3.4 Assurance

—

Over the history of the ride there had been a number of third-party reviews, inspections or audits carried out. Three years prior the incident a safety audit was conducted, and it recommended replacing the complicated emergency shut down process with a single emergency stop button, but this recommendation was never actioned [30].

There was also a legal requirement for the rides to have a "certificate to operate" and the Thunder River Rapids ride certificate was due to expire in January 2016. The company sought at extension as they were unable to find a "competent person" as defined under the legislation to inspect the ride [16]. Eight day before the incident the ride was inspected by a third-party engineer. He declared the ride as "structurally sound and properly functioning" even though he did not complete the full safety audit and merely did a visual inspection. He also did not inspect the log book as was required by law in a push to have the ride certified before it expired, though he admitted that he could have refused to certify the ride because of a "lack of documentation" [31, 32].

3.5 Human Factors

The control panel for the ride was confusing [3] and the operators needed to perform 36 different tasks in a one minute period, consistently [10]. This made for a system that encouraged human error, as it was not possible for the operators to be effective under these conditions. In addition to this, the engineering team believed the operators needed to simply follow procedures for the ride to be safe, not taking into account how difficult the procedures were or the actual workload. A senior engineer stated "In my view if procedures are followed the system is a safe system, you may as well not open the park. Procedures are there to be followed, if we don't follow procedures in everyday life, things happen." This blind focus on procedures does not take into account how the human interfaces with the machine and how the system advantages or disadvantages human performance.

3.6 Culture

A maintenance worker stated that he believed that the "safety culture at Dreamworld is very sound" [22] however this is in contrast to other workers who testified that familiarity had bred contempt with respect to the Thunder River Rapids ride [16], where new attractions were given more focus that existing ones [17]. Junior operators believed there was a culture of fear and had engaged with their union over their working conditions and the demands put on them regarding lack of rest breaks [33].

The culture also appeared to not learn from incidents, with the 2001 incident of flipping rafts said to be a one off and having "nil" chance of occurring with riders on board [6]. even though the consequence appeared to be understood at the time with an engineer writing in an email "I shudder when I think if there had been guests on the ride." [3].

The park relied on its apparent good safety history rather than challenging how and when an incident could occur [11]. They did not have an effective incident tracking system to

—

help with the learning process. They introduced a database but there was "general reluctance to use what was a fairly complex database" by the operators [34].

Safety did not appear to be a value in the company culture. The former safety manager stated that "everyone thought it was someone else's responsibility". He believed that deficiencies in the corporate structure led to a lot of problems. In 2018 his corporate safety role was made redundant.

Around five months prior to the incident, a memo was issued to staff to 'remain vigilant when monitoring conveyor movements" for items that might interfere with the rafts. This suggests there was some understanding of hazards and consequences, but it does to appear to have driven any improvement activities or modifications to prevent an incident [24].

4 Ramifications

The full ramifications of this incident are not yet clear. The Queensland government has initiated some legislative improvements to manage theme parks, but there is likely to be more recommendations coming from the Coronial Inquest when it releases its findings. Aside from the tragic loss of four lives, from the Dreamworld perspective, their guest numbers have remained low since the incident occurred, and it is struggling to attract patrons. This has led to their parent company suffering financial losses for the past two years and 50% decreases in the share price over the two year period [35].

The incident also provided an interesting lesson in crisis management. The company was scheduled to hold their annual general meeting in the days following the incident. The meeting went ahead and in it the CEO was awarded a performance bonus of $840,000 including both shares and a cash component despite the four deaths days before [36]. From a public relations perspective this was not a successful event. The CEO also stated they had been in contact with the victim's families, but this was publicly disputed by the families [37]. This further damaged the company reputation at a difficult time.

This incident will not only have ramifications for the families of the deceased, but for the survivors and the staff who responded due to the nature of injuries sustained by the victims. Four former staff members are suing Dreamworld for psychological damage [38].

These ramifications show how a process safety type incident can have significant impact on a business and its ability to continue operations in a positive way.

5 Could the Principles of Process Safety Have Helped?

As had been shown, there were clear and systematic failures across the whole organization that led to this tragedy. Starting with knowledge and competence, which relates to

—

organizational capability, had the operators been trained better in the safe operation of the ride as well as in emergency response, their performance that day may have been improved. From an engineering and design perspective, the design and installation of the ride was inadequate, it underwent undocumented changes which introduced hazards, and it was not adequately maintained. The systems and procedures for the operation and maintenance the ride was confusing and did not appear to contain necessary information. The assurance processes conducted did not seem to be thorough, rather they focused on keeping the ride operating and recommendations were not actioned. The ride was not designed with human factors in mind, from the high workload to the confusing operating panel. The organization culture did not seem to place a value on safety, it only focused on cost reduction. It also did not appear to be a learning culture, where lessons from the past are understood and embedded. Lastly at the time of crisis, the overall leadership displayed by the parent company was publicly found to be lacking.

There are transferrable learnings for the process industries from this incident, with some clear questions to ask. How are you managing your competency program to ensure you have the right capability? Are your designs adequate, maintained and any changes managed? Are your procedures thorough, identifying hazards, consequences and controls and easy to comprehend? How are you monitoring the operation of your facility? Does the design and operation of your facility enhance human performance or make it more difficult? Does your culture support safe operations?

6 Conclusion

A focus on the elements of process safety can help identify what is important in a business. In this instance there had been recommendations for a low-cost instrumented safety system to be installed that would have prevented the incident from occurring. The importance of this was not understood by the organization, and the opportunity to upgrade the ride during its recent refurbishment was ignored. It is clear that the hazard and potential consequences and controls were not understood. As is often the case, when the hazard is not understood and therefore not managed, people suffer tragic consequences.

7 References

[1] Kerin, T. *Managing Process Safety. In the Core Body of Knowledge for Generalist OHS Professionals.* Safety Institute of Australia, Tullamarine, VIC. 2017

[2] ISC (IChemE Safety Centre). Safety Centre Framework, Available at https://www.icheme.org/knowledge/safety-centre/framework/, Accessed on January 2, 2019

—

[3] ABC News, Available at https://www.abc.net.au/news/2018-06-18/dreamworld-fatal-accident-ride-had-broken-down-inquest-told/9871898, Accessed on January 2, 2019

[4] Australian Financial Review, Available at https://www.afr.com/news/dreamworlds-rapids-ride-different-from-others-around-the-world-20161101-gsfd07, Accessed on January 2, 2019

[5] 9 News, Video 11, Available at https://www.9news.com.au/2018/12/06/03/33/dreamworld-inquest-enters-final-days, Accessed on January 2, 2019

[6] news.com.au, Available at https://www.news.com.au/national/queensland/courts-law/coroner-james-mcdougall-requests-proof-all-dreamworld-rides-were-certified/news-story/e77275b97b0bbee9c1c541bda6df0035, Accessed on January 2, 2019

[7] ABC News, Available at https://www.abc.net.au/news/2018-06-19/dreamworld-operator-on-first-day-when-four-people-died-on-ride/9885214, Accessed on January 2, 2019

[8] ABC News, Available at https://www.abc.net.au/news/2018-06-27/dreamworld-trainer-promoted-since-four-killed-on-ride/9914188, Accessed on January 2, 2019

[9] 9 News, Video 5, Available at https://www.9news.com.au/2018/12/06/03/33/dreamworld-inquest-enters-final-days, Accessed on January 2, 2019

[10] ABC News, Available at https://www.abc.net.au/news/2018-06-20/dreamworld-operator-impossible-load-responsibilities-manage-ride/9889776, Accessed on January 2, 2019

[11] 9 News, Video 1, Available at https://www.9news.com.au/2018/12/06/03/33/dreamworld-inquest-enters-final-days, Accessed on January 2, 2019

[12] 9News, Available at https://www.9news.com.au/2018/11/20/03/32/dreamworld-ride-passed-wphs-safety-audit, Accessed on January 2, 2019

[13] ABC News, Available at https://www.abc.net.au/news/2018-10-08/dreamworld-inquest-hears-total-failure-to-ensure-safety/10350034, Accessed on January 2, 2019

[14] Intamin Worldwide, Available at https://www.intaminworldwide.com/infinity-falls-orlando/ , Accessed on December 12, 2018

—

[15] 9 News, Video 8, Available at https://www.9news.com.au/2018/12/06/03/33/dreamworld-inquest-enters-final-days, Accessed on January 2, 2019

[16] ABC News, Available at https://www.abc.net.au/news/2018-06-25/dreamworld-inquest-told-of-ride-maintenance-cutbacks/9906214, Accessed on January 2, 2019

[17] Australian Financial Review, Available at https://www.afr.com/real-estate/commercial/hotels-and-leisure/dreamworld-prioritised-new-attractions-over-maintenance-of-old-rides-inquest-20180626-h11urb, Accessed on January 2, 2019

[18] 9 News, Video 10, Available at https://www.9news.com.au/2018/12/06/03/33/dreamworld-inquest-enters-final-days, Accessed on January 2, 2019

[19] ABC News, Available at https://www.abc.net.au/news/2018-06-21/dreamworld-ride-operator-highly-distressed-give-evidence-inquest/9893308, Accessed on January 2, 2019

[20] news.com.au, Available at https://www.news.com.au/national/queensland/courts-law/dreamworld-engineer-tells-families-i-apologise-for-not-being-to-make-eye-contact-with-you/news-story/40545dc747c95eb96b0293c888d77d96, Accessed on January 2, 2019

[21] Sunshine Coast Daily, Available at https://www.sunshinecoastdaily.com.au/news/dreamworld-thunder-river-rapids-button-wasnt-regul/3543997/, Accessed on January 2, 2019

[22] ABC News, Available at https://www.abc.net.au/news/2018-06-26/dreamworld-ride-operator-says-conveyor-was-never-stopped/9910192, Accessed on January 2, 2019

[23] news.com.au, Available at https://www.news.com.au/national/queensland/courts-law/safety-report-shows-simple-change-on-dreamworld-ride-could-have-saved-lives/news-story/9c1fa2c195efd12f86a3e8ad516ea1b8, Accessed on January 2, 2019

[24] 9 News, Video 20, Available at https://www.9news.com.au/2018/12/06/03/33/dreamworld-inquest-enters-final-days, Accessed on January 2, 2019

[25] news.com.au, Available at https://www.news.com.au/travel/travel-updates/incidents/girl-11-dies-after-falling-into-the-water-on-splash-canyon-ride-during-school-trip-to-uk-theme-park/news-story/f83330b0ce12ea11df2d089ece676519, Accessed on January 2, 2019

—

[26] 9News, Available at https://www.9news.com.au/2018/12/06/03/33/dreamworld-inquest-enters-final-days, Accessed on January 2, 2019

[27] 9News, Available at https://www.9news.com.au/2018/10/16/14/43/dreamworld-inquest-thunder-river-rapids-ride-engineers-give-evidence, Accessed on January 2, 2019

[28] 9News, Available at https://www.9news.com.au/2018/10/17/10/49/dreamworld-inquest-engineer-to-appear-again, Accessed on January 2, 2019

[29] Australian Financial Review, Available at https://www.afr.com/news/dreamworld-ride-trainer-promoted-after-tragedy-staff-wanted-emergency-training-20180628-h11yke, Accessed on January 2, 2019

[30] 9 News, Video 12, Available at https://www.9news.com.au/2018/12/06/03/33/dreamworld-inquest-enters-final-days, Accessed on January 2, 2019

[31] ABC news, Available at https://www.abc.net.au/news/2018-10-09/dreamworld-ride-declared-safe-days-before-tragedy/10352164, Accessed on January 2, 2019

[32] ABC News, Available at https://www.abc.net.au/news/2018-10-17/dreamworld-supervisor-sympathy-has-families-in-tears-inquest/10387186, Accessed on January 2, 2019

[33] 9 News, Video 9, Available at https://www.9news.com.au/2018/12/06/03/33/dreamworld-inquest-enters-final-days, Accessed on January 2, 2019

[34] 9 News, Video 16, Available at https://www.9news.com.au/2018/12/06/03/33/dreamworld-inquest-enters-final-days, Accessed on January 2, 2019

[35] Australian Financial Review, Available at https://www.afr.com/research-tools/AAD/share-prices?days=1095, Accessed on November 26, 2018

[36] The Guardian, Available at https://www.theguardian.com/australia-news/2016/oct/27/deborah-thomas-set-to-receive-840000-in-bonuses-despite-dreamworld-deaths, Accessed on December 19, 2018

[37] ABC News, Available at https://www.abc.net.au/news/2016-10-27/dreamworlds-shareholders-ask-parent-company-to-support-victims/7970104, Accessed on January 2, 2019

—

[38] news.com.au, Available at https://www.news.com.au/lifestyle/real-life/news-life/dreamworld-first-responders-to-sue-theme-park-for-psychological-damage/news-story/e8fd7399eec9d165c657dbdb3ac35968, Accessed on January 2, 2019

Fatal Toxic Chemical Release at DuPont

Tamara Qureshi
U.S. Chemical Safety and Hazard Investigation Board
1750 Pennsylvania Ave, NW Suite 910 Washington, DC 20006
tamara.qureshi@csb.gov

Dan Tillema, Steve Cutchen, Lauren Grim, and Chris Lyon
U.S. Chemical Safety and Hazard Investigation Board
1750 Pennsylvania Ave, NW Suite 910 Washington, DC 20006

Prepared for Presentation at
American Institute of Chemical Engineers
2019 Spring Meeting and 15th Global Congress on Process Safety
New Orleans, LA
March 31 – April 3, 2019

Fatal Toxic Chemical Release at DuPont

Tamara Qureshi
U.S. Chemical Safety and Hazard Investigation Board
1750 Pennsylvania Ave, NW Suite 910 Washington, DC 20006
tamara.qureshi@csb.gov

Dan Tillema, Steve Cutchen, Lauren Grim, and Chris Lyon
U.S. Chemical Safety and Hazard Investigation Board

Keywords: DuPont, toxic release, gas detection, alarm

Abstract

On November 15, 2014, four workers were killed after approximately 24,000 pounds of highly toxic methyl mercaptan released from an insecticide production unit (Lannate® Unit) at the E.I. du Pont de Nemours chemical manufacturing facility in La Porte, Texas. This paper will discuss the events leading to the methyl mercaptan release, the deficiencies in DuPont La Porte's toxic gas detection system, and offer lessons on how to prevent some of the deficiencies in the toxic gas detection system from occurring. An effective toxic gas detection system can protect workers by alerting them of a potentially dangerous situation and preventing them from unknowingly entering a potential toxic chemical release.

1 Introduction

On November 15, 2014, approximately 24,000 pounds of highly toxic methyl mercaptan released from an insecticide production[1] unit (Lannate® Unit) at the E.I. du Pont de Nemours chemical manufacturing facility in La Porte, Texas. Four workers at the DuPont La Porte facility entered the Lannate® Unit's manufacturing building (manufacturing building) and the highly toxic methyl mercaptan release. Three of these workers were responding to a coworker's distress call from inside the manufacturing building. The fourth worker entered after all the workers inside the manufacturing building stopped communicating with workers inside the control room. The Lannate® Unit's methyl mercaptan toxic gas detection system did not effectively warn workers of the large chemical release occurring within the manufacturing building. Additionally, this toxic gas detection system did not provide emergency responders with sufficient information to effectively evaluate the potential harm to the public.

---

[1] The Insecticide Business Unit (IBU) was divided into two primary insecticide product lines: Lannate®, which is DuPont's trade name for methomyl; and Agricultural Products Intermediates (API), which produced insecticides and nematicides known as Vydate®, the trade name for oxamyl [3, p. 11].

The DuPont La Porte methyl mercaptan toxic gas detection system had several deficiencies. First, the methyl mercaptan detector system did not meet its design intent. The alarm set point for methyl mercaptan concentration and the method of conveying alarms did not forewarn workers of the immediately dangerous to life and health (IDLH) conditions they would encounter by entering the manufacturing building. The methyl mercaptan gas detectors only alarmed on the Lannate® control room operator's computer screen. There was no audible or visible alarm provided in the field for outside personnel. Second, the placement of the methyl mercaptan toxic gas detectors did not fully portray the extent of the November 15, 2014 methyl mercaptan release. Third, the DuPont La Porte site normalized the hazards associated with methyl mercaptan. Finally, the alarms did not prompt protective action or investigations into the sources of alarms. During the hours prior to the November 15, 2014 incident, multiple highly toxic chemical gas detectors alarmed (sounded). However, the DuPont emergency response team was not notified nor was the area cleared of personnel. In addition, methyl mercaptan releases on November 13 and 14, 2014, were identified by methyl mercaptan detectors, but were never reported as releases nor investigated as serious process safety incidents. Personnel believed they knew the source of the releases and normalized their responses to these alarms. Consequently, there was a delayed awareness of the release by operations, and workers who entered the manufacturing building after the release began were not alerted of nor prepared for the lethal concentrations of methyl mercaptan being released, resulting in the death of two of these workers.

The DuPont La Porte incident is the CSB's third investigation of a fatal DuPont incident in five years.[2] On September 30, 2015, the CSB issued a presentation, safety video, and an Interim Recommendations report [1],[3] which detailed needed safety improvements at the facility and issued recommendations to the DuPont La Porte facility to be implemented prior to resuming operations of the Lannate® Unit.[4] Due to the amount of methyl mercaptan stored at the DuPont La Porte site, the Lannate® process was covered by Occupational Safety Health Administration's (OSHA) Process Safety Management (PSM) program and the Environmental Protection Agency's (EPA) Risk Management Plan (RMP) Rule [2].

An effective toxic gas detection system can make the difference between personnel being prepared for a potentially dangerous situation or entering a potential toxic chemical release. To be effective, a toxic gas detection system needs to consider the chemical's hazards, the gas detector's

[2] The CSB also investigated an incident at a DuPont plant in Belle, West Virginia (three incidents that occurred in a 33-hour period over January 22-23, 2010, resulting in one fatality), and another in Buffalo, New York (a single incident that occurred on November 9, 2010, resulting in one fatality and one serious injury) [16], [17].

[3] In the Interim Recommendations, the CSB identified significant process safety management deficiencies at the DuPont La Porte facility, including delayed maintenance of safety-critical equipment, lack of written procedures, poor hazard analysis practices, lack of implementation of important inherently safer design concepts, and a lack of hazard recognition.

[4] In response to the CSB findings and recommendations, DuPont committed to making changes at the DuPont La Porte facility. DuPont La Porte, however, was not able to address all of these findings and recommendations, because in the spring of 2016 DuPont announced that it would not reopen the Lannate® Unit. The company had determined that "significant changes in market conditions during the period of the shutdown [would] persist over the long term [14]" and the cost required for the restart "[was] not a long-term viable and cost-efficient option for the DuPont Crop Protection business [15]." In 2017, DuPont dismantled and removed the buildings and equipment from the DuPont La Porte facility that were associated with its insecticide and herbicide business units, which included the Lannate® Unit.

capabilities, the location of the gas detectors, and the alarms associated with the gas detectors. This paper will discuss some of the events leading to the methyl mercaptan release, the deficiencies in DuPont La Porte's toxic gas detection system and offer lessons on how to prevent some of the deficiencies in the toxic gas detection system from occurring.

2 Events Leading to the Incident

On Monday, November 10, 2014, an inadvertent chemical dilution caused operating difficulties that forced a shutdown of DuPont La Porte's Lannate® Unit [3, pp. 20-21]. On Wednesday, November 12, 2014, operators tried to restart the Lannate® process. During the shutdown, the methyl mercaptan piping to the reaction section of the process plugged due to the formation of a clathrate hydrate (hydrate),[5] halting restart. At that time, operators did not know the source of the plugging and began troubleshooting the process to try to clear the piping, unaware that the solid hydrate had formed.

Two days later, on Friday, November 14, 2014, the troubleshooting efforts to clear the plugging were still ongoing. That morning, a technical team composed of engineers and other employees experienced in the Lannate® process, met with operations personnel to discuss troubleshooting options to clear the plugging. This meeting identified the likely scenario that water had entered the methyl mercaptan system and a solid hydrate formed as a result. Based on their understanding of the hydrate,[6] the technical team asked operators to put hot water on the outside of the methyl mercaptan piping, under the insulation, to warm the piping and its contents, to break-up (dissociate[7]) the plugging. The team also realized that when heated, methyl mercaptan would expand, requiring a safe place to vent to avoid over-pressuring the piping. To address this concern, operations personnel opened valves between the methyl mercaptan piping and a waste gas vent header, not recognizing that this alignment created a pathway for liquid methyl mercaptan to eventually release from the piping inside the manufacturing building when drain valves were opened. In addition, when operations personnel opened other valves while troubleshooting, such as drain valves, methyl mercaptan released to the atmosphere causing a strong methyl mercaptan odor[8] that could be smelled by site personnel.

---

[5] At low temperatures (≤ 52 °F) water and methyl mercaptan form a solid, ice-like material called methyl mercaptan clathrate hydrate [10]. Temperatures in the Houston area for the 24 hours preceding the incident averaged approximately 40 °F and had been consistently below 55 °F since Tuesday, November 11, 2014. Although the potential for methyl mercaptan hydrate formation had been identified in DuPont's methyl mercaptan technical standard and in a process hazard analysis years earlier, DuPont never implemented safeguards, such as heat tracing, or developed a procedure to dissociate the hydrate safely [3, p. 22].

[6] DuPont's methyl mercaptan technical standard identifies the potential to form a hydrate at low temperatures. The technical standard states, "[methyl mercaptan] will form a hydrate with water, which is a solid below 40 deg F [degrees Fahrenheit] per information provided by a [methyl mercaptan] supplier."

[7] Methyl mercaptan clathrate hydrates dissociate back to the liquid phases of methyl mercaptan and water at approximately 52 °F [10].

[8] Methyl mercaptan has a low odor threshold. It gives off a smell like rotten cabbage. It can be smelled at only two parts per billion (ppb) [9, p. 1]. OSHA's permissible exposure limit (PEL), however, is 10 parts per million (ppm), which is 5,000 times the minimum concentration it can be smelled. The low odor threshold of mercaptans makes them useful as fuel gas odorants [9, p. 2].

During this troubleshooting process, methyl mercaptan released both outside and inside the manufacturing building, triggering 32 methyl mercaptan gas alarms on the control panel throughout the 17 hours preceding the incident. While methyl mercaptan is a toxic and flammable chemical, and at high enough concentrations lethal, operations personnel did not respond in accordance with the nature of the emergency circumstances, because they associated the alarms with ongoing troubleshooting efforts. They did not perceive the methyl mercaptan alarms as signifying a serious hazard because they normalized[9] both the methyl mercaptan odor within the Lannate® Unit as well as the detector alarms associated with them.

Early in the morning on Saturday, November 15, 2014, the hot water from the hoses warmed the hydrate plugging, causing it to dissociate, clearing the plugging. Due to the valve alignment, liquid methyl mercaptan flowed into the waste gas vent header, located inside the manufacturing building [3, pp. 25-26]. This vent header, however, was not intended for liquid methyl mercaptan.

At 2:51 am alarms began to sound on the control system indicating high pressure in equipment located inside the manufacturing building. Operations personnel did not realize that liquid methyl mercaptan was causing the high pressure. Instead, they attributed the high pressure to a common, long-standing problem with process condensate (that DuPont La Porte personnel believed to be mostly water) accumulating in the vent header piping [3, p. 28]. The Lannate® control room operator (Board Operator) contacted the night shift supervisor (Shift Supervisor) and another operator (Operator 1) separately, asking them to help troubleshoot the high-pressure situation. The Shift Supervisor and Operator 1 each went separately to the manufacturing building, likely with the intent to drain liquid from the vent header piping manually.[10] Sometime between 3:01 am and 3:13 am, two sets of drain valves on the vent header piping, located on the third floor of the manufacturing building were opened (Figure 1). But instead of being comprised mostly of water as workers believed, liquid methyl mercaptan that had filled the piping escaped out of the valves, vaporized, and killed the Shift Supervisor.

Between 3:24 am and 3:26 am, three methyl mercaptan detectors inside the manufacturing building sensed at least 25 parts per million (ppm) methyl mercaptan[11] and triggered alarms at the control panel. The Board Operator—still focused on reducing the high pressure—did not realize a major chemical release inside the manufacturing building was occurring.

---

[9] Normalization of deviance is a phenomenon in which individuals or work teams gradually accept a lower standard of performance until the lower standard becomes the norm. It is typically the result of conditions slowly changing and eroding over time [11, p. 4].

[10] In 2011, DuPont installed a NOx (nitrogen oxides) reducing scrubbed (NRS) incinerator system. The waste gas vent header piping to the NRS Incinerator was designed without sufficient consideration of liquid accumulation. It contained low points where liquid could accumulate, and there was no engineered equipment provided, such as a knock out drum, to remove liquid safely from the waste gas vent header piping system. The liquid accumulation in the piping would create pressure in process equipment. Pressure buildup within the equipment was resolved by manually draining liquid from the waste gas vent header. See DuPont La Porte, Texas Chemical Facility Toxic Chemical Release: Interim Recommendations [3, p. 18].

[11] The methyl mercaptan detectors alarm at 25 ppm–well above the National Institute of Occupational Safety and Health (NIOSH) recommended exposure limit (REL) of 0.5 ppm and the OSHA enforceable (PEL) ceiling limit of 10 ppm. OSHA recommends that employers consider using the alternative occupational exposure limits because the agency believes that exposures above some of these alternative occupational exposure limits may be hazardous to workers, even when the exposure levels are in compliance with the relevant PELs [5]. The Cal-OSHA permissible exposure limit (PEL) is an 8-hour time-weighted average (TWA) of 0.5 ppm and NIOSH recommended exposure limit (REL) is a 15- minute ceiling limit of 0.5 ppm [7], [6].

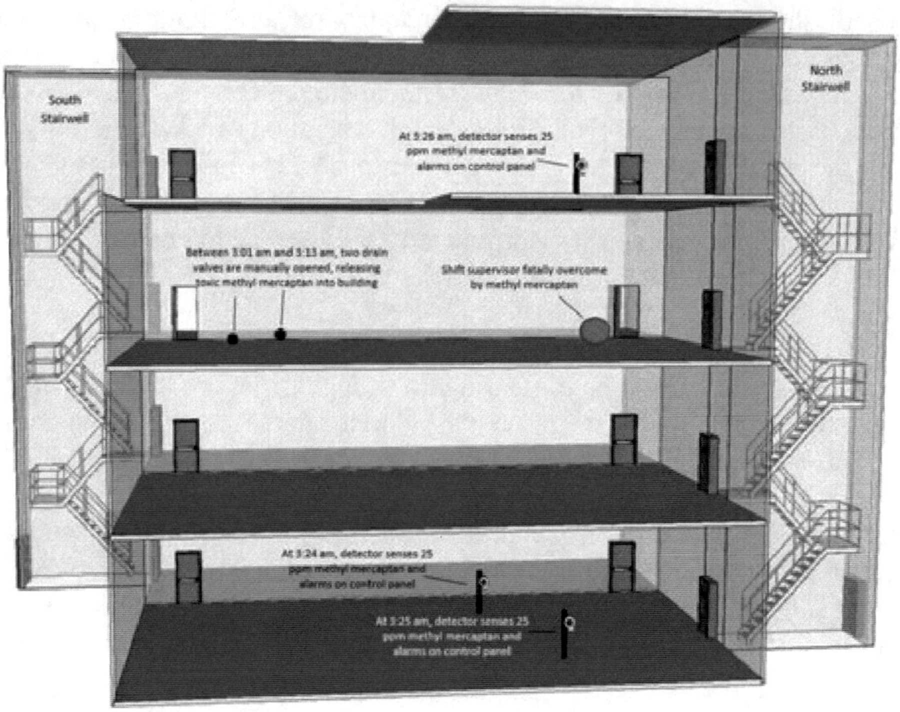

Figure 1. Depiction Showing Location Where Drain Valves Were Opened. These drain valves released toxic methyl mercaptan into the manufacturing building. Methyl mercaptan detectors on the first and fourth floors detected high concentrations of methyl mercaptan shortly after the release began. This depiction is a simplified graphic of the manufacturing building and does not show the location of Operator 1. Source: CSB

At about 3:30 am, Operator 1 made an urgent call for help over the radio. Various personnel interpreted her communication in different ways, generally as either "We need help!" or "I need help on the fourth floor!" The Board Operator tried to get more information through radio communication, but neither the Shift Supervisor nor Operator 1 ever responded [3, p. 29].
Two operators (Operator 2 and Operator 3) who were in the control room and heard the distress call ran to the manufacturing building to help. Another operator (Operator 4) saw Operator 2 and Operator 3 running to the manufacturing building, and he followed them. None of these operators knew there was a major release of toxic methyl mercaptan inside of the manufacturing building, and therefore they did not wear any respiratory protection when they ran into the manufacturing building. The manufacturing building lacked automatic visual or audible alarms to alert field workers or prevent them from entering a potentially toxic atmosphere.

Operator 2, Operator 3, and Operator 4 each entered the south stairway and took different routes inside the manufacturing building to try to find Operator 1. Operator 2 went to the third floor, where he was fatally overcome by methyl mercaptan [3, p. 29]. Operator 4 went to the second floor. He walked about ten feet and hit what he later described as a "wall" of methyl mercaptan, but he managed to retreat to the stairwell [3, p. 29]. Operator 3 went to the fourth floor but did not find anyone. He then announced on the Lannate® Unit's public-address system that he did not see anyone on the fourth floor. The Board Operator responded that the Shift Supervisor and Operator 1 may be on the third floor. Operator 3 then began to feel light headed. He made his way to the stairwell and lost consciousness while descending from the fourth floor [3, p. 29].

When Board Operator could not reach people inside the manufacturing building, an operator in the control room, Operator 6[12] (the brother of Operator 2), then grabbed three 5-minute escape respirators. Other operators in the control room warned him not to enter the manufacturing building because they did not know what was going on or where the non-responsive operators were. Operator 6, however, took the escape respirators and rushed to the manufacturing building [3, p. 29].

At about 3:40 am, Operator 6 encountered Operator 4 in the south stairway of the manufacturing building and put an escape respirator on him. The breathing air helped Operator 4 recover, and he exited the manufacturing building safely. A worker—likely Operator 6—then manually activated the manufacturing building fume release alarm, an alarm intended to alert area workers of a toxic chemical release in the building. Operator 6 then went to the third floor of the manufacturing building. At some point inside the manufacturing building after helping Operator 4, Operator 6 retrieved a self-contained breathing apparatus (SCBA). He tried to rescue his brother (Operator 2) by putting one of the escape respirators on his face. Operator 6 was then fatally overcome by methyl mercaptan before he could connect his own respirator mask to his SCBA tank.

3 Lessons from the DuPont La Porte Incident

3.1 Ensure Toxic Gas Detector Systems Operate as Intended

DuPont La Porte's methyl mercaptan technical standard stated, "[a]utomatic or continuous specific chemical detection systems *must* be installed where a [methyl mercaptan] release could have any off-site impact [...] [Emphasis in the original]." DuPont La Porte's methyl mercaptan toxic gas detection system consists of methyl mercaptan detectors, alarms, distributed control system (DCS), control panel, and a control room operator. The manufacturing building is equipped with gas detectors for methyl mercaptan in various locations. The third floor, where the release occurred, however, did not have any methyl mercaptan detectors. The detectors inside the manufacturing building were also set to alarm on the control panel when they detected a 25 ppm concentration of methyl mercaptan and were intended to protect workers from exposure to methyl mercaptan.[13]

DuPont used hydrogen sulfide (H_2S) detectors to identify methyl mercaptan. DuPont records indicate that the manufacturer has stated the detector should be capable of detecting methyl mercaptan within 1 ppm.[14] The alarms activated when the detectors sensed the hydrogen sulfide equivalent to 25 ppm methyl mercaptan.[15] The implementation of the methyl mercaptan toxic gas

[12] Operator 6 was a board operator for another process in the Lannate® Unit.

[13] The 25 ppm alarm set point is 50 times the NIOSH's REL of 0.5 ppm and more than double OSHA's ceiling limit of 10 ppm [7]. Therefore, alarming at this detection level could not protect workers from exposure to methyl mercaptan.

[14] DuPont La Porte's detectors for methyl mercaptan are calibrated for 0-50 ppm using a hydrogen sulfide (H_2S) calibration gas.

[15] The methyl mercaptan detectors were evaluated post-incident. Although the detectors inside the building have the ability to span 0-50 ppm, these detectors were set to span 0-25 ppm. Post-incident, the span was increased to 50 ppm. This configuration of the alarms limited the amount of information communicated to DuPont La Porte personnel.

detection system inside the manufacturing building, however, did not meet the design intent, federal safety regulations, or DuPont corporate standards.

The methyl mercaptan detectors outside and inside the manufacturing building were installed for two different purposes—the outside detectors were geared towards an offsite release and the inside detectors were meant to protect workers from a hazardous environment. DuPont La Porte documentation stated that the basis for the alarms outside the manufacturing building was to provide an early warning mechanism for a potential offsite methyl mercaptan release. The alarm setpoint was based on the ERPG (Emergency Response Planning Guidelines)-2[16] concentration[17] for methyl mercaptan which is 25 ppm. DuPont La Porte installed the methyl mercaptan detectors inside the manufacturing building after the methyl mercaptan detectors outside the building. Unlike the outside detectors, there was no documentation for the basis for setting the alarm setpoint to 25 ppm for the methyl mercaptan detectors inside the manufacturing building nor was there evidence that these setpoints were set at 25 ppm to prevent methyl mercaptan alarms from becoming a nuisance. It appears that the setpoints were copied from the outside methyl mercaptan detectors even though the inside detectors served a different purpose—worker protection. The methyl mercaptan detectors alarmed at 25 ppm—well above the NIOSH recommended exposure limit (REL) of 0.5 ppm and the OSHA enforceable PEL ceiling limit of 10 ppm.

Depending on the purpose of a toxic gas detection system alarm, there are different sets of values for facilities to consider. For instance, ERPG values are designed for the protection of the public during an emergency and not intended for the everyday routine protection of workers [4].[18] For worker protection, OSHA recommends that employers consider using the lower alternative occupational exposure limits because the agency believes that exposures above some of these alternative occupational exposure limits may be hazardous to workers, even when the exposure levels are in compliance with the relevant PELs [5]. For example, the Cal-OSHA permissible exposure limit (PEL) is an 8-hour time-weighted average (TWA) of 0.5 ppm and NIOSH's REL is a 15- minute ceiling limit of 0.5 ppm [6], [7].

Additionally, information from the methyl mercaptan detectors inside the manufacturing building only displayed on the DCS monitor and was not recorded by the site's DCS and data historian [3, p. 37]. CSB investigators, however, were able to obtain information from alarm logs about the detectors.

DuPont La Porte's management of change documentation for installation of methyl mercaptan detectors inside the manufacturing building stated the condition prompting the change was the

[16] ERPG-2 is the maximum airborne concentration below which nearly all individuals could be exposed for up to 1 hour without experiencing or developing irreversible or other serious health effects or symptoms which could impair an individual's ability to take protective action. At concentrations above ERPG-2, some members of the community may experience significant health effects which could impair them from taking protective action. Emergency responders often alert the community to take protective action in areas susceptible to the ERPG-2 concentration and above [19].

[17] The ERPG values for methyl mercaptan in parts per million (ppm) are: ERPG-1, 0.005; ERPG-2, 25; ERPG-3, 100 [18, p. 6].

[18] Emergency Response Planning Guidelines (ERPGs) are chemical concentration levels that are used during emergency response and when preparing for emergencies. They are designed to anticipate public health effects not occupational health effects from exposure to airborne chemicals [4].

possibility that personnel could enter a potentially hazardous location without knowing methyl mercaptan had been released:

> *Present Conditions and Description of Proposed Change:*
>
> *Personnel can enter a potentially hazardous location without knowing MeSH [methyl mercaptan] has been released to the environment.*

The methyl mercaptan gas detectors, however, only alarmed on the Lannate® control room operator's computer screen.

DuPont La Porte site did not have audible or visual alarms on or in the manufacturing building to alert field personnel of a methyl mercaptan release automatically. DuPont relied on verbal communication of alarms that automatically displayed on a continuously manned control board. Without being made aware of the major chemical release inside the building, Operator 2, Operator 3, and Operator 4 entered without respiratory protection. As a result, Operator 2 was killed from methyl mercaptan exposure, and Operator 3 and Operator 4 were exposed to dangerous concentrations of methyl mercaptan.

Facilities should make strategic decisions when choosing a toxic gas detector to ensure that a toxic gas detection system operates as intended. Considerations should include regulatory requirements, NIOSH recommendations, chemical interferences, and utility (worker protection, release modeling, etc.). The methyl mercaptan detectors at the DuPont La Porte facility had a span of 0-50 ppm. This range may be appropriate for early detection of a toxic leak when coupled with alarms and clear instructions to evacuate the premises at lower concentrations. If the goal is to understand other aspects of a release, like potential lethal concentration (150 ppm is methyl mercaptan's IDLH concentration) or modeling an offsite release, then a wider concentration range should be considered.

Additionally, facilities should also consider the use of automatic alarms designed to alert personnel of hazardous conditions—such as a release of a toxic chemical or explosive atmosphere—which can relay critical safety information immediately to personnel, without potential delays inherent in a system that solely relies on a control room operator to verbally communicate safety alarms to operators in the field.

3.2 Ensure Toxic Gas Detector Placement is Strategic

Although the DuPont La Porte site had 13[19] fixed methyl mercaptan detectors inside and outside the manufacturing building, there were no fixed methyl mercaptan detectors located on the third floor inside the manufacturing building where a majority of the November 15, 2014 methyl mercaptan release occurred. Additionally, no workers were provided with either portable or personal methyl mercaptan monitors. Part of the Lannate® process occurred inside the closed manufacturing building, a four-story building containing a large amount of piping and process equipment, dividing walls, multiple stairwells, and an elevator. The Lannate® Unit had methyl

---

[19] The site had a total of 15 methyl mercaptan detectors—there were two additional methyl mercaptan detectors installed in another process building adjacent to the manufacturing building.

mercaptan process piping both inside and outside the manufacturing building. A facility siting study and a process hazards analysis had identified the necessity of having methyl mercaptan detectors in the Lannate® manufacturing building. These studies identified the need for protecting workers from toxicity and explosivity hazards related to potential methyl mercaptan releases.

DuPont La Porte installed 10 methyl mercaptan detectors near process piping outside the manufacturing building. The design basis for the outside methyl mercaptan detectors was to provide an early warning of a methyl mercaptan release. DuPont La Porte placed these detectors around its methyl mercaptan storage tank and the area where methyl mercaptan could be unloaded. There were, however, no fixed methyl mercaptan detectors along the fence line that could relay whether there was an offsite release.

DuPont La Porte installed three methyl mercaptan detectors inside the manufacturing building. The stated design purpose for the methyl mercaptan detectors inside the building was worker protection. The inside detectors were located within the building on the first and fourth floors. These detector locations were determined based on potential release scenarios from a 2001 process hazard analysis (PHA).

The methyl mercaptan release occurred on the third floor where there were no methyl mercaptan detectors. Although the 2001 PHA did not identify methyl mercaptan detector locations on the third floor, DuPont La Porte routed the waste gas vent header piping along the third floor and this piping was directly connected to the top of the methyl mercaptan storage tank. During abnormal or upset conditions, the waste gas vent header could have been a potential source for a methyl mercaptan release. For example, ineffective hazard management while troubleshooting the plugged methyl mercaptan feed piping during the November 15th incident allowed liquid methyl mercaptan to flow into the waste gas vent header piping towards the NRS incinerator—a location where it was never intended to go.

Additionally, the routine draining of liquid from the waste gas vent header should have triggered a management of change review and considered whether additional safeguards like methyl mercaptan detectors were necessary. As discussed earlier, DuPont La Porte never fully resolved liquid accumulation in the waste gas vent header piping to the NRS incinerator with an engineered solution. Instead, operators were directed to drain the system by area management through written instructions in a daily instruction log book to operations staff [8, p. 18].[20] This routine draining presented potential dangers because DuPont La Porte personnel never sampled or analyzed the liquid workers routinely drained from this system [8, p. 18]. Daily instructions before the incident requested the operators to sample the uncharacterized liquid. DuPont La Porte management wanted samples of the liquid for a subsequent laboratory analysis to identify its chemical composition. Workers reported that the liquid had a strong, noxious odor of sulfur compounds [8, p. 18]. Even though DuPont La Porte was aware that the liquid likely contained more than just water and was unaware of the actual chemical composition [8, p. 18], instructions to workers did

[20] Beginning January 14, 2014, and continuing until the incident, the daily instructions directed operators to drain the methyl mercaptan waste gas vent header's "low point" on the 3rd floor once per shift. After the installation of a flexible hose from the waste gas vent header drain valve to the floor, the instructions changed to include directing the liquid through the hose under a running safety shower until the liquid stops flowing from the hose [8, p. 18].

not specify that additional precautions—like wearing additional PPE—were needed when draining the liquid from the waste gas vent header to the floor inside the manufacturing building.

Consequently, DuPont La Porte workers dealt with the common problem of liquid accumulation in the waste gas vent header on a routine basis by draining the liquid (line breaking) without engineered solutions or ensuring the use of safety procedures or personal protective equipment. On November 15, 2014, when the liquid drain valves were opened this time, flammable and highly toxic methyl mercaptan flowed onto the floor and filled the manufacturing building with toxic vapor and there no methyl mercaptan detectors on the third floor to alert personnel of the hazardous conditions that existed there.

As with toxic detector selection, the placement of toxic gas detectors must be strategic in order to capture the potential areas where a release could occur and the extent of the release. The methyl mercaptan detector placement both externally and internally at the DuPont La Porte facility did not operate as intended in part because the detectors did not fully cover the areas where the majority of the methyl mercaptan was released. For example, there was no real time data for emergency responders to use to evaluate the potential for offsite concentrations during the methyl mercaptan release. Although emergency responders conducted offsite monitoring hours after the initial release, the information obtained from air monitoring could not give an accurate portrayal of the methyl mercaptan release. Furthermore, the methyl mercaptan monitors inside the building did not provide accurate information about the release because they were not located near the open drain valves.

Facilities can adopt a variety of methods to select the location of toxic gas monitors with the understanding that they should be prepared for scenarios where the monitoring location may not be effective. One technique that a facility could use is to develop a series of release scenarios and based on those scenarios place monitors in the path of those releases (including by the potential sources of a leak). This technique, however, may not cover all potential release areas. Additionally, facilities should evaluate additional locations for toxic gas detectors if there is a change in process that creates new potential release scenarios. Another technique, if focusing on worker protection, is to give workers personal monitors when they enter toxic gas hazardous areas, especially if these areas are not normally occupied or congested with equipment. At DuPont La Porte, commercially available personal hydrogen sulfide monitors could have been issued to workers. Additionally, if using fixed toxic gas monitors for personnel protection, they should be installed near the breathing zone.

3.3 Ensure Personnel Understand Toxic Gas Hazards

DuPont La Porte personnel did not fully appreciate the effects of methyl mercaptan's physical characteristics. Although methyl mercaptan[21] is a highly toxic, flammable chemical, its low odor

[21] Personnel at DuPont La Porte would refer to methyl mercaptan as "MeSH." MeSH is shorthand based upon methyl mercaptan's chemical formula: CH_3SH. "Me" represents the methyl group (CH_3) and "SH" represents the mercaptan group (SH)—and when combined, these initialisms forms MeSH.

threshold[22] and ability to cause olfactory fatigue in personnel can generate a false sense of security when working with the chemical.

Although DuPont's methyl mercaptan technical standard states, "[a]ll DuPont sites *will* protect the site and neighboring businesses and communities from the escape of a highly toxic material [,]" methyl mercaptan detectors provided essentially no value to DuPont personnel [Emphasis in the original]. DuPont operations staff became accustomed to the odor of methyl mercaptan being present in their work environment. The low odor threshold of methyl mercaptan resulted in a false perception of safety. Consequently, the workers did not associate detecting the odor of methyl mercaptan as representing an immediate hazard.

> *"And I smell MeSH [methyl mercaptan] all the time ... but it never ... I smelled MeSH but I never felt I couldn't breathe. Like whether it's, there's -- whether it's a small leak or anything, it just never -- I didn't think MeSH was that dangerous. I thought it, you know, but it is."*

As seen by this comment from a DuPont La Porte worker, smelling methyl mercaptan at low concentrations without having any observable, adverse effects can potentially normalize the danger of a highly toxic chemical.

Furthermore, extended exposure to methyl mercaptan could potentially eliminate the perception of a methyl mercaptan release because olfactory fatigue will prevent workers from being able to smell the chemical. The odor threshold for methyl mercaptan is approximately two parts per billion (2 ppb). However, olfactory fatigue is known to occur with methyl mercaptan, resulting in the exposed person to no longer being able to detect the chemical by smell. As a result, using odor as an indicator of a release may not provide adequate warning of hazardous concentrations [9].

In the events leading to the major methyl mercaptan release, there were many smaller releases, filling the area with methyl mercaptan odor. In part because personnel believed they understood the source of releases and normalized the risk associated with these smaller releases, personnel did not apply additional safeguards like clearing the area or wearing the required personal protective equipment required by the site when dealing with a methyl mercaptan release.

Facilities should ensure that employees understand the hazards of chemicals that they are working with and may be potentially exposed to if there is a release. In cases where there are multiple hazardous chemicals, such as methyl isocyanate[23] (MIC) and methyl mercaptan (as was the case at DuPont La Porte), it is important that companies effectively train workers on the hazards of each chemical at their facility. Furthermore, companies should train workers that handle multiple highly toxic chemicals, such as methyl mercaptan and MIC, to treat these chemicals with equal

---

[22] Odor threshold, in general, is the lowest concentration of a gas or a material's vapor that can be detected by odor. Odor threshold values are not fixed physiological parameters or physical constants but are statistical points representing the best estimate value from a group of individual responses [12, p. 2].

[23] Methyl isocyanate (MIC) is a raw material used in the production of insecticides at La Porte including the Lannate® process. In December 3, 1984, a methyl isocyanate release at the Union Carbide insecticide plant in Bhopal, India resulted in an estimated 3,800 people that died within days, and tens of thousands that were injured. Eventually, the release killed tens of thousands of people. See CSB Safety Message on 30th anniversary of fatal chemical release that killed thousands in Bhopal, India [13].

importance. This training is important for workers to help ensure they have a working knowledge of those chemical hazards.

3.4 Ensure Response to a Toxic Gas Detector Alarm Is Appropriate

The toxic gas detection system at DuPont La Porte had a history of false readings. Because of these false readings, personnel supplanted their ability to observe and detect a leak over the toxic gas detection system. Personnel adapted their responses to a potential toxic gas release to include determining whether the toxic gas detectors were functioning properly by using their perception of smell before implementing the site's leak response procedures.

For example, the Lannate® Unit used chlorine in addition to methyl mercaptan. Workers could observe from a distance a chlorine leak by the green color of the gas or the immediate adverse effects (i.e., difficulty breathing) of the gas. Workers, similarly, would also use their sense of smell to detect methyl mercaptan. Methyl mercaptan gas, however, is colorless and the odor is not a reliable indicator of a release (for the reasons noted above). In both situations where workers used their physical observations to determine where there was a leak, workers were at risk for being overcome by a toxic chemical release because they were not using additional safeguards like personal protective equipment (PPE).

DuPont's highly toxic material standards require that toxic gas detection systems be treated the same as safety interlocks. DuPont defines a safety interlock as a system or function that detects an out-of-limits (abnormal) condition or improper sequence and brings it to a safe condition. In this case, an alarm should trigger a predetermined response from operators to bring the system to a safe condition. In practice, the methyl mercaptan detectors were not treated as a safety interlock.

Multiple DuPont employees informed the CSB that odor was the primary method used to locate a methyl mercaptan leak. DuPont personnel routinely responded to methyl mercaptan detector alarms without any additional respiratory protection. When they responded, if a significant leak was identified, these personnel consistently stated they would leave to get a SCBA and then return.

On the night of the incident, multiple methyl mercaptan detector alarms activated, but because the workers were aware that methyl mercaptan was being released from their activities associated with clearing the hydrate, these alarms were normalized and not treated as a safety interlock. The alarms were not treated as a near miss and there was no actual investigation into the source or how to prevent these releases. No additional personal protective equipment, such as air-purifying respirators or SCBAs was used. In addition, DuPont did not order a unit evacuation and did not notify the emergency response team until hours later, after Operator 1 made the distress call.

When a toxic detection system has a history of false readings it creates an opportunity for workers to normalize toxic gas detector readings and their response, making the toxic gas detection system ineffective. Facilities should take steps to address the potential for false readings and reduce opportunities for workers to normalize false readings and their alarms. One method is to treat toxic gas detector alarms as a near miss and investigate them even if there are no injuries or the readings are determined to be false. This method should trigger an effort to understand alarm management and safety management system issues and address them. Additionally, facilities should ensure that workers are using leak response procedures to prevent personnel from entering into potential hazardous situations without proper safeguards, like PPE.

4 Conclusions

The Lannate® Unit's toxic gas detection system did not effectively warn workers of the November 15, 2014 methyl mercaptan release occurring inside the manufacturing building. The manufacturing building, including the building's overall construction, detectors, and alarms, was not designed such that personnel could be readily notified of a major chemical release in the manufacturing building. Additionally, the methyl mercaptan toxic gas system did not provide real-time, accurate information to warn of potential offsite consequences.

There are several lessons that facilities can learn from the incident to help improve their toxic gas detection systems. Gas detectors must operate as intended to prevent workers from normalizing the hazard the detectors are protecting workers from. For toxic gas detection system to be effective, the detector selection and placement have to be strategic to identify and quantify releases. Finally, personnel have to understand and appreciate the hazards associated with the toxic gases they work with to help ensure they respond appropriately to potential toxic gas releases.

5 References

[1] U.S. Chemical Safety and Hazard Investigation Board (USCSB), "DuPont La Porte Facility Toxic Chemical Release," [Online]. Available: https://www.csb.gov/dupont-la-porte-facility-toxic-chemical-release-/. [Accessed 1 March 2019].

[2] U.S. Environmental Protection Agency (USEPA), "List of Regulated Substances under the Risk Management Plan (RMP) Program," [Online]. Available: https://www.epa.gov/rmp/list-regulated-substances-under-risk-management-plan-rmp-program. [Accessed 2017 August 28].

[3] U.S. Chemical Safety and Hazard Investigation Board (USCSB), "DuPont La Porte, Texas Chemical Facility Toxic Chemical Release: Interim Recommendations," 30 September 2015. [Online]. Available: https://www.csb.gov/assets/1/20/dupont_la_porte_interim_recommendations_2015-09-30_final1.pdf?15526. [Accessed 28 February 2019].

[4] National Oceanic and Atmospheric Administration (NOAA) Office of Response and Restoration, "Emergency Response Planning Guidelines (ERPGs)," [Online]. Available: https://response.restoration.noaa.gov/oil-and-chemical-spills/chemical-spills/resources/emergency-response-planning-guidelines-erpgs.html. [Accessed 10 March 2019].

[5] United States Department of Labor (USDOL) Occupational Safety and Health Administration (OSHA), "Permissible Exposure Limits - Annotated Tables," [Online]. Available: https://www.osha.gov/dsg/annotated-pels/. [Accessed 28 February 2019].

[6] United States Department of Labor (USDOL) Occupational Safety and Health Administration (OSHA), "Annotated Table Z-1," [Online]. Available: https://www.osha.gov/dsg/annotated-pels/tablez-1.html. [Accessed 28 February 2019].

[7] Centers for Diseases Control and Prevention National Institute of Occupational Safety and Health (NIOSH), "NIOSH Pocket Guide to Chemical Hazards - Methyl Mercaptan," 11 April 2016. [Online]. Available: https://www.cdc.gov/niosh/npg/npgd0425.html. [Accessed 28 February 2019].

[8] U.S. Chemical Safety and Hazard Investigation Board (USCSB), "DuPont La Porte, Texas Chemical Facility Toxic Chemical Release: Interim Recommendations," 30 September 2015. [Online]. Available: https://www.csb.gov/assets/1/20/dupont_la_porte_interim_recommendations_2015-09-30_final1.pdf?15526.

[9] Agency for Toxic Substances and Disease Registry (ATSDR), "Methyl Mercaptan," [Online]. Available: https://www.atsdr.cdc.gov/MHMI/mmg139.pdf. [Accessed 28 February 2019].

[10] A. H. Mohammadi and D. Richon, "Equilibrium Data of Sulfur Dioxide and Methyl Mercaptan Clathrate Hydrates," *Journal of Chemical & Engineering Data,* vol. 56, pp. 1666-1668, 2011.

[11] Center for Chemical Process Safety (CCPS), Recognizing Catastrophic Incident Warning Signs in the Process Industries, Hoboken: Wiley, 2012.

[12] A. H. Lehocky, S. S. Murnane and D. Patrick, Odor Thresholds for Chemicals with Established Occupational Health Standards, Falls Church: AIHA, 2013.

[13] USCSB, "On 30th Anniversary of Fatal Chemical Release that Killed Thousands in Bhopal, India, CSB Safety Message Warns it Could Happen Again," [Online]. Available: http://www.csb.gov/on-30th-anniversary-of-fatal-chemical-release-that-killed-thousands-in-bhopal-india-csb-safety-message-warns-it-could-happen-again-/?pg=4.

[14] K. Baker, "DuPont Closing Texas Plant Where 4 Died," The News Journal, 1 April 2016. [Online]. Available: https://www.usatoday.com/story/money/2016/04/01/dupont-closing-texas-plant-where-4-died/82512270/. [Accessed 29 June 2017].

[15] DuPont, "Form 8-K," 31 March 2016. [Online]. Available: https://www.sec.gov/Archives/edgar/data/30554/000003055416000115/dd33116form8-k.htm.

[16] U.S. Chemical Safety and Hazard Investigation Board (USCSB), "E.I. DuPont de Nemours & Co., Inc. Belle, West Virgina Investigative Report," September 2011. [Online]. Available: https://www.csb.gov/assets/1/20/csb%20final%20report.pdf?13966. [Accessed 1 March 2019].

[17] U.S. Chemical Safety and Hazard Investigation Board (USCSB), "E. I. DuPont De Nemours Co. Fatal Hotwork Explosion," [Online]. Available: https://www.csb.gov/e-i-dupont-de-nemours-co-fatal-hotwork-explosion/. [Accessed 1 March 2019].

[18] American Industrial Hygiene Association (AIHA), "Current ERPG Values," 2014. [Online]. Available: https://www.aiha.org/get-involved/AIHAGuidelineFoundation/EmergencyResponsePlanningGuidelines/Documents/2014%20ERPG%20Values.pdf.

[19] F. Cavender, S. Phillips and M. Holland, "Development of Emergency Response Planning Guidelines (ERPGs)," *Journal of Medical Toxicology,* vol. 4, no. 2, pp. 127-131, 2008.

553063 Hurricane Harvey 20 Months Later - Insights on the Arkema Crosby Flooding

On August 24, 2017, Hurricane Harvey, a Category 4 hurricane, made landfall in southeast Texas producing an unprecedented amount of rainfall. Hurricane Harvey killed 68 people and flooded over 300,000 structures, forcing roughly 40,000 people to evacuate their homes. Hurricane Harvey required a large emergency response effort supported by local, state, and Federal officials.

At Arkema's Crosby site, Harvey delivered an unprecedented amount of rainfall. The National Oceanic and Atmospheric Administration estimates show that the rainfall during Harvey had less than a one-in-one-thousand-year likelihood of occurring. Harris County has estimated that at Crosby, Harvey was a 5,000 year to 20,000 year event. The plant's hurricane preparedness procedure was followed and resulted in a safe shutdown and hurricane preparedness of the manufacturing units. However, finished goods inventory remaining in the warehouses was overwhelmed by high water and overcame multiple layers of protection used to safely store organic peroxide finished goods.

This presentation will explain the sites preparedness ahead of time, the events leading up to the flooding, the incident itself, and the emergency response.

The presentation will focus on the impacts, the main lessons learned and the key recommendations from the CSB report.

AUTHOR INDEX

AUTHOR INDEX

AUTHOR INDEX